Pure Mathematics 2

3, 8, 7

2, 9, 3

6, 4, 4

By the same authors

APPLIED MATHEMATICS 1
APPLIED MATHEMATICS 2
PURE MATHEMATICS 1

Pure Mathematics 2

L. Bostock, B.Sc.
Senior Mathematics Lecturer
Southgate Technical College

S. Chandler, B.Sc.

Stanley Thornes (Publishers) Ltd.

First published in 1979 by Stanley Thornes (Publishers) Ltd.,
Educa House, Old Station Drive, off Leckhampton Road,
CHELTENHAM GL53 0DN
Reprinted with minor correction 1980.
Reprinted with minor correction 1981.

ISBN 0 85950 097 7

Typeset at the Alden Press
Oxford London and Northampton
and printed in Great Britain at the Pitman Press, Bath

PREFACE

Pure Mathematics 2 is a continuation of the work covered in Volume 1 and is intended to complete a full two year course in Pure Mathematics. It caters for the Pure Mathematics content of such advanced level schemes as the University of London syllabus in Further Mathematics.

This book includes an extensive section on matrices and transformations, where the development assumes no prior knowledge of these topics. Another section is devoted to mathematical proof where we have introduced, in a simple way, some of the elementary logic concepts that provide a basis for appreciating sound proof. The text also examines, in general, functions and their properties.

Many worked examples are incorporated in the text to illustrate each main development of a topic and a set of straightforward problems follows each section. A selection of more challenging questions is given in the miscellaneous exercise at the end of most chapters, and, as in Volume 1, multiple choice exercises are included when appropriate.

We would like to express our sincere thanks to the many people whose suggestions and criticisms have been most useful. In particular we are grateful to Miss J. Broughton for her careful scrutiny of the text, Dr. C.P. Rourke for his invaluable help with the chapters on transformations and proof and Miss E. Clarke for typing the manuscript.

We are grateful to the following Examination Boards for permission to reproduce questions from past examination papers:
The Associated Examining Board (AEB)
University of London Entrance and School Examination Council (U of L)
University of Cambridge Local Examinations Syndicate (C)
Joint Matriculation Board (JMB)
Oxford Delegacy of Local Examinations (O)

L. Bostock
S. Chandler

CONTENTS

Preface		v
Notes on Use of the Book		xi
Chapter 1.	**Transformations, Matrices and Determinants**	1
	Column vectors. Transformations in two dimensions. Matrices. Multiplication of matrices. Determinants.	
Chapter 2.	**Coordinate Geometry I**	62
	Conic sections. Parabola. Ellipse. Diameters. Hyperbola. Asymptotes.	
Chapter 3.	**Coordinate Geometry II**	111
	Rectangular hyperbola. Live pair through the origin. Pole and Polar.	
Chapter 4.	**Three dimensional space – Coordinate methods**	139
	Cartesian frame of reference. Vectors in three dimensions. Cartesian base vectors. Direction ratios and direction cosines. Equations of a straight line. Equation of a plane.	
Chapter 5.	**Three dimensional space – Vector methods**	183
	Base vectors. Direction ratios and direction cosines. Unit vector. Equation of a straight line. Scalar product. Vector product. Equation of a plane.	
Chapter 6.	**Transformations and Linear Equations**	248
	Transformations of three dimensional space. Inverse transformations. Inverse matrix. Transpose matrix. Systems of equations and geometric interpretation in two and three dimensions. Solution by systematic elimination. Calculation of an inverse matrix by reduction.	

Chapter 7. **Mathematical Proof** 305

Statement. Negation. Conditional statement. Converse. Biconditional statement. Contrapositive. Direct proof. Proof by Induction. Indirect proof. Proof by contradiction. Counter example.

Chapter 8. **Summation of Series** 332

Number series. Use of partial fractions. Method of differences. Sequences.

Chapter 9. **Some Functions and their Properties** 364

Even and odd functions. Continuous functions. Discontinuity and its effect on differentiation and integration. Periodic functions. Logarithmic function. Hyperbolic functions. Inverse hyperbolic functions. General inverse functions.

Chapter 10. **Some Curves and their Properties** 410

Tangent at the origin. Inflexion. Inequalities. Graphical representation of inequalities. Simultaneous inequalities. Use of log–log and log–linear graph paper.

Chapter 11. **Polynomial Functions and Equations** 449

Remainder theorem. Repeated factors. Common factors. Homogeneous functions. Cyclic functions. Solution of polynomial equations. Relationship between roots and coefficients. Repeated roots. Number of real roots. Approximate solution of equations. Iterative methods.

Chapter 12. **Complex Number** 504

De Moivre's theorem. nth roots of a complex number. Exponential form. Relationship between hyperbolic, trigonometric and exponential functions. Transformations in the Argand diagram.

Chapter 13. **Further Integration and Differential Equations** 534

Reduction methods. First order differential equations. Integrating factor. Second order linear differential equations with constant coefficients. Particular integral.

Chapter 14. **Further Applications of Integration** 568

Length of arc. Area of surface of revolution. Volume of revolution. Root mean square value. Centroid and first moment. The theorems of Pappus. Second moment. Curvature. Radius of curvature.

Answers 609

Index 635

NOTES ON THE USE OF THE BOOK

Notation

$=$	is equal to
$\equiv$	is identical to
$\simeq$	is approximately equal to*
$>$	is greater than
$\geqslant$	is greater than or equal to
$<$	is less than
$\leqslant$	is less than or equal to
∞	infinitely large
$\rightarrow$	approaches
$\Rightarrow$	implies
$\Leftarrow$	is implied by
$\Leftrightarrow$	implies and is implied by

A stroke through any of the above symbols negates it. i.e. $\neq$ means 'is not equal to', $\not>$ means 'is not greater than'.

Abbreviations

$\parallel$	parallel
+ve	positive
−ve	negative
w.r.t.	with respect to

*Practical problems rarely have exact answers. Where numerical answers are given they are correct to two or three decimal places depending on their context, e.g. π is 3.142 correct to 3 d.p. and although we write $\pi = 3.142$ it is understood that this is not an exact value. We reserve the symbol $\simeq$ for those cases where the approximation being made is part of the method used.

Instructions for answering Multiple Choice Exercises

These exercises are at the end of most chapters. The questions are set in groups, each group representing one of the variations that may arise in examination papers. The answering techniques are different for each type of question and are classified as follows:

TYPE I

These questions consist of a problem followed by several alternative answers, only *one* of which is correct.

Write down the letter corresponding to the correct answer.

TYPE II

In this type of question some information is given and is followed by a number of possible responses. *One or more* of the suggested responses follow(s) directly and necessarily from the information given.

Write down the letter(s) corresponding to the correct response(s).

e.g. PQR is a triangle

(a) $\hat{P} + \hat{Q} + \hat{R} = 180°$

(b) $PQ + QR$ is less than PR

(c) if $\hat{P}$ is obtuse, $\hat{Q}$ and $\hat{R}$ must both be acute.

(d) $\hat{P} = 90°$, $\hat{Q} = 45°$, $\hat{R} = 45°$.

The correct responses are (a) and (c).

(b) is definitely incorrect and (d) may or may not be true of triangle PQR, i.e. it does not follow directly and necessarily from the information given. Responses of this kind should not be regarded as correct.

TYPE III

Each problem contains two independent statements (a) and (b).

1) If (a) always implies (b) but (b) does not always imply (a) write A.
2) If (b) always implies (a) but (a) does not always imply (b) write B.
3) If (a) always implies (b) *and* (b) always implies (a) write C.
4) If (a) denies (b) and (b) denies (a) write D.
5) If none of the first four relationships apply write E.

TYPE IV

A problem is introduced and followed by a number of pieces of information. You are not required to solve the problem but to decide whether:

1) the given information is *all* needed to solve the problem. In this case write A;

2) the total amount of information is insufficient to solve the problem. If so write I;
3) the problem can be solved without using one or more of the given pieces of information. In this case write down the letter(s) corresponding to the items not needed.

TYPE V

A single statement is made. Write T if it is true and F if it is false.

CHAPTER 1

TRANSFORMATIONS, MATRICES AND DETERMINANTS

COLUMN VECTORS

In Pure Mathematics I, Chapter 13, we saw that the position vector $\overrightarrow{OP}$ of a point $P(x, y)$ can be denoted either by the sum of its Cartesian components.

i.e. $$x\mathbf{i} + y\mathbf{j}$$

or by the column vector $$\begin{pmatrix} x \\ y \end{pmatrix}.$$

In this chapter we are going to use the column vector form of notation.

Hence if $\overrightarrow{OA}$ is the position vector of $A(1, 3)$ we write

$$\overrightarrow{OA} = \begin{pmatrix} 1 \\ 3 \end{pmatrix}$$

and if $\overrightarrow{OB}$ is the position vector of $B(-2, 3)$ we write

$$\overrightarrow{OB} = \begin{pmatrix} -2 \\ 3 \end{pmatrix}.$$

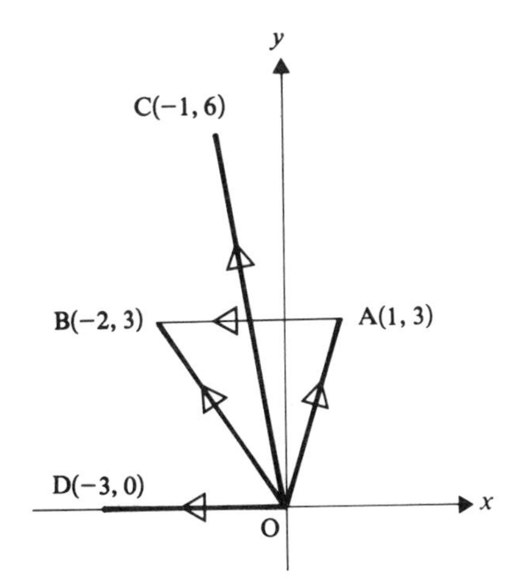

Using the triangle law for the addition of vectors gives

$$\overrightarrow{OA} + \overrightarrow{OB} = \begin{pmatrix} 1 \\ 3 \end{pmatrix} + \begin{pmatrix} -2 \\ 3 \end{pmatrix} = \begin{pmatrix} -1 \\ 6 \end{pmatrix}.$$

Hence, in the diagram,

$$\overrightarrow{OC} = \begin{pmatrix} -1 \\ 6 \end{pmatrix} \Rightarrow C \text{ is the point } (-1, 6).$$

Similarly $\overrightarrow{AB} = \overrightarrow{OB} - \overrightarrow{OA} = \begin{pmatrix} -2 \\ 3 \end{pmatrix} - \begin{pmatrix} 1 \\ 3 \end{pmatrix} = \begin{pmatrix} -3 \\ 0 \end{pmatrix}.$

Note that $\overrightarrow{AB}$ is *not* a position vector, and that in this case $\begin{pmatrix} -3 \\ 0 \end{pmatrix}$ can represent any vector equal to $\overrightarrow{AB}$. So if D is the point $(-3, 0)$ then $\overrightarrow{OD} = \overrightarrow{AB}$.

EXERCISE 1a

Given A(2, 5), B($-1, -3$), C(0, 4), D(-2, 3) find the column vector which represents:

1) $\overrightarrow{OA} + \overrightarrow{OC}$ 2) $\overrightarrow{OA} + \overrightarrow{OD}$ 3) $\overrightarrow{OC} - \overrightarrow{OB}$ 4) $\overrightarrow{OA} + \overrightarrow{OB} + \overrightarrow{OC}$

5) the position vector of the midpoint of AB

6) $\overrightarrow{AC}$ 7) $\overrightarrow{BD}$

8) the position vector of the midpoint of OC

9) the position vector of the centroid of $\triangle$ABC.

TRANSFORMATIONS IN TWO DIMENSIONS

Suppose that a rubber balloon has a picture printed on it. If the balloon is stretched, the picture is distorted. If the balloon is blown up the picture is enlarged. If the balloon is turned over, the picture changes position. Mathematically, such changes of position and distortions of shape are called transformations. In this section we are going to investigate transformations of the xy plane.

Suppose that the xy plane is rotated through an angle $\frac{\pi}{6}$ about the origin so that Ox and Oy are rotated to the position shown in diagram (ii).

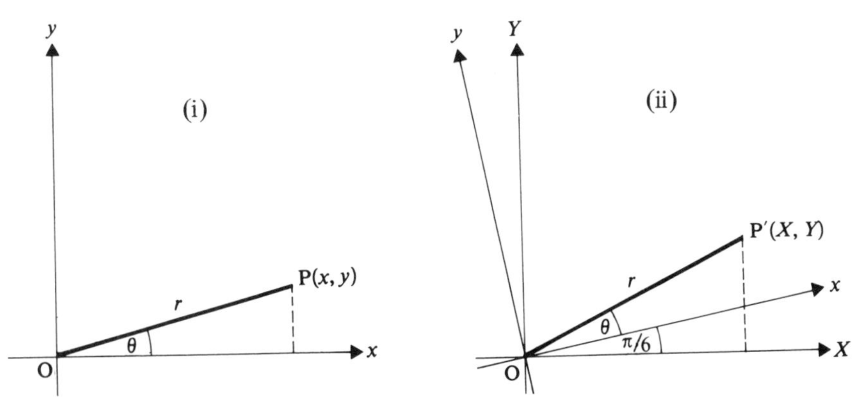

Now if $\overrightarrow{OP}$, where P is the point (x, y) becomes P′ in diagram (ii) then P′ is called the *image* of P and P is said to *map* to P′.
If we refer P′ to coordinate axes OX and OY then from the diagrams

$$\begin{cases} x = r\cos\theta \\ y = r\sin\theta \end{cases} \text{ and } \begin{cases} X = r\cos\left(\theta + \frac{\pi}{6}\right) = r\cos\theta\cos\frac{\pi}{6} - r\sin\theta\sin\frac{\pi}{6} \\ Y = r\sin\left(\theta + \frac{\pi}{6}\right) = r\sin\theta\cos\frac{\pi}{6} + r\cos\theta\sin\frac{\pi}{6} \end{cases}$$

$$\Rightarrow \begin{cases} X = x\cos\frac{\pi}{6} - y\sin\frac{\pi}{6} \\ Y = x\sin\frac{\pi}{6} + y\cos\frac{\pi}{6} \end{cases}$$

$$\Rightarrow \begin{cases} \frac{\sqrt{3}}{2}x - \frac{1}{2}y = X & [1] \\ \frac{1}{2}x + \frac{\sqrt{3}}{2}y = Y. & [2] \end{cases}$$

This pair of equations enables us to transform the coordinates of any point $P(x, y)$ to the coordinates of the point $P'(X, Y)$ where $\overrightarrow{OP'}$ is the vector obtained by rotating $\overrightarrow{OP}$ through $\frac{\pi}{6}$ about O.

We are now going to develop a notation to express the pair of scalar equations [1] and [2] as a single vector equation for transforming $\overrightarrow{OP}$ to $\overrightarrow{OP'}$,

i.e. $\begin{pmatrix} x \\ y \end{pmatrix}$ to $\begin{pmatrix} X \\ Y \end{pmatrix}$.

Equation (1) shows that X is a function of x and y, i.e. X is obtained by 'operating' on the position vector $\begin{pmatrix} x \\ y \end{pmatrix}$. Writing the coefficients of x and y in equation [1] in the form $(\frac{1}{2}\sqrt{3} \quad -\frac{1}{2})$ (known as a row vector), we can express equation [1] in the vector form

$$\left(\tfrac{1}{2}\sqrt{3} \quad -\tfrac{1}{2}\right)\begin{pmatrix} x \\ y \end{pmatrix} = X$$

where

$$\left(\tfrac{1}{2}\sqrt{3} \quad -\tfrac{1}{2}\right)\begin{pmatrix} x \\ y \end{pmatrix} = \frac{\sqrt{3}}{2}x - \frac{1}{2}y. \qquad [3]$$

Note that the RHS of [3] is obtained by *adding* the product of the first element in the row vector and the first element in the column vector *to* the product of the second element in the row vector and the second element in the column vector.

In general

$$(a \quad b)\begin{pmatrix} c \\ d \end{pmatrix} = ac + bd$$

e.g.

$$(3 \quad 1)\begin{pmatrix} -2 \\ 4 \end{pmatrix} = -6 + 4 = -2.$$

Note that $\begin{pmatrix} c \\ d \end{pmatrix}(a \quad b)$ is not defined at this stage.

Returning to equation [2], this may now be written in the vector form

$$\left(\tfrac{1}{2} \quad \tfrac{1}{2}\sqrt{3}\right)\begin{pmatrix} x \\ y \end{pmatrix} = Y.$$

So the pair of equations, [1] and [2], which transform $\begin{pmatrix} x \\ y \end{pmatrix}$ to $\begin{pmatrix} X \\ Y \end{pmatrix}$ under a rotation of $\frac{\pi}{6}$ can be written as

$$\left(\tfrac{1}{2}\sqrt{3} \quad -\tfrac{1}{2}\right)\begin{pmatrix} x \\ y \end{pmatrix} = X$$

$$\left(\tfrac{1}{2} \quad \tfrac{1}{2}\sqrt{3}\right)\begin{pmatrix} x \\ y \end{pmatrix} = Y.$$

These equations can now be included in a single equation by adjoining the row vectors to give a square array of numbers known as a *matrix*, i.e. $$\begin{pmatrix} \frac{\sqrt{3}}{2} & -\frac{1}{2} \\ \frac{1}{2} & \frac{\sqrt{3}}{2} \end{pmatrix}\begin{pmatrix} x \\ y \end{pmatrix} = \begin{pmatrix} X \\ Y \end{pmatrix}$$

where X is the product of the top row of the matrix and $\begin{pmatrix} x \\ y \end{pmatrix}$

and Y is the product of the bottom row of the matrix and $\begin{pmatrix} x \\ y \end{pmatrix}$

We may regard the matrix as an operator which rotates $\begin{pmatrix} x \\ y \end{pmatrix}$ through $\frac{\pi}{6}$ radians to $\begin{pmatrix} X \\ Y \end{pmatrix}$.

In general $$\begin{pmatrix} a & b \\ c & d \end{pmatrix}\begin{pmatrix} x \\ y \end{pmatrix} = \begin{pmatrix} ax + by \\ cx + dy \end{pmatrix},$$

e.g. $$\begin{pmatrix} 3 & 2 \\ 0 & 1 \end{pmatrix}\begin{pmatrix} 2 \\ 4 \end{pmatrix} = \begin{pmatrix} (3)(2) + (2)(4) \\ (0)(2) + (1)(4) \end{pmatrix} = \begin{pmatrix} 14 \\ 4 \end{pmatrix}.$$

Note that $\begin{pmatrix} x \\ y \end{pmatrix}\begin{pmatrix} a & b \\ c & d \end{pmatrix}$ is not defined and has no meaning.

Thus, if we want to rotate the square ABCD through $\frac{\pi}{6}$ about O, we can obtain the position vectors of its vertices after the rotation, by operating on $\overrightarrow{OA}, \overrightarrow{OB}, \overrightarrow{OC}, \overrightarrow{OD}$ with the matrix $\begin{pmatrix} \frac{\sqrt{3}}{2} & -\frac{1}{2} \\ \frac{1}{2} & \frac{\sqrt{3}}{2} \end{pmatrix}$ as follows.

If A, B, C, D are the points (2, 0), (4, 0) (4, 2) (2, 2) then

$$\begin{pmatrix} \frac{\sqrt{3}}{2} & -\frac{1}{2} \\ \frac{1}{2} & \frac{\sqrt{3}}{2} \end{pmatrix}\begin{pmatrix} 2 \\ 0 \end{pmatrix} = \begin{pmatrix} \sqrt{3} \\ 1 \end{pmatrix}$$

$$\begin{pmatrix} \frac{\sqrt{3}}{2} & -\frac{1}{2} \\ \frac{1}{2} & \frac{\sqrt{3}}{2} \end{pmatrix}\begin{pmatrix} 4 \\ 0 \end{pmatrix} = \begin{pmatrix} 2\sqrt{3} \\ 2 \end{pmatrix}$$

$$\begin{pmatrix} \frac{\sqrt{3}}{2} & -\frac{1}{2} \\ \frac{1}{2} & \frac{\sqrt{3}}{2} \end{pmatrix}\begin{pmatrix} 4 \\ 2 \end{pmatrix} = \begin{pmatrix} 2\sqrt{3}-1 \\ 2+\sqrt{3} \end{pmatrix}$$

$$\begin{pmatrix} \frac{\sqrt{3}}{2} & -\frac{1}{2} \\ \frac{1}{2} & \frac{\sqrt{3}}{2} \end{pmatrix}\begin{pmatrix} 2 \\ 2 \end{pmatrix} = \begin{pmatrix} \sqrt{3}-1 \\ 1+\sqrt{3} \end{pmatrix}.$$

The result of this transformation is shown in the diagram below.

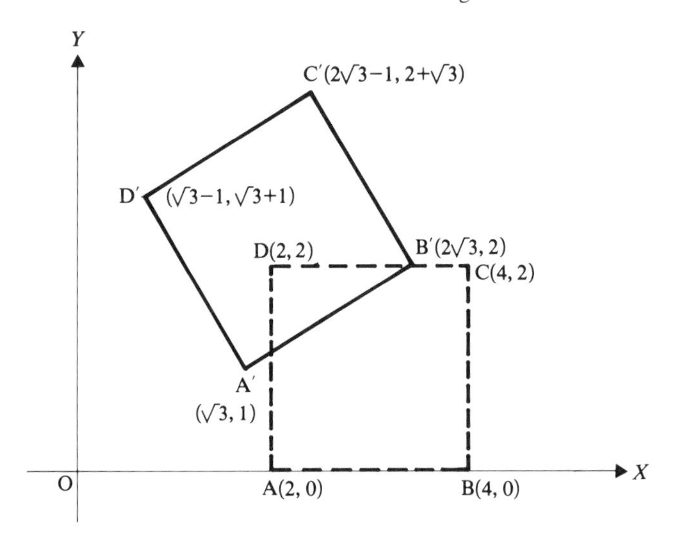

EXERCISE 1b

1) Evaluate

(a) $\begin{pmatrix} 1 & 3 \end{pmatrix}\begin{pmatrix} -1 \\ 4 \end{pmatrix}$ (b) $\begin{pmatrix} 2 & 7 \end{pmatrix}\begin{pmatrix} 3 \\ -4 \end{pmatrix}$ (c) $\begin{pmatrix} -1 & 0 \end{pmatrix}\begin{pmatrix} 3 \\ -1 \end{pmatrix}$.

2) Evaluate

(a) $\begin{pmatrix} 2 & 1 \\ 3 & -1 \end{pmatrix}\begin{pmatrix} 2 \\ 4 \end{pmatrix}$ (b) $\begin{pmatrix} -1 & 3 \\ 5 & 2 \end{pmatrix}\begin{pmatrix} -1 \\ -2 \end{pmatrix}$ (c) $\begin{pmatrix} -1 & 0 \\ -1 & 2 \end{pmatrix}\begin{pmatrix} 4 \\ 5 \end{pmatrix}$.

3) The points A(1, 0), B(2, 0), C(2, 3) form a triangle ABC. Use the matrix operator $\begin{pmatrix} 1 & 0 \\ 0 & -1 \end{pmatrix}$ to transform the position vectors $\overrightarrow{OA}, \overrightarrow{OB}, \overrightarrow{OC}$ to the position vectors $\overrightarrow{OA'}, \overrightarrow{OB'}, \overrightarrow{OC'}$. On the same diagram, draw triangle ABC and triangle A′B′C′ and hence describe the result of the transformation

$$\triangle ABC \rightarrow \triangle A'B'C'.$$

4) Repeat Question (3) with the matrix operator $\begin{pmatrix} -1 & 0 \\ 0 & 1 \end{pmatrix}$.

FURTHER TRANSFORMATIONS OF THE xy PLANE

In the previous section, the matrix which performed a rotation of $\pi/6$ radians about O was derived from the linear equations for the transformation.

Any transformation of $\begin{pmatrix} x \\ y \end{pmatrix}$ to $\begin{pmatrix} X \\ Y \end{pmatrix}$ that can be expressed by the linear equations $\begin{cases} ax + by = X \\ cx + dy = Y \end{cases}$ is called a *linear transformation* and is expressed by the matrix equation

$$\begin{pmatrix} a & b \\ c & d \end{pmatrix}\begin{pmatrix} x \\ y \end{pmatrix} = \begin{pmatrix} X \\ Y \end{pmatrix}$$

$\begin{pmatrix} a & b \\ c & d \end{pmatrix}$ is called the matrix of the transformation and we denote it by **M**.

Note that, under the operator **M**, $\begin{pmatrix} a & b \\ c & d \end{pmatrix}\begin{pmatrix} 0 \\ 0 \end{pmatrix} = \begin{pmatrix} 0 \\ 0 \end{pmatrix}$,

i.e. the origin does not change position.

So for any transformation matrix of the form $\begin{pmatrix} a & b \\ c & d \end{pmatrix}$ the origin is invariant.

Examples of such transformations are

1) *Rotation through any angle about the origin*

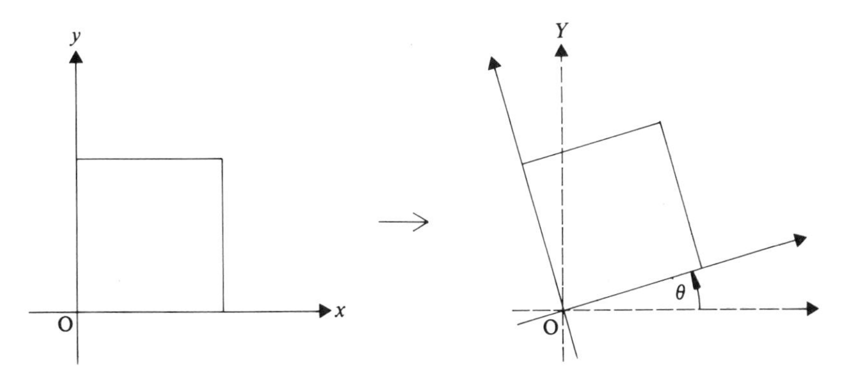

Note that rotation about any other point *does* change the position of O, as is seen in the following diagram,

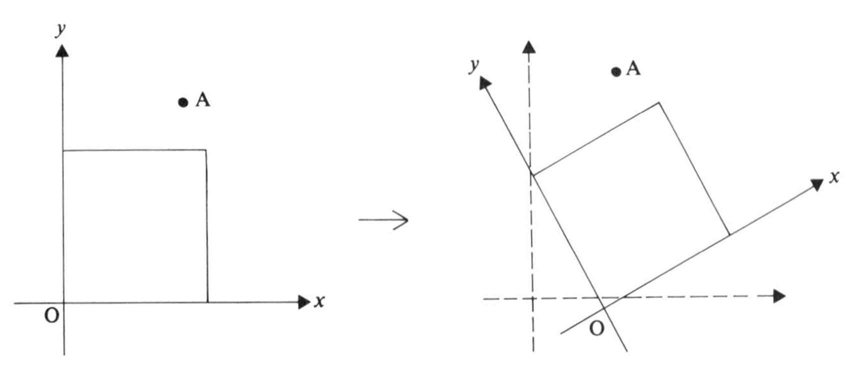

2) *Reflection in any line through* O

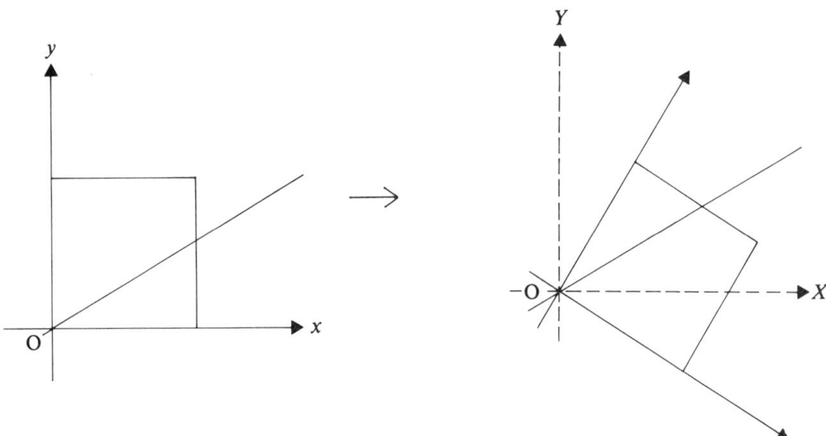

3) *Shearing parallel to the x axis (or y axis)*

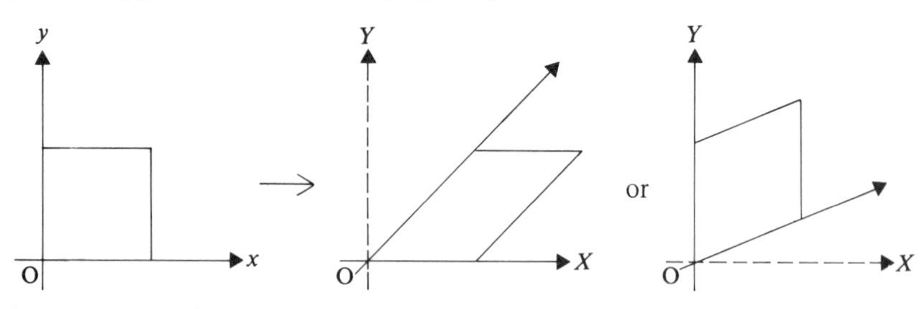

4) *Enlargement (or reduction)*

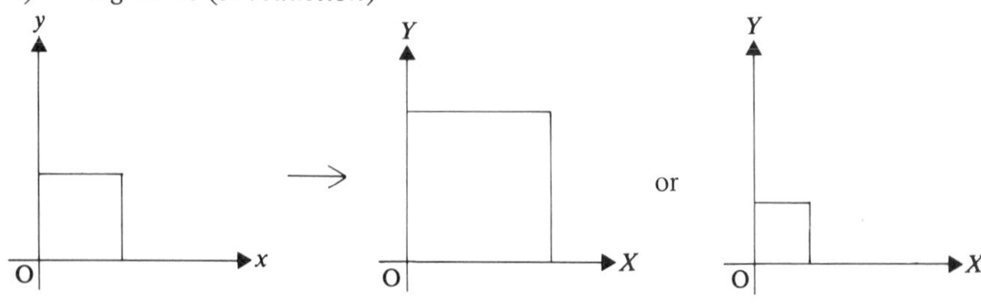

The matrix operator which produces each of these linear transformations can be found by the method used in the previous section, i.e. by first finding the appropriate pair of linear equations.
However, the algebra involved can be tedious, so we now develop a more direct method for obtaining a particular transformation matrix.
Suppose, that under a given linear transformation

$$\mathbf{i} \text{ maps to } \mathbf{p} \text{ where } \mathbf{p} = \begin{pmatrix} a \\ c \end{pmatrix} \quad \text{and} \quad \mathbf{j} \text{ maps to } \mathbf{q} \text{ where } \mathbf{q} = \begin{pmatrix} b \\ d \end{pmatrix}$$

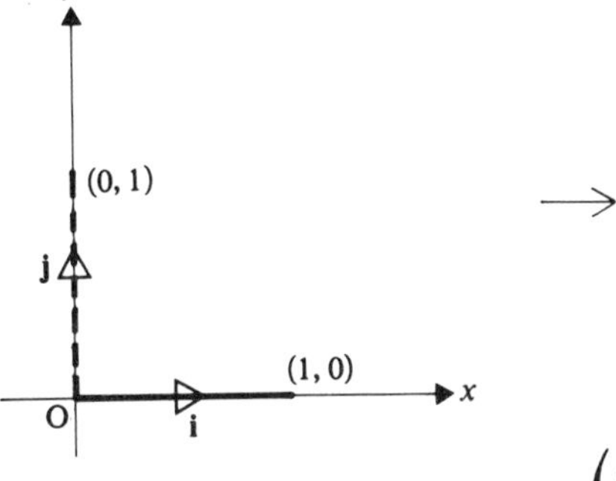

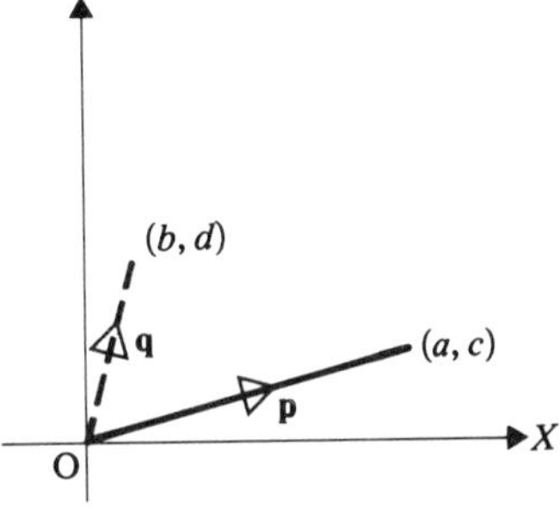

and $\quad \lambda\mathbf{i}$ maps to $\lambda\mathbf{p}$ where $\lambda\mathbf{p} = \begin{pmatrix} \lambda a \\ \lambda c \end{pmatrix}$

and $\quad \mu\mathbf{j}$ maps to $\mu\mathbf{q}$ where $\mu\mathbf{q} = \begin{pmatrix} \mu b \\ \mu d \end{pmatrix}$

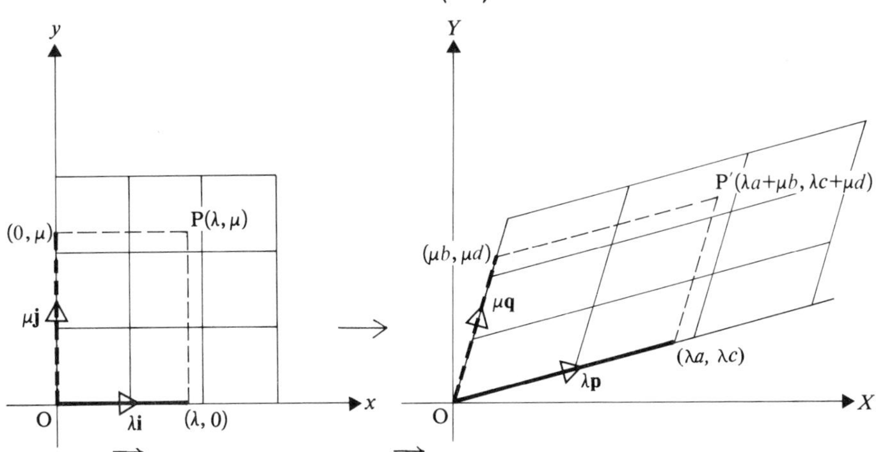

Hence $\overrightarrow{OP} = \lambda\mathbf{i} + \mu\mathbf{j}$ maps to $\overrightarrow{OP'} = \lambda\mathbf{p} + \mu\mathbf{q}$,

i.e. $\begin{pmatrix} \lambda \\ \mu \end{pmatrix}$ maps to $\begin{pmatrix} \lambda a + \mu b \\ \lambda c + \mu d \end{pmatrix}$

Now
$$\begin{pmatrix} \lambda a + \mu b \\ \lambda c + \mu d \end{pmatrix} = \begin{pmatrix} a & b \\ c & d \end{pmatrix}\begin{pmatrix} \lambda \\ \mu \end{pmatrix}$$

Hence the mapping of $\overrightarrow{OP}$ to $\overrightarrow{OP'}$ can be expressed by the equation

$$\begin{pmatrix} a & b \\ c & d \end{pmatrix}\overrightarrow{OP} = \overrightarrow{OP'}.$$

Noting that the first and second columns respectively of the matrix operator are

$\begin{pmatrix} a \\ c \end{pmatrix}$ which is the image of **i**

$\begin{pmatrix} b \\ d \end{pmatrix}$ which is the image of **j**

we see that

if for a particular linear transformation, we can find the image of **i** and the image of **j** we can *write down* the matrix operator for that transformation

Note that under the general linear transformation described, the images of the Cartesian base vectors **i**, **j** are **p**, **q** and that the image of the vector $\lambda\mathbf{i} + \mu\mathbf{j}$ in the Cartesian plane is of the form $\lambda\mathbf{p} + \mu\mathbf{q}$ in the transformed plane, i.e. the transformation changes the frame of reference from the Cartesian base vectors to a new set of base vectors.

We will now find the matrix operators for some common transformations.

1) *Reflection in the x-axis*

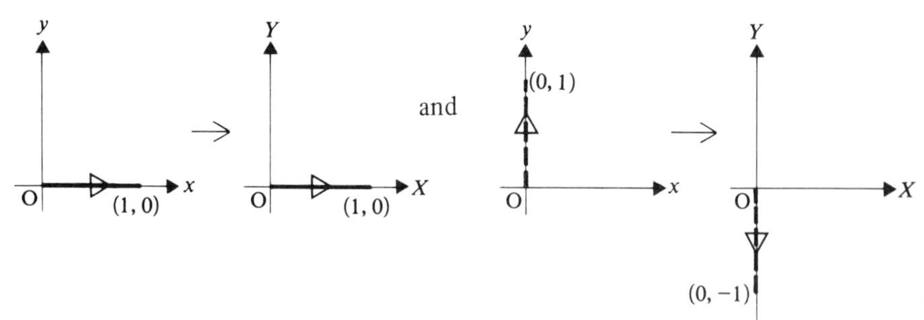

Under this transformation

$$\mathbf{i} \to \begin{pmatrix} 1 \\ 0 \end{pmatrix} \quad \text{and} \quad \mathbf{j} \to \begin{pmatrix} 0 \\ -1 \end{pmatrix}.$$

Therefore $\begin{pmatrix} 1 & 0 \\ 0 & -1 \end{pmatrix}$ is the matrix operator.

2) *Reflection in the line* $y = x$

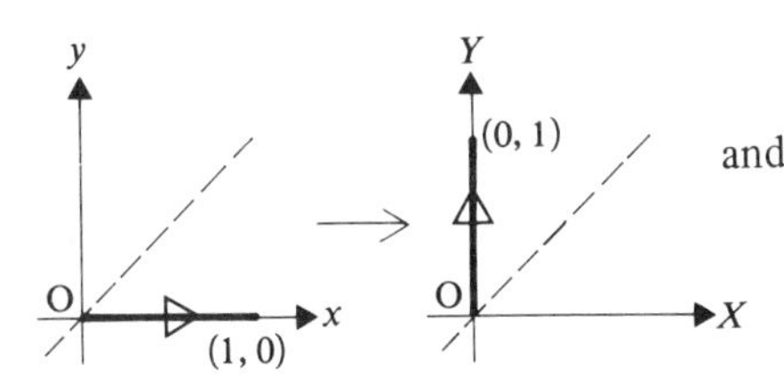

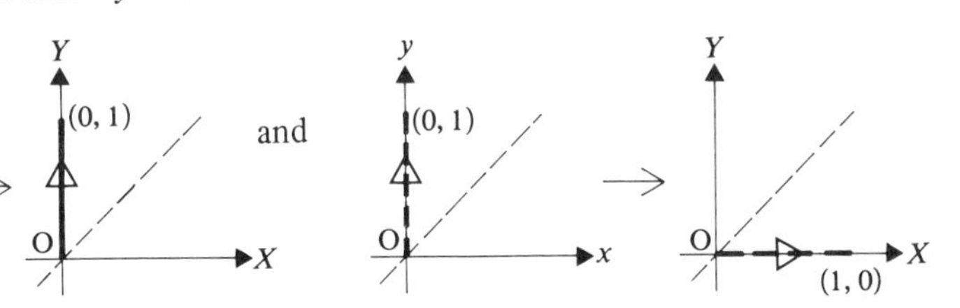

Under this transformation

$$\mathbf{i} \rightarrow \begin{pmatrix} 0 \\ 1 \end{pmatrix} \quad \text{and} \quad \mathbf{j} \rightarrow \begin{pmatrix} 1 \\ 0 \end{pmatrix}.$$

Therefore $\begin{pmatrix} 0 & 1 \\ 1 & 0 \end{pmatrix}$ is the matrix operator.

3) *Rotation through an angle* θ *about* O

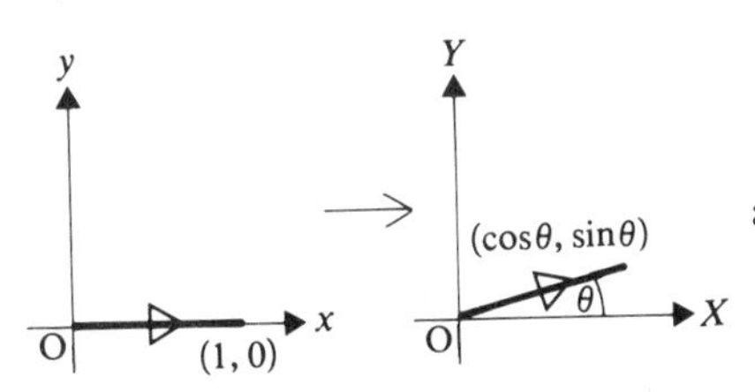

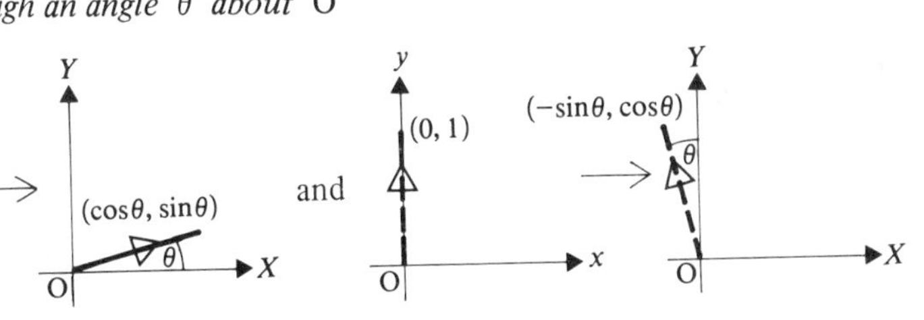

Under this transformation

$$\mathbf{i} \rightarrow \begin{pmatrix} \cos\theta \\ \sin\theta \end{pmatrix} \quad \text{and} \quad \mathbf{j} \rightarrow \begin{pmatrix} -\sin\theta \\ \cos\theta \end{pmatrix}.$$

Therefore $\begin{pmatrix} \cos\theta & -\sin\theta \\ \sin\theta & \cos\theta \end{pmatrix}$ is the matrix operator.

4) *Shear of 45° in the direction* Ox

The effect of this shear on a grid of squares is shown in the diagram below:

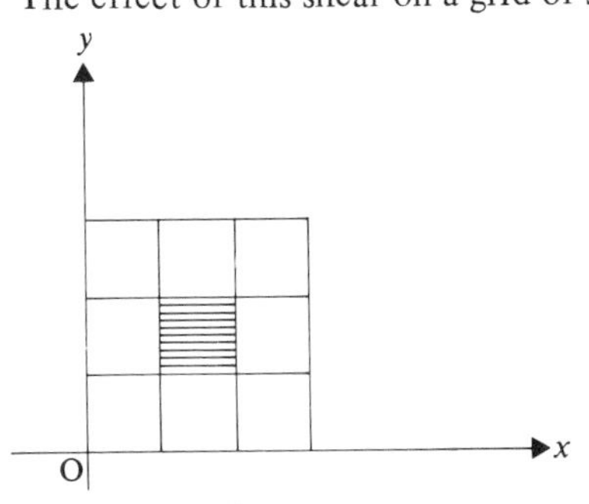

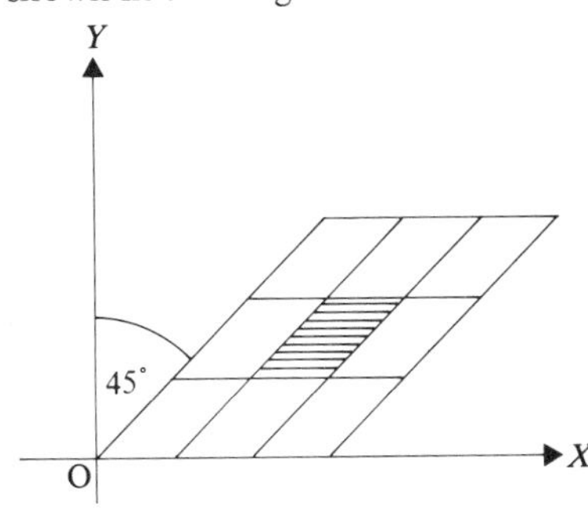

i.e. lines parallel to Oy are tilted at 45° and stretched to maintain height. (Lines parallel to Ox are not affected.)
Hence

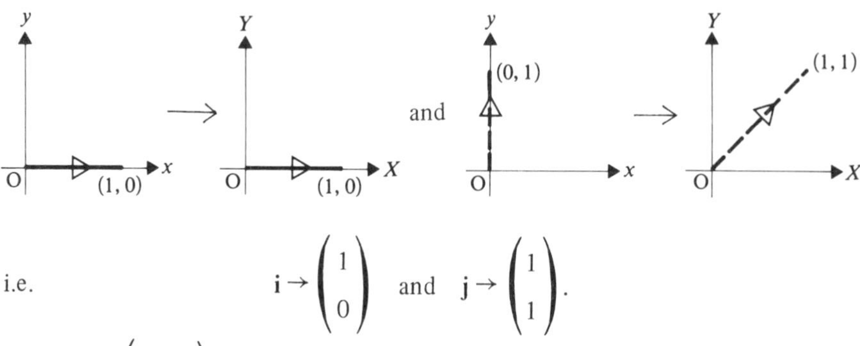

i.e. $$\mathbf{i} \to \begin{pmatrix} 1 \\ 0 \end{pmatrix} \quad \text{and} \quad \mathbf{j} \to \begin{pmatrix} 1 \\ 1 \end{pmatrix}.$$

Therefore $\begin{pmatrix} 1 & 1 \\ 0 & 1 \end{pmatrix}$ is the matrix operator.

EXAMPLES 1c

1) Find the images of the points $(3, 1), (3, 3), (6, 3), (6, 1)$ under the transformation $\begin{pmatrix} 1 & 0 \\ -2 & 1 \end{pmatrix}$.

Illustrate the effect of the transformation and describe it.

Now
$$\begin{pmatrix} 3 \\ 1 \end{pmatrix} \to \begin{pmatrix} 1 & 0 \\ -2 & 1 \end{pmatrix}\begin{pmatrix} 3 \\ 1 \end{pmatrix} = \begin{pmatrix} 3 \\ -5 \end{pmatrix}$$
$$\begin{pmatrix} 3 \\ 3 \end{pmatrix} \to \begin{pmatrix} 1 & 0 \\ -2 & 1 \end{pmatrix}\begin{pmatrix} 3 \\ 3 \end{pmatrix} = \begin{pmatrix} 3 \\ -3 \end{pmatrix}$$
$$\begin{pmatrix} 6 \\ 3 \end{pmatrix} \to \begin{pmatrix} 1 & 0 \\ -2 & 1 \end{pmatrix}\begin{pmatrix} 6 \\ 3 \end{pmatrix} = \begin{pmatrix} 6 \\ -9 \end{pmatrix}$$
$$\begin{pmatrix} 6 \\ 1 \end{pmatrix} \to \begin{pmatrix} 1 & 0 \\ -2 & 1 \end{pmatrix}\begin{pmatrix} 6 \\ 1 \end{pmatrix} = \begin{pmatrix} 6 \\ -11 \end{pmatrix}.$$

In the diagram on the next page, A, B, C, D are the given points and A′, B′, C′, D′ are their images under the transformation.

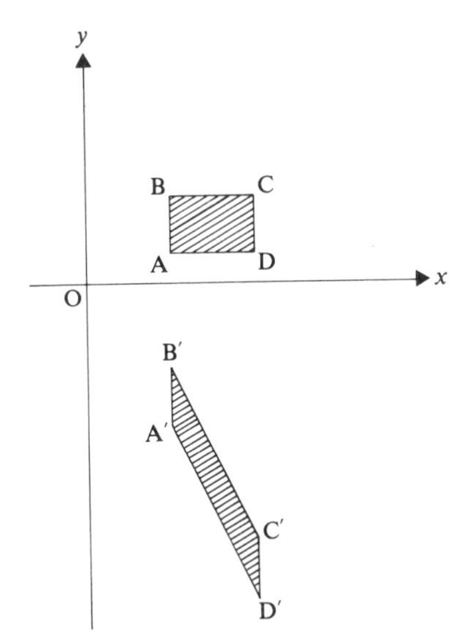

By inspection we see that the transformation is a shear in the direction of the negative y axis.

Alternatively, the effect of a transformation can be described from the matrix operator as follows:

$$\begin{pmatrix} 1 & 0 \\ -2 & 1 \end{pmatrix} \text{ maps } \mathbf{i} \to \begin{pmatrix} 1 \\ -2 \end{pmatrix} \text{ and } \mathbf{j} \to \begin{pmatrix} 0 \\ 1 \end{pmatrix}.$$

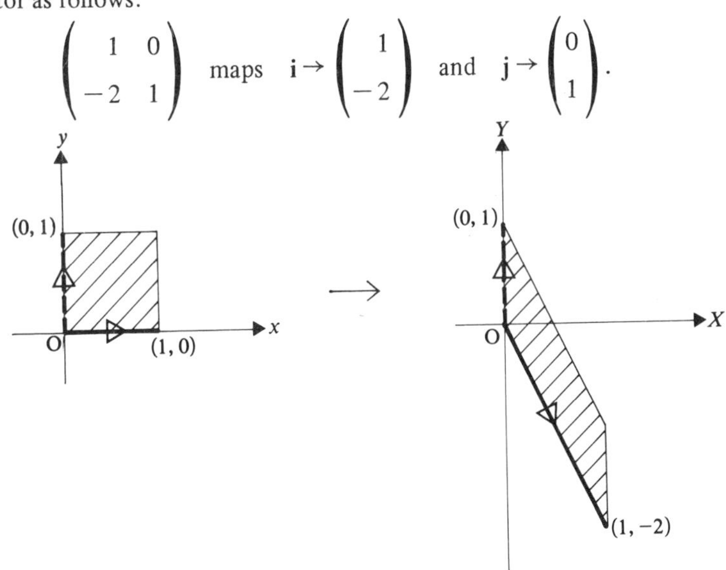

from which we see that the transformation is a shear.

2) Describe the transformation represented by the matrix $\begin{pmatrix} 1 & 1 \\ 1 & 1 \end{pmatrix}$

$$\begin{pmatrix} 1 & 1 \\ 1 & 1 \end{pmatrix} \text{ maps } \mathbf{i} \text{ to } \begin{pmatrix} 1 \\ 1 \end{pmatrix} \text{ and } \mathbf{j} \text{ to } \begin{pmatrix} 1 \\ 1 \end{pmatrix}.$$

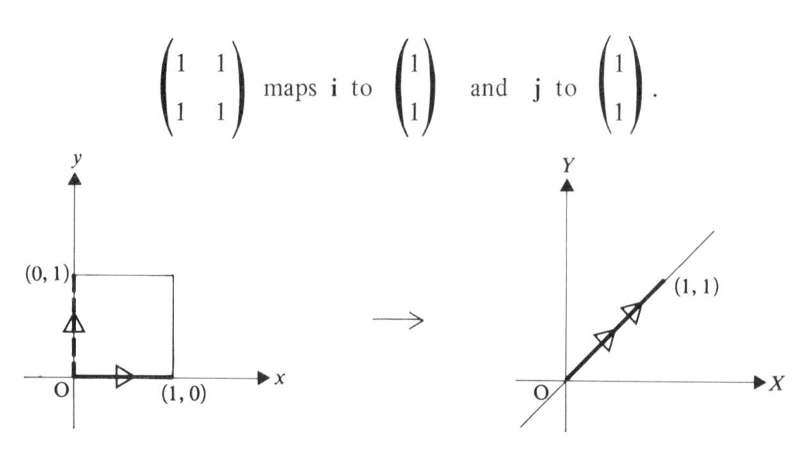

Also, for any point (x, y)

$$\begin{pmatrix} 1 & 1 \\ 1 & 1 \end{pmatrix}\begin{pmatrix} x \\ y \end{pmatrix} = \begin{pmatrix} x+y \\ x+y \end{pmatrix} = \begin{pmatrix} X \\ Y \end{pmatrix}$$

i.e. $\begin{pmatrix} 1 & 1 \\ 1 & 1 \end{pmatrix}$ maps any point to a point on the line $Y = X$.

This transformation is *singular* because it maps the plane to a line. Similarly any linear transformation which does not cover the whole plane is singular.

3) Find the equation of the image of the line $y = 2x - 1$ under the transformation $\begin{pmatrix} 2 & -1 \\ 0 & 2 \end{pmatrix}\begin{pmatrix} x \\ y \end{pmatrix} = \begin{pmatrix} X \\ Y \end{pmatrix}$.

Under this transformation,

$$\begin{pmatrix} x \\ y \end{pmatrix} \rightarrow \begin{pmatrix} 2 & -1 \\ 0 & 2 \end{pmatrix}\begin{pmatrix} x \\ y \end{pmatrix} = \begin{pmatrix} 2x - y \\ 2y \end{pmatrix} = \begin{pmatrix} X \\ Y \end{pmatrix}$$

i.e. a point $P(x, y)$ maps to the point $Q(X, Y)$

where
$$X = 2x - y \qquad [1]$$
$$Y = 2y \qquad [2]$$

But $P(x, y)$ is on the given line $\iff$ $y = 2x - 1$ (or $2x - y = 1$).
So from [1] we see that $X = 1$.
Therefore the image of $y = 2x - 1$ is the line $X = 1$.
In more general problems of this type, the equation of the locus of $Q(X, Y)$ is found by eliminating x and y from [1] and [2] and the equation of the locus of P.

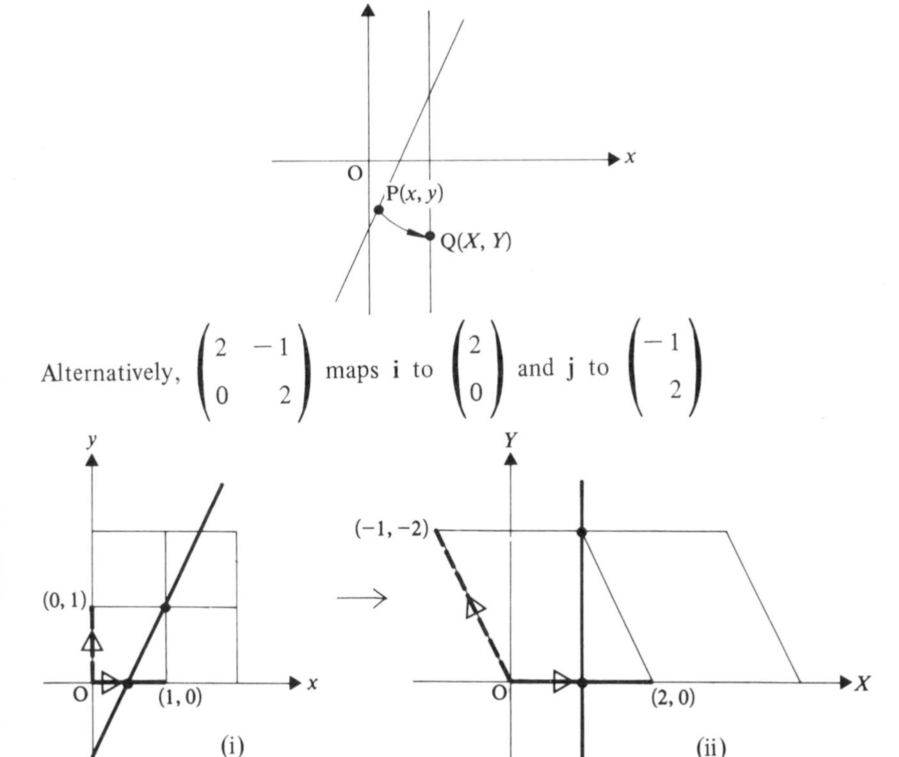

Alternatively, $\begin{pmatrix} 2 & -1 \\ 0 & 2 \end{pmatrix}$ maps **i** to $\begin{pmatrix} 2 \\ 0 \end{pmatrix}$ and **j** to $\begin{pmatrix} -1 \\ 2 \end{pmatrix}$

The line $y = 2x - 1$ in (i) and its image in (ii) are in the same relative positions on the grids, i.e. the line in (ii) passes through $(1, 0)$ and $(1, 2)$ and so its equation is $X = 1$.
Therefore the line $y = 2x - 1$ maps to the line $X = 1$.

4) Find the equations of the lines that are mapped on to themselves under the transformation

$$\begin{pmatrix} 2 & 1 \\ 3 & 0 \end{pmatrix}\begin{pmatrix} x \\ y \end{pmatrix} = \begin{pmatrix} X \\ Y \end{pmatrix}.$$

$\begin{pmatrix} x \\ y \end{pmatrix}$ is mapped to $\begin{pmatrix} X \\ Y \end{pmatrix}$ where

$$\begin{pmatrix} X \\ Y \end{pmatrix} = \begin{pmatrix} 2 & 1 \\ 3 & 0 \end{pmatrix}\begin{pmatrix} x \\ y \end{pmatrix} = \begin{pmatrix} 2x + y \\ 3x \end{pmatrix} \qquad [1]$$

If the line $y = mx + c$ in the xy plane maps to itself, then its equation in the transformed plane is $Y = mX + c$.

From [1] $\quad X = 2x + y$ and $Y = 3x$.

Hence $\quad Y = mX + c \Rightarrow 3x = m(2x+y) + c$

$$\Rightarrow \quad y = -\frac{3-2m}{m}x - \frac{c}{m}$$

So, if $y = mx + c$ maps to $Y = mX + c$ then $y = \frac{3-2m}{m}x - \frac{c}{m}$

Comparing coefficients we see that

$$m = \frac{(3-2m)}{m} \qquad [2] \quad \text{and} \quad c = -\frac{c}{m} \qquad [3]$$

From [2] $\quad m^2 = -2m + 3$

$\Rightarrow \quad m^2 + 2m - 3 = 0 \Rightarrow m = -3$ or 1.

From [3] $\quad c = -\frac{c}{m} \Rightarrow c(m+1) = 0 \Rightarrow c = 0$ or $m = -1$

but $\quad m \neq -1$; therefore $c = 0$.

Hence $\quad m = -3$ or $m = 1$ and $c = 0$.

Therefore the lines which map to themselves are $y = -3x$ and $y = x$.

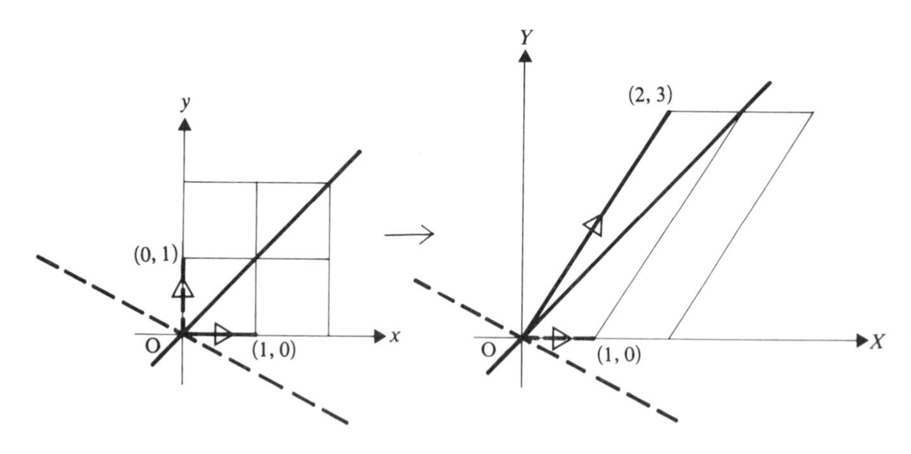

Note that although $y = x$ maps to $Y = X$, a particular point on $y = x$ does *not*, in general, map to the same point on $Y = X$, e.g. $\begin{pmatrix}1\\1\end{pmatrix} \rightarrow \begin{pmatrix}3\\3\end{pmatrix}$.

EXERCISE 1c

1) Find the matrices for the following transformations:
(a) reflection in the y axis,

(b) a rotation about the origin of 45°,
(c) a stretch by a factor of 2 parallel to Oy,
(d) reflection in the line $y = 2x$,
(e) a shear of 30° in the direction of Oy,
(f) a stretch by a factor of 3 parallel to Ox,
(g) an enlargement by a factor of 3,
(h) reflection in the x axis combined with a stretch by a factor of 2 parallel to the x axis.

2) Describe the transformations represented by the following matrices.

(a) $\begin{pmatrix} -1 & 0 \\ 0 & 1 \end{pmatrix}$ (b) $\begin{pmatrix} -1 & 0 \\ 0 & -1 \end{pmatrix}$ (c) $\begin{pmatrix} -1 & 1 \\ -1 & 1 \end{pmatrix}$

(d) $\begin{pmatrix} 1 & 0 \\ 0 & 1 \end{pmatrix}$ (e) $\begin{pmatrix} 2 & 1 \\ 0 & 2 \end{pmatrix}$ (f) $\begin{pmatrix} 1 & 1 \\ 0 & 0 \end{pmatrix}$

(g) $\begin{pmatrix} 0 & 0 \\ 0 & 0 \end{pmatrix}$ (h) $\begin{pmatrix} \sin\theta & -\cos\theta \\ \cos\theta & \sin\theta \end{pmatrix}$ (i) $\begin{pmatrix} 1 & 0 \\ -2 & 1 \end{pmatrix}$.

3) A triangle ABC has its vertices at the points A(1, 1), B(2, 1), C(2, 4). Draw diagrams to represent the image of triangle ABC under the transformations

(a) $\begin{pmatrix} 2 & 0 \\ 0 & 2 \end{pmatrix}$ (b) $\begin{pmatrix} -3 & 0 \\ 0 & -3 \end{pmatrix}$ (c) $\begin{pmatrix} -1 & 0 \\ 0 & 3 \end{pmatrix}$

(d) $\begin{pmatrix} 1 & 0 \\ 1 & 0 \end{pmatrix}$ (e) $\begin{pmatrix} 1 & 2 \\ 3 & 4 \end{pmatrix}$ (f) $\begin{pmatrix} 0 & 2 \\ 0 & -3 \end{pmatrix}$

4) Find the reflections of $\mathbf{i}$ and $\mathbf{j}$ in the line $y = x\tan\theta$.
Hence show that any matrix of the form $\begin{pmatrix} \cos 2\theta & \sin 2\theta \\ \sin 2\theta & -\cos 2\theta \end{pmatrix}$ represents a reflection in the line $y = mx$ where $m = \tan\theta$.
Write down the matrices representing a reflection in
(a) $3y = 4x$, (b) $y = -2x$.

5) Find the images of $\mathbf{i}$ and $\mathbf{j}$ under a rotation about the origin of θ radians.
Hence show that any matrix of the form $\begin{pmatrix} \cos\theta & -\sin\theta \\ \sin\theta & \cos\theta \end{pmatrix}$ represents a rotation about the origin.
Write down the matrices representing a rotation about the origin of

(a) $\frac{\pi}{3}$, (b) $-\frac{\pi}{4}$.

6) The image of the point (x, y) is (X, Y) under the transformation

$$\begin{pmatrix} 0 & -1 \\ -1 & 0 \end{pmatrix}\begin{pmatrix} x \\ y \end{pmatrix} = \begin{pmatrix} X \\ Y \end{pmatrix}.$$

Find the equation of the images under this transformation of the lines

(a) $2y = x$ (b) $x = 2$ (c) $x + y + 2 = 0$

Illustrate your results on a diagram.

7) The point $\begin{pmatrix} x \\ y \end{pmatrix}$ is mapped to the point $\begin{pmatrix} X \\ Y \end{pmatrix}$ under the transformation $\mathbf{M}\begin{pmatrix} x \\ y \end{pmatrix} = \begin{pmatrix} X \\ Y \end{pmatrix}$. Find the equations of the lines which are mapped on to themselves when $\mathbf{M}$ is

(a) $\begin{pmatrix} 0 & 1 \\ 5 & -4 \end{pmatrix}$ (b) $\begin{pmatrix} -2 & 2 \\ -2 & -3 \end{pmatrix}$ (c) $\begin{pmatrix} 0 & 3 \\ 1 & -2 \end{pmatrix}$.

Illustrate your results on diagrams.

MATRICES

In the previous section a form of notation was introduced to deal with the particular problem of linear transformations in the xy plane. This form of notation has far wider applications and may be used for a variety of problems involving linear relationships between any number of variables. In this section the notation is defined in more general terms, and the laws for operations between matrices are introduced; i.e. we develop a matrix algebra.

Definition of a Matrix

A matrix is an ordered rectangular array of numbers, e.g. $\begin{pmatrix} 1 & 3 & 7 \\ 2 & 0 & -1 \end{pmatrix}$.

A particular matrix is denoted by a bold capital letter,

e.g. $\mathbf{A}$ ($\underline{A}$ when handwritten).

The size of a matrix is described by the number of its rows and the number of its columns,

e.g. $\begin{pmatrix} 1 & 3 & 7 \\ 2 & 0 & -1 \end{pmatrix}$ has 2 rows and 3 columns

and is described as a 2 by 3 (or 2×3) matrix.

> In general if a matrix, $\mathbf{A}$, has m rows and n columns then $\mathbf{A}$ is called an $m \times n$ matrix.

A matrix with just one column, e.g. $\begin{pmatrix} 2 \\ -1 \\ 3 \end{pmatrix}$, is called a *column vector* and a matrix with just one row, e.g. $(-1 \quad 0 \quad 3 \quad -4)$, is called a *row vector*. Column vectors and row vectors are denoted by lower case bold letters, e.g.

$$\mathbf{a} = (1 \quad 2 \quad 3), \qquad \mathbf{b} = \begin{pmatrix} 2 \\ 5 \\ 1 \end{pmatrix}.$$

A matrix with an equal number of rows and columns is called a *square matrix*. The numbers in a matrix are called the *elements* of that matrix. A particular element can be identified by using the following notation.
A general 2×3 matrix is written as

$$\mathbf{A} = \begin{pmatrix} a_{11} & a_{12} & a_{13} \\ a_{21} & a_{22} & a_{23} \end{pmatrix}$$

where the suffixes attached to a particular element refer first to its row and second to its column.

In general, the element in the ith row and jth column of a matrix is denoted by a_{ij}.

Matrices of the Same Size

Two matrices, **A** and **B**, are the same size if the number of rows and columns in **A** is the same as the number of rows and columns in **B**.

Hence $\begin{pmatrix} 1 & 4 \\ 2 & 1 \\ 3 & -1 \end{pmatrix}$ and $\begin{pmatrix} 1 & 3 \\ 0 & 1 \\ -1 & 5 \end{pmatrix}$ are the same size

but $\begin{pmatrix} 2 & 1 \\ 3 & 4 \\ 1 & -1 \end{pmatrix}$ and $\begin{pmatrix} 1 & 0 & 1 \\ -2 & 5 & 0 \end{pmatrix}$ are not.

Equal Matrices

Two matrices **A** and **B** are equal if each element of **A** is equal to the corresponding element of **B**;

i.e. $\mathbf{A} = \mathbf{B} \iff a_{ij} = b_{ij}$ for all i and j.

So $\begin{pmatrix} 1 & 3 \\ 2 & 5 \end{pmatrix} = \begin{pmatrix} 1 & 3 \\ 2 & 5 \end{pmatrix}$ but $\begin{pmatrix} 1 & 3 \\ 2 & 5 \end{pmatrix} \neq \begin{pmatrix} 1 & 2 \\ 3 & 5 \end{pmatrix}$ although they are the same size

Addition of Matrices

If two matrices are the same size they can be added by summing the corresponding elements.

e.g. if $\mathbf{A} = \begin{pmatrix} 1 & 3 & 0 & 2 \\ -1 & 4 & 0 & -3 \end{pmatrix}$ and $\mathbf{B} = \begin{pmatrix} 0 & 1 & 4 & 3 \\ 7 & 4 & 9 & 10 \end{pmatrix}$

then

$$\mathbf{A}+\mathbf{B} = \begin{pmatrix} (1+0) & (3+1) & (0+4) & (2+3) \\ (-1+7) & (4+4) & (0+9) & (-3+10) \end{pmatrix} = \begin{pmatrix} 1 & 4 & 4 & 5 \\ 6 & 8 & 9 & 7 \end{pmatrix}.$$

Note that two matrices of equal size are said to be conformable for addition.
Note that if $\mathbf{A}$ and $\mathbf{B}$ are not the same size, then $\mathbf{A}+\mathbf{B}$ has no meaning.
Note also that if $\mathbf{A}+\mathbf{B}$ exists then $\mathbf{A}+\mathbf{B} = \mathbf{B}+\mathbf{A}$;
i.e., addition of matrices is commutative.

Multiplication by a Scalar

The matrix $2\mathbf{A}$ is defined as the matrix each of whose elements is twice the corresponding element of $\mathbf{A}$, e.g.

$$\text{if } \mathbf{A} = \begin{pmatrix} 1 & 3 \\ -2 & 4 \\ 0 & -1 \end{pmatrix} \text{ then } 2\mathbf{A} = \begin{pmatrix} 2\times 1 & 2\times 3 \\ 2\times(-2) & 2\times 4 \\ 2\times 0 & 2\times(-1) \end{pmatrix} = \begin{pmatrix} 2 & 6 \\ -4 & 8 \\ 0 & -2 \end{pmatrix}.$$

In general, if a_{ij} is a typical element of $\mathbf{A}$ then $\lambda\mathbf{A}$ is the matrix whose corresponding element is λa_{ij}.

EXERCISE 1d

Each of the questions below refers to the following matrices:

$\mathbf{A} = \begin{pmatrix} 1 & 6 & 0 \\ 2 & 1 & 5 \end{pmatrix}$ $\quad \mathbf{b} = (0 \quad 1 \quad 3)$ $\quad \mathbf{C} = \begin{pmatrix} -3 & 3 \\ 2 & 5 \\ 7 & 4 \end{pmatrix}$

$\mathbf{D} = \begin{pmatrix} 1 & 0 \\ -4 & 7 \\ 0 & -10 \end{pmatrix}$ $\quad \mathbf{E} = \begin{pmatrix} -3 & 0 & 10 \\ 12 & -9 & -1 \end{pmatrix}$ $\quad \mathbf{F} = \begin{pmatrix} 1 & 3 \\ 2 & 5 \end{pmatrix}$

$\mathbf{G} = \begin{pmatrix} -5 & 9 \\ 0 & -3 \end{pmatrix}$ $\quad \mathbf{h} = (-7 \quad 0 \quad -3).$

1) Describe the size of each of the matrices given above.

2) Write down the pairs of matrices which are conformable for addition.

3) Find, where it exists, $\mathbf{A}+\mathbf{E}$, $\mathbf{b}+\mathbf{h}$, $\mathbf{F}+\mathbf{G}$, $\mathbf{A}+\mathbf{D}$, $\mathbf{E}+\mathbf{F}$, $\mathbf{D}+\mathbf{C}$.

4) Find: (a) $4\mathbf{A}$, (b) $-2\mathbf{E}$, (c) $5\mathbf{G}$, (d) $\lambda\mathbf{b}$.

5) If $a_{ij}, b_{ij}, c_{ij}, \ldots$ are typical elements of $\mathbf{A}, \mathbf{b}, \mathbf{C}, \ldots$ write down the values of

$$a_{22}, b_{13}, c_{31}, d_{12}, e_{23}, f_{21}, g_{11}, h_{12}.$$

6) Find: (a) $2\mathbf{A}+3\mathbf{E}$, (b) $3\mathbf{C}-2\mathbf{D}$, (c) $\mathbf{F}-\mathbf{G}$.

Multiplication of Matrices

When the matrix notation was introduced to deal with linear transformations, the product of the row vector $\begin{pmatrix} a & b \end{pmatrix}$ and the column vector $\begin{pmatrix} x \\ y \end{pmatrix}$ was defined as

$$\begin{pmatrix} a & b \end{pmatrix}\begin{pmatrix} x \\ y \end{pmatrix} = ax + by$$

This can be extended to cover the product of any row vector and column vector *provided that the row vector and the column vector contain the same number of elements*.

e.g. $$\begin{pmatrix} 3 & 4 & -1 \end{pmatrix}\begin{pmatrix} 2 \\ -3 \\ 4 \end{pmatrix} = (3)(2) + (4)(-3) + (-1)(4) = -10$$

In general

$$\begin{pmatrix} a_{11} & a_{12} & a_{13} \end{pmatrix}\begin{pmatrix} b_{11} \\ b_{21} \\ b_{31} \end{pmatrix} = a_{11}b_{11} + a_{12}b_{21} + a_{13}b_{31}$$

Now consider the matrix operation

$$\begin{pmatrix} a & b \\ c & d \end{pmatrix}\begin{pmatrix} x \\ y \end{pmatrix} = \begin{pmatrix} ax + by \\ cx + dy \end{pmatrix}$$

which is equivalent to the two equations

$$\begin{pmatrix} a & b \end{pmatrix}\begin{pmatrix} x \\ y \end{pmatrix} = ax + by$$

and

$$\begin{pmatrix} c & d \end{pmatrix}\begin{pmatrix} x \\ y \end{pmatrix} = cx + dy.$$

This equivalence shows that the product $\begin{pmatrix} a & b \\ c & d \end{pmatrix}\begin{pmatrix} x \\ y \end{pmatrix}$ is defined as the column vector whose top element is $\begin{pmatrix} a & b \end{pmatrix}\begin{pmatrix} x \\ y \end{pmatrix}$ and whose bottom element is $\begin{pmatrix} c & d \end{pmatrix}\begin{pmatrix} x \\ y \end{pmatrix}$.

In general

$$\begin{pmatrix} a_{11} & a_{12} \\ a_{21} & a_{22} \end{pmatrix}\begin{pmatrix} b_{11} \\ b_{21} \end{pmatrix} = \begin{pmatrix} \begin{pmatrix} a_{11} & a_{12} \end{pmatrix}\begin{pmatrix} b_{11} \\ b_{21} \end{pmatrix} \\ \begin{pmatrix} a_{21} & a_{22} \end{pmatrix}\begin{pmatrix} b_{11} \\ b_{21} \end{pmatrix} \end{pmatrix} = \begin{pmatrix} a_{11}b_{11} + a_{12}b_{21} \\ a_{21}b_{11} + a_{22}b_{21} \end{pmatrix}$$

This definition may be extended to cover the product of any matrix and column vector *provided that the number of elements in each row of the matrix is the same as the number of elements in the column vector.*

e.g. $$\begin{pmatrix} 2 & -1 & 0 \\ 1 & 3 & 4 \end{pmatrix}\begin{pmatrix} 3 \\ -5 \\ 9 \end{pmatrix} = \begin{pmatrix} (2)(3)+(-1)(-5)+(0)(9) \\ (1)(3)+(-3)(-5)+(4)(9) \end{pmatrix} = \begin{pmatrix} 11 \\ 24 \end{pmatrix}$$

and

$$\begin{pmatrix} 1 & 3 & -2 & 4 \\ -1 & 0 & 2 & -1 \\ 3 & 1 & 0 & 2 \end{pmatrix}\begin{pmatrix} 2 \\ -1 \\ 0 \\ 2 \end{pmatrix} = \begin{pmatrix} 7 \\ -4 \\ 9 \end{pmatrix}$$

Note that the product $\begin{pmatrix} a \\ b \end{pmatrix}\begin{pmatrix} c & d \\ e & f \end{pmatrix}$ is *not* defined and has no meaning.

EXERCISE 1e

Find the following products:

1) $\begin{pmatrix} 1 & 0 & -3 \end{pmatrix}\begin{pmatrix} 2 \\ -4 \\ 1 \end{pmatrix}$

2) $\begin{pmatrix} -3 & 0 & 4 & 1 \end{pmatrix}\begin{pmatrix} 5 \\ 3 \\ 0 \\ 4 \end{pmatrix}$

3) $\begin{pmatrix} 5 & 7 & -10 \end{pmatrix}\begin{pmatrix} 4 \\ -1 \\ 0 \end{pmatrix}$

4) $\begin{pmatrix} p & q & r \end{pmatrix}\begin{pmatrix} 2p \\ -1 \\ 3r \end{pmatrix}$

5) $\begin{pmatrix} 3 & 2 \\ -1 & -2 \\ 4 & 0 \end{pmatrix}\begin{pmatrix} -3 \\ 0 \end{pmatrix}$

6) $\begin{pmatrix} -5 & 4 & -1 \\ 1 & 0 & -1 \end{pmatrix}\begin{pmatrix} -2 \\ 4 \\ -1 \end{pmatrix}$

7) $\begin{pmatrix} -1 & 3 & -1 \\ 4 & 0 & -2 \\ 1 & 3 & -2 \end{pmatrix}\begin{pmatrix} -3 \\ 5 \\ 3 \end{pmatrix}$

8) $\begin{pmatrix} 2 & -1 \\ -1 & 3 \\ -2 & 0 \\ 4 & -5 \end{pmatrix}\begin{pmatrix} -3 \\ 4 \end{pmatrix}$

9) $\begin{pmatrix} -5 & 0 & -1 \\ 4 & 0 & -2 \\ 0 & -1 & 1 \\ 1 & -2 & 1 \end{pmatrix}\begin{pmatrix} 1 \\ 2 \\ 3 \end{pmatrix}$

10) $\begin{pmatrix} \cos\theta & \sin\theta \\ -\sin\theta & \cos\theta \end{pmatrix}\begin{pmatrix} \cos\theta \\ \sin\theta \end{pmatrix}$

11) $\begin{pmatrix} x & 3x \\ 2y & y \end{pmatrix}\begin{pmatrix} 3x \\ 2y \end{pmatrix}$

12) $\begin{pmatrix} \cos\theta & \sin\theta & 1 \\ \sin\theta & \cos\theta & 2 \end{pmatrix}\begin{pmatrix} \cos\theta \\ \sin\theta \\ -1 \end{pmatrix}$

13) $\begin{pmatrix} t & t^2 \\ 2t & -t^2 \\ -t & 2t^2 \end{pmatrix}\begin{pmatrix} -t \\ 1 \end{pmatrix}$

14) $\begin{pmatrix} i & j \\ 1 & 1 \end{pmatrix}\begin{pmatrix} i \\ j \end{pmatrix}$

The Product of Two Matrices

Consider the line joining the points $A\begin{pmatrix} 2 \\ 1 \end{pmatrix}$ and $B\begin{pmatrix} -1 \\ 2 \end{pmatrix}$ which is transformed by the matrix $\begin{pmatrix} 3 & 4 \\ 5 & 6 \end{pmatrix}$ into the line joining A' and B' where A' and B' are the images of A and B.

The position vectors of A' and B' are given by

$$\begin{pmatrix} 3 & 4 \\ 5 & 6 \end{pmatrix}\begin{pmatrix} 2 \\ 1 \end{pmatrix} = \begin{pmatrix} 10 \\ 16 \end{pmatrix} \qquad [1]$$

and

$$\begin{pmatrix} 3 & 4 \\ 5 & 6 \end{pmatrix}\begin{pmatrix} -1 \\ 2 \end{pmatrix} = \begin{pmatrix} 5 \\ 7 \end{pmatrix} \qquad [2]$$

By adjoining the column vectors in [1] and [2], these two equations can be expressed as the single matrix equation

$$\begin{pmatrix} 3 & 4 \\ 5 & 6 \end{pmatrix}\begin{pmatrix} 2 & -1 \\ 1 & 2 \end{pmatrix} = \begin{pmatrix} 10 & 5 \\ 16 & 7 \end{pmatrix} \qquad [3]$$

As equation [3] is equivalent to equations [1] and [2], this equivalence defines the product

$$\begin{pmatrix} 3 & 4 \\ 5 & 6 \end{pmatrix}\begin{pmatrix} 2 & -1 \\ 1 & 2 \end{pmatrix}$$

as a 2×2 matrix whose first column is $\begin{pmatrix} 3 & 4 \\ 5 & 6 \end{pmatrix}\begin{pmatrix} 2 \\ 1 \end{pmatrix}$ and whose second column is $\begin{pmatrix} 3 & 4 \\ 5 & 6 \end{pmatrix}\begin{pmatrix} -1 \\ 2 \end{pmatrix}$.

So if $\mathbf{A} = \begin{pmatrix} a_{11} & a_{12} \\ a_{21} & a_{22} \end{pmatrix}$ and $\mathbf{B} = \begin{pmatrix} b_{11} & b_{12} \\ b_{21} & b_{22} \end{pmatrix}$

then $\mathbf{AB} = \begin{pmatrix} a_{11} & a_{12} \\ a_{21} & a_{22} \end{pmatrix}\begin{pmatrix} b_{11} & b_{12} \\ b_{21} & b_{22} \end{pmatrix} = \left(\mathbf{A}\begin{pmatrix} b_{11} \\ b_{21} \end{pmatrix} \quad \mathbf{A}\begin{pmatrix} b_{12} \\ b_{22} \end{pmatrix}\right)$

E.g. if $\mathbf{A} = \begin{pmatrix} -1 & 3 \\ -4 & 2 \end{pmatrix}$ and $\mathbf{B} = \begin{pmatrix} 0 & -1 \\ 3 & -2 \end{pmatrix}$

then $\mathbf{AB} = \begin{pmatrix} -1 & 3 \\ -4 & 2 \end{pmatrix}\begin{pmatrix} 0 & -1 \\ 3 & -2 \end{pmatrix} = \left(\mathbf{A}\begin{pmatrix} 0 \\ 3 \end{pmatrix} \quad \mathbf{A}\begin{pmatrix} -1 \\ -2 \end{pmatrix}\right)$

$$= \begin{pmatrix} 9 & -5 \\ 6 & 0 \end{pmatrix}.$$

Now $\mathbf{BA} = \begin{pmatrix} 0 & -1 \\ 3 & -2 \end{pmatrix}\begin{pmatrix} -1 & 3 \\ 4 & 2 \end{pmatrix} = \left(\mathbf{B}\begin{pmatrix} -1 \\ 4 \end{pmatrix} \quad \mathbf{B}\begin{pmatrix} 3 \\ 2 \end{pmatrix}\right)$

$$= \begin{pmatrix} -4 & -2 \\ -11 & 5 \end{pmatrix}$$

i.e. $\mathbf{BA} \neq \mathbf{AB}$

i.e. matrix multiplication is not commutative.

Now consider the more general case of a matrix $\mathbf{M}$ where

$$\mathbf{M} = \begin{pmatrix} a & b \\ c & d \end{pmatrix}$$

which maps $\begin{pmatrix} x_1 \\ y_1 \end{pmatrix}, \begin{pmatrix} x_2 \\ y_2 \end{pmatrix}, \begin{pmatrix} x_3 \\ y_3 \end{pmatrix}$ to $\begin{pmatrix} X_1 \\ Y_1 \end{pmatrix}, \begin{pmatrix} X_2 \\ Y_2 \end{pmatrix}, \begin{pmatrix} X_3 \\ Y_3 \end{pmatrix}$,

i.e. $$\begin{pmatrix} a & b \\ c & d \end{pmatrix}\begin{pmatrix} x_1 \\ y_1 \end{pmatrix} = \begin{pmatrix} X_1 \\ Y_1 \end{pmatrix} \qquad [1]$$

$$\begin{pmatrix} a & b \\ c & d \end{pmatrix}\begin{pmatrix} x_2 \\ y_2 \end{pmatrix} = \begin{pmatrix} X_2 \\ Y_2 \end{pmatrix} \qquad [2]$$

$$\begin{pmatrix} a & b \\ c & d \end{pmatrix}\begin{pmatrix} x_3 \\ y_3 \end{pmatrix} = \begin{pmatrix} X_3 \\ Y_3 \end{pmatrix}. \qquad [3]$$

These three equations can be expressed as a single matrix equation by adjoining the column vectors to give

$$\begin{pmatrix} a & b \\ c & d \end{pmatrix}\begin{pmatrix} x_1 & x_2 & x_3 \\ y_1 & y_2 & y_3 \end{pmatrix} = \begin{pmatrix} X_1 & X_2 & X_3 \\ Y_1 & Y_2 & Y_3 \end{pmatrix}. \qquad [4]$$

This defines the product $\begin{pmatrix} a & b \\ c & d \end{pmatrix}\begin{pmatrix} x_1 & x_2 & x_3 \\ y_1 & y_2 & y_3 \end{pmatrix}$ as a 2×3 matrix whose columns are

$$\begin{pmatrix} a & b \\ c & d \end{pmatrix}\begin{pmatrix} x_1 \\ y_1 \end{pmatrix}, \begin{pmatrix} a & b \\ c & d \end{pmatrix}\begin{pmatrix} x_2 \\ y_2 \end{pmatrix}, \begin{pmatrix} a & b \\ c & d \end{pmatrix}\begin{pmatrix} x_3 \\ y_3 \end{pmatrix}$$

respectively.

e.g. $\begin{pmatrix} -1 & 4 \\ 0 & 2 \end{pmatrix}\begin{pmatrix} 1 & 0 & 2 \\ 3 & 4 & 5 \end{pmatrix}$

$$= \left(\begin{pmatrix} -1 & 4 \\ 0 & 2 \end{pmatrix}\begin{pmatrix} 1 \\ 3 \end{pmatrix} \begin{pmatrix} -1 & 4 \\ 0 & 2 \end{pmatrix}\begin{pmatrix} 6 \\ 4 \end{pmatrix} \begin{pmatrix} -1 & 4 \\ 0 & 2 \end{pmatrix}\begin{pmatrix} 2 \\ 5 \end{pmatrix} \right)$$

$$= \begin{pmatrix} 11 & 16 & 18 \\ 6 & 8 & 10 \end{pmatrix}$$

This definition may now be extended to cover the product of any two matrices **AB**, provided that the number of columns in **A** is the same as the number of rows in **B**.

Hence if $\mathbf{A} = \begin{pmatrix} a_{11} & a_{12} \\ a_{21} & a_{22} \\ a_{31} & a_{32} \end{pmatrix}$ and $\mathbf{B} = \begin{pmatrix} b_{11} & b_{12} & b_{13} & b_{14} \\ b_{21} & b_{22} & b_{23} & b_{24} \end{pmatrix}$

then as **A** has two columns and **B** has two rows we define the product **AB** as

$$\begin{pmatrix} a_{11} & a_{12} \\ \boxed{a_{21} \quad a_{22}} \\ a_{31} & a_{32} \end{pmatrix}\begin{pmatrix} b_{11} & b_{12} & \boxed{\begin{matrix} b_{13} \\ b_{23} \end{matrix}} & b_{14} \\ b_{21} & b_{22} & & b_{24} \end{pmatrix} = \begin{pmatrix} c_{11} & c_{12} & c_{13} & c_{14} \\ c_{21} & c_{22} & \boxed{c_{23}} & c_{24} \\ c_{31} & c_{32} & c_{33} & c_{34} \end{pmatrix}$$

where c_{23} is the product of $\begin{cases}\text{the second row vector of } \mathbf{A} \\ \text{and the third column vector of } \mathbf{B}\end{cases}$

and we call this product 'row–column' multiplication.

Note that under this definition, **BA** has no meaning because **B** has four columns whereas **A** has only three rows.

EXAMPLE 1f

Find **AB** where $\mathbf{A} = \begin{pmatrix} 3 & -1 & 2 \\ 0 & 4 & -1 \end{pmatrix}$ and $\mathbf{B} = \begin{pmatrix} 2 & 0 \\ 4 & 3 \\ 1 & -1 \end{pmatrix}$.

Now **AB** exists because the number of columns in **A** and rows in **B** are the same.

Hence

$$\mathbf{AB} = \begin{pmatrix} 3 & -1 & 2 \\ 0 & 4 & -1 \end{pmatrix}\begin{pmatrix} 2 & 0 \\ 4 & 3 \\ 1 & -1 \end{pmatrix}$$

$$= \begin{pmatrix} (3)(2) + (-1)(4) + (2)(1) & (3)(0) + (-1)(3) + (2)(-1) \\ (0)(2) + (4)(4) + (-1)(1) & (0)(0) + (4)(3) + (-1)(-1) \end{pmatrix}$$

$$= \begin{pmatrix} 4 & -5 \\ 15 & 13 \end{pmatrix}$$

Now **BA** also exists as **B** has two columns and **A** has two rows.

Hence

$$\mathbf{BA} = \begin{pmatrix} 2 & 0 \\ 4 & 3 \\ 1 & -1 \end{pmatrix}\begin{pmatrix} 3 & -1 & 2 \\ 0 & 4 & -1 \end{pmatrix} = \begin{pmatrix} 6 & -2 & 4 \\ 12 & 8 & 5 \\ 3 & -5 & 3 \end{pmatrix}.$$

Note that $\mathbf{AB} \neq \mathbf{BA}$.

Because the order of the product matters, it is ambiguous to refer to 'the product of **A** and **B**' without specifying the order. For the order **AB** we say that **A** *premultiplies* **B** or that **B** *postmultiplies* **A**.

Note also that the product **AB** results in a matrix with the same number of rows as **A** and the same number of columns as **B**. Hence a 2×4 matrix premultiplying a 4×3 matrix results in a 2×3 matrix.

In general if **A** is of size $m \times n$, and **B** is of size $n \times p$ then **AB** is of size $m \times p$.

To summarize:

AB exists if the number of columns in **A** = the number of rows in **B**, i.e. if **A** is an $m \times n$ matrix and **B** is an $n \times t$ matrix **A** and **B** are then said

to be *compatible* for the product **AB**, and **AB** is a $m \times t$ matrix.
If **AB** exists, **BA** does not necessarily exist.
If **AB** exists and **BA** exists, then in general $\mathbf{AB} \neq \mathbf{BA}$.

In those cases where $\mathbf{AB} = \mathbf{BA}$, **A** and **B** are said to commute.

If **AB** exists and is equal to **C**, the element of the ith row and jth column of **C** (i.e. c_{ij}) is the product of the ith row of **A** and the jth column of **B**.

EXERCISE 1f

Express as a single matrix the product **AB**, and, where it exists, **BA**. State also any pair of matrices that commute.

1) $\mathbf{A} = \begin{pmatrix} 2 & -1 \\ 4 & 2 \end{pmatrix}$, $\mathbf{B} = \begin{pmatrix} -1 & 0 \\ 2 & -3 \end{pmatrix}$

2) $\mathbf{A} = \begin{pmatrix} 0 & -5 \\ 6 & -1 \end{pmatrix}$, $\mathbf{B} = \begin{pmatrix} 3 & -1 \\ -4 & 0 \end{pmatrix}$

3) $\mathbf{A} = \begin{pmatrix} -3 & 2 \\ 4 & 1 \end{pmatrix}$, $\mathbf{B} = \begin{pmatrix} 6 & -2 & 7 \\ 0 & 1 & -2 \end{pmatrix}$

4) $\mathbf{A} = \begin{pmatrix} 3 & -7 & 2 \\ -4 & 0 & -1 \\ 1 & -2 & 4 \end{pmatrix}$, $\mathbf{B} = \begin{pmatrix} 2 & 1 \\ -1 & 3 \\ 0 & -2 \end{pmatrix}$

5) $\mathbf{A} = \begin{pmatrix} 3 & -4 \\ 1 & 6 \\ 8 & 10 \end{pmatrix}$, $\mathbf{B} = \begin{pmatrix} 2 & -1 \\ 4 & -2 \end{pmatrix}$

6) $\mathbf{A} = \begin{pmatrix} -1 & 0 & 1 \\ -1 & 1 & 2 \\ 2 & 0 & 1 \end{pmatrix}$, $\mathbf{B} = \begin{pmatrix} 3 & -2 & 1 \\ 5 & 0 & 3 \\ 0 & -1 & 1 \end{pmatrix}$

7) $\mathbf{A} = \begin{pmatrix} 1 & 3 \\ 0 & -1 \end{pmatrix}$, $\mathbf{B} = \begin{pmatrix} 1 & 0 & -1 & 2 \\ 3 & -4 & 2 & 0 \end{pmatrix}$

8) $\mathbf{A} = \begin{pmatrix} 3 & 2 \\ 4 & -1 \\ 0 & 2 \end{pmatrix}$, $\mathbf{B} = \begin{pmatrix} 0 & 1 & 3 \\ -1 & 0 & 2 \end{pmatrix}$

9) $\mathbf{A} = \begin{pmatrix} 2 & -1 & 4 \\ 0 & -2 & 5 \\ 2 & 1 & 3 \end{pmatrix}$, $\mathbf{B} = \begin{pmatrix} 2 & 0 & 1 \\ 1 & 3 & -1 \\ 0 & 1 & 0 \end{pmatrix}$

10) $\mathbf{A} = \begin{pmatrix} 1 & 0 & 0 \\ 0 & 1 & 0 \\ 0 & 0 & 1 \end{pmatrix}$, $\mathbf{B} = \begin{pmatrix} 1 & 2 & 3 \\ 4 & 5 & 6 \\ 7 & 8 & 9 \end{pmatrix}$

11) $\mathbf{A} = \begin{pmatrix} 5 & 2 \\ 10 & 4 \end{pmatrix}$, $\mathbf{B} = \begin{pmatrix} 2 & 4 \\ -5 & -10 \end{pmatrix}$

12) $\mathbf{A} = \begin{pmatrix} 2 \\ 7 \end{pmatrix}$, $\mathbf{B} = (3 \quad 2)$

COMPOUND TRANSFORMATIONS

Consider the transformation, A,

a stretch by a factor of 2 in the direction of the x axis,

followed by the transformation, B,

a rotation of $\dfrac{\pi}{2}$ about the origin.

From the diagrams

for A

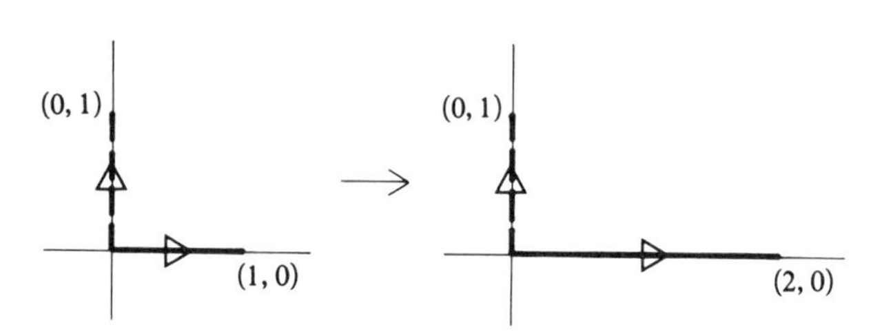

$\mathbf{i}$ maps to $\begin{pmatrix} 2 \\ 0 \end{pmatrix}$ and $\mathbf{j}$ maps to $\begin{pmatrix} 0 \\ 1 \end{pmatrix}$,

i.e. $\mathbf{A} = \begin{pmatrix} 2 & 0 \\ 0 & 1 \end{pmatrix}$ is the matrix operator;

for B

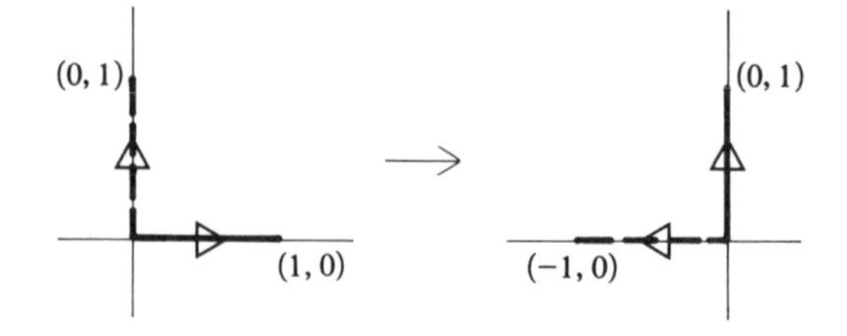

i maps to $\begin{pmatrix} 0 \\ 1 \end{pmatrix}$ and **j** to $\begin{pmatrix} -1 \\ 0 \end{pmatrix}$, i.e. $\mathbf{B} = \begin{pmatrix} 0 & -1 \\ 1 & 0 \end{pmatrix}$ is the matrix operator.

Now transformation A followed by transformation B maps

$$\mathbf{i} \to \begin{pmatrix} 2 \\ 0 \end{pmatrix} \to \begin{pmatrix} 0 & -1 \\ 1 & 0 \end{pmatrix}\begin{pmatrix} 2 \\ 0 \end{pmatrix} = \begin{pmatrix} 0 \\ 2 \end{pmatrix} \qquad [1]$$

and

$$\mathbf{j} \to \begin{pmatrix} 0 \\ 1 \end{pmatrix} \to \begin{pmatrix} 0 & -1 \\ 1 & 0 \end{pmatrix}\begin{pmatrix} 0 \\ 1 \end{pmatrix} = \begin{pmatrix} -1 \\ 0 \end{pmatrix}. \qquad [2]$$

Hence the combined effect of these two transformations is expressed by the operator

$$\mathbf{C} = \begin{pmatrix} 0 & -1 \\ 2 & 0 \end{pmatrix} \qquad [3]$$

But

$$\mathbf{BA} = \begin{pmatrix} 0 & -1 \\ 1 & 0 \end{pmatrix}\begin{pmatrix} 2 & 0 \\ 0 & 1 \end{pmatrix} = \begin{pmatrix} 0 & -1 \\ 2 & 0 \end{pmatrix}.$$

So when transformation B is carried out after the transformation A the compound transformation has as its matrix operator the product **BA** where **B** is the matrix operator for B and **A** is the matrix operator for A.

This result is confirmed by the diagram below

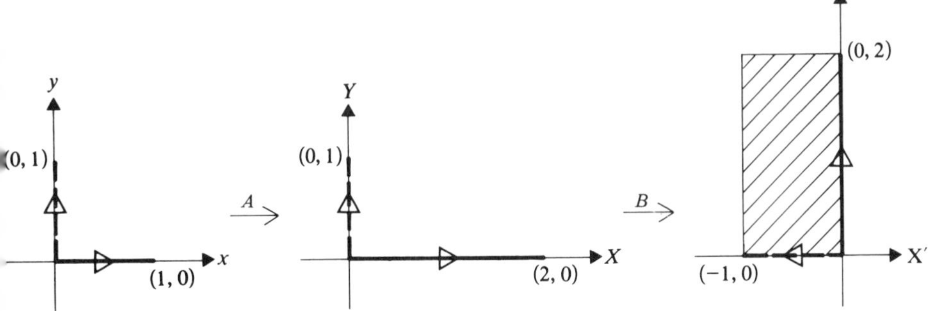

from which we see that
B carried out after **A** maps $\mathbf{i} \to \begin{pmatrix} 0 \\ 2 \end{pmatrix}$ and $\mathbf{j} \to \begin{pmatrix} -1 \\ 0 \end{pmatrix}$

$\Rightarrow$ the matrix operator $\mathbf{C} = \begin{pmatrix} 0 & -1 \\ 2 & 0 \end{pmatrix}$

If we reverse the order of the transformations, i.e. A carried out after B, to give the transformation D,

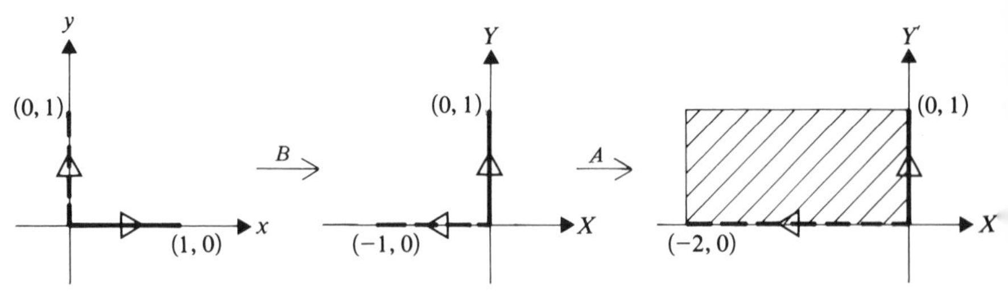

we find that we get a different result, for which the matrix operator is

$$\mathbf{D} = \begin{pmatrix} 0 & -2 \\ 1 & 0 \end{pmatrix} \quad \text{and that} \quad \mathbf{AB} = \begin{pmatrix} 2 & 0 \\ 0 & 1 \end{pmatrix}\begin{pmatrix} 0 & -1 \\ 1 & 0 \end{pmatrix} = \begin{pmatrix} 0 & -2 \\ 1 & 0 \end{pmatrix} = \mathbf{D}.$$

So the order in which the transformations are performed is important and must be stated.

EXAMPLE 1g

Find the matrix operator which transforms the xy plane by a reflection in the line $y = x$ followed by a stretch by a factor of 3 in the direction of Oy.

The matrix operator for a reflection in the line $y = x$ is $\mathbf{A} = \begin{pmatrix} 0 & 1 \\ 1 & 0 \end{pmatrix}$

i.e.

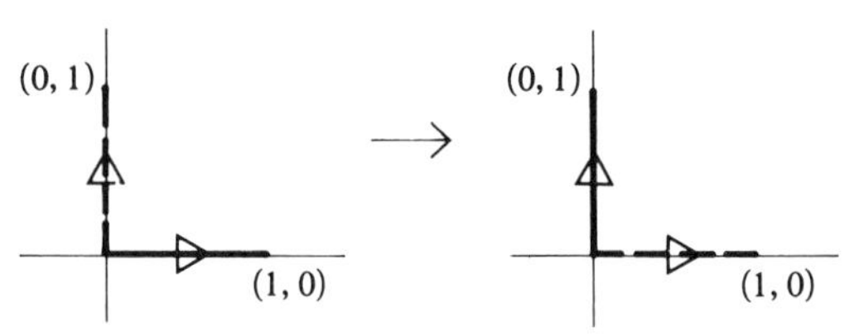

The matrix operator for a stretch by a factor of 3 in the direction Oy

is $\mathbf{B} = \begin{pmatrix} 1 & 0 \\ 0 & 3 \end{pmatrix}$

i.e.

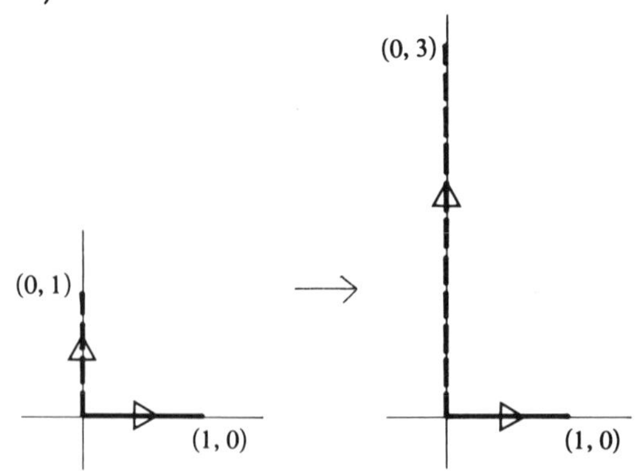

Hence the matrix operator for the combined transformation is

$$\mathbf{BA} = \begin{pmatrix} 1 & 0 \\ 0 & 3 \end{pmatrix}\begin{pmatrix} 0 & 1 \\ 1 & 0 \end{pmatrix} = \begin{pmatrix} 0 & 1 \\ 3 & 0 \end{pmatrix}$$

Note that the *first matrix* is the operator for the *second transformation*.

EXERCISE 1g

The following examples refer to the transformations

A, a reflection in Ox,
B, an enlargement by a factor of 2,
C, a rotation by $\dfrac{\pi}{2}$ about O,
D, a stretch by a factor of 3 in the direction Oy,
E, a reflection in the line $y = x$.

Find the matrix operators for

1) A followed by C
2) A following C
3) D followed by B
4) C carried out after E
5) A carried out after B
6) B followed by C
7) A followed by B followed by C.

The Meaning of $\mathbf{A}^n$ where n is a positive integer

For any matrix **A**, **AA** can exist only if the number of columns of **A** and the number of rows of **A** are the same; i.e. **AA** exists only if **A** is a square matrix. Hence

if $\mathbf{A} = \begin{pmatrix} 1 & 2 \\ 3 & 4 \end{pmatrix}$ then $\mathbf{A}^2 = \mathbf{AA} = \begin{pmatrix} 1 & 2 \\ 3 & 4 \end{pmatrix}\begin{pmatrix} 1 & 2 \\ 3 & 4 \end{pmatrix} = \begin{pmatrix} 7 & 10 \\ 15 & 22 \end{pmatrix}$.

Now $\mathbf{AA}^2 = \begin{pmatrix} 1 & 2 \\ 3 & 4 \end{pmatrix}\begin{pmatrix} 7 & 10 \\ 15 & 22 \end{pmatrix} = \begin{pmatrix} 37 & 54 \\ 81 & 118 \end{pmatrix}$

and $\mathbf{A}^2\mathbf{A} = \begin{pmatrix} 7 & 10 \\ 15 & 22 \end{pmatrix}\begin{pmatrix} 1 & 2 \\ 3 & 4 \end{pmatrix} = \begin{pmatrix} 37 & 54 \\ 81 & 118 \end{pmatrix}$

i.e. $\mathbf{AA}^2 = \mathbf{A}^2\mathbf{A}$.

This property can be shown to be true for any 2×2 matrix and for any 3×3 matrix.

Hence we define $\mathbf{A}^3$ as $\mathbf{AA}^2$ $(= \mathbf{A}^2\mathbf{A})$
and $\mathbf{A}^n$ as $\mathbf{AAA} \ldots \mathbf{A}$.

The Associative Law

We showed above that $\mathbf{A} \times (\mathbf{A} \times \mathbf{A}) = (\mathbf{A} \times \mathbf{A}) \times \mathbf{A}$.
In fact for *any* matrices **A**, **B** and **C** that are compatable for multiplication we find that $\mathbf{A} \times (\mathbf{B} \times \mathbf{C}) = (\mathbf{A} \times \mathbf{B}) \times \mathbf{C}$.

This is known as the associative law and it means that for a product of three matrices, *provided the order is not changed*, we can choose to multiply either the first pair or the second pair as a first step. It also means that we can write the product of three matrices as **ABC** without ambiguity.

For example, if $\mathbf{A} = \begin{pmatrix} 2 & 1 \\ 3 & 2 \end{pmatrix}$, $\mathbf{B} = \begin{pmatrix} 1 & 3 & -1 \\ 0 & 1 & 0 \end{pmatrix}$, $\mathbf{C} = \begin{pmatrix} -1 & 1 & 0 \\ -1 & 2 & 1 \\ 0 & 0 & 1 \end{pmatrix}$

$$\mathbf{A}(\mathbf{BC}) = \begin{pmatrix} 2 & 1 \\ 3 & 2 \end{pmatrix}\left[\begin{pmatrix} 1 & 3 & -1 \\ 0 & 1 & 0 \end{pmatrix}\begin{pmatrix} -1 & 1 & 0 \\ -1 & 2 & 1 \\ 0 & 0 & 1 \end{pmatrix}\right]$$

$$= \begin{pmatrix} 2 & 1 \\ 3 & 2 \end{pmatrix}\begin{pmatrix} -4 & 7 & 2 \\ -1 & 2 & 1 \end{pmatrix} = \begin{pmatrix} -9 & 16 & 5 \\ -14 & 25 & 8 \end{pmatrix}$$

$$(\mathbf{AB})\mathbf{C} = \left[\begin{pmatrix} 2 & 1 \\ 3 & 2 \end{pmatrix}\begin{pmatrix} 1 & 3 & -1 \\ 0 & 1 & 0 \end{pmatrix}\right]\begin{pmatrix} -1 & 1 & 0 \\ -1 & 2 & 1 \\ 0 & 0 & 1 \end{pmatrix}$$

$$= \begin{pmatrix} 2 & 7 & -2 \\ 3 & 11 & -3 \end{pmatrix}\begin{pmatrix} -1 & 1 & 0 \\ -1 & 2 & 1 \\ 0 & 0 & 1 \end{pmatrix} = \begin{pmatrix} -9 & 16 & 5 \\ -14 & 25 & 8 \end{pmatrix}$$

i.e. $\mathbf{A}(\mathbf{BC}) = (\mathbf{AB})\mathbf{C}$

THE UNIT MATRIX

Consider $\begin{pmatrix} 1 & 0 \\ 0 & 1 \end{pmatrix}\begin{pmatrix} a & b \\ c & d \end{pmatrix} = \begin{pmatrix} a & b \\ c & d \end{pmatrix} = \begin{pmatrix} a & b \\ c & d \end{pmatrix}\begin{pmatrix} 1 & 0 \\ 0 & 1 \end{pmatrix}$,

i.e. $\begin{pmatrix} a & b \\ c & d \end{pmatrix}$ is unchanged when premultiplied and when

postmultiplied by $\begin{pmatrix} 1 & 0 \\ 0 & 1 \end{pmatrix}$.

Hence $\begin{pmatrix} 1 & 0 \\ 0 & 1 \end{pmatrix}$ has the same effect in matrix multiplication that unity has in the multiplication of real numbers.

So $\begin{pmatrix} 1 & 0 \\ 0 & 1 \end{pmatrix}$ is called the *unit matrix* of size 2×2 and is denoted by **I**.

Similarly $\begin{pmatrix} 1 & 0 & 0 \\ 0 & 1 & 0 \\ 0 & 0 & 1 \end{pmatrix}$ when premultiplied or postmultiplied by any 3×3 matrix leaves that matrix unchanged, so $\begin{pmatrix} 1 & 0 & 0 \\ 0 & 1 & 0 \\ 0 & 0 & 1 \end{pmatrix}$ is the unit matrix of size 3×3 and is also denoted by **I**.

In fact, a unit matrix of any order is denoted by **I**, the order either being obvious from the context, or the order being stated, e.g. $\mathbf{I}_n$ denoting the $n \times n$ unit matrix.

Now consider **I** as a transformation matrix.

Under $\mathbf{I} = \begin{pmatrix} 1 & 0 \\ 0 & 1 \end{pmatrix}$, $\mathbf{i} \to \mathbf{i}$ and $\mathbf{j} \to \mathbf{j}$,

i.e. the xy plane and its image under **I** are identical, i.e. **I** represents the identity transformation, so **I** is also called an *identity matrix*.

It is interesting to note that if $\mathbf{C} = \begin{pmatrix} 1 & 2 & 3 \\ 4 & 5 & 6 \end{pmatrix}$

then $\mathbf{IC} = \begin{pmatrix} 1 & 0 \\ 0 & 1 \end{pmatrix}\begin{pmatrix} 1 & 2 & 3 \\ 4 & 5 & 6 \end{pmatrix} = \begin{pmatrix} 1 & 2 & 3 \\ 4 & 5 & 6 \end{pmatrix}$

That is, premultiplying any 2 row matrix by **I** leaves that matrix unchanged.

Note that **CI** has no meaning.

Also if $\mathbf{D} = \begin{pmatrix} 1 & 2 \\ 3 & 4 \\ 5 & 6 \end{pmatrix}$, $\mathbf{DI} = \begin{pmatrix} 1 & 2 \\ 3 & 4 \\ 5 & 6 \end{pmatrix}\begin{pmatrix} 1 & 0 \\ 0 & 1 \end{pmatrix} = \begin{pmatrix} 1 & 2 \\ 3 & 4 \\ 5 & 6 \end{pmatrix}$.

That is, postmultiplying any 2 column matrix by **I** leaves that matrix unchanged.

Similarly the unit matrix of order three is a premultiplying identity for any three row matrix and a postmultiplying identity for a three column matrix.

So for an $n \times p$ matrix, $\mathbf{I}_n$ is a premultiplying identity and $\mathbf{I}_p$ is a postmultiplying identity.

The Null Matrix

Any matrix, all of whose elements are zero, is called a *null or zero matrix*, and is denoted by **0**.

0 may be any size, i.e. **0** is not unique.

If $\mathbf{0} = \begin{pmatrix} 0 & 0 \\ 0 & 0 \end{pmatrix}$ then $\begin{pmatrix} 0 & 0 \\ 0 & 0 \end{pmatrix}\begin{pmatrix} 1 & 2 \\ 3 & 4 \end{pmatrix} = \begin{pmatrix} 0 & 0 \\ 0 & 0 \end{pmatrix} = \begin{pmatrix} 1 & 2 \\ 3 & 4 \end{pmatrix}\begin{pmatrix} 0 & 0 \\ 0 & 0 \end{pmatrix}$.

If $\mathbf{0} = \begin{pmatrix} 0 \\ 0 \end{pmatrix}$ then $\begin{pmatrix} 1 & 2 \\ 3 & 4 \end{pmatrix}\begin{pmatrix} 0 \\ 0 \end{pmatrix} = \begin{pmatrix} 0 \\ 0 \end{pmatrix}$.

In fact, if **0** is conformable for premultiplying **A**, $\mathbf{0A} = \mathbf{0}$,
and if **0** is conformable for postmultiplying **A**, $\mathbf{A0} = \mathbf{0}$.

For real numbers a and b, $ab=0 \Rightarrow a=0$ or $b=0$.
However, from Question 11 of Exercise 1e, it is seen that the product of two non-zero matrices can be a zero matrix

i.e. $$\mathbf{AB} = \mathbf{0} \nRightarrow \mathbf{A} = \mathbf{0} \text{ or } \mathbf{B} = \mathbf{0}.$$

Further for real numbers, a, b and c, $ab = ac \Rightarrow$ either $a = 0$ or $b = c$ but this is not true for matrix products as we see from the example below.
For the matrices $\mathbf{A}$, $\mathbf{B}$ and $\mathbf{C}$ as shown,

$$\mathbf{AB} = \begin{pmatrix} 1 & 1 \\ 1 & 1 \end{pmatrix}\begin{pmatrix} 1 & 0 \\ 0 & 1 \end{pmatrix} = \begin{pmatrix} 1 & 1 \\ 1 & 1 \end{pmatrix}$$

$$\mathbf{AC} = \begin{pmatrix} 1 & 1 \\ 1 & 1 \end{pmatrix}\begin{pmatrix} 0 & 1 \\ 1 & 0 \end{pmatrix} = \begin{pmatrix} 1 & 1 \\ 1 & 1 \end{pmatrix},$$

i.e. $\mathbf{AB} = \mathbf{AC}$ but $\mathbf{A} \neq \mathbf{0}$ and $\mathbf{B} \neq \mathbf{C}$,

i.e. $$\mathbf{AB} = \mathbf{AC} \nRightarrow \mathbf{A} = \mathbf{0} \text{ or } \mathbf{B} = \mathbf{C}.$$

EXAMPLES 1h

1) If $\mathbf{M} = \frac{1}{2}\begin{pmatrix} -1 & \sqrt{3} \\ -\sqrt{3} & -1 \end{pmatrix}$ find $\mathbf{M}^3$

and describe the transformation $\mathbf{M}\begin{pmatrix} x \\ y \end{pmatrix} = \begin{pmatrix} X \\ Y \end{pmatrix}$.

$$\mathbf{M}^2 = \tfrac{1}{4}\begin{pmatrix} -1 & \sqrt{3} \\ -\sqrt{3} & -1 \end{pmatrix}\begin{pmatrix} -1 & \sqrt{3} \\ -\sqrt{3} & -1 \end{pmatrix} = \tfrac{1}{4}\begin{pmatrix} -2 & -2\sqrt{3} \\ 2\sqrt{3} & -2 \end{pmatrix}$$

$$= \tfrac{1}{2}\begin{pmatrix} -1 & -\sqrt{3} \\ \sqrt{3} & -1 \end{pmatrix}$$

$$\mathbf{M}^3 = \mathbf{MM}^2 = \tfrac{1}{4}\begin{pmatrix} -1 & \sqrt{3} \\ -\sqrt{3} & -1 \end{pmatrix}\begin{pmatrix} -1 & -\sqrt{3} \\ \sqrt{3} & -1 \end{pmatrix} = \tfrac{1}{4}\begin{pmatrix} 4 & 0 \\ 0 & 4 \end{pmatrix} = \begin{pmatrix} 1 & 0 \\ 0 & 1 \end{pmatrix}$$

Hence $\mathbf{M}^3 = \mathbf{I}$.

Therefore $\mathbf{M}^3$ maps $\begin{pmatrix} x \\ y \end{pmatrix} \to \begin{pmatrix} x \\ y \end{pmatrix}$, i.e. $\mathbf{M}^3$ is the identity transformation.

Hence the transformation represented by $\mathbf{M}$, when repeated three times, maps the xy plane on to itself.
One transformation that would give such a result is a rotation through a third of a revolution about O.

i.e. through $\pm \frac{2\pi}{3}$ radians, as shown below.

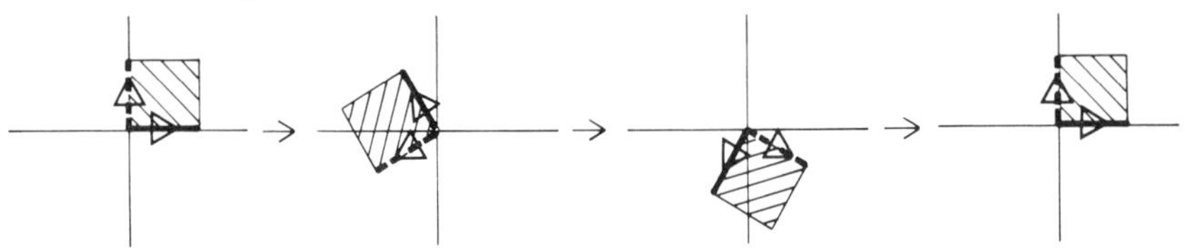

The transformation matrix for a rotation about O of $\pm \frac{2\pi}{3}$ is

$$\begin{pmatrix} \cos\left(\pm\frac{2\pi}{3}\right) & -\sin\left(\pm\frac{2\pi}{3}\right) \\ \sin\left(\pm\frac{2\pi}{3}\right) & \cos\left(\pm\frac{2\pi}{3}\right) \end{pmatrix}$$ and, comparing with **M**, we see that

$$\mathbf{M} = \begin{pmatrix} \cos\left(-\frac{2\pi}{3}\right) & -\sin\left(-\frac{2\pi}{3}\right) \\ \sin\left(-\frac{2\pi}{3}\right) & \cos\left(-\frac{2\pi}{3}\right) \end{pmatrix} = \begin{pmatrix} -\frac{1}{2} & \frac{\sqrt{3}}{2} \\ -\frac{\sqrt{3}}{2} & -\frac{1}{2} \end{pmatrix}$$

That is, $\mathbf{M}\begin{pmatrix} x \\ y \end{pmatrix} = \begin{pmatrix} X \\ Y \end{pmatrix}$ rotates $\begin{pmatrix} x \\ y \end{pmatrix}$ about O by $-\frac{2\pi}{3}$ radians.

EXERCISE 1h

1) If $\mathbf{A} = \begin{pmatrix} 1 & 0 \\ 2 & -1 \end{pmatrix}$ find $\mathbf{A}^2$ and $\mathbf{A}^3$.

2) If $\mathbf{B} = \begin{pmatrix} 1 & 0 & -1 \\ 0 & 1 & 1 \\ -1 & 0 & 0 \end{pmatrix}$ find $\mathbf{B}^2$ and $\mathbf{B}^3$.

3) Find a 2×2 matrix, **B**, such that $\mathbf{B}^2 = \mathbf{I}$.
(*Hint*: Think in terms of a double transformation that maps $\mathbf{i} \to \mathbf{i}$ and $\mathbf{j} \to \mathbf{j}$.)

4) Find a 2×2 matrix, **C**, such that $\mathbf{C}^4 = \mathbf{I}$ and $\mathbf{C}^2 \neq \mathbf{I}$.

5) If $\mathbf{A} = \begin{pmatrix} i & 0 \\ 0 & i \end{pmatrix}$ where $i = \sqrt{(-1)}$, find $\mathbf{A}^2$ and $\mathbf{A}^4$.

6) If $\mathbf{D} = \begin{pmatrix} \cos\theta & 0 & 0 \\ 0 & \sin\theta & 0 \\ 0 & 0 & -\cos\theta \end{pmatrix}$ find $\mathbf{D}^2$.

7) If **A** and **B** are two matrices such that **AB** and **BA** exist, what condition is imposed on the size of **A** and the size of **B**? If, also $\mathbf{AB} = \mathbf{BA}$, what further condition is imposed on the sizes of **A** and **B**?

8) If $\mathbf{M} = \begin{pmatrix} 1 & 0 \\ 1 & 0 \end{pmatrix}$, describe the transformation $\mathbf{M}\begin{pmatrix} x \\ y \end{pmatrix} = \begin{pmatrix} X \\ Y \end{pmatrix}$ and the transformation $\mathbf{M}^2\begin{pmatrix} x \\ y \end{pmatrix} = \begin{pmatrix} X' \\ Y' \end{pmatrix}$.

9) If $\mathbf{A} = \begin{pmatrix} 1 & 1 \\ 2 & 2 \end{pmatrix}$ and $\mathbf{B} = \begin{pmatrix} 1 & -3 \\ -1 & 3 \end{pmatrix}$, describe, with the help of a diagram, the transformations

(a) $\mathbf{AB}\begin{pmatrix} x \\ y \end{pmatrix} = \begin{pmatrix} X \\ Y \end{pmatrix}$, (b) $\mathbf{BA}\begin{pmatrix} x \\ y \end{pmatrix} = \begin{pmatrix} X \\ Y \end{pmatrix}$.

Discuss the statement $\mathbf{AB} = \mathbf{0} \Rightarrow \mathbf{BA} = \mathbf{0}$.

10) If $\mathbf{A} = \begin{pmatrix} 1 & 0 \\ 2 & -1 \end{pmatrix}$ and $\mathbf{B} = \begin{pmatrix} -2 & 1 \\ 3 & 0 \end{pmatrix}$, find

(a) $(\mathbf{A} + \mathbf{B})^2$, (b) $\mathbf{A}^2 + \mathbf{AB} + \mathbf{BA} + \mathbf{B}^2$.

Hence show that $(\mathbf{A} + \mathbf{B})^2 = \mathbf{A}^2 + \mathbf{AB} + \mathbf{BA} + \mathbf{B}^2$.

11) The matrix $\mathbf{B}$ is said to be a real square root of the matrix $\mathbf{A}$ if $\mathbf{A} = \mathbf{B}^2$ and all the elements of $\mathbf{B}$ are real. Find two real and different square roots of $\mathbf{A} = \begin{pmatrix} 9 & 0 \\ 0 & 9 \end{pmatrix}$.

DETERMINANTS

Consider the effect that the transformation matrix $\mathbf{M} = \begin{pmatrix} a & b \\ c & d \end{pmatrix}$ has on the area of a figure.

$$\mathbf{M} \text{ maps } \mathbf{i} \rightarrow \begin{pmatrix} a \\ c \end{pmatrix} \text{ and } \mathbf{j} \rightarrow \begin{pmatrix} b \\ d \end{pmatrix}.$$

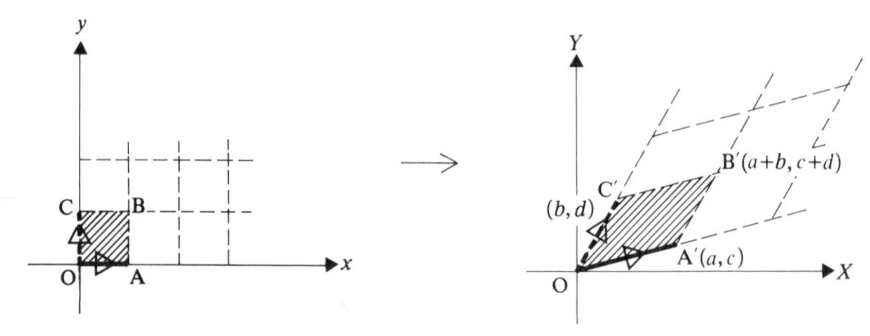

From the diagram, the area of OABC is 1 square unit
and the area of $OA'B'C'$ is $(ad - bc)$ square units.

Hence **M** maps an area of 1 square unit in the xy-plane
to an area of $(ad - bc)$ square units in the transformed plane.

Thus an area A in the xy plane is mapped by $\begin{pmatrix} a & b \\ c & d \end{pmatrix}$ to an area $(ad - bc)A$ in the transformed plane,

i.e. $\begin{pmatrix} a & b \\ c & d \end{pmatrix}$ changes area by a factor $(ad - bc)$.

This factor, $(ad - bc)$, is a value obviously related to the matrix $\begin{pmatrix} a & b \\ c & d \end{pmatrix}$, and it is called the *determinant* of the matrix **M**.
The determinant of the matrix **M** is denoted by $|\mathbf{M}|$, or by $\det \mathbf{M}$, and we write

$$|\mathbf{M}| = \begin{vmatrix} a & b \\ c & d \end{vmatrix} = ad - bc.$$

In general, if $\mathbf{A} = \begin{pmatrix} a_{11} & a_{12} \\ a_{21} & a_{22} \end{pmatrix}$

then $$|\mathbf{A}| = \begin{vmatrix} a_{11} & a_{12} \\ a_{21} & a_{22} \end{vmatrix} = a_{11}a_{22} - a_{12}a_{21}.$$

Note that $|\mathbf{A}|$ is defined for a 2×2 matrix *only*, at this stage, and that the value of $|\mathbf{A}|$ is the product of the elements of **A** on the leading diagonal (top left to lower right) *minus* the product of the elements on the other diagonal.

e.g. $$\begin{vmatrix} \boxed{4} & 2 \\ -1 & \boxed{3} \end{vmatrix} = \boxed{4 \times 3} - 2 \times (-1) = 14$$

and $$\begin{vmatrix} -7 & -4 \\ 3 & 6 \end{vmatrix} = -42 - (-12) = -30$$

EXAMPLES 1i

1) The triangle OAB is mapped by **M** to the triangle $OA'B'$ where A is the point $(0, 2)$, B is the point $(3, 0)$ and $\mathbf{M} = \begin{pmatrix} 3 & 2 \\ 5 & 8 \end{pmatrix}$.

Find the area of triangle $OA'B'$.

From the diagram, the area of triangle OAB = 3 square units.

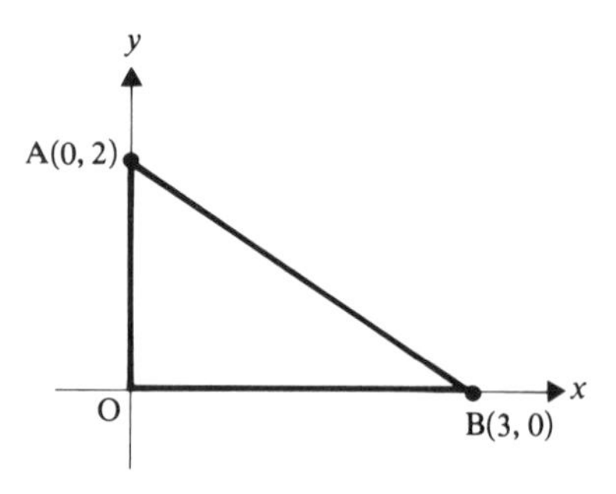

$$|\mathbf{M}| = \begin{vmatrix} 3 & 2 \\ 5 & 8 \end{vmatrix} = 14$$

Therefore the area of triangle OA′B′ is $14 \times 3 = 42$ square units.

2) The square, whose vertices are the points A(0, 3), B(1, 1), C(3, 2), D(2, 4) is mapped to the plane figure A′B′C′D′ by $\mathbf{M} = \begin{pmatrix} 1 & 2 \\ 1 & -2 \end{pmatrix}$

Find the area of A′B′C′D′.

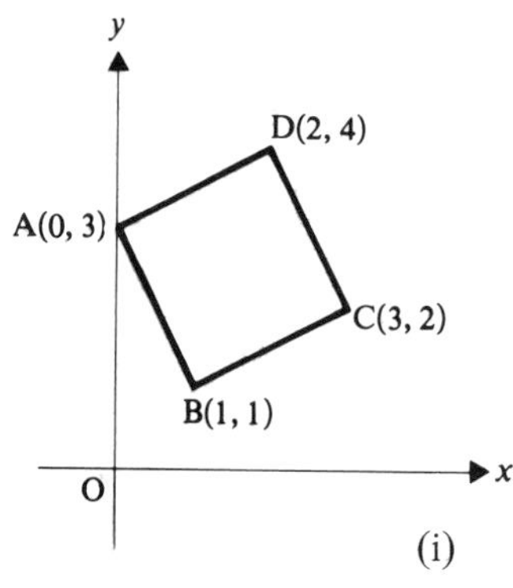

(i)

From the diagram, the area of ABCD is 5 square units. Now **M** changes area by a factor $|\mathbf{M}|$.

$$|\mathbf{M}| = \begin{vmatrix} 1 & 2 \\ 1 & -2 \end{vmatrix} = -4.$$

To interpret the meaning of the negative sign we look at the image figure A′B′C′D′. Under **M**, the images of A, B, C, D are $\begin{pmatrix} 6 \\ -6 \end{pmatrix}$, $\begin{pmatrix} 3 \\ -1 \end{pmatrix}$, $\begin{pmatrix} 7 \\ -1 \end{pmatrix}$, $\begin{pmatrix} 10 \\ -6 \end{pmatrix}$ respectively.

Comparing ABCD, diagram (i), with its image A′B′C′D′, diagram (ii), we see that not only is the shape of ABCD distorted and rotated under **M**, it is also *turned over*; i.e. in (ii) we are looking at an area which is the image of the *reverse side* of the area in (i).

So the negative sign in $|\mathbf{M}| = -4$ indicates that a reflection (among other transformations) is involved in the transformation under **M**.

So **M** changes area by a factor 4.

Hence the area of A′B′C′D′ is $5 \times 4 = 20$ square units.

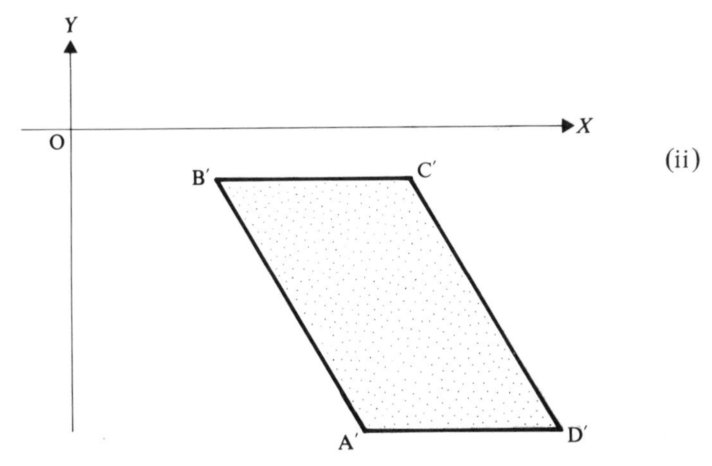

In general, if $|\mathbf{M}|$ is negative, the transformation $\begin{pmatrix} x \\ y \end{pmatrix} \rightarrow \mathbf{M}\begin{pmatrix} x \\ y \end{pmatrix}$ turns an area over.

On page 14, a matrix, **M**, is called singular if the transformation

$\mathbf{M}\begin{pmatrix} x \\ y \end{pmatrix} = \begin{pmatrix} X \\ Y \end{pmatrix}$ maps the xy plane on to a line or on to the origin. Under such a transformation, area is destroyed, i.e. any area in the xy plane maps to zero area in the transformed plane. So a singular matrix **M** changes area by a factor of zero.

i.e. $\qquad \mathbf{M}$ is singular $\Rightarrow$ $|\mathbf{M}| = 0$.

EXERCISE 1i

Evaluate

1) $\begin{vmatrix} 3 & 2 \\ 1 & 5 \end{vmatrix}$, 2) $\begin{vmatrix} -1 & 3 \\ 2 & -1 \end{vmatrix}$, 3) $\begin{vmatrix} 2 & -1 \\ 4 & -2 \end{vmatrix}$, 4) $\begin{vmatrix} 2 & -1 \\ -4 & -1 \end{vmatrix}$

5) If each of the determinants in Questions 1 to 4 is a determinant of a transformation matrix, describe the effect that the transformation has on the square OABC where $\overrightarrow{OA} = \mathbf{i}$ and $\overrightarrow{OC} = \mathbf{j}$. Illustrate your answers with a diagram.

6) If $\mathbf{A} = \begin{pmatrix} 1 & 3 \\ -2 & 1 \end{pmatrix}$ and $\mathbf{B} = \begin{pmatrix} 2 & 4 \\ -1 & 5 \end{pmatrix}$ find

(a) **AB**, (b) $|\mathbf{A}|$, (c) $|\mathbf{B}|$, (d) $|\mathbf{AB}|$.
Is $|\mathbf{AB}|$ equal to $|\mathbf{A}||\mathbf{B}|$?

7) If $\mathbf{A} = \begin{pmatrix} 3 & 0 \\ -1 & 4 \end{pmatrix}$ and $\mathbf{B} = \begin{pmatrix} -2 & 1 \\ 0 & -3 \end{pmatrix}$ find $|\mathbf{AB}|$ and $|\mathbf{BA}|$.

8) If $\mathbf{A}$ and $\mathbf{B}$ are any 2×2 matrices, determine which of the following are true.
(a) $|\mathbf{AB}| = |\mathbf{A}|\,|\mathbf{B}|$. (b) $|\mathbf{AB}| = |\mathbf{BA}|$.
(c) $|\mathbf{AB}| = 0 \quad \Rightarrow \quad |\mathbf{A}| = 0 \quad \text{or} \quad |\mathbf{B}| = 0$.
(d) $|\mathbf{AB}| = |\mathbf{BA}| \quad \Rightarrow \quad \mathbf{AB} = \mathbf{BA}$.

9) For those statements in Question 8 which you decide are true, write down a general proof.

10) For those statements in Question 8 which you decide are false, find examples that prove them to be false.

11) Expand and simplify

(a) $\begin{vmatrix} \cos\theta & \sin\theta \\ -\sin\theta & \cos\theta \end{vmatrix}$, (b) $\begin{vmatrix} x & y \\ x^2 & y^2 \end{vmatrix}$, (c) $\begin{vmatrix} a & b \\ b & a \end{vmatrix}$,

(d) $\begin{vmatrix} a & a+b \\ b & a \end{vmatrix}$, (e) $\begin{vmatrix} \ln 2 & \ln 4 \\ \ln 5 & \ln 6 \end{vmatrix}$, (f) $\begin{vmatrix} \cos 2\theta & \sin\theta \\ -\sin 2\theta & \cos\theta \end{vmatrix}$.

12) Solve for x and y the equations
$$a_1x + b_1y = c_1$$
$$a_2x + b_2y = c_2.$$

Hence show that $x = \dfrac{\begin{vmatrix} c_1 & b_1 \\ c_2 & b_2 \end{vmatrix}}{\begin{vmatrix} a_1 & b_1 \\ a_2 & b_2 \end{vmatrix}}$ and $y = \dfrac{\begin{vmatrix} a_1 & c_1 \\ a_2 & c_2 \end{vmatrix}}{\begin{vmatrix} a_1 & b_1 \\ a_2 & b_2 \end{vmatrix}}$.

13) Use the general result obtained in Question 12 to solve the following equations simultaneously for x and y:
(a) $3x - 2y = 1$, $x + 3y = 2$ (b) $3x + 4y = 3$, $2x - 7y = -1$ (c) $7x - 3y = -2$, $4x + 2y = 3$.

THE DETERMINANT OF A 3×3 MATRIX

Any square matrix has an associated value which is called its determinant and is denoted by $|\mathbf{A}|$, e.g. if $\mathbf{A}$ is a 3×3 matrix where

$$\mathbf{A} = \begin{pmatrix} 1 & 2 & 3 \\ 4 & 5 & 6 \\ 7 & 8 & 9 \end{pmatrix}, \quad \text{then} \quad |\mathbf{A}| = \begin{vmatrix} 1 & 2 & 3 \\ 4 & 5 & 6 \\ 7 & 8 & 9 \end{vmatrix}.$$

The value of $|\mathbf{A}|$ is found, basically, by obtaining 2×2 determinants from the 3×3 determinant.
If the row and column through a particular element are crossed out, this leaves four elements which form a 2×2 determinant,
e.g. if, in $|\mathbf{A}|$, we cross out the row and column through the element 4,

i.e. $\begin{vmatrix} 1 & 2 & 3 \\ \textcircled{4} & 5 & 6 \\ 7 & 8 & 9 \end{vmatrix}$,

this leaves the determinant $\begin{vmatrix} 2 & 3 \\ 8 & 9 \end{vmatrix}$.

This determinant is known as the *minor* of the element 4.
In general, for a 3×3 determinant, Δ, where

$$\Delta = \begin{vmatrix} a_{11} & a_{12} & a_{13} \\ a_{21} & a_{22} & \textcircled{a_{23}} \\ a_{31} & a_{32} & a_{33} \end{vmatrix}, \quad \text{the minor of } a_{23} \text{ is } \begin{vmatrix} a_{11} & a_{12} \\ a_{31} & a_{32} \end{vmatrix}.$$

These minors have associated signs, + or −, depending on the position in the determinant of the element of which they are a minor.
These associated signs are shown in this diagram.

$$\begin{vmatrix} + & - & + \\ - & + & - \\ + & - & + \end{vmatrix}$$

So the associated sign for the minor of a_{23} is −, that for the minor of a_{31} is +, and so on.

The minor of a particular element, together with its associated sign, is called the cofactor of that element.

e.g. if $\Delta = \begin{vmatrix} 1 & 2 & 3 \\ 4 & 5 & 6 \\ 7 & 8 & 9 \end{vmatrix}$, the cofactor of 4 is $-\begin{vmatrix} 2 & 3 \\ 8 & 9 \end{vmatrix}$

and the cofactor of 3 is $+\begin{vmatrix} 4 & 5 \\ 7 & 8 \end{vmatrix}$.

In general, if $\Delta = \begin{vmatrix} a_{11} & a_{12} & a_{13} \\ a_{21} & a_{22} & a_{23} \\ a_{31} & a_{32} & a_{33} \end{vmatrix}$

the cofactor of a_{21}, say, is $-\begin{vmatrix} a_{12} & a_{13} \\ a_{32} & a_{33} \end{vmatrix}$

and it is denoted by A_{21}.
We now define the value of Δ as the sum of the products of the elements of the first row with their respective cofactors.

e.g., if $\Delta = \begin{vmatrix} 1 & 2 & 3 \\ 4 & 5 & 6 \\ 7 & 8 & 9 \end{vmatrix}$

the cofactor of 1 is $+\begin{vmatrix} 5 & 6 \\ 8 & 9 \end{vmatrix}$

the cofactor of 2 is $-\begin{vmatrix} 4 & 6 \\ 7 & 9 \end{vmatrix}$

the cofactor of 3 is $+\begin{vmatrix} 4 & 5 \\ 7 & 8 \end{vmatrix}$.

Hence

$$\begin{aligned}\Delta &= (1)\begin{vmatrix} 5 & 6 \\ 8 & 9 \end{vmatrix} - (2)\begin{vmatrix} 4 & 6 \\ 7 & 9 \end{vmatrix} + (3)\begin{vmatrix} 4 & 5 \\ 7 & 8 \end{vmatrix} \\ &= (1)(-3)-(2)(-6)+(3)(-3) \\ &= -3+12-9 = 0.\end{aligned}$$

In general,

$$\begin{aligned}\Delta &= \begin{vmatrix} a_1 & b_1 & c_1 \\ a_2 & b_2 & c_2 \\ a_3 & b_3 & c_3 \end{vmatrix} \\ &= a_1A_1 + b_1B_1 + c_1C_1 \\ &= a_1\begin{vmatrix} b_2 & c_2 \\ b_3 & c_3 \end{vmatrix} - b_1\begin{vmatrix} a_2 & c_2 \\ a_3 & c_3 \end{vmatrix} + c_1\begin{vmatrix} a_2 & b_2 \\ a_3 & b_3 \end{vmatrix} \\ &= a_1b_2c_3 - a_1c_2b_3 - b_1a_2c_3 + b_1c_2a_3 + c_1a_2b_3 - c_1b_2a_3. \qquad [1]\end{aligned}$$

Clearly the determinant notation $\begin{vmatrix} a_1 & b_1 & c_1 \\ a_2 & b_2 & c_2 \\ a_3 & b_3 & c_3 \end{vmatrix}$ expresses the full expansion [1] much more neatly. So a 3×3 determinant may be considered as a shorthand form for expressions like [1]. If the elements of a determinant are real, the determinant itself has a real value and so is a member of the set of real numbers.

At the beginning of this section, a 3×3 determinant was introduced as the associated value of a 3×3 matrix and an apparantly arbitrary definition was given for its value.

However expansions of this form occur frequently in many branches of mathematics. An example arises in coordinate geometry.

Consider the area of the triangle whose vertices are the points (x_1, y_1), (x_2, y_2), (x_3, y_3).

Using the trapeziums SACT, TCBU and SABU, we find the area of triangle ABC to be

$$\tfrac{1}{2}(x_2y_3 - y_2x_3 - x_1y_3 + x_3y_1 + x_1y_2 - x_2y_1).$$

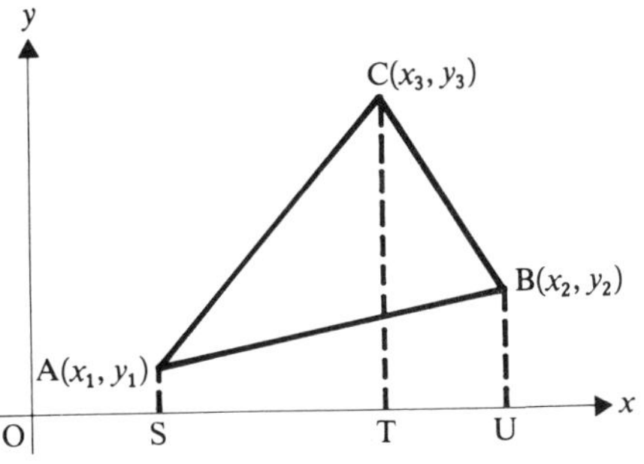

Inserting a factor of one into each term we have

$$\text{Area of } \triangle\text{ABC} = \tfrac{1}{2}(1x_2y_3 - 1y_2x_3 - 1x_1y_3 + 1x_3y_1 + 1x_1y_2 - 1x_2y_1).$$

Comparing with the expansion of the general determinant we see that the expression obtained for the area of $\triangle$ABC can be written $\tfrac{1}{2}\begin{vmatrix} 1 & 1 & 1 \\ x_1 & x_2 & x_3 \\ y_1 & y_2 & y_3 \end{vmatrix}$.

i.e. the area of the triangle whose vertices are the points $(x_1, y_1), (x_2, y_2), (x_3, y_3)$ is

$$\tfrac{1}{2}\begin{vmatrix} 1 & 1 & 1 \\ x_1 & x_2 & x_3 \\ y_1 & y_2 & y_3 \end{vmatrix}.$$

If the area of the triangle is zero, then the three points are collinear. Hence the condition for three points, $(x_1, y_1), (x_2, y_2), (x_3, y_3)$ to be collinear is

$$\begin{vmatrix} 1 & 1 & 1 \\ x_1 & x_2 & x_3 \\ y_1 & y_2 & y_3 \end{vmatrix} = 0.$$

EXERCISE 1j

1) Find the cofactors of each term of the determinants

(a) $\begin{vmatrix} 3 & 2 & -1 \\ 0 & 4 & -6 \\ 2 & -1 & 3 \end{vmatrix}$, (b) $\begin{vmatrix} -1 & 0 & 2 \\ -4 & 1 & -3 \\ 7 & 0 & -2 \end{vmatrix}$.

2) Evaluate

(a) $\begin{vmatrix} -2 & 1 & 4 \\ 3 & -2 & 5 \\ 0 & 1 & 3 \end{vmatrix}$, (b) $\begin{vmatrix} 0 & 1 & 3 \\ 0 & -1 & 4 \\ 2 & 6 & -2 \end{vmatrix}$, (c) $\begin{vmatrix} -2 & 0 & 1 \\ 3 & -4 & 5 \\ -7 & -3 & 2 \end{vmatrix}$.

3) Find the area of the triangle ABC where A, B and C are the points
(a) $(1,3), (2,-4)$ and $(5,7)$, (b) $(-6,2), (-1,-1)$ and $(-3,5)$.

4) Determine which of the following sets of points are collinear.

(a) $(0, \frac{1}{3}), (-1, 0), (5, 2)$, (b) $(0, 1), (1, 0), (1, -1)$,
(c) $(0, -6), (1, -3), (3, 3)$.

5) Expand and simplify the determinants

(a) $\begin{vmatrix} 1 & 1 & 1 \\ \cos\theta & \cos^2\theta & 1 \\ \sin\theta & \sin^2\theta & 1 \end{vmatrix}$, (b) $\begin{vmatrix} a & b & c \\ a^2 & b^2 & c^2 \\ a^3 & b^3 & c^3 \end{vmatrix}$

(c) $\begin{vmatrix} \cos\theta & 1 & 1 \\ \sin\theta & 1 & 1 \\ 1 & \sin\theta & \cos\theta \end{vmatrix}$, (d) $\begin{vmatrix} 1 & n & n^2 \\ n & n^2 & n^3 \\ 1/n & 1/n^2 & 1/n^3 \end{vmatrix}$.

6) Verify, by expansion that $\begin{vmatrix} a_1 & a_2 & a_3 \\ b_1 & b_2 & b_3 \\ c_1 & c_2 & c_3 \end{vmatrix}$

$$= a_1A_1 + a_2A_2 + a_3A_3 = b_1B_1 + b_2B_2 + b_3B_3 = c_1C_1 + c_2C_2 + c_3C_3.$$

SIMPLIFICATION OF DETERMINANTS

If the elements of a determinant are large numbers or complicated algebraic expressions, the evaluation of the determinant can involve some tedious work. There are, however, various properties of determinants which can be used to reduce the complexity of the elements without altering the value of the determinant. These properties are set out below and, in all cases, Δ refers to the

general 3×3 determinant $\begin{vmatrix} a_1 & a_2 & a_3 \\ b_1 & b_2 & b_3 \\ c_1 & c_2 & c_3 \end{vmatrix}$.

(1) The value of a determinant is unaltered when the rows and columns are completely interchanged.

Interchanging the rows and columns of Δ results in the determinant

$$\begin{vmatrix} a_1 & b_1 & c_1 \\ a_2 & b_2 & c_2 \\ a_3 & b_3 & c_3 \end{vmatrix} = \Delta'.$$

(i.e. Δ' is the determinant whose rows are the columns of Δ.)

$$\begin{aligned} \Delta' &= a_1(b_2c_3 - b_3c_2) - b_1(a_2c_3 - a_3c_2) + c_1(a_2b_3 - a_3b_2) \\ &= a_1(b_2c_3 - b_3c_2) - a_2(b_1c_3 - b_3c_1) + a_3(b_2c_1 - b_1c_2) \\ &= \Delta. \end{aligned}$$

This property does not help to reduce the complexity of the elements but it does mean that

any property proved for rows is also valid for columns.

Question 6 in Exercise 1h showed that a determinant can be expanded by using any row and its cofactors. As $\Delta' = \Delta$ it therefore follows that a determinant can also be expanded using any column and its cofactors,

e.g. if $\Delta = \begin{vmatrix} 2 & -1 & 7 \\ 0 & 8 & -2 \\ 0 & -4 & 2 \end{vmatrix}$, we see that column 1 contains two zeros, so it is sensible to use this column and its cofactors to evaluate Δ,

i.e.

$$\Delta = 2\begin{vmatrix} 8 & -2 \\ -4 & 2 \end{vmatrix} + 0 + 0$$
$$= 16.$$

(2) If any row (or column) is added to or subtracted from any other row (or column), the value of Δ is not changed.

If Δ_1 is the determinant formed by adding the first row to the second row of Δ, then

$$\Delta_1 = \begin{vmatrix} a_1 & a_2 & a_3 \\ b_1 + a_1 & b_2 + a_2 & b_3 + a_3 \\ c_1 & c_2 & c_3 \end{vmatrix}$$
$$= a_1[(b_2 + a_2)c_3 - (b_3 + a_3)c_2] - a_2[(b_1 + a_1)c_3 - (b_3 + a_3)c_1] + a_3[(b_1 + a_1)c_2 - (b_2 + a_2)c_1]$$
$$= a_1(b_2c_3 - b_3c_2) - a_2(b_1c_3 - b_3c_1) + a_3(b_1c_2 - b_2c_1)$$
$$= \Delta.$$

This result is a simplified form of the next property which, together with the remaining properties, is stated without proof. Any of the following properties can be proved from first principles, adapting the approach used in (1) and (2).

(3) The value of Δ is unaltered if a multiple of any row (or column) is added to any other row (or column).

For example, if λ times row one is added to row two, then

$$\begin{vmatrix} a_1 & a_2 & a_3 \\ \lambda a_1 + b_1 & \lambda a_2 + b_2 & \lambda a_3 + b_3 \\ c_1 & c_2 & c_3 \end{vmatrix} = \begin{vmatrix} a_1 & a_2 & a_3 \\ b_1 & b_2 & b_3 \\ c_1 & c_2 & c_3 \end{vmatrix}.$$

This is a most useful property for simplifying determinants before expansion, as the following examples illustrate.

(a)

$$\Delta = \begin{vmatrix} 4 & 6 & 2 \\ 3 & 7 & 1 \\ 3 & 5 & 2 \end{vmatrix}$$

Subtracting row 3 from row 1 gives

$$\Delta = \begin{vmatrix} 1 & 1 & 0 \\ 3 & 7 & 1 \\ 3 & 5 & 2 \end{vmatrix}$$

Subtracting column 1 from column 2 gives

$$\Delta = \begin{vmatrix} 1 & 0 & 0 \\ 3 & 4 & 1 \\ 3 & 2 & 2 \end{vmatrix}$$

which may now be evaluated easily as follows:

$$\Delta = 1\begin{vmatrix} 4 & 1 \\ 2 & 2 \end{vmatrix} + 0 + 0 = 6.$$

(b)

$$\Delta = \begin{vmatrix} 10 & 42 & -7 \\ 2 & 10 & 1 \\ -3 & -9 & 4 \end{vmatrix}$$

Subtracting 3 times column 1 from column 2 gives

$$\Delta = \begin{vmatrix} 10 & 12 & -7 \\ 2 & 4 & 1 \\ -3 & 0 & 4 \end{vmatrix}$$

Subtracting 3 times row 2 from row 1 gives

$$\Delta = \begin{vmatrix} 4 & 0 & -10 \\ 2 & 4 & 1 \\ -3 & 0 & 4 \end{vmatrix} = 4(16-30) = -56.$$

Note that the aim in simplifying a determinant is to obtain as many zeros as possible in one row or column. However, it is easy to get carried away; the evaluation of a determinant is usually a small part of any problem and simplification is meant to minimize the risk of arithmetic mistakes. Mistakes are just as likely to occur when combining multiples of rows or columns as they are when expanding the determinant, so in general only straightforward combinations are worth using. For example, in the determinant above, we could have started by adding $4 \times$ (row one) to $3 \times$ (row three) to obtain a zero in the first element of the bottom row, but the mental arithmetic involved is more than likely (in the authors' experience) to give rise to a mistake **Note** also that when solving problems, the simplification of a determinant can be written down directly without explaining the combination of rows or columns used.

(4) If two rows (or columns) are interchanged the determinant changes sign.

e.g.
$$\begin{vmatrix} a_1 & a_2 & a_3 \\ b_1 & b_2 & b_3 \\ c_1 & c_2 & c_3 \end{vmatrix} = -\begin{vmatrix} b_1 & b_2 & b_3 \\ a_1 & a_2 & a_3 \\ c_1 & c_2 & c_3 \end{vmatrix}$$

(5) A determinant, Δ, may be expressed as the sum or difference of two determinants Δ_1 and Δ_2 where two columns (or rows) of Δ, Δ_1 and Δ_2 are identical and the elements in the remaining column of Δ are the sum (or difference) of the corresponding elements in Δ_1 and Δ_2.

e.g.
$$\begin{vmatrix} a_1 & b_1 & c_1 + d_1 \\ a_2 & b_2 & c_2 + d_2 \\ a_3 & b_3 & c_3 + d_3 \end{vmatrix} = \begin{vmatrix} a_1 & b_1 & c_1 \\ a_2 & b_2 & c_2 \\ a_3 & b_3 & c_3 \end{vmatrix} + \begin{vmatrix} a_1 & b_1 & d_1 \\ a_2 & b_2 & d_2 \\ a_3 & b_3 & d_3 \end{vmatrix}$$

FACTORIZATION OF DETERMINANTS

There are some further properties of determinants that are useful for factorization, particularly when the elements are algebraic expressions.

(6) If one row (or column) of Δ is multiplied by λ, the resulting determinant is equal to $\lambda\Delta$.

e.g.
$$\begin{vmatrix} a_1 & a_2 & a_3 \\ \lambda b_1 & \lambda b_2 & \lambda b_3 \\ c_1 & c_2 & c_3 \end{vmatrix} = \lambda \begin{vmatrix} a_1 & a_2 & a_3 \\ b_1 & b_2 & b_3 \\ c_1 & c_2 & c_3 \end{vmatrix}$$

It follows that if all three rows are multiplied by λ, the determinant is multiplied by a factor λ^3,

i.e.
$$\begin{vmatrix} \lambda a_1 & \lambda a_2 & \lambda a_3 \\ \lambda b_1 & \lambda b_2 & \lambda b_3 \\ \lambda c_1 & \lambda c_2 & \lambda c_3 \end{vmatrix} = \lambda^3 \begin{vmatrix} a_1 & a_2 & a_3 \\ b_1 & b_2 & b_3 \\ c_1 & c_2 & c_3 \end{vmatrix}$$

It also follows that a common factor of the elements of one row (or column) is a factor of the determinant.

For example,
$$\Delta = \begin{vmatrix} x^3 & x^2y & xy^2 \\ x^2 & xy^2 & y^3 \\ x & y^3 & 1 \end{vmatrix}$$

has common factors x in the first column and y in the second column.

Therefore
$$\Delta = xy \begin{vmatrix} x^2 & x^2 & xy^2 \\ x & xy & y^3 \\ 1 & y^2 & 1 \end{vmatrix}$$

Also the top row has a common factor x

so
$$\Delta = x^2y \begin{vmatrix} x & x & y^2 \\ x & xy & y^3 \\ 1 & y^2 & 1 \end{vmatrix}$$

(7) If all the elements in one row (or column) of a determinant are zero, the determinant is zero.
Also, if two rows (or columns) of a determinant are identical, it follows from property (2) that a row (or column) of zeros can be obtained.

Hence if two rows (or columns) of a determinant are identical then $\Delta = 0$.
This property, in conjunction with the factor theorem, is very useful for factorizing determinants.

e.g. if, $f(x) \equiv \begin{vmatrix} x & a & b \\ x^2 & a^2 & b^2 \\ x^3 & a^3 & b^3 \end{vmatrix}$,

then $f(a) = \begin{vmatrix} a & a & b \\ a^2 & a^2 & b^2 \\ a^3 & a^3 & b^3 \end{vmatrix} = 0$ as two columns are identical.

So $x - a$ is a factor of $f(x)$.
Similarly $f(b) = 0 \Rightarrow (x - b)$ is a factor of $f(x)$.
The following examples illustrate how these properties can be used to evaluate

or factorize a determinant. It should be noted that the first step in the simplification of a determinant should be the removal of any common factors.

EXAMPLES 1k

1) Evaluate $\begin{vmatrix} -7 & 14 & 7 \\ 2 & -8 & 6 \\ 9 & -3 & 12 \end{vmatrix}$.

Removing common factors from the three rows gives

$$\begin{vmatrix} -7 & 14 & 7 \\ 2 & -8 & 6 \\ 9 & -3 & 12 \end{vmatrix} = (7)(2)(3)\begin{vmatrix} -1 & 2 & 1 \\ 1 & -4 & 3 \\ 3 & -1 & 4 \end{vmatrix}.$$

Adding row 2 to row 1 gives

$$42\begin{vmatrix} 0 & -2 & 4 \\ 1 & -4 & 3 \\ 3 & -1 & 4 \end{vmatrix} = 42\left\{2\begin{vmatrix} 1 & 3 \\ 3 & 4 \end{vmatrix} + 4\begin{vmatrix} 1 & -4 \\ 3 & -1 \end{vmatrix}\right\}$$

$$= 1428.$$

2) Factorize $\begin{vmatrix} x & 1 & 2 \\ x^2 & 1 & 4 \\ x^3 & 1 & 8 \end{vmatrix}$.

Removing common factors from the first and last columns gives

$$\begin{vmatrix} x & 1 & 2 \\ x^2 & 1 & 4 \\ x^3 & 1 & 8 \end{vmatrix} = 2x\begin{vmatrix} 1 & 1 & 1 \\ x & 1 & 2 \\ x^2 & 1 & 4 \end{vmatrix}.$$

If $f(x) \equiv \begin{vmatrix} 1 & 1 & 1 \\ x & 1 & 2 \\ x^2 & 1 & 4 \end{vmatrix}$ then $f(1) = \begin{vmatrix} 1 & 1 & 1 \\ 1 & 1 & 2 \\ 1 & 1 & 4 \end{vmatrix} = 0$ (two columns identical).

So $(x - 1)$ is a factor.

Also $f(2) = \begin{vmatrix} 1 & 1 & 1 \\ 2 & 1 & 2 \\ 4 & 1 & 4 \end{vmatrix} = 0$. So $(x-2)$ is a factor.

By inspection (i.e. without expanding) it is clear that the given determinant is a polynomial of degree 3. As we have found three linear factors, viz., $2x, (x - 1), (x - 2)$, the only other possible factor is a constant, K say.

Then
$$\begin{vmatrix} x & 1 & 2 \\ x^2 & 1 & 4 \\ x^3 & 1 & 8 \end{vmatrix} \equiv 2Kx(x-1)(x-2).$$

K can be evaluated by comparing the coefficients of a particular power of x. For instance, in the expansion of the determinant the coefficient of x^3 is 2 while the corresponding coefficient in the factorized form is $2K$,

i.e.

$$2 = 2K \Rightarrow K = 1 \quad \text{and} \quad \begin{vmatrix} x & 1 & 2 \\ x^2 & 1 & 4 \\ x^3 & 1 & 8 \end{vmatrix} \equiv 2x(x-1)(x-2).$$

Alternatively we may proceed as follows:

$$f(x) \equiv \begin{vmatrix} x & 1 & 2 \\ x^2 & 1 & 4 \\ x^3 & 1 & 8 \end{vmatrix} \equiv 2x \begin{vmatrix} 1 & 1 & 1 \\ x & 1 & 2 \\ x^2 & 1 & 4 \end{vmatrix}.$$

$$f(1) = 2 \begin{vmatrix} 1 & 1 & 1 \\ 1 & 1 & 2 \\ 1 & 1 & 4 \end{vmatrix} \Rightarrow (x-1) \text{ is a factor.}$$

Subtracting column 2 from column 1 gives

$$f(x) \equiv 2x \begin{vmatrix} 0 & 1 & 1 \\ x-1 & 1 & 2 \\ x^2-1 & 1 & 4 \end{vmatrix}$$

(As $x-1$ is a known factor we have looked for a combination of rows or columns to give $x-1$ as a common factor of a row or column.)
Removing $(x-1)$ from column 1 gives

$$f(x) \equiv 2x(x-1) \begin{vmatrix} 0 & 1 & 1 \\ 1 & 1 & 2 \\ x+1 & 1 & 4 \end{vmatrix} \equiv 2x(x-1) \begin{vmatrix} 0 & 1 & 0 \\ 1 & 1 & 1 \\ x+1 & 1 & 3 \end{vmatrix}$$

The remaining determinant is now easily expanded to give

$$f(x) \equiv 2x(x-1)\left[(-1)\begin{vmatrix} 1 & 1 \\ x+1 & 3 \end{vmatrix}\right] = 2x(x-1)(x-2)$$

This method has the advantage that all the factors (including K) are found directly. It has the disadvantage that it is not always easy to see the combination or rows (or columns) that will produce a common factor.
Note that it is also possible to expand Δ without any preliminary work and to factorize the result by using the factor theorem.

EXERCISE 1k

Evaluate the following determinants.

1) $\begin{vmatrix} 2 & -7 & 12 \\ 9 & -3 & 21 \\ 2 & 4 & 6 \end{vmatrix}$ 2) $\begin{vmatrix} 1 & 8 & -10 \\ 2 & 4 & 15 \\ 1 & 12 & 5 \end{vmatrix}$

3) $\begin{vmatrix} 150 & 200 & -100 \\ 80 & -90 & 50 \\ 70 & 10 & -20 \end{vmatrix}$ 4) $\begin{vmatrix} -5 & 15 & 7 \\ 6 & 9 & 2 \\ -3 & 8 & -5 \end{vmatrix}$

Factorize the following determinants.

5) $\begin{vmatrix} x & x^2 & 1 \\ x^2 & x & 1 \\ x^3 & x^3 & 1 \end{vmatrix}$ 6) $\begin{vmatrix} x-1 & 1 & x+1 \\ -1 & 1 & 1 \\ x+1 & 1 & x-1 \end{vmatrix}$

7) $\begin{vmatrix} \sin\theta & \cos\theta & 1 \\ \sin^2\theta & \cos^2\theta & 1 \\ \sin^3\theta & \cos^3\theta & 1 \end{vmatrix}$ 8) $\begin{vmatrix} 1 & a & a+1 \\ a+1 & 1 & a \\ a & a+1 & 1 \end{vmatrix}$

9) $\begin{vmatrix} 1 & 1 & 1 \\ x^2+4 & x^2+9 & x^2+16 \\ 2 & 3 & 4 \end{vmatrix}$

10) Solve the equation $\begin{vmatrix} 1 & 1 & 1 \\ x & x+1 & x-1 \\ x-1 & 2x & x+1 \end{vmatrix} = 0.$

SUMMARY

If $\mathbf{M} = \begin{pmatrix} a & b \\ c & d \end{pmatrix}$ and $\mathbf{M}$ maps (x, y) to (X, Y)

then $\mathbf{M}$ maps $\mathbf{i}$ to $\begin{pmatrix} a \\ c \end{pmatrix}$ and $\mathbf{j}$ to $\begin{pmatrix} b \\ d \end{pmatrix}$.

$\mathbf{M}$ alters area by a factor $|\mathbf{M}| = ad - bc$.

$\mathbf{M} = \begin{pmatrix} \cos\theta & -\sin\theta \\ \sin\theta & \cos\theta \end{pmatrix}$ represents a rotation about the origin through an angle θ.

$\mathbf{M} = \begin{pmatrix} \cos 2\theta & \sin 2\theta \\ \sin 2\theta & -\cos 2\theta \end{pmatrix}$ represents a reflection in the line $y = x\tan\theta$.

If $|\mathbf{M}| = 0$, $\mathbf{M}$ is singular and maps two dimensional space to a line or a point.

$\mathbf{A} = \mathbf{B} \iff a_{ij} = b_{ij}$ for all i and j.

$\mathbf{A} = \mathbf{0} \iff a_{ij} = 0$ for all i and j.

$\mathbf{A}$ and $\mathbf{B}$ are the same size if both $\mathbf{A}$ and $\mathbf{B}$ are $n \times m$ matrices.

$\mathbf{A} + \mathbf{B}$ exists if $\mathbf{A}$ and $\mathbf{B}$ are the same size, when $\mathbf{A} + \mathbf{B}$ is given by adding the corresponding elements of $\mathbf{A}$ and $\mathbf{B}$.

$\mathbf{A} + \mathbf{B} = \mathbf{B} + \mathbf{A}$.

$$\lambda\mathbf{A} = \begin{pmatrix} \lambda a_{11} & \dots & \\ & \dots & \lambda a_{ij} \end{pmatrix}$$

$\mathbf{AB}$ is defined if $\mathbf{A}$ is $m \times n$ and $\mathbf{B}$ is $n \times p$ and then $\mathbf{AB}$ is $m \times p$.

If $\mathbf{AB} = \mathbf{C}$, $\quad c_{ij} = (i\text{th row of } \mathbf{A}) \begin{pmatrix} j\text{th} \\ \text{column} \\ \text{of } \mathbf{B} \end{pmatrix}$.

In general $\mathbf{AB} \neq \mathbf{BA}$

$$\mathbf{A} \times (\mathbf{B} \times \mathbf{C}) = (\mathbf{A} \times \mathbf{B}) \times \mathbf{C}.$$

$\mathbf{A}^n$ exists if $\mathbf{A}$ is square, when $\mathbf{A}^n = \mathbf{A} \times \mathbf{A} \times \dots \mathbf{A}$.

The unit, or identity, matrix $\mathbf{I}$ is square with unit elements in the leading diagonal and zeros elsewhere.

$$\Delta = \begin{vmatrix} a_1 & a_2 \\ b_1 & b_2 \end{vmatrix} = a_1 b_2 - a_2 b_1$$

$$\Delta = \begin{vmatrix} a_1 & a_2 & a_3 \\ b_1 & b_2 & b_3 \\ c_1 & c_2 & c_3 \end{vmatrix} = a_1 A_1 + a_2 A_2 + a_3 A_3$$

where A_1, A_2, A_3 are the cofactors of a_1, a_2, a_3

and $A_1 = +\begin{vmatrix} b_2 & b_3 \\ c_2 & c_3 \end{vmatrix}$, $A_2 = -\begin{vmatrix} b_1 & b_3 \\ c_1 & c_3 \end{vmatrix}$, $A_3 = +\begin{vmatrix} b_1 & b_2 \\ c_1 & c_2 \end{vmatrix}$

$$\Delta' = \begin{vmatrix} a_1 & b_1 & c_1 \\ a_2 & b_2 & c_2 \\ a_3 & b_3 & c_3 \end{vmatrix}$$

If a multiple of any row (column) is added to any other row (column), the value of Δ is unaltered.

A common factor of any row (column) is a factor of the determinant.

MULTIPLE CHOICE EXERCISE 1

(Instructions for answering these questions are given on page xii.)

TYPE I

1) $(2 \quad 4 \quad 6)\begin{pmatrix} 3 \\ 2 \\ -1 \end{pmatrix} =$

(a) 20 (b) -8 (c) 8 (d) $\begin{pmatrix} 6 \\ 8 \\ -6 \end{pmatrix}$ (e) none of these.

2) Under the transformation $\begin{pmatrix} 1 & 2 \\ 1 & 2 \end{pmatrix}\begin{pmatrix} x \\ y \end{pmatrix} = \begin{pmatrix} X \\ Y \end{pmatrix}$, the area of a unit square is mapped to an area
(a) twice the size,
(b) the same size but which has been reflected in a line through O,
(c) which is destroyed, (d) none of these.

3) $\begin{pmatrix} 3 & 7 \\ -1 & 4 \end{pmatrix}\begin{pmatrix} 2 & 7 & -3 \\ 1 & 0 & 1 \end{pmatrix} =$

(a) $\begin{pmatrix} 13 & 21 & -2 \\ 2 & -7 & 7 \end{pmatrix}$ (b) $\begin{pmatrix} 15 & 2 \\ 21 & -7 \\ -2 & 7 \end{pmatrix}$

(c) $\begin{pmatrix} 1 & 0 \\ 0 & 1 \end{pmatrix}$ (d) $\begin{pmatrix} -1 & 21 & -16 \\ -6 & -7 & -1 \end{pmatrix}$.

4) $(3 \quad 1 \quad 2) + (2 \quad 1 \quad 3) =$

(a) $\begin{pmatrix} 3 & 1 & 2 \\ 2 & 1 & 3 \end{pmatrix}$ (b) $2(3 \quad 1 \quad 2)$ (c) 12 (d) $(5 \quad 2 \quad 5)$
(e) has no meaning.

5) $\begin{vmatrix} 1 & 0 & 1 \\ 0 & 1 & 0 \\ 1 & 0 & 1 \end{vmatrix} =$

(a) 0 (b) 1 (c) 2 (d) -1 (e) 5.

6) If $\mathbf{A} = \begin{pmatrix} 1 & 0 \\ -1 & 1 \end{pmatrix}$, $\mathbf{b} = \begin{pmatrix} 1 \\ 2 \end{pmatrix}$, then $\mathbf{bA} =$:

(a) $\begin{pmatrix} 1 \\ -1 \end{pmatrix}$ (b) $\begin{pmatrix} 0 \\ 0 \end{pmatrix}$ (c) 2 (d) $(1 \quad 1)$ (e) has no meaning.

7) The transformation represented by $\mathbf{M} = \begin{pmatrix} 1 & -1 \\ 1 & 1 \end{pmatrix}$ is:

(a) a rotation of $\frac{\pi}{4}$ about O, (b) a reflection in the y axis,

(c) a rotation of $\frac{\pi}{4}$ together with an enlargement by a factor $\sqrt{2}$,

(d) a shear parallel to Ox, (e) none of these.

8) If $\mathbf{A} = \begin{pmatrix} 1 & 2 \\ 3 & 4 \end{pmatrix}$, then $\mathbf{A}^2 =$

(a) $\begin{pmatrix} 1 & 4 \\ 9 & 16 \end{pmatrix}$ (b) $\begin{pmatrix} 5 & 5 \\ 25 & 25 \end{pmatrix}$ (c) $\begin{pmatrix} 7 & 10 \\ 15 & 22 \end{pmatrix}$ (d) $\begin{pmatrix} 7 & 10 \\ 3 & 5 \end{pmatrix}$

(e) $\mathbf{A}^2$ has no meaning.

9) The cofactor of the element 6 in $\begin{vmatrix} 1 & 2 & 3 \\ 4 & 5 & 6 \\ 7 & 8 & 9 \end{vmatrix}$ is:

(a) $\begin{vmatrix} 1 & 2 \\ 7 & 8 \end{vmatrix}$ (b) $\begin{vmatrix} 7 & 8 \\ -1 & -2 \end{vmatrix}$ (c) $\begin{vmatrix} -1 & -2 \\ 7 & 8 \end{vmatrix}$ (d) $\begin{vmatrix} -1 & -2 \\ -7 & -8 \end{vmatrix}$.

10) $\begin{vmatrix} 2 & -1 & 4 \\ -2 & 1 & 4 \\ 2 & 1 & 0 \end{vmatrix} =$

(a) $\begin{vmatrix} -3 & -1 & 4 \\ -3 & 1 & 3 \\ 3 & 1 & 0 \end{vmatrix}$ (b) $\begin{vmatrix} 0 & -1 & 4 \\ 0 & 1 & 4 \\ 4 & 1 & 0 \end{vmatrix}$ (c) $\begin{vmatrix} 1 & -1 & 2 \\ -1 & 1 & 2 \\ 1 & 1 & 0 \end{vmatrix}$

(d) $\begin{vmatrix} 0 & 0 & 0 \\ -2 & 1 & 4 \\ 2 & 1 & 0 \end{vmatrix}$.

TYPE II

11) The area of the triangle ABC, where A, B, C are the points (1, 2), (3, 1), (−2, 1) is given by:

(a) $\frac{1}{2}\begin{vmatrix} 1 & 1 & 1 \\ 1 & 3 & -2 \\ 2 & 1 & 1 \end{vmatrix}$ (b) $\frac{1}{2}\begin{vmatrix} 1 & 1 & 2 \\ 1 & 3 & 1 \\ 1 & -2 & 1 \end{vmatrix}$ (c) $\frac{1}{2}\begin{vmatrix} 1 & 1 & 1 \\ 2 & 1 & 1 \\ 1 & 3 & -2 \end{vmatrix}$.

12) $\mathbf{M} = \begin{pmatrix} \cos\theta & -\sin\theta \\ \sin\theta & \cos\theta \end{pmatrix}$.

(a) $\mathbf{M}$ is singular.
(b) $\mathbf{M}$ represents a reflection in $y = mx$, where $m = \tan\theta$.
(c) $\mathbf{M}$ represents a rotation of θ about O.

13) $\mathbf{I} = \begin{pmatrix} 1 & 0 \\ 0 & 1 \end{pmatrix}$ and $\mathbf{A} = \begin{pmatrix} 1 & 1 & 1 \\ 2 & 2 & 1 \end{pmatrix}$.

(a) $\mathbf{IA} = \mathbf{A}$.
(b) $\mathbf{AI} = \mathbf{0}$.
(c) $\mathbf{A}^2 = \begin{pmatrix} 1 & 1 & 1 \\ 4 & 4 & 1 \end{pmatrix}$.

14) $\mathbf{A} = \begin{pmatrix} a_1 & a_2 & a_3 \\ b_1 & b_2 & b_3 \\ c_1 & c_2 & c_3 \end{pmatrix}$.

(a) $|\mathbf{A}| = \begin{vmatrix} a_1 & b_1 & c_1 \\ a_2 & b_2 & c_2 \\ a_3 & b_3 & c_3 \end{vmatrix}$. (b) $\mathbf{A}^2$ exists. (c) $|\mathbf{A}| = \begin{vmatrix} c_1 & c_2 & c_3 \\ a_1 & a_2 & a_3 \\ b_1 & b_2 & b_3 \end{vmatrix}$.

15) $\mathbf{A} = \begin{pmatrix} a_{11} & a_{12} \\ a_{21} & a_{22} \end{pmatrix}$, $\mathbf{B} = \begin{pmatrix} b_{11} & b_{12} \\ b_{21} & b_{22} \end{pmatrix}$.

(a) $\mathbf{AB} = \mathbf{BA}$. (b) $|\mathbf{AB}| = |\mathbf{BA}|$. (c) $\mathbf{A}^2\mathbf{B}^2 = (\mathbf{AB})^2$.

TYPE III

16) (a) $\mathbf{AB} = \mathbf{BA}$.
(b) $\mathbf{A} = \mathbf{I}$ or $\mathbf{B} = \mathbf{I}$.

17) (a) **A** and **B** are the same size.
(b) $\mathbf{A} + \mathbf{B} = \mathbf{B} + \mathbf{A}$.

18) (a) $\mathbf{A} = \mathbf{0}$ or $\mathbf{B} = \mathbf{0}$.
(b) $\mathbf{AB} = \mathbf{0}$.

19) (a) **A** is an $n \times p$ matrix, **B** is an $m \times n$ matrix.
(b) **AB** exists.

20) (a) $\Delta = \begin{vmatrix} 1 & 2 \\ 3 & 4 \end{vmatrix}$

(b) $\dfrac{1}{\Delta} = \begin{vmatrix} 1 & \frac{1}{3} \\ \frac{1}{4} & \frac{1}{10} \end{vmatrix}$.

21) (a) $\mathbf{IA} = \mathbf{A}$.
(b) **A** is square.

22) (a) The matrix **M** maps **i** to $\begin{pmatrix} 1 \\ 2 \end{pmatrix}$ and **j** to $\begin{pmatrix} 2 \\ 1 \end{pmatrix}$.

(b) The matrix **M** maps $\begin{pmatrix} 2 \\ 2 \end{pmatrix}$ to $\begin{pmatrix} 6 \\ 6 \end{pmatrix}$.

23) (a) $\mathbf{A} = \begin{pmatrix} \cos\theta & \cos\theta \\ \sin\theta & -\sin\theta \end{pmatrix}$

(b) $|\mathbf{A}| = 0$.

TYPE IV

24) Find the equations of the lines which map to themselves under the transformation $\mathbf{M}\begin{pmatrix} x \\ y \end{pmatrix} = \begin{pmatrix} X \\ Y \end{pmatrix}$.

(a) $\mathbf{M}\begin{pmatrix} 0 \\ 0 \end{pmatrix} = \begin{pmatrix} 0 \\ 0 \end{pmatrix}$. (b) $\mathbf{M}$ maps $\mathbf{i}$ to $\begin{pmatrix} 1 \\ 1 \end{pmatrix}$.

(c) $\mathbf{M}$ maps $\mathbf{j}$ to $\begin{pmatrix} 0 \\ -1 \end{pmatrix}$.

25) Find the factor by which the transformation $\mathbf{M}\begin{pmatrix} x \\ y \end{pmatrix} = \begin{pmatrix} X \\ Y \end{pmatrix}$ changes area.

(a) $\mathbf{M}$ represents a rotation of $\dfrac{\pi}{3}$ about O.

(b) $\mathbf{M}$ maps $\mathbf{i}$ to $\frac{1}{2}\begin{pmatrix} 1 \\ \sqrt{3} \end{pmatrix}$.

(c) The origin is invariant under $\mathbf{M}$.

26) Find a square root of the matrix $\mathbf{A}$.

(a) $\mathbf{A}$ is a 2×2 matrix. (b) $(a_{11} \quad a_{12}) = (1 \quad 2)$.

(c) $\begin{pmatrix} a_{11} \\ a_{21} \end{pmatrix} = \begin{pmatrix} 1 \\ 3 \end{pmatrix}$.

TYPE V

27) $\lambda|\mathbf{A}| = |\lambda\mathbf{A}|$.

28) $\begin{vmatrix} a_1 & a_2 \\ b_1 & b_2 \end{vmatrix} = \begin{vmatrix} b_1 & b_2 \\ a_1 & a_2 \end{vmatrix}$.

29) $\begin{vmatrix} a_1 & a_2 \\ b_1 & b_2 \end{vmatrix} = \begin{vmatrix} a_1 & b_1 \\ a_2 & b_2 \end{vmatrix}$.

30) If $\mathbf{AB} = \mathbf{BA}$ then $\mathbf{A}$ and $\mathbf{B}$ must both be square.

31) If $\mathbf{AB}$ and $\mathbf{BA}$ both exist then $\mathbf{A}$ and $\mathbf{B}$ must both be square.

32) If the plane defined by the base vectors $\mathbf{i}$ and $\mathbf{j}$ is transformed to the plane defined by the base vectors $\mathbf{p}$ and $\mathbf{q}$ then the vector $\lambda\mathbf{i} + \mu\mathbf{j}$ is transformed to the vector $\lambda\mathbf{p} + \mu\mathbf{q}$.

33) If the xy plane is rotated about the point $(1, 1)$, the image of any point (x, y) may be obtained from an equation of the form $\begin{pmatrix} a & b \\ c & d \end{pmatrix}\begin{pmatrix} x \\ y \end{pmatrix} = \begin{pmatrix} X \\ Y \end{pmatrix}$.

MISCELLANEOUS EXERCISE 1

1) Show that the determinant

$$\begin{vmatrix} 1 & 1 & 1 \\ x & y & z \\ yz & zx & xy \end{vmatrix} = (x-y)(y-z)(z-x).$$

(U of L)p

2) If two dimensional space is transformed by

$$\begin{pmatrix} x \\ y \end{pmatrix} \to \begin{pmatrix} 4 & -1 \\ 6 & -3 \end{pmatrix}\begin{pmatrix} x \\ y \end{pmatrix}$$

find the equations of the straight lines which are mapped onto themselves.

(U of L)p

3) Show that the transformation

$$\begin{pmatrix} x_2 \\ y_2 \end{pmatrix} = \begin{pmatrix} \cos\theta & -\sin\theta \\ \sin\theta & \cos\theta \end{pmatrix}\begin{pmatrix} x_1 \\ y_1 \end{pmatrix}$$

represents a rotation about the origin.
Show also that the transformation

$$\begin{pmatrix} x_2 \\ y_2 \end{pmatrix} = \begin{pmatrix} -\frac{3}{5} & \frac{4}{5} \\ \frac{4}{5} & \frac{3}{5} \end{pmatrix}\begin{pmatrix} x_1 \\ y_1 \end{pmatrix}$$

represents a reflection in a fixed line through the origin. (U of L)p

4) Show, with the help of a diagram, that the matrix **P** of the linear transformation which rotates the plane in the counter-clockwise sense through an angle θ about the origin is

$$\begin{pmatrix} \cos\theta & -\sin\theta \\ \sin\theta & \cos\theta \end{pmatrix}.$$

Find the matrix **Q** of the linear transformation which reflects the points of the plane in the line $x = y$. Find the values of θ for which $\mathbf{PQ} = \mathbf{QP}$.

(U of L)p

5) Find the 2×2 matrices corresponding to:

(a) the reflection in the line through the origin making an angle of 60° with the positive x axis,

(b) the rotation about the origin through an angle of 90°,

(c) the reflection in the line through the origin making an angle of 120° with the positive x axis.

(All angles are measured anti-clockwise.)

Describe geometrically the resultant of the three transformations, taken in the given order. (O)

6) Show that, if $\mathbf{A} = \begin{pmatrix} 1 & -1 \\ 2 & -1 \end{pmatrix}$ and $\mathbf{B} = \begin{pmatrix} 1 & 1 \\ 4 & -1 \end{pmatrix}$ then

$$(\mathbf{A}+\mathbf{B})^2 = \mathbf{A}^2 + \mathbf{B}^2$$

(U of L)p

7) Show that $(a-b)$ is a factor of the determinant

$$\begin{vmatrix} 1+a^2 & a & 1 \\ 1+b^2 & b & 1 \\ 1+c^2 & c & 1 \end{vmatrix}$$

and express the determinant in a completely factorized form. (JMB)

8) Simplify the determinant

$$\begin{vmatrix} 1 & z & z+1 \\ z+1 & 1 & z \\ z & z+1 & 1 \end{vmatrix}.$$

Hence prove that the value of the determinant is a real number if z^3 is real.

(JMB)

9) Show that $\sin 4\theta - \sin\theta$ is a factor of the determinant

$$D = \begin{vmatrix} 1 & 1 & 1 \\ \frac{1}{2} & \sin\theta & \sin 4\theta \\ \frac{1}{4} & \sin^2\theta & \sin^2 4\theta \end{vmatrix}$$

and express D as the product of three factors, each of which depends on θ. Find all the values of θ for which $D = 0$. (JMB)

10) Prove that the value of a 3×3 determinant is unaltered if λ times the first column is added to the last column.

11) If $\mathbf{A}$ and $\mathbf{B}$ are matrices of order 3, prove that $|\mathbf{AB}| = |\mathbf{BA}|$.

12) The point (x, y) is transformed to the point (X, Y) under the transformation

$$\begin{pmatrix} X \\ Y \end{pmatrix} = \begin{pmatrix} 2 \\ 1 \end{pmatrix} + \begin{pmatrix} -1 & 0 \\ 0 & 1 \end{pmatrix}\begin{pmatrix} x \\ y \end{pmatrix}.$$

Find the images of A(1, 2) and O under this transformation and draw a diagram to illustrate the effect of the transformation on AO. Is the origin invariant under this transformation?

13) Show that the transformation of the plane given by the matrix $\mathbf{S} = \begin{pmatrix} 1 & 0 \\ 0 & -1 \end{pmatrix}$ is a reflection in the x axis.

Show also that the transformation of the plane given by the matrix
$\mathbf{R}_\alpha = \begin{pmatrix} \cos\alpha & -\sin\alpha \\ \sin\alpha & \cos\alpha \end{pmatrix}$ is a rotation about the origin through an angle α.
Form the product $\mathbf{R}_\alpha \mathbf{S} \mathbf{R}_{-\alpha}$ and show that the transformation of the plane given by this matrix is a reflection in the line $y = x \tan\alpha$. (U of L)

14) Find the two numerical values of λ such that

$$\begin{pmatrix} 4 & 3 \\ 1 & 2 \end{pmatrix}\begin{pmatrix} u \\ 1 \end{pmatrix} = \lambda\begin{pmatrix} u \\ 1 \end{pmatrix}.$$

Hence, or otherwise, find the equations of the two lines through the origin which are invariant under the transformation of the plane defined by

$$\begin{pmatrix} x' \\ y' \end{pmatrix} = \begin{pmatrix} 4 & 3 \\ 1 & 2 \end{pmatrix}\begin{pmatrix} x \\ y \end{pmatrix}.$$

(C)

15) $\mathbf{M}$ is the matrix $\begin{pmatrix} a & b \\ c & d \end{pmatrix}$, and $\mathbf{i}, \mathbf{j}$ are the vectors $\begin{pmatrix} 1 \\ 0 \end{pmatrix}, \begin{pmatrix} 0 \\ 1 \end{pmatrix}$ respectively. Write down the vectors $\mathbf{u}$ and $\mathbf{v}$, where $\mathbf{u} = \mathbf{Mi}$ and $\mathbf{v} = \mathbf{Mj}$, and show that:
(a) if $\mathbf{u}$ and $\mathbf{v}$ are perpendicular then $ab + cd = 0$,
(b) if $\mathbf{u}$ and $\mathbf{v}$ have equal magnitudes then $a^2 + c^2 = b^2 + d^2$.
$\mathbf{r}_1 = x_1\mathbf{i} + y_1\mathbf{j}$ and $\mathbf{r}_2 = x_2\mathbf{i} + y_2\mathbf{j}$ are any two vectors, and when multiplied by $\mathbf{M}$ they are transformed to $\mathbf{s}_1$ and $\mathbf{s}_2$ respectively. Given that a, b, c, d satisfy $ab + cd = 0$ and $a^2 + c^2 = b^2 + d^2$, show that the angle between $\mathbf{s}_1$ and $\mathbf{s}_2$ is the same as that between $\mathbf{r}_1$ and $\mathbf{r}_2$. (C)

16) It is known that three non-null 2×2 matrices, $\mathbf{P}, \mathbf{Q}, \mathbf{R}$ satisfy the equation $\mathbf{PQ} = \mathbf{RQ}$. State whether the deduction that $\mathbf{P} = \mathbf{R}$ is true or false. If you think the deduction is true, prove it, if you think it is false, given an example of three non-null matrices $\mathbf{P}, \mathbf{Q}$ and $\mathbf{R}$ which disproves it. (C)p

17) Show that the determinant

$$\begin{vmatrix} 1 & 1 & 1 \\ \cos^2 a & \cos^4 a & \sec^2 a \\ \sin^2 a & \sin^4 a & \tan^2 a \end{vmatrix}$$

is equal to $2\sin^4 a \cos^2 a$. (JMB)

18) Evaluate

$$\begin{vmatrix} x & x-y & x+y \\ x-y & x+y & x \\ x+y & x & x-y \end{vmatrix}.$$

Hence, or otherwise, show that

$$\begin{vmatrix} 4 & 11 & -3 \\ 11 & -3 & 4 \\ -3 & 4 & 11 \end{vmatrix} = \begin{vmatrix} 4 & -3 & 11 \\ -3 & 11 & 4 \\ 11 & 4 & -3 \end{vmatrix}.$$ (JMB)

19) Express $\begin{vmatrix} 1 & 1 & n \\ n+1 & n-1 & 2 \\ n(n-1) & n(n+1) & 0 \end{vmatrix}$ as the product of factors linear in n.

Hence show that for all integer values of n, the determinant is divisible by 24. (JMB)

20) Let $\mathbf{A}$ be the matrix $\begin{pmatrix} a & b \\ c & d \end{pmatrix}$, where no one of a, b, c, d is zero.
It is required to find a non-zero 2×2 matrix $\mathbf{X}$ such that $\mathbf{AX} + \mathbf{XA} = \mathbf{0}$, where $\mathbf{0}$ is the zero 2×2 matrix. Prove that either
(a) $a + d = 0$, in which case the general solution for $\mathbf{X}$ depends on two parameters, or
(b) $ad - bc = 0$, in which case the general solution for $\mathbf{X}$ depends on one parameter. (O)

21) Write down the matrices corresponding to:
(a) rotation about the origin O through an angle of 30°,
(b) reflection in the line through O that makes an angle of 120° with Ox,
(c) rotation about O through an angle of 210°.
(All angles are measured in the anticlockwise sense.)
Give the complete geometrical description of the single transformation of the plane that is the resultant of (a), (b), (c) in that order. (O)

22) $\mathbf{A}$ and $\mathbf{X}$ are the matrices $\begin{pmatrix} a & b \\ c & d \end{pmatrix}$ and $\begin{pmatrix} x & y \\ u & v \end{pmatrix}$ respectively, where b is not equal to zero. Prove that if $\mathbf{AX} = \mathbf{XA}$ then $u = \dfrac{cy}{b}$ and $v = x + \dfrac{(d-a)y}{b}$.

Hence prove that if $\mathbf{AX} = \mathbf{XA}$ then there are numbers p and q such that $\mathbf{X} = p\mathbf{A} + q\mathbf{I}$, where $\mathbf{I}$ is the unit matrix $\begin{pmatrix} 1 & 0 \\ 0 & 1 \end{pmatrix}$, and find p and q in terms of a, b, x, y. (O)

23) When are two matrices conformable for multiplication?

Express $(1,\ 2,\ -3)\begin{pmatrix} 2, & 1, & 4 \\ 1, & 0, & 3 \\ 4, & 3, & 5 \end{pmatrix}\begin{pmatrix} 1 \\ 2 \\ -3 \end{pmatrix}$ as a single matrix. (U of L)p

24) If $\mathbf{A}$ is the matrix $\begin{pmatrix} \cos\frac{\pi}{n} & -\sin\frac{\pi}{n} \\ \sin\frac{\pi}{n} & \cos\frac{\pi}{n} \end{pmatrix}$, where n is a positive integer, prove that $\mathbf{A}^{2n} = \mathbf{I}$, where $\mathbf{I}$ is the identity (or unit) matrix of order 2.

(U of L)p

25) Find the most general form for the matrix $\mathbf{P}$ if $\mathbf{PQ} = \mathbf{QP}$, where $\mathbf{Q}$ is the matrix $\begin{pmatrix} 2 & 1 \\ 4 & 5 \end{pmatrix}$.

If the non-zero column vectors $\mathbf{x}$ and $\mathbf{y}$ are such that $\mathbf{Qx} = \mathbf{x}$ and $\mathbf{Qy} = 6\mathbf{y}$, obtain the particular matrix $\mathbf{P}$ such that $\mathbf{PQ} = \mathbf{QP}$, $\mathbf{Px} = -\mathbf{x}$ and $\mathbf{Py} = 4\mathbf{y}$.

(U of L)p

26) The point (x, y) is transformed to the point (x', y') by means of the transformation

$$\begin{pmatrix} x' \\ y' \end{pmatrix} = \begin{pmatrix} 3 & 0 \\ 0 & 4 \end{pmatrix}\begin{pmatrix} x \\ y \end{pmatrix} + \begin{pmatrix} 1 \\ 1 \end{pmatrix}$$

Find the image of the line $y = 2x$ under this transformation. (U of L)

CHAPTER 2

COORDINATE GEOMETRY I

CONIC SECTIONS

The circle, the parabola and the ellipse were introduced in Volume 1, Chapter 11.
These curves, together with the hyperbola and a pair of straight lines, are collectively known as conic sections. The reason for this is seen by considering a double cone (obtained by rotating a straight line through one revolution about an axis that intersects the line at an angle α say). When a plane cuts this double cone, the shape of the section formed depends upon the inclination θ of the plane to the axis.

If $\theta = \dfrac{\pi}{2}$, the plane cuts only one half of the double cone and the cross-section is a circle.

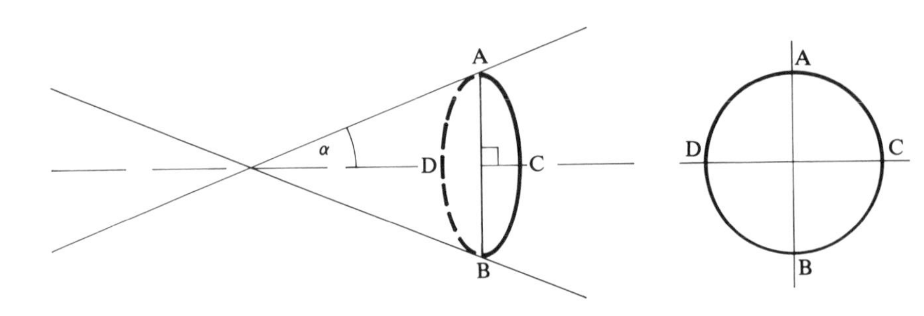

If $\alpha < \theta < \frac{\pi}{2}$, the plane again cuts only one half of the cone and the cross-section is an ellipse.

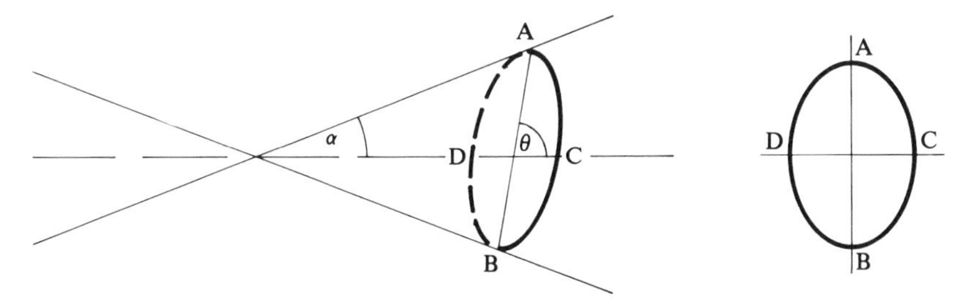

If $\theta = \alpha$, the plane is parallel to a generator of the cone and hence cuts only one half of the double cone in a section which is open-ended. This section is a parabola.

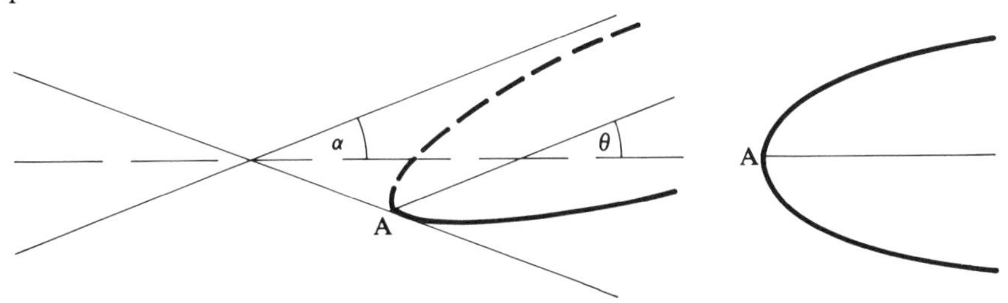

If $\theta < \alpha$, the plane cuts into both halves of the double cone, producing (unless the plane passes through the vertex) a section comprising two open-ended curves. This section is called a hyperbola.

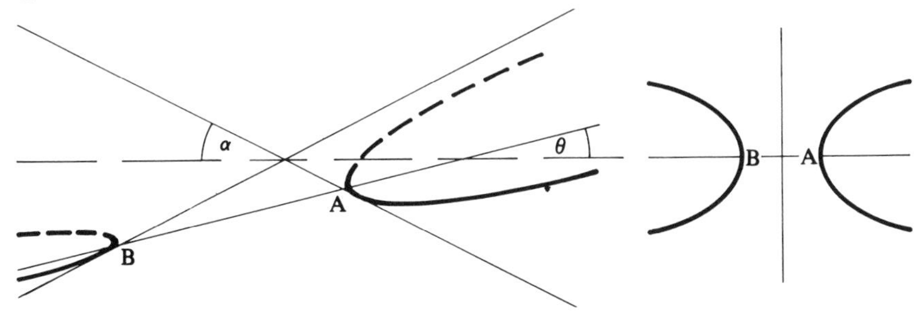

If $\theta < \alpha$ and the plane *does* pass through the vertex, the section is a pair of straight lines.
We now have the complete set of conic sections (often referred to simply as *the conics*).

Conic Sections as Loci

Another factor linking the members of this group of curves is that each is the locus of a set of points, P, obeying the same basic restriction, which is that:

the distances of P from a fixed point (the focus) and a fixed straight line (the directrix) are in a constant ratio e (the eccentricity).
Different values of e correspond to different conics.
We saw in Volume 1 that if $e = 1$ the locus of P is a parabola.

We also saw that if the point $(ae, 0)$ is a focus, and the line $x = \frac{a}{e}$ is a directrix, the locus of P has a Cartesian equation

$$\frac{x^2}{a^2} + \frac{y^2}{a^2(1 - e^2)} = 1. \qquad [1]$$

If $0 < e < 1$, $a^2(1 - e^2) > 0$ and can be replaced by b^2 to give

$$\frac{x^2}{a^2} + \frac{y^2}{b^2} = 1$$

which is the equation of an ellipse.
If $e = 0$, equation [1] becomes $x^2 + y^2 = a^2$ which is the equation of a circle.
If $e > 1$, $a^2(1 - e^2) < 0$ and so can be replaced by $-b^2$ to give

$$\frac{x^2}{a^2} - \frac{y^2}{b^2} = 1$$

which is the equation of a hyperbola.
So there is both a geometric and a definitive relationship linking the conic sections. But each member of the set has certain unique characteristics and properties. Some of these were dealt with in Volume 1, Chapter 11 and these the reader is recommended to revise at this stage.
We will now investigate individual conics.

THE PARABOLA

As all parabolas have the same geometric properties, we usually study one with a simple Cartesian equation. This standard parabola has its vertex at the origin, its focus, S, on the x-axis at a point $(a, 0)$ and the line $x = -a$ as its directrix. We saw in Volume 1 that:

(a) its Cartesian equation is $y^2 = 4ax$,
(b) its parametric equations are $x = at^2$, $y = 2at$,
(c) the gradient at the point $(at^2, 2at)$ is $\frac{1}{t}$,
(d) the parametric equation of the tangent at the point $(at^2, 2at)$ is $ty = x + at^2$.

Optical Property of a Parabola

Any ray of light parallel to the axis of a parabolic mirror is reflected through the focus. This property, which is of considerable practical use in optics, can be proved by showing that the normal at any point P on a parabola, bisects the angle between PS and the line PQ that is parallel to the axis of the parabola (the angle of incidence and the angle of reflection are equal).

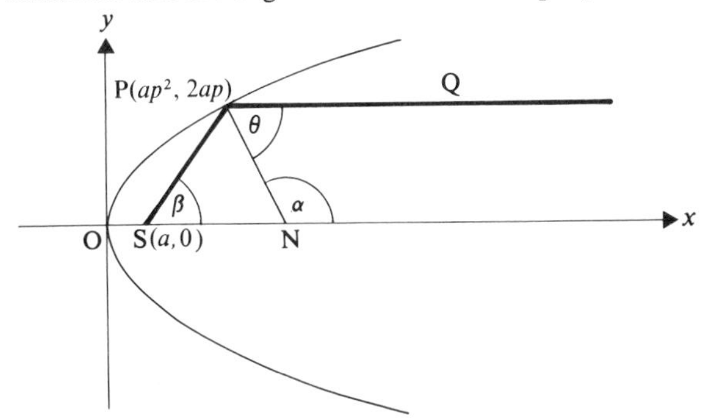

In the diagram PN is the normal at P(ap^2, $2ap$),

so $\tan\alpha = -p$

and $\tan\theta = \tan(\pi - \alpha) = p.$

Also $\tan\beta = \dfrac{2ap}{ap^2 - a} = \dfrac{2p}{p^2 - 1}$

so $\tan \text{QPS} = \tan(\pi - \beta) = \dfrac{2p}{1 - p^2} = \tan 2\theta.$

Hence $\angle\text{QPS} = 2(\angle\text{QPN}),$

i.e. PN bisects $\angle$QPS.

Focal Chord Properties

A line joining two points P and Q on a parabola is a focal chord if PQ passes through the focus S. The focal chord that is perpendicular to the axis of the parabola is called the *latus rectum*.

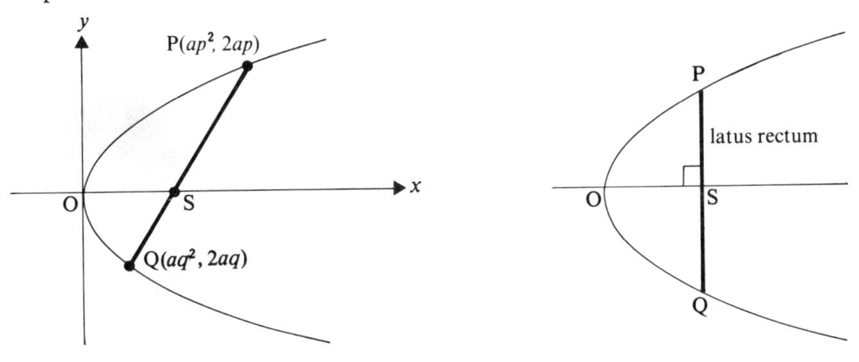

If PSQ is a straight line, the gradients of QS and PS are equal. So, taking $(ap^2, 2ap)$ and $(aq^2, 2aq)$ as the coordinates of P and Q, we have

$$\frac{2aq}{aq^2 - a} = \frac{2ap}{ap^2 - a}$$

$$\Rightarrow \quad q(p^2 - 1) = p(q^2 - 1)$$

$$\Rightarrow \quad pq(p - q) = -(p - q)$$

$$\Rightarrow \quad pq = -1 \quad (\text{as } p \neq q).$$

This property of the parametric values at the ends of a focal chord is quotable.

Now consider the tangents at P and at Q. Their gradients are $\frac{1}{p}$ and $\frac{1}{q}$ respectively.

But if $pq = -1$, then $\left(\frac{1}{p}\right)\left(\frac{1}{q}\right) = -1$ showing that the tangents at P and Q are perpendicular to each other.

Further, the point R where these tangents intersect can be found by solving simultaneously the equations

$$\begin{cases} py = x + ap^2 & \text{(tangent at P)} \\ qy = x + aq^2 & \text{(tangent at Q)} \end{cases}$$

giving $\{apq, a(p + q)\}$ as the coordinates of R.

But if $pq = -1$, $apq = -a$ showing that R is a point on the directrix $(x = -a)$.

So we see that, if PQ is a focal chord,

(1) $pq = -1$,

(2) the tangents at P and Q are perpendicular,

(3) the tangents at P and Q meet on the directrix.

Note. The tangents at *any* two points P and Q on the parabola meet where

$$x = apq \quad \text{and} \quad y = a(p + q).$$

The reader may observe that these coordinates are the geometric and arithmetic mean respectively of the coordinates of P and Q, which provides an interesting aid to memory.

Diameter of a Parabola

We usually think of a diameter as existing only in a closed curve (e.g. circle or ellipse) but, defining a diameter as 'the locus of midpoints of a set of parallel chords when this locus is a straight line' we find that a parabola has diameters.

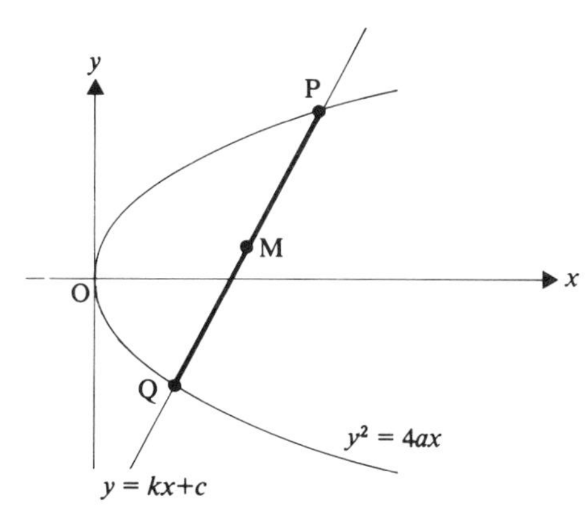

Consider any chord of fixed gradient k and thus with equation

$$y = kx + c.$$

This chord meets the parabola $y^2 = 4ax$ at points P and Q whose y-coordinates are the roots of the equation

$$y = k\left(\frac{y^2}{4a}\right) + c$$

or $$ky^2 - 4ay + 4ac = 0. \quad [1]$$

If these roots are y_P and y_Q then, at the midpoint, M, of PQ,

$$y_M = \tfrac{1}{2}(y_P + y_Q),$$

i.e. $$y_M = \tfrac{1}{2}(\text{sum of roots of equation } [1])$$

$$= \tfrac{1}{2}\left(\frac{4a}{k}\right)$$

$$= \frac{2a}{k}.$$

But k is constant, so the y-coordinate of M is fixed.
That is, as P and Q move on the parabola, the gradient of PQ remaining constant, the locus of the midpoint of PQ is a fixed horizontal line with equation $y = \dfrac{2a}{k}$.

Thus we see that diameters of a parabola are straight lines parallel to the axis of the parabola.

The following exercise makes use of the work on tangents and normals covered in Volume 1 together with the methods used and properties derived in this chapter.

EXERCISE 2a

1) The tangent at a point P on the parabola $y^2 = 4ax$ meets the directrix at Q. The line through Q parallel to the axis of the parabola meets the normal at P at the point R. Find the equation of the locus of R. Show that the locus is another parabola and find its vertex.

2) Prove that the line $x - ty + at^2 = 0$ touches the parabola $y^2 = 4ax$ for all values of t. Find the coordinates of the point of contact.

3) The tangent at P to the parabola $y^2 = 4ax$ meets the x-axis at T. Prove that PS = TS where S is the focus.

4)

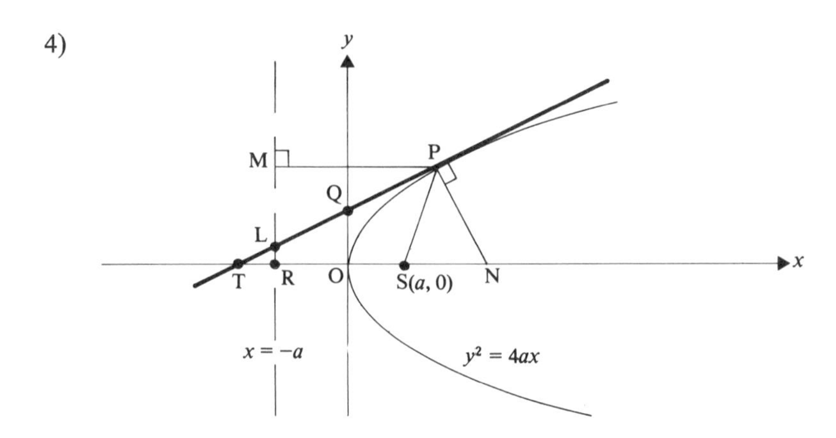

In the diagram, PT and PN are the tangent and normal respectively to the parabola $y^2 = 4ax$. MR is the directrix and S is the focus. Prove that

(a) PT bisects $\angle$MPS,
(b) SPMT is a parallelogram,
(c) the perpendicular from S to the tangent PT meets PT at Q,
(d) the line through L, perpendicular to PL, touches the parabola,
(e) $(\text{SQ})^2 = (\text{SO})(\text{ST})$,
(f) if a line PV is drawn so that $\angle\text{NPS} = \angle\text{SPV}$, then MPV is a straight line.

5) PQ is a focal chord of the parabola $y^2 = 4ax$, and O is the origin. Using parametric coordinates $(ap^2, 2ap)$ for P, find the coordinates of the centroid, G, of triangle OPQ and hence find the equation of the locus of G as PQ varies.

6) If $\text{A}(h, k)$ lies on the normal at $\text{P}(at^2, 2at)$ to the parabola $y^2 = 4ax$, show that $at^3 + (2a - h)t - k = 0$.
By considering the possible number of real roots of this cubic equation for t, show that, for certain points A, three normals can be drawn to the parabola, while from a different set of points A, only one normal can be drawn. Illustrate your deduction on a diagram.

7) If the normal at $\text{P}(at^2, 2at)$ to the parabola $y^2 = 4ax$ meets the curve again at $\text{Q}(aq^2, 2aq)$, prove that $p^2 + pq + 2 = 0$. The tangents at P and Q intersect at R. If the line through R parallel to the axis of the parabola meets the parabola at T, show that PT is a focal chord.

8) PQ is a focal chord of the parabola with parametric equations $x = at^2$, $y = 2at$.
PR is another chord that meets the x axis at the point $(ka, 0)$. If RQ produced meets the axis at T, prove that $\text{RT} = k\text{QT}$.

9) $\text{P}(ap^2, 2ap)$ and $\text{Q}(aq^2, 2aq)$ are two variable points on the parabola $y^2 = 4ax$. If PQ subtends a right angle at the origin prove that $pq = -4$.

(a) Show that PQ passes through a fixed point on the axis of the parabola.
(b) The tangents at P and Q meet at R. Find the equation of the locus of R.

10) A variable tangent to the parabola $y^2 = 4x$ meets the parabola $y^2 = 8x$ at P and Q. Find the equation of the locus of the midpoint of PQ.

11) Find the equation of the locus of the midpoints of focal chords of the parabola with parametric equations $x = at^2$, $y = 2at$.

12) P and Q are the points of contact of the tangents drawn from a point R to a parabola with focus S and M is the midpoint of PQ.
Prove that RP.RQ = 2RS.RM.

THE ELLIPSE

We saw in Volume 1 that an ellipse has two foci, two directrices and two axes of symmetry. The standard ellipse whose centre is at the origin, whose foci are the points $(\pm ae, 0)$ and whose directrices are $x = \pm\dfrac{a}{e}$ has a Cartesian equation

$$\frac{x^2}{a^2}+\frac{y^2}{b^2} = 1$$

$$\text{where } b^2 = a^2(1-e^2).$$

Suitable parametric equations are

$$x = a\cos\theta \qquad y = b\sin\theta.$$

Such an ellipse has a *major axis* of length $2a$ and a *minor axis* of length $2b$.

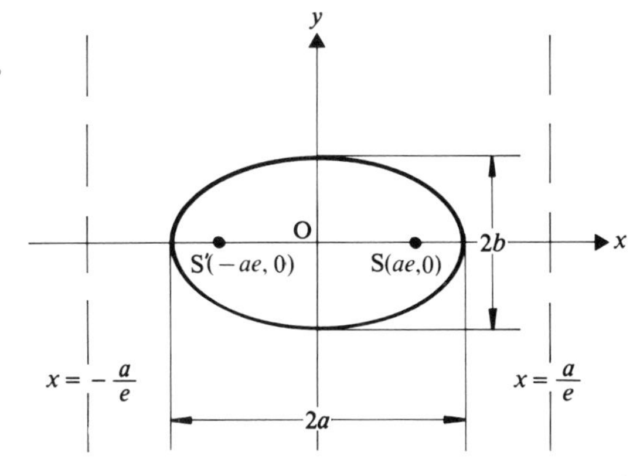

The major axis is a segment of the line through the two foci which can, alternatively, lie on the y axis. In this case the major axis is vertical and the Cartesian equation is $\dfrac{x^2}{b^2}+\dfrac{y^2}{a^2} = 1$.

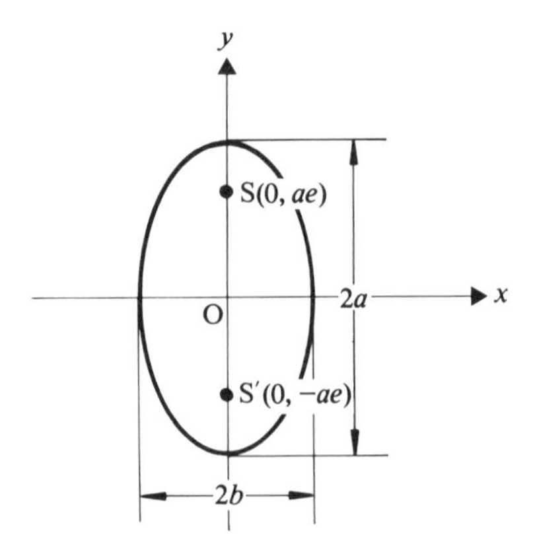

Note. If $a = b$ the equation of the ellipse becomes $x^2 + y^2 = a^2$ which is the equation of a circle.
In this case the eccentricity is given by $a^2 = a^2(1 - e^2) \Rightarrow e = 0$.
This confirms that the conic section with zero eccentricity is a circle.
The centre, foci and axes lengths of any ellipse with horizontal and vertical axes can be identified by comparing its equation with that of a standard ellipse

$$\frac{X^2}{a^2} + \frac{Y^2}{b^2} = 1$$

whose centre is given by $X = 0$, $Y = 0$
and whose axes lengths are $2a$ and $2b$.
For example, given an ellipse with equation

$$\frac{(x-1)^2}{25} + \frac{y^2}{16} = 1$$

comparison with

$$\frac{X^2}{a^2} + \frac{Y^2}{b^2} = 1$$

shows that $X = x - 1, \quad Y = y, \quad a = 5, \quad b = 4.$

So the centre $(X = 0, Y = 0)$ is $(1, 0)$.

The major axis is horizontal and of length 10 and the minor axis is of length 8.
From $b^2 = a^2(1 - e^2)$ we see that $e = \frac{3}{5}$. The foci, which are always on the major axis, are each distant $ae\ (= 3)$ from the centre, i.e. at $(4, 0)$ and $(-2, 0)$.
The directrices are perpendicular to the major axis and distant $\frac{a}{e}\left(= \frac{25}{3}\right)$ from the centre. So their equations are $x = \frac{28}{3}$ and $x = -\frac{22}{3}$.

The given ellipse can now be sketched.

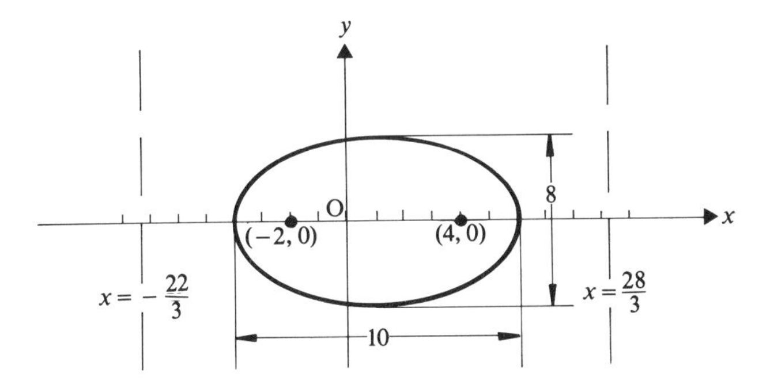

Note that $\dfrac{(x-1)^2}{25}+\dfrac{y^2}{16}=1$ is transformed into $\dfrac{X^2}{25}+\dfrac{Y^2}{16}=1$ by moving the origin to the point $(1, 0)$ and using a Y-axis moved to the position $x=1$.

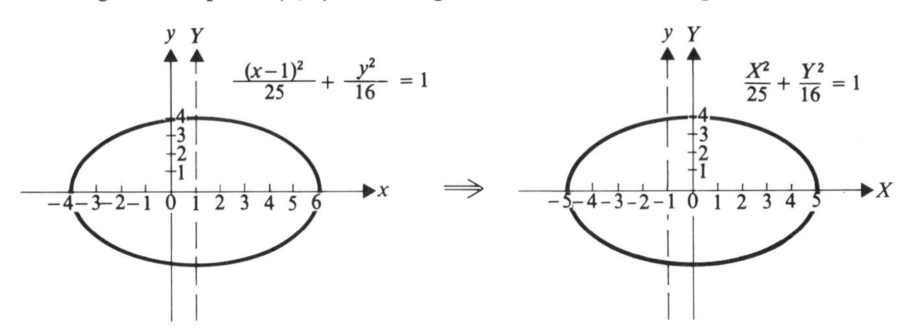

In general, the equation $\dfrac{(x-p)^2}{a^2}+\dfrac{(y-q)^2}{b^2}=1$ can be transformed into the form $\dfrac{X^2}{a^2}+\dfrac{Y^2}{b^2}=1$ by moving the axes to $x=p$ and $y=q$.

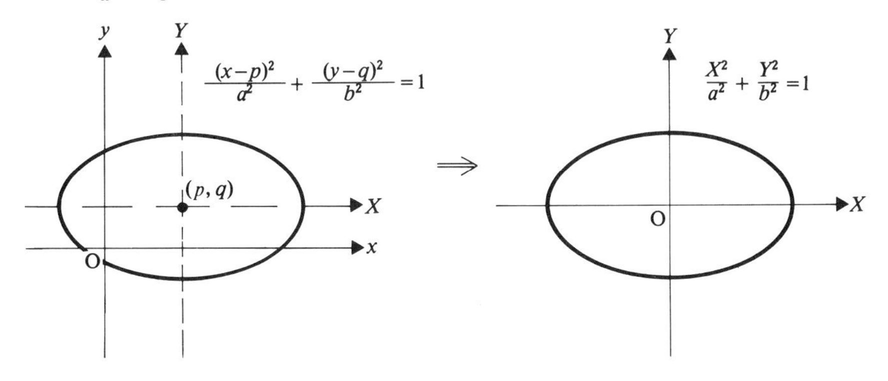

EXAMPLES 2b

1) An ellipse has its foci at the points $(-1, 0)$ and $(7, 0)$ and its eccentricity is $\frac{1}{2}$. Find its equation in Cartesian and in parametric form.

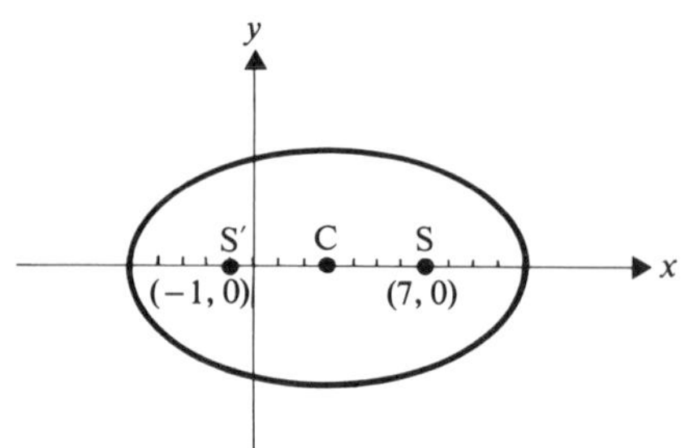

The centre, C, is midway between the foci,

i.e. C is $(3, 0)$.

The distance between the centre and each focus is ae,

i.e. $\quad CS = 4 = \frac{1}{2}a$

$\Rightarrow \quad a = 8.$

Now using $b^2 = a^2(1-e^2)$, we have

$$b^2 = 64(1-\tfrac{1}{4}) = 48$$

so the equation of the ellipse is

$$\frac{(x-3)^2}{64}+\frac{y^2}{48} = 1.$$

Comparing this equation with the standard equation

$$\frac{X^2}{a^2}+\frac{Y^2}{b^2} = 1 \quad \text{where} \quad \begin{cases} X = a\cos\theta \\ Y = b\sin\theta \end{cases}$$

gives $\quad x-3 = 8\cos\theta \qquad y = 4\sqrt{3}\sin\theta.$

So the parametric equations of the given ellipse are

$$x = 3+8\cos\theta \qquad y = 4\sqrt{3}\sin\theta.$$

2) An ellipse has its foci on the y-axis and its centre at the origin. The distance between the foci is 6 and the major axis is of length 10. Find the eccentricity, the Cartesian equation and the equations of the directrices of this ellipse.

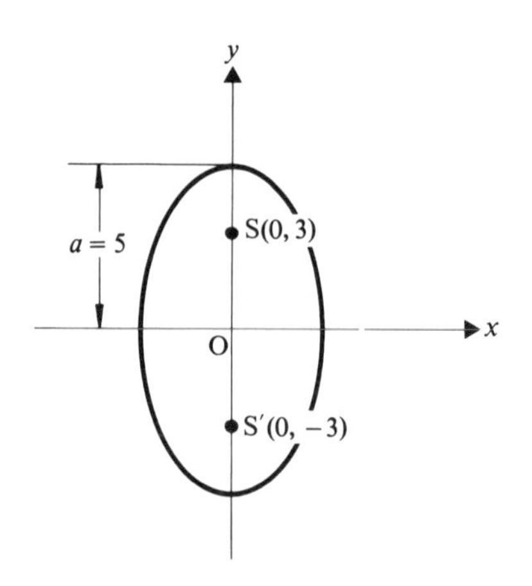

The centre of the ellipse is the origin so the foci are the points $(0, 3)$ and $(0, -3)$ and $ae = 3$.
But $a = 5 \Rightarrow e = \frac{3}{5}$.
Using $b^2 = a^2(1-e^2)$ gives $b^2 = 25(1-\frac{9}{25}) = 16$.

As the major axis is vertical, the Cartesian equation is of the form $\dfrac{x^2}{b^2}+\dfrac{y^2}{a^2} = 1$,

i.e. $$\frac{x^2}{16}+\frac{y^2}{25} = 1.$$

The directrices are horizontal and distant $\dfrac{a}{e}$ from O.

So their equations are $y = \pm\frac{25}{3}$.

EXERCISE 2b

1) Write down the equations of the following ellipses.
(a) Centre 0, major axis horizontal and of length 6, minor axis of length 4.
(b) Centre $(0, 1)$, major axis vertical and of length 8, minor axis of length 4.
(c) Centre $(2, 0)$, major axis horizontal and of length $4a$, eccentricity $\frac{1}{2}$.
(d) Centre $(3, 4)$, minor axis horizontal and of length 6, eccentricity $\frac{1}{4}\sqrt{7}$.

2) Find the coordinates of the centre, the lengths of the axes and the eccentricity of the following ellipses.

(a) $\dfrac{x^2}{9}+\dfrac{y^2}{16} = 1$ (b) $\dfrac{x^2}{4}+(y-1)^2 = 1$

(c) $16x^2+25y^2 = 400$ (d) $16x^2+25y^2 = 16$

(e) $x^2-4x+4y^2 = 12$ (*Hint*: Complete the square on the x terms.)

(f) $b^2(x-a)^2+a^2(y-b)^2 = a^2b^2$, $a > b$

(g) $\dfrac{(x+3)^2}{3}+\dfrac{(y-4)^2}{2} = 1.$

3) Find the coordinates of the foci and the equations of the directrices of each of the ellipses in Question 1. Sketch each ellipse.

4) An ellipse has its centre at the origin and its foci on the x axis. The distance between the foci is 3 cm and the distance between the directrices is 12 cm. Find the eccentricity and the length of the major axis. Hence find the Cartesian equation of the ellipse.

5) Use the locus definition of an ellipse to find the equations of the following ellipses.
(a) A focus is at $(3, 4)$, the corresponding directrix is the line $x+y = 1$ and the eccentricity is $\frac{1}{2}$.
(b) The set of points that are three times as far from the x axis as they are from the point $(0, 2)$.

The Focal Distance Property

If S and S' are the foci of an ellipse whose major axis is of length $2a$ and P is any point on the ellipse, then

$$PS + PS' = 2a.$$

This property provides the following simple method of drawing an ellipse. Take a piece of string of length $2a$ and fix its ends to two points (S and S′) whose distance apart is less than $2a$. A pencil (P) placed in the loop of string, keeping the string taut, now satisfies the condition $PS + PS' = 2a$. So, as the pencil moves, it traces out an ellipse.

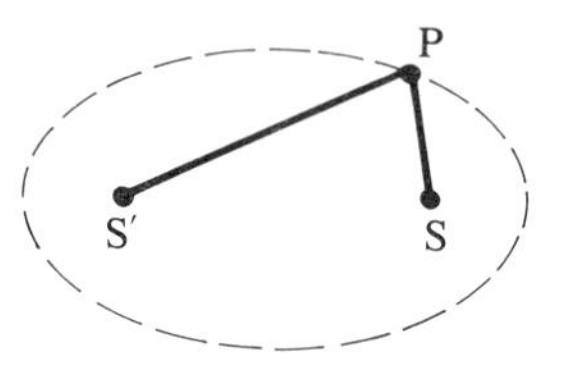

Note that it is not necessary to use this method to *sketch* an ellipse.
This property can be proved from the locus definition of an ellipse as follows:

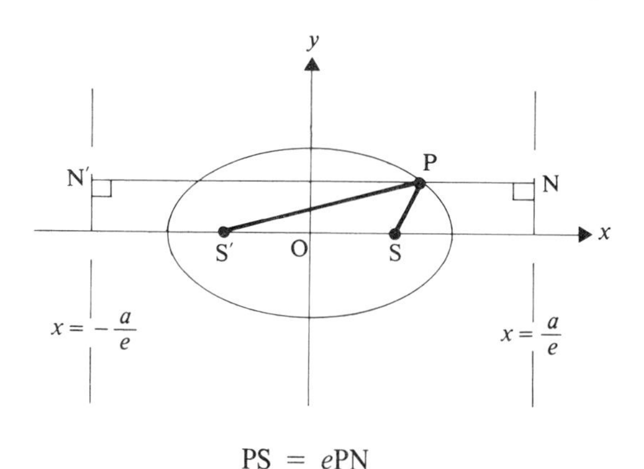

$$\begin{aligned} PS &= ePN \\ PS' &= ePN' \\ PS + PS' &= e(PN + PN') \\ &= e(NN'). \end{aligned}$$

But NN′ is the distance between the directrices,

i.e.
$$NN' = \frac{2a}{e}.$$

Hence
$$PS + PS' = 2a.$$

Note that the locus definition of an ellipse can be applied to derive a variety of its geometric properties.

Diameters

A diameter of any conic is the locus of the midpoints of a set of parallel chords.
In the case of an ellipse we find that each of its diameters passes through the centre of the ellipse.

Consider the ellipse with equation $\frac{x^2}{a^2}+\frac{y^2}{b^2}=1$ and a set of chords with constant gradient k.
One such chord with equation $y = kx + c$ cuts the ellipse at A and B.

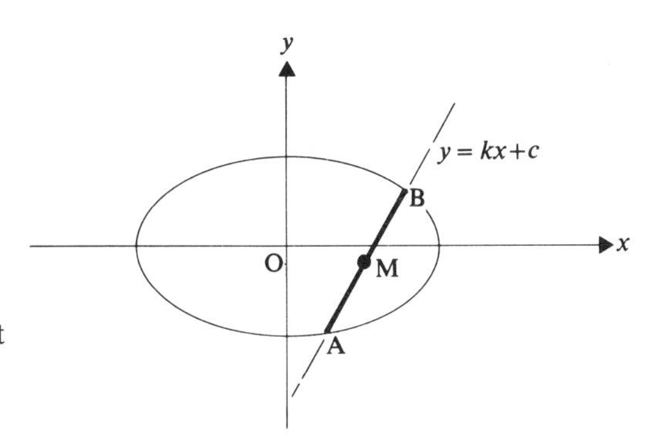

The x coordinates of A and B are the roots of the equation

$$\frac{x^2}{a^2}+\frac{(kx+c)^2}{b^2} = 1$$

or $$(b^2+a^2k^2)x^2+2a^2kcx+a^2(c^2-b^2) = 0. \qquad [1]$$

At the midpoint, $M(X, Y)$, of AB,

$$X = \tfrac{1}{2}(x_A + x_B)$$
$$= \tfrac{1}{2}(\text{sum of roots of equation [1]})$$
$$= -\frac{a^2kc}{b^2+a^2k^2}. \qquad [2]$$

Also, as M is a point on the line with equation $y = kx + c$, we have

$$Y = kX + c. \qquad [3]$$

Equations [2] and [3] are valid for all chords with gradient k, so the Cartesian equation of the locus of M is given by eliminating c (which varies as A and B move on the ellipse) from these two equations;

i.e. $$Y = kX - \left(\frac{b^2+a^2k^2}{a^2k}\right)X$$

$\Rightarrow$ $$a^2kY + b^2X = 0.$$

This line passes through O for all values of k, so all diameters of an ellipse pass through its centre.

Note that for a set of chords of the ellipse $\frac{x^2}{a^2}+\frac{y^2}{b^2}=1$ with gradient k the diameter has gradient $-\frac{b^2}{a^2k}$.

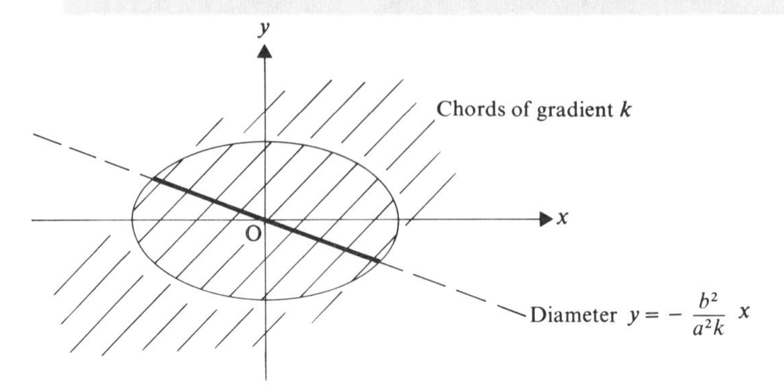

If AB is a diameter and if CD is the diameter containing the midpoints of the set of chords parallel to AB, then AB and CD are *conjugate diameters*.

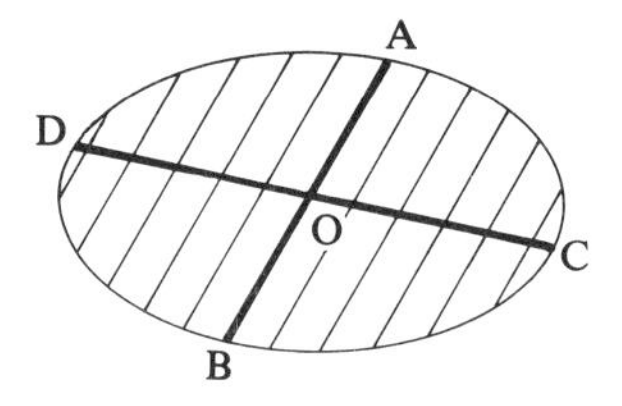

Now if the set of chords parallel to AB has a gradient m, we have seen that the diameter CD has a gradient $-\frac{b^2}{a^2m}$. Hence the products of the gradients of a pair of conjugate diameters is

$$(m)\left(-\frac{b^2}{a^2m}\right) = -\frac{b^2}{a^2}.$$

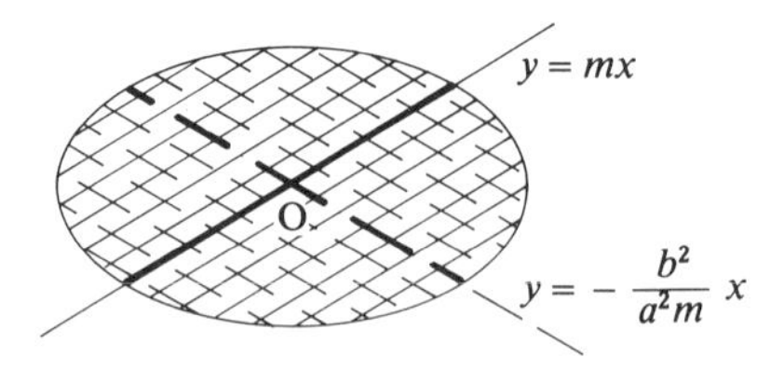

Note that, as the product of the gradients of conjugate diameters is $-\frac{b^2}{a^2}$, they are not perpendicular (unless $m=0$ or ∞).
The tangent at the end of a diameter is the limiting member of the set of parallel

chords. So the tangents at the ends of the diameter $y = -\dfrac{b^2}{a^2m}x$ have gradient m, i.e. they are parallel to the conjugate diameter.

Condition for a Line to Touch an Ellipse

In Volume 1 Chapter 11 the problem of tangency was dealt with by first considering the possible intersection of a line and a curve. We shall now apply this method when the curve is an ellipse.

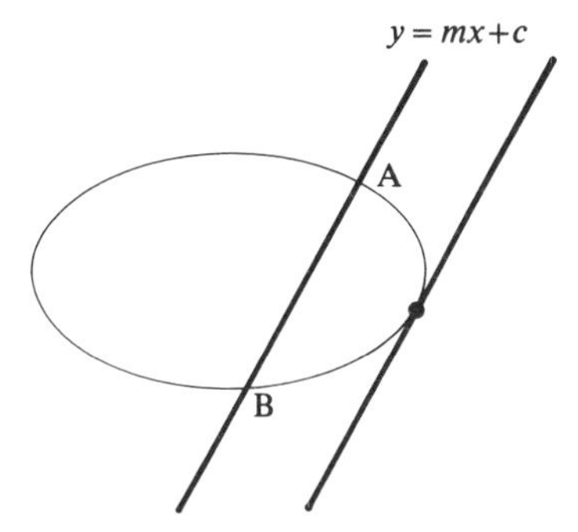

Consider a line $y = mx + c$ and the ellipse $\dfrac{x^2}{a^2} + \dfrac{y^2}{b^2} = 1$.

If the line meets the curve at A and B, the x coordinates of A and B are given by the roots of the equation

$$\frac{x^2}{a^2} + \frac{(mx+c)^2}{b^2} = 1$$

or

$$(b^2 + a^2m^2)x^2 + 2a^2mcx + a^2(c^2 - b^2) = 0.$$

If the line touches the ellipse, A and B coincide, so this equation has equal roots, i.e.

$$(2a^2mc)^2 = 4a^2(b^2 + a^2m^2)(c^2 - b^2),$$

which simplifies to give

$$c^2 = b^2 + a^2m^2.$$

This condition should not be regarded as quotable. Instead, the method used to derive it should be adapted to each individual problem on tangency.
However, it is useful to remember that, from this condition we can deduce that the line

$$y = mx \pm \sqrt{(a^2m^2 + b^2)}$$

always touches the ellipse $\dfrac{x^2}{a^2} + \dfrac{y^2}{b^2} = 1$.

This provides a useful approach to problems involving two tangents to an ellipse from a given point, a technique that is illustrated in the following paragraph.

The Director Circle

If two perpendicular tangents are drawn from a point $P(x, y)$ to an ellipse, $\dfrac{x^2}{a^2} + \dfrac{y^2}{b^2} = 1$, the locus of P as the points of contact vary, is a circle called the *director circle* and its equation is

$$x^2 + y^2 = a^2 + b^2.$$

This can be proved by considering the equations of the two tangents from P(x, y) to the ellipse. These equations can be written in the form

$$y = mx \pm \sqrt{(a^2m^2 + b^2)}$$

or $$(x^2 - a^2)m^2 - 2xym + (y^2 - b^2) = 0.$$

The roots of this equation, m_1 and m_2, are the gradients of the two tangents from P.

So when the tangents are at right angles, $m_1 m_2 = -1$.

In this case the coordinates of P satisfy the relationship

$$\frac{y^2 - b^2}{x^2 - a^2} = -1$$

$\Rightarrow$ $$x^2 + y^2 = a^2 + b^2.$$

EXAMPLES 2c

1) The normal at a point P(x_1, y_1) on an ellipse of eccentricity e, meets the major axis at G. Prove that GS $= e$PS where S is a focus.

Using the standard ellipse with equation

$$\frac{x^2}{a^2} + \frac{y^2}{b^2} = 1$$

differentiating gives $$\frac{2x}{a^2} + \frac{2y}{b^2}\frac{dy}{dx} = 0,$$

i.e. $$\frac{dy}{dx} = -\frac{b^2 x_1}{a^2 y_1} \quad \text{at P.}$$

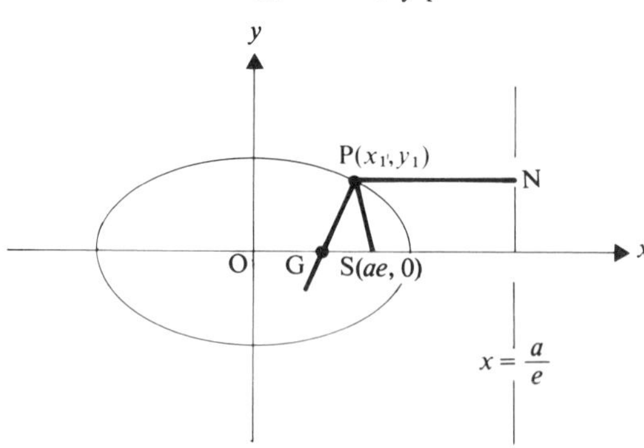

The normal at P has equation

$$y - y_1 = \frac{a^2 y_1}{b^2 x_1}(x - x_1).$$

At G, $y = 0$, so $$a^2y_1x = (a^2 - b^2)x_1y_1$$

$$\Rightarrow \qquad x = \frac{(a^2 - b^2)}{a^2}x_1.$$

But $$b^2 = a^2(1 - e^2)$$

or $$a^2 - b^2 = a^2e^2.$$

So at G, $$x = e^2x_1.$$

Hence $$SG = ae - e^2x_1.$$

Now by definition $$PS = ePN$$

$$= e\left(\frac{a}{e} - x_1\right)$$

$$= a - ex_1$$

So $$ePS = ae - e^2x_1.$$

Hence $$SG = ePS.$$

2) The line $y = 2x$ is a diameter of the ellipse $4x^2 + 9y^2 = 36$. Find the equation of the tangents at the ends of this diameter. Show that the area enclosed by the tangent at the ends of this diameter and its conjugate, is 24 square units.

Writing the equation of the ellipse as

$$\frac{x^2}{9} + \frac{y^2}{4} = 1$$

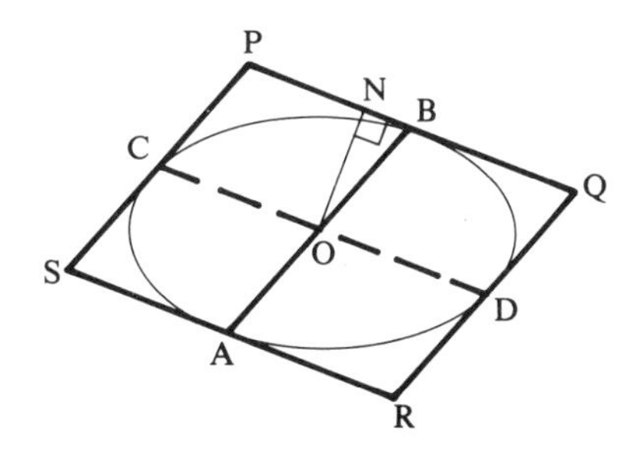

we see that

$a^2 = 9$ and $b^2 = 4$.

The conjugate of the diameter $y = 2x$ has gradient

$$-\frac{b^2}{2a^2} = -\frac{2}{9}.$$

The tangents at the ends of the diameter $y = 2x$ therefore have a gradient $-\frac{2}{9}$ because they are parallel to the conjugate diameter.

At A and B, $$4x^2 + 9(4x^2) = 36 \quad \Rightarrow \quad x = \pm\frac{3}{\sqrt{10}}.$$

So the equations of the tangents at A and B are

$$\begin{cases} y - \dfrac{6}{\sqrt{10}} = -\dfrac{2}{9}\left(x - \dfrac{3}{\sqrt{10}}\right) \\ y + \dfrac{6}{\sqrt{10}} = -\dfrac{2}{9}\left(x + \dfrac{3}{\sqrt{10}}\right) \end{cases}$$

i.e. $$9y + 2x = \pm 6\sqrt{10}.$$

To find the area of the parallelogram PQRS we will find the length of CD (the conjugate diameter) and the perpendicular distance from O to PQ.
The equation of CD is $y = -\frac{2}{9}x$.
Therefore at C and D

$$4x^2 + 9(-\tfrac{2}{9}x)^2 = 36$$

$\Rightarrow$ $$x = \pm\frac{9}{\sqrt{10}}.$$

Therefore at D, $$x = \frac{9}{\sqrt{10}} \quad \text{and} \quad y = -\frac{2}{\sqrt{10}}$$

$\Rightarrow$ $$\text{OD} = \sqrt{\frac{85}{10}} = \text{OC}.$$

Also, ON is the perpendicular distance from $(0, 0)$ to the line $9y + 2x - 6\sqrt{10} = 0$,

i.e. $$\text{ON} = \left|\frac{-6\sqrt{10}}{\sqrt{(81+4)}}\right| = 6\sqrt{\frac{10}{85}}.$$

Then
$$\begin{aligned} \text{area PQRS} &= 4 \times \text{area OCPB} \\ &= 4 \times \text{OC} \times \text{ON} \\ &= 4 \times \sqrt{\frac{85}{10}} \times 6 \times \sqrt{\frac{10}{85}} \\ &= 24 \text{ square units.} \end{aligned}$$

Note. In general such a parallelogram is of area $4ab$.

EXERCISE 2c

1) Find the values of c if $y = 3x + c$ is a tangent to the ellipse $x^2 + 4y^2 = 4$.
Find the coordinates of the points of contact.

2) Prove that the line $3y = 2x + 5$ touches the ellipse $\dfrac{x^2}{4} + y^2 = 1$.

3) Determine whether the line $y = x + 4$ meets the ellipse $\frac{x^2}{n^2} + y^2 = 1$ in two distinct points, touches the ellipse or misses it completely, when
(a) $n = 2$, (b) $n = 4$, (c) $n = \sqrt{15}$.

4) Find the equation of the locus of the midpoints of chords with gradient $\frac{1}{2}$ of the ellipse $4x^2 + 9y^2 = 36$. What is the equation of the conjugate diameter?
Find the equations of the tangents at the ends of this conjugate diameter.

5) Prove that the line $y = mx \pm \sqrt{(1 + 4m^2)}$ is a tangent to the ellipse $x^2 + 4y^2 = 4$, for all values of m.

6) Find the distances from the foci of the point $P(x_1, y_1)$ on the ellipse $\frac{x^2}{9} + \frac{y^2}{25} = 1$. (*Hint*: use the focal distance property.)

7) $P(x, y)$ is a point on the ellipse $\frac{x^2}{36} + \frac{y^2}{25} = 1$. If P is distant 5 units from one focus, how far is it from the other?

8) The foci of an ellipse are S, S′ and P is any point on the curve. The normal at P meets SS′ at G. Prove that

$$PG^2 = (1 - e^2)PS \,.\, PS'$$

where e is the eccentricity of the ellipse.

9) An ellipse with major axis horizontal and centre at the origin has foci that are 8 units apart and directrices that are 18 units apart. Find its Cartesian equation. A tangent to this ellipse is equally inclined to the axes. If the tangent cuts the axes at A and B prove that $AB = 4\sqrt{7}$.

10) Prove that the line $lx + my + n = 0$ is a tangent to the ellipse $\frac{x^2}{a^2} + \frac{y^2}{b^2} = 1$ if $a^2l^2 + b^2m^2 = n^2$.
Tangents are drawn from a point P to the ellipse. Lines from the origin are drawn perpendicular to these two tangents. If these lines are conjugate diameters prove that the equation of the locus of P is

$$a^2x^2 + b^2y^2 = a^4 + b^4.$$

11) Two conjugate diameters of the ellipse $\frac{x^2}{a^2} + \frac{y^2}{b^2} = 1$ are drawn to meet a directrix at P and Q. PM is perpendicular to OQ and QN is perpendicular to OP.
Prove that PM and QN meet at a fixed point and state its coordinates.

12) Find (i.e. do not quote) the equation of the director circle of the ellipse

$$\frac{x^2}{9}+\frac{y^2}{16} = 1.$$

PARAMETRIC COORDINATES

For the standard ellipse with Cartesian equation

$$\frac{x^2}{a^2}+\frac{y^2}{b^2} = 1$$

suitable parametric equations are

$$x = a\cos\theta, \quad y = b\sin\theta.$$

Thus the coordinates of a point P on this ellipse can be given in the form $(a\cos\theta, b\sin\theta)$ where each value of θ corresponds to one and only one point on the ellipse.

The Eccentric Angle

The geometric significance of the parameter θ is explained in the following diagram.

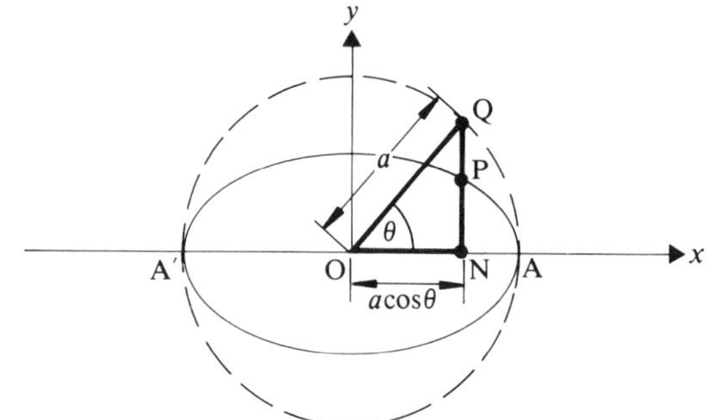

A circle is drawn on AA′ as diameter. For a point P on the ellipse, the ordinate NP is produced to cut the circle at Q.

Now since $OQ = a$ and $ON = OQ\cos\angle QON$ we see that

the x coordinate of P $=$ ON $= a\cos\angle QON$.

But the x coordinate of P $= a\cos\theta$.

So θ is the angle QON and it is sometimes called the *eccentric angle* for the point P on the ellipse. The circle on AA′ as diameter is referred to as the *eccentric circle or auxiliary circle.*

Parametric Analysis of the Ellipse

For each value of the parameter θ, there is one and only one point P on the ellipse, so the behaviour of the curve at P is uniquely related to θ. The equations of the tangent and normal at P can therefore be found in terms of θ.

The Parametric Equation of the Tangent

At P, $x = a\cos\theta$ and $y = b\sin\theta$ so the gradient of the tangent at P is the value of $\dfrac{dy}{dx}$ at P.

$$\frac{dy}{dx} = \frac{dy}{d\theta}\Big/\frac{dx}{d\theta} = -\frac{b\cos\theta}{a\sin\theta}.$$

The equation of the tangent at P is therefore

$$y - b\sin\theta = -\frac{b\cos\theta}{a\sin\theta}(x - a\cos\theta)$$

$$\Rightarrow \quad bx\cos\theta + ay\sin\theta = ab(\sin^2\theta + \cos^2\theta)$$

$$\Rightarrow \quad bx\cos\theta + ay\sin\theta = ab$$

or
$$\frac{x}{a}\cos\theta + \frac{y}{b}\sin\theta = 1.$$

Note. The trig simplification used above is not so obvious if $\dfrac{dy}{dx}$ is simplified to $-\dfrac{b}{a}\cot\theta$. In parametric work on an ellipse it is usually better to use $\cos\theta$ and $\sin\theta$ rather than $\tan\theta$ or $\cot\theta$.

The Equation of a Chord Joining Two Points on the Ellipse

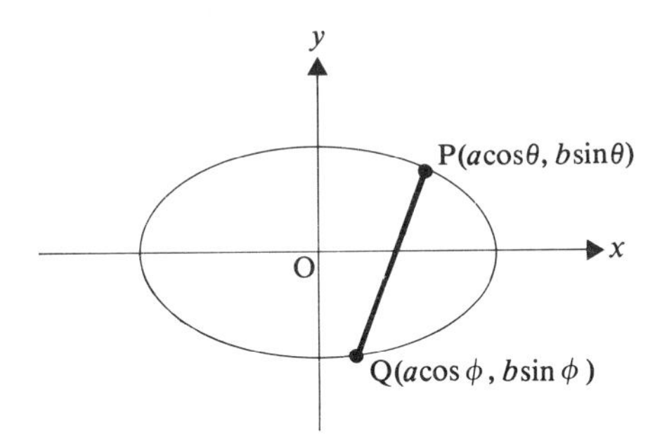

For two different points on the ellipse we use two different eccentric angles, θ and ϕ in this case.
The equation of PQ is

$$\frac{y - b\sin\theta}{x - a\cos\theta} = \frac{b(\sin\theta - \sin\phi)}{a(\cos\theta - \cos\phi)}$$

$$\Rightarrow \quad ay(\cos\theta - \cos\phi) - ab\sin\theta(\cos\theta - \cos\phi)$$
$$= bx(\sin\theta - \sin\phi) - ab\cos\theta(\sin\theta - \sin\phi)$$

$\Rightarrow$ $$ay(\cos\theta - \cos\phi) - bx(\sin\theta - \sin\phi) = ab(\sin\phi\cos\theta - \sin\theta\cos\phi).$$

But $$\cos\theta - \cos\phi \equiv 2\sin\frac{\theta+\phi}{2}\sin\frac{\phi-\theta}{2},$$

$$\sin\theta - \sin\phi \equiv 2\cos\frac{\theta+\phi}{2}\sin\frac{\theta-\phi}{2},$$

and $$\sin\phi\cos\theta - \sin\theta\cos\phi \equiv \sin(\phi-\theta) \equiv 2\sin\frac{\phi-\theta}{2}\cos\frac{\phi-\theta}{2}.$$

So the equation of the chord becomes

$$ay\sin\frac{\theta+\phi}{2} + bx\cos\frac{\theta+\phi}{2} = ab\cos\frac{\theta-\phi}{2}.$$

Note. This particular piece of work is included primarily to indicate that the parametric analysis of an ellipse requires familiarity with standard trig identities which should be revised at this stage.

From the equation of a chord we can deduce the equation of a tangent by considering the result when $Q \to P$, i.e. when $\phi \to \theta$.
The equation of the chord then becomes

$$ay\sin\theta + bx\cos\theta = ab.$$

Clearly this is not a good method for *finding* the equation of the tangent at P but it is an interesting deduction from the equation of a chord.
Note that a chord through a focus that is perpendicular to the major axis is called the *latus rectum* (see p. 65).

Conjugate Diameters and Related Parameters

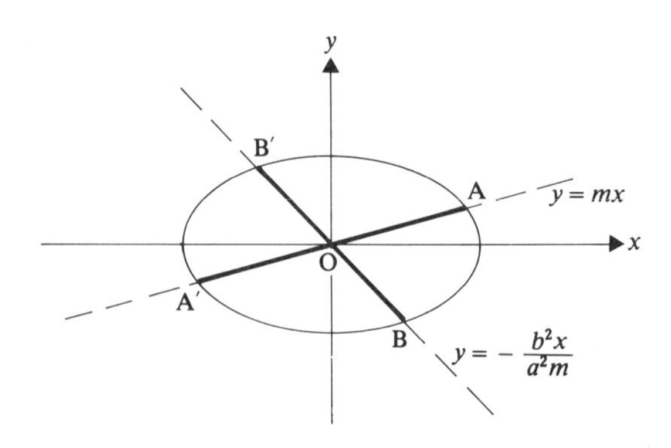

It has already been proved that the product of the gradients of conjugate diameters of the ellipse $\dfrac{x^2}{a^2}+\dfrac{y^2}{b^2}=1$ is $-\dfrac{b^2}{a^2}$.

AA′ and BB′ are a pair of conjugate diameters where A is the point $(a\cos\theta, b\sin\theta)$ and B is the point $(a\cos\phi, b\sin\phi)$.

The gradient of OA is $\dfrac{b\sin\theta}{a\cos\theta}$, i.e. $\dfrac{b}{a}\tan\theta$

and the gradient of OB is $\dfrac{b}{a}\tan\phi$.

Hence
$$\left(\frac{b}{a}\tan\theta\right)\left(\frac{b}{a}\tan\phi\right) = -\frac{b^2}{a^2}$$

$$\Rightarrow \quad \tan\theta\tan\phi = -1$$

$$\Rightarrow \quad \tan\theta = -\cot\phi = \tan\left(\phi+\frac{\pi}{2}\right)$$

$$\Rightarrow \quad \theta = \phi+\frac{\pi}{2}.$$

So we see that the eccentric angles at the ends of a pair of conjugate diameters differ by $\dfrac{\pi}{2}$.

EXAMPLES 2d

1) Show that the curve with parametric equations $x = 1+4\cos\theta$, $y = 2+3\sin\theta$ is an ellipse. State the coordinates of the centre and the lengths of the semi-axes. Find the equation of the tangent to the ellipse at the point $(1+4\cos\theta, 2+3\sin\theta)$.

Rearranging the parametric equations in the form $\cos\theta = \dfrac{x-1}{4}$, $\sin\theta = \dfrac{y-2}{3}$ and using $\cos^2\theta+\sin^2\theta\equiv 1$, we have

$$\left(\frac{x-1}{4}\right)^2+\left(\frac{y-2}{3}\right)^2 = 1$$

which is the Cartesian equation of an ellipse with centre $(1, 2)$ and semi-axes of lengths 4 and 3.

The gradient at any point is given by

$$\frac{dy}{dx} = \frac{dy}{d\theta}\Bigg/\frac{dx}{d\theta} = -\frac{3\cos\theta}{4\sin\theta}.$$

So the equation of the tangent at the point $(1+4\cos\theta, 2+3\sin\theta)$ is

$$y-(2+3\sin\theta) = -\frac{3\cos\theta}{4\sin\theta}\{x-(1+4\cos\theta)\},$$

i.e. $\quad 4y\sin\theta+3x\cos\theta = 3\cos\theta(1+4\cos\theta)+4\sin\theta(2+3\sin\theta),$

i.e. $\quad 4y\sin\theta+3x\cos\theta = 3\cos\theta+8\sin\theta+12.$

2) Show that the tangent at the point $A(a\cos\theta, b\sin\theta)$ on the ellipse $b^2x^2+a^2y^2=a^2b^2$ meets the tangent at $B(a\cos\theta, a\sin\theta)$ on the circle $x^2+y^2=a^2$ at a point on the x axis.

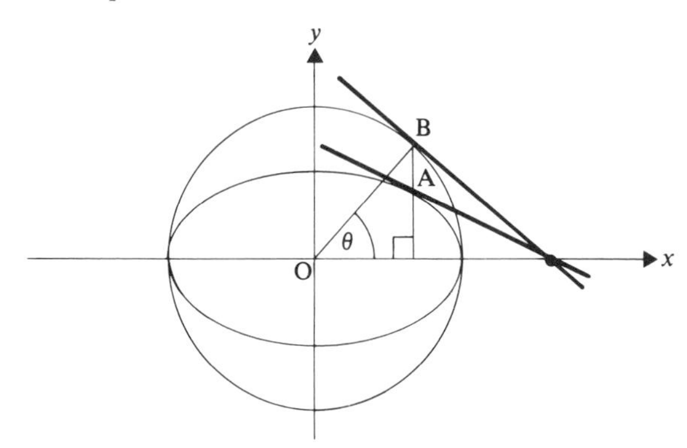

The equation of the tangent at A to the ellipse is

$$y-b\sin\theta = -\frac{b\cos\theta}{a\sin\theta}(x-a\cos\theta). \qquad [1]$$

The equation of the tangent at B to the circle is

$$y-a\sin\theta = -\frac{a\cos\theta}{a\sin\theta}(x-a\cos\theta). \qquad [2]$$

[1] becomes $\quad \dfrac{ay\sin\theta}{b}-a\sin^2\theta = -x\cos\theta+a\cos^2\theta.$

[2] becomes $\quad y\sin\theta-a\sin^2\theta = -x\cos\theta+a\cos^2\theta.$

$$(1)-(2) \Rightarrow \left(\frac{a}{b}-1\right)y\sin\theta = 0 \Rightarrow y = 0.$$

So the two tangents meet on the x axis.

3) The normal at a point $P(4\cos\theta, 3\sin\theta)$ on the ellipse $\dfrac{x^2}{16}+\dfrac{y^2}{9}=1$ meets the x and y axes at A and B. Show that the locus of M, the midpoint of AB, is an ellipse with the same eccentricity as the given ellipse. Sketch the two ellipses on the same diagram.

For the given ellipse, $a = 4$ and $b = 3$, so the major axis is horizontal and

$$e^2 = 1 - \frac{b^2}{a^2} = 1 - \frac{9}{16},$$

i.e. $e = \frac{1}{4}\sqrt{7}$.

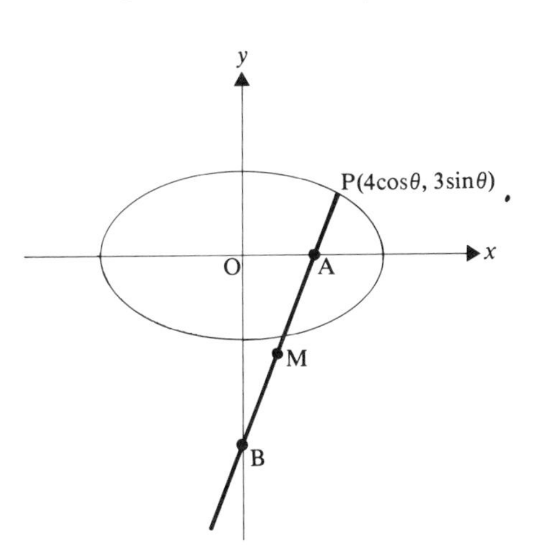

The gradient of the tangent at P is given by

$$\frac{dy}{dx} = -\frac{3\cos\theta}{4\sin\theta}$$

so the equation of the normal at P is

$$y - 3\sin\theta = \frac{4\sin\theta}{3\cos\theta}(x - 4\cos\theta),$$

i.e. $4x\sin\theta - 3y\cos\theta = 7\sin\theta\cos\theta$.

Therefore at A, $y = 0$ and $x = \frac{7}{4}\cos\theta$

and at B, $x = 0$ and $y = -\frac{7}{3}\sin\theta$.

The coordinates of M, the midpoint of AB, are therefore

$$x = \tfrac{7}{8}\cos\theta, y = -\tfrac{7}{6}\sin\theta$$

and, using $\cos^2\theta + \sin^2\theta \equiv 1$, the equation of the locus of M becomes

$$\left(\frac{8x}{7}\right)^2 + \left(-\frac{6y}{7}\right)^2 = 1,$$

i.e.

$$\frac{x^2}{(\frac{7}{8})^2} + \frac{y^2}{(\frac{7}{6})^2} = 1.$$

This is the equation of an ellipse with semi-axes of lengths $\frac{7}{8}$ and $\frac{7}{6}$, the major axis being vertical $(\frac{7}{6} > \frac{7}{8})$. So the eccentricity of this ellipse is given by

$$e^2 = 1 - (\tfrac{7}{8})^2/(\tfrac{7}{6})^2 = \tfrac{7}{16},$$

i.e. $e = \frac{1}{4}\sqrt{7}$ which is equal to the eccentricity of the given ellipse.
The two ellipses can now be sketched as shown

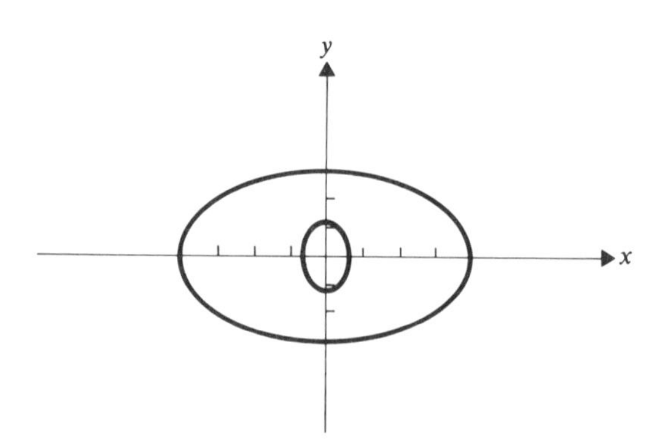

EXERCISE 2d

1) Sketch the loci of the following points as θ varies from 0 to 2π:

(a) $(4\cos\theta, 3\sin\theta)$ (b) $(\cos\theta, 2\sin\theta)$

(c) $(5+4\cos\theta, 2-3\sin\theta)$ (d) $(\cos\theta-1, 2\sin\theta)$.

2) Find the eccentricity, the coordinates of the centre and the coordinates of the foci of the ellipses in Question 1.

3) Obtain the Cartesian equations of the ellipses in Question 1.

4) Obtain parametric equations for the following ellipses:

(a) $\dfrac{(x-1)^2}{9}+\dfrac{(y-3)^2}{4}=1$ (b) $3(x+2)^2+2(y+1)^2=6$

(c) $16x^2+9y^2=144$ (d) $(x+1)^2+4(y-1)^2=4$.

5) Find the equations of the tangents to the following ellipses at the specified points:

(a) $x=4\cos\theta,\quad y=3\sin\theta,$ at the point where $\theta=\dfrac{\pi}{3}$,

(b) $x=\cos\theta,\quad y=2\sin\theta,$ at the point where $\theta=-\dfrac{\pi}{6}$,

(c) $x=4\cos\theta,\quad y=\sin\theta,$ at the point where $\theta=\dfrac{3\pi}{4}$,

(d) $x=5\cos\theta,\quad y=3\sin\theta,$ at the point where $\theta=-\dfrac{2\pi}{3}$.

6) Find the equation of the tangent and the normal at the point $P(2\cos\theta, \sin\theta)$ on the ellipse $\dfrac{x^2}{4}+y^2=1$.

7) Find an equation of the chord joining the points A and B if OA and OB are conjugate semi-diameters of the ellipse $\dfrac{x^2}{9}+\dfrac{y^2}{4}=1$.
(A and B both have positive ordinates.)

8) Find the coordinates of the points where
(a) the tangent (b) the normal
at the point $P(a\cos\theta, b\sin\theta)$ on the ellipse $\dfrac{x^2}{a^2}+\dfrac{y^2}{b^2}=1$ meet:
(i) the x axis, (ii) the y axis.

9) (a) Show that the point with parametric coordinates $(1-3\cos\theta,$ $2+\sin\theta)$ always lies on the ellipse $\dfrac{(x-1)^2}{9}+(y-2)^2=1$.

(b) Find a suitable pair of parametric equations for the ellipse $4(x+2)^2+(y-5)^2=4$.

10) S is the focus, on the positive x axis, of the ellipse $\dfrac{x^2}{9}+\dfrac{y^2}{4}=1$ and $P(3\cos\theta, 2\sin\theta)$ is a variable point on the ellipse. If SP is produced to Q so that PQ = 2PS, find the equation of the locus of Q as P moves on the ellipse.

11) The tangent at P on the ellipse $\dfrac{x^2}{a^2}+\dfrac{y^2}{b^2}=1$ meets the x and y axes at A and B.
Find, in terms of the eccentric angle of P, the ratio of the lengths of AP and BP.

12) Repeat Question 6 using the normal at P.

13) The tangent at $P(4\cos\theta, 3\sin\theta)$ on the ellipse $9x^2+16y^2=144$ meets the tangent at the positive end of the major axis at Q and meets the tangent at the positive end of the minor axis at R. Find
(a) the ratio of the lengths PQ and PR,
(b) the parametric equations of the locus of the midpoint of QR.

14) The tangents at the ends of a pair of conjugate diameters of the ellipse $x^2+4y^2=16$ are inclined to each other at an angle $\arctan\dfrac{5}{3}$. Find the coordinates of the extremities of the diameters.

15) AB is a diameter of the ellipse $x^2+9y^2=25$ and the eccentric angle of A is $\dfrac{\pi}{6}$. Find
(a) the eccentric angle of B,
(b) the equations of the tangents at A and B,
(c) the equation of the conjugate diameter.

16) Prove that the line $3x\cos\phi + 4y\sin\phi = 12$ is a tangent to the ellipse $\dfrac{x^2}{16}+\dfrac{y^2}{9}=1$ for all values of ϕ.

17) P is a point on an ellipse whose major axis is AB and the tangent at P meets the minor axis at Q. PA and PB cut the minor axis at R and T. Prove that Q bisects RT.

THE HYPERBOLA

The definition of a hyperbola, in common with the other conic sections, is the locus of points whose distances from a focus and a directrix are in a fixed ratio e (>1).

A convenient focus and directrix are the point $(ae, 0)$ and the line $x=\dfrac{a}{e}$.

The reader will recognise these as being a focus and directrix of the standard ellipse but, in the case of the hyperbola $ae>\dfrac{a}{e}$.

The Cartesian equation of this standard hyperbola can be found as follows.

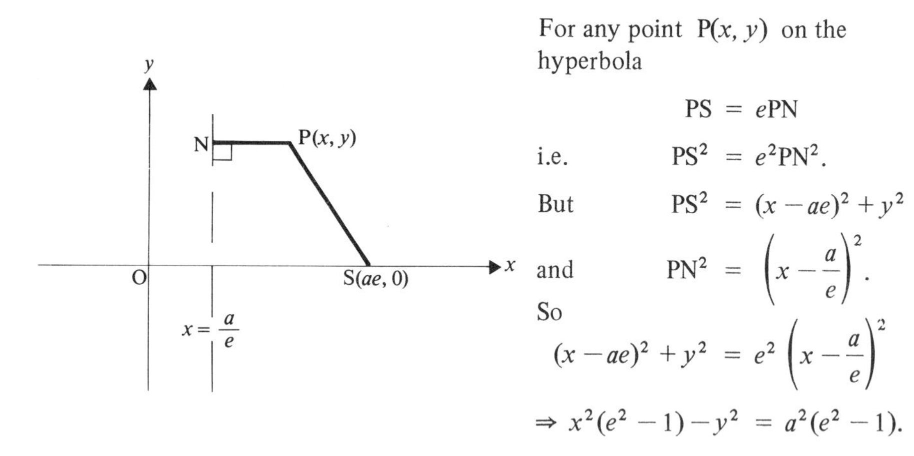

For any point $P(x, y)$ on the hyperbola

$$PS = ePN$$

i.e. $$PS^2 = e^2PN^2.$$

But $$PS^2 = (x-ae)^2+y^2$$

and $$PN^2 = \left(x-\frac{a}{e}\right)^2.$$

So

$$(x-ae)^2+y^2 = e^2\left(x-\frac{a}{e}\right)^2$$

$$\Rightarrow x^2(e^2-1)-y^2 = a^2(e^2-1).$$

Now $a^2(e^2-1)$ is always positive and so can be replaced by b^2 giving the Cartesian equation of the hyperbola in the form

$$\frac{x^2}{a^2}-\frac{y^2}{b^2}=1 \quad \text{where} \quad b^2 = a^2(e^2-1).$$

The general shape of this curve can be deduced by noting that

(a) There is symmetry about both the x axis and the y axis (as the equation contains only even powers of both x and y).

(b) $\frac{y^2}{b^2} = \frac{x^2}{a^2} - 1 \quad \Rightarrow \quad \frac{x^2}{a^2} - 1 \geqslant 0 \quad \Rightarrow \quad |x| \geqslant a.$

(c) $\begin{cases} \text{As} \quad x \to \infty, & y \to \pm\infty \\ \text{As} \quad x \to -\infty, & y \to \pm\infty. \end{cases}$

(d) $\frac{dy}{dx} = \frac{b^2x}{a^2y}$ so $\frac{dy}{dx}$ is infinite when $y = 0$.

From these properties the possible shape of the hyperbola is deduced.

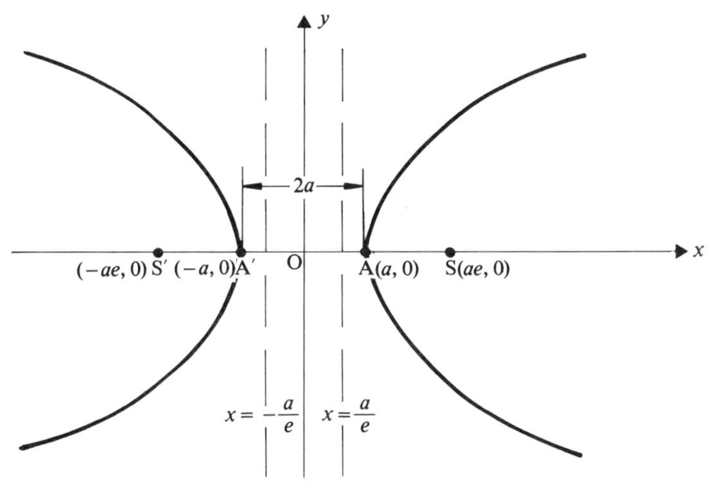

The hyperbola has two foci and two directrices (from symmetry).
AA′ is called the major axis, the points A, A′ are the vertices, and the midpoint of AA′ is the centre. The foci are always on the major axis produced.

If the major axis is vertical the equation of the hyperbola becomes

$$\frac{y^2}{a^2} - \frac{x^2}{b^2} = 1.$$

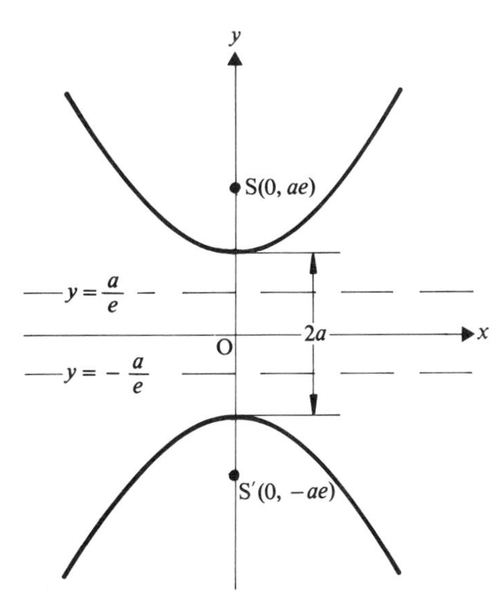

Note that the major axis is not necessarily given by the larger denominator in the Cartesian equation. It is the *positive term* whose denominator gives this axis.

For example, $\dfrac{x^2}{9} - \dfrac{y^2}{64} = 1$ has a horizontal major axis of length 6,

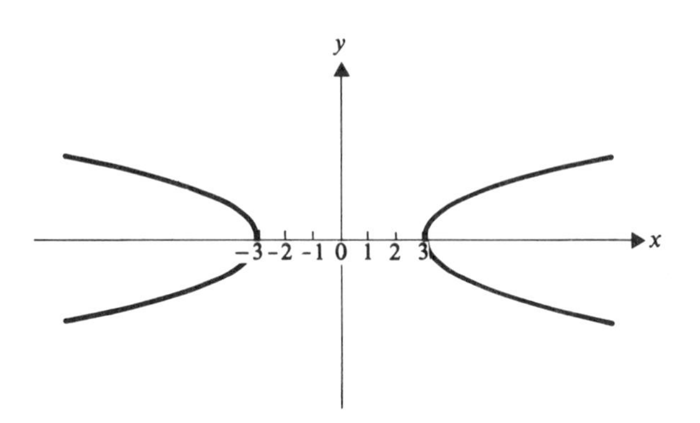

while $\dfrac{y^2}{25} - \dfrac{x^2}{16} = 1$ has a vertical major axis of length 10.

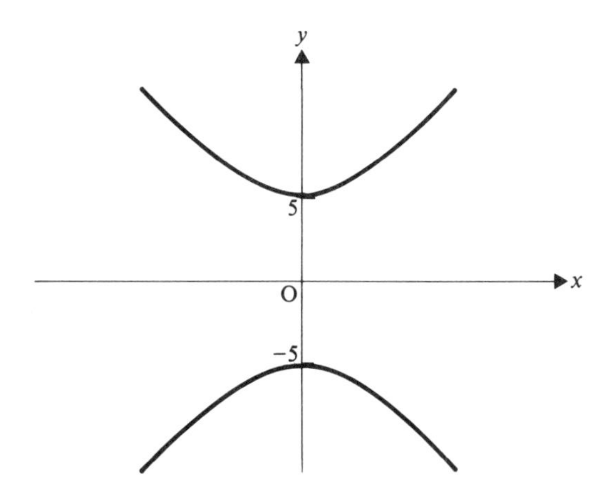

The centre, foci and axes lengths of any hyperbola with a horizontal or vertical major axis can be identified by comparing its equation with that of a standard hyperbola

$$\frac{X^2}{a^2} - \frac{Y^2}{b^2} = 1$$

whose centre is given by $X = 0, \quad Y = 0$
and whose major axis is of length $2a$.

For example, comparing the hyperbola

$$\frac{(x-3)^2}{16} - \frac{(y+2)^2}{9} = 1$$

with

$$\frac{X^2}{a^2} - \frac{Y^2}{b^2} = 1,$$

(i.e. by moving the axes to $x = 3$, $y = -2$)
we see that $x - 3 = X$, $y + 2 = Y$, $a = 4$, $b = 3$.
So the centre $(X = 0,\ Y = 0)$ is the point $(3, -2)$.
The major axis is horizontal (as the x^2 term is positive) and of length 8.
From $b^2 = a^2(e^2 - 1)$ we see that $e = \frac{5}{4}$.
The foci, which are on the major axis and distant ae from the centre, are the points $(3 \pm 5, -2)$, i.e. $(8, -2)$ and $(-2, -2)$.
The directrices (distant $\dfrac{a}{e}$ from the centre) are the lines $x = 3 \pm \frac{16}{5}$, i.e. $x = \frac{31}{5}$ and $x = -\frac{1}{5}$.

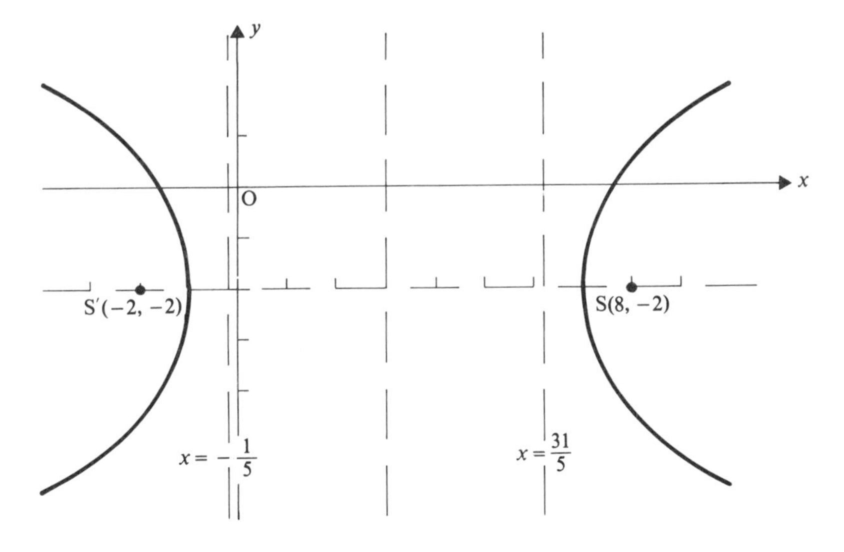

EXERCISE 2e

1) Find the equations of the following hyperbolas:
(a) foci at $(\pm 6, 0)$ and eccentricity 2;
(b) centre at the origin, one directrix $x = 2$, distance between vertices 12;
(c) centre $(2, 0)$, eccentricity 3, directrices vertical, one focus at $(4, 0)$;
(d) foci at $(0, 4)$ and $(0, -2)$, eccentricity $\frac{3}{2}$.

2) Find the centre and the distance between the vertices of each of the following hyperbolas:

(a) $\dfrac{x^2}{4} - y^2 = 1$ (b) $\dfrac{x^2}{16} - \dfrac{y^2}{25} = 1$

(c) $9(x-1)^2 - y^2 = 9$ (d) $\dfrac{(y+1)^2}{4} - \dfrac{(x+2)^2}{9} = 1$

(e) $x^2 - y^2 = 1$ (f) $y^2 - x^2 = 4$.

3) Find the coordinates of the foci and the equations of the directrices of each of the hyperbolas in Question 2.

4) Find, from the first principles, the equation of a hyperbola with a focus at $(2, 1)$ a directrix $y = x$, and eccentricity 2.

5) A set of points is such that each point is three times as far from the y axis as it is from the point $(4, 0)$. Find the equation of the locus of P and sketch the locus.

The Focal Distance Property

The distances of any point P on a hyperbola from the foci S and S′ are such that

$$|PS - PS'| = 2a.$$

This property can be proved by using the definition of a hyperbola as a locus, as follows:

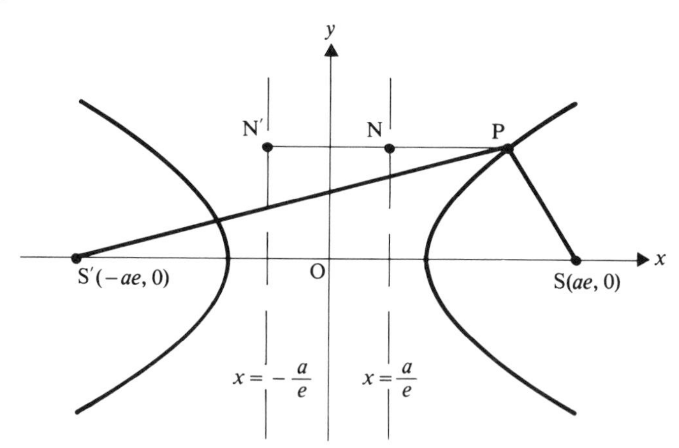

For any point P on the hyperbola,

$$PS = e PN$$

$$PS' = e PN'$$

so $$PS' - PS = e(PN' - PN).$$

But $$\text{PN}' - \text{PN} = \text{NN}'$$
$$= 2\left(\frac{a}{e}\right).$$

Hence $$\text{PS}' - \text{PS} = 2a.$$

If P is on the branch of the hyperbola with focus S′, we find that
$$\text{PS} - \text{PS}' = 2a.$$

So, in general, $|\text{PS} - \text{PS}'| = 2a.$

ASYMPTOTES

When x and y both become very large, the equation $\dfrac{x^2}{a^2} - \dfrac{y^2}{b^2} = 1$ approximates to $\dfrac{x^2}{a^2} - \dfrac{y^2}{b^2} = 0$ (since 1 is negligible compared with x^2 and y^2).

But $\dfrac{x^2}{a^2} - \dfrac{y^2}{b^2} = 0$ is the equation of two straight lines $\dfrac{x}{a} - \dfrac{y}{b} = 0$ and $\dfrac{x}{a} + \dfrac{y}{b} = 0.$

So, for large values of x and y, the hyperbola $\dfrac{x^2}{a^2} - \dfrac{y^2}{b^2} = 1$ approximates to a pair of straight lines.

Hence the pair of lines $y = \dfrac{b}{a}x$ and $y = -\dfrac{b}{a}x$ are the asymptotes of the hyperbola.

The hyperbola $\dfrac{x^2}{a^2} - \dfrac{y^2}{b^2} = 1$ can now be sketched more accurately,

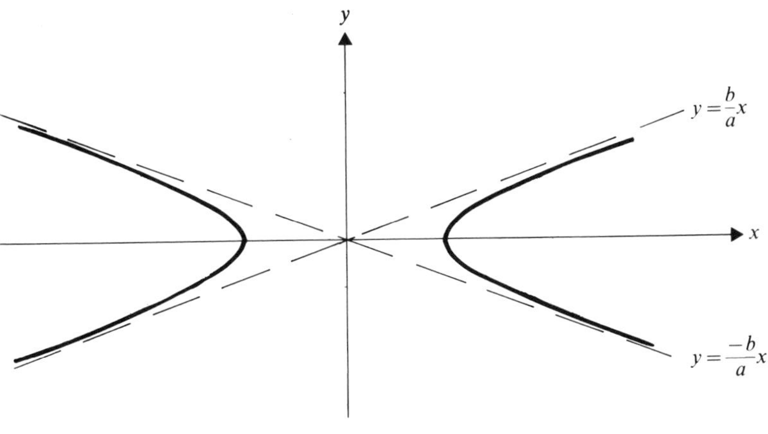

The ratio $b:a$ determines the shape of the hyperbola.

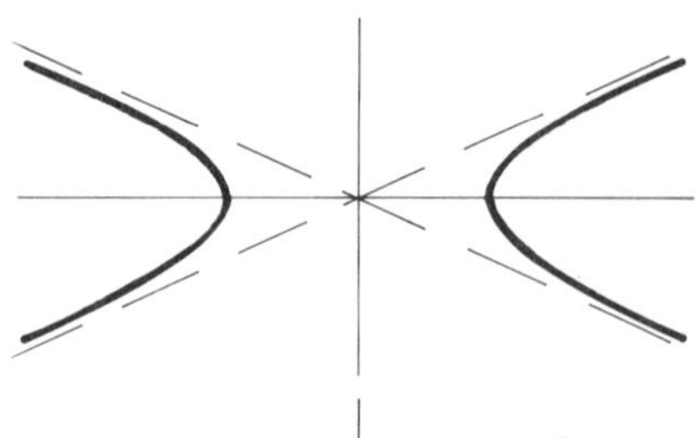

If b is small compared with a the hyperbola is sharp.

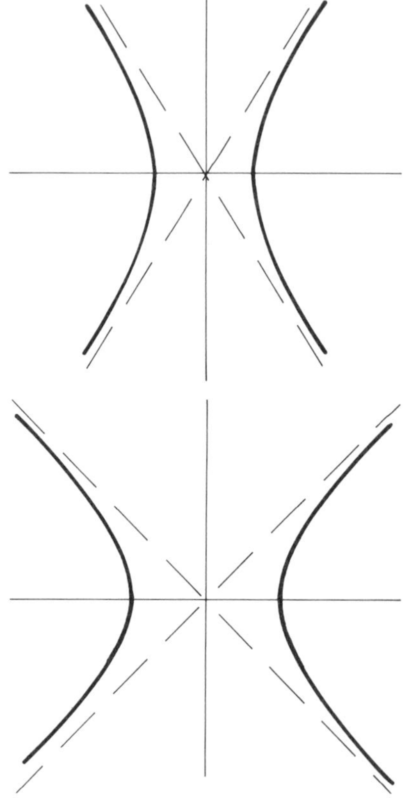

whereas the curve is more rounded when b is large compared with a.

The special case when a and b are equal gives a hyperbola whose asymptotes are perpendicular to each other.

This curve is called a *rectangular hyperbola* and its standard equation is

$$x^2 - y^2 = a^2.$$

The eccentricity of a rectangular hyperbola $(a = b)$ is given by
$a^2 = a^2(e^2 - 1)$

i.e. $$e = \sqrt{2}.$$

We will investigate the rectangular hyperbola later, but first we will complete an analysis of the general hyperbola in its standard position. As most of the techniques and methods used for a curve with Cartesian equation $\frac{x^2}{a^2} - \frac{y^2}{b^2} = 1$

are clearly similar to those used for the ellipse with equation $\dfrac{x^2}{a^2}+\dfrac{y^2}{b^2}=1$, no worked examples involving Cartesian methods are given but problems are set in Exercise 2f. Problems solved by parametric analysis, although based on principles used already in the case of the ellipse, involve different trigonometric detail, so this section is now covered more fully.

PARAMETRIC ANALYSIS

The similarity of the equation $\dfrac{x^2}{a^2}-\dfrac{y^2}{b^2}=1$ to the trig identity $\sec^2\theta - \tan^2\theta \equiv 1$ suggests the parametric equations that are usually used,

i.e. $$x = a\sec\theta, \qquad y = b\tan\theta.$$

Note that the parameter, θ, has no geometric significance in this case.

Gradient

At any point $P(a\sec\theta, b\tan\theta)$, the gradient is given by

$$\frac{dy}{dx} = \frac{dy}{d\theta}\Big/\frac{dx}{d\theta} = \frac{b\sec^2\theta}{a\sec\theta\tan\theta} = \frac{b\sec\theta}{a\tan\theta}$$

(although a further simplification gives $\dfrac{dy}{dx}=\dfrac{b}{a\sin\theta}$, this form is not recommended if the gradient is going to be used to find the equations of the tangent and the normal).

The Equation of the Tangent at $P(a\sec\theta, b\tan\theta)$

Using the gradient in the form $\dfrac{b\sec\theta}{a\tan\theta}$, the equation of the tangent is

$$y - b\tan\theta = \frac{b\sec\theta}{a\tan\theta}(x - a\sec\theta),$$

i.e. $$bx\sec\theta - ay\tan\theta = ab(\sec^2\theta - \tan^2\theta) = ab.$$

The Equation of the Normal at $P(a\sec\theta, b\tan\theta)$

The gradient of the normal is $-\dfrac{a\tan\theta}{b\sec\theta}$ so the equation of the normal is

$$y - b\tan\theta = -\frac{a\tan\theta}{b\sec\theta}(x - a\sec\theta),$$

i.e. $$by\sec\theta + ax\tan\theta = (a^2+b^2)\sec\theta\tan\theta.$$

The Equation of the Chord Joining P($a \sec \theta, b \tan \theta$) and Q($a \sec \phi, b \tan \phi$)

The equation of PQ is

$$y - b \tan \theta = \left(\frac{b \tan \theta - b \tan \phi}{a \sec \theta - a \sec \phi}\right)(x - a \sec \theta) \quad \text{or}$$

$$\frac{x}{a}(\tan \theta - \tan \phi) - \frac{y}{b}(\sec \theta - \sec \phi) = \sec \theta(\tan \theta - \tan \phi) - \tan \theta(\sec \theta - \sec \phi)$$

$$= \tan \theta \sec \phi - \sec \theta \tan \phi.$$

Multiplying throughout by $\cos \theta \cos \phi$ gives

$$\frac{x}{a}(\sin \theta \cos \phi - \cos \theta \sin \phi) - \frac{y}{b}(\cos \phi - \cos \theta) = \sin \theta - \sin \phi.$$

Then, using compound angle and factor formulae, we get

$$\frac{x}{a}\left(2 \sin \frac{\theta - \phi}{2} \cos \frac{\theta - \phi}{2}\right) - \frac{y}{b}\left(2 \sin \frac{\theta + \phi}{2} \sin \frac{\theta - \phi}{2}\right) = 2 \cos \frac{\theta + \phi}{2} \sin \frac{\theta - \phi}{2}$$

and finally the equation of the chord PQ can be written

$$\frac{x}{a} \cos \frac{\theta - \phi}{2} - \frac{y}{b} \sin \frac{\theta + \phi}{2} = \cos \frac{\theta + \phi}{2}.$$

Note that any problem involving two points on a hyperbola given in parametric form will necessarily contain some demanding trig manipulation.
Note that, if $Q \to P$
so that $\phi \to \theta$ and the chord PQ $\to$ the tangent at P,
the equation of the chord becomes

$$\frac{x}{a} - \frac{y}{b} \sin \theta = \cos \theta$$

or $\dfrac{x}{a} \sec \theta - \dfrac{y}{b} \tan \theta = 1$, which is the equation of the tangent at P.

The Tangent-Intercept Property

If the tangent at a point P on a hyperbola cuts the asymptotes at A and B, then

$$\text{PA} = \text{PB}.$$

This property can be proved as follows.

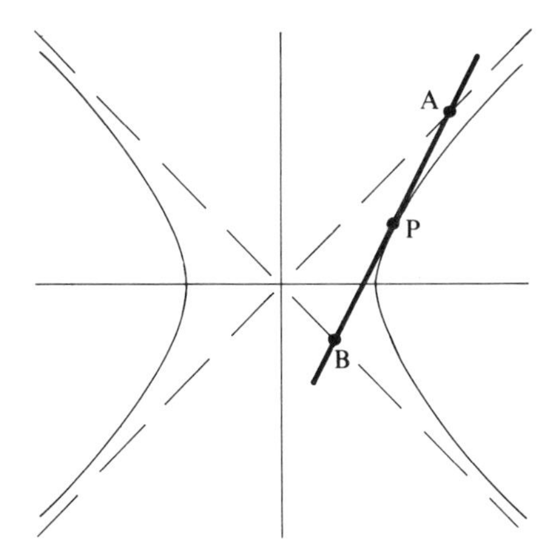

The tangent at P has equation

$$bx \sec\theta - ay \tan\theta = ab$$

and the equations of the asymptotes are

$$y = \pm\frac{b}{a}x.$$

So, at A and B

$$bx \sec\theta - (\pm bx) \tan\theta = ab$$

$\Rightarrow$
$$x = \frac{a}{\sec\theta \mp \tan\theta}.$$

At the midpoint of AB

$$x = \frac{1}{2}\left\{\frac{a}{\sec\theta + \tan\theta} + \frac{a}{\sec\theta - \tan\theta}\right\}$$

$$= \frac{1}{2}\left\{\frac{2a\sec\theta}{\sec^2\theta - \tan^2\theta}\right\}$$

$$= a\sec\theta$$

$=$ the x coordinate of P.

So P is the midpoint of the segment of the tangent at P cut off by the asymptotes.

EXAMPLES 2f

1) The tangent and normal at a point on the hyperbola $\dfrac{x^2}{a^2} - \dfrac{y^2}{b^2} = 1$ cut the

y axis at A and B. Prove that the circle on AB as diameter passes through the foci of the hyperbola.

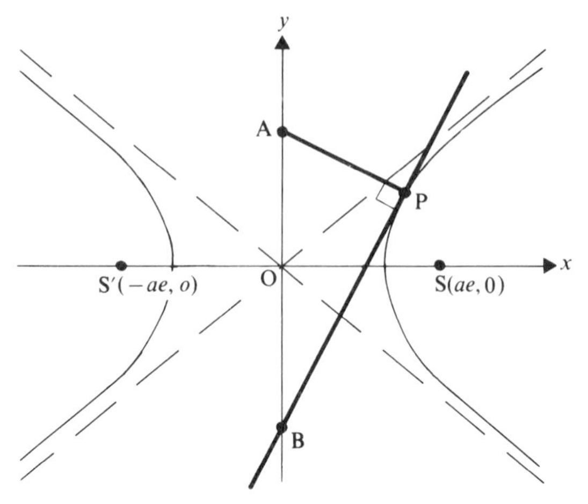

If P is the point $(a \sec\theta, b \tan\theta)$ then the equation of the tangent is

$$\frac{x \sec\theta}{a} - \frac{y \tan\theta}{b} = 1$$

And the equation of the normal (treating it as a line through P perpendicular to the tangent) is

$$\frac{x}{b}\tan\theta + \frac{y}{a}\sec\theta = \frac{a\sec\theta}{b}\tan\theta + \frac{b\tan\theta}{a}\sec\theta$$

$$= \left(\frac{a^2+b^2}{ab}\right)\sec\theta\tan\theta.$$

Hence at A $$y = \left(\frac{a^2+b^2}{b}\right)\tan\theta = \frac{a^2e^2}{b}\tan\theta$$

and at B $$y = -\frac{b}{\tan\theta}.$$

The gradient of AS is

$$\left(\frac{a^2e^2}{b}\tan\theta\right) \Big/ (-ae) = -\frac{ae}{b}\tan\theta$$

and the gradient of BS is

$$\left(\frac{b}{\tan\theta}\right) \Big/ ae = \frac{b}{ae\tan\theta}.$$

The product of these gradients is -1, showing that $\angle$ASB is a right angle.

So S is on the circle with AB as diameter.
From symmetry, S′ also lies on this circle.

2) Prove that all diameters of a hyperbola pass through the centre.

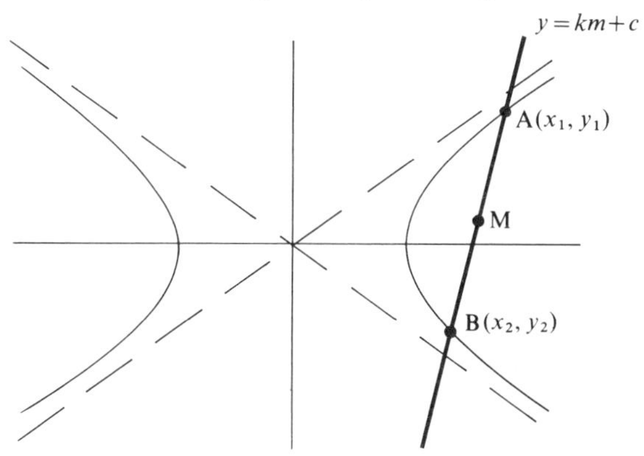

Consider a set of chords of the hyperbola $\dfrac{x^2}{a^2} - \dfrac{y^2}{b^2} = 1$, that have a constant gradient k.
One such chord, with equation $y = kx + c$, meets the hyperbola at A and B where

$$\frac{x^2}{a^2} - \frac{(kx+c)^2}{b^2} = 1,$$

i.e. $$(b^2 - a^2k^2)x^2 - 2kca^2x - a^2(b^2 + c^2) = 0. \qquad [1]$$

The x coordinate of M, the midpoint of AB, is

$$\tfrac{1}{2}(x_1 + x_2) = \tfrac{1}{2}(\text{sum of roots of equation [1]})$$
$$= \frac{1}{2}\left(\frac{2kca^2}{b^2 - a^2k^2}\right).$$

So, at M $$x = \frac{kca^2}{b^2 - a^2k^2}$$

and $$y = kx + c.$$

Eliminating c, which varies as A and B move on the hyperbola while k remains constant, gives

$$y = kx + \frac{(b^2 - a^2k^2)}{ka^2}x$$

$\Rightarrow$ $$y = \frac{b^2x}{a^2k}.$$

This is the equation of the locus of the midpoints of a set of chords with gradient k, so it is the equation of a diameter of the hyperbola. As it is satisfied by $x = 0$, $y = 0$, it passes through the origin which is the centre of the hyperbola, for any value of k.

Note that this proof is equally true for chords that join two points on different branches of the hyperbola.

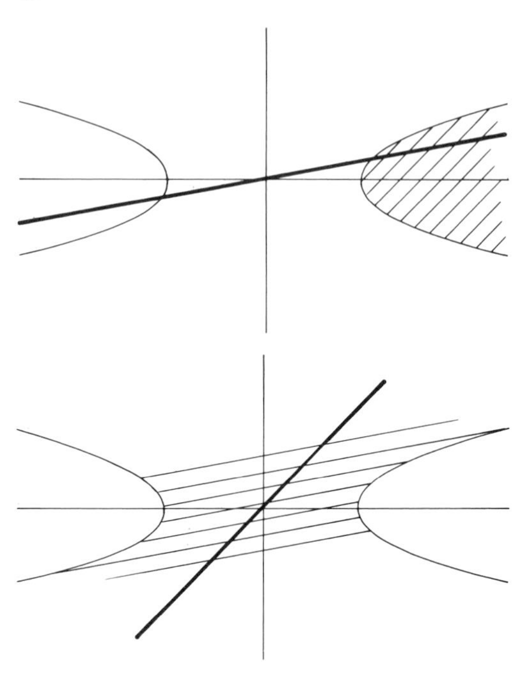

3) Find the equations of the tangents to the hyperbola $x^2 - 9y^2 = 9$ that are drawn from the point $(3, 2)$ and find the points of contact.
Find the area of the triangle that these tangents form with their chord of contact.

Consider any line with equation $y = mx + c$. This line meets the hyperbola $x^2 - 9y^2 = 9$ at points whose x coordinates are given by

$$x^2 - 9(mx + c)^2 = 9$$

i.e.
$$(9m^2 - 1)x^2 + 18mcx + 9(1 + c^2) = 0. \qquad [1]$$

If the line touches the curve, the roots of this equation are equal.
In this case

$$(18mc)^2 = 36(9m^2 - 1)(1 + c^2)$$

$\Rightarrow$
$$9m^2 = 1 + c^2. \qquad [2]$$

But the point $(3, 2)$ is on the line $y = mx + c$

so
$$2 = 3m + c. \qquad [3]$$

From [2] and [3],

$$9m^2 = 1 + (2 - 3m)^2$$

$$\Rightarrow \qquad m = \tfrac{5}{12} \quad \text{or } m \text{ is infinte,}$$

i.e.

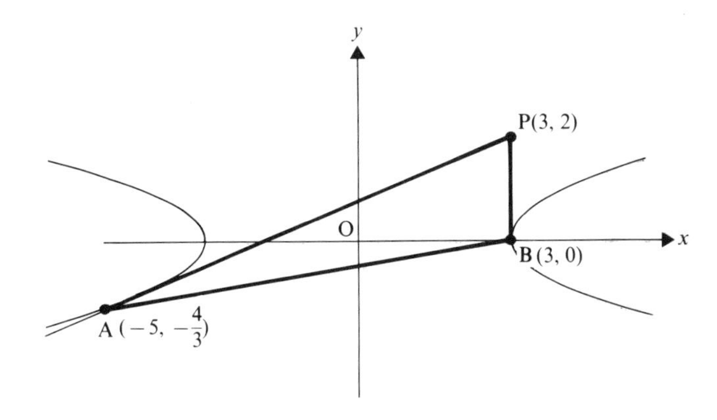

If $m = \frac{5}{12}$, then $c = 2 - 3m = \frac{3}{4}$, so the equation of one of the tangents is

$$12y = 5x + 9.$$

If $m \to \infty$, the tangent is parallel to the y axis, and as it passes through the point $(3, 2)$ its equation is

$$x = 3.$$

Now when the roots of equation [1] are equal, $x = -\dfrac{18mc}{2(9m^2 - 1)}$, and this is the x coordinate of the point of contact.
So, when $m = \frac{5}{12}$, $x = -5$, showing that the tangent $12y = 5x + 9$ touches the hyperbola at the point $(-5, -\frac{4}{3})$.
By observation, the other tangent touches the hyperbola at the vertex $(3, 0)$.
The triangle formed by the two tangents, PA and PB, and the chord of contact, AB, has vertices $(3, 2), (-5, -\frac{4}{3}), (3, 0)$.

The area, Δ, of the triangle APB is given by

$$2\Delta = \begin{vmatrix} 1 & 1 & 1 \\ x_A & x_P & x_B \\ y_A & y_P & y_B \end{vmatrix} = \begin{vmatrix} 1 & 1 & 1 \\ 3 & -5 & 3 \\ 2 & -\frac{4}{3} & 0 \end{vmatrix}$$

$$\Rightarrow \qquad \Delta = 8 \text{ square units.}$$

EXERCISE 2f

1) *Write down* the equation of the following tangents:

(a) to $\dfrac{x^2}{4}-\dfrac{y^2}{3}=1$ at the point $\left(\dfrac{4}{\sqrt{3}}, 1\right)$,

(b) to $x^2-y^2=9$ at the point $(4, 5)$,

(c) to $x^2-\dfrac{y^2}{3}=1$ at the point $(2, 3)$.

2) State the equations of the asymptotes of each of the hyperbolas in Question 1.

3) If the line $y = 2x + 1$ is a tangent to the hyperbola $x^2-ky^2=1$, find the value of k.

4) Find the condition that the line $y = mx + c$ shall touch the hyperbola $\dfrac{x^2}{a^2}-\dfrac{y^2}{b^2}=1$. Hence find the equations of the tangents from the origin to the hyperbola.

5) Prove that the line $y = x + \sqrt{3}$ is a tangent to the hyperbola $\dfrac{x^2}{4}-y^2=1$ and find the coordinates of the point of contact.

6) For what value of m does the line $y = mx + 1$ touch the hyperbola $9x^2-4y^2=36$?

7) Prove that the lines $y = mx \pm a\sqrt{(m^2-1)}$ are tangents to the hyperbola $x^2-y^2=a^2$ for all values of m.

8) Find the equation of the locus of midpoints of the chords of the hyperbola $x^2-4y^2=4$ that have a gradient of 2.

9) Prove that the product of the distances from the foci of the hyperbola $\dfrac{x^2}{a^2}-\dfrac{y^2}{b^2}=1$ to any tangent is b^2.

10) The asymptotes to the hyperbola $b^2x^2-a^2y^2=a^2b^2$ are inclined to one another at an angle α.

Prove that $\tan\alpha = \dfrac{2ab}{a^2-b^2}$.

11) $P(a\sec\theta, b\tan\theta)$ is a point on the hyperbola $\dfrac{x^2}{a^2}-\dfrac{y^2}{b^2}=1$. The ordinate at P and the tangent at P meet one asymptote at Q and R respectively. If the normal at P meets the x axis at N prove that RQ and QN are perpendicular.

12) PQ is a diameter of the hyperbola $\frac{x^2}{4} - y^2 = 1$, whose foci are S and S′. If ∠PSQ and ∠PS′Q are both right angles find the area of PSQS′.

13) $P(a \sec\theta, b\tan\theta)$ is a point on the hyperbola $x^2 - y^2 = a^2$, and A is the fixed point $(2a, 0)$. Show that the locus of M, the midpoint of AP, is another hyperbola. Sketch this locus, indicating the coordinates of the centre and the vertices, and the gradients of its asymptotes.

14) Find the equations of the tangents to the hyperbola $\frac{x^2}{3} - \frac{y^2}{2} = 1$, that are parallel to the line $y = x$. State the coordinates of the points of contact and find the area of the triangle formed by one of these tangents and the asymptotes.

MISCELLANEOUS EXERCISE 2

1) Prove that the equation of the chord joining the points $P(ap^2, 2ap)$ and $Q(aq^2, 2aq)$ of the parabola $y^2 = 4ax$ is

$$2x - (p+q)y + 2apq = 0.$$

S is the focus of the parabola and M is the midpoint of PQ.
The line through S perpendicular to PQ meets the directrix at R.
Prove that

$$2RM = SP + SQ.$$

(U of L)

2) Find the equation of the tangent at $P(at^2, 2at)$ to the parabola $y^2 = 4ax$. The tangent and the normal to the parabola at P meet the x axis at T and G respectively and M is the foot of the perpendicular from P to the line $x = -a$; S is the point $(a, 0)$.
Show that ST = PM = SP and deduce that PT bisects the angle SPM. Show also that PG bisects the angle between PS and MP produced. (U of L)

3) Find the equation of the chord joining the points $P(at_1^2, 2at_1)$ and $Q(at_2^2, 2at_2)$ on the parabola $y^2 = 4ax$.
If the chord PQ passes through the focus of the parabola and the normals at P and Q meet at R, show that the locus of R is given by $y^2 + 3a^2 = ax$.
(U of L)

4) P and Q are the points $(ap^2, 2ap)$, $(aq^2, 2aq)$ respectively on the parabola $y^2 = 4ax$, and M is the midpoint of the chord PQ.
(a) Show that the area, A, enclosed by the curve and the chord PQ is given by $9A^2 = a^4(p-q)^6$.
(b) If $q = p - 4$, give the coordinates of M in terms of p only, and find the equation of the locus of M as the value of p varies continuously. Sketch the two curves on the same diagram.

5) A point P on the parabola $(x-a)^2 = 4ay$ has coordinates $x = a + 2at$, $y = at^2$. Find the equations of the tangent and the normal to the parabola at P.
If the tangent and normal cut the x axis at the points T and N respectively, prove that

$$PT^2/TN = at.$$

Find the coordinates of the point Q in which the normal at P intersects the parabola again. (U of L)

6) Find the equation of the tangent to the parabola $y^2 = \frac{1}{2}x$ at the point $(2t^2, t)$.
Hence, or otherwise, find the equations of the common tangents of the parabola $y^2 = \frac{1}{2}x$ and the ellipse $5x^2 + 20y^2 = 4$.
Draw a sketch to illustrate the results. (O)

7) The tangent at $P(a\cos\theta, b\sin\theta)$ to the ellipse

$$b^2x^2 + a^2y^2 = a^2b^2$$

cuts the y axis at Q. The normal at P is parallel to the line joining Q to one focus S′. If S is the other focus, show that PS is parallel to the y axis. (U of L)

8) Show that for all values of m the straight lines with equation $y = mx \pm \sqrt{(b^2 + a^2m^2)}$ are tangents to the ellipse

$$\frac{x^2}{a^2} + \frac{y^2}{b^2} = 1.$$

Hence show that, if the tangents from an external point P to the ellipse meet at right angles, the locus of P is a circle and find its equation. (U of L)

9) Show that the equation of the normal to the ellipse $x^2/a^2 + y^2/b^2 = 1$ at the point $P(a\cos\theta, b\sin\theta)$ is

$$ax\sin\theta - by\cos\theta = (a^2 - b^2)\sin\theta\cos\theta.$$

If the normal to the ellipse $x^2/25 + y^2/9 = 1$ at a point Q meets the coordinate axes at A and B respectively, show that, as Q varies, the locus of the midpoint of AB is another ellipse, and give the coordinates of the foci of this second ellipse. (U of L)

10) It is given that the line $y = mx + c$ is a tangent to the ellipse

$$\frac{x^2}{a^2} + \frac{y^2}{b^2} = 1 \quad \text{if} \quad a^2m^2 = c^2 - b^2.$$

Show that if the line $y = mx + c$ passes through the point $(\frac{5}{4}, 5)$ and is a tangent to the ellipse $8x^2 + 3y^2 = 35$, then $c = \frac{35}{3}$ or $\frac{35}{9}$.
Find the coordinates of the points of contact of the tangents from the point $(\frac{5}{4}, 5)$ to the curve $8x^2 + 3y^2 = 35$.

(U of L)

11) Determine the equation of the tangent and of the normal at a point $P(a\cos\theta, b\sin\theta)$ on the ellipse $\dfrac{x^2}{a^2}+\dfrac{y^2}{b^2}=1$.
If the normal at P meets the x axis at A and the y axis at B, prove that the locus of the midpoint of AB is an ellipse.
If the centre of the given ellipse is O and the tangent at P meets the x axis at C, determine the equation of the circle passing through the points P, O, B and C. (AEB, '66)

12) Show that the equation of the normal at $P(a\cos\theta, b\sin\theta)$ to the ellipse $b^2x^2+a^2y^2=a^2b^2$ is

$$ax\sin\theta - by\cos\theta = (a^2-b^2)\sin\theta\cos\theta.$$

The normal at P meets the x axis at A and the y axis at B.
Show that the area of the triangle OAB, where O is the origin, cannot exceed $(a^2-b^2)^2/(4ab)$. Find the equation of the locus of the centroid of the triangle OAB. (U of L)

13) Prove that the equation of the chord of the ellipse $\dfrac{x^2}{a^2}+\dfrac{y^2}{b^2}=1$ joining the points $(a\cos\theta, b\sin\theta)$ and $(a\cos\phi, b\sin\phi)$ is

$$\frac{x}{a}\cos\tfrac{1}{2}(\theta+\phi)+\frac{y}{b}\sin\tfrac{1}{2}(\theta+\phi) = \cos\tfrac{1}{2}(\theta-\phi).$$

Prove that, if this chord touches the ellipse

$$\frac{x^2}{a^2}+\frac{y^2}{b^2}=\frac{1}{2},$$

θ and ϕ differ by an odd multiple of $\dfrac{\pi}{2}$. (AEB, '71)

14) The line $y=mx+c$ cuts the ellipse $x^2+4y^2=16$ in the points P and Q. Show that the coordinates of M, the midpoint of PQ, are

$$x = -4mc/(4m^2+1), \quad y = c/(4m^2+1).$$

If the chord PQ passes through the point $(2, 0)$, show that M lies on the ellipse $x^2+4y^2=2x$. Sketch the two ellipses in the same diagram. (U of L)

15) The tangent and the normal at a point $P(3\sqrt{2}\cos\theta, 3\sin\theta)$ on the ellipse $x^2/18+y^2/9=1$ meet the y axis at T and N respectively. If O is the origin, prove that OT . ON is independent of the position of P. Find the coordinates of X, the centre of the circle through P, T and N. Find also the equation of the locus of the point Q on PX produced such that X is the midpoint of PQ. (AEB, '72)

16) Show that the equation of the chord joining the points $P(a\cos\phi, b\sin\phi)$ and $Q(a\cos\theta, b\sin\theta)$ on the ellipse $b^2x^2+a^2y^2=a^2b^2$ is

$$bx \cos \tfrac{1}{2}(\theta + \phi) + ay \sin \tfrac{1}{2}(\theta + \phi) = ab \cos \tfrac{1}{2}(\theta - \phi).$$

Prove that, if the chord PQ subtends a right angle at the point $(a, 0)$, then PQ passes through a fixed point on the x axis. (U of L)

17) Prove that the equation of the tangent to the ellipse $\dfrac{x^2}{a^2} + \dfrac{y^2}{b^2} = 1$ at the point $P_1(x_1, y_1)$ is $\dfrac{xx_1}{a^2} + \dfrac{yy_1}{b^2} = 1$. The tangent at P_1 meets the tangent at $P_2(x_2, y_2)$ at T.
Show that the line

$$\frac{xx_1}{a^2} + \frac{yy_1}{b^2} = \frac{xx_2}{a^2} + \frac{yy_2}{b^2}$$

passes through T and through the midpoint of P_1P_2.
Prove that if P_1TP_2 is a right angle, then

$$\frac{x_1x_2}{a^4} + \frac{y_1y_2}{b^4} = 0. \quad \text{(U of L)}$$

18) A point P moves so that its distances from $A(a, 0)$, $A'(-a, 0)$, $B(b, 0)$, $B'(-b, 0)$ are related by the equation $AP.PA' = BP.PB'$.
Show that the locus of P is a hyperbola and find the equations of its asymptotes. (U of L)

19) Two diameters of the ellipse $b^2x^2 + a^2y^2 - a^2b^2 = 0$ are *conjugate* if each bisects all chords parallel to the other. If the gradients of two conjugate diameters are m and m', prove that $a^2mm' + b^2 = 0$.
If the conjugate diameters are AB, CD prove that

$$AB^2 + CD^2 = 4(a^2 + b^2).$$

Prove that the area of the parallelogram formed by tangents to the ellipse at A, B, C, D is $4ab$. (U of L)

20) The tangents to the hyperbola $b^2x^2 - a^2y^2 = a^2b^2$ at points A and B on the curve meet at T. If M is the midpoint of AB, prove that TM passes through the centre of the hyperbola. Prove that the product of the gradients of AB and TM is constant. (U of L)

21) Show that the equation of the tangent to the ellipse $x^2/a^2 + y^2/b^2 = 1$ at the point $(a \cos \theta, b \sin \theta)$ is

$$\frac{x \cos \theta}{a} + \frac{y \sin \theta}{b} = 1.$$

P is any point on the ellipse and the tangent at P meets the coordinate axes at Q, R. If P is the midpoint of QR, show that P lies on a diagonal of the rectangle which circumscribes the ellipse and has its sides parallel to the axes of coordinates.
Find the equation of the locus of the midpoint of QR. (U of L)

22) Write down the equations of the two asymptotes of the hyperbola $x^2/9-y^2/16=1$.
The tangent to the hyperbola at the point $P(3\sec\theta, 4\tan\theta)$ meets the asymptotes at X and Y. Show that
(a) P is the midpoint of XY,
(b) If O is the origin, the area of the $\triangle XOY$ is independent of θ. (U of L)

23) Obtain the equations of the tangent and the normal to the ellipse $b^2x^2+a^2y^2=a^2b^2$ at the point $P(a\cos\phi, b\sin\phi)$.
The normal at P meets the axes at Q and R, and the midpoint of QR is M. Show that the locus of M is an ellipse.
Sketch the two ellipses and show that they have the same eccentricity. (U of L)

24) Show that the straight line

$$x\cos\alpha+y\sin\alpha = p$$

is a tangent to the hyperbola

$$\frac{x^2}{a^2}-\frac{y^2}{b^2} = 1$$

if

$$a^2\cos^2\alpha-b^2\sin^2\alpha = p^2.$$

Find the coordinates of the point of contact.
Obtain the equations of the tangents to the hyperbola $9x^2-16y^2=144$ which touch the circle $x^2+y^2=9$. (U of L)

25) Prove that the ellipse $4x^2+9y^2=36$ and the hyperbola $4x^2-y^2=4$ have the same foci, and that they intersect at right angles.
Find the equation of the circle through the points of intersection of the two conics. (U of L)

26) Prove that the gradient of the tangent at an extremity of a latus rectum of the hyperbola

$$\frac{x^2}{9}-\frac{y^2}{7} = 1$$

is equal to the eccentricity of the hyperbola. (AEB, '67)p

27) Find the coordinates of the points of intersection of the ellipse $\dfrac{x^2}{a^2}+\dfrac{y^2}{b^2}=1\ (a>b)$ and the hyperbola $x^2-y^2=c^2$.
If the curves cut at right angles at each of these points, find the relation between a, b and c. (O)

28) Find the equation of the tangent to the hyperbola $b^2x^2-a^2y^2=a^2b^2$ at the point $(a\sec t, b\tan t)$ and show that the equation of the normal to the curve at this point is

$$ax\sin t+by = (a^2+b^2)\tan t.$$

Show that the product of the areas of the two triangles formed by each of these lines and the coordinate axes is independent of t.
Find the locus of the circumcentre of the triangle formed by the tangent and the coordinate axes. (U of L)

29) A hyperbola of the form $x^2/\alpha^2 - y^2/\beta^2 = 1$ has asymptotes $y^2 = m^2x^2$ and passes through the point $(a, 0)$. Find the equation of the hyperbola in terms of x, y, a and m.
A point P on this hyperbola is equidistant from one of its asymptotes and the x-axis. Prove that, for all values of m, P lies on the curve

$$(x^2 - y^2)^2 = 4x^2(x^2 - a^2).$$

(U of L)

30) Find the equation of the normal to the hyperbola $x^2/a^2 - y^2/b^2 = 1$ at the point $P(a \sec \theta, b \tan \theta)$.
If there is a value of θ such that the normal at P passes through the point $(2a, 0)$, show that the eccentricity of the hyperbola cannot be greater than $\sqrt{2}$.
Show that in this case, the parameter ϕ of the point on the hyperbola where the normal passes through the point $(0, -2b)$, is such that $-1 \leqslant \tan \phi < 0$.
If the normal at any point P meets the y axis at L, find the locus, as θ varies, of the midpoint of PL. (AEB '73)

31) If a given line l meets the hyperbola

$$\frac{x^2}{a^2} - \frac{y^2}{b^2} = \lambda$$

at P_1 and P_2, show that the midpoint M of P_1P_2 is independent of λ.
If the coordinates of M are (x_0, y_0), find the equation of l. (O)

32) Find the gradient of the tangent to the hyperbola $x^2/a^2 - y^2/b^2 = 1$ at the point $(a \sec \theta, b \tan \theta)$ and deduce that the equation of this tangent is $bx \sec \theta - ay \tan \theta = ab$.
If this tangent passes through a focus of the ellipse $x^2/a^2 + y^2/b^2 = 1$, show that it is parallel to one of the lines $y = x$, $y = -x$ and that its point of contact with the hyperbola lies on a directrix of the ellipse. (U of L)

33) Show that the equation

$$\frac{x^2}{29 - c} + \frac{y^2}{4 - c} = 1$$

represents:
(a) an ellipse if c is any constant less than 4,
(b) a hyperbola if c is any constant between 4 and 29.
Show that the foci of each ellipse in (a) and each hyperbola in (b) are independent of the value of c.
If $c = 13$ find the coordinates of the vertices A and B of the hyperbola.
If P and Q are points on this hyperbola such that PQ is a double ordinate, prove that the locus of the intersection of AP and BQ is an ellipse. (AEB '72)

CHAPTER 3

COORDINATE GEOMETRY II

THE RECTANGULAR HYPERBOLA

A hyperbola is said to be rectangular if its asymptotes are perpendicular to each other.
We saw in Chapter 2 that the eccentricity of a rectangular hyperbola is $\sqrt{2}$ and the standard equation is $x^2 - y^2 = a^2$.
If this hyperbola is rotated through an angle $\frac{\pi}{4}$ about the origin, its asymptotes become a suitable pair of Cartesian axes.

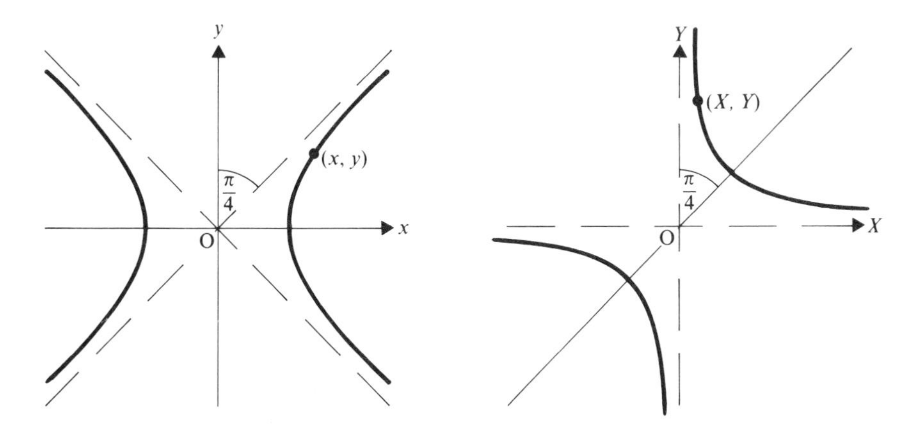

To carry out this transformation, a point (x, y) is mapped to a point (X, Y) by the matrix

$$\begin{pmatrix} \cos\frac{\pi}{4} & -\sin\frac{\pi}{4} \\ \sin\frac{\pi}{4} & \cos\frac{\pi}{4} \end{pmatrix}, \quad \text{i.e.} \quad \begin{pmatrix} \frac{1}{\sqrt{2}} & -\frac{1}{\sqrt{2}} \\ \frac{1}{\sqrt{2}} & \frac{1}{\sqrt{2}} \end{pmatrix}.$$

So
$$\begin{pmatrix} X \\ Y \end{pmatrix} = \begin{pmatrix} \frac{1}{\sqrt{2}} & -\frac{1}{\sqrt{2}} \\ \frac{1}{\sqrt{2}} & \frac{1}{\sqrt{2}} \end{pmatrix} \begin{pmatrix} x \\ y \end{pmatrix} = \begin{pmatrix} \frac{x}{\sqrt{2}} - \frac{y}{\sqrt{2}} \\ \frac{x}{\sqrt{2}} + \frac{y}{\sqrt{2}} \end{pmatrix}.$$

Hence
$$\begin{cases} X = \frac{1}{\sqrt{2}}(x - y) \\ Y = \frac{1}{\sqrt{2}}(x + y). \end{cases}$$

Multiplying these equations together gives

$$XY = \tfrac{1}{2}(x^2 - y^2).$$

But
$$x^2 - y^2 = a^2.$$

So
$$XY = \frac{a^2}{2} = c^2 \quad \text{where} \quad a = \sqrt{2}c.$$

Thus the Cartesian equation of a rectangular hyperbola referred to its asymptotes as axes is $xy = c^2$.

Note. In the term xy, the *combined* power of x and y is 2, so this is a second order equation. A general second order equation in x and y can thus include terms of the type x^2, y^2, xy, x, y and a constant and is usually written

$$ax^2 + 2hxy + by^2 + 2gx + 2fy + c = 0.$$

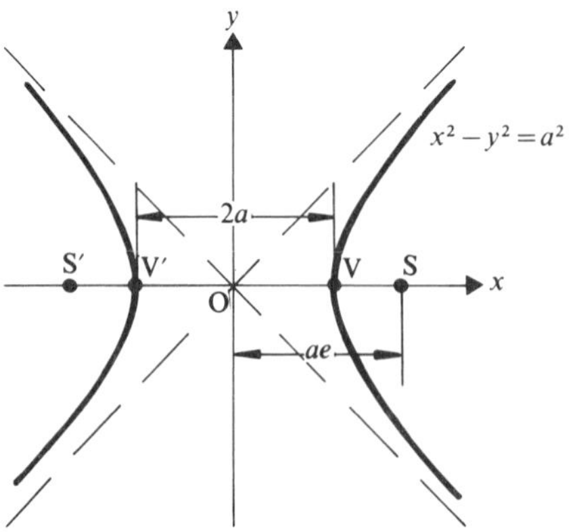

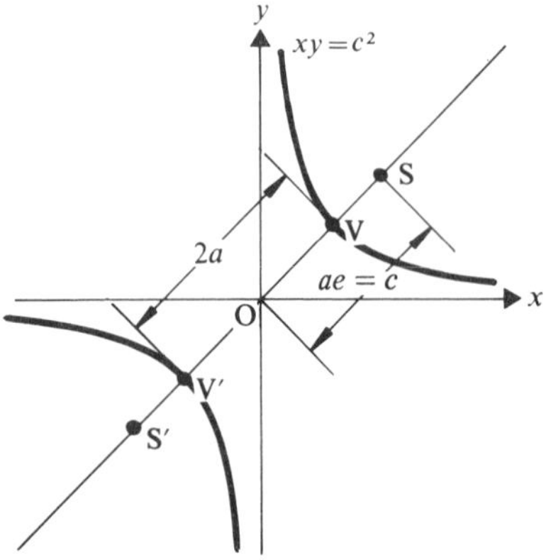

The distance between the centre and each vertex is a, i.e. $\sqrt{2}c$. So the coordinates of the vertices of $xy = c^2$ are (c, c) and $(-c, -c)$. The distance OS between the centre and a focus is ae. But $e = \sqrt{2}$ and $a = \sqrt{2}c$, so $\text{OS} = \text{OS}' = 2c$. Thus the coordinates of the foci of $xy = c^2$ are $(\sqrt{2}c, \sqrt{2}c)$ and $(-\sqrt{2}c, -\sqrt{2}c)$.

Note that the equation of a rectangular hyperbola located in the second and fourth quadrants is $xy = -c^2$.

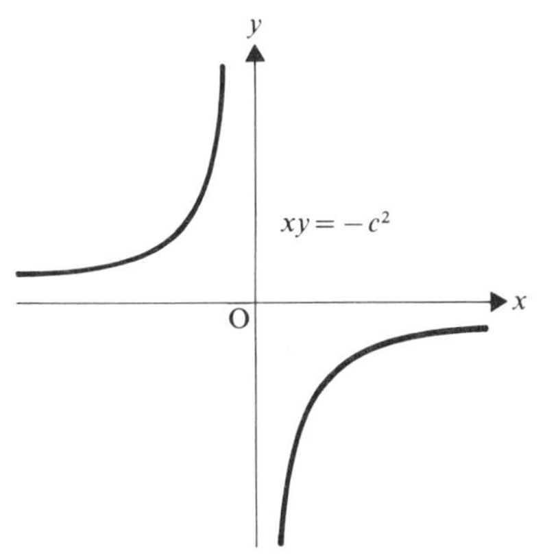

PARAMETRIC ANALYSIS

There are several suitable pairs of parametric equations that can be used for the hyperbola $xy = c^2$ but the commonest are

$$x = ct, \qquad y = \frac{c}{t}.$$

Gradient

At a point $P\left(ct, \frac{c}{t}\right)$ the gradient is given by

$$\frac{dy}{dx} = \frac{dy}{dt} \Big/ \frac{dx}{dt} = -\frac{c}{t^2} \Big/ c$$

i.e.

$$\frac{dy}{dx} = -\frac{1}{t^2}.$$

Note that this result confirms that the gradient of $xy = c^2$ is always negative, a property that can be seen from the shape of the curve.

The Parametric Equation of the Tangent is

$$y - \frac{c}{t} = -\frac{1}{t^2}(x - ct)$$

i.e. $$t^2y + x = 2ct.$$

The Parametric Equation of the Normal is

$$y - \frac{c}{t} = t^2(x - ct),$$

i.e. $$ty - t^3x = c(1 - t^4).$$

The Equation of a Chord Joining $P\left(cp, \frac{c}{p}\right)$ and $Q\left(cq, \frac{c}{q}\right)$

The gradient of PQ is

$$\frac{\frac{c}{p} - \frac{c}{q}}{cp - cq} = \frac{(q-p)}{pq(p-q)} = -\frac{1}{pq}.$$

So the equation of PQ is

$$y - \frac{c}{p} = -\frac{1}{pq}(x - cp),$$

i.e. $$pqy + x = c(p+q).$$

As we have seen before, the equation of the tangent at P can be deduced from the equation of the chord through P and Q by allowing Q to approach P so that $q \to p$ and the equation of PQ $\to$ the equation of the tangent at P.
Note that a chord may join either two points on one branch of the curve or two points, one on each branch of the curve.

The Condition for a Line to Touch a Rectangular Hyperbola

A general line $y = mx + k$ meets the rectangular hyperbola $xy = c^2$ at points whose x coordinates are given by

$$\frac{c^2}{x} = mx + k \qquad [1]$$

so the line is a tangent if the roots of this equation are equal,

i.e. if $$k^2 + 4mc^2 = 0.$$

(**Note** that, as c is a standard symbol in the equation of the rectangular hyperbola, it cannot be used for the constant term in the equation of the line, so k is used instead.)
Alternatively, if a line *with a specified gradient* is to touch the rectangular hyperbola $xy = c^2$, the problem can be dealt with parametrically using the gradient of a tangent in the form $-\frac{1}{t^2}$.

For example, if a line with gradient -4 is to touch $xy = c^2$, then the value of t at the point(s) of contact must satisfy

$$-\frac{1}{t^2} = -4 \quad \Rightarrow \quad t = \pm\frac{1}{2}.$$

Thus the points of contact are $\left(\frac{c}{2}, 2c\right), \left(-\frac{c}{2}, -2c\right)$

and the equations of the required lines are

$$y - c = -4\left(x - \frac{c}{2}\right)$$

and $$y + c = -4\left(x + \frac{c}{2}\right).$$

The Number of Normals from a Given Point

If a normal to the rectangular hyperbola is drawn from a point $A(h, k)$, not on the hyperbola, then the equation of the normal is of the form

$$ty - t^3x = c(1 - t^4)$$

where t is the value of the parameter of the point, P, where the normal meets the hyperbola. Also this equation must be satisfied by $x = h$, $y = k$,

i.e. $$tk - t^3h = c(1 - t^4). \qquad [1]$$

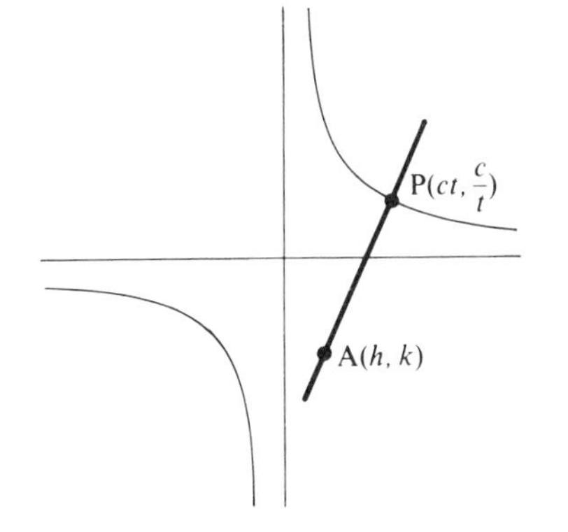

So the position of P is determined by finding the value of t that satisfies equation [1].

But equation [1] is a quartic equation for t,

i.e. $$ct^4 - ht^3 + kt - c = 0$$

and so may have either 4 real roots
or 2 real roots
or no real roots

(since complex roots always occur in conjugate pairs).

Thus we deduce that P is not unique and that from certain points it is possible to draw four normals to the hyperbola while from other points only two normals can be drawn (clearly there is no point A for which equation [1] has no real roots).
e.g.

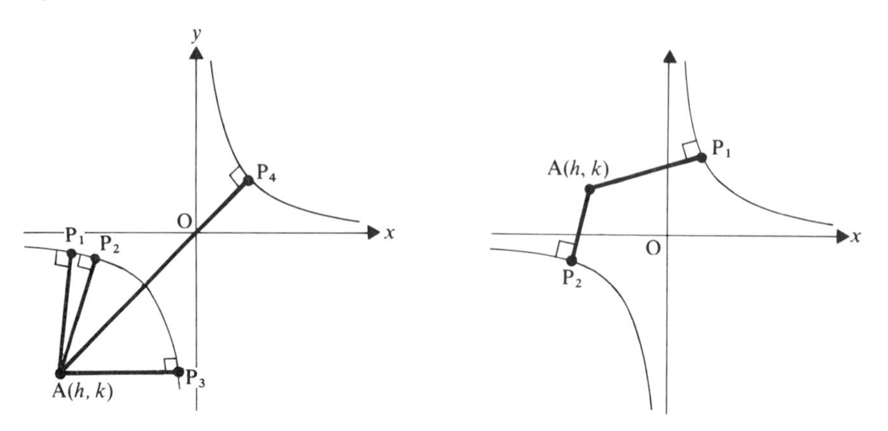

Note that the quartic equation for t may have roots that are real *and equal* and, if so, the corresponding normals coincide.

EXAMPLES 3a

1) A chord PQ of the rectangular hyperbola $xy = c^2$ subtends a right angle at another point R on the hyperbola. Prove that the normal at R is parallel to PQ.

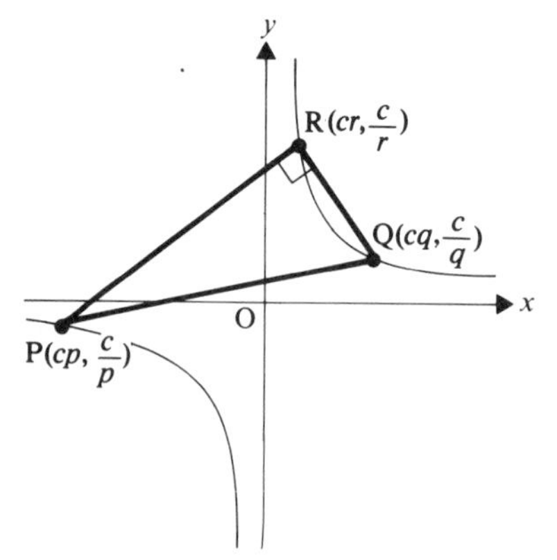

The gradient of PR is

$$\left(\frac{c}{r}-\frac{c}{p}\right)\bigg/(cr-cp) = -\frac{1}{pr}.$$

Similarly the gradient of QR is $-\dfrac{1}{qr}$.

But PR is perpendicular to QR

so $$\left(-\frac{1}{pr}\right)\left(-\frac{1}{qr}\right) = -1,$$

i.e. $$\frac{1}{pqr^2} = -1. \qquad [1]$$

The gradient of the normal at R is r^2

and the gradient of PQ is $-\dfrac{1}{pq}$.

But, from [1], $\dfrac{1}{pq} = -r^2$, so the gradient of PQ is r^2.

Thus the normal at R is parallel to PQ.

2) A chord PQ of the rectangular hyperbola $xy = c^2$ meets the asymptotes at L and M.
Prove that PL = QM.
A tangent to the parabola $y^2 = 4x$ meets the hyperbola $xy = 9$ at two points P and Q.
Find, and sketch, the equation of the locus of the midpoint of PQ.

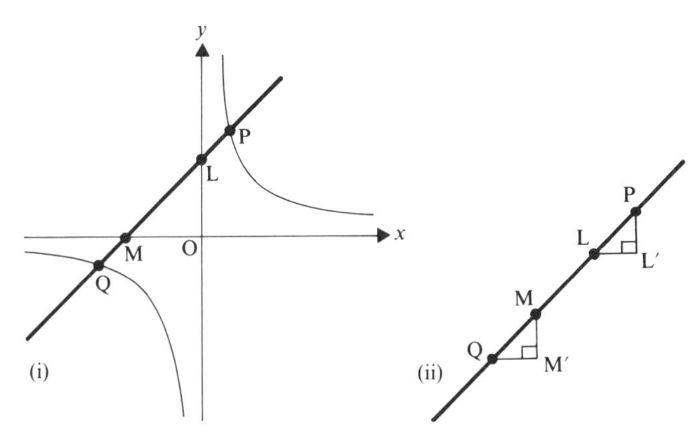

The chord joining $P\left(cp, \dfrac{c}{p}\right)$ and $Q\left(cq, \dfrac{c}{q}\right)$ has equation

$$y - \frac{c}{p} = -\frac{1}{pq}(x - cp)$$

i.e.
$$pqy + x = c(p+q).$$

The asymptotes of the rectangular hyperbola $xy = c^2$ are the x and y axes. Therefore

at L:
$$x = 0, \quad y = \frac{c(p+q)}{pq}$$

at M:
$$y = 0, \quad x = c(p+q).$$

So, in diagram (ii),

$$\begin{aligned} LL' &= x_P - x_L &&= cp - 0 \\ QM' &= x_{M'} - x_Q &&= c(p+q) - cq. \\ & &&= cp. \end{aligned}$$

Therefore, since triangles QM′M and LL′P are similar and have equal bases,

$$QM = LP.$$

For the hyperbola $xy = 9$, $c = 3$.
So the equation of PQ is

$$pqy + x = 3(p + q)$$

If PQ touches the parabola $y^2 = 4x$, then

$$pqy + \frac{y^2}{4} = 3(p + q)$$

gives equal values for y.

i.e. $\quad y^2 + 4pqy - 12(p + q) = 0$

has equal roots.

So
$$(4pq)^2 + 48(p + q) = 0,$$

i.e.
$$(pq)^2 + 3(p + q) = 0. \qquad [1]$$

The coordinates of N, the midpoint of PQ, and therefore the midpoint of LM (since $\text{QM} = \text{LP}$), are

$$x = \frac{3}{2}(p + q) \qquad [2]$$

$$y = \frac{3}{2pq}(p + q). \qquad [3]$$

The equation of the locus of N is given by eliminating p and q from equations [1], [2] and [3].

From [1] and [2]

$$(pq)^2 + 2x = 0.$$

Then, squaring [3], we have

$$4(pq)^2 y^2 = 9(p + q)^2$$

$$\Rightarrow \qquad 4(-2x)y^2 = 9\left(\frac{2x}{3}\right)^2.$$

So the equation of the locus of N is

$$-2y^2 = x$$

which is a parabola.

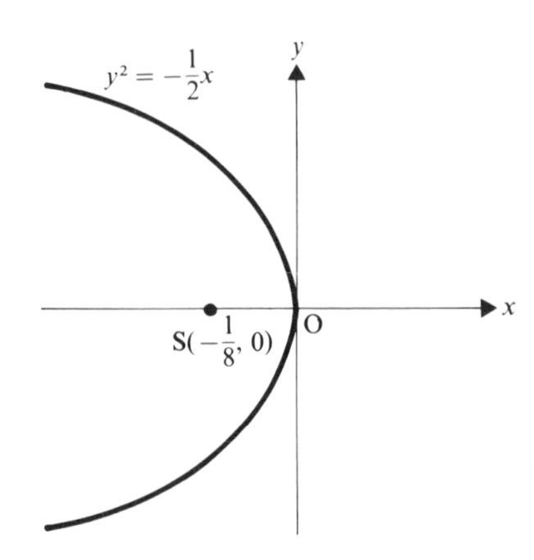

3) The normal at $P\left(ct, \frac{c}{t}\right)$ on the rectangular hyperbola $xy = c^2$ meets the curve again at Q. Find the coordinates of Q.
Hence find the equation of the locus of midpoints of normal chords of the given hyperbola. (A normal chord is normal to the hyperbola at one end.)

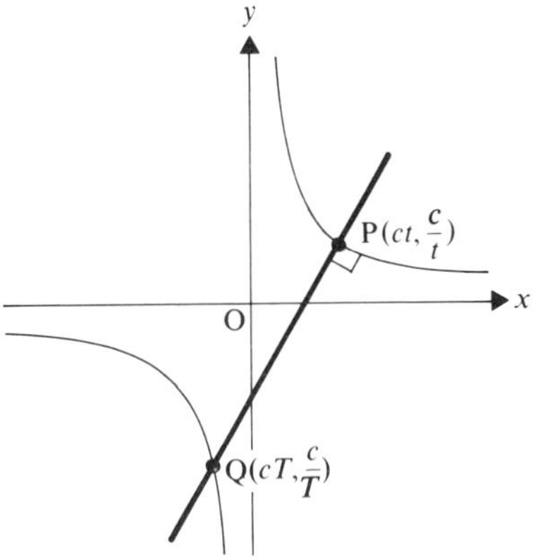

The equation of the normal at P is

$$ty - t^3x = c(1 - t^4).$$

Q is on the rectangular hyperbola, so its coordinates can be used in the form $\left(cT, \frac{c}{T}\right)$.

Q is also on the normal at P

so $$\frac{tc}{T} - ct^3T = c(1 - t^4)$$

i.e. $$t^3T^2 + (1 - t^4)T - t = 0.$$

One solution of this equation is $T = t$ (since P is on both the hyperbola and the normal).

The other solution is therefore $T = -\frac{t}{t^3} \div t$ $\left(\text{using } \alpha\beta = \frac{c}{a}\right)$.

So at Q, $T = -\frac{1}{t^3}$ and therefore the coordinates of Q are $\left(-\frac{c}{t^3}, -ct^3\right)$.

As PQ is a normal chord, we require the locus of its midpoint as P varies.
The midpoint, M, of PQ has coordinates

$$x = \frac{c}{2}\left(t - \frac{1}{t^3}\right), \quad y = \frac{c}{2}\left(\frac{1}{t} - t^3\right)$$

i.e. $$2x = \frac{c}{t^3}(t^4 - 1), \quad 2y = \frac{c}{t}(1 - t^4).$$

To find the equation of the locus of M, t must be eliminated from these two equations. It is impossible to eliminate t in one step (as t appears in the forms t, t^3 and t^4), so we first eliminate $(t^4 - 1)$, which is a factor in both equations,

i.e. $$t^4 - 1 = \frac{2xt^3}{c} = -\frac{2yt}{c} \Rightarrow t^2 = -\frac{y}{x}.$$

Now we use $t^2 = -\frac{y}{x}$ in one of the parametric equations, e.g. $$2y = \frac{c}{t}(1-t^4) \Rightarrow 4y^2 = \frac{c^2}{t^2}(1-t^4)^2$$

$$\Rightarrow 4y^2 = c^2\left(-\frac{x}{y}\right)\left(1-\frac{y^2}{x^2}\right)^2.$$

In this way t is completely eliminated and the equation of the locus of M becomes

$$4x^3y^3 + c^2(x^2-y^2)^2 = 0.$$

EXERCISE 3a

1) Write down parametric equations for the following rectangular hyperbolas:

(a) $xy = 16$ (b) $xy + 25 = 0$ (c) $(x-2)y = 1$

(d) $y = \dfrac{9}{x}$ (e) $x - 1 = \dfrac{4}{y}$ (f) $4xy = 1$.

2) Find the Cartesian equation of each of the following loci:

(a) $\left(2t, \dfrac{2}{t}\right)$ (b) $\left(3t, -\dfrac{3}{t}\right)$ (c) $\left(\dfrac{1}{t}, t\right)$

(d) $\left(1+t, \dfrac{1}{t}\right)$ (e) $\left(4t, 1-\dfrac{4}{t}\right)$ (f) $\left(-2t, \dfrac{2}{t}\right)$.

3) Find the coordinates of the centre, the foci and the vertices of each of the rectangular hyperbolas given in Questions 1 and 2 and sketch each curve.

4) Find the equation of the tangent and the normal to each of the following hyperbolas at the specified points:

(a) $xy = 16; \left(4t, \dfrac{4}{t}\right)$ (b) $xy = 1; (2, \frac{1}{2})$

(c) $xy = -9; \left(-3t, \dfrac{3}{t}\right)$ (d) $xy = 4$; each vertex.

5) Find the equations of the tangents to the rectangular hyperbola $xy = 4$ that

(a) have a gradient of $-\frac{1}{2}$,
(b) pass through the point $(2, 0)$.

6) The normal at the point $A(6, \frac{3}{2})$ on the rectangular hyperbola $x = 3t$, $y = \dfrac{3}{t}$ meets the curve again at B. Find the coordinates of B and the length of AB.

7) The tangent at $P(5, 2)$ on the curve $xy = 10$ meets the asymptotes at Q and R. Find the length of QR.

8) Repeat Question 7 for the normal at P.

9) Prove that the line $y + m^2x = 2cm$ touches the rectangular hyperbola $xy = c^2$ for all values of m. Find the coordinates of the point of contact.

10) A normal chord AB is drawn at the point A(4, 1) on the rectangular hyperbola $xy = 4$. Find its length.

11) Prove that the area of the triangle bounded by the asymptotes and a tangent to the rectangular hyperbola $xy = c^2$ is constant.

12) The tangent at a point on the curve $x = ct$, $y = \frac{c}{t}$ meets the x and y axes at P and Q respectively. The normal at the same point meets the lines $y = x$ and $y = -x$ at R and S respectively. Prove that PRQS is a rhombus unless $t^2 = 1$.

13) P and Q are variable points on the hyperbola $xy = 9$. The tangents at P and Q meet at R. If PQ passes through the point (6, 2), find the equation of the locus of R.

14) Find the condition that the line $lx + my + n = 0$ shall be a tangent to the hyperbola $xy = c^2$.

15) The tangent at a point P on the hyperbola $xy = c^2$ meets the asymptotes at A and B. Prove that P bisects AB.

16) $A(x_1, y_1)$ and $B(x_2, y_2)$ are the ends of a diameter of a circle. Show that the equation of the circle is

$$(x - x_1)(x - x_2) + (y - y_1)(y - y_2) = 1.$$

If A and B are also points on opposite branches of the hyperbola $xy = c^2$ and the circle on AB as diameter cuts the hyperbola again at C and D, prove that CD is a diameter of the hyperbola.

17) A line drawn from the centre of a rectangular hyperbola, perpendicular to the tangent at $P\left(ct, \frac{c}{t}\right)$, meets the tangent at T. As P moves on the hyperbola, find the equation of the locus of T.

18) The tangents at two points P and Q on a rectangular hyperbola, cut one asymptote at A and B and the other asymptote at C and D. Prove that PQ bisects both AB and CD.

THE LINE PAIR

Each of the conic sections we have analysed so far has had a Cartesian equation of the second degree. The last conic section, a pair of straight lines, also has a second order equation as can be seen by considering two lines with equations

$$m_1x - y + c_1 = 0 \quad \text{and} \quad m_2x - y + c_2 = 0.$$

Separately, these equations are linear, but they can be combined to form a single equation

$$(m_1x - y + c_1)(m_2x - y + c_2) = 0.$$

This second order equation is satisfied by any point on either line and therefore represents the pair of lines.
Conversely, an equation of order two represents a line pair if it is made up of two linear factors,
e.g. $x^2 + 3y^2 + 4xy + x + 3y = 0$ can be factorised to give
$(x + y + 1)(x + 3y) = 0$ and therefore is the equation of the pair of lines

$$x + y + 1 = 0 \quad \text{and} \quad x + 3y = 0.$$

THE LINE PAIR THROUGH THE ORIGIN

The equations of any two lines through the origin can be given in the form

$$m_1x - y = 0 \quad \text{and} \quad m_2x - y = 0.$$

When these equations are combined, the corresponding line pair becomes

$$(m_1x - y)(m_2x - y) = 0,$$

i.e. $$m_1m_2x^2 - (m_1 + m_2)xy + y^2 = 0. \qquad [1]$$

As *every term* in this equation is of the same degree, two, it is called a *homogeneous equation of the second order*.
In general, such a homogeneous equation can be expressed in the form

$$ax^2 + 2hxy + by^2 = 0$$

or $$\frac{a}{b}x^2 + \frac{2h}{b}xy + y^2 = 0 \qquad [2]$$

and represents a pair of lines through the origin provided that it has real linear factors,

i.e. if $$\left(\frac{2h}{b}\right)^2 - 4\left(\frac{a}{b}\right) \geqslant 0$$

$\Rightarrow$ $$h^2 - ab \geqslant 0.$$

Comparing equations [1] and [2] it can be seen that,
when $ax^2 + 2hxy + by^2 = 0$ represents a line pair through O, then the gradients of these lines, m_1 and m_2, are such that

$$m_1 + m_2 = -\frac{2h}{b} \quad \text{and} \quad m_1 m_2 = \frac{a}{b}.$$

These relationships can be used to find various properties of pairs of lines. For example, the two lines are

(a) perpendicular if m_1 and m_2 are real and $m_1 m_2 = -1$,

i.e. if $\quad h^2 - ab \geqslant 0$ and $\frac{a}{b} = -1 \Rightarrow a + b = 0.$

In this case equation [2] is of the form

$$x^2 + kxy - y^2 = 0 \qquad \left(k = \frac{2h}{a}\right).$$

(b) identical if m_1 and m_2 are real and equal,
i.e. if $h^2 = ab$.
In this case the LHS of equation [2] is a perfect square,

i.e. $$ax^2 + 2\sqrt{(ab)}\,xy + by^2 = 0$$

$\Rightarrow$ $$(x\sqrt{a} + y\sqrt{b})^2 = 0.$$

The Angle Contained by the Line Pair

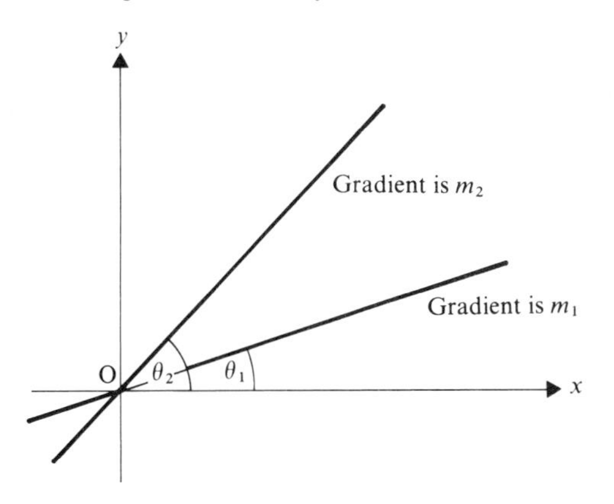

If the two lines are inclined at angles θ_1, θ_2 to the x axis, then

$$\tan\theta_1 = m_1$$

$$\tan\theta_2 = m_2$$

and $$\alpha = \theta_2 - \theta_1.$$

Thus $$\tan\alpha = \tan(\theta_2 - \theta_1),$$

i.e. $$\tan\alpha = \frac{m_2 - m_1}{1 + m_1 m_2}$$

which can be expressed in terms of a, b and h.

ASYMPTOTES AS LINE PAIRS

Consider the hyperbola $4x^2 - 9y^2 = 36$.
As x and y both become very large, the terms $4x^2$ and $9y^2$ are so large that

the term 36 is negligible and the equation of the hyperbola approximates to

$$4x^2 - 9y^2 = 0.$$

This equation has real factors

$$(2x-3y)(2x+3y) = 0$$

so the hyperbola approximates to the pair of lines $2x-3y=0$ and $2x+3y=0$ for large values of both x and y. These lines are therefore the asymptotes of the hyperbola.

This method can be applied to any second order equation that approximates to a homogeneous equation of the second degree when x and y are both very large, in order to determine whether it has real asymptotes through the origin.

A general second order equation can be written in the form

$$ax^2 + 2hxy + by^2 + 2gx + 2fy + c = 0.$$

Then if $x \to \infty$ and $y \to \infty$, the terms $2gx$, $2fy$ and c are negligible compared with ax^2, $2hxy$ and by^2.

So the general equation approximates to $ax^2 + 2hxy + by^2 = 0$.

If the curve has real asymptotes through O, this equation must represent a line pair. Therefore it must have real factors. Thus the general second order curve has real asymptotes through the origin if $h^2 - ab \geqslant 0$.

EXAMPLES 3b

1) Find the equations of the lines with equation $2x^2 + 5xy - 12y^2 = 0$, and find the angle between these lines.

$$2x^2 + 5xy - 12y^2 \equiv (2x-3y)(x+4y)$$

so $2x^2 + 5xy - 12y^2 = 0$ is the combined equation of the lines

$$2x - 3y = 0 \quad \text{and} \quad x + 4y = 0.$$

The gradients of these lines are $\frac{2}{3}$ and $-\frac{1}{4}$.

The angle, α, between them is given by

$$\tan\alpha = \frac{\frac{2}{3}-(-\frac{1}{4})}{1+(\frac{2}{3})(-\frac{1}{4})} = \frac{11}{10} \Rightarrow \alpha = \arctan\frac{11}{10}.$$

2) Find the value(s) of λ for which $6x^2 - xy + \lambda y^2 = 0$ represents

(a) two perpendicular lines,

(b) two distinct lines,

(c) two lines inclined at an angle $\dfrac{\pi}{4}$.

Comparing $\quad 6x^2 - xy + \lambda y^2 = 0$

with $\quad ax^2 + 2hxy + by^2 = 0$

we see that $a = 6$, $h = -\frac{1}{2}$ and $b = \lambda$.
The equation represents a pair of real lines if

$$h^2 - ab \geqslant 0,$$

i.e.
$$\tfrac{1}{4} - 6\lambda \geqslant 0$$

$\Rightarrow$
$$\lambda \leqslant \tfrac{1}{24}.$$

(a) The two lines are perpendicular if *also* the product of their gradients $\left(\frac{a}{b}\right)$ is -1,

i.e. if
$$\frac{6}{\lambda} = -1 \quad \Rightarrow \quad \lambda = -6.$$

This value of λ is within the range for which the given equation represents real lines, so $\lambda = -6$ gives a perpendicular line pair.

(b) The lines are distinct if the given equation has real different factors,

i.e. if
$$h^2 - ab > 0 \quad \Rightarrow \quad \lambda < \tfrac{1}{24}.$$

(c) If the lines are inclined at $\frac{\pi}{4}$, then their gradients, m_1 and m_2, satisfy

$$\tan\frac{\pi}{4} = 1 = \left|\frac{m_1 - m_2}{1 + m_1 m_2}\right|.$$

But
$$\begin{aligned}(m_1 - m_2)^2 &\equiv (m_1 + m_2)^2 - 4m_1 m_2 \\ &= \left(-\frac{2h}{b}\right)^2 - 4\left(\frac{a}{b}\right) \\ &= \left(\frac{1}{\lambda}\right)^2 - 4\left(\frac{6}{\lambda}\right) \\ &= \frac{1}{\lambda^2}(1 - 24\lambda).\end{aligned}$$

So
$$\left|\frac{m_1 - m_2}{1 + m_1 m_2}\right| = \left|\frac{\sqrt{(1 - 24\lambda)}}{\lambda\left(1 + \frac{6}{\lambda}\right)}\right| = 1,$$

i.e. $$1-24\lambda = \lambda^2+12\lambda+36$$

$$\Rightarrow \qquad \lambda^2+36\lambda+35 = 0$$

$$\Rightarrow \qquad (\lambda+35)(\lambda+1) = 0$$

$$\Rightarrow \qquad \lambda = -35 \quad \text{or} \quad \lambda = -1.$$

Both of these values of λ are less than $\frac{1}{24}$, so both correspond to real pairs of lines inclined at $\dfrac{\pi}{4}$ to each other.

3) A curve has equation $x^2+kxy+9y^2+3x-4y+2=0$.
(a) Determine the value(s) of k for which it has real asymptotes through O.
(b) If $k=10$ find the equations of its asymptotes.
(c) Comment on the case when $k=6$.

(a) When $x\to\infty$ and $y\to\infty$, the equation of the given curve approximates to

$$x^2+kxy+9y^2 = 0.$$

This equation represents a line pair through the origin if

$$\left(\frac{k}{2}\right)^2-(1)(9) \geqslant 0,$$

i.e. if $k^2 \geqslant 36$.
Therefore if k has any value *except* $-6<k<6$, the given curve has real asymptotes through O.

(b) If $k=10$ (a value for which there are real asymptotes), the asymptotes form the line pair $x^2+10xy+9y^2=0$

$$\Rightarrow \qquad (x+9y)(x+y) = 0.$$

So the asymptotes are the lines

$$x+9y = 0 \quad \text{and} \quad x+y = 0.$$

(c) If $k=6$ the asymptotes are real and their combined equation is

$$x^2+6xy+9y^2 = 0$$

$$\Rightarrow \qquad (x+3y)^2 = 0.$$

So for this value of k we have a pair of coincident lines, and there is only one asymptote, the line $x+3y=0$.

EXERCISE 3b

1) Which of the following equations represent a pair of straight lines, distinct or identical, through the origin?

(a) $x^2 + 2xy + y^2 = 0$ (b) $2x^2 - xy + 3y^2 = 0$
(c) $x^2 - y^2 = 0$ (d) $x^2 + y^2 = 0$
(e) $2x^2 + 3xy + y^2 = 0$ (f) $x^2 + 4xy + 3y^2 = 0$.

2) In those parts of Question 1 where a pair of straight lines is represented, write down the equations of these lines.

3) Find the equation of each of the following line pairs through O:
(a) lines with gradients 2 and 3,
(b) lines through $(2, 3)$ and $(1, -4)$ respectively,
(c) lines inclined at $\pm\frac{\pi}{3}$ to the x axis.

4) Determine which of the following pairs of lines are
(a) perpendicular (b) coincident
(c) equally inclined to Ox (but not coincident):
(i) $x^2 - y^2 = 0$ (ii) $y^2 - 4x^2 = 0$
(iii) $x^2 + 4xy + 4y^2 = 0$ (iv) $x^2 + 9y^2 = 0$
(v) $3x^2 + 10xy + 3y^2 = 0$.

5) Find the angle between the following pairs of lines:
(a) $2x^2 + 5xy + 2y^2 = 0$ (b) $x^2 - 9y^2 = 0$
(c) $x^2 + 2xy - 3y^2 = 0$ (d) $x^2 + pxy + qy^2 = 0$.

6) Find the value of n if $nx^2 + 2xy + y^2 = 0$ represents
(a) two identical lines (b) two perpendicular lines
(c) two lines inclined at 60°.

7) Find the condition(s) that the equation

$$px^2 + qxy + ry^2 = 0$$

shall represent (a) two distinct lines,
(b) a pair of perpendicular lines.
If $2p = 2r = q$, what does the equation represent?

8) By considering the set of points $P(x, y)$ that are equidistant from the two lines represented by the equation $ax^2 + 2hxy + by^2 = 0$, show that the equation of the pair of angle bisectors of these lines is

$$hx^2 + (b - a)xy - hy^2 = 0.$$

9) Show that, if α is the acute angle between the lines that form the line pair

$$ax^2 + 2hxy + by^2 = 0,$$

then $\tan\alpha = \left|\dfrac{2\sqrt{(h^2-ab)}}{a+b}\right|$. Deduce the condition that the lines shall be (a) perpendicular, (b) coincident.

10) Find the equations of the asymptotes of the hyperbola $\dfrac{x^2}{a^2}-\dfrac{y^2}{b^2}=1$, and find an expression for the tangent of the angle between them. Hence find the condition for the hyperbola to be rectangular.

11) Without assuming any knowledge of its shape, prove that the ellipse $\dfrac{x^2}{a^2}+\dfrac{y^2}{b^2}=1$ has no real asymptotes.

12) Determine whether the following curves have real asymptotes through the origin. If they have, find the equations of the asymptotes

(a) $3x^2-xy+y^2-x+2y=3$ (b) $x^2+2xy-3y^2-2=0$

(c) $4x^2+xy-5y^2+2y=x$ (d) $(x-2)^2+(y-3)^2=1$

(e) $y^2=4(x^2+1)$ (f) $(2x-1)^2+(3y+2)^2=7$.

THE LINE PAIR THROUGH THE POINTS COMMON TO A GIVEN LINE AND A GIVEN CURVE

Consider a line with equation $L=0$ [1]

and a curve with equation $C=0$ [2]

where L is a linear function of x and y, and C is of second degree in x and y.
The coordinates of any point on the line satisfy [1]
and the coordinates of any point on the curve satisfy [2].
So the coordinates of a point common to the line and the curve will give a zero value to any combination of L and C.
The particular combination we are going to consider is the one that uses equation [1] to make equation [2] into a homogeneous equation.
Suppose, for example, that the equations of a given line and curve are

$$2x+y-4=0 \quad [1]$$

and

$$x^2+3y^2-2x-y-1=0. \quad [2]$$

From [1] we use $1=\dfrac{2x+y}{4}$ so that any term in [2] which is not already of degree two, can be multiplied by $\left(\dfrac{2x+y}{4}\right)$ or $\left(\dfrac{2x+y}{4}\right)^2$ to give a homogeneous equation which is satisfied by the coordinates of A and B, the points where the given line and curve meet.

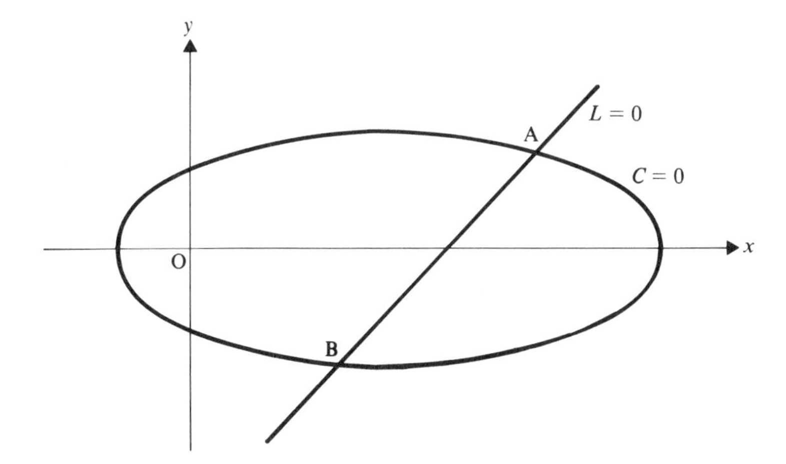

So A and B are points on the conic

$$x^2 + 3y^2 - 2x\left(\frac{2x+y}{4}\right) - y\left(\frac{2x+y}{4}\right) - 1\left(\frac{2x+y}{4}\right)^2 = 0,$$

i.e.

$$4x^2 + 20xy - 43y^2 = 0. \qquad [3]$$

But this equation is homogeneous of degree two and satisfies the condition $h^2 - ab \geqslant 0$, so it represents a pair of lines through O.
We already know that A and B lie on the conic section with equation [3].
Therefore equation [3] represents the line pair OA and OB.

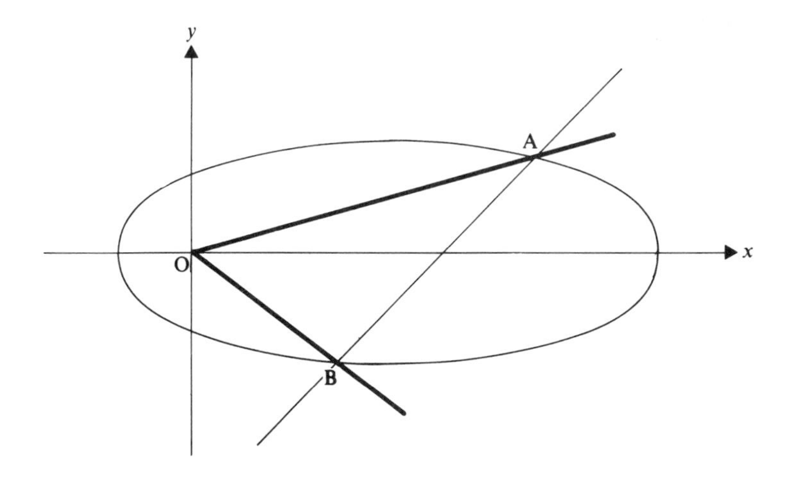

In general, if the line $lx + my + n = 0$ meets the curve $ax^2 + 2hxy + by^2 + 2gx + 2fy + c = 0$ at real points A and B, then the equation of the line pair OA and OB is given by

$$ax^2 + 2hxy + by^2 + (2gx + 2fy)\left(\frac{lx + my}{-n}\right) + c\left(\frac{lx + my}{-n}\right)^2 = 0.$$

Note that

(a) if this equation does not satisfy the condition for real factors, we deduce that the given line and curve *do not intersect*.

(b) if the *given line touches the given curve*, then A and B coincide. In this case the homogeneous equation [3], above, represents *two identical lines* and so satisfies the condition for *equal factors*.

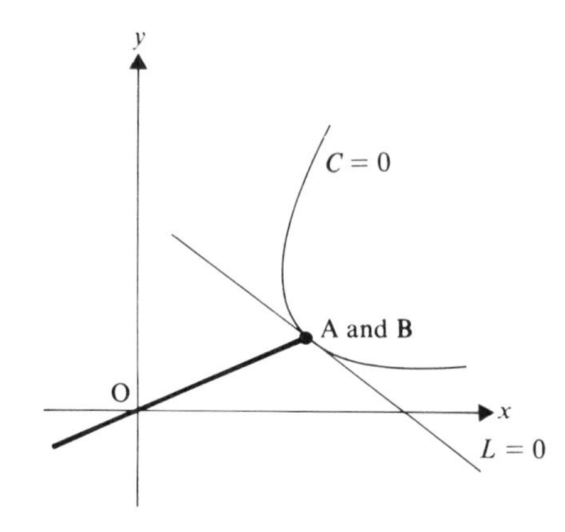

This property provides an alternative method for handling problems on tangency.

POLE AND POLAR

Suppose that, from a point $P(p, q)$ two tangents are drawn to a conic, touching it at $A(x_1, y_1)$ and $B(x_2, y_2)$.

Then AB is the *chord of contact* of the tangents from P.
AB is also called the *polar of P* with respect to the given conic and the point P is called the *pole*.

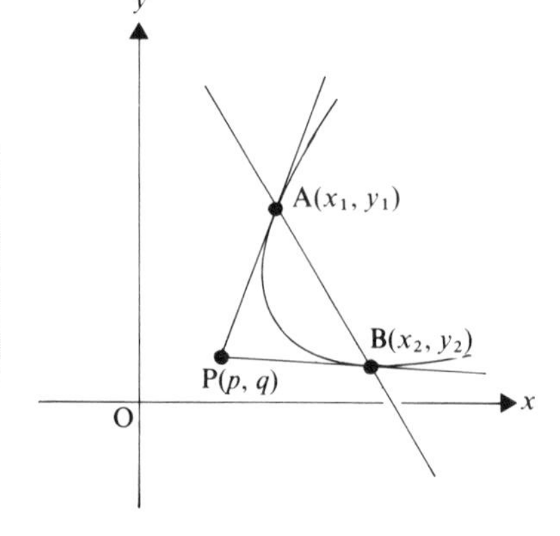

Attempting to find the equation of AB by determining the actual coordinates of A and B in each individual problem is tedious. The equation of the chord of contact is better found by the deductive method illustrated in the following example.

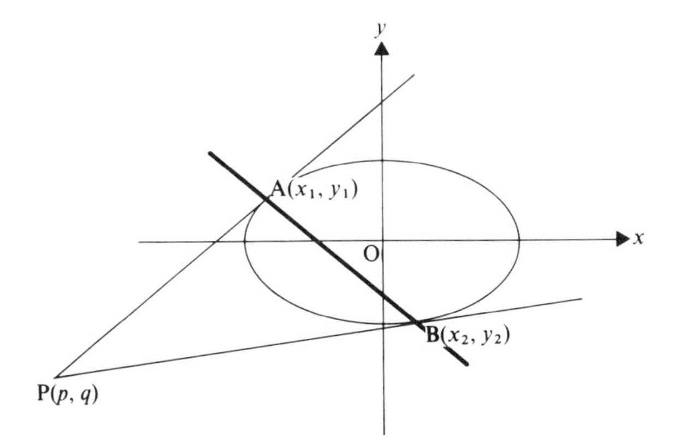

Suppose that the given conic is the ellipse $\dfrac{x^2}{a^2}+\dfrac{y^2}{b^2}=1.$

Then at $A(x_1, y_1)$ the equation of the tangent PA is

$$\frac{xx_1}{a^2}+\frac{yy_1}{b^2} = 1.$$

As $P(p, q)$ is on this tangent,

$$\frac{px_1}{a^2}+\frac{qy_1}{b^2} = 1. \qquad [1]$$

Similarly, considering the tangent PB, we have

$$\frac{px_2}{a^2}+\frac{qy_2}{b^2} = 1. \qquad [2]$$

Now observe the equation

$$\frac{px}{a^2}+\frac{qy}{b^2} = 1 \qquad [3]$$

and note that

(a) it is the equation of a line

(b) it is satisfied by (x_1, y_1) (see equation [1])

(c) it is satisfied by (x_2, y_2) (see equation [2])

(d) it is *not* satisfied by (p, q) $\dfrac{p^2}{a^2}+\dfrac{q^2}{b^2}\neq 1$ as (p, q) is not on the ellipse.

So equation [3] represents a line through A and B and is therefore the equation of AB,
i.e. the equation of the *polar* of $P(p, q)$ with respect to the ellipse $\frac{x^2}{a^2}+\frac{y^2}{b^2}=1$ is

$$\frac{px}{a^2}+\frac{qy}{b^2} = 1.$$

Extending this method to a general conic (i.e. a general second order curve) we find that the polar of $P(p, q)$ to the conic

$$ax^2 + 2hxy + by^2 + 2gx + 2fy + c = 0$$

is $$apx + h(py + qx) + bqy + g(p + x) + f(q + y) + c = 0.$$

Note that although this equation has the *same form* as that of a tangent at a given point on the conic, *the pole, P(p, q) is not on the conic.*
The equation of the polar of a given point with respect to a given conic can be quoted unless derivation is required;
e.g. the polar of the point $(2, 3)$ with respect to the hyperbola $4x^2 - y^2 = 4$ is given by

$$4xp - yq - 4 = 0 \quad \Rightarrow \quad 8x - 3y = 4.$$

EXAMPLES 3c

1) Find the equation of the line pair through the origin and the points of intersection of the line $y = mx + 2$ and the parabola $y^2 = 4(x - 1)$. Hence find the value(s) of m for which the given line touches the parabola.

If the line and the parabola meet at A and B, then at A and B

$$\frac{y - mx}{2} = 1$$

and $$y^2 - 4x + 4 = 0.$$

So the coordinates of A and B satisfy

$$y^2 - 4x\left(\frac{y - mx}{2}\right) + 4\left(\frac{y - mx}{2}\right)^2 = 0,$$

i.e. $$(2m + m^2)x^2 - 2(m + 1)xy + 2y^2 = 0.$$

As this is a homogeneous equation of order two, it can represent a line pair through O.
Thus it represents the pair of lines OA and OB, provided that A and B are real.
If the line $y = mx + 2$ touches the parabola $y^2 = 4(x - 1)$ so that the points A and B coincide, then the lines OA and OB also coincide. In this case

$$(2m + m^2)x^2 - 2(m + 1)xy + 2y^2 = 0$$

has equal factors (i.e. $h^2 = ab$).

Thus $$(m + 1)^2 = (2m + m^2)(2)$$

$\Rightarrow$ $$m^2 + 2m - 1 = 0$$

$\Rightarrow$ $$m = -1 \pm \sqrt{2}.$$

2) Write down the equation of the polar of the point (p, q) with respect to the circle $x^2 + y^2 - 6x - 4y + 12 = 0$. Use your result to derive the equation of the pair of tangents from the origin to this circle.

The polar of (p, q) with respect to the given circle is

$$px + qy - 3(x + p) - 2(y + q) + 12 = 0.$$

This line is the chord of contact of the tangents drawn to the circle from the point (p, q), so, for tangents from the origin we use $p = 0$, $q = 0$ giving the equation of the chord of contact, AB, as $3x + 2y - 12 = 0$.

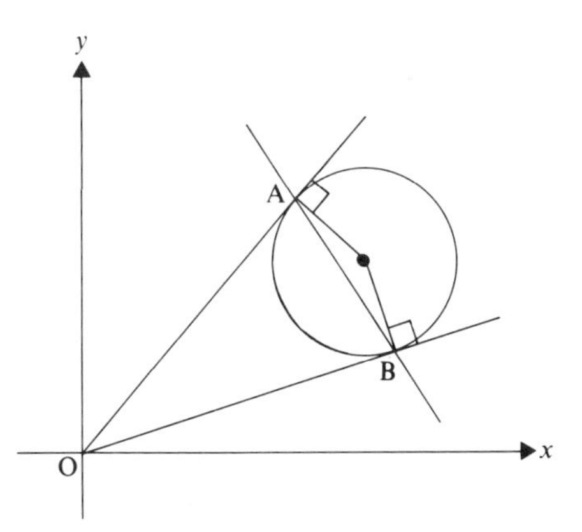

Now the equation of the line pair OA, OB is given by using $\dfrac{3x + 2y}{12} = 1$ to make the equation of the circle into a homogeneous equation,

i.e. $$x^2 + y^2 - 6x\left(\frac{3x + 2y}{12}\right) - 4y\left(\frac{3x + 2y}{12}\right) + 12\left(\frac{3x + 2y}{12}\right)^2 = 0$$

$\Rightarrow$ $$3x^2 - 12xy + 8y^2 = 0.$$

This equation is therefore the equation of the tangents from O to the circle.

EXERCISE 3c

Find the equation of the line pair from the origin through the points of intersection of the following lines and curves:

1) $x + y = 3, \quad 16x^2 + 25y^2 = 400.$

2) $2x - 1 = 4y, \quad x^2 + y^2 = 9.$

3) $4x + 3y = 5, \quad x^2 + y^2 - 4x - 6y = 0.$

4) $y = 2x - 5, \quad y^2 = 9x.$

5) $y = mx + c, \quad x^2 + y^2 = a^2.$

Write down the polar with respect to the given curve of the given point in each of the following cases:

6) $\frac{x^2}{a^2} + \frac{y^2}{b^2} = 1, \quad (2a, 2b).$

7) $x^2 + y^2 - 2x - 4y + 4 = 0, \quad (3, 5).$

8) $y^2 = 4x, \quad (-2, 1).$

9) $xy = 16, \quad (1, 1).$

10) $x^2 + 2y^2 + 3xy - 4x = 5, \quad (p, q).$

By considering the equation of the pair of lines from the origin through the possible points of intersection of the given line and the given curve, determine in the following questions whether the given line crosses, touches or misses the given curve:

11) $x^2 + y^2 = 4, \quad y = 2x + 3.$

12) $x^2 + 2y^2 = 4, \quad 2y = 5x + 9.$

13) $\frac{x^2}{4} + \frac{y^2}{9} = 1; \quad x + y = 7.$

14) $xy = 16; \quad 2x + 3y = 4.$

15) $x^2 - y^2 = 9; \quad y = 3x - 6\sqrt{2}$

SUMMARY

When a rectangular hyperbola is referred to its asymptotes as axes

(a) its equation is $xy = \pm c^2$,
(b) its eccentricity is $\sqrt{2}$,
(c) suitable parametric equations are

$$\begin{cases} x = ct \\ y = \dfrac{c}{t}. \end{cases}$$

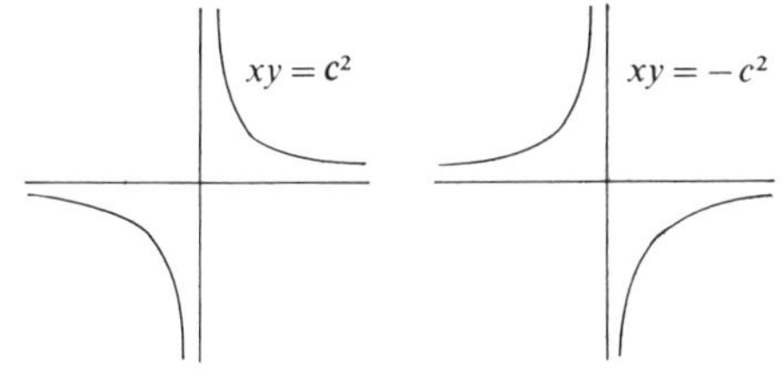

The equation $ax^2 + 2hxy + by^2 = 0$ represents a pair of straight lines through O if $h^2 - ab \geqslant 0$.
The gradients m_1 and m_2 of these lines satisfy

$$m_1 + m_2 = -\frac{2h}{b} \quad \text{and} \quad m_1 m_2 = \frac{a}{b}.$$

The lines $\begin{cases} \text{coincide if } h^2 = ab \\ \text{are perpendicular if } a + b = 0. \end{cases}$

The polar of a point (x_1, y_1) with respect to a given conic is given by replacing

$$x^2 \text{ by } xx_1, \qquad y^2 \text{ by } yy_1,$$

$$2x \text{ by } (x + x_1), \qquad 2y \text{ by } (y + y_1)$$

and

$$2xy \text{ by } (xy_1 + x_1 y)$$

in the equation of the given curve.

MISCELLANEOUS EXERCISE 3

1) Find the equation of the normal to the hyperbola $xy = c^2$ at the point P, whose coordinates are $(ct, c/t)$. Find the coordinates of the point P′ in which the normal at P cuts the hyperbola again, and write down the coordinates of the point P″ in which the normal at P′ cuts the hyperbola again.
Find the equation of the locus of the midpoint of PP′. (U of L)

2) Prove that the chord joining the points $P(cp, c/p)$ and $Q(cq, c/q)$ on the rectangular hyperbola $xy = c^2$ has the equation

$$x + pqy = c(p + q).$$

Three points P, Q, R are given on the rectangular hyperbola $xy = c^2$. Prove that
(a) if PQ and PR are equally inclined to the axes of coordinates, then QR passes through the origin O,
(b) if angle QPR is a right angle, then QR is perpendicular to the tangent at P. (C)

3) Two points $P(4p, 4/p)$ and $Q(4q, 4/q)$ lie on one branch of the rectangular

hyperbola $xy = 16$. If the line LPQM meets the axes at L and M, show that $LP = QM$.
The tangent at a point T on the other branch of the rectangular hyperbola meets the axes at R and S. Prove that $TR = TS$.
The tangents to the hyperbola at P and Q meet at U. Show that if PQ and RS are parallel, the points T and U are collinear with the origin. (U of L)

4) The tangent and normal at the point $P(ct, c/t)$ of the hyperbola $xy = c^2$ meet the x axis at Q and R. Find the equation of the circle on QR as diameter (which, of course, passes through P).
If this circle passes through the point of the hyperbola with parameter $-t$, prove that $t^4 = \frac{1}{3}$, and that the circle then passes also through the points with parameters $1/t$ and $-1/t$. (O)

5) The line $y = mx$ meets the hyperbola $xy = c^2$, where $c > 0$, at the points R and S. Prove that the tangents to the hyperbola at R and S are parallel. Find the distance between the parallel tangents and show that, as m varies, the maximum distance between them is $2c\sqrt{2}$.
The tangents and normals at R and S together form a rectangle.
Find the area of the rectangle and show that, when $m = 3$, the area is $6.4c^2$. (U of L)

6) Prove that the equation of the chord joining the points $P(cp, c/p)$ and $Q(cq, c/q)$ on the rectangular hyperbola $xy = c^2$ is

$$x + pqy = c(p+q).$$

If this chord is also the normal at P, prove that $p^3q + 1 = 0$.
If, in this case, the normal at Q cuts the hyperbola again at R, prove that PR has the equation

$$x + p^{10}y = cp(1+p^8).$$

(C)

7) Write down the coordinates of the centre and the equations of the asymptotes of the rectangular hyperbola

$$(x-h)(y-k) = c^2.$$

Sketch the hyperbolae $2x(y-2) = 3$ and $2y(x-1) = 3$ and find the coordinates of the points P and Q in which they intersect.
Show that the tangents to the hyperbolae at P and Q form a parallelogram. (U of L)

8) Find the equation of the tangent to the curve $xy = c^2$ at the point $P(cp, c/p)$.

Show that the tangents at P and Q $\left(cq, \dfrac{c}{q}\right)$ can never be perpendicular and that when they are parallel the line PQ passes through the origin, O.
Find the coordinates of R, the point of intersection of the tangents at P and Q, and show that the line RO, produced if necessary, passes through the midpoint of PQ. (AEB, '78)

9) Find the equation of the normal to the rectangular hyperbola $xy = c^2$ at the point $P\left(ct, \dfrac{c}{t}\right)$.
The normal at P meets the curve again at Q. Determine the coordinates of Q. QR is the diameter through Q of the hyperbola. Show that the locus of the midpoint of PR as P varies is

$$4x^3y^3 = c^2(x^2 + y^2)^2.$$ (U of L)

10) The straight line $y = mx + b$ meets the coordinate axes at P_1 and Q_1; it also meets the rectangular hyperbola $xy = c^2$ at P and Q. Prove that P_1Q_1 and PQ have the same midpoint.
If a set of parallel lines is drawn to cut the hyperbola, prove that the midpoints of the chords so formed lie on a straight line through the origin. (C)

11) The foot of the perpendicular from a point P to the straight line

$$x + y = \sqrt{2}$$

is the point R, and Q is the point with coordinates $(\sqrt{2}, \sqrt{2})$. If P varies in such a way that $PQ^2 = 2PR^2$, show that its locus is the rectangular hyperbola

$$xy = 1.$$

Find the equation of the tangent to this hyperbola at the point $(t, 1/t)$.
This tangent cuts the x axis at A and the y axis at B, and C is the point on AB such that $AC : CB = a : b$. Show that the locus of C as t varies is the rectangular hyperbola

$$xy = \frac{4ab}{(a+b)^2}.$$

Determine the two possible values of the ratio $a : b$ such that the straight line

$$x + y = \sqrt{2}$$

is a tangent to the locus of C. (JMB)

12) The chord through two variable points P and Q on the rectangular hyperbola $xy = c^2$ cuts the x axis at R. If S is the mid-point of PQ and O is the origin, prove that the triangle OSR is isosceles. Show that, if OP, OQ and OS make angles θ_1, θ_2 and θ_3 respectively with OR, then

$$\tan^2\theta_3 = \tan\theta_1 \tan\theta_2.$$ (AEB, '71)

13) The line of gradient m $(\neq 0)$ through the point $A(a, 0)$ is a tangent to the rectangular hyperbola $xy = c^2$ at the point P. Find m in terms of a and c, and show that the coordinates of P are $(\frac{1}{2}a, 2c^2/a)$.
The line through A parallel to the y axis meets the hyperbola at Q, and the line joining Q to the origin O intersects AP at R. Given that OQ and AP are perpendicular to each other, find the numerical value of c^2/a^2 and the numerical value of the ratio AR : RP. (JMB)

14) Find the equation of the pair of lines joining the origin to the points where the line $x + 2y - 5 = 0$ meets the pair of lines

$$4x^2 - 15xy - 4y^2 + 39x + 65y - 169 = 0.$$

Find also the angle between the pair of lines through the origin.

15) Write down the equation of the chord of contact of the tangents from the origin to the circle

$$x^2 + y^2 + 6x + 4y + 4 = 0.$$

Prove that the area of the triangle formed by these tangents and the chord of contact is $\frac{24}{13}$ square units.

16) Prove that the line $lx + my + n = 0$ touches the parabola $y^2 = 4ax$ if $am^2 = nl$. If the polar of a point P with respect to the ellipse $4(x+1)^2 + y^2 = 4$ touches the parabola $y^2 = 4x$, show that the equation of the locus of P is $16x(x+1) = y^2$.

17) Find the equation of the pair of lines that pass through the origin and through the points of intersection of the line $y = \lambda(x+1)$ and the circle $(x-2)^2 + (y-2)^2 = 9$. If these lines are perpendicular, find λ.
For what value of λ do these lines coincide? Deduce the equations of the tangents to the circle from the point $(1, 0)$.

18) Show that the equation of the pair of straight lines joining the origin to the points A and B where the line $lx + my = 1$ meets the conic $ax^2 + by^2 = 1$, is

$$(a - l^2)x^2 - 2lmxy + (b - m^2)y^2 = 0.$$

If angle AOB is a right angle, show that AB touches the circle

$$(a + b)(x^2 + y^2) = 1.$$

19) Find the equation of the line pair obtained by rotating the pair of lines with equation $x^2 + 3xy + 2y^2 = 0$ through an angle of $\dfrac{\pi}{3}$ about the origin.

20) One of the medians of the triangle formed by the line pair $ax^2 + 2hxy + by^2 = 0$ and the line $px + qy = r$ lies along the y axis. If neither a nor r is zero, prove that $bp + hq = 0$. (U of L)

21) Form the equation of the straight lines joining the origin to the points of intersection of $ax^2 + 2hxy + by^2 + 2gx + 2fy + c = 0$ and $px + qy + r = 0$, and write down the condition that these lines should be at right angles.
If this condition is satisfied, show that the equation of the locus of the foot of the perpendicular from the origin to the line $px + qy + r = 0$ is

$$(a + b)(x^2 + y^2) + 2gx + 2fy + c = 0.$$

(U of L)

CHAPTER 4

THREE DIMENSIONAL SPACE: COORDINATE METHODS

All analysis so far has been concerned with problems involving two variables. Any two variable problem can be represented visually in two dimensional space, i.e. in a plane. Similarly any three variable problem can be represented in three-dimensional space, so in this chapter we extend the methods of working in a plane to working in three dimensions, i.e. we develop coordinate geometry in three dimensions.

THE CARTESIAN FRAME OF REFERENCE

The unambiguous location of a particular point in space requires coordinates referred to a fixed frame of reference, the most convenient being the Cartesian frame of reference. This consists of a fixed point O, the origin, and three mutually perpendicular axes, Ox, Oy and Oz.

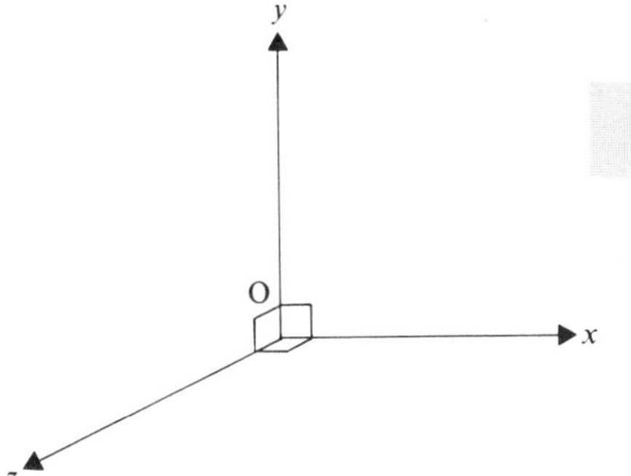

The axes are placed in such a way that they form a right-handed set.
This means that if a screw, placed at the origin, is turned in the sense from the positive x axis to the positive y axis, it moves in the direction of the positive z axis.

A point is located within this frame by giving its directed distances from O in the directions of the positive x axis, y axis and z axis. These coordinates are written as the ordered set (x, y, z).
Hence the point P whose coordinates are (a, b, c) is

a units from O in the direction Ox,
b units from O in the direction Oy,
c units from O in the direction Oz.

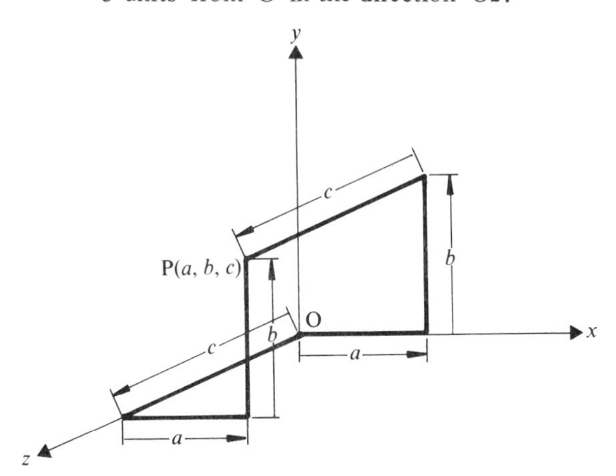

The distance of $P(a, b, c)$ from the origin can be found by Pythagoras' Theorem from the diagram as follows:

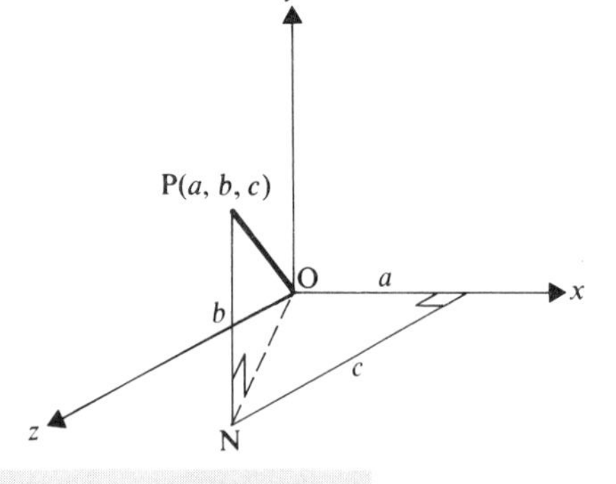

$$ON^2 = a^2 + c^2$$

$$OP^2 = ON^2 + b^2 = a^2 + c^2 + b^2$$

$$\Rightarrow OP = \sqrt{(a^2 + b^2 + c^2)}.$$

Hence the distance from O of any point $P(x, y, z)$ is given by

$$OP = \sqrt{(x^2 + y^2 + z^2)}.$$

VECTORS IN THREE DIMENSIONS

Vector methods lend themselves particularly well to three dimensional analysis. The basic properties of vectors are discussed in Chapter 13, Volume 1 and are summarised below.

1) A vector can be represented in magnitude and direction by a segment of a line.

2) The modulus, $|\mathbf{a}|$, of a vector $\mathbf{a}$ is the magnitude of $\mathbf{a}$, i.e. the length of the line representing $\mathbf{a}$.

3) A unit vector is *any* vector that has a magnitude of one unit.

4) A *free* vector has no specific location and can be represented by any one of a set of equal and parallel line segments. A *tied* vector has a particular location in space.

5) If $\mathbf{a}$ and $\mathbf{b}$ are two vectors such that $\mathbf{a} = \lambda\mathbf{b}$, then $\mathbf{a}$ and $\mathbf{b}$ are parallel.
When λ is positive $\mathbf{a}$ and $\mathbf{b}$ have the same direction.
When λ is negative $\mathbf{a}$ and $\mathbf{b}$ are opposite in direction.
When $\lambda = 1$, $\mathbf{a}$ and $\mathbf{b}$ are *equal*.
When $\lambda = -1$, $\mathbf{a}$ and $\mathbf{b}$ are *equal and opposite*.

6) If lines representing vectors in magnitude and direction are drawn consecutively, the line which completes the polygon represents the resultant vector.
This property can be expressed as a vector equation.

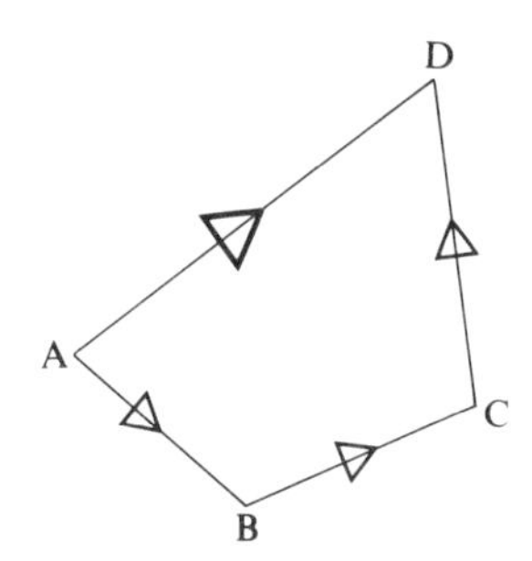

$$\overrightarrow{AD} = \overrightarrow{AB} + \overrightarrow{BC} + \overrightarrow{CD}$$

(A, B, C and D are not necessarily coplanar.)

$\overrightarrow{AD}$ represents the resultant vector.
$\overrightarrow{AB}$, $\overrightarrow{BC}$ and $\overrightarrow{CD}$ represent the components of $\overrightarrow{AD}$.
Thus *a vector is equal to the vector sum of its components.*

7) If $\mathbf{a}$ and $\mathbf{b}$ are the position vectors of the points A and B, the position vector of the point D dividing AB in the ratio $m:n$ is $\dfrac{n\mathbf{a} + m\mathbf{b}}{n + m}$.

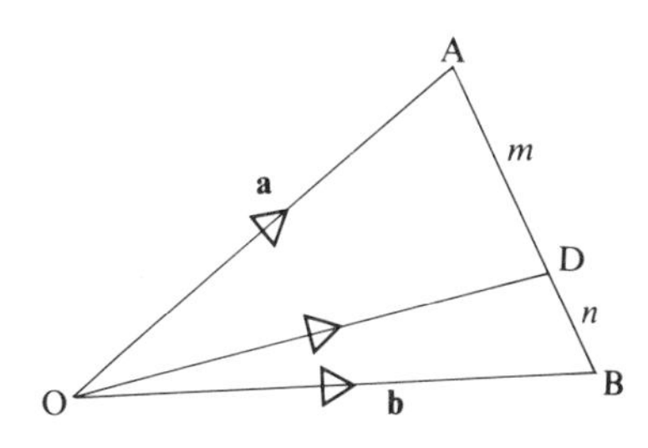

CARTESIAN BASE VECTORS

For work in coordinate geometry we take a set of mutually perpendicular unit vectors, $\mathbf{i}, \mathbf{j}$, and $\mathbf{k}$ and relate them formally to the Cartesian frame of reference as follows:

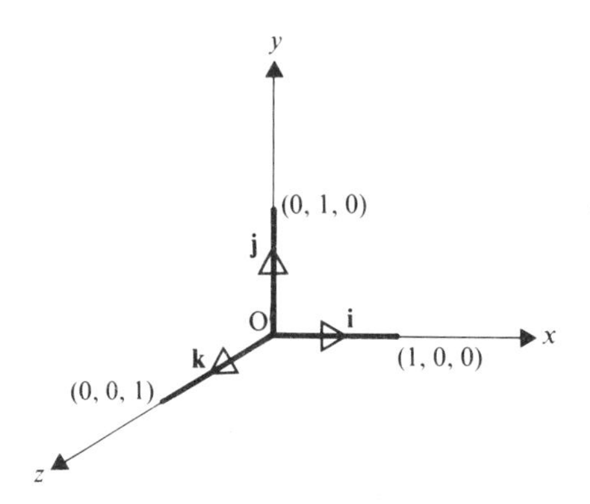

$\mathbf{i}$ is a unit vector in the direction Ox,

$\mathbf{j}$ is a unit vector in the direction Oy,

$\mathbf{k}$ is a unit vector in the direction Oz.

If $P(a, b, c)$ is any point and $\mathbf{r}$ is the position vector of P then $\mathbf{r}$ has components

$a\mathbf{i}$ in the direction Ox,

$b\mathbf{j}$ in the direction Oy,

$c\mathbf{k}$ in the direction Oz.

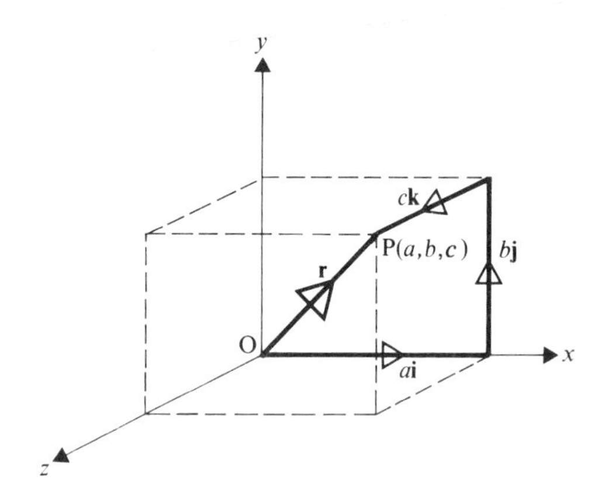

That is, $\overrightarrow{OP} = \mathbf{r} = a\mathbf{i} + b\mathbf{j} + c\mathbf{k}$.

In general, for any point $P(x, y, z)$, the position vector, $\mathbf{r}$, of P is given by $\mathbf{r} = x\mathbf{i} + y\mathbf{j} + z\mathbf{k}$.

Hence the position vector of the point $(2, -1, 3)$ is $2\mathbf{i} - \mathbf{j} + 3\mathbf{k}$ and conversely the coordinates of the point whose position vector is $-4\mathbf{i} + \mathbf{j} - 2\mathbf{k}$ are $(-4, 1, -2)$.

However, given a vector such as $\mathbf{V} = 2\mathbf{i} + \mathbf{j} + 3\mathbf{k}$, then without further information we may assume that $\mathbf{V}$ is a free vector. Thus although we can represent $\mathbf{V}$ by the position vector $\overrightarrow{OP}$ of the point $P(2, 1, 3)$, we can also represent $\mathbf{V}$ by any other line of the same length and direction as $\overrightarrow{OP}$.

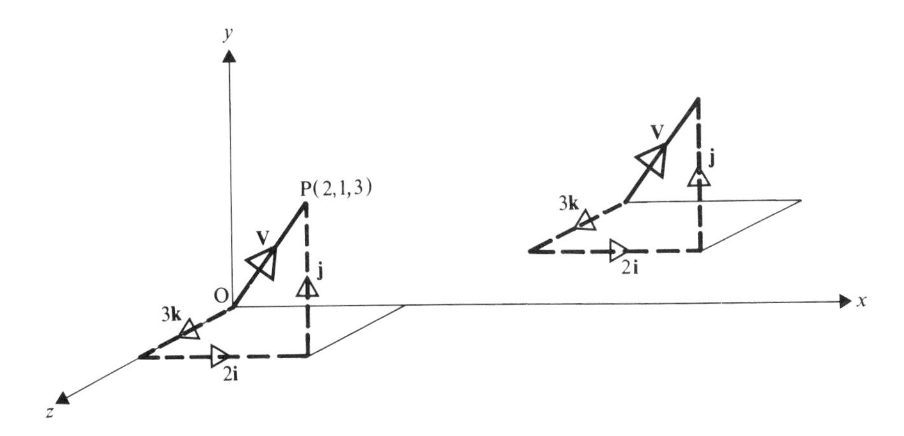

Note. Wherever we choose to represent $\mathbf{V}$ the components of $\mathbf{V}$ are $2\mathbf{i}, \mathbf{j}, 3\mathbf{k}$.

Modulus

The magnitude or *modulus* of $\mathbf{r}$ is written as $|\mathbf{r}|$ or r and is represented by the length d of OP.

Hence $$|\mathbf{r}| = d = \sqrt{(a^2+b^2+c^2)}.$$

Therefore, by its definition, $|\mathbf{r}|$ is always positive,

e.g. if $\mathbf{r} = 2\mathbf{i}-\mathbf{j}+3\mathbf{k}$, $|\mathbf{r}| = \sqrt{[2^2+(-1)^2+3^2]} = \sqrt{14}$.

The free vector $\mathbf{V} = a\mathbf{i}+b\mathbf{j}+c\mathbf{k}$ may also be represented by $\overrightarrow{OP}$, hence for any vector $\mathbf{V}$

$$|\mathbf{V}| = \sqrt{(a^2+b^2+c^2)}.$$

Resultant Vectors

Consider the vectors

$$\mathbf{V}_1 = 3\mathbf{i}+2\mathbf{j}+2\mathbf{k}$$

$$\mathbf{V}_2 = \mathbf{i}+2\mathbf{j}+\mathbf{k}$$

If $$\mathbf{V}_1 = \overrightarrow{OA} \quad \text{and} \quad \mathbf{V}_2 = \overrightarrow{AB}$$

then $$\mathbf{V}_1+\mathbf{V}_2 = \overrightarrow{OB}.$$

The components of $\overrightarrow{OB}$ are

$$3\mathbf{i}+\mathbf{i},\quad 2\mathbf{j}+2\mathbf{j},\quad 2\mathbf{k}+\mathbf{k},$$

i.e. $$\mathbf{V}_1+\mathbf{V}_2 = 4\mathbf{i}+4\mathbf{j}+3\mathbf{k}.$$

In general if
$$\mathbf{V}_1 = x_1\mathbf{i} + y_1\mathbf{j} + z_1\mathbf{k}$$
$$\mathbf{V}_2 = x_2\mathbf{i} + y_2\mathbf{j} + z_2\mathbf{k}$$
then
$$\mathbf{V}_1 + \mathbf{V}_2 = (x_1 + x_2)\mathbf{i} + (y_1 + y_2)\mathbf{j} + (z_1 + z_2)\mathbf{k}.$$

EXAMPLE 4a

A triangle ABC has its vertices at the points A(2, − 1, 4), B(3, − 2, 5), C(− 1, 6, 2).
Find in the form $a\mathbf{i} + b\mathbf{j} + c\mathbf{k}$, the vectors $\overrightarrow{AB}, \overrightarrow{BC}, \overrightarrow{CA}$ and hence find the lengths of the sides of the triangle.

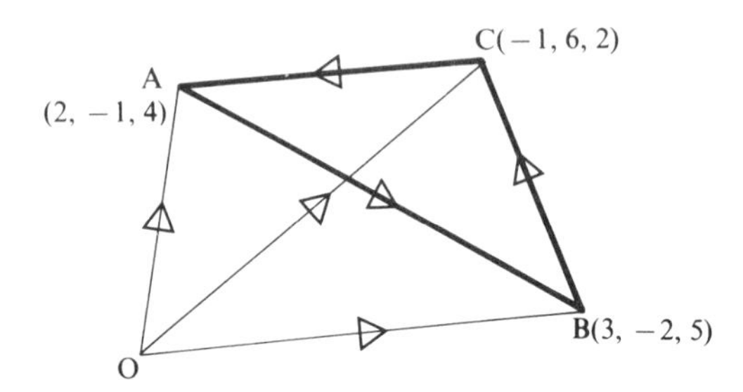

Note. The coordinate axes are not drawn in this diagram. When two, or more, points are illustrated, the presence of the axes tends to reduce the clarity of a diagram and so, in general, they are not introduced. However, the origin should be marked, as it provides a reference point.

Now
$$\begin{aligned}
\overrightarrow{AB} &= \overrightarrow{OB} - \overrightarrow{OA} \\
&= (3\mathbf{i} - 2\mathbf{j} + 5\mathbf{k}) - (2\mathbf{i} - \mathbf{j} + 4\mathbf{k}) \\
&= \mathbf{i} - \mathbf{j} + \mathbf{k} \\
\overrightarrow{BC} &= \overrightarrow{OC} - \overrightarrow{OB} \\
&= (-\mathbf{i} + 6\mathbf{j} + 2\mathbf{k}) - (3\mathbf{i} - 2\mathbf{j} + 5\mathbf{k}) \\
&= -4\mathbf{i} + 8\mathbf{j} - 3\mathbf{k} \\
\overrightarrow{CA} &= \overrightarrow{OA} - \overrightarrow{OC} \\
&= (2\mathbf{i} - \mathbf{j} + 4\mathbf{k}) - (-\mathbf{i} + 6\mathbf{j} + 2\mathbf{k}) \\
&= 3\mathbf{i} - 7\mathbf{j} + 2\mathbf{k}.
\end{aligned}$$

Hence
$$\begin{aligned}
AB &= |\overrightarrow{AB}| = \sqrt{[(1)^2 + (-1)^2 + (1)^2]} = \sqrt{3} \\
BC &= |\overrightarrow{BC}| = \sqrt{[(-4)^2 + (8)^2 + (-3)^2]} = \sqrt{89} \\
CA &= |\overrightarrow{CA}| = \sqrt{[(3)^2 + (-7)^2 + (2)^2]} = \sqrt{62}
\end{aligned}$$

EXERCISE 4a

1) Write down, in the form $a\mathbf{i} + b\mathbf{j} + c\mathbf{k}$, the vector represented by $\overrightarrow{OP}$ if P is a point with coordinates
(a) $(3, 6, 4)$ (b) $(1, -2, -7)$ (c) $(1, 0, -3)$.

2) $\overrightarrow{OP}$ represents a vector $\mathbf{r}$. Write down the coordinates of P if
(a) $\mathbf{r} = 5\mathbf{i} - 7\mathbf{j} + 2\mathbf{k}$ (b) $\mathbf{r} = \mathbf{i} + 4\mathbf{j}$ (c) $\mathbf{r} = \mathbf{j} - \mathbf{k}$.

3) Find the length of the line OP if P is the point
(a) $(2, -1, 4)$ (b) $(3, 0, 4)$ (c) $(-2, -2, 1)$.

4) Find the modulus of the vector $\mathbf{V}$ if
(a) $\mathbf{V} = 2\mathbf{i} - 4\mathbf{j} + 4\mathbf{k}$ (b) $\mathbf{V} = 6\mathbf{i} + 2\mathbf{j} - 3\mathbf{k}$ (c) $\mathbf{V} = 11\mathbf{i} - 7\mathbf{j} - 6\mathbf{k}$.

5) If $\mathbf{a} = \mathbf{i} + \mathbf{j} + \mathbf{k}$, $\mathbf{b} = 2\mathbf{i} - \mathbf{j} + 3\mathbf{k}$, $\mathbf{c} = -\mathbf{i} + 3\mathbf{j} - \mathbf{k}$ find
(a) $\mathbf{a} + \mathbf{b}$ (b) $\mathbf{a} - \mathbf{c}$ (c) $\mathbf{a} + \mathbf{b} + \mathbf{c}$ (d) $\mathbf{a} - 2\mathbf{b} + 3\mathbf{c}$.

6) The triangle ABC has its vertices at the points $A(-1, 3, 0)$, $B(-3, 0, 7)$, $C(-1, 2, 3)$. Find in the form $a\mathbf{i} + b\mathbf{j} + c\mathbf{k}$ the vectors representing
(a) $\overrightarrow{AB}$ (b) $\overrightarrow{AC}$ (c) $\overrightarrow{CB}$.

7) Find the lengths of the sides of the triangle described in Question 6.

8) Find $|\mathbf{a} - \mathbf{b}|$ where $\mathbf{a} = \mathbf{i} - \mathbf{j} + 2\mathbf{k}$, $\mathbf{b} = 2\mathbf{i} - \mathbf{j}$.

9) A, B, C and D are the points $(0, 0, 2)$, $(-1, 3, 2)$, $(1, 0, 4)$ and $(-1, 2, -2)$ respectively. Find the vectors representing $\overrightarrow{AB}$, $\overrightarrow{BD}$, $\overrightarrow{CD}$, $\overrightarrow{AD}$.

DIRECTION RATIOS AND DIRECTION COSINES OF A LINE

Consider the line OP where P is the point (a, b, c). The coordinates of P determine the direction of $\overrightarrow{OP}$ relative to the axes and the ratios $a:b:c$ are called the direction ratios of the line OP.

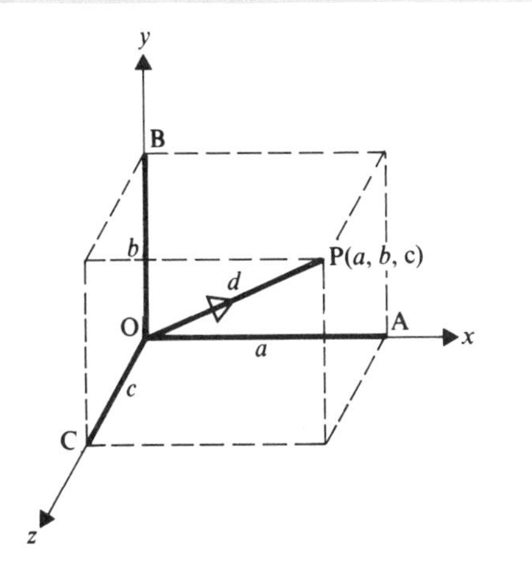

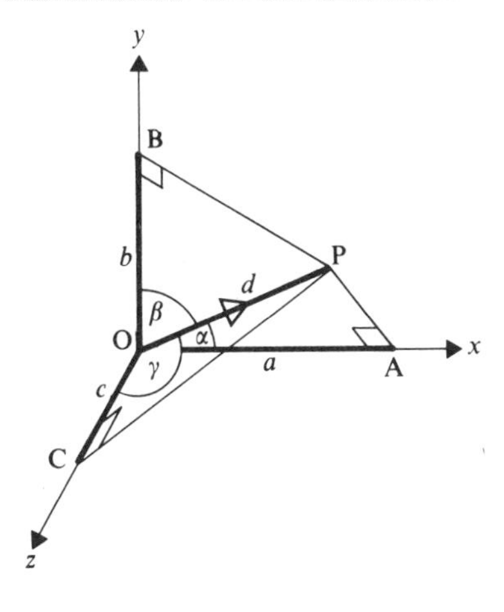

In the diagram, lines PA, PB and PC are drawn from P perpendicular to Ox, Oy and Oz respectively, so that angles PAO, PBO and PCO are each $90°$.
But $OA = a$, $OB = b$, $OC = c$ and $OP = \sqrt{(a^2+b^2+c^2)} = d$.
So if OP makes angles α, β and γ with Ox, Oy and Oz respectively, then

$$\cos\alpha = \frac{a}{d} \qquad \cos\beta = \frac{b}{d} \qquad \cos\gamma = \frac{c}{d}$$

But the angles α, β and γ determine the direction of OP, so their cosines are called the *direction cosines* of OP and are often given the symbols l, m and n, so that

$$\cos\alpha = \frac{a}{d} = l$$

$$\cos\beta = \frac{b}{d} = m$$

$$\cos\gamma = \frac{c}{d} = n.$$

So if the direction ratios of a line are $a:b:c$
the direction cosines of that line are

$$\frac{a}{\sqrt{(a^2+b^2+c^2)}}, \quad \frac{b}{\sqrt{(a^2+b^2+c^2)}}, \quad \frac{c}{\sqrt{(a^2+b^2+c^2)}}.$$

From these three relationships it follows that

1) $$l^2+m^2+n^2 = \left(\frac{a}{d}\right)^2 + \left(\frac{b}{d}\right)^2 + \left(\frac{c}{d}\right)^2$$

$$= \frac{a^2+b^2+c^2}{d^2}$$

But $$a^2+b^2+c^2 = d^2$$

Hence $$l^2+m^2+n^2 = 1$$

Thus *the sum of the squares of the direction cosines of any line is unity.*

2) $l:m:n = a:b:c$.

3) As $a = dl$, $b = dm$, $c = dn$

the coordinates of $P(a, b, c)$ may be written (dl, dm, dn).

Hence $$\overrightarrow{OP} = dl\mathbf{i} + dm\mathbf{j} + dn\mathbf{k}$$

$$= d(l\mathbf{i} + m\mathbf{j} + n\mathbf{k})$$

If $$\overrightarrow{OQ} = l\mathbf{i} + m\mathbf{j} + n\mathbf{k},$$

Q is the point (l, m, n) and $OQ = \sqrt{(l^2 + m^2 + n^2)} = 1$,
i.e. $\overrightarrow{OQ}$ is a unit vector in the direction $\overrightarrow{OP}$.

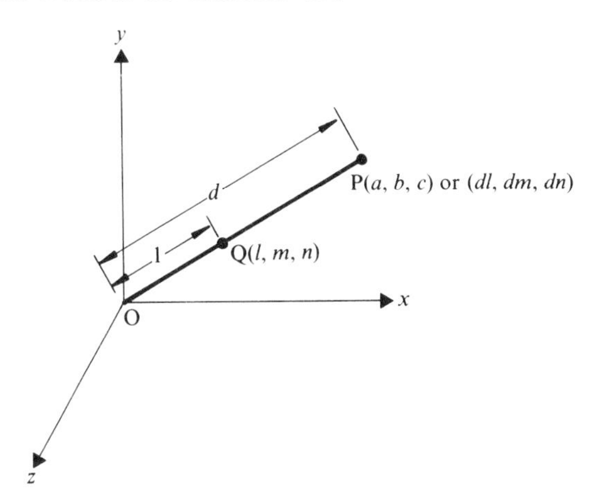

EXAMPLES 4b

1) Find the direction cosines of the line OP where P is the point $(2, 3, -6)$.

$$\mathbf{OP} = 2\mathbf{i} + 3\mathbf{j} - 6\mathbf{k}$$

$$OP = \sqrt{[2^2 + 3^2 + (-6)^2]} = 7$$

OP has direction ratios $2:3:-6$

Therefore OP has direction cosines $\frac{2}{7}, \frac{3}{7}, -\frac{6}{7}$.

2) A line OP is inclined to Ox at 45° and to Oy at 60°. Find its inclination to Oz. If the length of OP is 12 units find the coordinates of P.

The direction cosines of OP are $\cos 45^\circ$, $\cos 60^\circ$, $\cos\gamma$,

so $$l = 1/\sqrt{2}, \quad m = \tfrac{1}{2}, \quad n = \cos\gamma$$

But $$l^2 + m^2 + n^2 = 1 \quad \Rightarrow \quad n^2 = 1 - \tfrac{1}{2} - \tfrac{1}{4} = \tfrac{1}{4}$$

$$\Rightarrow \quad n = \pm\tfrac{1}{2}.$$

Therefore $\cos\gamma = \pm\frac{1}{2}$
and so OP is inclined to Oz at 60° or 120°.
Now the coordinates of P are (dl, dm, dn), and $d = 12$ (given),

so P is the point $\left(\dfrac{12}{\sqrt{2}}, 6, \pm 6\right)$, i.e. $(6\sqrt{2}, 6, \pm 6)$.

3) Find the coordinates of P if OP is of length 5 units and is in the direction $\overrightarrow{OR}$ where R is the point $(2, -1, 4)$.

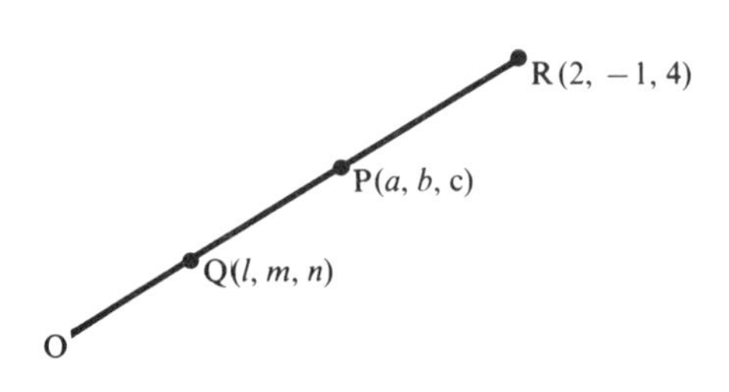

The direction cosines of $\overrightarrow{OR}$ are

$$l = \frac{2}{\sqrt{21}}, \quad m = -\frac{1}{\sqrt{21}}, \quad n = \frac{4}{\sqrt{21}}.$$

Hence P is the point (dl, dm, dn),

i.e. $\left(\frac{10}{\sqrt{21}}, -\frac{5}{\sqrt{21}}, \frac{20}{\sqrt{21}}\right)$.

EXERCISE 4b

1) Find the direction cosines of the line OP, where P is the point
(a) $(2, 2, -1)$, (b) $(6, -2, -3)$, (c) $(3, 0, 4)$, (d) $(1, 8, 4)$.

2) Find the inclination to Oy of OP, where P is the point
(a) $(1, 1, 1)$, (b) $(1, 0, 1)$, (c) $(0, 1, 1)$, (d) $(2, -2, 1)$.

3) Find the coordinates of Q if $|\overrightarrow{OQ}| = 1$ and $\overrightarrow{OQ}$ is in the direction of
(a) $\mathbf{i} + 2\mathbf{j} - 2\mathbf{k}$, (b) $-3\mathbf{i} + 2\mathbf{j} + 6\mathbf{k}$.

4) Find the coordinates of P, where
(a) $|\overrightarrow{OP}| = 6$ and $\overrightarrow{OP}$ is in the direction of $2\mathbf{i} - 3\mathbf{j} + 6\mathbf{k}$,
(b) $|\overrightarrow{OP}| = 2$ and $\overrightarrow{OP}$ is in the direction of $8\mathbf{i} + \mathbf{j} - 4\mathbf{k}$,
(c) $|\overrightarrow{OP}|$ is inclined at equal acute angles to Ox, Oy and Oz and $|\overrightarrow{OP}| = 4$.

5) A line OP is inclined as 60° to Ox and Oz. If $|\overrightarrow{OP}| = 3$, find the coordinates of P.

6) P is the point $(1, a, b)$ where $|\overrightarrow{OP}| = 5$ and OP is inclined at 45° to Oz. Find a and b.

PROPERTIES OF A LINE JOINING TWO POINTS

Consider the line joining the two points $A(x_1, y_1, z_1)$ and $B(x_2, y_2, z_2)$.

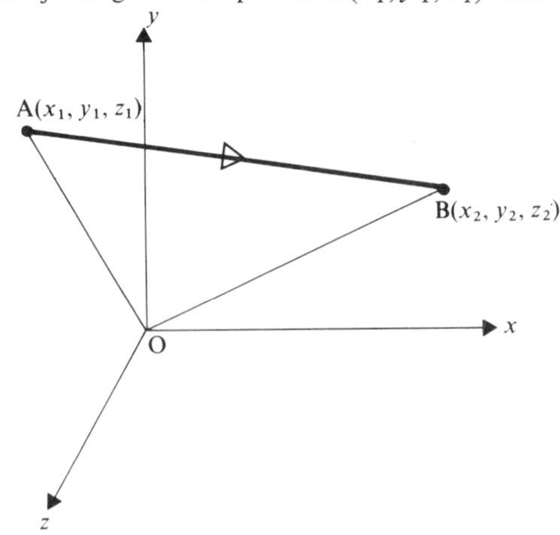

Length of AB

Now $\quad \overrightarrow{OA} \equiv x_1\mathbf{i}+y_1\mathbf{j}+z_1\mathbf{k} \quad$ and $\quad \overrightarrow{OB} = x_2\mathbf{i}+y_2\mathbf{j}+z_2\mathbf{k}$.

Hence
$$\overrightarrow{AB} = \overrightarrow{OB}-\overrightarrow{OA}$$
$$= (x_2-x_1)\mathbf{i}+(y_2-y_1)\mathbf{j}+(z_2-z_1)\mathbf{k}$$

Now the length of AB $= |\overrightarrow{AB}|$
$$= \sqrt{[(x_2-x_1)^2+(y_2-y_1)^2+(z_2-z_1)^2]},$$

i.e. the length of the line joining $A(x_1,y_1,z_1)$ to $B(x_2,y_2,z_2)$ is
$$\sqrt{[(x_2-x_1)^2+(y_2-y_1)^2+(z_2-z_1)^2]}$$

Direction Ratios of AB

$$\overrightarrow{AB} = (x_2-x_1)\mathbf{i}+(y_2-y_1)\mathbf{j}+(z_2-z_1)\mathbf{k},$$

so the direction ratios of $\overrightarrow{AB}$ are $\quad (x_2-x_1):(y_2-y_1):(z_2-z_1)$

and the direction cosines of $\overrightarrow{AB}$ are $\dfrac{(x_2-x_1)}{AB}, \; \dfrac{(y_2-y_1)}{AB}, \; \dfrac{(z_2-z_1)}{AB}$

Now
$$\overrightarrow{BA} = (x_1-x_2)\mathbf{i}+(y_1-y_2)\mathbf{j}+(z_1-z_2)\mathbf{k}.$$

So the direction *ratios* of $\overrightarrow{AB}$ and $\overrightarrow{BA}$ are the same.
Thus we can refer to the direction ratios of the line AB without ambiguity.
But the *direction cosines of* $\overrightarrow{BA}$ *are opposite in sign to those of* $\overrightarrow{AB}$ and so when giving the direction cosines of a line we must specify which of the two possible directions along the line (i.e. $\overrightarrow{AB}$ or $\overrightarrow{BA}$) the values refer to.

Midpoint of AB

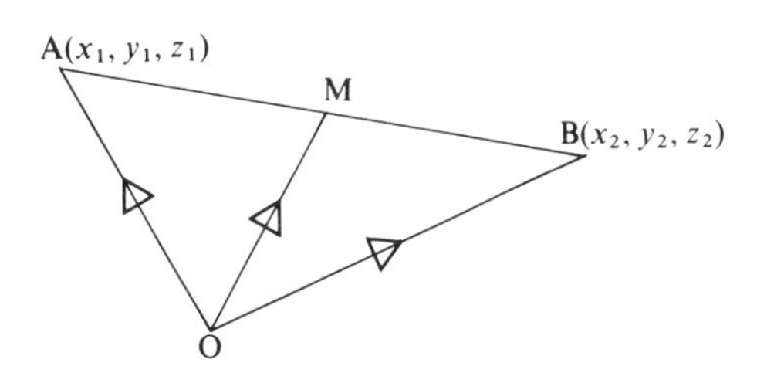

If M is the midpoint of AB

then
$$\overrightarrow{OM} = \overrightarrow{OA}+\overrightarrow{AM}$$
$$= \overrightarrow{OA}+\tfrac{1}{2}(\overrightarrow{AB})$$
$$= \overrightarrow{OA}+\tfrac{1}{2}(\overrightarrow{OB}-\overrightarrow{OA})$$
$$= \tfrac{1}{2}(\overrightarrow{OA}+\overrightarrow{OB})$$

i.e.
$$\overrightarrow{OM} = \tfrac{1}{2}[(x_1+x_2)\mathbf{i}+(y_1+y_2)\mathbf{j}+(z_1+z_2)\mathbf{k}]$$

and $\quad$ M has coordinates $\left(\dfrac{x_1+x_2}{2}, \dfrac{y_1+y_2}{2}, \dfrac{z_1+z_2}{2}\right)$

Point Dividing AB in a Given Ratio

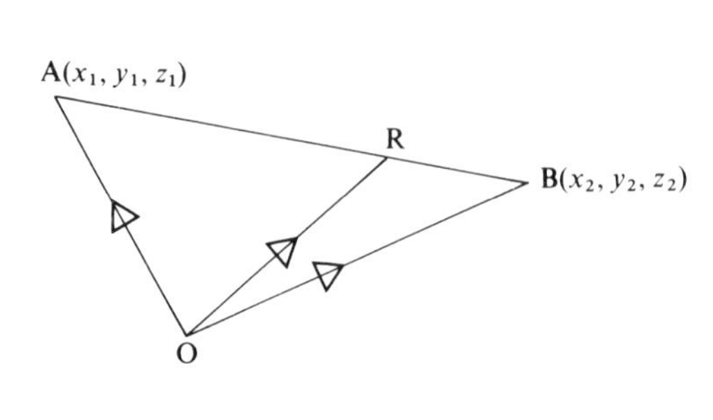

If R is the point dividing AB internally in the ratio $\lambda:\mu$ then $\dfrac{AR}{RB} = \dfrac{\lambda}{\mu}$ and

$$\overrightarrow{OR} = \overrightarrow{OA} + \overrightarrow{AR}$$

$$= \overrightarrow{OA} + \frac{\lambda}{\lambda+\mu}\overrightarrow{AB}$$

$$= \overrightarrow{OA} + \frac{\lambda}{\lambda+\mu}(\overrightarrow{OB} - \overrightarrow{OA})$$

$$= \frac{\lambda\overrightarrow{OB} + \mu\overrightarrow{OA}}{\lambda+\mu},$$

i.e.
$$\overrightarrow{OR} = \frac{\lambda(x_2\mathbf{i} + y_2\mathbf{j} + z_2\mathbf{k}) + \mu(x_1\mathbf{i} + y_1\mathbf{j} + z_1\mathbf{k})}{\lambda+\mu}$$

$\Rightarrow$ R has coordinates $\left(\dfrac{\lambda x_2 + \mu x_1}{\lambda+\mu}, \dfrac{\lambda y_2 + \mu y_1}{\lambda+\mu}, \dfrac{\lambda z_1 + \mu z_2}{\lambda+\mu}\right)$.

EXAMPLES 4c

1) Find the length of the median through O of the triangle OAB, where A is the point $(2, 7, -1)$ and B is the point $(4, 1, 2)$.

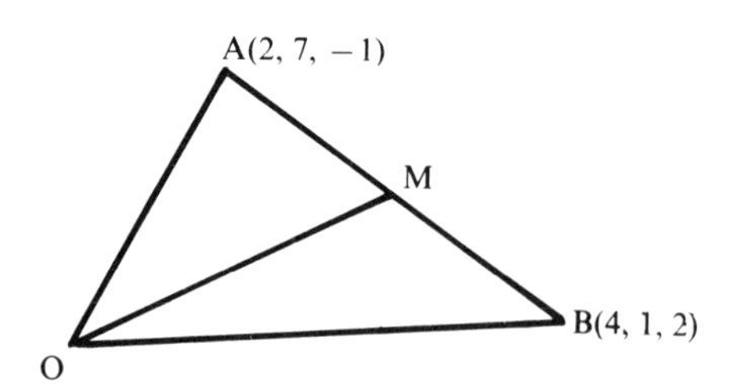

The coordinates of M, the midpoint of AB, are

$$\left(\frac{4+2}{2}, \frac{1+7}{2}, \frac{2-1}{2}\right) = (3, 4, \tfrac{1}{2})$$

So the length of $\text{OM} = \sqrt{[3^2 + 4^2 + (\tfrac{1}{2})^2]} = \dfrac{\sqrt{101}}{2}$

2) A and B are two points with position vectors $\mathbf{i}-\mathbf{j}+4\mathbf{k}$ and $7\mathbf{i}-\mathbf{j}-2\mathbf{k}$ respectively. Find the coordinates of points P and Q which divide AB
(a) internally in the ratio 5 : 1,
(b) externally in the ratio 3 : 2.

(a)

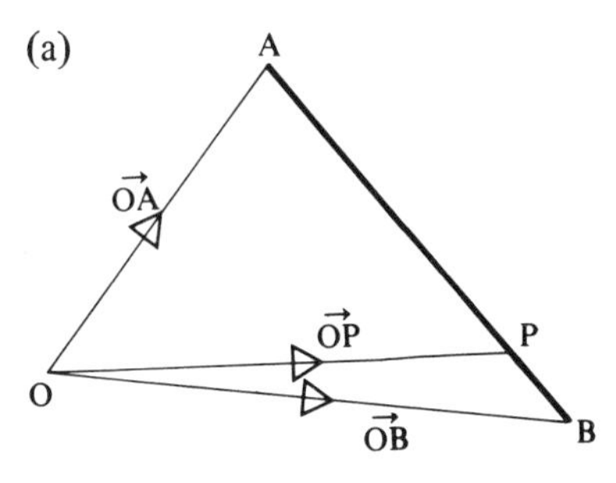

Since $\quad AP:PB = 5:1$

$$\overrightarrow{OP} = \frac{5\overrightarrow{OB} + \overrightarrow{OA}}{5+1}$$

$$= \frac{5(7\mathbf{i}-\mathbf{j}-2\mathbf{k}) + (\mathbf{i}-\mathbf{j}+4\mathbf{k})}{6}$$

$$= 6\mathbf{i}-\mathbf{j}-\mathbf{k}$$

Hence P is the point $(6, -1, -1)$.

(b)

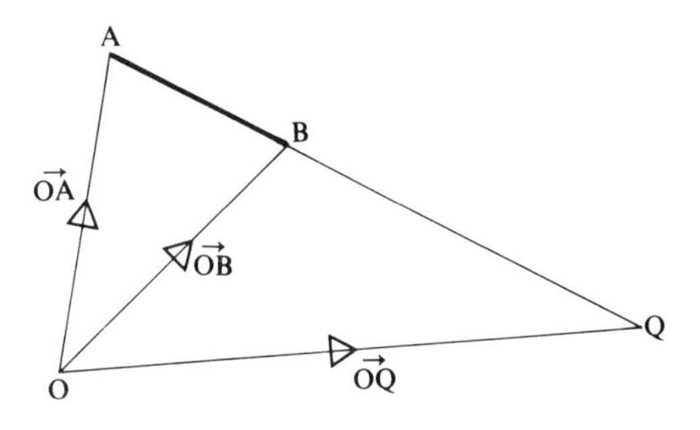

Since $\quad AQ:QB = 3:-2$

$$\overrightarrow{OQ} = \frac{3\overrightarrow{OB} - 2\overrightarrow{OA}}{3-2}$$

$$= \frac{3(7\mathbf{i}-\mathbf{j}-2\mathbf{k}) - 2(\mathbf{i}-\mathbf{j}+4\mathbf{k})}{3-2}$$

$$= 19\mathbf{i}-\mathbf{j}-14\mathbf{k}.$$

Hence Q is the point $(19, -1, -14)$.

3) Find the length and direction cosines of the line LM where L is the midpoint of AB, M is the midpoint of BC, and A, B, C are the points $(3, -1, 5)$, $(7, 1, 3)$, $(-5, 9, -1)$ respectively.

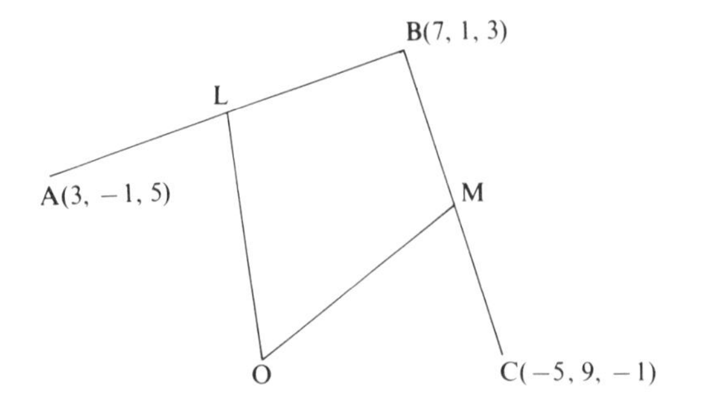

L is the point

$$\left(\frac{7+3}{2}, \frac{-1+1}{2}, \frac{5+3}{2}\right)$$

i.e. $(5, 0, 4)$

M is the point

$$\left(\frac{7-5}{2}, \frac{1+9}{2}, \frac{3-1}{2}\right)$$

i.e. $(1, 5, 1)$

Hence $\quad LM = \sqrt{[(5-1)^2 + (0-5)^2 + (4-1)^2]} = 5\sqrt{2}$

Now $\quad \overrightarrow{LM} = (1-5)\mathbf{i} + (5-0)\mathbf{j} + (1-4)\mathbf{k}$

$$= -4\mathbf{i} + 5\mathbf{j} - 3\mathbf{k}$$

Hence the direction cosines of $\overrightarrow{LM}$ are $-\dfrac{4}{5\sqrt{2}}, \dfrac{5}{5\sqrt{2}}, -\dfrac{3}{5\sqrt{2}}$

i.e. $-\dfrac{2\sqrt{2}}{5}, \dfrac{\sqrt{2}}{2}, -\dfrac{3\sqrt{2}}{10}$

4) The points OABC form a tetrahedron, where A, B, and C are the points $(4, 1, -2)$, $(3, 5, -1)$ and $(-1, 2, 4)$ respectively.
Show that the line joining the midpoint of OA to the midpoint of OB is parallel to the line joining the midpoint of AC to the midpoint of CB.

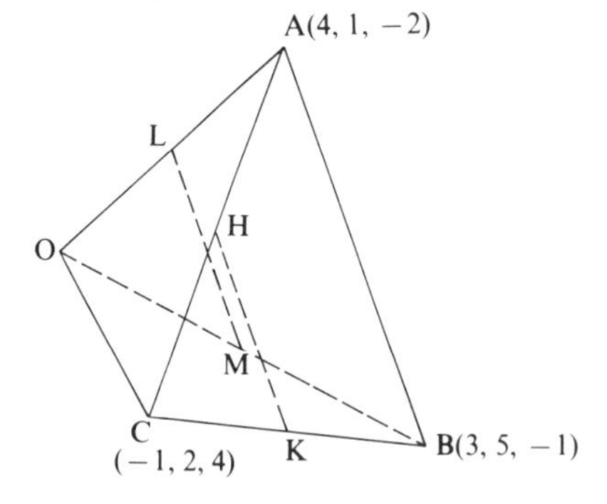

If L and M are the midpoints of OA and OB then

$$\text{L is the point } (2, \tfrac{1}{2}, -1)$$

$$\text{M is the point } (\tfrac{3}{2}, \tfrac{5}{2}, -\tfrac{1}{2}).$$

If H and K are the midpoints of AC and CB, then

$$\text{H is the point } (\tfrac{3}{2}, \tfrac{3}{2}, 1)$$

$$\text{K is the point } (1, \tfrac{7}{2}, \tfrac{3}{2}).$$

The direction ratios of LM are

$$(2-\tfrac{3}{2}):(\tfrac{1}{2}-\tfrac{5}{2}):(-1+\tfrac{1}{2}) = 1:-4:-1$$

The direction ratios of HK are

$$(\tfrac{3}{2}-1):(\tfrac{3}{2}-\tfrac{7}{2}):(1-\tfrac{3}{2}) = 1:-4:-1$$

Therefore LM and HK are parallel.

EXERCISE 4c

In Questions 1–6, A, B, C and D are points whose coordinates are $(1, 1, 0)$, $(3, -2, 1)$, $(-3, -3, 0)$, and $(7, 2, 1)$ respectively.

1) Find the modulus and direction cosines of $\overrightarrow{AB}$, $\overrightarrow{AC}$ and $\overrightarrow{BD}$.

2) Find the coordinates of a point which
(a) divides AD internally in the ratio $1:2$,
(b) bisects AB,
(c) divides BC externally in the ratio $3:1$.

3) Check whether the following pairs of lines are parallel:
(a) AB and CD, (b) AC and BD, (c) AD and BC.

4) Find out whether
(a) $AC = BC$, (b) $AC = BD$, (c) $AD = BC$, (d) $AB = DC$.

5) If L and M are the midpoints of AD and BD respectively find the position vectors of L and M and show that LM is parallel to AB.

6) If H and K are the midpoints of AC and CD show that $\overrightarrow{HK} = \frac{1}{2}\overrightarrow{AD}$.

The remainder of this chapter develops methods for three dimensional work based on the use of coordinates. The *alternative* vector approach is developed in Chapter 5.

THE EQUATIONS OF A STRAIGHT LINE

A particular line is uniquely located in space if
(a) it has a known direction and passes through a known point,
(b) it passes through two known points.

(a) Consider a line whose direction ratios are $a:b:c$ and which passes through the point $A(x_1, y_1, z_1)$.

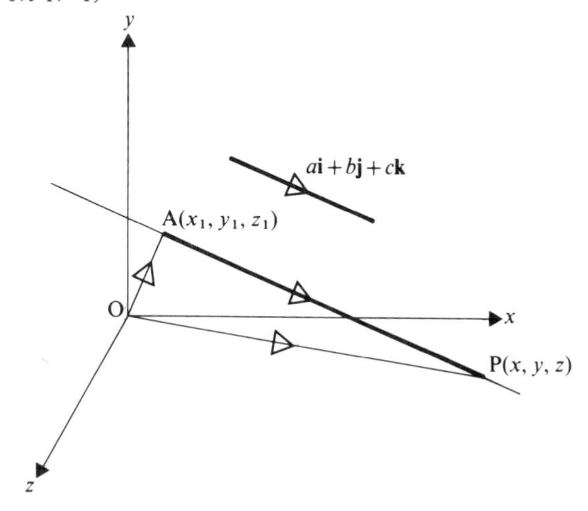

As the line has direction ratios $a:b:c$ it is parallel to the vector $a\mathbf{i}+b\mathbf{j}+c\mathbf{k}$.
Hence $P(x, y, z)$ is a point on the line $\iff$ $\overrightarrow{AP}=\lambda(a\mathbf{i}+b\mathbf{j}+c\mathbf{k})$ where λ is a scalar variable.

Now $$\overrightarrow{AP} = \overrightarrow{OP}-\overrightarrow{OA}.$$

Hence $$\lambda(a\mathbf{i}+b\mathbf{j}+c\mathbf{k}) = \overrightarrow{OP}-(x_1\mathbf{i}+y_1\mathbf{j}+z_1\mathbf{k})$$

i.e. $$\overrightarrow{OP} = x_1\mathbf{i}+y_1\mathbf{j}+z_1\mathbf{k}+\lambda(a\mathbf{i}+b\mathbf{j}+c\mathbf{k}) \qquad [1]$$

or $$x\mathbf{i}+y\mathbf{j}+z\mathbf{k} = (x_1+\lambda a)\mathbf{i}+(y_1+\lambda b)\mathbf{j}+(z_1+\lambda c)\mathbf{k}$$

$\Rightarrow$ $$x = x_1+\lambda a, \quad y = y_1+\lambda b, \quad z = z_1+\lambda c$$

$\Rightarrow$ $$\frac{x-x_1}{a}=\frac{y-y_1}{b}=\frac{z-z_1}{c} \quad (=\lambda) \qquad [2]$$

Equation [1] is known as the vector equation of the straight line and equations [2] are called the Cartesian equations of the line.
For example, the equations of a line passing through $(2, -1, 3)$ and with direction ratios $5:-2:4$ may be written

$$\frac{x-2}{5}=\frac{y+1}{-2}=\frac{z-3}{4} \quad (=\lambda)$$

Note that $(2, -1, 3)$ is only one of an infinite set of fixed points that could have been chosen on this line and so the set of equations which represent a particular line is not unique.

Also the equations $\dfrac{x-2}{5}=\dfrac{y+1}{-2}=\dfrac{z-3}{4}$ $(=\lambda)$ may be presented in several ways,

e.g., in parametric form $$\begin{cases} x = 5\lambda+2 \\ y = -2\lambda-1 \\ z = 4\lambda+3 \end{cases}$$

or, eliminating λ, by the pair of equations

$$\begin{cases} 2x+5y = -1 \\ 2x+y-2z = -3 \end{cases}$$

As there are several ways of eliminating λ, this pair of equations is only one of several pairs that can be obtained.
Note that the direction ratios of the line appear as the denominators in the standard Cartesian equations and as the coefficients of λ in the parametric equations.

(b) Now consider the definition of a particular line in space as the line passing through two fixed points $A(x_1, y_1, z_1)$ and $B(x_2, y_2, z_2)$.

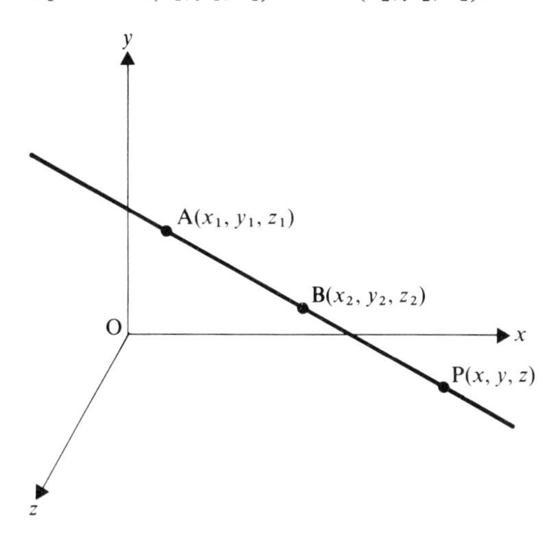

The direction ratios of AB, and hence of the line, are

$$(x_2 - x_1):(y_2 - y_1):(z_2 - z_1)$$

Therefore the Cartesian equations of this line may be written

$$\frac{x - x_1}{x_2 - x_1} = \frac{y - y_1}{y_2 - y_1} = \frac{z - z_1}{z_2 - z_1} \quad (= \lambda)$$

Note that the corresponding vector equation is

$$\overrightarrow{OP} = x_1\mathbf{i} + y_1\mathbf{j} + z_1\mathbf{k} + \lambda[(x_2 - x_1)\mathbf{i} + (y_2 - y_1)\mathbf{j} + (z_2 - z_1)\mathbf{k}]$$

EXAMPLES 4d

1) Find Cartesian equations for the line with direction ratios $2:-1:-4$ which passes through the point $(3, -1, \frac{1}{2})$.

By using $\dfrac{x - x_1}{a} = \dfrac{y - y_1}{b} = \dfrac{z - z_1}{c}$, the equations of this line may be written

$$\frac{x - 3}{2} = \frac{y + 1}{-1} = \frac{z - \frac{1}{2}}{-4}$$

$$\Rightarrow \qquad \frac{x - 3}{2} = \frac{y + 1}{-1} = \frac{2z - 1}{-8}$$

Note that the equations above may be written with positive denominators,

i.e. $$\frac{x - 3}{2} = \frac{-y - 1}{1} = \frac{1 - 2z}{8}$$

2) Find the direction cosines of the line whose equations are
(a) $x = 2y - 4 = 3 - 4z$ (b) $2x - y + 3z = 4$
$x + 4y - 2z = 6.$

(a) Writing the given equations in standard form

i.e. in the form $\dfrac{x - x_1}{a} = \dfrac{y - y_1}{b} = \dfrac{z - z_1}{c}$

$$\frac{x}{1} = \frac{2y-4}{1} = \frac{3-4z}{1}$$

$$\Rightarrow \qquad \frac{x}{1} = \frac{y-2}{\frac{1}{2}} = \frac{z-\frac{3}{4}}{-\frac{1}{4}}$$

we see that the direction ratios of the line are

$$1:\tfrac{1}{2}:-\tfrac{1}{4} = 4:2:-1.$$

Hence the direction cosines are $\dfrac{4}{\sqrt{21}}, \dfrac{2}{\sqrt{21}}, \dfrac{-1}{\sqrt{21}}.$

(b) The equations
$$2x - y + 3z = 4 \qquad [1]$$
$$3x + 4y - z = 6 \qquad [2]$$

may be converted to standard form as follows.
Eliminating z from [1] and [2] gives

$$11x + 11y = 22 \Rightarrow x = 2 - y \qquad [3]$$

Eliminating y from [1] and [2] gives

$$11x + 11z = 22 \Rightarrow x = 2 - z \qquad [4]$$

Thus equations [3] and [4] may be written

$$x = 2 - y = 2 - z$$

$$\Rightarrow \qquad \frac{x}{1} = \frac{y-2}{-1} = \frac{z-2}{-1}.$$

Hence the direction ratios are $1:-1:-1$

and the direction cosines are $\dfrac{1}{\sqrt{3}}, -\dfrac{1}{\sqrt{3}}, \dfrac{-1}{\sqrt{3}}$

3) Find the equations of the line through $(2, 0, 4)$ which is parallel to the line whose equations are $x - 1 = 3y + 2, \quad z = 1.$

The line whose equations are

$$\frac{x-1}{1} = \frac{y+\frac{2}{3}}{\frac{1}{3}}, \quad z = 1$$

or

$$\frac{x-1}{3} = \frac{y+\frac{2}{3}}{1}, \quad z = 1$$

is parallel to the xy plane and at a constant distance of one unit from the xy plane (as the z coordinate of every point on this line is unity).
Any parallel line will also be at a constant distance from the xy plane so one of its equations is $z = c$. Therefore as the required line passes through $(2, 0, 4)$, one of its equations is $z = 4$.
If the given line makes angles α and β with Ox and Oy then

$$\cos\alpha : \cos\beta = 3:1$$

Thus any parallel line has as one of its equations $\dfrac{x-x_1}{3} = \dfrac{y-y_1}{1}$

So the equations of the required line are $\dfrac{x-2}{3} = \dfrac{y}{1}, \quad z = 4.$

Note that, as the given line is perpendicular to Oz, $\cos\gamma = 0$, but to say that the directions *ratios* are $3:1:0$ is meaningless as the ratio $3:1:0$ is undefined.
However we *can* say that $l:m = 3:1$ and $n = 0$

$$\Rightarrow \qquad l = \frac{3}{\sqrt{10}}, \quad m = \frac{1}{\sqrt{10}}, \quad n = 0$$

Note also that the direction cosines of the x axis are $1, 0, 0$.
Hence the equations of the x axis are $y = z = 0$.
Similarly the equations of the y axis are $x = z = 0$
and the equations of the z axis are $x = y = 0$.

EXERCISE 4d

1) Write down the equations of the line which
(a) has direction ratios $1:2:3$ and passes through $(2, -1, 4)$,
(b) passes through $(3, -1, 5)$ and $(7, -2, 8)$,
(c) passes through $(4, -2, 1)$ and is parallel to the line

$$\frac{x-1}{3} = \frac{y+2}{5} = \frac{z-1}{7}$$

2) Find the direction ratios and direction cosines of the following lines:

(a) $\dfrac{x-3}{2} = \dfrac{2y+1}{3} = \dfrac{3z-1}{4}$ (b) $2x-1 = 3y+2 = 4-3z$

(c) $x+3 = 2y-1, \quad z = 3$ (d) $2x-3 = z+4, \quad y = 2$

(e) $x+3y-z = 4, \quad 2x-y+z = 2.$

3) Find the equations of the line which

(a) has direction cosines $0, \frac{1}{\sqrt{2}}, \frac{1}{\sqrt{2}}$ and which passes through $(3, -1, 4)$,

(b) passes through $(3, 2, 7)$, $(4, 2, 3)$,

(c) is parallel to Oz and passes through $(1, 2, 0)$.

4) Determine which of the following sets of points are collinear:

(a) $(1, -1, 2)$, $(4, 1, 7)$, $(-2, -3, -2)$.

(b) $(2, 4, 2)$, $(3, 6, 2)$, $(1, 2, -1)$.

(c) $(2, 6, -6)$, $(3, 2, 0)$, $(4, -2, 6)$.

5) Write the equations $\frac{x-2}{1} = \frac{1-3y}{2} = \frac{4-2z}{1}$ in parametric form. Hence, or otherwise, find the coordinates of the point where the line cuts

(a) the xy plane, (b) the xz plane, (c) the yz plane,

(d) the plane parallel to the yz plane wich cuts Ox at $(2, 0, 0)$.

PAIRS OF LINES

The location of two lines in space may be such that

(a) the lines are parallel,

(b) the lines are not parallel and intersect,

(c) the lines are not parallel and do not intersect. Such lines are called *skew* lines.

Parallel Lines

We have already seen that if two lines are parallel they have equal direction ratios.
So if two lines are parallel, this property can be observed from their equations.

Non-parallel Lines

Consider two lines whose equations are

$$\frac{x-x_1}{a_1} = \frac{y-y_1}{b_1} = \frac{z-z_1}{c_1} = \lambda \quad \text{and} \quad \frac{x-x_2}{a_2} = \frac{y-y_2}{b_2} = \frac{z-z_2}{c_2} = \mu.$$

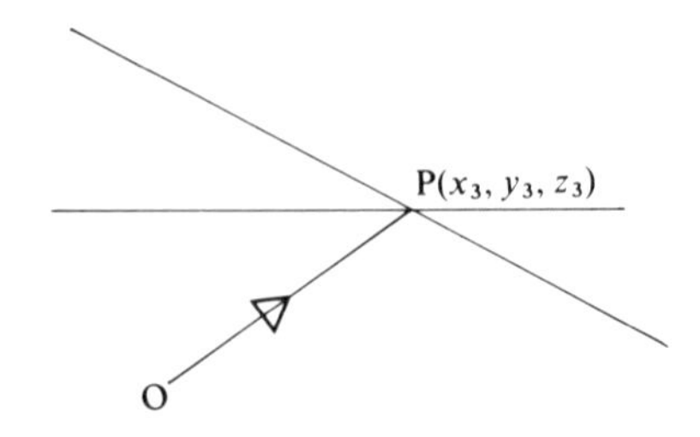

In order that these lines shall intersect there must exist unique values of λ and μ such that $x = x_3$, $y = y_3$, $z = z_3$ satisfy both sets of equations. If no such values of λ and μ can be found, the lines are skew.

EXAMPLES 4e

1) Determine whether the following pairs of lines are parallel, intersecting or skew:

(a) $\dfrac{x-1}{3} = \dfrac{y-1}{-2} = \dfrac{z-2}{4}, \quad \dfrac{x-2}{-6} = \dfrac{y+1}{4} = \dfrac{z-3}{-8}$

(b) $x-1 = 1-y = z-3, \quad \dfrac{x-2}{2} = y-4 = -z$

(c) $x-1 = \dfrac{y}{3} = \dfrac{z-1}{4}, \quad \dfrac{x-2}{4} = \dfrac{y-3}{-1} = z.$

(a) The first line has direction ratios $3:-2:4$
The second line has direction ratios $-6:4:-8 = 3:-2:4$
so these two lines are parallel.

(b) The direction ratios of this pair of lines are $1:-1:1$ and $2:1:-1$ so the lines are not parallel.

$$x-1 = 1-y = z-3 = \lambda \quad \text{and} \quad \frac{x-2}{2} = y-4 = -z = \mu$$

may be written in the parametric forms

$$\left.\begin{aligned} x &= \lambda+1 \\ y &= -\lambda+1 \\ z &= \lambda+3 \end{aligned}\right\} \quad \text{and} \quad \left\{\begin{aligned} x &= 2\mu+2 \\ y &= \mu+4 \\ z &= -\mu \end{aligned}\right.$$

Equating x and y gives

$$\left.\begin{aligned} \lambda+1 &= 2(\mu+1) \\ -\lambda+1 &= \mu+4 \end{aligned}\right\} \Rightarrow \lambda = -\tfrac{5}{3}, \quad \mu = -\tfrac{4}{3}$$

With these values of λ and μ, the values of z are

$$\left.\begin{aligned} &\text{first line} & \lambda+3 &= \tfrac{4}{3} \\ &\text{second line} & -\mu &= \tfrac{4}{3} \end{aligned}\right\} \text{equal values.}$$

So this pair of lines intersect.
The point of intersection (which is given by using $\lambda = -\frac{5}{3}$ in the equations of the first line or by using $\mu = -\frac{4}{3}$ in the equations of the second line) is $(-\frac{1}{3}, \frac{7}{3}, \frac{4}{3})$.

(c) The direction ratios of this pair of lines are $1:3:4$ and $4:-1:1$ so the lines are not parallel.

Now $$x-1 = \frac{y}{3} = \frac{z-1}{4} = \lambda \quad \text{and} \quad \frac{x-2}{4} = \frac{y-3}{-1} = z = \mu$$

may be written in parametric form to give

$$\left.\begin{aligned} x &= \lambda+1 \\ y &= 3\lambda \\ z &= 4\lambda+1 \end{aligned}\right\} \quad \text{and} \quad \left\{\begin{aligned} x &= 4\mu+2 \\ y &= -\mu+3 \\ z &= \mu \end{aligned}\right.$$

Equating x and y gives

$$\left.\begin{aligned} \lambda+1 &= 2+4\mu \\ 3\lambda &= 3-\mu \end{aligned}\right\} \Rightarrow \quad \mu = 0 \quad \text{and} \quad \lambda = 1.$$

With these values of λ and μ, the values of z are

$$\left.\begin{aligned} \text{first line} \quad 4\lambda+1 &= 5 \\ \text{second line} \quad \mu &= 0 \end{aligned}\right\} \text{these values are unequal.}$$

So this pair of lines is skew.

The Angle Between a Pair of Lines

Consider two lines L_1 and L_2.
If OA and OB are drawn parallel to L_1 and L_2 then the angle between L_1 and L_2 is defined as the angle θ between OA and OB.

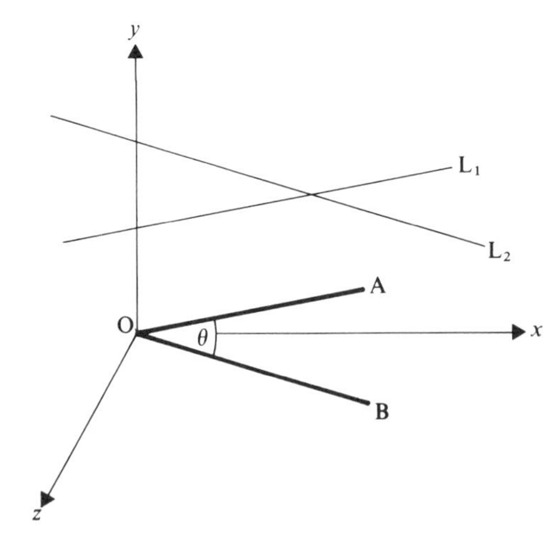

So the angle between two lines depends on their directions and not on their positions, i.e. the angle between skew lines, as well as intersecting lines, is defined.

If L_1 has direction cosines l_1, m_1, n_1
L_2 has direction cosines l_2, m_2, n_2

and OA, OB are drawn such that $|\overrightarrow{OA}| = |\overrightarrow{OB}| = 1$
then A is the point (l_1, m_1, n_1) and B is the point (l_2, m_2, n_2).

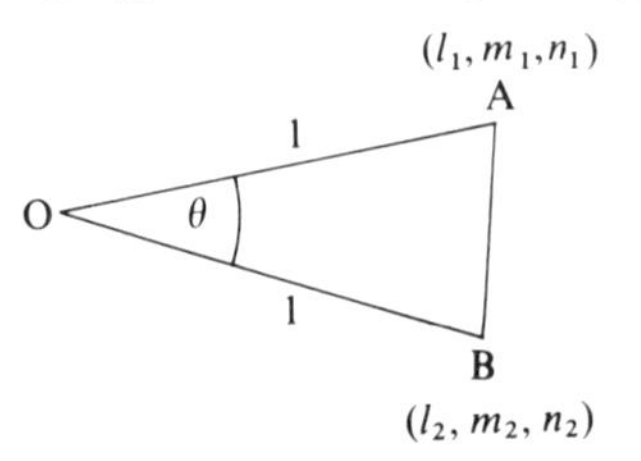

So $\quad AB^2 = (l_1 - l_2)^2 + (m_1 - m_2)^2 + (n_1 - n_2)^2$

$$= l_1^2 - 2l_1l_2 + l_2^2 + m_1^2 - 2m_1m_2 + m_2^2 + n_1^2 - 2n_1n_2 + n_2^2$$

But $\quad l_1^2 + m_1^2 + n_1^2 = 1$ and $l_2^2 + m_2^2 + n_2^2 = 1$

So $\quad AB^2 = 2 - 2(l_1l_2 + m_1m_2 + n_1n_2)$

Using the cosine formula in ΔOAB gives

$$\cos\theta = \frac{OA^2 + OB^2 - AB^2}{2\,OA.OB}$$

$$= \frac{2 - [2 - 2(l_1l_2 + m_1m_2 + n_1n_2)]}{2}$$

i.e. $\quad \cos\theta = l_1l_2 + m_1m_2 + n_1n_2$

For perpendicular lines, $\cos\theta = 0$

therefore

$$L_1 \text{ and } L_2 \text{ are perpendicular} \iff l_1l_2 + m_1m_2 + n_1n_2 = 0$$

Further, as $a = ld$, $b = lm$, $c = ln$,

$$a_1a_2 + b_1b_2 + c_1c_2 = 0$$

where $a_1{:}b_1{:}c_1$ and $a_2{:}b_2{:}c_2$ are the direction ratios of L_1 and L_2.

EXAMPLES 4e (continued)

2) Find the angle between the lines

$$\frac{x-3}{2} = \frac{y+4}{-1} = \frac{z-3}{2} \quad \text{and} \quad \frac{x+1}{6} = \frac{y-2}{-3} = \frac{z-3}{-2}$$

The first line has direction ratios $2{:}-1{:}2$
and hence has direction cosines $\frac{2}{3}, -\frac{1}{3}, \frac{2}{3}$.
The second line has direction ratios $6{:}-3{:}-2$
and direction cosines $\frac{6}{7}, -\frac{3}{7}, -\frac{2}{7}$.

Hence the angle θ between these lines is given by

$$\cos\theta = (\tfrac{2}{3})(\tfrac{6}{7}) + (-\tfrac{1}{3})(-\tfrac{3}{7}) + (\tfrac{2}{3})(-\tfrac{2}{7})$$
$$= \tfrac{11}{21}$$

so $$\theta = \arccos\tfrac{11}{21}$$

3) Find the equations of the line through $A(2, -1, 5)$ which is perpendicular to, and intersects, the line whose equations are

$$\frac{x-3}{1} = \frac{y-1}{2} = \frac{z+1}{2}$$

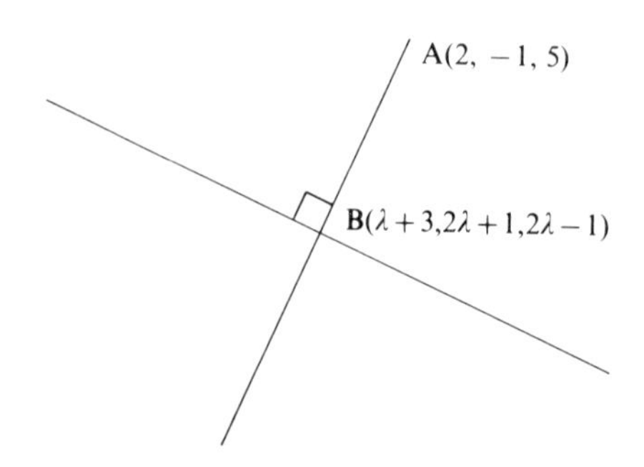

B is the point of intersection of the required line and the given line L whose equations are

$$\frac{x-3}{1} = \frac{y-1}{2} = \frac{z+1}{2} = \lambda$$

As B is on L, its coordinates can be written

$$(\lambda+3, 2\lambda+1, 2\lambda-1)$$

So the direction ratios of AB are $(\lambda+1):(2\lambda+2):(2\lambda-6)$
Also the direction ratios of L are $1:2:2$
As AB is perpendicular to L

$$1(\lambda+1) + 2(2\lambda+2) + 2(2\lambda-6) = 0$$

$$\Rightarrow \qquad \lambda = \tfrac{7}{9}$$

Hence the direction ratios of AB are $\tfrac{16}{9}:\tfrac{32}{9}:-\tfrac{40}{9}$

$$= 2:4:-5$$

So the equations of the line through A and B are

$$\frac{x-2}{2} = \frac{y+1}{4} = \frac{z-5}{-5}$$

Note that in three dimensional work it is not possible to *write down* a general form for the equations of a line which is perpendicular to a given line.

EXERCISE 4e

1) Find out whether the following pairs of lines are parallel, intersecting or skew. If they intersect, give the point of intersection.

(a) $\dfrac{x-1}{3} = \dfrac{y+1}{-4} = z-1$ and $\dfrac{x}{-3} = \dfrac{y}{4} = \dfrac{z}{-1}$

(b) $\dfrac{x-4}{1} = \dfrac{y-8}{2} = \dfrac{z-3}{1}$ and $\dfrac{x-7}{6} = \dfrac{y-6}{4} = \dfrac{z-5}{5}$

(c) $\dfrac{x-1}{2} = y = z-3$ and $x-2 = \dfrac{y+1}{-2}$, $z = 1$

2) The lines $x-2 = \dfrac{y-9}{2} = \dfrac{z-13}{3}$ and $\dfrac{x-a}{-1} = \dfrac{y-7}{2} = \dfrac{z+2}{-3}$ intersect. Find a and the point of intersection.

3) A(4, 7, 1), B(1, 2, 3) and C(−2, 0, 5) are the vertices of a triangle. Find the equations of all three medians.
Use these equations to prove that the medians are concurrent and find the coordinates of the centroid of triangle ABC.

4) Find the angle between each of the following pairs of lines:

(a) $x-3 = \dfrac{y-2}{2} = \dfrac{z+4}{2}$ and $\dfrac{x}{3} = \dfrac{y-5}{2} = \dfrac{z+2}{6}$

(b) a line with direction ratios 4:4:2 and the line through (3, 1, 4) and (7, 2, 12),

(c) $\dfrac{x+4}{3} = \dfrac{y-1}{5} = \dfrac{z+3}{4}$ and $\dfrac{x+1}{1} = \dfrac{y-4}{1} = \dfrac{z-5}{2}$

(d) $\dfrac{x-2}{3} = \dfrac{y+1}{-2}$, $z = 2$ and $\dfrac{x-1}{1} = \dfrac{2y+3}{3} = \dfrac{z+5}{2}$

5) Show that the line $x-1 = \dfrac{y-2}{-7} = \dfrac{z+1}{3}$ is perpendicular to the line

$x+1 = y-3 = \dfrac{z+2}{2}$

6) C is the foot of the perpendicular from A(1, 1, 1) to
$\dfrac{x-1}{2} = \dfrac{y-1}{1} = \dfrac{z-2}{1}$
Find the coordinates of C and hence the distance of A from the line.

7) Find the equations of the line through $A(1, 0, 1)$ which is perpendicular to and intersects the line $\dfrac{x-2}{3} = \dfrac{y}{2} = \dfrac{z+1}{-1}$

8) The points $A(1, 1, 1)$, $B(1, -1, 3)$, $C(2, -1, 0)$ form a triangle. Find the perpendicular distance of A from the line through B and C. Hence find the area of $\triangle ABC$.

9) Show that the lines $\dfrac{x-1}{1} = \dfrac{y-3}{2} = \dfrac{z+1}{3}$

and $\dfrac{x+2}{1} = \dfrac{y-7}{-1} = \dfrac{z-2}{3}$ are skew.

Find the angle between each of these lines and a line whose direction ratios are $a:b:c$.
Find the values of $a:b:c$ such that each of the given lines is perpendicular to the third line.

THE EQUATION OF A PLANE

A particular plane can be located in space if one point on the plane is known and if the plane is known to be *perpendicular* to a given direction.
For example, consider the plane which contains the point $A(2, -1, 4)$ and which is perpendicular to any line with direction ratios $3:5:7$.

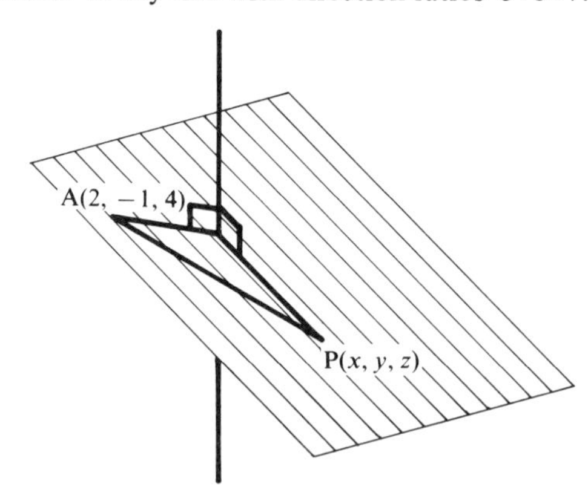

If L is any line with direction ratios $3:5:7$ then *any* line in the plane is perpendicular to L.
Hence $P(x, y, z)$ is a point on the plane $\iff$ AP is perpendicular to L.
Now the direction ratios of AP are $(x-2):(y+1):(z-4)$.
For AP to be perpendicular to L $\quad 3(x-2)+5(y+1)+7(z-4) = 0$

i.e. $$3x+5y+7z = 29$$

So $P(x, y, z)$ is a point on the plane $\iff$ $3x + 5y + 7z = 29$

Thus $3x + 5y + 7z = 29$ is the equation of this plane.
Note that the direction ratios of L, the *normal* to the plane, appear as the coefficients of x, y and z.
In general, if a plane contains $A(x_1, y_1, z_1)$ and is normal to any line with direction ratios $a:b:c$, then

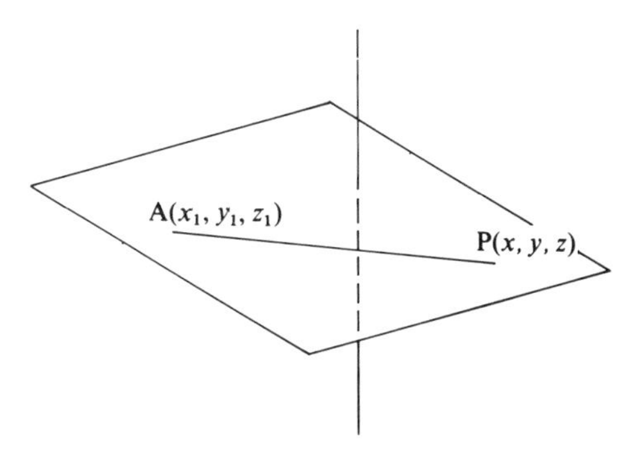

$P(x, y, z)$ is on the plane $\iff$ AP is perpendicular to a line with direction ratios $a:b:c$.

$$\iff a(x - x_1) + b(y - y_1) + c(z - z_1) = 0$$

$$\iff ax + by + cz = (ax_1 + by_1 + cz_1)$$

Thus any equation of the form $ax + by + cz = D$ is the equation of a plane. Further the coefficients of x, y, z (i.e. a, b, c) are the direction ratios of any *normal* to the plane.

If a plane is normal to a line with direction ratios $2:-1:3$ and contains the point $(3, -1, 0)$ its equation may be written down directly as

$$2x - y + 3z = (2)(3) + (-1)(-1) + (3)(0)$$

i.e. $$2x - y + 3z = 7$$

The location of a particular plane can be given in many ways. For example, three non-collinear fixed points locate a particular plane. The following examples show how we can find the equation of a particular plane defined in a variety of ways.

EXAMPLES 4f

1) Find the equation of the plane which contains the points A(3, 1, 2), B(2, −1, 0), C(1, 3, −1).

Any plane has an equation of the form $ax + by + cz = D$. If A, B, C are points on the plane, their coordinates satisfy its equation.

So
$$\begin{aligned} 3a + b + 2c &= D \\ 2a - b &= D \\ a + 3b - c &= D \end{aligned}$$

Solving these equations for a, b, c in terms of D gives

$$a = \frac{10D}{19}, \quad b = \frac{D}{19}, \quad c = \frac{-6D}{19}$$

So the plane through A, B and C has equation

$$\frac{10D}{19}x + \frac{D}{19}y - \frac{6D}{19}z = D$$

i.e.
$$10x + y - 6z = 19$$

Note that a, b, and c do not have unique values, this is because they are direction ratios.
Note that if a plane contains the origin, then $(0, 0, 0)$ satisfies the equation $ax + by + cz = D \quad \Rightarrow \quad D = 0$,
i.e. any plane through the origin has an equation of the form

$$ax + by + cz = 0.$$

If the method of Example 1 is used on such a plane, an unusual situation arises. For example, the plane containing the points $(2, 1, -1)$, $(1, -4, -2)$, $(0, 3, 1)$ gives the equations

$$\begin{aligned} 2a + b - c &= D \\ a - 4b - 2c &= D \\ 3b + c &= D \end{aligned}$$

Eliminating a from the first pair of equations reduces the equations to

$$\begin{aligned} 9b + 3c &= -D \\ 3b + c &= D \end{aligned}$$

for which the only possible solution is $D = 0 \quad \Rightarrow$ plane contains O.

2) Find the equation of the plane which contains the lines L_1 and L_2 whose equations are

$$\frac{x-3}{5} = \frac{y+1}{2} = \frac{z-3}{1} \quad \text{and} \quad \frac{x-3}{2} = \frac{y+1}{4} = \frac{z-3}{3}$$

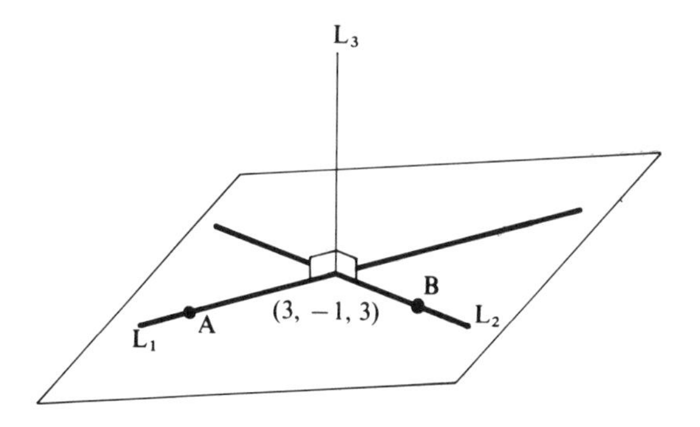

Two lines can be contained in a plane only if they are parallel or intersecting. In either case, since three noncollinear points define a plane uniquely, any two points on L_1 and one point on L_2 are sufficient to define the plane containing L_1 and L_2.
L_1 has equations

$$\frac{x-3}{5} = \frac{y+1}{2} = \frac{z-3}{1} = \lambda$$

where each value of λ corresponds to one (and only one) point on L_1.
So using $\lambda = 0$, $\lambda = -1$ we get the two points $(3, -1, 3)$, $(-2, -3, 2)$ on L_1
For L_2,

$$\frac{x-3}{2} = \frac{y+1}{4} = \frac{z-3}{3} = \mu$$

and $\mu = -1$ gives the point $(1, -5, 0)$ on L_2.
(**Note** that $\mu = 0$ gives $(3, -1, 3)$ which is also on L_1.)
The equation of the plane through these three points can now be found as in Example 1.
Note: Values of λ and μ should be chosen to give points with simple coordinates (preferably integers).

Alternatively we can proceed as follows.
Let L_3, with direction ratios $a:b:c$, be a normal to the plane.
Then L_3 is perpendicular to L_1 so $5a + 2b + c = 0$
Also L_3 is perpendicular to L_2 so $2a + 4b + 3c = 0$
Eliminating c from these equations gives

$$13a + 2b = 0$$

$$\Rightarrow \qquad a:b = -2:13 = -1:\tfrac{13}{2}$$

Eliminating b from these equations gives

$$8a - c = 0 \Rightarrow a:c = 1:8$$

Therefore
$$a:b:c = 1:-\frac{13}{2}:8$$
$$= 2:-13:16$$

Using $(3,-1,3)$ as the fixed point on the plane we can now write its equation as

$$2x - 13y + 16z = (2)(3) + (-13)(-1) + (16)(3)$$

$$\Rightarrow \quad 2x - 13y + 16z = 67.$$

Note that in Example 1 the direction ratios of AB and AC can be found and hence the direction ratios of a line normal to AB and to AC. Then the plane can be found by the method used in Example 2.

3) Show that the line $\dfrac{x-2}{3} = \dfrac{y-4}{2} = \dfrac{z-3}{2}$ is parallel to the plane $2x - 7y + 4z = 0$.

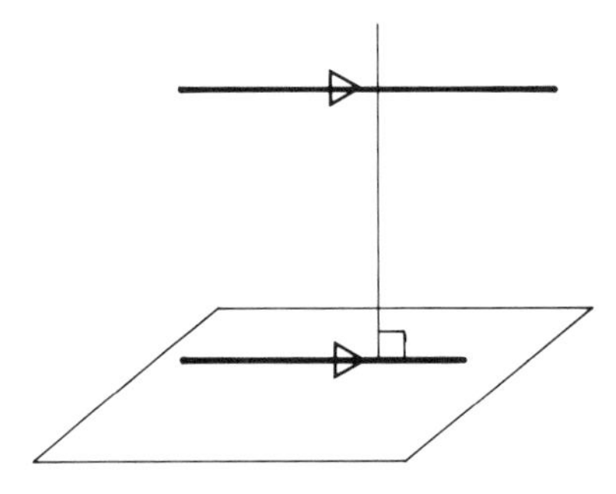

Any line in the plane or parallel to the plane is perpendicular to the normal, L, to the plane.

The given line has direction ratios $3:2:2$
The normal to the plane has direction ratios $2:-7:4$
Now $(3)(2) + (2)(-7) + (2)(4) = 0$.
So the given line is parallel to the plane.

4) Show that the line $\dfrac{x-2}{1} = \dfrac{y-3}{4} = \dfrac{z+1}{1}$ is contained in the plane $3x - y + z = 2$.

A plane can be shown to contain a line if any two points on the line satisfy the equation of the plane.

If $\dfrac{x-2}{1} = \dfrac{y-3}{4} = \dfrac{z+1}{1} = \lambda$ then $\lambda = 0$ and $\lambda = 1$ gives $(2,3,-1)$ and $(3,7,0)$ as two points on the line.
Substituting $x = 2$, $y = 3$, $z = -1$ in the LHS of the equation of the plane gives

$$3(2) - 3 + (-1) = 2 = \text{RHS}.$$

Substituting $x = 3$, $y = 7$, $z = 0$ in the LHS of the equation of the plane gives

$$3(3) - 7 + 0 = 2 = \text{RHS}.$$

Thus $(2, 3, -1)$ and $(3, 7, 0)$ are contained in the plane, so the given line is contained in the plane.

5) Find the point of intersection of the line $\dfrac{x-1}{4} = \dfrac{y+2}{3} = \dfrac{z-1}{2}$ with the plane $2x - y + 2z = 5$.

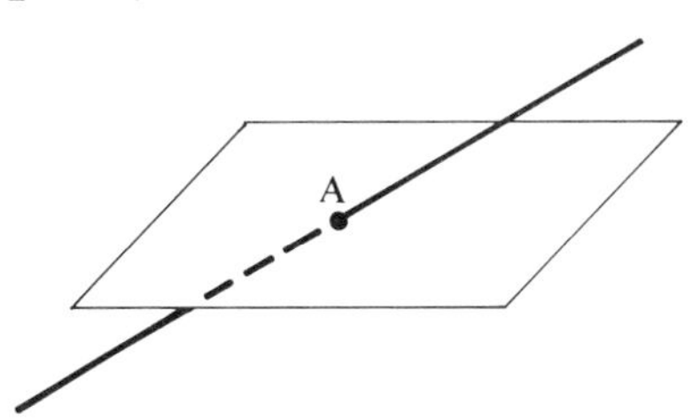

Writing the equations of the line in parametric form

i.e. $$x = 4\lambda + 1, \quad y = 3\lambda - 2, \quad z = 2\lambda + 1$$

and substituting in the equation of the plane gives

$$2(4\lambda + 1) - (3\lambda - 2) + 2(2\lambda + 1) = 5$$

$$\Rightarrow \qquad \lambda = -\tfrac{1}{9}$$

The coordinates of A, the point of intersection, are therefore

$$\left(\tfrac{5}{9}, -\tfrac{7}{3}, \tfrac{7}{9}\right) \qquad \text{(using } \lambda = -\tfrac{1}{9} \text{ in [1].)}$$

EXERCISE 4f

1) Find the equations of the planes passing through the given points and whose normals have the given direction ratios
(a) $(2, -1, 3)$, $1:2:-1$ (b) $(3, 0, -2)$, $1:3:4$
(c) $(0, 0, 0)$, $2:-1:2$ (d) $(1, 1, 0)$, $1:2:-3$.

2) Find the equations of the planes passing through the points
(a) $(2, -1, 4), (3, 2, -6), (4, 1, 5)$,
(b) $(1, 0, 2), (0, 1, 3), (2, -1, 4)$,
(c) $(0, 0, 0), (1, 1, 1), (-1, 2, -4)$.

3) Find the equations of the planes containing the given lines

(a) $\dfrac{x-1}{3} = \dfrac{y+1}{2} = \dfrac{z-4}{5}$, $\dfrac{x-1}{6} = \dfrac{y+1}{1} = \dfrac{z-4}{3}$

(b) $\dfrac{x}{2} = \dfrac{y}{3} = \dfrac{z}{4}$, $\dfrac{x-2}{1} = \dfrac{y-3}{7} = \dfrac{z-4}{-1}$

4) Find the equation of the plane through $(2, 1, 5)$ which is parallel to the plane $2x - y + 5z = 4$.

5) Find the equation of the plane which has positive intercepts $2, 2$ and 4 respectively on the x, y and z axes.

6) Show that the line $\dfrac{x-2}{1} = \dfrac{y-3}{1} = \dfrac{2z}{-1}$ is contained in the plane $3x - y + 2z = 3$.

7) Show that the points $(1, 1, 2)$, $(2, 3, 3)$, $(9, 2, 5)$ and $(3, 2, 3)$ are coplanar.

8) Show that the line $\dfrac{x-1}{3} = \dfrac{y-2}{-9} = \dfrac{z-3}{-1}$ is parallel to the plane $2x + y - 3z = 4$.

9) A plane contains the points $(2, -1, 0)$ and $(5, 2, 4)$ and is perpendicular to the plane $3x - 2y + 7z = 0$. Find its equation.

10) Find the point of intersection of the line $\dfrac{x}{4} = \dfrac{y}{7} = \dfrac{z}{2}$ and the plane $x + 2y - z = 3$.

11) Find the point of intersection of the line $\dfrac{x-1}{3} = \dfrac{y+2}{2} = \dfrac{z-1}{5}$ and the plane $2x - y + 4z = 6$.

Distance of a Plane from the Origin

Consider the plane Π whose equation is $ax + by + cz = D$.

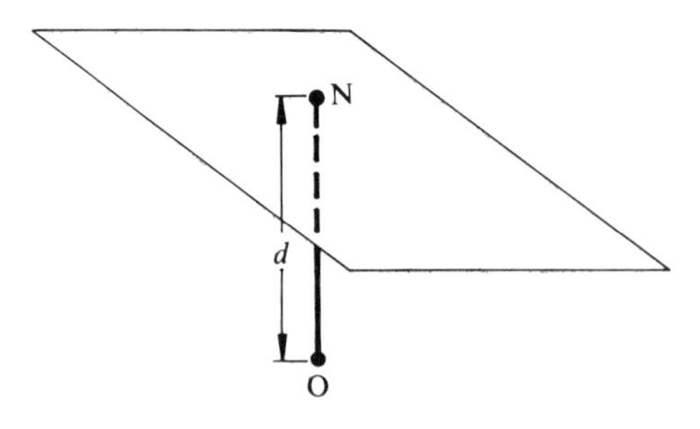

If ON is drawn perpendicular to Π to cut the plane at N, then ON has direction ratios $a:b:c$

and direction cosines $l = \dfrac{a}{r}$, $m = \dfrac{b}{r}$, $n = \dfrac{c}{r}$

where $r = \sqrt{(a^2 + b^2 + c^2)}$.

If ON is of length d, the coordinates of N are (dl, dm, dn),

i.e. $$\frac{da}{r}, \frac{db}{r}, \frac{dc}{r}$$

But N is a point on Π, so the coordinates of N satisfy the equation of Π,

i.e. $$a\left(\frac{da}{r}\right) + b\left(\frac{db}{r}\right) + c\left(\frac{dc}{r}\right) = D$$

$\Rightarrow$ $$d(a^2+b^2+c^2) = D\sqrt{(a^2+b^2+c^2)}$$

$\Rightarrow$ $$d = \frac{D}{\sqrt{(a^2+b^2+c^2)}}$$

So when the equation $ax + by + cz = D$ is divided by $\sqrt{(a^2+b^2+c^2)}$ it becomes $lx + my + nz = d$
where l, m, n are the direction cosines of a normal to the plane and d is the distance of the plane from the origin.

Distance of a Point from a Plane

If the equation of a plane Π is $ax + by + cz = D$ and $A(x_1, y_1, z_1)$ is a point not on Π, then $|\overrightarrow{AN}|$ is the distance of A from Π.

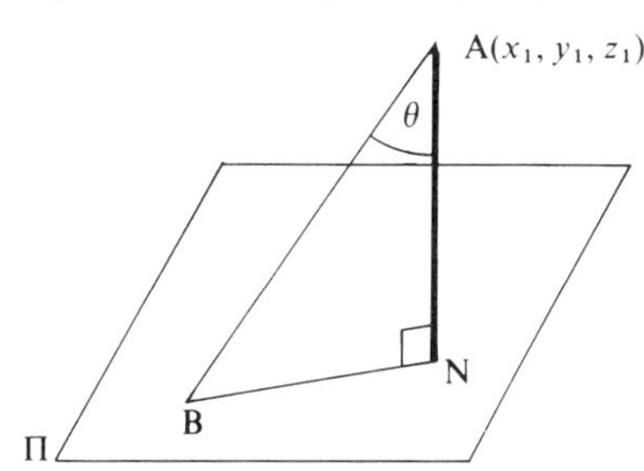

As AN is perpendicular to Π it has direction ratios $a:b:c$.
If B is any point on Π,
then $AN = AB\cos\theta$.

For example, if Π is the plane $2x + 2y + z = 4$ and A is the point $(-1, -2, 7)$, a point B on Π can be found by giving x and y any convenient values (zero say) and finding the corresponding value of z.
So we take B as the point $(0, 0, 4)$
Then AN has direction ratios $2:2:1$ and direction cosines $\frac{2}{3}, \frac{2}{3}, \frac{1}{3}$
AB has direction ratios $-1:-2:3$ and direction cosines

$$-\frac{1}{\sqrt{14}}, -\frac{2}{\sqrt{14}}, \frac{3}{\sqrt{14}}$$

and $$AB = \sqrt{14}$$

Hence $$AN = (\sqrt{14})\left[\left(\frac{2}{3}\right)\left(-\frac{1}{\sqrt{14}}\right) + \left(\frac{2}{3}\right)\left(-\frac{2}{\sqrt{14}}\right) + \left(\frac{1}{3}\right)\left(\frac{3}{\sqrt{14}}\right)\right]$$
$$= 1.$$

So the distance of A from the plane is 1 unit.

Distance Between Two Parallel Planes

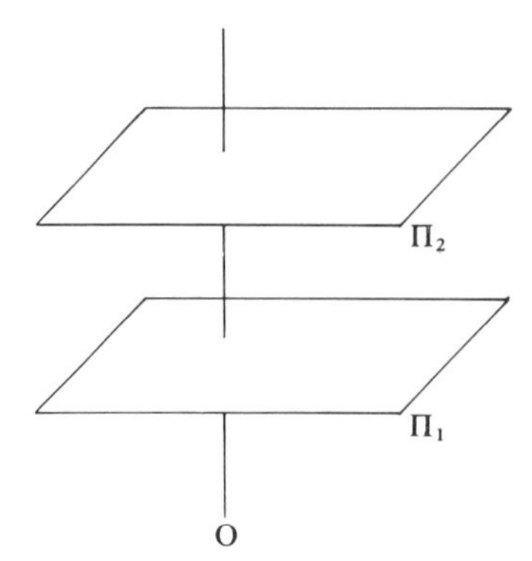

If Π_1 has equation $ax + by + cz = D_1$ then the normal to Π_1 will also be a normal to any parallel plane Π_2.
So Π_2 has an equation of the form

$$ax + by + cz = D_2$$

Π_1 is at a distance $\dfrac{D_1}{\sqrt{(a^2 + b^2 + c^2)}}$ from O,

Π_2 is at a distance $\dfrac{D_2}{\sqrt{(a^2 + b^2 + c^2)}}$ from O,

so the distance between Π_1 and Π_2 is $\dfrac{D_2 - D_1}{\sqrt{(a^2 + b^2 + c^2)}}$

For example, the distance between the pair of parallel planes $2x + y - 2z = 1$ and $2x + y - 2z = 5$ is given by

$$\frac{5 - 1}{\sqrt{[2^2 + 1^2 + (-2)^2]}} = \frac{4}{3}$$

The Angle Between a Line and a Plane

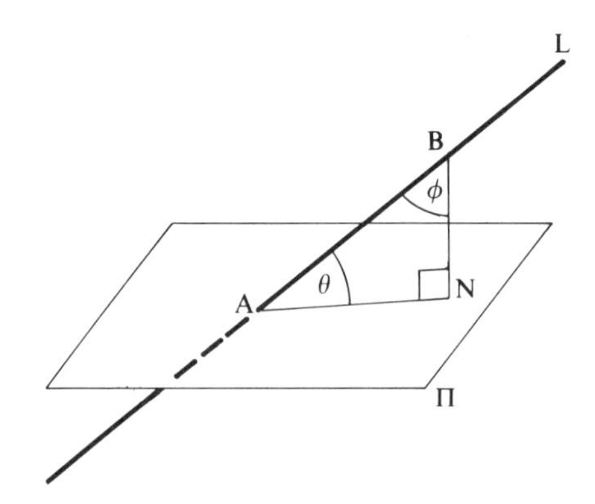

Unless a line L is parallel to a plane Π it will cut the plane at a point A. From any other point B on the line L, BN is drawn perpendicular to Π. Then the angle θ between L and Π is the angle between AB and AN.

Now $\theta = \dfrac{\pi}{2} - \phi$ where ϕ is the angle between AB and BN.

So $\sin\theta = \cos\phi$.

If Π is the plane $2x - y + 2z = 5$

and L is the line $\dfrac{x-1}{4} = \dfrac{y+2}{3} = \dfrac{z-1}{2} = \lambda$

BN is normal to Π.

So BN has direction ratios $2:-1:2$ and direction cosines $\frac{2}{3}, -\frac{1}{3}, \frac{2}{3}$.

AB has direction ratios $4:3:2$ and direction cosines $\dfrac{4}{\sqrt{29}}, \dfrac{3}{\sqrt{29}}, \dfrac{2}{\sqrt{29}}$

So
$$\cos\phi = \left(\frac{2}{3}\right)\left(\frac{4}{\sqrt{29}}\right) + \left(-\frac{1}{3}\right)\left(\frac{3}{\sqrt{29}}\right) + \left(\frac{2}{3}\right)\left(\frac{2}{\sqrt{29}}\right)$$
$$= \frac{3}{\sqrt{29}} = \sin\theta.$$

Hence θ, the angle between AB and the plane, is given by

$$\sin\theta = \frac{3}{\sqrt{29}}$$

The Angle Between Two Planes

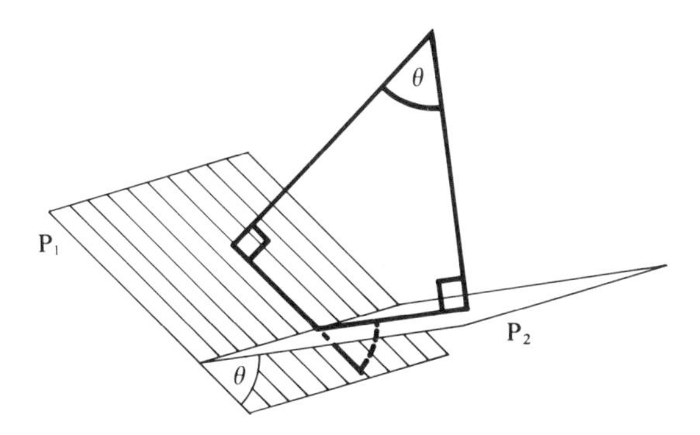

Consider two planes P_1 and P_2 with equations

$$a_1x + b_1y + c_1z = D_1 \quad \text{and} \quad a_2x + b_2y + c_2z = D_2$$

The acute angle θ between P_1 and P_2 is equal to the acute angle between the normals to P_1 and P_2,

i.e. θ is the angle between two lines with direction ratios $a_1:b_1:c_1$ and $a_2:b_2:c_2$.

Thus the angle between the planes

$$2x - y + 2z = 4$$

and

$$3x - 2y + 6z = 3$$

is equal to the angle θ between lines with direction ratios $2:-1:2$ and $3:-2:6$

so
$$\cos\theta = (\tfrac{2}{3})(\tfrac{3}{7}) + (-\tfrac{1}{3})(-\tfrac{2}{7}) + (\tfrac{2}{3})(\tfrac{6}{7}) = \tfrac{20}{21}$$

INTERSECTION OF TWO PLANES

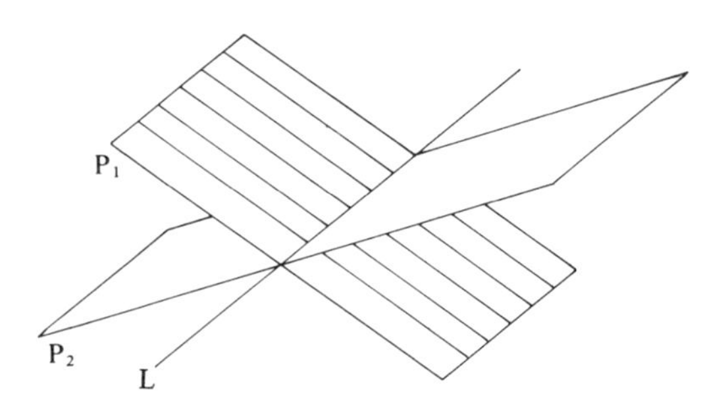

If two planes are not parallel they intersect in a line, i.e. they have a line in common.

If P_1 is the plane $x - 2y + z = 3$
and P_2 is the plane $2x + y - 3z = 1$

then the pair of equations
$$\begin{cases} x - 2y + z = 3 \\ 2x + y - 3z = 1 \end{cases}$$

are the equations of the line of intersection of the planes.

Writing this pair of equations in standard form (by eliminating z and solving for x and then eliminating y and solving for x) gives

$$\frac{x}{1} = \frac{y+2}{1} = \frac{z+1}{1}$$

Now consider the equations of P_1 and P_2 in the form

$$x - 2y + z - 3 = 0 \qquad [1]$$

$$2x + y - 3z - 1 = 0 \qquad [2]$$

The coordinates of any point on L, the line of intersection of P_1 and P_2, satisfy both [1] and [2],
i.e. for any point on L,

$$x - 2y + z - 3 = 2x + y - 3z - 1 = 0 \qquad [3]$$

Now consider the equation formed by adding [1] to a multiple of [2],

i.e. $$(x-2y+z-3)+k(2x+y-3z-1) = 0.$$

For any value of k this is the equation of a plane P_3 and from [3] the coordinates of any point on L satisfy the equation of P_3,
i.e. P_3 also contains L.

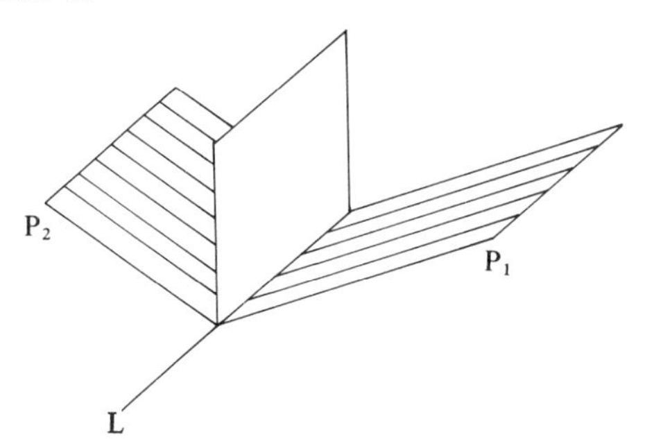

Therefore any plane through the line of intersection of P_1 and P_2 has an equation of the form

$$(x-2y+z-3)+k(2x+y-3z-1) = 0$$

The particular plane containing l which also contains the origin can now be found because $(0, 0, 0)$ satisfies the equation of this plane,

i.e. $$(-3)+k(-1) = 0 \quad \Rightarrow \quad k = -3$$

Thus $$(x-2y+z-3)-3(2x+y-3z-1) = 0$$

$\Rightarrow$ $$-5x-5y+10z = 0$$

or $$x+y-2z = 0$$

is the equation of the plane which contains the line of intersection of P_1 and P_2 and the origin.
In general if $E_1 = 0$ and $E_2 = 0$ are the equations of two members of a family of curves, the equation $E_1 + kE_2 = 0$ represents, for all values of k, another member of the family that contains the points of intersection of $E_1 = 0$ and $E_2 = 0$.

EXERCISE 4g

1) Find the distance of the following planes from the origin:
(a) $2x-y+5z = 2$ (b) $x-2y+2z = 6$
(c) $2x+3y-6z = 14$ (d) $3x+y-2z = 4$.

2) Determine whether the following pairs of planes are parallel or intersecting. If they are parallel, find the distance between them. If they intersect find the direction ratios of the line of intersection.

(a) $2x - y + 3z = 7$ and $-4x + 2y - 6z = 3$,
(b) $x - y + 2z = 1$ and $x + y - 3z = 0$,
(c) $x + 3y - z = 4$ and $2x + 6y - 2z = 3$,
(d) $x + 4y - 3z = 4$ and $x - 3y + 3z = 2$.

3) Find the angle between the pairs of planes (a) and (b), (a) and (c) in Question 1.

4) Find the distance of the point $(-1, 3, 2)$ from the plane
(a) $x - 3y + z = 1$, (b) $2x + y - 3z = 4$.

5) Find the angle between
(a) the line $\dfrac{x-1}{3} = \dfrac{y+2}{5} = \dfrac{z-1}{2}$ and the plane $x - 3y + z = 2$
(b) the line $\dfrac{x+2}{3} = \dfrac{y-2}{2} = \dfrac{z+3}{7}$ and the plane $2x + y - 2z = 1$.

The examples that follow illustrate how the various properties of lines and planes can be used in particular problems.

EXAMPLES 4h

1) A point A has coordinates $(2, -1, 3)$ and a plane Π has equation

$$x - 2y + z - 4 = 0.$$

Show that the point $B(1, 1, 5)$ is contained in the plane Π and find the equations of the line L through A and B.
Find the image of the point A in the plane Π and hence find the equations of the image of the line L in the plane Π.

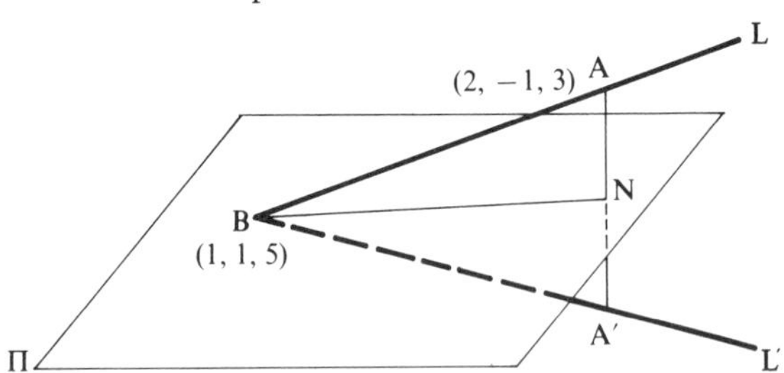

Substituting the coordinates of the point B into the LHS of the equation of the plane Π gives $1 - 2(1) + 5 - 4 = 0$.
Therefore B is contained in Π.
The line L has direction ratios

$$(2-1):(-1-1):(3-5) = 1:-2:-2$$

and hence has equations $\dfrac{x-1}{1} = \dfrac{y-1}{-2} = \dfrac{z-5}{-2}$

If A$'$ is the image of A in Π then AA$'$ is perpendicular to Π and thus parallel to a normal to Π.
Therefore AN has direction ratios $1:-2:1$ and hence equations

$$\frac{x-2}{1} = \frac{y+1}{-2} = \frac{z-3}{1} = \lambda$$

or
$$x = 2+\lambda, \quad y = -1-2\lambda, \quad z = 3+\lambda.$$

At N, since it lies in the plane Π,

$$(2+\lambda)-2(-1-2\lambda)+(3+\lambda)-4 = 0 \quad \Rightarrow \quad \lambda = -\tfrac{1}{2}.$$

Therefore N is the point $(\frac{3}{2}, 0, \frac{5}{2})$.
As N is the midpoint of AA$'$, A$'$ is the point $(1, 1, 2)$.
The line L$'$ through A$'$B, i.e. through $(1, 1, 5)$ and $(1, 1, 2)$ has direction cosines 0, 0, 1. (It is meaningless to say that the direction ratios are $0:0:3$.)
So L$'$ is parallel to the z axis and perpendicular to the x and y axes.
Hence the equations of L$'$ are $x = y = 1$.

2) A sphere of radius 6 units has its centre at the point R$(1, -1, 2)$. Show that the points A$(1, -1, 8)$, B$(7, -1, 2)$ and C$(5, 3, 0)$ lie on the surface of the sphere.
Show further that the equation of the surface of the sphere is

$$(x-1)^2+(y+1)^2+(z-2)^2 = 36.$$

Find the equation of the plane containing A, B and C and hence find the centre of the circle of intersection of this plane and the sphere.

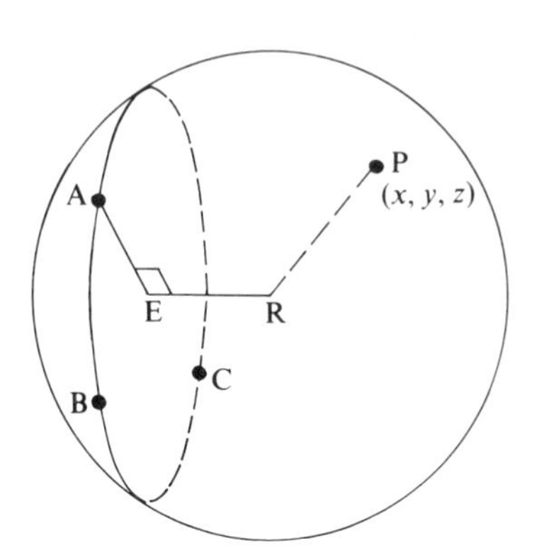

A, B and C lie on the surface of the sphere if AR = BR = CR = 6.

Now
$$\text{AR} = \sqrt{[(1-1)^2+(-1+1)^2+(8-2)^2]} = 6$$
$$\text{BR} = \sqrt{[(7-1)^2+(-1+1)^2+(2-2)^2]} = 6$$
$$\text{CR} = \sqrt{[(5-1)^2+(3+1)^2+(0-2)^2]} = 6$$

So A, B and C lie on the surface of the sphere.
Any point $P(x, y, z)$ lies on the surface of the sphere if and only if $PR = 6$,

i.e. $$PR^2 = 36,$$

i.e. $$(x-1)^2 + (y+1)^2 + (z-2)^2 = 36$$

and this is the equation of the surface of the sphere.

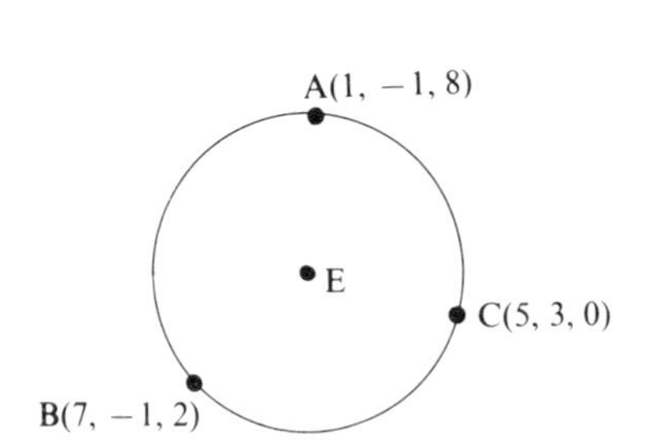

If $ax + by + cz = D$ is the equation of the plane containing A, B and C

then
$$a - b + 8c = D$$
$$5a + 3b = D$$
$$7a - b + 2c = D$$
$$\Rightarrow \quad a = \frac{D}{8}, \quad b = \frac{D}{8}, \quad c = \frac{D}{8}$$

Hence the equation of the plane containing A, B and C is

$$x + y + z = 8.$$

If E is the centre of the circle through A, B and C then ER is normal to this plane and so has direction ratios $1:1:1$.
Using the coordinates of R, the equations of the line ER are

$$x - 1 = y + 1 = z - 2 = \lambda$$

and any point on this line has coordinates $(\lambda + 1, \lambda - 1, \lambda + 2)$.
The line ER cuts the plane $x + y + z = 8$

where $$(\lambda + 1) + (\lambda - 1) + (\lambda + 2) = 8$$

$$\Rightarrow \qquad \lambda = 2.$$

Hence E is the point $(3, 1, 4)$.

MISCELLANEOUS EXERCISE 4

1) A line has the Cartesian equations $x + 4 = \frac{1}{4}(1 - y) = \frac{1}{3}z$.
(a) Find the coordinates of the foot of the perpendicular from the origin to the line.
(b) Show that the plane Π whose equation is $x + 4y + 5z = 0$ contains the line.
(c) Find the equations of the line L which is parallel to the plane Π, passes through the point $(-3, 2, 1)$ and meets the z axis. (JMB)

2) Prove that the lines $$\frac{x+2}{2} = \frac{y-5}{-1} = \frac{z+3}{2} \qquad (L_1)$$

and $$\frac{x-1}{1} = \frac{y-1}{2} = \frac{z+2}{3} \qquad (L_2)$$

intersect and find the coordinates of the point of intersection.
Find
(a) the equation of the plane Π_1 containing L_1 and L_2,
(b) the equation of the plane Π_2 which passes through the line L_1 and which is perpendicular to the plane Π_1,
(c) the equation of the plane which passes through the origin and contains the line of intersection of Π_1 and Π_2. (JMB)

3) The coordinates of the points A, B, and C are $(0, 1, 0)$, $(1, 1, 1)$ and $(2, -3, 1)$ respectively.
Find
(a) the equations of the line joining A and C,
(b) the equation of the plane p containing the point C and the z axis,
(c) the cosine of the angle between the plane p and the line AC,
(d) the perpendicular distance of the origin from the plane

$$4x + y - 4z - 1 = 0. \qquad \text{(AEB '74)}$$

4) Obtain the equation of the straight line which passes through the point $A(-1, 2, 3)$ and which is normal to the plane $2x - 3y + 4z + 8 = 0$.
Calculate the coordinates of P the point of intersection of this line and the plane.
If the point $B(a, 2a, 3)$ lies on the plane, find the value of a and calculate the angle between AP and AB. (AEB '75)p

5) Prove that the points A, B and C whose coordinates are respectively $(2, -1, 5)$, $(3, 1, -2)$ and $(1, -3, 12)$ are collinear. Find
(a) the sine of the angle between the line ABC and the plane $7x + 2y + z = 9$,
(b) the coordinates of the point of intersection of ABC and this plane.
Find the mirror image in the plane of the point $(3, 1, -2)$ (AEB '75)

6) Obtain the equations of the straight line through the origin parallel to the straight line through the points $P(3, 2, 3)$ and $Q(-1, -2, 1)$. Find the equation of the plane which passes through both these two lines.
The line of intersection of a plane through P and a plane through Q is the y axis. Find the angle between these two planes. (U of L)

7) The point $P(14 + 2\lambda, 5 + 2\lambda, 2 - \lambda)$ lies on a fixed straight line for all values of λ. Find Cartesian equations for this line and find the cosine of the acute angle between this line and the line $x = z = 0$.
Show that the line $2x = -y = -z$ is perpendicular to the locus of P.
Hence, or otherwise, find the equation of the plane containing the origin and all possible positions of P. (U of L)

8) Verify that the points $A(2, 1, -2)$, $B(-1, 4, 6)$ and $C(4, 1, 2)$ lie on the surface of a sphere of centre $R(-1, 1, 2)$. Find the equation of the circle of intersection of the sphere with the xy plane, and the equations of the tangent planes parallel to this plane. (JMB)

9) Determine the coordinates of the point B where the straight line

$$\frac{x-2}{-1} = \frac{y+1}{3} = \frac{z+4}{2}$$

meets the xy plane and of the point C where the straight line

$$\frac{x-2}{1} = \frac{y-4}{2} = \frac{z}{2}$$

meets the xz plane.
Prove that the lines intersect and find the coordinates of A, their point of intersection. Verify that both lines lie in the plane $2x + 4y - 5z = 20$.
Find the volume of the tetrahedron OABC, where O is the origin, and show that the area of the triangle ABC is $\frac{3}{2}\sqrt{5}$. (JMB)

10) Find the direction cosines of the line L common to the planes

$$x + y + z - 3 = 0, \quad x + y + 2z + 1 = 0.$$

Show that for all values of the parameter λ the equation

$$(x + y + z - 3) + \lambda(x + y + 2z + 1) = 0$$

represents a plane passing through the line L. Find the equation of the plane P which passes through the line L and through the origin.
Obtain the equations of the straight line through the origin which lies in the plane P and is perpendicular to the line L. (U of L)

11) The coordinates of the vertices of a tetrahedron ABCD are as follows:

$$A(2, 3, 4),\ B(1, -1, 2),\ C(0, 4, 5),\ D(-2, 3, -4).$$

Find the equations of the line AB and the equation of the plane ABC.
Find, in degrees and minutes, the acute angles between
(a) the lines AB and CD,
(b) the line AD and the plane ABC.
Find the perpendicular distance from D to the plane ABC. (AEB '76)

12) A right circular cone has vertex at the point $V(0, 0, 16)$ and the centre of its base at the origin O. The points A, B, C lie on the circumference of the base, which is of radius 12 units. A is on the positive x-axis and B is on the positive y-axis. The point C is on the minor arc AB and AOC is an equilateral triangle. Find the equations of
(a) the line VA, (b) the line VC, (c) the plane VAC.
A sphere has its centre on the z axis and the cone touches the sphere along the circle ACB. Find the radius of the sphere and the coordinates of its centre. (U of L)

13) The line L_1 is the line of intersection of the planes $2x-y-3=0$ and $x+z-3=0$. The equations of the line L_2 are $x-2=y+2=z+3$.
Find the equation of the plane P which passes through L_1 and is parallel to L_2.
Hence, or otherwise, find the shortest distance between L_1 and L_2. Show that the point $Q(1,1,-2)$ lies in the plane P and determine the equations of the straight line in this plane which is perpendicular to L_2 and which passes through Q. (AEB '71)

14) Find the equation of the locus of points which are equidistant from the points $A(0,1,4)$ and $B(1,-3,2)$ and hence show that the locus is a plane p. Find the equations of the line AB and verify that AB is perpendicular to p. Find the equation of the plane perpendicular to p which contains A and which passes through the point $(2,1,3)$. (AEB '71)

15) A regular tetrahedron ABCO has vertices A, B and O which lie in the plane $z=0$, where O is the origin and A is the point $(4,0,0)$.
If the coordinates of C are all positive find the coordinates of B and C. Determine the direction cosines of the straight line through B and C and write down equations for this line. Find also the equation for the plane which contains BC and is parallel to OA. (U of L)

16) In a tetrahedron ABCD the coordinates of the vertices B, C, D are respectively $(1,2,3)$, $(2,3,3)$, $(3,2,4)$. Find
(a) the equation of the plane BCD,
(b) the sine of the angle between BC and the plane $x+2y+3z=4$.
If AC and AD are perpendicular to BD and BC respectively and if $AB=\sqrt{26}$, find the coordinates of the two possible positions of A. (U of L)

17) The corners O, A, B, C of the base of a rectangular box are at the points $(0,0,0)$, $(1,0,0)$, $(0,2,0)$ and $(1,2,0)$ respectively. One corner D of the top of the box is at the point $(1,0,3)$. Find
(a) The equations of the diagonal of the box passing through A,
(b) the direction ratios of the diagonal of the box passing through D,
(c) the angles between the diagonals of the box passing through A and D,
(d) the equation of the plane passing through the points O, C and D,
(e) the perpendicular distance of A from this plane. (AEB '73)

18) The cross-section of a right prism is an equilateral triangle. The rectangular face ABCD of the prism lies in the plane $x+y+5z=\sqrt{2}$, where A and B are the points $(0,\sqrt{2},0)$ and $(\sqrt{2},0,0)$ respectively. EF is the edge in which the other two rectangular faces ADEF and BCEF meet.
Prove that the equation of the plane containing A, B, F is

$$5x+5y-2z = 5\sqrt{2}.$$

If the origin lies inside the prism, determine the equations of the line EF. (U of L)

19) A right circular cone has its vertex at the point $(4, -5, 3)$ and the centre of its base at the point $(0, 1, -1)$. Write down
(a) equations of the axis of the cone,
(b) the equation of the plane containing the base of the cone.
If the line $\dfrac{x-4}{3} = \dfrac{y+5}{-8} = \dfrac{z-3}{2}$ is a generator of the cone, find the coordinates of the point where this generator meets the base, and deduce that the volume of the cone is $6\pi\sqrt{17}$. (U of L)

20) A tetrahedron has one vertex at the origin O and the other three vertices at the points A$(0, 2, 2)$, B$(1, 2, 3)$, C$(3, 1, 6)$. Find
(a) the equation of the plane ABC,
(b) the coordinates of the foot of the perpendicular from O to the plane ABC,
(c) the volume of the tetrahedron.
(U of L)

CHAPTER 5

THREE DIMENSIONAL SPACE VECTOR METHODS

As the title of this chapter suggests, the three dimensional analysis that follows covers the same ground as did the previous chapter, but uses vector methods rather than coordinate methods. It should therefore be regarded as an *alternative* approach to the same work.

BASE VECTORS FOR THREE DIMENSIONS

Let $\mathbf{a}$, $\mathbf{b}$ and $\mathbf{c}$ be any three *non-parallel* and *non-coplanar* vectors. If O is a fixed point in space and if P is *any* other point in space then we can draw the closed polygon OPQR such that

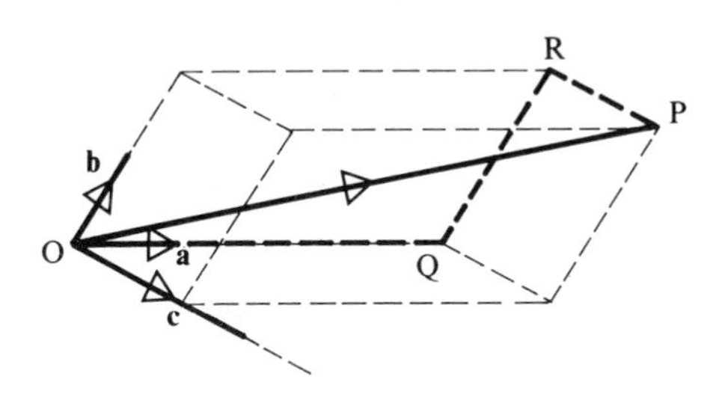

$\overrightarrow{OQ}$ is parallel to $\mathbf{a}$, i.e. $\overrightarrow{OQ} = \lambda\mathbf{a}$

$\overrightarrow{QR}$ is parallel to $\mathbf{b}$, i.e. $\overrightarrow{QR} = \mu\mathbf{b}$

$\overrightarrow{RP}$ is parallel to $\mathbf{c}$, i.e. $\overrightarrow{RP} = \eta\mathbf{c}$.

Now
$$\overrightarrow{OP} = \overrightarrow{OQ} + \overrightarrow{QR} + \overrightarrow{RP}$$
$$= \lambda \mathbf{a} + \mu \mathbf{b} + \eta \mathbf{c}.$$

Thus the position vector of any point in space can be expressed in terms of **a**, **b** and **c**,
i.e. **a**, **b** and **c** together with O form a frame of reference for three dimensional space.
It also follows that any free vector equal to $\overrightarrow{OP}$ can be expressed in the form

$$\lambda \mathbf{a} + \mu \mathbf{b} + \eta \mathbf{c} \quad \text{where} \quad \lambda, \mu, \eta \text{ are scalar.}$$

a, **b** and **c** are known as the base, or basis, vectors of the frame of reference.

Cartesian Base Vectors

Working with general base vectors that are neither unit vectors nor perpendicular to each other would involve tedious calculations on scalene triangles. To avoid this we work with the Cartesian base vectors, **i**, **j** and **k**. These are introduced and developed in Chapter 4 on pages 142–144.

DIRECTION RATIOS AND DIRECTION COSINES OF A VECTOR

In Chapter 4 we saw that if **r** is the position vector of $P(a, b, c)$,

i.e.
$$\mathbf{r} = a\mathbf{i} + b\mathbf{j} + c\mathbf{k}$$

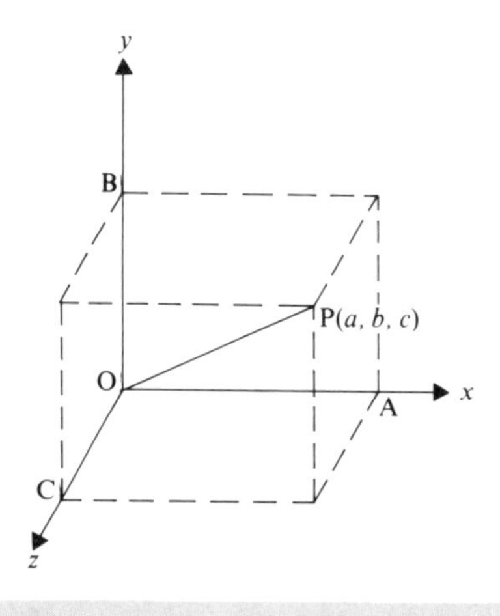

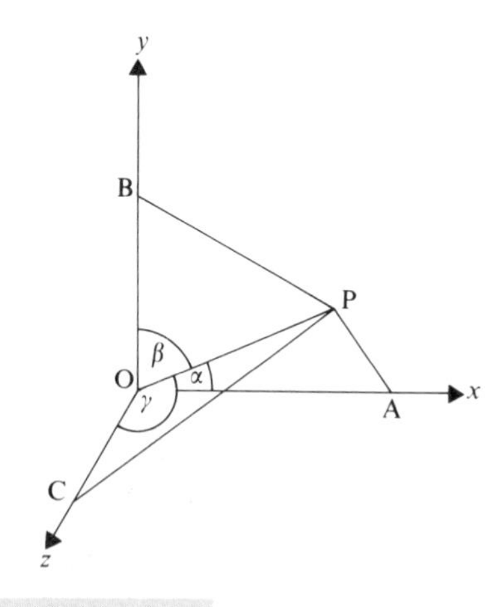

then $a:b:c$ are the direction ratios of $\mathbf{r} = a\mathbf{i} + b\mathbf{j} + c\mathbf{k}$

and if $\overrightarrow{OP}$ makes angles α, β and γ with Ox, Oy, Oz then

$$\cos\alpha = \frac{a}{|\overrightarrow{OP}|}, \quad \cos\beta = \frac{b}{|\overrightarrow{OP}|}, \quad \cos\gamma = \frac{c}{|\overrightarrow{OP}|}$$

so the direction cosines of $\mathbf{r}$ are

$$l = \cos\alpha = \frac{a}{|\mathbf{r}|}$$

$$m = \cos\beta = \frac{b}{|\mathbf{r}|}$$

$$n = \cos\gamma = \frac{c}{|\mathbf{r}|}$$

where $l^2 + m^2 + n^2 = 1$.
The free vector $\mathbf{v} = a\mathbf{i} + b\mathbf{j} + c\mathbf{k}$ has the same direction as the position vector $\mathbf{r} = a\mathbf{i} + b\mathbf{j} + c\mathbf{k}$ and may also be represented by $\overrightarrow{OP}$.

Hence for *any* vector $\mathbf{v} = a\mathbf{i} + b\mathbf{j} + c\mathbf{k}$
the direction ratios are $a:b:c$
and the direction cosines are $l = \frac{a}{|\mathbf{v}|}$, $m = \frac{b}{|\mathbf{v}|}$, $n = \frac{c}{|\mathbf{v}|}$
where $l^2 + m^2 + n^2 = 1$.

Direction Ratios and Direction Cosines of Parallel Vectors

Consider two parallel vectors $\mathbf{v}_1$ and $\mathbf{v}_2$,

i.e. $\mathbf{v}_2 = \lambda\mathbf{v}_1$

So if $\mathbf{v}_1 = a\mathbf{i} + b\mathbf{j} + c\mathbf{k}$

then $\mathbf{v}_2 = \lambda a\mathbf{i} + \lambda b\mathbf{j} + \lambda c\mathbf{k}$

Now the direction ratios of $\mathbf{v}_1$ are $a:b:c$
and the direction ratios of $\mathbf{v}_2$ are $\lambda a:\lambda b:\lambda c$.
But $\lambda a:\lambda b:\lambda c = a:b:c$ whatever the value of λ.
Hence parallel vectors have equal direction ratios.
Now considering the direction cosines of $\mathbf{v}_1$ and $\mathbf{v}_2$ we see that:

$\mathbf{v}_1$ has direction cosines $\frac{a}{|\mathbf{v}_1|}, \frac{b}{|\mathbf{v}_1|}, \frac{c}{|\mathbf{v}_1|}$,

$\mathbf{v}_2$ has direction cosines $\frac{\lambda a}{|\lambda\mathbf{v}_1|}, \frac{\lambda b}{|\lambda\mathbf{v}_1|}, \frac{\lambda c}{|\lambda\mathbf{v}_1|}$.

As $|\lambda\mathbf{v}_1|$ is always positive, $\mathbf{v}_1$ and $\mathbf{v}_2$ have the same direction cosines when λ is positive. But if λ is negative the direction cosines of $\mathbf{v}_2$ are opposite in sign to those of $\mathbf{v}_1$.

Hence *parallel vectors have equal direction ratios.*
Whereas *like parallel vectors have equal direction cosines* but *unlike parallel vectors have direction cosines equal in magnitude but opposite in sign.*

It follows that the direction ratios of a vector are not unique but its direction cosines are unique.

Unit Vectors

Any vector of magnitude 1 unit is a unit vector ($\mathbf{i}, \mathbf{j}$ and $\mathbf{k}$ are all unit vectors).
Consider a line OPQ where the position vector of P is $\mathbf{r}$ and OQ is of length 1 unit.

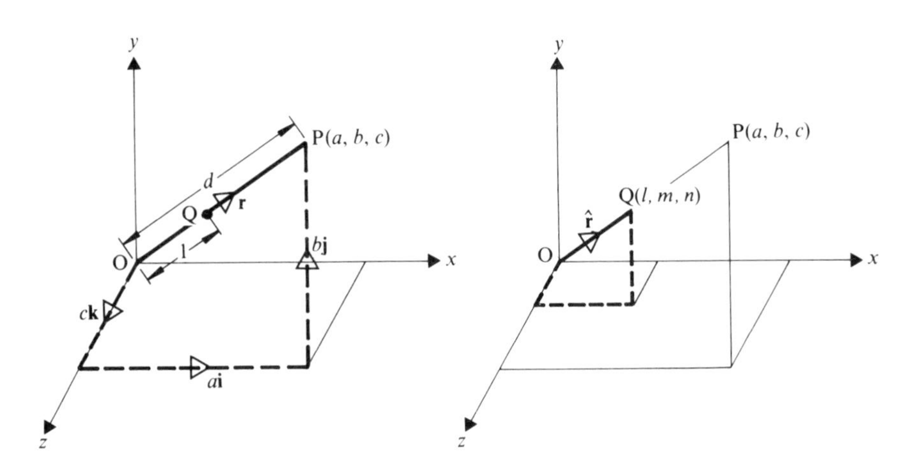

Then $\overrightarrow{OQ}$ represents a unit vector in the direction of $\mathbf{r}$.

Such a unit vector is written $\hat{\mathbf{r}}$.

Now if $OP = d$ then as $OQ = 1$

$$\overrightarrow{OP} = d\,\overrightarrow{OQ}.$$

But $d = |\mathbf{r}|$.

Hence $$\mathbf{r} = |\mathbf{r}|\hat{\mathbf{r}},$$

i.e. any vector, $\mathbf{v}$, can be expressed as the product of its magnitude and the unit vector in the same direction,

i.e. $$\mathbf{v} = |\mathbf{v}|\hat{\mathbf{v}}.$$

If P is the point (a, b, c), we saw in Chapter 4 that Q is the point (l, m, n),

i.e. $$\hat{\mathbf{r}} = l\mathbf{i} + m\mathbf{j} + n\mathbf{k}.$$

Direction Vectors

A vector which is used to specify the direction of another vector can be called a *direction vector*.
For example if we are told that a vector $\mathbf{V}$ of magnitude 14 units is in the direction of the vector $3\mathbf{i} + 6\mathbf{j} + 2\mathbf{k}$ then $3\mathbf{i} + 6\mathbf{j} + 2\mathbf{k}$ is the direction vector for $\mathbf{V}$.

The unit direction vector $\hat{\mathbf{V}} = \frac{1}{7}(3\mathbf{i} + 6\mathbf{j} + 2\mathbf{k})$,

so $\qquad \mathbf{V} = |\mathbf{V}| \times (\text{unit direction vector})$

$$\Rightarrow \qquad \mathbf{V} = 14\left(\frac{3\mathbf{i} + 6\mathbf{j} + 2\mathbf{k}}{7}\right) = 6\mathbf{i} + 12\mathbf{j} + 4\mathbf{k}.$$

Summarising, we have

if $\qquad \mathbf{V} = a\mathbf{i} + b\mathbf{j} + c\mathbf{k}$.

$$|\mathbf{V}| = \sqrt{(a^2 + b^2 + c^2)} = d.$$

$\mathbf{V}$ has direction ratios $a:b:c$.

$\mathbf{V}$ has direction cosines $\dfrac{a}{d}, \dfrac{b}{d}, \dfrac{c}{d}$

or l, m, n.

$$\hat{\mathbf{V}} = \mathbf{V}/d = l\mathbf{i} + m\mathbf{j} + n\mathbf{k}.$$

$$l^2 + m^2 + n^2 = 1.$$

EXAMPLE 5a

A vector $\mathbf{V}$ is inclined to Ox at $45°$ and to Oy at $60°$. Find its inclination to Oz.
If the magnitude of $\mathbf{V}$ is 12 units, express $\mathbf{V}$ in the form $a\mathbf{i} + b\mathbf{j} + c\mathbf{k}$.

The direction cosines of $\mathbf{V}$ are $\cos 45°, \cos 60°, \cos\gamma$.

So $\qquad l = \dfrac{1}{\sqrt{2}}, \quad m = \dfrac{1}{2}, \quad n = \cos\gamma.$

But $\qquad l^2 + m^2 + n^2 = 1.$

Hence $\qquad n^2 = 1 - \frac{1}{2} - \frac{1}{4} = \frac{1}{4}$

$\Rightarrow \qquad n = \pm\frac{1}{2} = \cos\gamma.$

Therefore $\mathbf{V}$ is inclined to Oz either at $60°$ or at $120°$.
Now $l\mathbf{i}, m\mathbf{j}$ and $n\mathbf{k}$ are the components of $\hat{\mathbf{V}}$,

i.e. $\qquad \hat{\mathbf{V}} = \dfrac{1}{\sqrt{2}}\mathbf{i} + \dfrac{1}{2}\mathbf{j} \pm \dfrac{1}{2}\mathbf{k}.$

But $\qquad \mathbf{V} = |\mathbf{V}|\hat{\mathbf{V}}.$

Hence $\qquad \mathbf{V} = 12\left(\dfrac{1}{\sqrt{2}}\mathbf{i} + \dfrac{1}{2}\mathbf{j} \pm \dfrac{1}{2}\mathbf{k}\right)$

$\Rightarrow \qquad \mathbf{V} = 6\sqrt{2}\mathbf{i} + 6\mathbf{j} \pm 6\mathbf{k}.$

EXERCISE 5a

1) Find the direction cosines of the vectors
(a) $2\mathbf{i}+2\mathbf{j}-\mathbf{k}$, (b) $6\mathbf{i}-2\mathbf{j}-3\mathbf{k}$, (c) $3\mathbf{i}+4\mathbf{k}$, (d) $\mathbf{i}+8\mathbf{j}+4\mathbf{k}$.

2) Find the unit vectors in the direction of the vectors in Question 1.

3) Find the coordinates of Q if $|\overrightarrow{OQ}|=1$ and $\overrightarrow{OQ}$ is in the direction of
(a) $\mathbf{i}+2\mathbf{j}-2\mathbf{k}$, (b) $3\mathbf{i}+2\mathbf{j}+6\mathbf{k}$.

4) Find the angle at which the following vectors are inclined to each of the coordinate axes:
(a) $\mathbf{i}-\mathbf{j}+\mathbf{k}$, (b) $4\mathbf{i}+8\mathbf{j}+\mathbf{k}$, (c) $\mathbf{j}-\mathbf{k}$.

5) Find the vector $\mathbf{V}$ if

(a) $\mathbf{V}=\overrightarrow{OP}$ and P is the point $(0,4,5)$,
(b) $|\mathbf{V}|=24$ units and $\hat{\mathbf{V}}=\frac{2}{3}\mathbf{i}-\frac{2}{3}\mathbf{j}-\frac{1}{3}\mathbf{k}$,
(c) $\mathbf{V}$ is inclined at $60°$ to Oy and at $60°$ to Oz and is of magnitude 8 units,
(d) $\mathbf{V}$ is parallel to the vector $8\mathbf{i}+\mathbf{j}+4\mathbf{k}$ and is equal in magnitude to the vector $\mathbf{i}-2\mathbf{j}+2\mathbf{k}$.

6) Find the magnitude and the inclination to each of the coordinate axes of a vector $\mathbf{V}$ if
(a) $\mathbf{V}=3\mathbf{i}+4\mathbf{j}+5\mathbf{k}$,
(b) $\mathbf{V}=-\mathbf{i}+\mathbf{j}-\mathbf{k}$,

(c) $\mathbf{V}$ is represented by $\overrightarrow{OP}$ where P is the point $(5,1,4)$.

7) Find $\hat{\mathbf{r}}$ in the form $a\mathbf{j}+b\mathbf{j}+c\mathbf{k}$ if
(a) $\mathbf{r}=\mathbf{i}-\mathbf{j}+\mathbf{k}$,

(b) $\mathbf{r}=\overrightarrow{OP}$ and P is the point $(3,2,-6)$.

8) A vector $\mathbf{V}$ is inclined at equal acute angles to Ox, Oy and Oz. If the magnitude of $\mathbf{V}$ is 6 units, find $\mathbf{V}$.

9) If $\mathbf{r}_1=2\mathbf{i}-\mathbf{j}+\mathbf{k}$ and $\mathbf{r}_2=\mathbf{i}+3\mathbf{j}+2\mathbf{k}$ find the modulus and direction cosines of $\mathbf{r}_1+\mathbf{r}_2$, $\mathbf{r}_1-\mathbf{r}_2$.

PROPERTIES OF A LINE JOINING TWO POINTS

Consider the line joining $A(x_1,y_1,z_1)$ and $B(x_2,y_2,z_2)$ whose position vectors are $\mathbf{a}$ and $\mathbf{b}$ respectively. In Chapter 4 we saw that

The length of $\overrightarrow{AB}$ is

$$|\overrightarrow{AB}| = |\mathbf{b}-\mathbf{a}| = \sqrt{[(x_2-x_1)^2+(y_2-y_1)^2+(z_2-z_1)^2]}.$$

The direction ratios of $\overrightarrow{AB}$ are

$$(x_2 - x_1):(y_2 - y_1):(z_2 - z_1).$$

The direction cosines of $\overrightarrow{AB}$ are

$$l = \frac{x_2 - x_1}{AB}, \quad m = \frac{y_2 - y_1}{AB}, \quad n = \frac{z_2 - z_1}{AB}.$$

The midpoint of $\overrightarrow{AB}$ is M, where

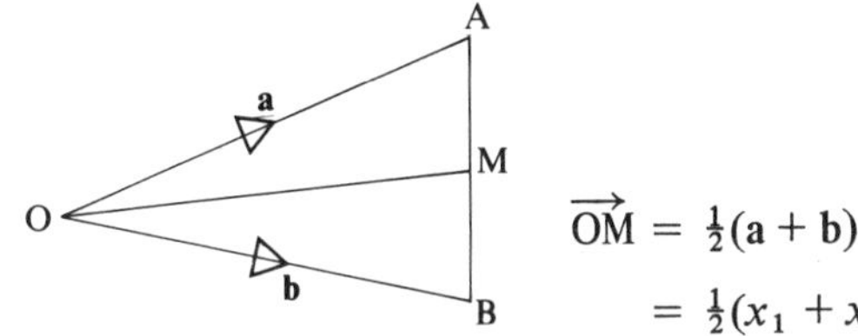

$$\overrightarrow{OM} = \tfrac{1}{2}(\mathbf{a} + \mathbf{b})$$
$$= \tfrac{1}{2}(x_1 + x_2)\mathbf{i} + \tfrac{1}{2}(y_1 + y_2)\mathbf{j} + \tfrac{1}{2}(z_1 + z_2)\mathbf{k}.$$

The position vector of the point P dividing AB in the ratio $\boldsymbol{\lambda} : \boldsymbol{\mu}$ is $\overrightarrow{OP}$ where

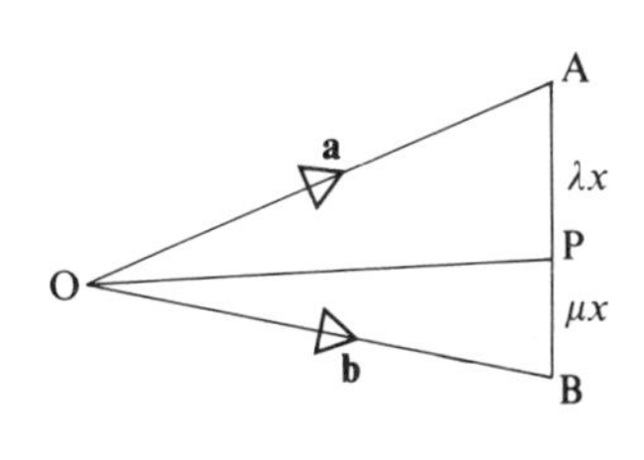

$$\overrightarrow{OP} = \frac{\mu\mathbf{a} + \lambda\mathbf{b}}{\mu + \lambda}$$
$$= \left(\frac{\mu x_1 + \lambda x_2}{\mu + \lambda}\right)\mathbf{i} + \left(\frac{\mu y_1 + \lambda y_2}{\mu + \lambda}\right)\mathbf{j}$$
$$+ \left(\frac{\mu z_1 + \lambda z_2}{\mu + \lambda}\right)\mathbf{k}.$$

EXAMPLES 5b

1) A and B are two points whose position vectors are $3\mathbf{i} + \mathbf{j} - 2\mathbf{k}$ and $\mathbf{i} - 3\mathbf{j} - \mathbf{k}$ respectively. Find the position vectors of the points dividing AB
(a) internally in the ratio $1:3$,
(b) externally in the ratio $3:1$.

(a)

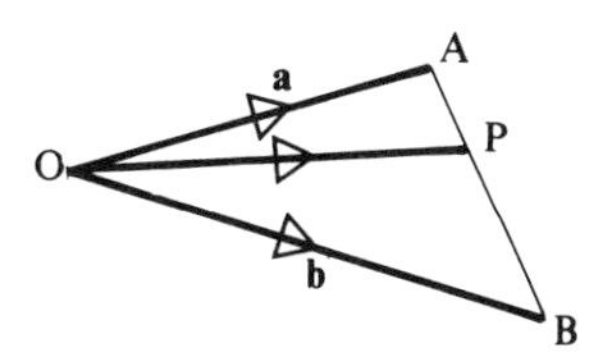

If P is the point dividing AB internally in the ratio $1:3$, then

$$AP:PB = 1:3.$$

So $$\overrightarrow{OP} = \frac{3\mathbf{a} + \mathbf{b}}{3 + 1}$$

$$= \frac{3(3\mathbf{i}+\mathbf{j}-2\mathbf{k})+(\mathbf{i}-3\mathbf{j}-\mathbf{k})}{4}$$

$$= \tfrac{5}{2}\mathbf{i}-\tfrac{7}{4}\mathbf{k}.$$

(b)

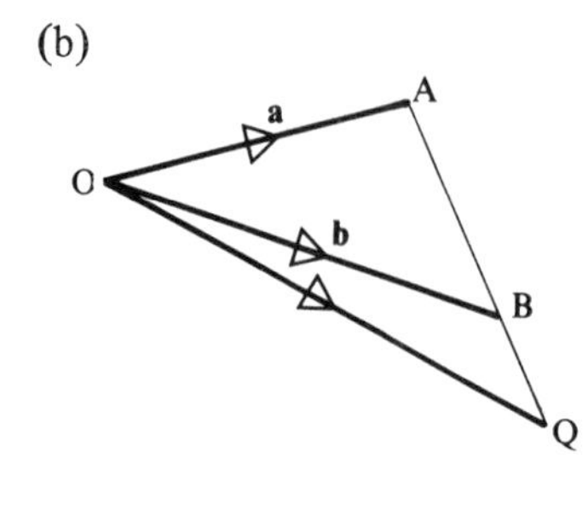

If Q is the point dividing AB externally in the ratio $3:1$, then

$$AQ:QB = 3:-1.$$

So $$\overrightarrow{OQ} = \frac{-\mathbf{a}+3\mathbf{b}}{-1+3}$$

$$= \frac{-(3\mathbf{i}+\mathbf{j}-2\mathbf{k})+3(\mathbf{i}-3\mathbf{j}-\mathbf{k})}{2}$$

$$= -5\mathbf{j}-\tfrac{1}{2}\mathbf{k}$$

2) A, B and C are the points whose position vectors are $2\mathbf{i}-\mathbf{j}+5\mathbf{k}$, $\mathbf{i}-2\mathbf{j}+\mathbf{k}$, $3\mathbf{i}+\mathbf{j}-2\mathbf{k}$ respectively. L and M are the midpoints of AC and CB. Show that LM is parallel to BA.

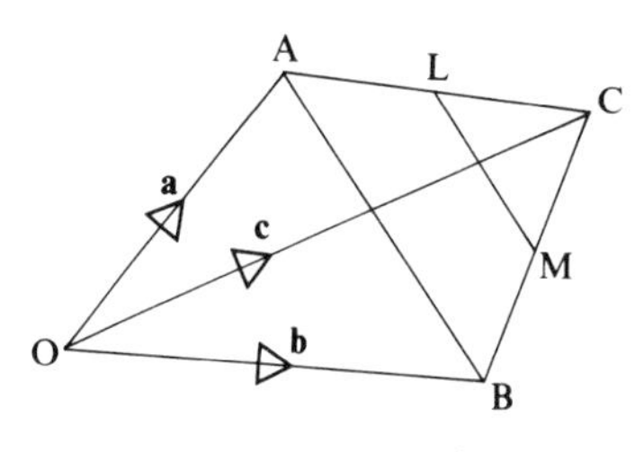

$$\overrightarrow{OL} = \tfrac{1}{2}(\mathbf{a}+\mathbf{c})$$

$$= \tfrac{1}{2}(5\mathbf{i}+3\mathbf{k})$$

$$\overrightarrow{OM} = \tfrac{1}{2}(\mathbf{b}+\mathbf{c})$$

$$= \tfrac{1}{2}(4\mathbf{i}-\mathbf{j}-\mathbf{k}).$$

So $$\overrightarrow{LM} = \overrightarrow{OM}-\overrightarrow{OL} = \tfrac{1}{2}(-\mathbf{i}-\mathbf{j}-4\mathbf{k})$$

Now $$\overrightarrow{BA} = \mathbf{a}-\mathbf{b} \quad = \mathbf{i}+\mathbf{j}+4\mathbf{k}$$

Therefore $$\overrightarrow{BA} = -2\overrightarrow{LM}$$

Hence BA and LM are parallel.

EXERCISE 5b

In Questions 1–9, A, B, C and D are the points with position vectors $\mathbf{i}+\mathbf{j}-\mathbf{k}$, $\mathbf{i}-\mathbf{j}+2\mathbf{k}$, $\mathbf{j}+\mathbf{k}$ and $2\mathbf{i}+\mathbf{j}$ respectively.

1) Find $|\overrightarrow{AB}|$ and $|\overrightarrow{BD}|$.

2) Find the direction cosines of $\overrightarrow{CD}$ and $\overrightarrow{AC}$.

3) Find the position vector of the point which
(a) divides BC internally in the ratio $3:2$,
(b) divides AC externally in the ratio $3:2$.

4) Determine whether any of the following pairs of lines are parallel:
(a) AB and CD, (b) AC and BD, (c) AD and BC.

5) If L and M are the position vectors of the midpoints of AD and BD respectively, show that $\overrightarrow{LM}$ is parallel to $\overrightarrow{AB}$.

6) If H and K are the midpoints of AC and CD show that $\overrightarrow{HK} = \frac{1}{2}\overrightarrow{AD}$.

7) If L, M, N and P are the midpoints of AD, BD, BC and AC respectively, show that $\overrightarrow{LM}$ is parallel to $\overrightarrow{NP}$.

THE EQUATION OF A STRAIGHT LINE

A particular line is uniquely located in space if
(a) it has a known direction and passes through a known fixed point, or
(b) it passes through two known fixed points.

(a) Consider a line which is parallel to a vector $\mathbf{m}$ and which passes through the fixed point A with position vector $\mathbf{a}$.

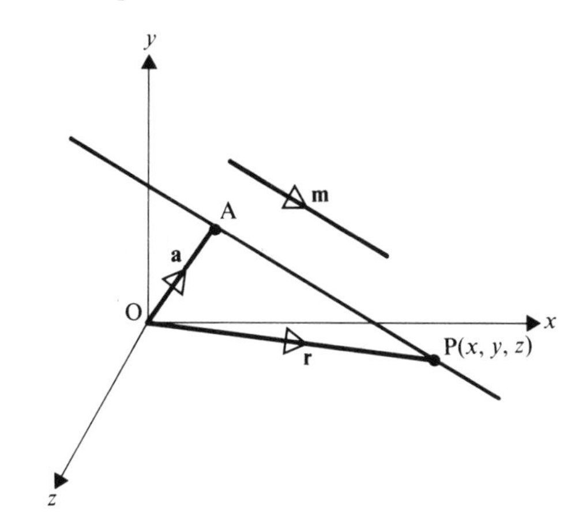

If $\mathbf{r}$ is the position vector, $\overrightarrow{OP}$, of a point P then
P is a point on this line $\iff$ $\overrightarrow{AP} = \lambda\mathbf{m}$
where λ is a variable scalar, i.e. a parameter.

Now $\overrightarrow{AP} = \overrightarrow{OP} - \overrightarrow{OA}$

i.e. $\lambda\mathbf{m} = \mathbf{r} - \mathbf{a}$,

i.e. P is on the line $\iff$ $\mathbf{r} = \mathbf{a} + \lambda\mathbf{m}$.

For each value of the parameter λ this equation gives the position vector of a point on the line and is called the vector equation of the line.

For example, the vector equation of the line which is parallel to the vector $2\mathbf{i}-\mathbf{j}+3\mathbf{k}$ and which passes through the point $(5,-2,4)$ is

$$\mathbf{r} = (5\mathbf{i}-2\mathbf{j}+4\mathbf{k})+\lambda(2\mathbf{i}-\mathbf{j}+3\mathbf{k}). \qquad [1]$$

Now $\mathbf{r}$ is the position vector of any point $P(x,y,z)$ on the line,

i.e.
$$\begin{aligned} x\mathbf{i}+y\mathbf{j}+z\mathbf{k} &= 5\mathbf{i}-2\mathbf{j}+4\mathbf{k}+\lambda(2\mathbf{i}-\mathbf{j}+3\mathbf{k}) \\ &= (5+2\lambda)\mathbf{i}+(-2-\lambda)\mathbf{j}+(4+3\lambda)\mathbf{k} \end{aligned}$$

$$\Rightarrow \quad \left.\begin{aligned} x &= 5+2\lambda \\ y &= -2-\lambda \\ z &= 4+3\lambda. \end{aligned}\right\} \qquad [2]$$

Solving each of these equations for λ gives

$$\frac{x-5}{2} = \frac{y+2}{-1} = \frac{z-4}{3} \quad (=\lambda) \qquad [3]$$

So there are three ways of expressing the relationships between the coordinates of any point P on this line:
[1] is the vector equation of the line,
[2] are the parametric equations of the line,
[3] are the Cartesian equations of the line.
Note that the line is parallel to $2\mathbf{i}-\mathbf{j}+3\mathbf{k}$ so the direction ratios of the line are $2:-1:3$. These direction ratios appear in all three forms of the equations of the line.
In general, if a line passes through $A(x_1, y_1, z_1)$ and is parallel to $a\mathbf{i}+b\mathbf{j}+c\mathbf{k}$ its equations may be written as

$$\mathbf{r} = (x_1\mathbf{i}+y_1\mathbf{j}+z_1\mathbf{k})+\lambda(a\mathbf{i}+b\mathbf{j}+c\mathbf{k})$$

or
$$\begin{cases} x = x_1+\lambda a \\ y = y_1+\lambda b \\ z = z_1+\lambda c \end{cases}$$

or
$$\frac{x-x_1}{a} = \frac{y-y_1}{b} = \frac{z-z_1}{c} \quad (=\lambda).$$

Note that the direction ratios of the line, $a:b:c$, appear in all three forms.
Note also that although the direction ratios of a particular line are unique, the point (x_1, y_1, z_1) is only one of an infinite set of fixed points on the line. Hence the equations representing a particular line are not unique.

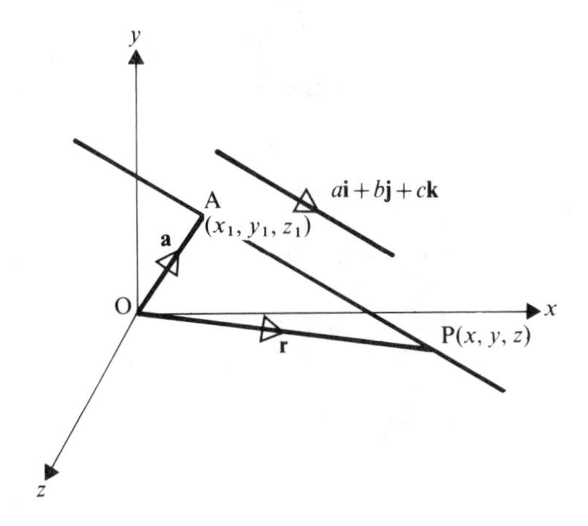

(b) Now consider the definition of a particular line in space as the line passing through two fixed points $A(x_1, y_1, z_1)$ and $B(x_2, y_2, z_2)$ whose position vectors are **a** and **b** respectively.

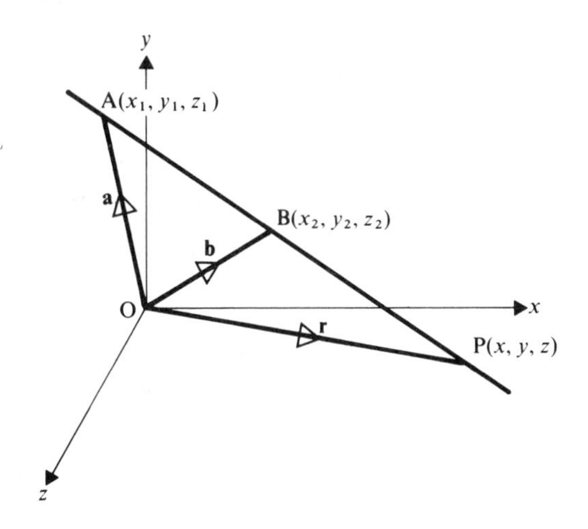

If **r** is the position vector, $\overrightarrow{OP}$, of the point P then

$$\text{P is a point on the line} \iff \overrightarrow{AP} = \lambda\overrightarrow{AB}$$

Now $\overrightarrow{AB} = \mathbf{b} - \mathbf{a}$ and $\overrightarrow{AP} = \mathbf{r} - \mathbf{a}$.

Therefore P is on the line $\iff \mathbf{r} - \mathbf{a} = \lambda(\mathbf{b} - \mathbf{a})$,

i.e.
$$\mathbf{r} = \mathbf{a} + \lambda(\mathbf{b} - \mathbf{a}). \qquad [1]$$

Using the Cartesian components of **a** and **b**, [1] becomes

$$\mathbf{r} = x_1\mathbf{i} + y_1\mathbf{j} + z_1\mathbf{k} + \lambda[(x_2 - x_1)\mathbf{i} + (y_2 - y_1)\mathbf{j} + (z_2 - z_1)\mathbf{k}].$$

EXAMPLES 5c

1) (a) The Cartesian equations of a line are $\dfrac{x-5}{3}=\dfrac{y+4}{7}=\dfrac{z-6}{2}$

Find a vector equation for the line.

(b) If the vector equation of a line is $\mathbf{r}=\mathbf{i}-3\mathbf{j}+2\mathbf{k}+\lambda(5\mathbf{i}+2\mathbf{j}-\mathbf{k})$, express the equation of the line in parametric form and hence find the coordinates of the point where the line crosses the xy plane.

(a) Comparing $$\frac{x-5}{3}=\frac{y+4}{7}=\frac{z-6}{2}$$

with $$\frac{x-x_1}{a}=\frac{y-y_1}{b}=\frac{z-z_1}{c}$$

we see that this line has direction ratios $3:7:2$. Hence its direction vector is $3\mathbf{i}+7\mathbf{j}+2\mathbf{k}$.
One point on the line has coordinates $(5,-4,6)$.
The position vector of this point is $5\mathbf{i}-4\mathbf{j}+6\mathbf{k}$.
Hence a vector equation of the line is

$$\mathbf{r} = 5\mathbf{i}-4\mathbf{j}+6\mathbf{k}+\lambda(3\mathbf{i}+7\mathbf{j}+2\mathbf{k}).$$

(b) $$\mathbf{r} = \mathbf{i}-3\mathbf{j}+2\mathbf{k}+\lambda(5\mathbf{i}+2\mathbf{j}-\mathbf{k})$$

$$\Rightarrow \quad \mathbf{r} = (1+5\lambda)\mathbf{i}+(-3+2\lambda)\mathbf{j}+(2-\lambda)\mathbf{k}.$$

A general point on the line has coordinates

$$x = 1+5\lambda, \quad y = -3+2\lambda, \quad z = 2-\lambda$$

and these are the parametric equations of the line.
At the point where the line crosses the xy plane, $z=0$,
so $2-\lambda=0$.
When $\lambda=2$, $x=11$ and $y=1$.
Therefore the line crosses the xy plane at the point $(11, 1, 0)$.

2) A line passes through the point with position vector $2\mathbf{i}-\mathbf{j}+4\mathbf{k}$ and is in the direction of $\mathbf{i}+\mathbf{j}-2\mathbf{k}$. Find equations for the line in vector and in Cartesian form.

The equation of the line in vector form is

$$\mathbf{r} = 2\mathbf{i}-\mathbf{j}+4\mathbf{k}+\lambda(\mathbf{i}+\mathbf{j}-2\mathbf{k}).$$

This shows that the coordinates of any point P on the line are

$$[(2+\lambda), (-1+\lambda), (4-2\lambda)].$$

Hence the Cartesian equations are

$$x-2 = y+1 = \frac{z-4}{-2} \quad (= \lambda).$$

3) Find a vector equation for the line through the points A(3, 4, −7) and B(1, −1, 6).

$$\overrightarrow{OA} = \mathbf{a} = 3\mathbf{i}+4\mathbf{j}-7\mathbf{k},$$

$$\overrightarrow{OB} = \mathbf{b} = \mathbf{i}-\mathbf{j}+6\mathbf{k}.$$

For any point P on the line, $\overrightarrow{OP} = \mathbf{r}$,

so
$$\mathbf{r} = \mathbf{a}+\lambda(\mathbf{b}-\mathbf{a})$$
$$= 3\mathbf{i}+4\mathbf{j}-7\mathbf{k}+\lambda(-2\mathbf{i}-5\mathbf{j}+13\mathbf{k}).$$

4) Show that the line through the points $\mathbf{i}+\mathbf{j}-3\mathbf{k}$ and $4\mathbf{i}+7\mathbf{j}+\mathbf{k}$ is parallel to the line $\mathbf{r} = \mathbf{i}-\mathbf{k}+\lambda(\frac{3}{2}\mathbf{i}+3\mathbf{j}+2\mathbf{k})$.

The line through the two given points has direction ratios

$$(1-4):(1-7):(-3-1) = 3:6:4.$$

The given line has direction ratios

$$\tfrac{3}{2}:3:2 = 3:6:4.$$

Hence the two lines are parallel.

5) Find the coordinates of the point where the line through A(3, 4, 1) and B(5, 1, 6) crosses the xy plane.

The vector equation of the line through A and B is

$$\mathbf{r} = 3\mathbf{i}+4\mathbf{j}+\mathbf{k}+\lambda(2\mathbf{i}-3\mathbf{j}+5\mathbf{k}).$$

The line crosses the xy plane where $z = 0$.
Any point on the line has coordinates $[(3+2\lambda), (4-3\lambda), (1+5\lambda)]$.
If $z = 0$ then $1+5\lambda = 0 \Rightarrow \lambda = -\frac{1}{5}$

Hence $x = 3-\frac{2}{5} = \frac{13}{5}$; $y = 4+\frac{3}{5} = \frac{23}{5}$

6) Write down the direction cosines of the line

$$\mathbf{r} = 3\mathbf{i}+3\mathbf{j}-\mathbf{k}+\lambda(4\mathbf{i}+3\mathbf{k})$$

and describe its position relative to the x, y, and z axes.

From its equation, the line is seen to be parallel to $4\mathbf{i}+0\mathbf{j}+3\mathbf{k}$.
Hence the direction cosines of the line are $\frac{4}{5}, 0, \frac{3}{5}$

so
$$\cos\beta = 0 \Rightarrow \beta = \frac{\pi}{2}$$

Therefore the line is perpendicular to Oy and hence parallel to the xz plane. As the point $(3, 3, -1)$ lies on the line, the line lies in the plane parallel to the xz plane cutting Oy at $y = 3$, and is inclined at $\arccos\frac{4}{5}$ to Ox and $\arccos\frac{3}{5}$ to Oz.

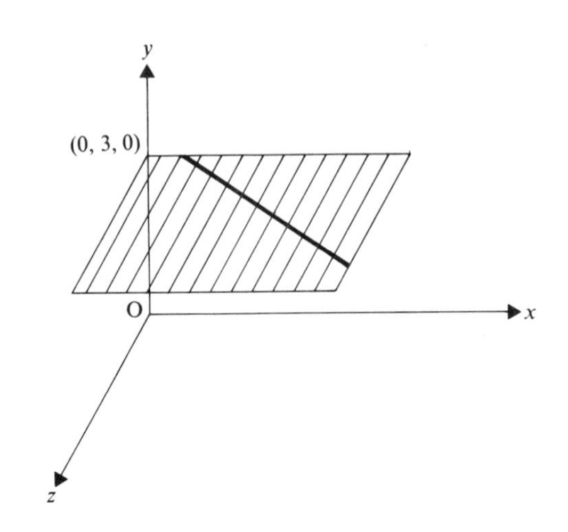

EXERCISE 5c

1) Convert the following vector equations to Cartesian form:
(a) $\mathbf{r} = 2\mathbf{i} + 3\mathbf{j} - \mathbf{k} + \lambda(\mathbf{i} + \mathbf{j} + \mathbf{k})$,
(b) $\mathbf{r} = 4\mathbf{j} + \lambda(3\mathbf{i} + 5\mathbf{k})$,
(c) $\mathbf{r} = \lambda(2\mathbf{i} + 3\mathbf{j} + 4\mathbf{k})$.

2) Convert to vector form, the following equations:

(a) $\dfrac{x-3}{4} = \dfrac{y-1}{2} = \dfrac{z-7}{6}$,

(b) $x = 3\lambda + 2, \quad y = \lambda - 5, \quad z = 4\lambda + 1$,

(c) $\dfrac{1-x}{3} = \dfrac{y}{5} = z$.

3) Write down equations, in vector and in Cartesian form, for the line through a point A with position vector $\mathbf{a}$ and with a direction vector $\mathbf{b}$ if:
(a) $\mathbf{a} = \mathbf{i} - 3\mathbf{j} + 2\mathbf{k}$ $\quad \mathbf{b} = 5\mathbf{i} + 4\mathbf{j} - \mathbf{k}$,
(b) $\mathbf{a} = 2\mathbf{i} + \mathbf{j}$ $\quad \mathbf{b} = 3\mathbf{j} - \mathbf{k}$,
(c) A is the origin $\quad \mathbf{b} = \mathbf{i} - \mathbf{j} - \mathbf{k}$.

4) State whether or not the following pairs of lines are parallel:
(a) $\mathbf{r} = \mathbf{i} + \mathbf{j} - \mathbf{k} + \lambda(2\mathbf{i} - 3\mathbf{j} + \mathbf{k})$
$\mathbf{r} = 2\mathbf{i} - 4\mathbf{j} + 5\mathbf{k} + \lambda(\mathbf{i} + \mathbf{j} - \mathbf{k})$

(b) $\dfrac{x-1}{2}=\dfrac{y-4}{3}=\dfrac{z+1}{-4}$

$\dfrac{x}{4}=\dfrac{y+5}{6}=\dfrac{3-z}{8}$,

(c) $\mathbf{r}=2\mathbf{i}-\mathbf{j}+4\mathbf{k}+\lambda(\mathbf{i}+\mathbf{j}+3\mathbf{k})$

$x-4=y+7=\dfrac{z}{3}$,

(d) $\mathbf{r}=\lambda(3\mathbf{i}-3\mathbf{j}+6\mathbf{k})$
$\mathbf{r}=4\mathbf{j}+\lambda(-\mathbf{i}+\mathbf{j}-2\mathbf{k})$,

(e) $\mathbf{r}=3\mathbf{i}+\mathbf{k}+\lambda(\mathbf{i}-\mathbf{j}-2\mathbf{k})$

$\dfrac{x-3}{1}=\dfrac{y}{1}=\dfrac{z-1}{2}$

5) The points A(4, 5, 10), B(2, 3, 4) and C(1, 2, −1) are three vertices of a parallelogram ABCD. Find vector and Cartesian equations for the sides AB and BC and find the coordinates of D.

6) Write down a vector equation for the line through A and B if

(a) $\overrightarrow{OA}$ is $3\mathbf{i}+\mathbf{j}-4\mathbf{k}$ and $\overrightarrow{OB}$ is $\mathbf{i}+7\mathbf{j}+8\mathbf{k}$,
(b) A and B have coordinates (1, 1, 7) and (3, 4, 1).
Find, in each case, the coordinates of the points where the line crosses the *xy* plane, the *yz* plane and the *zx* plane.

7) A line has Cartesian equations $\dfrac{x-1}{3}=\dfrac{y+2}{4}=\dfrac{z-4}{5}$

Find a vector equation for a parallel line passing through the point with position vector $5\mathbf{i}-2\mathbf{j}-4\mathbf{k}$ and find the coordinates of the point on this line where $y=0$.

8) The Cartesian equations of a line are $x-2=2y+1=3z-2$.
Find the direction ratios of the line and write down the vector equation of the line through (2, −1, −1) which is parallel to the given line.

PAIRS OF LINES

The location of two lines in space may be such that:
(a) the lines are parallel,
(b) the lines are not parallel and intersect,
(c) the lines are not parallel and do not intersect. Such lines are called *skew*.

Parallel Lines

We have already seen that parallel lines have equal direction ratios. So if two lines are parallel, this property can be observed from their equations.

Non-parallel Lines

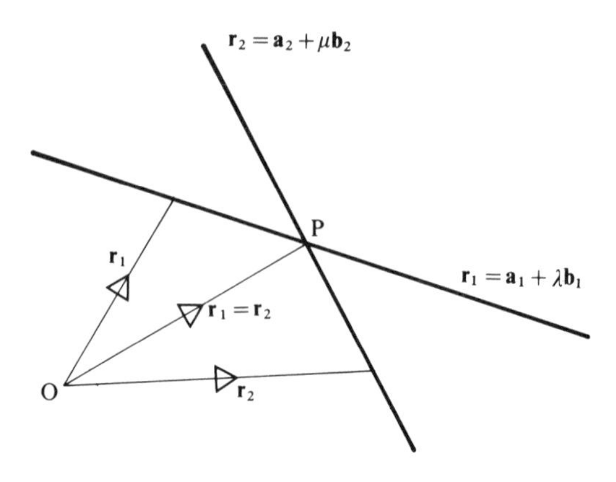

Consider two lines whose vector equations are $\mathbf{r}_1 = \mathbf{a}_1 + \lambda\mathbf{b}_1$ and $\mathbf{r}_2 = \mathbf{a}_2 + \mu\mathbf{b}_2$.
In order that these lines shall intersect there must be unique values of λ and μ such that

$$\mathbf{a}_1 + \lambda\mathbf{b}_1 = \mathbf{a}_2 + \mu\mathbf{b}_2.$$

If no such values can be found, the lines do not intersect.

EXAMPLE 5d

Find out whether the following pairs of lines are parallel, non-parallel and intersecting, or non-parallel and non-intersecting:

(a) $\mathbf{r}_1 = \mathbf{i} + \mathbf{j} + 2\mathbf{k} + \lambda(3\mathbf{i} - 2\mathbf{j} + 4\mathbf{k})$
$\mathbf{r}_2 = 2\mathbf{i} - \mathbf{j} + 3\mathbf{k} + \mu(-6\mathbf{i} + 4\mathbf{j} - 8\mathbf{k})$,

(b) $\mathbf{r}_1 = \mathbf{i} - \mathbf{j} + 3\mathbf{k} + \lambda(\mathbf{i} - \mathbf{j} + \mathbf{k})$
$\mathbf{r}_2 = 2\mathbf{i} + 4\mathbf{j} + 6\mathbf{k} + \mu(2\mathbf{i} + \mathbf{j} + 3\mathbf{k})$,

(c) $\mathbf{r}_1 = \mathbf{i} + \mathbf{k} + \lambda(\mathbf{i} + 3\mathbf{j} + 4\mathbf{k})$
$\mathbf{r}_2 = 2\mathbf{i} + 3\mathbf{j} + \mu(4\mathbf{i} - \mathbf{j} + \mathbf{k})$.

(a) Checking first whether the lines are parallel we compare the direction ratios of the two lines.
The first line has direction ratios $3:-2:4$.
The second line has direction ratios $-6:4:-8 = 3:-2:4$.
Therefore these two lines are parallel.

(b) In this case the two sets of direction ratios are $1:-1:1$ and $2:1:3$.
These are not equal, so these two lines are not parallel.
Now if the lines intersect it will be at a point where $\mathbf{r}_1 = \mathbf{r}_2$,
i.e. where

$$(1 + \lambda)\mathbf{i} - (1 + \lambda)\mathbf{j} + (3 + \lambda)\mathbf{k} = 2(1 + \mu)\mathbf{i} + (4 + \mu)\mathbf{j} + (6 + 3\mu)\mathbf{k}.$$

Equating the coefficients of **i** and **j**, we have

$$1 + \lambda = 2(1 + \mu)$$

$$-(1 + \lambda) = 4 + \mu.$$

Hence $\mu = -2, \quad \lambda = -3$

With these values for λ and μ, the coefficients of **k** become

$$\left.\begin{array}{lrl} \text{first line} & 3 + \lambda &= 0 \\ \text{second line} & 6 + 3\mu &= 0 \end{array}\right\} \text{equal values.}$$

So $\mathbf{r}_1 = \mathbf{r}_2$ when $\lambda = -3$ and $\mu = -2$.

Therefore the lines *do* intersect at the point with position vector

$$(1-3)\mathbf{i} - (1-3)\mathbf{j} + (3-3)\mathbf{k}, \qquad (\lambda = -3 \text{ in } \mathbf{r}_1)$$

i.e. $$-2\mathbf{i} + 2\mathbf{j}.$$

(c) The direction ratios of these two lines are not equal so the lines are not parallel.
If the lines intersect it will be where $\mathbf{r}_1 = \mathbf{r}_2$,
i.e. where

$$(1 + \lambda)\mathbf{i} + 3\lambda\mathbf{j} + (1 + 4\lambda)\mathbf{k} = (2 + 4\mu)\mathbf{i} + (3 - \mu)\mathbf{j} + \mu\mathbf{k}.$$

Equating the coefficients of **i** and **j** we have

$$1 + \lambda = 2 + 4\mu$$

$$3\lambda = 3 - \mu.$$

Hence $\mu = 0, \quad \lambda = 1.$

With these values of λ and μ, the coefficients of **k** become

$$\left.\begin{array}{lrl} \text{first line} & 1 + 4\lambda &= 5 \\ \text{second line} & \mu &= 0 \end{array}\right\} \text{unequal values.}$$

So there are no values of λ and μ for which $\mathbf{r}_1 = \mathbf{r}_2$ and these lines do not intersect and are skew.

EXERCISE 5d

1) Find whether the following pairs of lines are parallel, intersecting or skew. In the case of intersection state the position vector of the common point.

(a) $\mathbf{r} = \mathbf{i} - \mathbf{j} + \mathbf{k} + \lambda(3\mathbf{i} - 4\mathbf{j} + \mathbf{k})$
$\mathbf{r} = \mu(-9\mathbf{i} + 12\mathbf{j} - 3\mathbf{k}),$

(b) $\dfrac{x-4}{1}=\dfrac{y-8}{2}=\dfrac{z-3}{1}$

$\dfrac{x-7}{6}=\dfrac{y-6}{4}=\dfrac{z-5}{5}$,

(c) $\mathbf{r}=\mathbf{i}+3\mathbf{k}+\lambda(2\mathbf{i}+\mathbf{j}+\mathbf{k})$
$\mathbf{r}=2\mathbf{i}-\mathbf{j}+\mathbf{k}+\mu(\mathbf{i}-2\mathbf{j})$.

2) Two lines have equations

$$\mathbf{r} = 2\mathbf{i}+9\mathbf{j}+13\mathbf{k}+\lambda(\mathbf{i}+2\mathbf{j}+3\mathbf{k})$$

$$\mathbf{r} = a\mathbf{i}+7\mathbf{j}-2\mathbf{k}+\mu(-\mathbf{i}+2\mathbf{j}-3\mathbf{k}).$$

If they intersect, find the value of a and the position vector of the point of intersection.

3) Show that the lines

$$\mathbf{r} = 2\mathbf{i}-\mathbf{j}+\mathbf{k}+\lambda(\mathbf{i}-2\mathbf{j}+2\mathbf{k})$$

$$\mathbf{r} = \mathbf{i}-3\mathbf{j}+4\mathbf{k}+\lambda(2\mathbf{i}+3\mathbf{j}-6\mathbf{k})$$

are skew.
$\overrightarrow{OQ}$ is the unit vector in the direction of the first line and $\overrightarrow{OR}$ is the unit vector in the direction of the second line. Write down the coordinates of Q and R. By using the cosine formula in triangle OQR find the angle between $\overrightarrow{OQ}$ and $\overrightarrow{OR}$ and hence the angle between the given lines.

4) If $\overrightarrow{OA}$ is the unit vector $l_1\mathbf{i}+m_1\mathbf{j}+n_1\mathbf{k}$ and $\overrightarrow{OB}$ is the unit vector $l_2\mathbf{i}+m_2\mathbf{j}+n_2\mathbf{k}$, by using the cosine formula in triangle OAB show that θ, the angle between $\overrightarrow{OA}$ and $\overrightarrow{OB}$, is given by $\cos\theta = l_1l_2+m_1m_2+n_1n_2$.

SCALAR PRODUCT

Geometric analysis often involves the use of expressions containing the sine or the cosine of an angle (for example, in the solution of triangles).
So we now look at a vector operation between two vectors **a** and **b** which, among other things, gives rise to $\cos\theta$ where θ is the angle between **a** and **b**.
Later in this chapter a second vector operation is introduced which involves $\sin\theta$.
These two operations are both called products but, being vector operations they are in no way related to the product of real numbers.
The first of these results in a scalar quantity and it is therefore known as the *scalar product* of two vectors.
The second operation, which is referred to as the *vector product* of two vectors, produces a vector quantity.

Distinction is drawn between the two processes by using the multiplication 'dot' symbol $(\mathbf{a}.\mathbf{b})$ exclusively for the scalar product and the multiplication 'cross' symbol $(\mathbf{a} \times \mathbf{b})$ exclusively for the vector product.
(This can cause some difficulty in vector work since the 'dot' and 'cross' can cause confusion if used to represent multiplication of numbers in the context of vector analysis. This problem can be avoided by the use of brackets.)
The scalar product was introduced in Pure Mathematics I and the results obtained are summarised below and then extended.
For two vectors $\mathbf{a}$ and $\mathbf{b}$

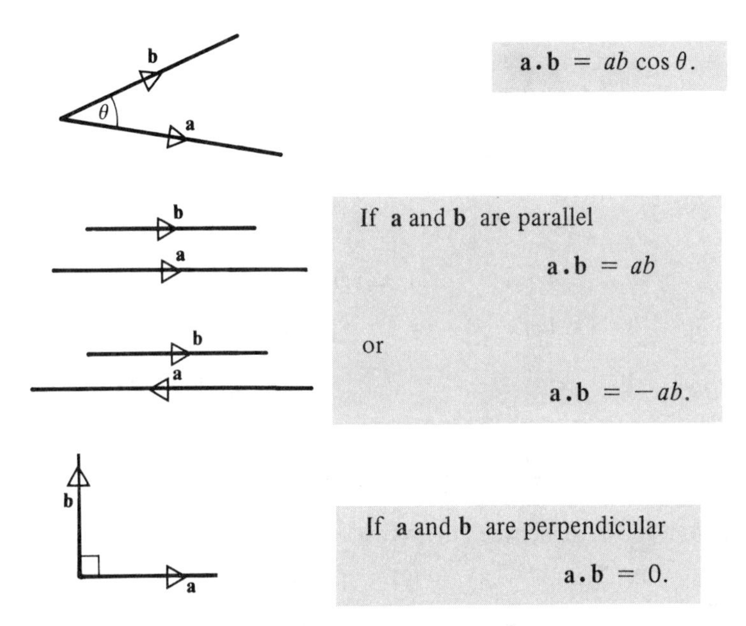

$$\mathbf{a}.\mathbf{b} = ab \cos\theta.$$

If $\mathbf{a}$ and $\mathbf{b}$ are parallel

$$\mathbf{a}.\mathbf{b} = ab$$

or

$$\mathbf{a}.\mathbf{b} = -ab.$$

If $\mathbf{a}$ and $\mathbf{b}$ are perpendicular

$$\mathbf{a}.\mathbf{b} = 0.$$

In the special case when $\mathbf{a} = \mathbf{b}$, $\mathbf{a}.\mathbf{b} = \mathbf{a}.\mathbf{a} = a^2$ (sometimes $\mathbf{a}.\mathbf{a}$ is written as $\mathbf{a}^2$).
In the case of the unit vectors $\mathbf{i}, \mathbf{j}$ and $\mathbf{k}$ these results give

$$\mathbf{i}.\mathbf{i} = \mathbf{j}.\mathbf{j} = \mathbf{k}.\mathbf{k} = 1$$

$$\mathbf{i}.\mathbf{j} = \mathbf{j}.\mathbf{k} = \mathbf{k}.\mathbf{i} = 0.$$

Scalar product is commutative, i.e. $\mathbf{a}.\mathbf{b} = \mathbf{b}.\mathbf{a}$.
Scalar product is distributive across addition,

i.e. $$\mathbf{a}.(\mathbf{b} + \mathbf{c}) = \mathbf{a}.\mathbf{b} + \mathbf{a}.\mathbf{c}.$$

This property can be proved as follows.

$$\begin{aligned} \mathbf{a}.(\mathbf{b} + \mathbf{c}) &= |\mathbf{a}||\mathbf{b} + \mathbf{c}| \cos\theta \\ &= (\text{OA})(\text{OC}) \cos\theta \end{aligned}$$

$= (OA)(ON)$

$= (OA)(OM + MN)$

$= (OA)(OB \cos \phi) + (OA)(BC \cos \psi)$

$= \mathbf{a}.\mathbf{b} + \mathbf{a}.\mathbf{c}.$

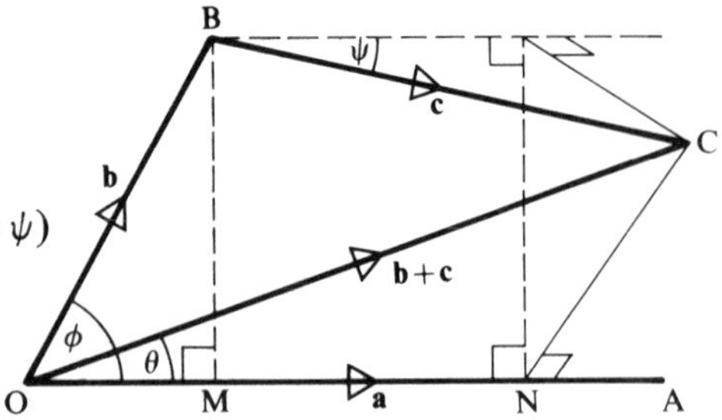

Hence if $\mathbf{a} = x_1\mathbf{i} + y_1\mathbf{j} + z_1\mathbf{k}$

and $\mathbf{b} = x_2\mathbf{i} + y_2\mathbf{j} + z_2\mathbf{k}$

then

$$\begin{aligned} \mathbf{a}.\mathbf{b} &= (x_1\mathbf{i} + y_1\mathbf{j} + z_1\mathbf{k}).(x_2\mathbf{i} + y_2\mathbf{j} + z_2\mathbf{k}) \\ &= (x_1x_2\mathbf{i}.\mathbf{i} + y_1y_2\mathbf{j}.\mathbf{j} + z_1z_2\mathbf{k}.\mathbf{k}) \\ &\quad + (x_1y_2\mathbf{i}.\mathbf{j} + y_1z_2\mathbf{j}.\mathbf{k} + z_1x_2\mathbf{k}.\mathbf{i}) \\ &\quad + (y_1x_2\mathbf{j}.\mathbf{i} + z_1y_2\mathbf{k}.\mathbf{j} + x_1z_2\mathbf{i}.\mathbf{k}) \\ &= (x_1x_2 + y_1y_2 + z_1z_2) + (0) + (0), \end{aligned}$$

i.e.

$$(x_1\mathbf{i} + y_1\mathbf{j} + z_1\mathbf{k}).(x_2\mathbf{i} + y_2\mathbf{j} + z_2\mathbf{k}) = x_1x_2 + y_1y_2 + z_1z_2.$$

For example,

$$(2\mathbf{i} - 3\mathbf{j} + 4\mathbf{k}).(\mathbf{i} + 3\mathbf{j} - 2\mathbf{k}) = (2)(1) + (-3)(3) + (4)(-2) = -15$$

Further, if θ is the angle between these two vectors

$$\begin{aligned} (2\mathbf{i} - 3\mathbf{j} + 4\mathbf{k}).(\mathbf{i} + 3\mathbf{j} - 2\mathbf{k}) &= |2\mathbf{i} - 3\mathbf{j} + 4\mathbf{k}||\mathbf{i} + 3\mathbf{j} - 2\mathbf{k}|\cos\theta \\ &= (\sqrt{29})(\sqrt{14})\cos\theta, \end{aligned}$$

i.e.

$$\theta = \arccos\left(-\frac{15}{\sqrt{406}}\right)$$

So we can use the scalar product to find the angle between a pair of vectors.

Note that $\begin{pmatrix} x_1 & y_1 & z_1 \end{pmatrix}\begin{pmatrix} x_2 \\ y_2 \\ z_2 \end{pmatrix} = x_1x_2 + y_1y_2 + z_1z_2$

So the scalar product $(x_1\mathbf{i} + y_1\mathbf{j} + z_1\mathbf{k}).(x_2\mathbf{i} + y_1\mathbf{j} + z_1\mathbf{k})$ may be written $\begin{pmatrix} x_1 & y_1 & z_1 \end{pmatrix}\begin{pmatrix} x_2 \\ y_2 \\ z_2 \end{pmatrix}$

Angle Between a Pair of Lines

If two lines have equations

$$\begin{cases} \mathbf{r} = \mathbf{a}_1 + \lambda \mathbf{b}_1 \\ \mathbf{r} = \mathbf{a}_2 + \mu \mathbf{b}_2 \end{cases}$$

they are in the directions of $\mathbf{b}_1$ and $\mathbf{b}_2$ respectively.

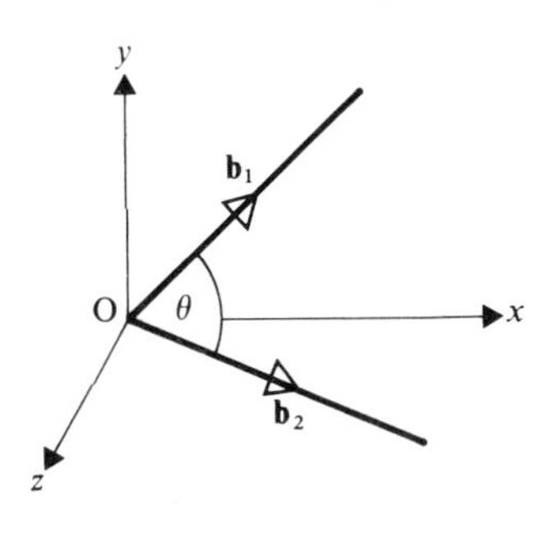

Drawing $\mathbf{b}_1$ and $\mathbf{b}_2$ through the origin we define the angle between the lines at θ, the angle between the direction vectors, $\mathbf{b}_1$ and $\mathbf{b}_2$, of the lines.
So the angle between a pair of lines depends only on their directions and not on their positions.

As $\mathbf{b}_1 . \mathbf{b}_2 = b_1 b_2 \cos\theta$

$$\cos\theta = \frac{\mathbf{b}_1 . \mathbf{b}_2}{b_1 b_2}$$

This applies to a pair of skew lines as well as to a pair of intersecting lines.
It follows that if the lines are perpendicular

$$\cos\theta = 0 \quad \Rightarrow \quad \mathbf{b}_1 . \mathbf{b}_2 = 0.$$

If two lines have equations

$$\mathbf{r} = x_1\mathbf{i} + y_1\mathbf{j} + z_1\mathbf{k} + \lambda(a_1\mathbf{i} + b_1\mathbf{j} + c_1\mathbf{k})$$

and

$$\mathbf{r} = x_2\mathbf{i} + y_2\mathbf{j} + z_2\mathbf{k} + \mu(a_2\mathbf{i} + b_2\mathbf{j} + c_2\mathbf{k})$$

respectively, the angle θ between these lines is the angle between their direction vectors, $\mathbf{v}_1 = a_1\mathbf{i} + b_1\mathbf{j} + c_1\mathbf{k}$ and $\mathbf{v}_2 = a_2\mathbf{i} + b_2\mathbf{j} + c_2\mathbf{k}$.
Now

$$(a_1\mathbf{i} + b_1\mathbf{j} + c_1\mathbf{k}).(a_2\mathbf{i} + b_2\mathbf{j} + c_2\mathbf{k}) = a_1a_2 + b_1b_2 + c_1c_2 = |\mathbf{v}_1||\mathbf{v}_2|\cos\theta.$$

Hence

$$\cos\theta = \frac{a_1a_2 + b_1b_2 + c_1c_2}{|\mathbf{v}_1||\mathbf{v}_2|}$$

But $a_1 : b_1 : c_1$ are the direction ratios of the first line

and $a_2 : b_2 : c_2$ are the direction ratios of the second line,

so $\dfrac{a_1}{|\mathbf{v}_1|}, \dfrac{b_1}{|\mathbf{v}_1|}, \dfrac{c_1}{|\mathbf{v}_1|}$ are the direction cosines l_1, m_1, n_1 of the first line,

and $\dfrac{a_2}{|\mathbf{v}_2|}, \dfrac{b_2}{|\mathbf{v}_2|}, \dfrac{c_2}{|\mathbf{v}_2|}$ are the direction cosines, l_2, m_2, n_2 of the second line.

Hence the angle θ between two lines is given by

$$\cos\theta = l_1 l_2 + m_1 m_2 + n_1 n_2.$$

EXAMPLES 5e

1) Find the scalar product of $\mathbf{a} = 2\mathbf{i} - 3\mathbf{j} + 5\mathbf{k}$ and $\mathbf{b} = \mathbf{i} - 3\mathbf{j} + \mathbf{k}$ and hence find the cosine of the angle between $\mathbf{a}$ and $\mathbf{b}$.

$$\mathbf{a}.\mathbf{b} = (2)(1) + (-3)(-3) + (5)(1) = 16$$

But $$\mathbf{a}.\mathbf{b} = |\mathbf{a}||\mathbf{b}|\cos\theta$$

$$|\mathbf{a}| = \sqrt{(4 + 9 + 25)} = \sqrt{38}$$

$$|\mathbf{b}| = \sqrt{(1 + 9 + 1)} = \sqrt{11}$$

Hence $$\cos\theta = \frac{\mathbf{a}.\mathbf{b}}{|\mathbf{a}||\mathbf{b}|} = \frac{16}{\sqrt{11}\sqrt{38}} = \frac{16}{\sqrt{418}}$$

2) If $\mathbf{a} = 10\mathbf{i} - 3\mathbf{j} + 5\mathbf{k}$, $\mathbf{b} = 2\mathbf{i} + 6\mathbf{j} - 3\mathbf{k}$ and $\mathbf{c} = \mathbf{i} + 10\mathbf{j} - 2\mathbf{k}$, verify that $\mathbf{a}.\mathbf{b} + \mathbf{a}.\mathbf{c} = \mathbf{a}.(\mathbf{b} + \mathbf{c})$.

$$\mathbf{a}.\mathbf{b} = (10)(2) + (-3)(6) + (5)(-3) = -13$$

$$\mathbf{a}.\mathbf{c} = (10)(1) + (-3)(10) + (5)(-2) = -30$$

$$\mathbf{b} + \mathbf{c} = 3\mathbf{i} + 16\mathbf{j} - 5\mathbf{k}$$

Hence $$\mathbf{a}.(\mathbf{b} + \mathbf{c}) = (10)(3) + (-3)(16) + (5)(-5) = -43$$

But $$\mathbf{a}.\mathbf{b} + \mathbf{a}.\mathbf{c} = -13 - 30 = -43$$

Therefore $$\mathbf{a}.\mathbf{b} + \mathbf{a}.\mathbf{c} = \mathbf{a}.(\mathbf{b} + \mathbf{c})$$

3) The resultant of two vectors $\mathbf{a}$ and $\mathbf{b}$ is perpendicular to $\mathbf{a}$. If $|\mathbf{b}| = \sqrt{2}|\mathbf{a}|$, show that the resultant of $2\mathbf{a}$ and $\mathbf{b}$ is perpendicular to $\mathbf{b}$.

The resultant of $\mathbf{a}$ and $\mathbf{b}$ is $\mathbf{a} + \mathbf{b}$.
Since $\mathbf{a} + \mathbf{b}$ is perpendicular to $\mathbf{a}$,

$$(\mathbf{a} + \mathbf{b}).\mathbf{a} = 0$$

Hence $$\mathbf{a}.\mathbf{a} + \mathbf{b}.\mathbf{a} = 0$$

But $$\mathbf{a}.\mathbf{a} = |\mathbf{a}|^2$$

Hence $$\mathbf{b}.\mathbf{a} = -|\mathbf{a}|^2$$

Now the resultant of $2\mathbf{a}$ and $\mathbf{b}$ is $2\mathbf{a} + \mathbf{b}$

and $$\mathbf{b}.(2\mathbf{a} + \mathbf{b}) = 2\mathbf{b}.\mathbf{a} + \mathbf{b}.\mathbf{b}$$

But $$\mathbf{b}.\mathbf{b} = |\mathbf{b}|^2 = b^2$$

and $$2\mathbf{a}.\mathbf{b} = -2|\mathbf{a}|^2 = -2a^2$$

Hence $\quad \mathbf{b}.(2\mathbf{a}+\mathbf{b}) = -2a^2 + b^2.$

But $\quad b = a\sqrt{2}$ (given)

so $\quad \mathbf{b}.(2\mathbf{a}+\mathbf{b}) = 0.$

Hence $2\mathbf{a}+\mathbf{b}$ is perpendicular to $\mathbf{b}$.

4) Find the angle between the lines

$$\mathbf{r}_1 = \mathbf{i}-2\mathbf{j}+3\mathbf{k}+\lambda(2\mathbf{i}-3\mathbf{j}+6\mathbf{k}) \qquad [1]$$

$$\mathbf{r}_2 = 2\mathbf{i}-7\mathbf{j}+10\mathbf{k}+\mu(\mathbf{i}+2\mathbf{j}+2\mathbf{k}). \qquad [2]$$

The angle between the lines depends only upon their directions.
Line [1] has direction cosines $\frac{2}{7}, -\frac{3}{7}, \frac{6}{7}$.
Line [2] has direction cosines $\frac{1}{3}, \frac{2}{3}, \frac{2}{3}$.
The angle θ between the lines is given by

$$\cos\theta = (\tfrac{2}{7})(\tfrac{1}{3}) + (-\tfrac{3}{7})(\tfrac{2}{3}) + (\tfrac{6}{7})(\tfrac{2}{3})$$

$$\Rightarrow \qquad \theta = \arccos\tfrac{8}{21}.$$

5) Find a unit vector which is perpendicular to AB and AC if $\overrightarrow{AB} = \mathbf{i}+2\mathbf{j}+3\mathbf{k}$ and $\overrightarrow{AC} = 4\mathbf{i}-\mathbf{j}+2\mathbf{k}$.

Let $a\mathbf{i}+b\mathbf{j}+c\mathbf{k}$ be a vector perpendicular both to AB and to AC.

It is perpendicular to AB so $\quad (a\mathbf{i}+b\mathbf{j}+c\mathbf{k}).(\mathbf{i}+2\mathbf{j}+3\mathbf{k}) = 0.$

It is perpendicular to AC so $\quad (a\mathbf{i}+b\mathbf{j}+c\mathbf{k}).(4\mathbf{i}-\mathbf{j}+2\mathbf{k}) = 0.$

Therefore
$$\begin{cases} a+2b+3c = 0 \\ 4a-b+2c = 0. \end{cases}$$

Eliminating b gives $\quad a = -\frac{7}{9}c.$

Eliminating a gives $\quad b = -\frac{10}{9}c.$

Hence
$$a\mathbf{i}+b\mathbf{j}+c\mathbf{k} = -\tfrac{7}{9}c\mathbf{i}-\tfrac{10}{9}c\mathbf{j}+c\mathbf{k}$$
$$= \tfrac{1}{9}c(-7\mathbf{i}-10\mathbf{j}+9\mathbf{k}).$$

Thus $-7\mathbf{i}-10\mathbf{j}+9\mathbf{k}$ is perpendicular to both AB and AC.
A unit vector perpendicular to AB and AC is therefore

$$(-7\mathbf{i}-10\mathbf{j}+9\mathbf{k})/\sqrt{230}.$$

(An alternative method for finding a common perpendicular to two vectors is given on p. 210, using the vector product.)

EXERCISE 5e

1) Calculate $\mathbf{a.b}$ if
(a) $\mathbf{a} = 2\mathbf{i} - 4\mathbf{j} + 5\mathbf{k}$, $\mathbf{b} = \mathbf{i} + 3\mathbf{j} + 8\mathbf{k}$,
(b) $\mathbf{a} = 3\mathbf{i} - 7\mathbf{j} + 2\mathbf{k}$, $\mathbf{b} = 5\mathbf{i} + \mathbf{j} - 4\mathbf{k}$,
(c) $\mathbf{a} = 2\mathbf{i} - 3\mathbf{j} + 6\mathbf{k}$, $\mathbf{b} = \mathbf{i} + \mathbf{j}$.
What conclusion can you draw in (b)?

2) Find $\mathbf{p.q}$ and the cosine of the angle between $\mathbf{p}$ and $\mathbf{q}$ if
(a) $\mathbf{p} = 2\mathbf{i} + 4\mathbf{j} + \mathbf{k}$, $\mathbf{q} = \mathbf{i} + \mathbf{j} + \mathbf{k}$,
(b) $\mathbf{p} = -\mathbf{i} + 3\mathbf{j} - 2\mathbf{k}$, $\mathbf{q} = \mathbf{i} + \mathbf{j} - 6\mathbf{k}$.

3) The angle between two vectors $\mathbf{v}_1$ and $\mathbf{v}_2$ is $\arccos \frac{4}{21}$.
If $\mathbf{v}_1 = 6\mathbf{i} + 3\mathbf{j} - 2\mathbf{k}$ and $\mathbf{v}_2 = -2\mathbf{i} + \lambda\mathbf{j} - 4\mathbf{k}$, find the positive value of λ.

4) If $\mathbf{a} = 3\mathbf{i} + 4\mathbf{j} - \mathbf{k}$, $\mathbf{b} = \mathbf{i} - \mathbf{j} + 3\mathbf{k}$ and $\mathbf{c} = 2\mathbf{i} + \mathbf{j} - 5\mathbf{k}$, find
(a) $\mathbf{a.b}$ (b) $\mathbf{a.c}$ (c) $\mathbf{a.(b+c)}$ (d) $(2\mathbf{a} + 3\mathbf{b}).\mathbf{c}$ (e) $(\mathbf{a} - \mathbf{b}).\mathbf{c}$.

5) In a triangle ABC, $\overrightarrow{AB} = \mathbf{i} + 2\mathbf{j} + 3\mathbf{k}$ and $\overrightarrow{BC} = -\mathbf{i} + 4\mathbf{j}$. Find the cosine of angle ABC.
Find the vector $\overrightarrow{AC}$ and use it to calculate the angle BAC.

6) A, B and C are points with position vectors $\mathbf{a}$, $\mathbf{b}$ and $\mathbf{c}$ respectively, relative to the origin O. AB is perpendicular to OC and BC is perpendicular to OA.
Show that AC is perpendicular to OB.

7) Given two vectors $\mathbf{a}$ and $\mathbf{b}$ $(\mathbf{a} \neq 0,\ \mathbf{b} \neq 0)$, show that
(a) if $\mathbf{a} + \mathbf{b}$ and $\mathbf{a} - \mathbf{b}$ are perpendicular then $|\mathbf{a}| = |\mathbf{b}|$,
(b) if $|\mathbf{a} + \mathbf{b}| = |\mathbf{a} - \mathbf{b}|$ then $\mathbf{a}$ and $\mathbf{b}$ are perpendicular.

8) Three vectors $\mathbf{a}$, $\mathbf{b}$ and $\mathbf{c}$ are such that $\mathbf{a} \neq \mathbf{b} \neq \mathbf{c} \neq 0$.
(a) If $\mathbf{a}.(\mathbf{b} + \mathbf{c}) = \mathbf{b}.(\mathbf{a} - \mathbf{c})$ prove that $\mathbf{c}.(\mathbf{a} + \mathbf{b}) = 0$.
(b) If $(\mathbf{a.b})\mathbf{c} = (\mathbf{b.c})\mathbf{a}$ show that $\mathbf{c}$ and $\mathbf{a}$ are parallel.

9) Find the angle between each of the following pairs of lines:
(a) $\mathbf{r}_1 = 3\mathbf{i} + 2\mathbf{j} - 4\mathbf{k} + \lambda(\mathbf{i} + 2\mathbf{j} + 2\mathbf{k})$,
$\mathbf{r}_2 = 5\mathbf{j} - 2\mathbf{k} + \mu(3\mathbf{i} + 2\mathbf{j} + 6\mathbf{k})$.
(b) A line with direction ratios 2:2:1,
A line joining (3, 1, 4) to (7, 2, 12).
(c) $\dfrac{x+4}{3} = \dfrac{y-1}{5} = \dfrac{z+3}{4}$,

$\dfrac{x+1}{1} = \dfrac{y-4}{1} = \dfrac{z-5}{2}$

10) Find the angle between the following pairs of lines:

(a) $\dfrac{x-2}{3} = \dfrac{y+1}{-2}, \quad z = 2$

$\dfrac{x-1}{1} = \dfrac{2y+3}{3} = \dfrac{z+5}{2}$

or

(b) $\mathbf{r} = 4\mathbf{i} - \mathbf{j} + \lambda(\mathbf{i} + 2\mathbf{j} - 2\mathbf{k})$
$\mathbf{r} = \mathbf{i} - \mathbf{j} + 2\mathbf{k} - \mu(2\mathbf{i} + 4\mathbf{j} - 4\mathbf{k})$.

11) Show that $\mathbf{i} + 7\mathbf{j} + 3\mathbf{k}$ is perpendicular to both $\mathbf{i} - \mathbf{j} + 2\mathbf{k}$ and $2\mathbf{i} + \mathbf{j} - 3\mathbf{k}$.

12) Show that $13\mathbf{i} + 23\mathbf{j} + 7\mathbf{k}$ is perpendicular to both $2\mathbf{i} + \mathbf{j} - 7\mathbf{k}$ and $3\mathbf{i} - 2\mathbf{j} + \mathbf{k}$.

13) Find a unit vector which is perpendicular to **a** and **b** if
(a) $\mathbf{a} = 6\mathbf{i} + \mathbf{j} + 3\mathbf{k}$, $\quad \mathbf{b} = 5\mathbf{i} + \mathbf{k}$
(b) $\mathbf{a} = \mathbf{i} - \mathbf{j} - \mathbf{k}$, $\quad \mathbf{b} = 7\mathbf{i} - 2\mathbf{j} + 3\mathbf{k}$.

VECTOR PRODUCT

The vector product of two vectors **a** and **b** which are inclined at an angle θ is written as $\mathbf{a} \times \mathbf{b}$ (or sometimes $\mathbf{a} \wedge \mathbf{b}$) and is defined as a vector of magnitude $ab \sin\theta$ in a direction perpendicular to the plane containing **a** and **b** in the sense of a right-handed screw turned from **a** to **b**.

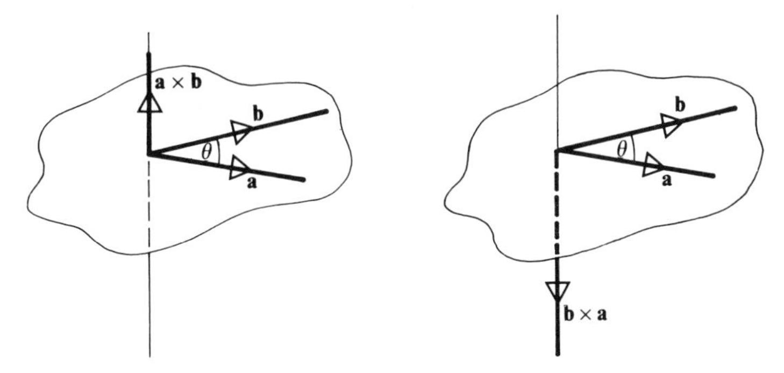

From this definition it follows that the direction of $\mathbf{b} \times \mathbf{a}$ is in the sense of a right-handed screw turned from **b** to **a**. So the direction of $\mathbf{b} \times \mathbf{a}$ is opposite to that of $\mathbf{a} \times \mathbf{b}$, but the magnitude of $\mathbf{b} \times \mathbf{a}$ (i.e. $ba \sin\theta$) is the same as the magnitude of $\mathbf{a} \times \mathbf{b}$.

Therefore $$\mathbf{a} \times \mathbf{b} = -\mathbf{b} \times \mathbf{a}$$

Thus vector product is *not* commutative.

Vector Product of Parallel Vectors

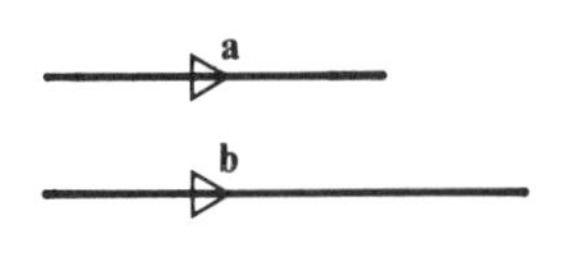

If $\mathbf{a}$ and $\mathbf{b}$ are parallel vectors

$$|\mathbf{a} \times \mathbf{b}| = ab \sin\theta.$$

But $\quad \sin\theta = 0.$

Therefore $\quad \mathbf{a} \times \mathbf{b} = \mathbf{0}$

Vector Product of Perpendicular Vectors

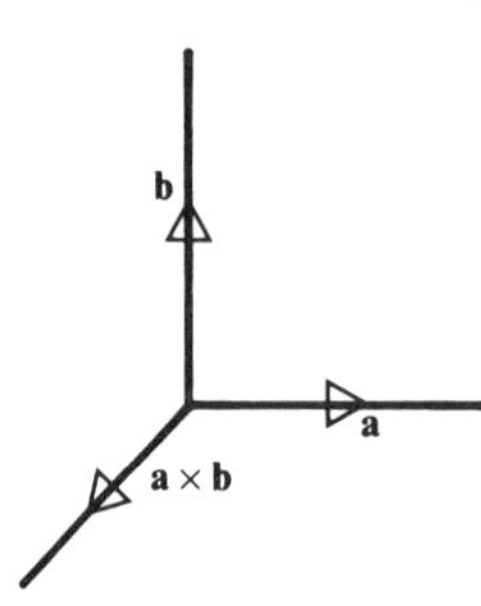

If $\mathbf{a}$ and $\mathbf{b}$ are perpendicular vectors, $\sin\theta = 1$ and $|\mathbf{a} \times \mathbf{b}| = ab$.
In this case $\mathbf{a}, \mathbf{b}$ and $\mathbf{a} \times \mathbf{b}$ form a right-handed set of three mutually perpendicular vectors as shown in the diagram.

This result is particularly important in the case of the unit vectors $\mathbf{i}, \mathbf{j}$ and $\mathbf{k}$.

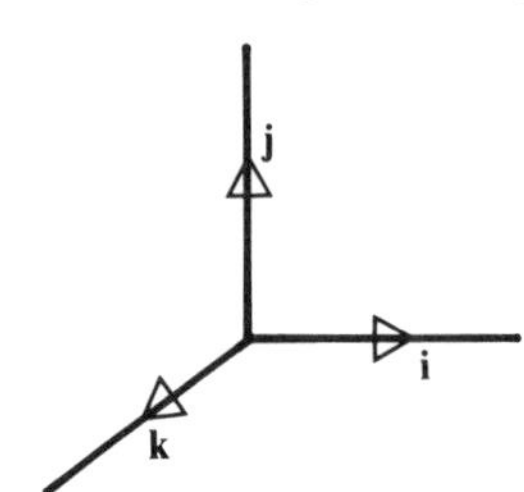

Thus $\quad \mathbf{i} \times \mathbf{j} = \mathbf{k}$ and $\mathbf{j} \times \mathbf{i} = -\mathbf{k}$

$\mathbf{j} \times \mathbf{k} = \mathbf{i}$ and $\mathbf{k} \times \mathbf{j} = -\mathbf{i}$

$\mathbf{k} \times \mathbf{i} = \mathbf{j}$ and $\mathbf{i} \times \mathbf{k} = -\mathbf{j}$

Also $\quad \mathbf{i} \times \mathbf{i} = \mathbf{j} \times \mathbf{j} = \mathbf{k} \times \mathbf{k} = 0$

Vector Product of Vectors in Cartesian Component Form

Vector product is distributive: i.e. $\mathbf{a} \times (\mathbf{b} + \mathbf{c}) = \mathbf{a} \times \mathbf{b} + \mathbf{a} \times \mathbf{c}$
(This property is proved on page 214.)
Therefore if

$$\mathbf{a} = x_1\mathbf{i} + y_1\mathbf{j} + z_1\mathbf{k} \quad \text{and} \quad \mathbf{b} = x_2\mathbf{i} + y_2\mathbf{j} + z_2\mathbf{k}$$

$$\begin{aligned}
\mathbf{a} \times \mathbf{b} &= (x_1\mathbf{i} + y_1\mathbf{j} + z_1\mathbf{k}) \times (x_2\mathbf{i} + y_2\mathbf{j} + z_2\mathbf{k}) \\
&= x_1x_2(\mathbf{i} \times \mathbf{i}) + x_1y_2(\mathbf{i} \times \mathbf{j}) + x_1z_2(\mathbf{i} \times \mathbf{k}) + y_1x_2(\mathbf{j} \times \mathbf{i}) + y_1y_2(\mathbf{j} \times \mathbf{j}) \\
&\quad + y_1z_2(\mathbf{j} \times \mathbf{k}) + z_1x_2(\mathbf{k} \times \mathbf{i}) + z_1y_2(\mathbf{k} \times \mathbf{j}) + z_1z_2(\mathbf{k} \times \mathbf{k}) \\
&= x_1y_2\mathbf{k} - x_1z_2\mathbf{j} - y_1x_2\mathbf{k} + y_1z_2\mathbf{i} + z_1x_2\mathbf{j} - z_1y_2\mathbf{i}
\end{aligned}$$

(using the results above)

$$= (y_1z_2 - z_1y_2)\mathbf{i} - (x_1z_2 - z_1x_2)\mathbf{j} + (x_1y_2 - y_1x_2)\mathbf{k}$$

This expression is the expansion of the determinant $\begin{vmatrix} \mathbf{i} & \mathbf{j} & \mathbf{k} \\ x_1 & y_1 & z_1 \\ x_2 & y_2 & z_2 \end{vmatrix}$

$$\text{Therefore} \quad (x_1\mathbf{i}+y_1\mathbf{j}+z_1\mathbf{k}) \times (x_2\mathbf{i}+y_2\mathbf{j}+z_2\mathbf{k}) = \begin{vmatrix} \mathbf{i} & \mathbf{j} & \mathbf{k} \\ x_1 & y_1 & z_1 \\ x_2 & y_2 & z_2 \end{vmatrix}$$

For example,
$$(2\mathbf{i}+\mathbf{j}-2\mathbf{k}) \times (\mathbf{j}+3\mathbf{k}) = \begin{vmatrix} \mathbf{i} & \mathbf{j} & \mathbf{k} \\ 2 & 1 & -2 \\ 0 & 1 & 3 \end{vmatrix}$$
$$= 5\mathbf{i}-6\mathbf{j}+2\mathbf{k}.$$

Some calculations involve a mixture of vector and scalar product, e.g. $\mathbf{a}\times\mathbf{b}\,.\,\mathbf{c}$. Brackets are unnecessary in expressions of this type as the cross product *must* be calculated first. If $\mathbf{b}\,.\,\mathbf{c}$ were worked first this would lead to the vector product of $\mathbf{a}$ and a scalar quantity, which is meaningless.

To summarize:

$\mathbf{a}\times\mathbf{b} = ab\sin\theta\,\hat{\mathbf{n}}$ where $\hat{\mathbf{n}}$ is a unit vector perpendicular to $\mathbf{a}$ and $\mathbf{b}$ in the direction of a right-handed screw turned from $\mathbf{a}$ to $\mathbf{b}$.

$\mathbf{a}\times\mathbf{b} = -\mathbf{b}\times\mathbf{a}$ so the order of the vectors in the product is important.

If $\mathbf{a}$ and $\mathbf{b}$ are parallel $\mathbf{a}\times\mathbf{b} = \mathbf{0}$.

If $\mathbf{a}$ and $\mathbf{b}$ are perpendicular $\mathbf{a}, \mathbf{b}$ and $\mathbf{a}\times\mathbf{b}$ form a right-handed set of mutually perpendicular vectors.

EXAMPLES 5f

1) Simplify (a) $\mathbf{a}\times(\mathbf{a}-\mathbf{b})$, (b) $\mathbf{a}\times\mathbf{b}\,.\,\mathbf{a}$.

(a)
$$\mathbf{a}\times(\mathbf{a}-\mathbf{b}) = \mathbf{a}\times\mathbf{a}-\mathbf{a}\times\mathbf{b}$$
$$= \mathbf{0}-\mathbf{a}\times\mathbf{b}$$
$$= \mathbf{b}\times\mathbf{a}$$

(b) By definition $\mathbf{a}\times\mathbf{b}$ is perpendicular to $\mathbf{a}$ and the scalar product of perpendicular vectors is zero.
Therefore $\mathbf{a}\times\mathbf{b}\,.\,\mathbf{a} = 0$.

2) If $\mathbf{a} = (2\mathbf{i}+\mathbf{j}-\mathbf{k})$, $\mathbf{b} = (3\mathbf{i}-\mathbf{j}+\mathbf{k})$ and $\mathbf{c} = (\mathbf{i}+2\mathbf{j})$ find $\mathbf{a}\times(\mathbf{b}\times\mathbf{c})$.

$$\mathbf{b}\times\mathbf{c} = \begin{vmatrix} \mathbf{i} & \mathbf{j} & \mathbf{k} \\ 3 & -1 & 1 \\ 1 & 2 & 0 \end{vmatrix} = -2\mathbf{i}+\mathbf{j}+7\mathbf{k}.$$

Therefore $$\mathbf{a}\times(\mathbf{b}\times\mathbf{c}) = \begin{vmatrix} \mathbf{i} & \mathbf{j} & \mathbf{k} \\ 2 & 1 & -1 \\ -2 & 1 & 7 \end{vmatrix} = 8\mathbf{i}-12\mathbf{j}+4\mathbf{k}$$

$$= 4(2\mathbf{i}-3\mathbf{j}+\mathbf{k}).$$

3) Find the sine of the angle between AB and BC where A, B and C are the points $(0, 1, 3), (-1, 0, 1)$ and $(1, -1, -2)$ respectively. Find also a unit vector which is perpendicular to the plane containing A, B and C.

$$\left.\begin{aligned} \overrightarrow{OA} &= \mathbf{j}+3\mathbf{k} \\ \overrightarrow{OB} &= -\mathbf{i}+\mathbf{k} \\ \overrightarrow{OC} &= \mathbf{i}-\mathbf{j}-2\mathbf{k} \end{aligned}\right\} \Rightarrow \begin{aligned} \overrightarrow{BA} &= \overrightarrow{OA}-\overrightarrow{OB} = \mathbf{i}+\mathbf{j}+2\mathbf{k} \\ \overrightarrow{BC} &= \overrightarrow{OC}-\overrightarrow{OB} = 2\mathbf{i}-\mathbf{j}-3\mathbf{k} \end{aligned}$$

Now $$|\overrightarrow{BA}\times\overrightarrow{BC}| = (BA)(BC)\sin\theta.$$

Therefore $$\sin\theta = \frac{|\overrightarrow{BA}\times\overrightarrow{BC}|}{(BA)(BC)}$$

But $$\overrightarrow{BA}\times\overrightarrow{BC} = \begin{vmatrix} \mathbf{i} & \mathbf{j} & \mathbf{k} \\ 1 & 1 & 2 \\ 2 & -1 & -3 \end{vmatrix} = -\mathbf{i}+7\mathbf{j}-3\mathbf{k}$$

$\Rightarrow$ $$|\overrightarrow{BA}\times\overrightarrow{BC}| = \sqrt{59}$$

$$BA = |\mathbf{i}+\mathbf{j}+2\mathbf{k}| = \sqrt{6}$$

$$BC = |2\mathbf{i}-\mathbf{j}-3\mathbf{k}| = \sqrt{14}$$

Therefore $$\sin\theta = \sqrt{59}/(\sqrt{6}\sqrt{14}) = \sqrt{(59/84)}$$

$\overrightarrow{BA}\times\overrightarrow{BC}$ is a vector which is perpendicular to both BA and BC and is therefore perpendicular to the plane ABC.

Therefore the unit vector perpendicular to the plane ABC is $\dfrac{\overrightarrow{BA}\times\overrightarrow{BC}}{|\overrightarrow{BA}\times\overrightarrow{BC}|}$

$$= \frac{1}{\sqrt{59}}(-\mathbf{i}+7\mathbf{j}-3\mathbf{k})$$

4) $\mathbf{a}, \mathbf{b}$ and $\mathbf{c}$ are three vectors such that $\mathbf{a} \times \mathbf{b} = \mathbf{c} \times \mathbf{a}$, $\mathbf{a} \neq \mathbf{0}$.
Find a linear relationship between $\mathbf{a}, \mathbf{b}$ and $\mathbf{c}$

As $\quad \mathbf{a} \times \mathbf{b} = \mathbf{c} \times \mathbf{a}$

$\mathbf{a} \times \mathbf{b} = -\mathbf{a} \times \mathbf{c}$

Therefore $\quad \mathbf{a} \times \mathbf{b} + \mathbf{a} \times \mathbf{c} = \mathbf{0}$.

Therefore $\quad \mathbf{a} \times (\mathbf{b} + \mathbf{c}) = \mathbf{0}$ (distributive law)

so either $\mathbf{a}$ and $\mathbf{b} + \mathbf{c}$ are parallel vectors,
or $\mathbf{b} + \mathbf{c} = \mathbf{0}$,

i.e. either $\mathbf{a} = k(\mathbf{b} + \mathbf{c})$ where k is a scalar quantity
or $\mathbf{b} = -\mathbf{c}$

EXERCISE 5f

1) Simplify the following:
(a) $(\mathbf{a} + \mathbf{b}) \times \mathbf{b}$ (b) $(\mathbf{a} + \mathbf{b}) \times (\mathbf{a} + \mathbf{b})$
(c) $(\mathbf{a} - \mathbf{b}) \times (\mathbf{a} + \mathbf{b})$ (d) $\mathbf{a} \times (\mathbf{b} + \mathbf{c}) \cdot \mathbf{b}$
(e) $\mathbf{a} \cdot (\mathbf{b} + \mathbf{c}) \times \mathbf{a}$ (f) $\mathbf{a} \times \mathbf{b} \cdot \mathbf{a} + \mathbf{b} \cdot \mathbf{a} \times \mathbf{b}$

2) If $\mathbf{a} = \mathbf{i} + \mathbf{j} - \mathbf{k}$ and $\mathbf{b} = 2\mathbf{i} - \mathbf{j} + \mathbf{k}$ find
(a) $\mathbf{a} \times \mathbf{b}$ (b) $\mathbf{a} \times (\mathbf{a} + \mathbf{b})$
and verify that $\mathbf{a} \cdot \mathbf{a} \times \mathbf{b} = 0$.

3) If $\mathbf{a} = \mathbf{i} + 2\mathbf{j} - \mathbf{k}$ and $\mathbf{b} = \mathbf{j} + \mathbf{k}$ find the unit vector perpendicular to both $\mathbf{a}$ and $\mathbf{b}$. Calculate also the sine of the angle between $\mathbf{a}$ and $\mathbf{b}$.

4) If $\mathbf{a} = \mathbf{i} + \mathbf{j} - \mathbf{k}$, $\mathbf{b} = \mathbf{i} - \mathbf{j}$ and $\mathbf{c} = 2\mathbf{i} + \mathbf{k}$, find $(\mathbf{a} \times \mathbf{b}) \times \mathbf{c}$ and $\mathbf{a} \times (\mathbf{b} \times \mathbf{c})$.
Verify also that $\mathbf{a} \cdot \mathbf{b} \times \mathbf{c} = \mathbf{a} \times \mathbf{b} \cdot \mathbf{c}$

5) If $\mathbf{a} = \mathbf{i} + \mathbf{j}$ and $\mathbf{b} = 2\mathbf{i} + \mathbf{k}$ find the sine of the angle between $\mathbf{a}$ and $\mathbf{b}$, and the unit vector perpendicular to both $\mathbf{a}$ and $\mathbf{b}$.

6) A, B and C are the points $(0, 1, 2)$, $(3, 2, 1)$ and $(1, -1, 0)$ respectively. Find the unit vector which is perpendicular to the plane ABC.

7) Three vectors $\mathbf{a}, \mathbf{b}$ and $\mathbf{c}$ are such that $\mathbf{a} \times \mathbf{b} = \mathbf{a} \times \mathbf{c}$ $(\mathbf{a} \neq \mathbf{0})$.
Show that $\mathbf{b} - \mathbf{c} = k\mathbf{a}$ where k is a scalar.

8) Three vectors $\mathbf{a}, \mathbf{b}$ and $\mathbf{c}$ are such that $\mathbf{a} \times 3\mathbf{b} = 2\mathbf{a} \times \mathbf{c}$. Find a linear relationship between $\mathbf{a}, \mathbf{b}$ and $\mathbf{c}$.

9) Show that $\mathbf{u} = (\mathbf{i} + \mathbf{j})$ is a solution of the equation

$$\mathbf{u} \times (\mathbf{i} + 4\mathbf{j}) = 3\mathbf{k}$$

Show also that the general solution to this equation is

$$\mathbf{u} = -3\mathbf{j} + t(\mathbf{i} + 4\mathbf{j})$$

10) Find a general solution to the equation

$$\mathbf{u} \times (\mathbf{i} - 3\mathbf{k}) = 2\mathbf{j}$$

APPLICATIONS OF THE VECTOR PRODUCT

1. Area of a Parallelogram

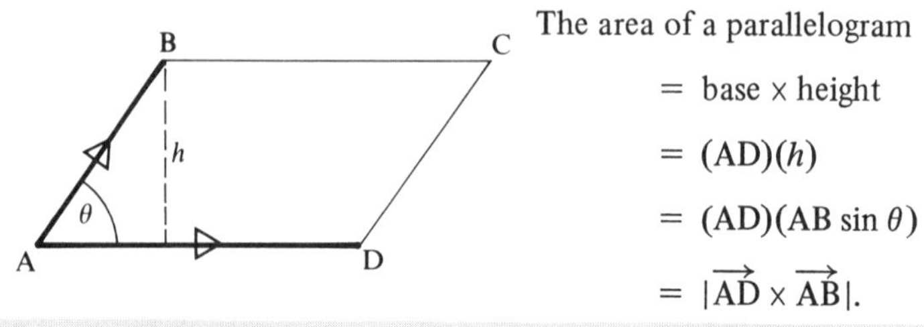

The area of a parallelogram

$= \text{base} \times \text{height}$

$= (\text{AD})(h)$

$= (\text{AD})(\text{AB} \sin \theta)$

$= |\overrightarrow{\text{AD}} \times \overrightarrow{\text{AB}}|.$

Therefore the area of a parallelogram is the magnitude of the vector product of two adjacent sides.

2. Area of a Triangle

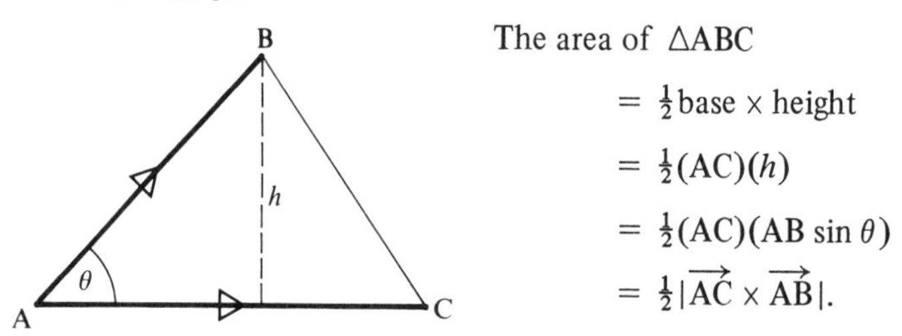

The area of $\triangle$ABC

$= \frac{1}{2}\text{base} \times \text{height}$

$= \frac{1}{2}(\text{AC})(h)$

$= \frac{1}{2}(\text{AC})(\text{AB} \sin \theta)$

$= \frac{1}{2}|\overrightarrow{\text{AC}} \times \overrightarrow{\text{AB}}|.$

Therefore the area of a triangle is half the magnitude of the vector product of two sides.

EXAMPLES 5g

1) A triangle ABC has its vertices at the points A(1, 2, 1), B(1, 0, 3), C(−1, 2, −1).
Find the area of $\triangle$ABC.

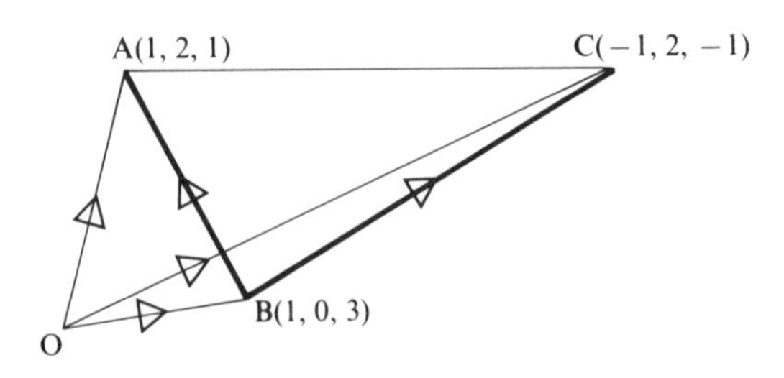

If O is the origin then

$$\vec{OA} = \mathbf{i} + 2\mathbf{j} + \mathbf{k}$$

$$\vec{OB} = \mathbf{i} + 3\mathbf{k}$$

$$\vec{OC} = -\mathbf{i} + 2\mathbf{j} - \mathbf{k}.$$

Therefore

$$\vec{BA} = \vec{OA} - \vec{OB} = 2\mathbf{j} - 2\mathbf{k}$$

$$\vec{BC} = \vec{OC} - \vec{OB} = -2\mathbf{i} + 2\mathbf{j} - 4\mathbf{k}.$$

$$\text{Area of } \triangle ABC = \tfrac{1}{2}|\vec{BA} \times \vec{BC}|$$

$$\vec{BA} \times \vec{BC} = \begin{vmatrix} \mathbf{i} & \mathbf{j} & \mathbf{k} \\ 0 & 2 & -2 \\ -2 & 2 & -4 \end{vmatrix} = -4\mathbf{i} + 4\mathbf{j} + 4\mathbf{k}.$$

Therefore area of $\triangle ABC = \tfrac{1}{2}|-4\mathbf{i} + 4\mathbf{j} + 4\mathbf{k}| = 2\sqrt{3}$.

3. Volume of a Parallelepiped

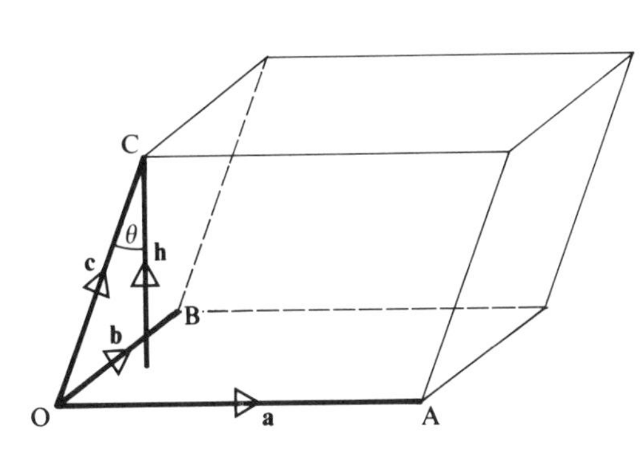

The volume of a parallelepiped

$$= \text{area of base} \times \text{height}$$

$$= |\mathbf{a} \times \mathbf{b}|h$$

$$= |\mathbf{a} \times \mathbf{b}||\mathbf{c}| \cos\theta.$$

θ is the angle between $\mathbf{c}$ and $\mathbf{h}$ and as $\mathbf{h}$ is perpendicular to $\mathbf{a}$ and $\mathbf{b}$, θ is also the angle between $\mathbf{a} \times \mathbf{b}$ and $\mathbf{c}$.
$\therefore$ $|\mathbf{a} \times \mathbf{b}||\mathbf{c}| \cos\theta$ is the scalar product of the vectors $\mathbf{a} \times \mathbf{b}$ and $\mathbf{c}$.

Therefore the volume of the parallelepiped is $|\mathbf{a} \times \mathbf{b} \cdot \mathbf{c}|$

4. Volume of a Tetrahedron

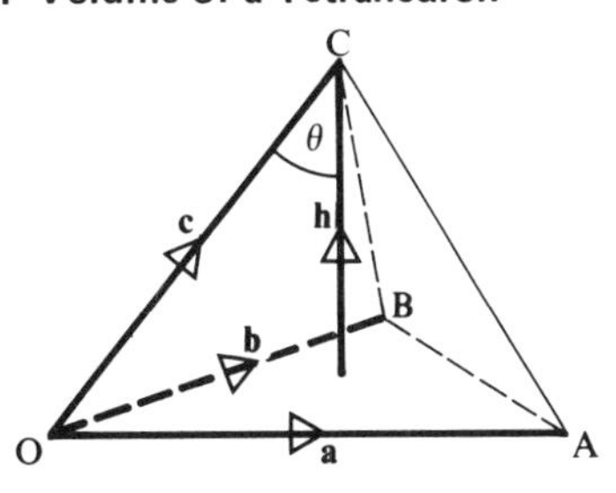

The volume of a tetrahedron

$$= \tfrac{1}{3}\,\text{area of base} \times \text{height}$$

$$= \tfrac{1}{3}(\tfrac{1}{2}|\mathbf{a} \times \mathbf{b}|)|\mathbf{c}| \cos\theta$$

$$= \tfrac{1}{6}|\mathbf{a} \times \mathbf{b} \cdot \mathbf{c}|$$

EXAMPLES 5g (continued)

2) Find the volume of the tetrahedron OABC where O is the origin and A, B, C are the points $(2, 1, 1), (0, -1, 1), (-1, 3, 0)$.

The volume of OABC is $\frac{1}{6}|\overrightarrow{OA} \times \overrightarrow{OB} . \overrightarrow{OC}|$

$$= \tfrac{1}{6}|(2\mathbf{i}+\mathbf{j}+\mathbf{k}) \times (-\mathbf{j}+\mathbf{k}) . (-\mathbf{i}+3\mathbf{j})|$$

$$= \tfrac{1}{6}|(2\mathbf{i}-2\mathbf{j}-2\mathbf{k}) . (-\mathbf{i}+3\mathbf{j})|$$

$$= 1\tfrac{1}{3}.$$

Triple Scalar Product and the Proof of the Distributive Law for Vector Products

Expressions of the form $\mathbf{a} \times \mathbf{b} . \mathbf{c}$ are called triple scalar products.
The volume of the parallelepiped above could have been obtained by considering the side defined by $\mathbf{b}$ and $\mathbf{c}$ as the base, in which case the result would have been in the form $\mathbf{b} \times \mathbf{c} . \mathbf{a}$,

i.e. $$\mathbf{a} \times \mathbf{b} . \mathbf{c} = \mathbf{b} \times \mathbf{c} . \mathbf{a}$$

The order in which a scalar product is performed does not matter

i.e. $$\mathbf{b} \times \mathbf{c} . \mathbf{a} = \mathbf{a} . \mathbf{b} \times \mathbf{c}$$

Therefore $$\mathbf{a} \times \mathbf{b} . \mathbf{c} = \mathbf{a} . \mathbf{b} \times \mathbf{c}$$

Thus in a triple scalar product, the 'cross' and 'dot' may be interchanged without altering the value of the expression. This property will be referred to as the *triple scalar product property*.
(But if the order of the vectors is altered, the expression may not remain the same: e.g. $\mathbf{a} \times \mathbf{b} . \mathbf{c} = -\mathbf{b} \times \mathbf{a} . \mathbf{c}$.)
We will now use this property of a triple scalar product to prove that the vector product is distributive.

Consider $\mathbf{d} . (\mathbf{a} \times \mathbf{b} + \mathbf{a} \times \mathbf{c})$

$= \mathbf{d} . \mathbf{a} \times \mathbf{b} + \mathbf{d} . \mathbf{a} \times \mathbf{c}$ (scalar product is distributive)

$= \mathbf{d} \times \mathbf{a} . \mathbf{b} + \mathbf{d} \times \mathbf{a} . \mathbf{c}$ (triple scalar product property)

$= \mathbf{d} \times \mathbf{a} . (\mathbf{b} + \mathbf{c})$ (scalar product is distributive)

$= \mathbf{d}\ \ \mathbf{a} \times (\mathbf{b} + \mathbf{c})$ (triple scalar product property),

i.e. $$\mathbf{d} . (\mathbf{a} \times \mathbf{b} + \mathbf{a} \times \mathbf{c}) = \mathbf{d} . \mathbf{a} \times (\mathbf{b} + \mathbf{c})$$

Therefore $$\mathbf{a} \times \mathbf{b} + \mathbf{a} \times \mathbf{c} = \mathbf{a} \times (\mathbf{b} + \mathbf{c}),$$

i.e. the vector product is distributive.

5. Distance of a Point from a Line

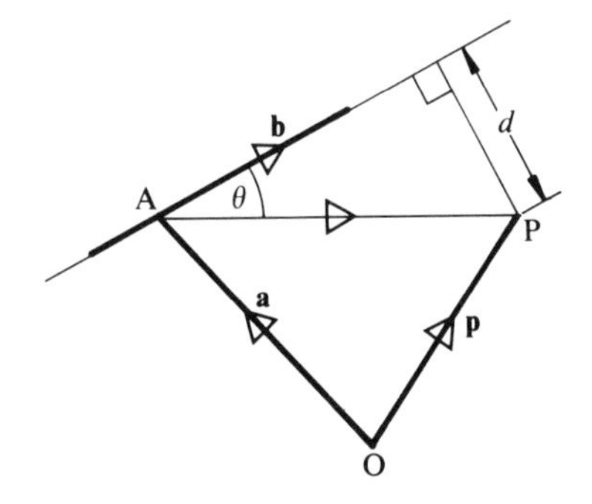

Consider the line whose vector equation is $\mathbf{r} = \mathbf{a} + \lambda\mathbf{b}$ and the point P whose position vector is $\mathbf{p}$. If A is the point on the line with position vector $\mathbf{a}$ then $\overrightarrow{AP} = \mathbf{p} - \mathbf{a}$.

If θ is the angle between $\overrightarrow{AP}$ and the line (which is parallel to $\mathbf{b}$)

then
$$d = AP \sin\theta$$
$$= \frac{|\mathbf{b} \times \overrightarrow{AP}|}{|\mathbf{b}|} = \frac{|\mathbf{b} \times (\mathbf{p} - \mathbf{a})|}{|\mathbf{b}|}$$

Therefore the perpendicular distance of a point P whose position vector is $\mathbf{p}$ from the line whose vector equation is $\mathbf{r} = \mathbf{a} + \lambda\mathbf{b}$ is

$$\frac{|\mathbf{b} \times (\mathbf{p} - \mathbf{a})|}{|\mathbf{b}|}$$

For example, the distance of the point P(1, 3, 2) from the line $\mathbf{r} = (1 - t)\mathbf{i} + (2 - 3t)\mathbf{j} + 2\mathbf{k}$ can be found as follows.

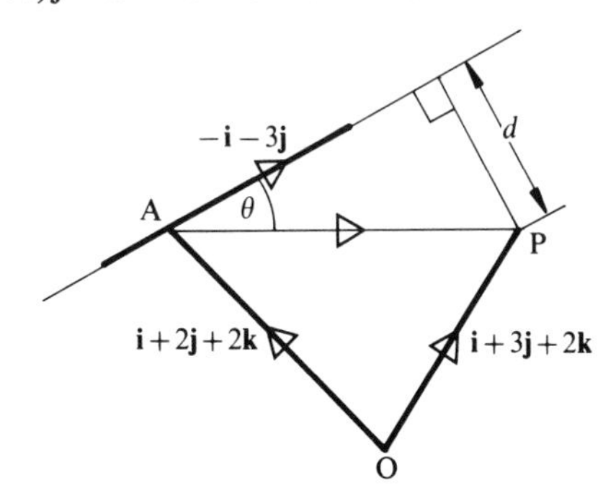

Rewriting the equation of the line in the form

$$\mathbf{r} = \mathbf{i} + 2\mathbf{j} + 2\mathbf{k} + t(-\mathbf{i} - 3\mathbf{j})$$

we see that $\mathbf{i} + 2\mathbf{j} + 2\mathbf{k}$ is the position vector of a point on the line (A say),

therefore
$$\overrightarrow{AP} = (\mathbf{i} + 3\mathbf{j} + 2\mathbf{k}) - (\mathbf{i} + 2\mathbf{j} + 2\mathbf{k})$$
$$= \mathbf{j}.$$

If d is the distance of P from the line

$$d = \frac{|\mathbf{j}\times(-\mathbf{i}-3\mathbf{j})|}{|-\mathbf{i}-3\mathbf{j}|} = \frac{1}{\sqrt{10}}\begin{vmatrix} \mathbf{i} & \mathbf{j} & \mathbf{k} \\ 0 & 1 & 0 \\ -1 & -3 & 0 \end{vmatrix} = \frac{1}{\sqrt{10}}|\mathbf{k}|$$

$$= \frac{1}{\sqrt{10}}$$

6. Shortest Distance Between Two Skew Lines

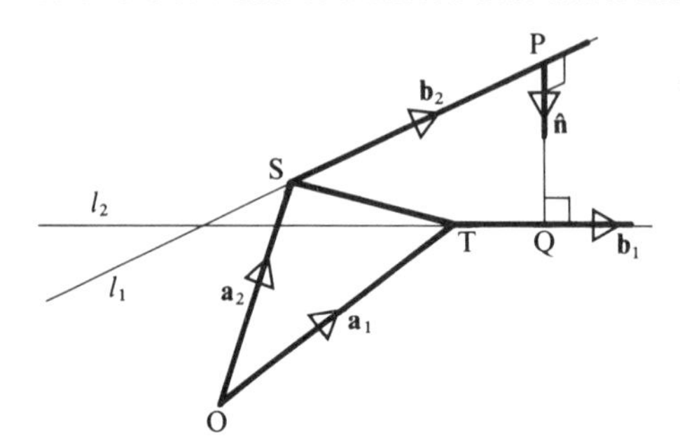

Consider two lines l_1 and l_2 whose vector equations are

$$\mathbf{r}_1 = \mathbf{a}_1 + \lambda\mathbf{b}_1$$

$$\mathbf{r}_2 = \mathbf{a}_2 + \mu\mathbf{b}_2.$$

If PQ is the shortest distance between the lines then PQ is perpendicular to both lines, i.e. $\overrightarrow{PQ}$ is parallel to $\mathbf{b}_1 \times \mathbf{b}_2$.

Therefore the unit vector $\hat{\mathbf{n}}$ in the direction of $\overrightarrow{PQ}$ is $\dfrac{\mathbf{b}_1\times\mathbf{b}_2}{|\mathbf{b}_1\times\mathbf{b}_2|}$,

and $\mathbf{a}_1 - \mathbf{a}_2$ represents the displacement from a point S on the first line to a point T on the second line.

If θ is the angle between $\overrightarrow{ST}$ and $\overrightarrow{PQ}$

then $$PQ = ST\cos\theta$$

$$= |\overrightarrow{ST}.\hat{\mathbf{n}}|,$$

i.e. $$d = \left|\frac{(\mathbf{a}_1 - \mathbf{a}_2).\mathbf{b}_1\times\mathbf{b}_2}{|\mathbf{b}_1\times\mathbf{b}_2|}\right|$$

EXAMPLES 5g (continued)

3) Find the shortest distance between the lines whose vector equations are $\mathbf{r}_1 = \mathbf{i}+\mathbf{j}+\lambda(2\mathbf{i}-\mathbf{j}+\mathbf{k})$ and $\mathbf{r}_2 = 2\mathbf{i}+\mathbf{j}-\mathbf{k}+\mu(3\mathbf{i}-5\mathbf{j}+2\mathbf{k})$.

A vector which is perpendicular to both lines is

$$(2\mathbf{i}-\mathbf{j}+\mathbf{k})\times(3\mathbf{i}-5\mathbf{j}+2\mathbf{k}) = 3\mathbf{i}-\mathbf{j}-7\mathbf{k}.$$

Therefore a unit vector perpendicular to both lines is $(3\mathbf{i}-\mathbf{j}-7\mathbf{k})/\sqrt{59}$.
A displacement vector from a point on one line to a point on the other is $(\mathbf{i}+\mathbf{j})-(2\mathbf{i}+\mathbf{j}-\mathbf{k}) = -\mathbf{i}+\mathbf{k}$.

Thus the least distance between the two lines is given by

$$\frac{1}{\sqrt{59}}|(-\mathbf{i}+\mathbf{k}).(3\mathbf{i}-\mathbf{j}-7\mathbf{k})| = 10/\sqrt{59}.$$

Condition That Two Lines Intersect

Two lines with equations $\mathbf{r}_1 = \mathbf{a}_1 + \lambda\mathbf{b}_1$, $\mathbf{r}_2 = \mathbf{a}_2 + \mu\mathbf{b}_2$ intersect if the least distance between them is zero. i.e. if $(\mathbf{a}_1 - \mathbf{a}_2).\mathbf{b}_1 \times \mathbf{b}_2 = 0$.
(This is the *condition* that two lines shall intersect. It does not give the position vector of the point of intersection. If this is required use the method given on page 198.)

EXERCISE 5g

1) The vectors $(2\mathbf{i}+3\mathbf{j}-\mathbf{k})$ and $(\mathbf{i}+2\mathbf{j}+\mathbf{k})$ represent two sides of a triangle. Find the area of the triangle.

2) The triangle ABC has its vertices at the points A(0, 0, 1), B(1, 0, 1), C(2, 1, 3). Find the area of the triangle ABC.

3) The vertices of a triangle are at the points with position vectors **a**, **b** and **c**. Prove that the area of the triangle is $\frac{1}{2}|\mathbf{a}\times\mathbf{b}+\mathbf{b}\times\mathbf{c}+\mathbf{c}\times\mathbf{a}|$.

4) A parallelogram OABC has one vertex O at the origin and the vertices A and B at the points (0, 1, 3), (0, 2, 5). Find the area of OABC.

5) The parallelogram ABCD has three of its vertices A, B and C at the points (1, 2, − 1), (1, 3, 2) and (− 1, 3, − 1). Find the area of the parallelogram.

6) The vectors $\overrightarrow{OA}, \overrightarrow{OB}, \overrightarrow{OC}$ are three edges of a parallelepiped where O is the origin and A, B and C are the points (2, 1, 0), (− 1, − 1, 1), (0, 2, − 1). Find the volume of the parallelepiped.

7) Find the volume of the tetrahedron OABC where O is the origin and A, B and C are the points (2, 0, 1), (3, 1, 2) and (− 1, 3, 0).

8) ABCD is a tetrahedron and A, B, C and D are the points (0, 1, 0), (0, 0, 4), (1, 1, 1) and (− 1, 3, 2). Find the volume of the tetrahedron.

9) The four vertices of a tetrahedron are at the points with position vectors **a**, **b**, **c**, **d**. Find the volume of the tetrahedron.

10) Find the perpendicular distance of the given point from the given line in the following cases:
(a) (1, 0, 2), $\mathbf{l} = \mathbf{i}+\mathbf{j}+\lambda(\mathbf{i}-\mathbf{j}+\mathbf{k})$,
(b) (0, 0, 0), $\mathbf{l} = (1-\lambda)\mathbf{i}+(2\lambda-1)\mathbf{j}+\lambda\mathbf{k}$,
(c) point with position vector **a**, $\mathbf{l} = \mathbf{a}-\mathbf{b}+\lambda\mathbf{c}$,
(d) point with position vector **r**, $\mathbf{l} = (\lambda-1)\mathbf{r}+(2\lambda+1)\mathbf{p}$.

11) The vector equations of two lines are

$$\mathbf{l}_1 = \mathbf{i}+2\mathbf{j}+\mathbf{k}+\lambda(\mathbf{i}-\mathbf{j}+\mathbf{k})$$
$$\mathbf{l}_2 = 2\mathbf{i}-\mathbf{j}-\mathbf{k}+\mu(2\mathbf{i}+\mathbf{j}+2\mathbf{k}).$$

Find the shortest distance between these lines.

12) The vector equations of two lines are

$$\mathbf{r}_1 = (1-t)\mathbf{i}+(t-2)\mathbf{j}+(3-2t)\mathbf{k}$$
$$\mathbf{r}_2 = (s+1)\mathbf{i}+(2s-1)\mathbf{j}-(2s+1)\mathbf{k}.$$

Find the shortest distance between these lines.

13) Determine whether the following pairs of lines intersect:
(a) $\mathbf{r}=\mathbf{i}-\mathbf{j}+\lambda(2\mathbf{i}+\mathbf{k})$; $\mathbf{r}=2\mathbf{i}-\mathbf{j}+\mu(\mathbf{i}+\mathbf{j}-\mathbf{k})$,
(b) $\mathbf{r}=\mathbf{i}+\mathbf{j}-\mathbf{k}+\lambda(3\mathbf{i}-\mathbf{j})$; $\mathbf{r}=4\mathbf{i}-\mathbf{k}+\mu(2\mathbf{i}+3\mathbf{k})$,
(c) $\dfrac{x-1}{2}=\dfrac{y+1}{3}=z$; $\dfrac{x+1}{5}=\dfrac{y-2}{1}$, $z=2$.

EQUATION OF A PLANE

A particular plane can be specified in several ways, for example:

(a) one and only one plane can be drawn through three non-collinear points, therefore three given points specify a particular plane;

(b) one and only one plane can be drawn to contain two concurrent lines, therefore two given concurrent lines specify a particular plane;

(c) one and only one plane can be drawn perpendicular to a given direction at a given distance from the origin, therefore the normal to a plane and the distance of the plane from the origin specify a particular plane;

(d) one and only one plane can be drawn through a given point and perpendicular to a given direction, therefore a point on the plane and a normal to the plane specify a particular plane.

There are many other ways of specifying a particular plane but the most useful are described in (c) and (d) as these lead to the same, very simple, form for the vector equation of a plane.
Consider the plane which is at a distance d from the origin and which is perpendicular to the unit vector $\hat{\mathbf{n}}$ ($\hat{\mathbf{n}}$ being directed *away* from O).

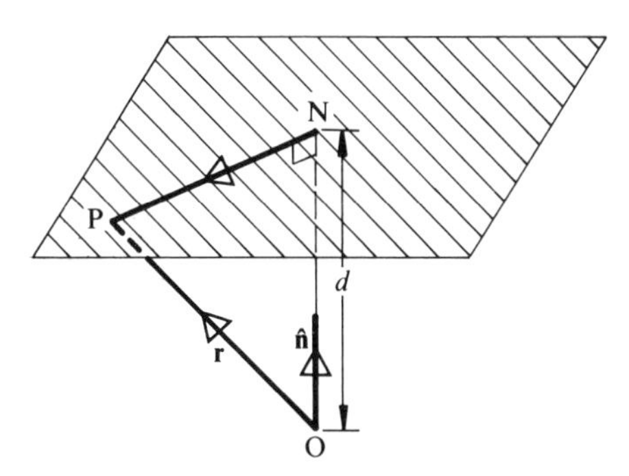

If ON is the perpendicular from the origin to the plane then $\overrightarrow{ON} = d\hat{\mathbf{n}}$.
If P is any point on the plane, NP is perpendicular to ON, and conversely if P is not on the plane, NP is not perpendicular to ON.

Therefore P is a point on the plane $\iff \overrightarrow{NP}.\overrightarrow{ON} = 0$. [1]

This equation is called the scalar product form of the vector equation of the plane.
If $\mathbf{r}$ is the position vector of P, $\overrightarrow{NP} = \mathbf{r} - d\hat{\mathbf{n}}$.

Therefore [1] becomes $(\mathbf{r} - d\hat{\mathbf{n}}).d\hat{\mathbf{n}} = 0$

$\Rightarrow \qquad \mathbf{r}.\hat{\mathbf{n}} - d\hat{\mathbf{n}}.\hat{\mathbf{n}} = 0$

$\Rightarrow \qquad \mathbf{r}.\hat{\mathbf{n}} = d \qquad$ as $\hat{\mathbf{n}}.\hat{\mathbf{n}} = 1$.

The equation $\mathbf{r}.\hat{\mathbf{n}} = d$ is the standard form of the vector equation of the plane, where
$\mathbf{r}$ is the position vector of any point on the plane,
$\hat{\mathbf{n}}$ is the unit vector perpendicular to the plane,
d is the distance of the plane from the origin.

Now consider the plane which contains the point A whose position vector is $\mathbf{a}$ and which is perpendicular to the unit vector $\hat{\mathbf{n}}$.

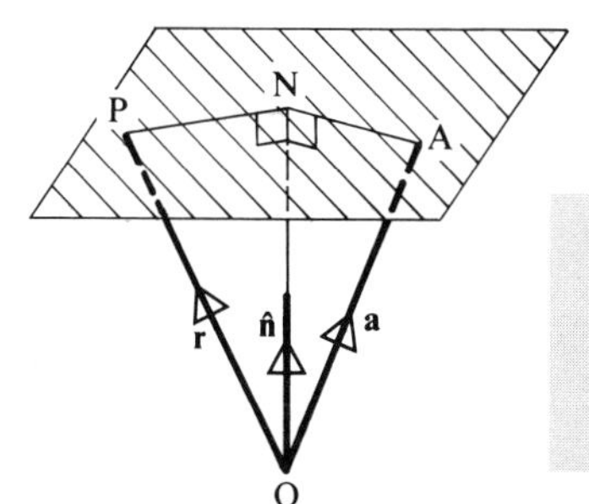

The distance of the plane from the origin is ON and $ON = \mathbf{a}.\hat{\mathbf{n}}$.
Therefore the equation of the plane is

$$\mathbf{r}.\hat{\mathbf{n}} = \mathbf{a}.\hat{\mathbf{n}}$$

where $\mathbf{r}$ is the position vector of any point P on the plane.

The standard form of the vector equation of a plane can be multiplied by any scalar quantity, thus
any equation of the form $\mathbf{r}.\mathbf{n} = D$ represents a plane perpendicular to $\mathbf{n}$.

Converting to standard form we have $\mathbf{r}.\hat{\mathbf{n}} = \dfrac{D}{|\mathbf{n}|}$ where $\dfrac{D}{|\mathbf{n}|}$ is the distance of the plane from the origin.

The Cartesian Equation of a Plane

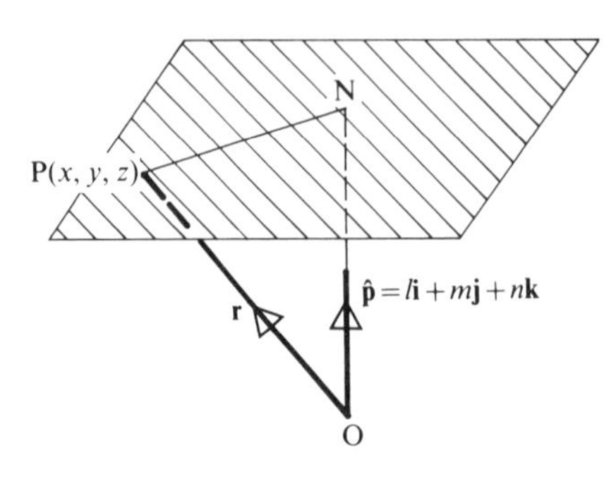

The standard form of the vector equation of the plane shown in the diagram is

$$\mathbf{r}.\hat{\mathbf{p}} = d. \qquad [1]$$

If $P(x,y,z)$ is any point on the plane

$$\mathbf{r} = x\mathbf{i} + y\mathbf{j} + z\mathbf{k}$$

and if l, m, n are the direction cosines of $\hat{\mathbf{p}}$ then $\hat{\mathbf{p}} = l\mathbf{i} + m\mathbf{j} + n\mathbf{k}$.

Substituting in [1]

$$(x\mathbf{i} + y\mathbf{j} + z\mathbf{k}).(l\mathbf{i} + m\mathbf{j} + n\mathbf{k}) = d$$

$$\Rightarrow \qquad lx + my + nz = d$$

which is the Cartesian equation of the plane.
Therefore if $\mathbf{r}.(A\mathbf{i} + B\mathbf{j} + C\mathbf{k}) = D$ is the vector equation of a plane then $Ax + By + Cz = D$ is the Cartesian equation of the same plane where A, B, C are the direction ratios of the normal to the plane.
(So there is an easy transfer between these two forms for the equation of a plane.)

The Parametric Form for the Vector Equation of a Plane

Consider the plane which is parallel to the vectors **d** and **e** (**d** not parallel to **e**) and which also contains the point A whose position vector is **a**.

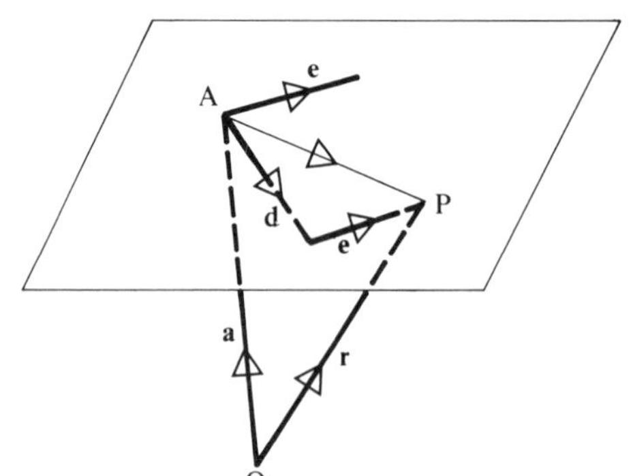

The vectors **d** and **e** determine the orientation of the plane and **a** fixes it in space, so these vectors specify a particular plane.
If P is any point on this plane
$\overrightarrow{AP} = \lambda\mathbf{d} + \mu\mathbf{e}$ where λ and μ are independent parameters.

If $\mathbf{r}$ is the position vector of P
$$\mathbf{r} = \mathbf{a} + \overrightarrow{AP} = \mathbf{a} + \lambda\mathbf{d} + \mu\mathbf{e}.$$

Thus any equation of the form $\mathbf{r} = \mathbf{a} + \lambda\mathbf{d} + \mu\mathbf{e}$, where λ and μ are independent parameters, represents the plane parallel to the vectors $\mathbf{d}$ and $\mathbf{e}$ and containing the point $\mathbf{a}$.

It should be noted that the parametric form is not a unique equation for a particular plane: $\mathbf{a}$ is any one of an infinite number of points on the plane, and $\mathbf{d}$ and $\mathbf{e}$ are only one pair of the infinite set of vectors parallel to the plane. An interesting variation of the parametric form is found by considering the plane passing through three given points.

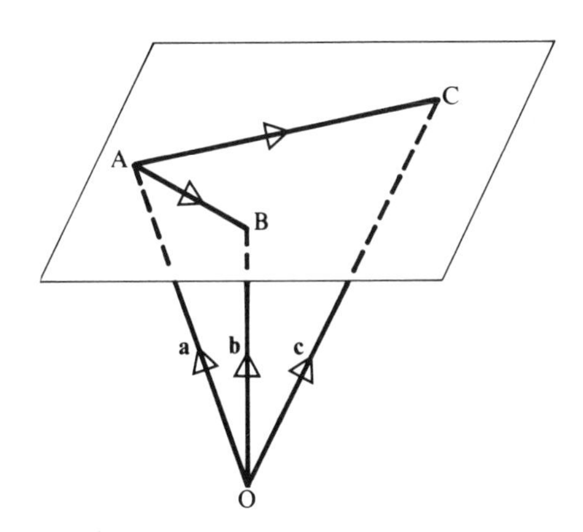

The vector equation of the plane in the diagram can be written

$$\begin{aligned}\mathbf{r} &= \mathbf{a} + t\overrightarrow{AB} + u\overrightarrow{AC}\\ &= \mathbf{a} + t(\mathbf{b} - \mathbf{a}) + u(\mathbf{c} - \mathbf{a})\\ &= (1 - t - u)\mathbf{a} + t\mathbf{b} + u\mathbf{c}\\ &= s\mathbf{a} + t\mathbf{b} + u\mathbf{c}\end{aligned}$$

where $s = 1 - t - u$.

Therefore any equation of the form $\mathbf{r} = s\mathbf{a} + t\mathbf{b} + u\mathbf{c}$, where $s + t + u = 1$, represents the plane passing through the points $\mathbf{a}, \mathbf{b}, \mathbf{c}$.
(**Note**: In this form, s, t and u are *not* independent parameters.)

To summarize:

The *scalar product form* of the vector equation of a plane is

$$\mathbf{r}.\mathbf{n} = D$$

where $\mathbf{n}$ is a vector perpendicular to the plane and $\dfrac{D}{|\mathbf{n}|}$ is the distance of the plane from the origin.

The *parametric form* of the vector equation of a plane is

$$\mathbf{r} = \mathbf{a} + \lambda\mathbf{b} + \mu\mathbf{c}$$

where $\mathbf{a}$ is a point on the plane and $\mathbf{b}$ and $\mathbf{c}$ are parallel to the plane.

The *Cartesian equation* of a plane is

$$Ax + By + Cz = D$$

where A, B, C are the direction ratios of $\mathbf{n}$, the normal to the plane.

EXAMPLES 5h

1) Find the vector equation of the plane containing the points A(0, 1, 1), B(2, 1, 0), C(−2, 0, 3), (a) in parametric form, (b) in scalar product form.

(a) The parametric equation of this plane is

$$\mathbf{r} = \lambda(\mathbf{j}+\mathbf{k})+\mu(2\mathbf{i}+\mathbf{j})+\eta(-2\mathbf{i}+3\mathbf{k})$$

where $\lambda+\mu+\eta = 1$.
Replacing η by $1-(\lambda+\mu)$ and simplifying gives

$$\mathbf{r} = (-2+2\lambda+4\mu)\mathbf{i}+(\lambda+\mu)\mathbf{j}+(3-2\lambda-3\mu)\mathbf{k}. \qquad [1]$$

(b) If $P(x,y,z)$ is any point on this plane, then from equation [1] we have

$$x = -2+2\lambda+4\mu$$

$$y = \lambda+\mu$$

$$z = 3-2\lambda-3\mu.$$

Eliminating λ and μ gives $x+2y+2z = 4$ which is the Cartesian equation of the plane.
Therefore the scalar product form of the equation is

$$\mathbf{r}.(\mathbf{i}+2\mathbf{j}+2\mathbf{k}) = 4.$$

2) Show that the line L whose vector equation is

$$\mathbf{r} = 2\mathbf{i}-2\mathbf{j}+3\mathbf{k}+\lambda(\mathbf{i}-\mathbf{j}+4\mathbf{k})$$

is parallel to the plane Π whose vector equation is

$$\mathbf{r}.(\mathbf{i}+5\mathbf{j}+\mathbf{k}) = 5$$

and find the distance between them.

L is parallel to $\mathbf{i}-\mathbf{j}+4\mathbf{k}$ and Π is normal to $\mathbf{i}+5\mathbf{j}+\mathbf{k}$.
Now $(\mathbf{i}-\mathbf{j}+4\mathbf{k}).(\mathbf{i}+5\mathbf{j}+\mathbf{k}) = 0$.
So L is perpendicular to the normal to Π and hence parallel to Π.

The distance, d_1, of Π from the origin is $\dfrac{5}{\sqrt{27}}$.

As $2\mathbf{i}-2\mathbf{j}+3\mathbf{k}$ is a point on L we may write the equation of the plane which is parallel to Π and which contains L as

$$\mathbf{r}.(\mathbf{i}+5\mathbf{j}+\mathbf{k}) = (2\mathbf{i}-2\mathbf{j}+3\mathbf{k}).(\mathbf{i}+5\mathbf{j}+\mathbf{k})$$

$$\Rightarrow \qquad \mathbf{r}.(\mathbf{i}+5\mathbf{j}+\mathbf{k}) = -5 \quad \text{or} \quad \mathbf{r}.(-\mathbf{i}-5\mathbf{j}-\mathbf{k}) = 5.$$

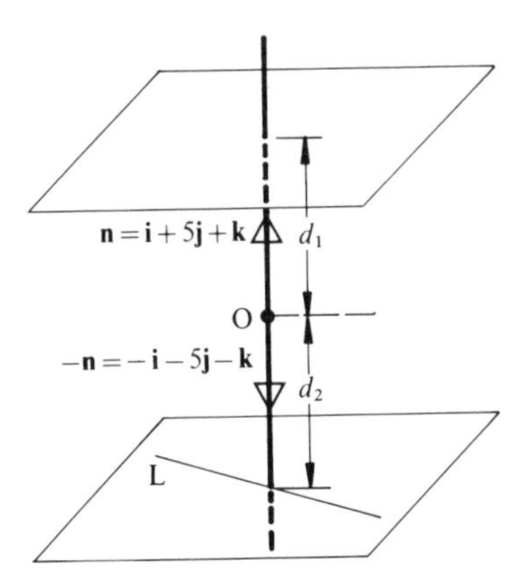

Since the normal vectors to these planes are in opposite directions we deduce that this plane and Π are on opposite sides of the origin.

The distance of this plane from the origin is given by $d_2 = \dfrac{5}{\sqrt{27}}$.

Hence the distance between the two planes, and hence between L and Π is

$$d_1 + d_2 = \frac{10}{\sqrt{27}}.$$

3) Show that the plane whose vector equation is $\mathbf{r}.(\mathbf{i} + 2\mathbf{j} - \mathbf{k}) = 3$ contains the line whose vector equation is $\mathbf{r} = \mathbf{i} + \mathbf{j} + \lambda(2\mathbf{i} + \mathbf{j} + 4\mathbf{k})$.

The line is contained in the plane if any two points on the line are on the plane. Taking $\lambda = 0$ and $\lambda = 1$ we find that $\mathbf{i} + \mathbf{j}$ and $3\mathbf{i} + 2\mathbf{j} + 4\mathbf{k}$ are two points on the line.
If $\mathbf{r} = \mathbf{i} + \mathbf{j}$ then $\mathbf{r}.(\mathbf{i} + 2\mathbf{j} - \mathbf{k}) = (\mathbf{i} + \mathbf{j}).(\mathbf{i} + 2\mathbf{j} - \mathbf{k}) = 3$.
Therefore $\mathbf{i} + \mathbf{j}$ is a point on the plane.
If $\mathbf{r} = 3\mathbf{i} + 2\mathbf{j} + 4\mathbf{k}$ then $\mathbf{r}.(\mathbf{i} + 2\mathbf{j} - \mathbf{k}) = (3\mathbf{i} + 2\mathbf{j} + 4\mathbf{k}).(\mathbf{i} + 2\mathbf{j} - \mathbf{k}) = 3$.
Therefore $3\mathbf{i} + 2\mathbf{j} + 4\mathbf{k}$ is a point on the plane.
Therefore the line is contained in the plane.

4) Find the vector equation of line passing through the point $(3, 1, 2)$ and perpendicular to the plane $\mathbf{r}.(2\mathbf{i} - \mathbf{j} + \mathbf{k}) = 4$. Find also the point of intersection of this line and the plane.

As $\mathbf{r}.(2\mathbf{i} - \mathbf{j} + \mathbf{k}) = 4$ is the equation of the plane, $2\mathbf{i} - \mathbf{j} + \mathbf{k}$ is perpendicular to the plane and therefore parallel to the required line.
As this line passes through the point $3\mathbf{i} + \mathbf{j} + 2\mathbf{k}$ its equation is

$$\mathbf{r} = (3\mathbf{i} + \mathbf{j} + 2\mathbf{k}) + \lambda(2\mathbf{i} - \mathbf{j} + \mathbf{k}).$$

Writing the equations of the line in parametric form

$$x = 3 + 2\lambda, \quad y = 1 - \lambda, \quad z = 2 + \lambda$$

we see that this line meets the plane, where

$$[(3+2\lambda)\mathbf{i}+(1-\lambda)\mathbf{j}+(2+\lambda)\mathbf{k}].(2\mathbf{i}-\mathbf{j}+\mathbf{k}) = 4$$

$$\Rightarrow \quad 2(3+2\lambda)-(1-\lambda)+(2+\lambda) = 4$$

$$\Rightarrow \quad \lambda = -\tfrac{1}{2}$$

$$\Rightarrow \quad x = 2, \quad y = \tfrac{3}{2}, \quad z = \tfrac{3}{2}.$$

So the point of intersection of the line and the plane is $(2, \frac{3}{2}, \frac{3}{2})$.

EXERCISE 5h

1) Find a vector equation of the plane containing the points A, B and C in parametric form and in scalar product form where
(a) A, B and C are the points $(1, 2, -1), (1, 3, 2), (0, 2, 1)$ respectively,
(b) the position vectors of A, B and C are $\mathbf{i}+\mathbf{j}-2\mathbf{k}$, $\mathbf{i}+\mathbf{k}$, $-2\mathbf{i}+\mathbf{j}-3\mathbf{k}$.

2) Find the vector equation of the following planes in scalar product form:
(a) $\mathbf{r} = \mathbf{i}-\mathbf{j}+\lambda(\mathbf{i}+\mathbf{j}+\mathbf{k})+\mu(\mathbf{i}-2\mathbf{j}+3\mathbf{k})$,
(b) $\mathbf{r} = 2\mathbf{i}-\mathbf{k}+\lambda(\mathbf{i})+\mu(\mathbf{i}-2\mathbf{j}-\mathbf{k})$,
(c) $\mathbf{r} = (1+s-t)\mathbf{i}+(2-s)\mathbf{j}+(3-2s+2t)\mathbf{k}$.

3) Find the Cartesian equations of the following planes:
(a) $\mathbf{r}.(\mathbf{i}+\mathbf{j}-\mathbf{k}) = 2$,
(b) $\mathbf{r}.(2\mathbf{i}+3\mathbf{j}-4\mathbf{k}) = 1$,
(c) $\mathbf{r} = (s-2t)\mathbf{i}+(3-t)\mathbf{j}+(2s+t)\mathbf{k}$.

4) Find the vector equation in scalar product form of the plane that contains the lines $\mathbf{r} = (\mathbf{i}+\mathbf{j})+s(\mathbf{i}+2\mathbf{j}-\mathbf{k})$ and $\mathbf{r} = (\mathbf{i}+\mathbf{j})+t(-\mathbf{i}+\mathbf{j}-2\mathbf{k})$.

5) Find the vector equation in parametric form of the plane that contains the lines $\mathbf{r} = -3\mathbf{i}-2\mathbf{j}+t(\mathbf{i}-2\mathbf{j}+\mathbf{k})$, $\mathbf{r} = \mathbf{i}-11\mathbf{j}+4\mathbf{k}+s(2\mathbf{i}-\mathbf{j}+2\mathbf{k})$.

6) Find the vector equation in parametric form of the plane that goes through the point with position vector $\mathbf{i}+\mathbf{j}$ and which is parallel to the lines $\mathbf{r}_1 = 2\mathbf{i}-\mathbf{j}+\lambda(\mathbf{i}+\mathbf{k})$ and $\mathbf{r}_2 = 2\mathbf{j}-\mathbf{k}+\mu(\mathbf{i}-\mathbf{j}+\mathbf{k})$. Is either of these lines contained in the plane?

7) A plane goes through the three points whose position vectors are $\mathbf{a}$, $\mathbf{b}$ and $\mathbf{c}$ where

$$\mathbf{a} = \mathbf{i}+\mathbf{j}+2\mathbf{k}$$
$$\mathbf{b} = 2\mathbf{i}-\mathbf{j}+3\mathbf{k}$$
$$\mathbf{c} = -\mathbf{i}+2\mathbf{j}-2\mathbf{k}.$$

Find the vector equation of this plane in scalar product form and hence find the distance of the plane from the origin.

8) A plane goes through the points whose position vectors are $\mathbf{i}-2\mathbf{j}+\mathbf{k}$ and $2\mathbf{i}-\mathbf{j}-\mathbf{k}$ and is parallel to the line $\mathbf{r}=\mathbf{i}-\mathbf{j}+\lambda(3\mathbf{i}+\mathbf{j}-2\mathbf{k})$. Find the distance of this plane from the origin.

9) Two planes Π_1 and Π_2 have vector equations $\mathbf{r}.(2\mathbf{i}+\mathbf{j}-2\mathbf{k})=3$ and $\mathbf{r}.(2\mathbf{i}+\mathbf{j}-2\mathbf{k})=9$. Explain why Π_1 and Π_2 are parallel and hence find the distance between them.

10) Find the vector equation of the line through the origin which is perpendicular to the plane $\mathbf{r}.(\mathbf{i}-2\mathbf{j}+\mathbf{k})=3$.

11) Find the vector equation of the line through the point $(2, 1, 1)$ which is perpendicular to the plane $\mathbf{r}.(\mathbf{i}+2\mathbf{j}-3\mathbf{k})=6$.

12) Find the vector equation of the plane which goes through the point $(0, 1, 6)$ and is parallel to the plane $\mathbf{r}.(\mathbf{i}-2\mathbf{j})=3$.

13) Find the vector equation of the plane which goes through the origin and which contains the line $\mathbf{r}=2\mathbf{i}+\lambda(\mathbf{j}+\mathbf{k})$.

14) Find the point of intersection of the line $\mathbf{r}=(\mathbf{i}+\mathbf{j}-2\mathbf{k})+\lambda(\mathbf{i}-\mathbf{j}+\mathbf{k})$ and the plane $\mathbf{r}.(\mathbf{i}+2\mathbf{j}-\mathbf{k})=2$.

15) Find the point of intersection of the line $x-2=2y+1=3-z$ and the plane $x+2y+z=3$.

16) Show that the line $x+1 = y = \dfrac{z-3}{2}$ is parallel to the plane $\mathbf{r}.(\mathbf{i}+\mathbf{j}-\mathbf{k})=3$ and find the distance between them.

17) Determine whether the given lines are parallel to, contained in, or intersect the plane $\mathbf{r}.(2\mathbf{i}+\mathbf{j}-3\mathbf{k})=5$:
(a) $\mathbf{r}=3\mathbf{i}-\mathbf{j}+\mathbf{k}+\lambda(-2\mathbf{i}+\mathbf{j}-3\mathbf{k})$,
(b) $\mathbf{r}=\mathbf{i}-\mathbf{j}+\mu(2\mathbf{i}+\mathbf{j}-3\mathbf{k})$,
(c) $x=y=z$,
(d) $\mathbf{r}=(2\mathbf{i}+\mathbf{j})+s(3\mathbf{i}+2\mathbf{k})$.

THE ANGLE BETWEEN TWO PLANES

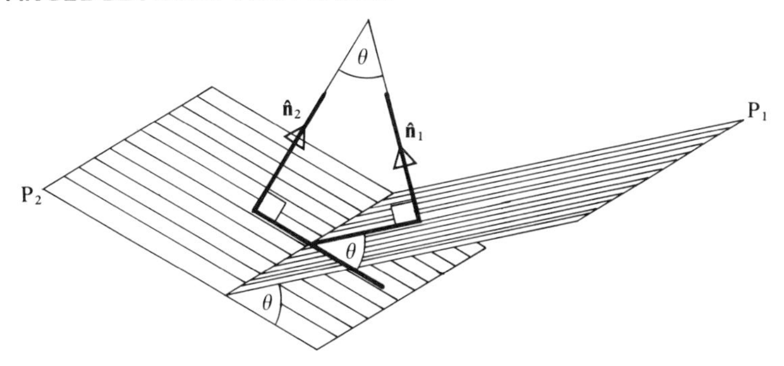

Consider two planes P_1 and P_2 whose vector equations are

$$\mathbf{r}.\hat{\mathbf{n}}_1 = d_1 \quad \text{and} \quad \mathbf{r}.\hat{\mathbf{n}}_2 = d_2.$$

The angle between P_1 and P_2 is equal to the angle between the normals to P_1 and P_2, i.e. the angle between $\hat{\mathbf{n}}_1$ and $\hat{\mathbf{n}}_2$.
Therefore if θ is the angle between P_1 and P_2,

$$\cos\theta = \hat{\mathbf{n}}_1 . \hat{\mathbf{n}}_2 \qquad [1]$$

e.g. the angle between the planes whose vector equations are
$\mathbf{r}.(\mathbf{i}+\mathbf{j}-2\mathbf{k}) = 3$ and $\mathbf{r}.(2\mathbf{i}-2\mathbf{j}+\mathbf{k}) = 2$ is given by

$$\cos\theta = \frac{(\mathbf{i}+\mathbf{j}-2\mathbf{k})}{\sqrt{6}} . \frac{(2\mathbf{i}-2\mathbf{j}+\mathbf{k})}{3} = -\frac{\sqrt{6}}{9}$$

This is the cosine of the obtuse angle between the planes.
The acute angle is $\arccos\sqrt{6}/9$.

Two planes are perpendicular if $\hat{\mathbf{n}}_1 . \hat{\mathbf{n}}_2 = 0$
and two planes are parallel if $\hat{\mathbf{n}}_1 = \hat{\mathbf{n}}_2$.

THE ANGLE BETWEEN A LINE AND A PLANE

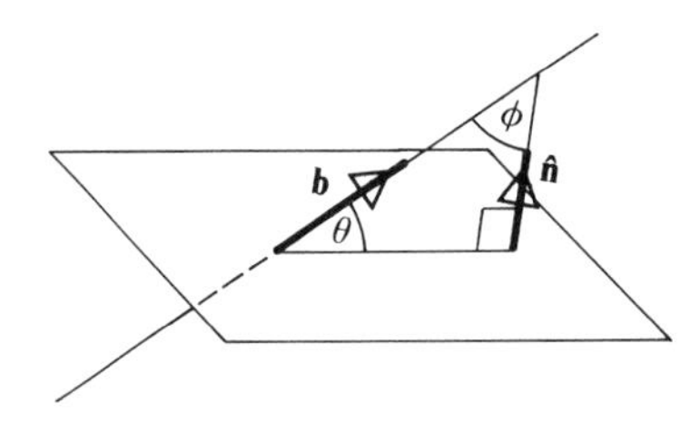

Consider the line $\mathbf{r} = \mathbf{a} + \lambda\mathbf{b}$ and the plane $\mathbf{r}.\hat{\mathbf{n}} = d$.
The angle ϕ between the line and the normal to the plane is given by

$$\cos\phi = \frac{\mathbf{b}.\hat{\mathbf{n}}}{|\mathbf{b}|}.$$

If θ is the angle between the line and the plane then $\theta = \frac{\pi}{2} - \phi$

i.e. $\sin\theta = \cos\phi.$

Therefore the angle between the line $\mathbf{r} = \mathbf{a} + \lambda\mathbf{b}$ and the plane $\mathbf{r}.\hat{\mathbf{n}} = d$ is given by

$$\sin\theta = \frac{\mathbf{b}.\hat{\mathbf{n}}}{|\mathbf{b}|}$$

e.g. the angle θ between the line $\mathbf{r} = (\mathbf{i} + 2\mathbf{j} - \mathbf{k}) + \lambda(\mathbf{i} - \mathbf{j} + \mathbf{k})$ and the plane $\mathbf{r}.(2\mathbf{i} - \mathbf{j} + \mathbf{k}) = 4$ is given by

$$\sin\theta = \frac{(\mathbf{i} - \mathbf{j} + \mathbf{k})}{\sqrt{3}} \cdot \frac{(2\mathbf{i} - \mathbf{j} + \mathbf{k})}{\sqrt{6}} = \frac{2\sqrt{2}}{3}$$

Therefore $$\theta = \arcsin\frac{2\sqrt{2}}{3}$$

THE DISTANCE OF A POINT FROM A PLANE

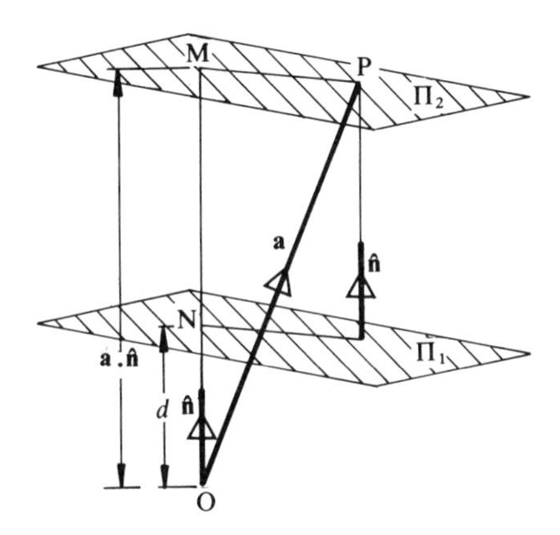

Consider a point P with position vector $\mathbf{a}$ and a plane Π_1 whose equation is

$$\mathbf{r}.\hat{\mathbf{n}} = d.$$

The equation of the plane Π_2 through P parallel to the plane Π_1 is

$$\mathbf{r}.\hat{\mathbf{n}} = \mathbf{a}.\hat{\mathbf{n}},$$

i.e. the distance OM of this plane from the origin is $\mathbf{a}.\hat{\mathbf{n}}$.
Therefore (assuming that P and O are on opposite sides of Π_1) the distance MN of P from the plane Π_1 is $\mathbf{a}.\hat{\mathbf{n}} - d$.
(If P and O are on the same side of the plane the use of this formula will give a negative result.)
Thus the distance of the point with position vector $\mathbf{i} - 3\mathbf{j} + 3\mathbf{k}$ from the plane Π with equation $\mathbf{r}.(2\mathbf{i} + 3\mathbf{j} - 6\mathbf{k}) = 9$ is given by

$$(\mathbf{i} - 3\mathbf{j} + 3\mathbf{k}).\frac{(2\mathbf{i} + 3\mathbf{j} - 6\mathbf{k})}{7} - \frac{9}{7} = -\frac{25}{7} - \frac{9}{7} = -\frac{34}{7}$$

The negative sign indicates that the point and the origin are on the same side of the plane.

THE INTERSECTION OF TWO PLANES

Unless two planes are parallel they will contain a common line which is the line of intersection of the two planes.

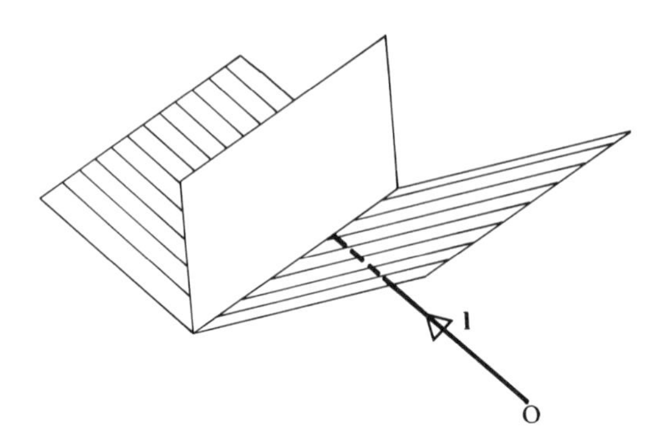

If Π_1 and Π_2 are planes with equations $\mathbf{r}.\hat{\mathbf{n}}_1 = d_1$ and $\mathbf{r}.\hat{\mathbf{n}}_2 = d_2$ respectively, the position vector of any point on the line of intersection must satisfy both equations.
If $\mathbf{l}$ is the position vector of a point on this line then
$\mathbf{l}.\hat{\mathbf{n}}_1 = d_1$ and $\mathbf{l}.\hat{\mathbf{n}}_2 = d_2$.
Therefore, for any value of k,

$$\mathbf{l}.\hat{\mathbf{n}}_1 - k\mathbf{l}.\hat{\mathbf{n}}_2 = d_1 - kd_2$$

or

$$\mathbf{l}.(\hat{\mathbf{n}}_1 - k\hat{\mathbf{n}}_2) = d_1 - kd_2.$$

But the equation $\mathbf{r}.(\hat{\mathbf{n}}_1 - k\hat{\mathbf{n}}_2) = d_1 - kd_2$ represents a plane Π_3 which is such that if any vector $\mathbf{r}$ satisfies the equations of Π_1 and Π_2 it also satisfies the equation of Π_3.

i.e. any plane passing through the intersection of the planes $\mathbf{r}.\hat{\mathbf{n}}_1 = d_1$ and $\mathbf{r}.\hat{\mathbf{n}}_2 = d_2$ has an equation

$$\mathbf{r}.(\hat{\mathbf{n}}_1 - k\hat{\mathbf{n}}_2) = d_1 - kd_2.$$

Conversely, for all real values of k the equation $\mathbf{r}.(\hat{\mathbf{n}}_1 - k\hat{\mathbf{n}}_2) = d_1 - kd_2$ represents the family of planes passing through the line of intersection of the planes $\mathbf{r}.\hat{\mathbf{n}}_1 = d_1$ and $\mathbf{r}.\hat{\mathbf{n}}_2 = d_2$.
This is a particular case of a more general result, viz, if $E_1 = 0$ and $E_2 = 0$ are the equations of two members of a family of curves (or surfaces) then the equation

$$E_1 = kE_2$$

represents, for all real values of k, those members of that family that contain the point (or points) of intersection of E_1 and E_2.

The Line of Intersection of Two Planes

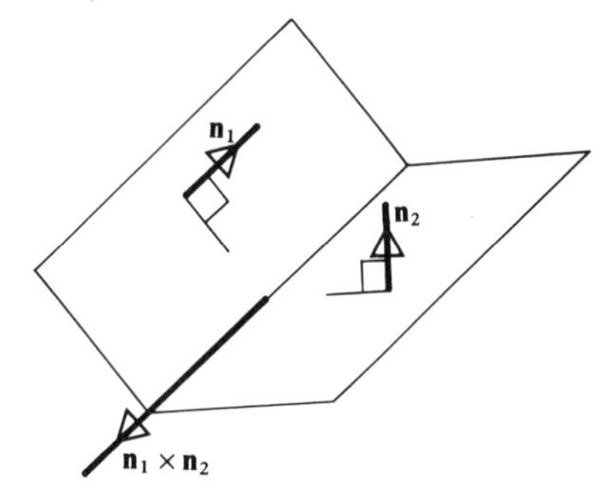

As the line of intersection of two planes

$$\mathbf{r}.\mathbf{n}_1 = D_1$$

$$\mathbf{r}.\mathbf{n}_2 = D_2$$

is contained in both planes it is perpendicular to both $\mathbf{n}_1$ and $\mathbf{n}_2$, i.e. parallel to $\mathbf{n}_1 \times \mathbf{n}_2$.

To find the equation of the line of intersection of two planes, consider, as an example, the planes

$$\left.\begin{aligned}\mathbf{r}.(\mathbf{i}+\mathbf{j}-3\mathbf{k}) &= 6\\ \mathbf{r}.(2\mathbf{i}-\mathbf{j}+\mathbf{k}) &= 4\end{aligned}\right\} \Rightarrow \begin{cases} x+y-3z = 6 & [1]\\ 2x-y+z = 4. & [2]\end{cases}$$

These planes meet where $\quad 3x-2z = 10 \qquad [1]+[2]$

and $\quad 7x-2y = 18, \qquad [1]+3[2]$

i.e. where $\quad x = \dfrac{2z+10}{3} = \dfrac{2y+18}{7} = \lambda.$

These are the Cartesian equations of the line.
So any point on the line has coordinates

$$x = \lambda, \quad y = \frac{7\lambda-18}{2}, \quad z = \frac{3\lambda-10}{2}$$

or $\quad x = 2s, \; y = 7s-9 \quad z = 3s-5 \quad (\lambda = 2s).$

Hence the position vector of any point on the line is

$$\mathbf{r} = -9\mathbf{j}-5\mathbf{k}+s(2\mathbf{i}+7\mathbf{j}+3\mathbf{k}).$$

This is the vector equation of the line.
If the equations of the planes are in parametric form it is not necessary to convert these equations into Cartesian form. The line of intersection may be found as follows.

Consider the planes

$$\mathbf{r} = \mathbf{i}+\mathbf{j}+\lambda(2\mathbf{i}-\mathbf{k})+\mu(\mathbf{i}-\mathbf{j}+\mathbf{k}) \qquad [1]$$

$$\mathbf{r} = 3\mathbf{i}-\mathbf{k}+s(\mathbf{i}-\mathbf{j}+2\mathbf{k})+t(\mathbf{i}+2\mathbf{j}-\mathbf{k}). \qquad [2]$$

Rearranging gives

$$\mathbf{r} = (1+2\lambda+\mu)\mathbf{i}+(1-\mu)\mathbf{j}+(-\lambda+\mu)\mathbf{k}$$

$$\mathbf{r} = (3+s+t)\mathbf{i}+(-s+2t)\mathbf{j}+(-1+2s-t)\mathbf{k}.$$

These planes meet where

$$1+2\lambda+\mu = 3+s+t$$

$$1-\mu = -s+2t$$

$$-\lambda+\mu = -1+2s-t.$$

Eliminating λ and μ from these equations gives $3 = 2s+5t$,

i.e. the planes meet at all points where $s = \dfrac{3-5t}{2}$.

Therefore substituting $\dfrac{3-5t}{2}$ for s in equation [2] gives the vector equation of the line of intersection of the planes,

i.e. $$\mathbf{r} = \tfrac{1}{2}[(9\mathbf{i}-3\mathbf{j}+4\mathbf{k})+t(-3\mathbf{i}+9\mathbf{j}-12\mathbf{k})].$$

EXERCISE 5i

1) Find the cosine of the angle between the two planes whose equations are
(a) $\mathbf{r}.(\mathbf{i}-\mathbf{j}+3\mathbf{k}) = 3$, $\mathbf{r}.(2\mathbf{i}-\mathbf{j}+2\mathbf{k}) = 5$;
(b) $\mathbf{r} = (\mathbf{i}+\mathbf{j})+\lambda(\mathbf{i}+\mathbf{j}-\mathbf{k})+\mu(2\mathbf{i}-\mathbf{j}+3\mathbf{k})$,
$\mathbf{r} = (\mathbf{i}-2\mathbf{j}+\mathbf{k})+s(2\mathbf{i}+\mathbf{k})+t(\mathbf{i}-2\mathbf{j}-\mathbf{k})$;
(c) $2x+2y-3z = 3$, $x+3y-4z = 6$.

2) Find the sine of the angle between the line and plane whose equations are
(a) $\mathbf{r} = \mathbf{i}-\mathbf{j}+\lambda(\mathbf{i}+\mathbf{j}+\mathbf{k})$, $\mathbf{r}.(\mathbf{i}-2\mathbf{j}+2\mathbf{k}) = 4$;
(b) $\mathbf{r} = \mathbf{i}-2\mathbf{j}+\mathbf{k}+\lambda(2\mathbf{i}-\mathbf{j})$, $\mathbf{r} = \mathbf{i}-\mathbf{j}+s(\mathbf{i}+\mathbf{k})+t(\mathbf{j}-\mathbf{k})$;
(c) $\dfrac{x-2}{2} = \dfrac{y+1}{6} = \dfrac{z+3}{3}$, $2x-y-2z = 4$.

3) Find the distance of the point $(1, 3, 2)$ from the following planes:
(a) $\mathbf{r}.(7\mathbf{i}+4\mathbf{j}+4\mathbf{k}) = 9$;
(b) $6x+6y+3z = 8$;
(c) $\mathbf{r} = \mathbf{i}-\mathbf{j}+\mathbf{k}+\lambda(\mathbf{i})+\mu(\mathbf{j}-2\mathbf{k})$.

4) Find the vector equation of the line of intersection of the following pairs of planes :
(a) $\mathbf{r}.(\mathbf{i}-2\mathbf{j}+\mathbf{k}) = 3$, $\mathbf{r}.(3\mathbf{i}+\mathbf{j}-2\mathbf{k}) = 4$;
(b) $\mathbf{r} = (1-\lambda+\mu)\mathbf{i}+(\lambda-\mu)\mathbf{j}+(2-\mu)\mathbf{k}$,
$\mathbf{r} = (2-s)\mathbf{i}+(1-3t)\mathbf{j}+(2s-3t)\mathbf{k}$.

5) Prove that the line $\mathbf{r} = \mathbf{i}-2\mathbf{j}+\lambda(\mathbf{i}-3\mathbf{j}-\mathbf{k})$ is parallel to the intersection of the planes $\mathbf{r}.(\mathbf{i}+\mathbf{j}-2\mathbf{k}) = 2$ and $\mathbf{r}.(2\mathbf{i}+\mathbf{j}-\mathbf{k}) = 0$.

The results given in this Chapter contain several formulae which must be used with caution. The use of a formula, particularly in the case of finding an area or volume, is not always the simplest method of solving a geometric problem. Consideration should first be given to the particular information provided in that problem so that full use can be made of special properties.

EXAMPLES 5j

1) Show that the points $P(3, 0, 1)$, $Q(2, 1, -2)$ lie on opposite sides of the plane Π whose equation is $\mathbf{r}.(2\mathbf{i}-\mathbf{j}+\mathbf{k}) = 3$. Find the coordinates of the point of intersection of the plane Π and the line PQ.

Rewriting the equation of Π in the form $\mathbf{r}.\hat{\mathbf{n}} = d$ gives

$$\mathbf{r}.\frac{1}{\sqrt{6}}(2\mathbf{i}-\mathbf{j}+\mathbf{k}) = \frac{3}{\sqrt{6}}.$$

The distance of P from Π is

$$\overrightarrow{OP}.\hat{\mathbf{n}} - d = (3\mathbf{i}+\mathbf{k}).\frac{1}{\sqrt{6}}(2\mathbf{i}-\mathbf{j}+\mathbf{k}) - \frac{3}{\sqrt{6}} = \frac{4}{\sqrt{6}},$$

i.e. P and O are on opposite sides of Π.
The distance of Q from Π is

$$\overrightarrow{OQ}.\hat{\mathbf{n}} - d = (2\mathbf{i}+\mathbf{j}-2\mathbf{k}).\frac{1}{\sqrt{6}}(2\mathbf{i}-\mathbf{j}+\mathbf{k}) - \frac{3}{\sqrt{6}} = -\frac{2}{\sqrt{6}},$$

i.e. Q and O are on the same side of Π.
Therefore P and Q are on opposite sides of Π.

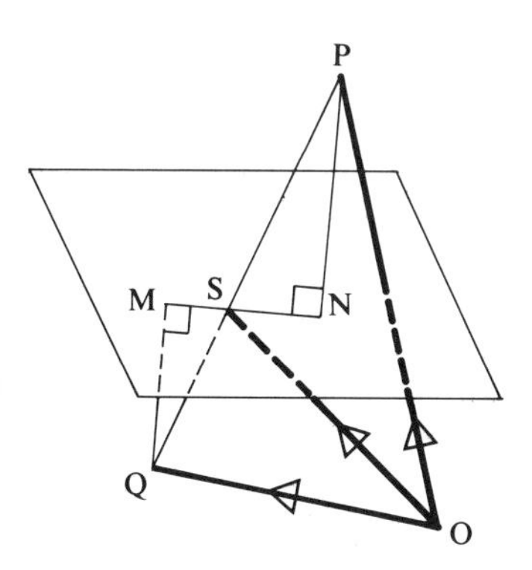

If S is the point of intersection of PQ and Π, $PS:SQ = PN:MQ = 2:1$.
Therefore the position vector of S is given by $\frac{1}{3}(\overrightarrow{OP} + 2\overrightarrow{OQ})$.
Therefore the coordinates of S are $(\frac{7}{3}, \frac{2}{3}, -1)$.
(Alternatively $\overrightarrow{OS}$ can be found by solving simultaneously the equation of Π and the equation of PQ, but this method is longer.)

2) A right circular cone has its vertex at the point $(2, 1, 3)$ and the centre of its plane face at the point $(1, -1, 2)$. A generator of the cone has equation $\mathbf{r} = (2\mathbf{i} + \mathbf{j} + 3\mathbf{k}) + \lambda(\mathbf{i} - \mathbf{j} - \mathbf{k})$. Find the radius of the base of the cone and hence its volume.

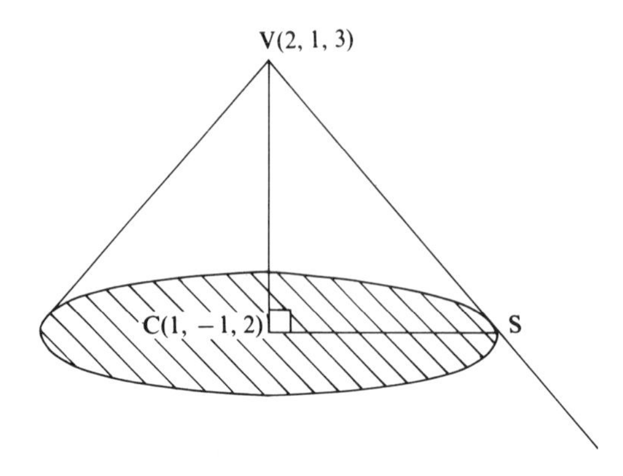

To find the radius of the base we need to find S, the point where the generator VS meets the plane Π containing the base.
The equation of the plane Π is found as follows.

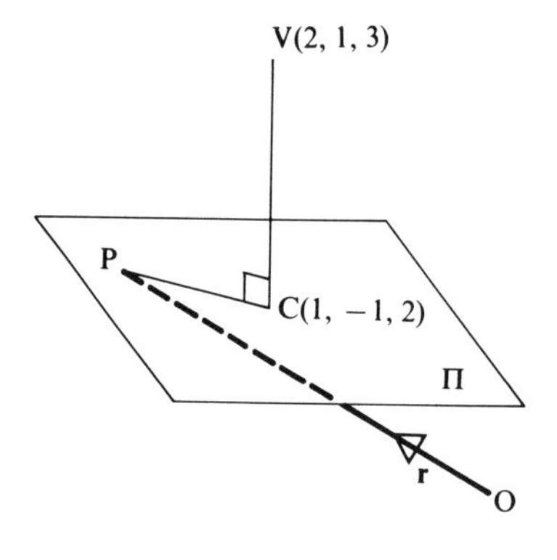

If $\mathbf{r}$ is the position vector of any point P in the plane Π then $\overrightarrow{PC} \cdot \overrightarrow{VC} = 0$,

i.e. $[\mathbf{r} - (\mathbf{i} - \mathbf{j} + 2\mathbf{k})] \cdot (\mathbf{i} + 2\mathbf{j} + \mathbf{k}) = 0$

$\Rightarrow \qquad \mathbf{r} \cdot (\mathbf{i} + 2\mathbf{j} + \mathbf{k}) = 1.$

Any point on the given generator has coordinates $[(2 + \lambda), (1 - \lambda), (3 - \lambda)]$.

The coordinates of S also satisfy the equation of Π. So at S
$[(2 + \lambda)\mathbf{i} + (1 - \lambda)\mathbf{j} + (3 - \lambda)\mathbf{k}] \cdot (\mathbf{i} + 2\mathbf{j} + \mathbf{k}) = 1 \quad \Rightarrow \quad \lambda = 3.$
Therefore the coordinates of S are $(5, -2, 0)$.
The radius of the base is CS, where
$CS = \sqrt{[(1 - 5)^2 + (-1 + 2)^2 + (2 - 0)^2]} = \sqrt{21}.$
The volume of the cone is $\frac{1}{3}\pi r^2 h = \frac{1}{3}\pi(21)\sqrt{(1^2 + 2^2 + 1^2)}$
$= 7\pi\sqrt{6}.$

3) A tetrahedron has one vertex at O and the other vertices at the points A(2, 0, 0), B(0, 3, 0), C(0, 0, 1). Find the volume of the tetrahedron. This tetrahedron is divided into two parts by a plane Π which contains the line OB and which is inclined at 60° to the face OAB. Find the vector equation of Π in scalar product form and the ratio in which it divides the volume of the tetrahedron OABC.

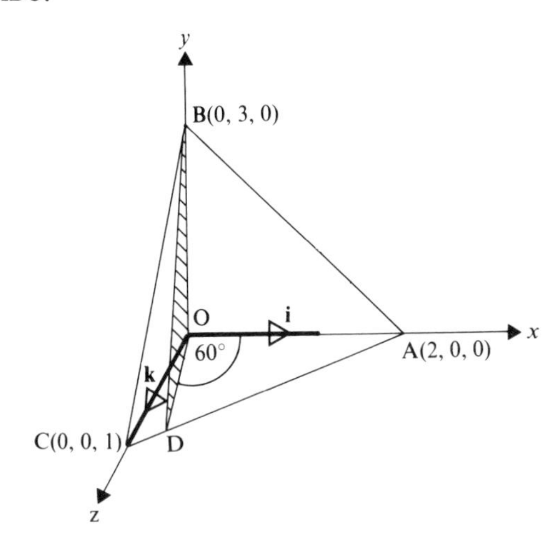

$$\text{The volume of OABC} = (\tfrac{1}{3}\text{ area }\Delta\text{OAC})(\text{OB})$$
$$= \tfrac{1}{3}(1)(3) = 1.$$

The plane Π contains
the line OB (which is parallel to $\mathbf{j}$)
the line OD (which is parallel to $\cos 60^\circ\mathbf{i} + \cos 30^\circ\mathbf{k}$, or $\mathbf{i} + \sqrt{3}\mathbf{k}$)
and the origin.

Therefore the equation of Π in parametric form is

$$\mathbf{r} = \lambda\mathbf{j} + \mu(\mathbf{i} + \sqrt{3}\mathbf{k}).$$

If (x, y, z) is any point on Π:

$$\left.\begin{aligned} x &= \mu \\ y &= \lambda \\ z &= \sqrt{3}\mu \end{aligned}\right\} \Rightarrow \sqrt{3}x - z = 0.$$

Therefore the equation of Π in scalar product form is

$$\mathbf{r}\cdot(\sqrt{3}\mathbf{i} - \mathbf{k}) = 0.$$

If D is the point of intersection of Π and AC, OBCD and OABD are tetrahedrons with a common base OBD; therefore their volumes are in the ratio of the distances of C and A from the plane containing OBD (i.e. Π).

The distance of C from Π is $|\mathbf{k}.\frac{1}{2}(\sqrt{3}\mathbf{i}-\mathbf{k})-0| = \frac{1}{2}$.
The distance of A from Π is $|2\mathbf{i}.\frac{1}{2}(\sqrt{3}\mathbf{i}-\mathbf{k})-0| = \sqrt{3}$.
Therefore Π divides the volume of OABC in the ratio

$$\sqrt{3}:\tfrac{1}{2} = 2\sqrt{3}:1.$$

4) Find the reflection P′ of the point P whose position vector is $2\mathbf{i}+\mathbf{j}-2\mathbf{k}$ in the plane $\mathbf{r}.(\mathbf{i}-\mathbf{j}+4\mathbf{k}) = 0$.

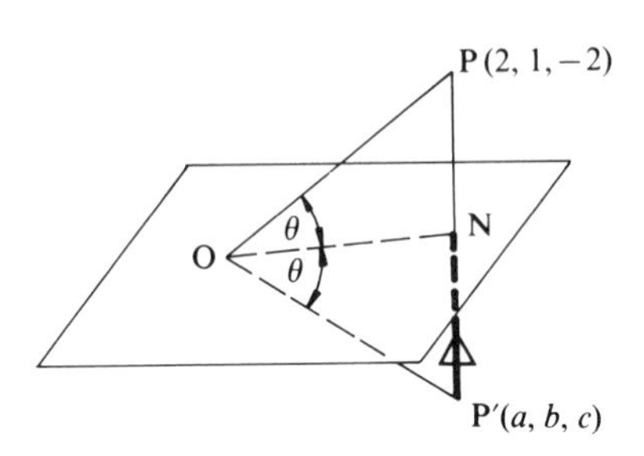

The given plane, Π, $\mathbf{r}.(\mathbf{i}-\mathbf{j}+4\mathbf{k}) = 0$, contains the origin.
So if P′ is the reflection of P(2, 1, − 2) in the plane, then PP′ is perpendicular to the plane and hence parallel to $\mathbf{i}-\mathbf{j}+4\mathbf{k}$.
So the equation of PP′ is

$$\begin{aligned}\mathbf{r} &= 2\mathbf{i}+\mathbf{j}-2\mathbf{k}+\lambda(\mathbf{i}-\mathbf{j}+4\mathbf{k})\\ &= (2+\lambda)\mathbf{i}+(1-\lambda)\mathbf{j}+(4\lambda-2)\mathbf{k}.\end{aligned}$$

PP′ cuts the plane at N, where

$$[(2+\lambda)\mathbf{i}+(1-\lambda)\mathbf{j}+(4\lambda-2)\mathbf{k}].[\mathbf{i}-\mathbf{j}+4\mathbf{k}] = 0$$

$$\Rightarrow \qquad \lambda = \tfrac{7}{18}$$

Hence N is the point $(\frac{43}{18}, \frac{11}{18}, -\frac{8}{18})$
As N is the midpoint of PP′, P′ is the point $(\frac{25}{9}, \frac{2}{9}, \frac{10}{9})$

5) Π is the plane whose equation is $\mathbf{r}.(\mathbf{i}+\mathbf{j}-3\mathbf{k}) = 5$.
Find the radius of the circle of intersection of the plane Π and the sphere whose centre is $(1, 1, 1)$ and whose radius is 2.

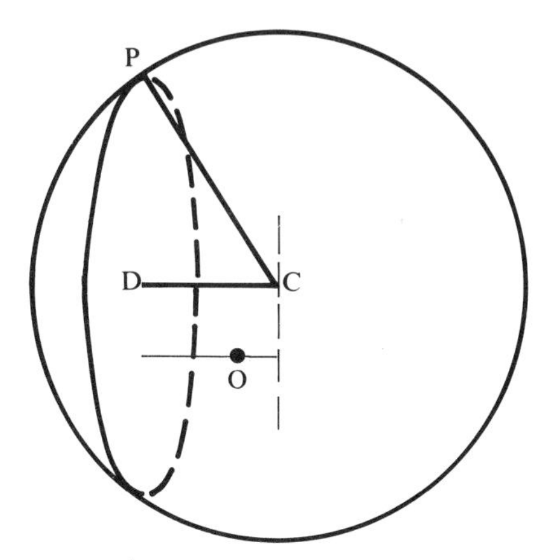

Let C be the centre of the sphere and D be the centre of the circle of intersection. The equation of the plane through C parallel to Π is

$$\mathbf{r}.(\mathbf{i}+\mathbf{j}-3\mathbf{k}) = (\mathbf{i}+\mathbf{j}+\mathbf{k}).(\mathbf{i}+\mathbf{j}-3\mathbf{k}) = -1$$

i.e.
$$\mathbf{r}.(-\mathbf{i}-\mathbf{j}+3\mathbf{k}) = 1$$

Hence Π and this plane are on opposite sides of O and the distance between Π and this plane is

$$\frac{5}{\sqrt{11}}+\frac{1}{\sqrt{11}} = \frac{6}{\sqrt{11}}$$

If P is any point on the circumference of the circle of intersection then

$$\text{PC} = 2 \quad \text{and} \quad \angle\text{PDC} = 90°.$$

From Pythagoras' Theorem

$$\begin{aligned}\text{PD}^2 &= \text{PC}^2 - \text{DC}^2\\ &= 4-\frac{36}{11} = \frac{8}{11}\end{aligned}$$

Hence the radius of the circle is given by $\text{PD} = 2\sqrt{\tfrac{2}{11}}$

6) Show that the vectors $\mathbf{a} = \mathbf{i}-\mathbf{j}+2\mathbf{k}$, $\mathbf{b} = \mathbf{i}+\mathbf{j}+\mathbf{k}$, $\mathbf{c} = 2\mathbf{i}-\mathbf{j}+\mathbf{k}$ form a set of base vectors for three dimensions. Express the vector $\mathbf{d} = 3\mathbf{i}-2\mathbf{j}+4\mathbf{k}$ in terms of **a**, **b** and **c**.

a, **b** and **c** form a set of base vectors for three dimensions provided
(a) they are not parallel (which from their direction ratios they clearly are not),
(b) they are not coplanar.

Now
$$\mathbf{a}\times\mathbf{b} = \begin{vmatrix}\mathbf{i} & \mathbf{j} & \mathbf{k}\\ 1 & -1 & 2\\ 1 & 1 & 1\end{vmatrix} = -3\mathbf{i}+\mathbf{j}+2\mathbf{k}$$

and $$\mathbf{a}\times\mathbf{c} = \begin{vmatrix} \mathbf{i} & \mathbf{j} & \mathbf{k} \\ 1 & -1 & 2 \\ 2 & -1 & 1 \end{vmatrix} = \mathbf{i}+3\mathbf{j}+\mathbf{k}.$$

So the plane containing $\mathbf{a}$ and $\mathbf{b}$ is not parallel to the plane containing $\mathbf{a}$ and $\mathbf{c}$, i.e. $\mathbf{a}$, $\mathbf{b}$ and $\mathbf{c}$ are not coplanar.
So $\mathbf{a}$, $\mathbf{b}$ and $\mathbf{c}$ *do* form a set of base vectors for three dimensions.
Now any other vector $\mathbf{d}$ can be expressed as $\lambda\mathbf{a}+\mu\mathbf{b}+\eta\mathbf{c}$.
Hence

$$\begin{aligned} 3\mathbf{i}-2\mathbf{j}+4\mathbf{k} &= \lambda(\mathbf{i}-\mathbf{j}+2\mathbf{k})+\mu(\mathbf{i}+\mathbf{j}+\mathbf{k})+\eta(2\mathbf{i}-\mathbf{j}+\mathbf{k}) \\ &= (\lambda+\mu+2\eta)\mathbf{i}+(-\lambda+\mu-\eta)\mathbf{j}+(2\lambda+\mu+\eta)\mathbf{k} \end{aligned}$$

$$\Rightarrow \quad \left.\begin{aligned} \lambda+\mu+2\eta &= 3 \\ -\lambda+\mu-\eta &= -2 \\ 2\lambda+\mu+\eta &= 4 \end{aligned}\right\} \Rightarrow \lambda = \tfrac{8}{5}, \quad \mu = \tfrac{1}{5}, \quad \eta = \tfrac{3}{5}$$

Hence $\mathbf{d} = \frac{8}{5}\mathbf{a}+\frac{1}{5}\mathbf{b}+\frac{3}{5}\mathbf{c}$.

EXERCISE 5j

1) A tetrahedron has one vertex at O and the other vertices at the points $A(1,3,2)$, $B(1,-1,0)$, $C(2,3,1)$. Find the distance of O from the face ABC.

2) Show that the points $P(3,2,-2)$, $Q(1,2,1)$ are on opposite sides of the plane $\mathbf{r}.(\mathbf{i}-\mathbf{j}-\mathbf{k}) = 2$. Find the position vector of the point of intersection of the line PQ with the plane.

3) A tetrahedron has vertices at the points $A(2,-1,0)$, $B(3,0,1)$, $C(1,-1,2)$, $D(-1,3,0)$. Find the cosine of the angle between the faces ABC and ABD.

4) OABC is one face of a cube, where A and C are the points $(1,4,-1)$, $(3,0,3)$ respectively. Find the coordinates of B. Find also the vector equation of the plane containing the other face of the cube of which AB is an edge.

5) A tetrahedron is bounded by the planes $\mathbf{r}.\mathbf{i} = 0$, $\mathbf{r}.\mathbf{j} = 0$, $\mathbf{r}.\mathbf{k} = 0$, $\mathbf{r}.(2\mathbf{i}-\mathbf{j}+\mathbf{k}) = 4$. Find the coordinates of the vertices and the volume of this tetrahedron.

6) A right circular cylinder has its plane faces contained in the planes $\mathbf{r}.(2\mathbf{i}-\mathbf{j}+2\mathbf{k}) = 10$, $\mathbf{r}.(-2\mathbf{i}+\mathbf{j}-2\mathbf{k}) = 6$. Find the height of the cylinder. The lines $\mathbf{r} = (2\mathbf{i}+\mathbf{j})+\lambda(2\mathbf{i}-\mathbf{j}+2\mathbf{k})$, $\mathbf{r} = (\mathbf{i}-\mathbf{j}+\mathbf{k})+\mu(2\mathbf{i}-\mathbf{j}+2\mathbf{k})$ are generators of the curved surface of the cylinder, passing through opposite ends of a diameter of its plane face. Find the radius of the cylinder and hence its volume.

7) Find the radius of the circle in which the plane $\mathbf{r}.(2\mathbf{i}+\mathbf{j}-2\mathbf{k})=9$ cuts the sphere of radius 5 and centre the origin. Find the volume of the cone of which this circle is the base and whose vertex is the origin.

8) Show that the point $C(2,0,1)$ lies in the plane Π whose equation is $\mathbf{r}.(\mathbf{i}-2\mathbf{j}+2\mathbf{k})=4$.
A right circular cone has its plane face lying in the plane Π and its centre is at C. Find the vector equation of the axis of the cone.
The line $\mathbf{r}=4\mathbf{i}-3\mathbf{j}+5\mathbf{k}+\lambda(\mathbf{i}+\mathbf{j}+2\mathbf{k})$ is a generator of the cone. Find the coordinates of the vertex of the cone and the point where this generator meets the plane Π. Hence find the volume of the cone.

9) A tetrahedron has three of its vertices at the points $A(3,2,0)$, $B(1,3,-1)$, $C(0,2,0)$. Find the unit vector perpendicular to the face ABC. The fourth vertex D is such that $\overrightarrow{DA}.\overrightarrow{AB}=\overrightarrow{DA}.\overrightarrow{AC}=0$. Find the vector equation of AD.
If the volume of the tetrahedron is $3\sqrt{2}$ cubic units and if D is on the same side of the face ABC as the origin, find the coordinates of D.

10) Show that the lines

$$\begin{aligned}\mathbf{r} &= (\mathbf{i}+\mathbf{j})+\lambda(3\mathbf{i}-\mathbf{j}+5\mathbf{k})\\ \mathbf{r} &= (3\mathbf{i}+2\mathbf{j}+5\mathbf{k})+\mu(\mathbf{i}-2\mathbf{j})\\ \mathbf{r} &= (2\mathbf{i}-\mathbf{j})+\eta(2\mathbf{i}+\mathbf{j}+5\mathbf{k})\end{aligned}$$

are coplanar and find in parametric form the vector equation of the plane containing them.

11) A right prism has a triangular cross-section. Two of its rectangular faces are contained in the planes

$$\begin{aligned}\mathbf{r}.(\mathbf{i}-2\mathbf{j}) &= 0,\\ \mathbf{r}.(3\mathbf{i}-\mathbf{j}+\mathbf{k}) &= 4.\end{aligned}$$

The two edges of the prism which are parallel to the intersection of these two planes pass through the origin and the point $(1,2,-1)$ respectively. Find the vector equations of these edges.
Find also the equation of the plane which contains the cross-section of this prism, one of whose vertices is the origin. Find the area of this cross-section.

THE USE OF VECTORS IN GENERAL GEOMETRIC PROBLEMS

Many of the theorems of Euclidean geometry can be proved easily and quickly using vector methods, as illustrated in the following examples.

EXAMPLES 5k

1) Prove that the diagonals of a parallelogram bisect each other.

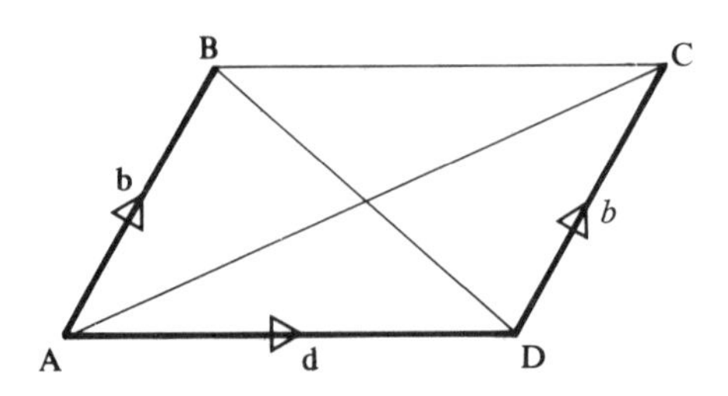

Taking one vertex A as origin and the position vectors of B and D as **b** and **d** respectively gives

$$\overrightarrow{AC} = \overrightarrow{AD} + \overrightarrow{DC} = \mathbf{d} + \mathbf{b}.$$

The position vector of the midpoint of $BD = \frac{1}{2}(\mathbf{b} + \mathbf{d}) = \frac{1}{2}\overrightarrow{AC}$.
Therefore the diagonals bisect each other.

2) Prove that in any triangle ABC,

$$\frac{\sin A}{a} = \frac{\sin B}{b} = \frac{\sin C}{c}$$

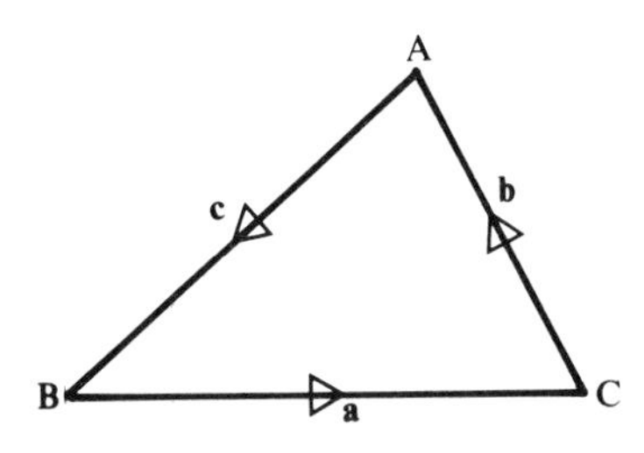

With the notation in the diagram

$$\begin{aligned} 2\triangle ABC &= |\mathbf{b} \times \mathbf{c}| = bc \sin(180° - A) = bc \sin A \\ &= |\mathbf{a} \times \mathbf{c}| = ac \sin(180° - B) = ac \sin B \\ &= |\mathbf{a} \times \mathbf{b}| = ab \sin(180° - C) = ab \sin C. \end{aligned}$$

Therefore $bc \sin A = ac \sin B = ab \sin C$

$$\frac{\sin A}{a} = \frac{\sin B}{b} = \frac{\sin C}{c}$$

3) Prove that the altitudes of any triangle are concurrent.

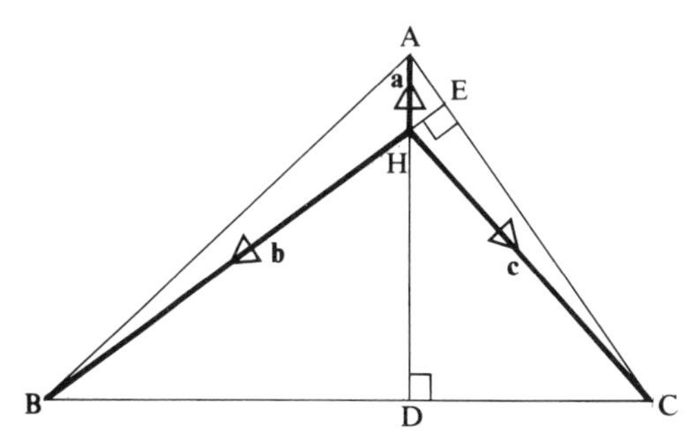

Let the altitudes AD and BE intersect at H.
Taking H as origin, let the position vectors of A, B, C be **a**, **b**, **c** respectively.

Now $\overrightarrow{HA} \cdot \overrightarrow{BC} = 0$ i.e. $\mathbf{a} \cdot (\mathbf{c} - \mathbf{b}) = 0$ [1]

and $\overrightarrow{HB} \cdot \overrightarrow{AC} = 0$ i.e. $\mathbf{b} \cdot (\mathbf{c} - \mathbf{a}) = 0$ [2]

$\Rightarrow$ $(\mathbf{a} \cdot \mathbf{c} - \mathbf{a} \cdot \mathbf{b}) - (\mathbf{b} \cdot \mathbf{c} - \mathbf{b} \cdot \mathbf{a}) = 0$ ([1]−[2])

$\Rightarrow$ $\mathbf{a} \cdot \mathbf{c} - \mathbf{b} \cdot \mathbf{c} = 0$

$\Rightarrow$ $\mathbf{c} \cdot (\mathbf{a} - \mathbf{b}) = 0.$

Therefore **c** is perpendicular to $\mathbf{a} - \mathbf{b}$, or $\overrightarrow{HC}$ is perpendicular to $\overrightarrow{BA}$ and so $\overrightarrow{HC}$ is the third altitude.
Therefore the three altitudes are concurrent.

4) Prove that in a skew quadrilateral the joins of the mid points of opposite sides bisect each other.

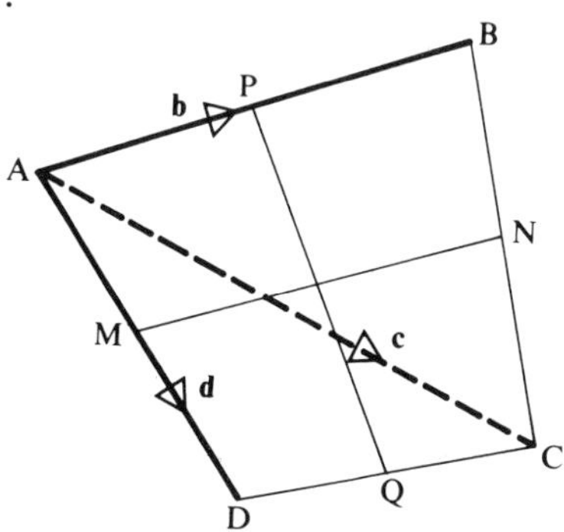

Taking A as origin, the position vectors of B, C and D as **b**, **c**, **d**, and P, N, Q, M as the midpoints of AB, BC, CD, DA respectively we have

$$\overrightarrow{BC} = \mathbf{c} - \mathbf{b}, \qquad \overrightarrow{DC} = \mathbf{c} - \mathbf{d}.$$

Therefore $\overrightarrow{AM} = \frac{1}{2}\mathbf{d}$ and $\overrightarrow{AN} = \frac{1}{2}(\mathbf{b} + \mathbf{c})$.

Therefore the position vector of the midpoint of MN is

$$\tfrac{1}{2}(\overrightarrow{AM} + \overrightarrow{AN}) = \tfrac{1}{4}\mathbf{d} + \tfrac{1}{4}\mathbf{b} + \tfrac{1}{4}\mathbf{c}.$$

Also $\overrightarrow{AP} = \tfrac{1}{2}\mathbf{b}$, $\overrightarrow{AQ} = \tfrac{1}{2}(\mathbf{d} + \mathbf{c})$.
Therefore the position vector of the midpoint of PQ is

$$\tfrac{1}{2}(\overrightarrow{AP} + \overrightarrow{AQ}) = \tfrac{1}{4}\mathbf{d} + \tfrac{1}{4}\mathbf{b} + \tfrac{1}{4}\mathbf{c}.$$

As the midpoints of PQ and MN have the same position vector, these lines bisect each other.

MULTIPLE CHOICE EXERCISE 5

(The instructions for answering these questions are on p. xii.)

TYPE I

1) The modulus of the vector $6\mathbf{i} - 2\mathbf{j} - 3\mathbf{k}$ is
(a) $\sqrt{23}$ (b) 7 (c) 1 (d) 49 (e) $\sqrt{11}$.

2) The direction cosines of the vector $\mathbf{i} + \mathbf{j} + \mathbf{k}$ are
(a) 1, 1, 1 (b) $\tfrac{1}{3}, \tfrac{1}{3}, \tfrac{1}{3}$ (c) $\sqrt{\tfrac{1}{3}}, \sqrt{\tfrac{1}{3}}, \sqrt{\tfrac{1}{3}}$ (d) $\sqrt{\tfrac{1}{2}}, \sqrt{\tfrac{1}{2}}, \sqrt{\tfrac{1}{2}}$
(e) $-\sqrt{\tfrac{1}{3}}, -\sqrt{\tfrac{1}{3}}, -\sqrt{\tfrac{1}{3}}$.

3) If $\mathbf{a} = \mathbf{i} - \mathbf{j} + \mathbf{k}$ and $\mathbf{b} = \mathbf{i} + \mathbf{j} - \mathbf{k}$ then $\mathbf{a} \cdot \mathbf{b}$ is
(a) $2\mathbf{i}$ (b) -1 (c) $-2\mathbf{j} + 2\mathbf{k}$ (d) 2 (e) 3.

4) If $\mathbf{a} = \mathbf{i} + \mathbf{j}$ and $\mathbf{b} = 2\mathbf{i} - \mathbf{j}$ then $\mathbf{a} \times \mathbf{b}$ is
(a) $2\mathbf{i} - \mathbf{j}$ (b) $\mathbf{0}$ (c) $\mathbf{i}$ (d) $\sqrt{10}$ (e) $-3\mathbf{k}$.

5) $\mathbf{a}$, $\mathbf{b}$ and $\mathbf{c}$ are the position vectors of the vertices of a triangle. The area of the triangle is
(a) $\mathbf{a} \times \mathbf{b}$ (b) $\tfrac{1}{2}|\mathbf{b} \times \mathbf{c}|$ (c) $\tfrac{1}{2}|(\mathbf{a} - \mathbf{b}) \times (\mathbf{b} - \mathbf{c})|$
(d) $(\mathbf{b} - \mathbf{a}) \times (\mathbf{a} - \mathbf{c})$ (e) none of these.

6) The line whose equation is $\mathbf{r} = \lambda(2\mathbf{i} - \mathbf{j} + \mathbf{k})$ has direction ratios
(a) 0, 0, 0 (b) $-2:1:-1$ (c) $2:1:1$ (d) $\tfrac{1}{3}:-\tfrac{1}{6}:\tfrac{1}{6}$.

7) The plane whose equation is $\mathbf{r} \cdot (\mathbf{i} - \mathbf{j} + \mathbf{k}) = 2$ contains the point
(a) $(1, -1, 1)$ (b) $(-1, 1, 0)$ (c) $(0, 1, 1)$ (d) $(2, 0, 0)$
(e) $(0, 0, 0)$.

8) The angle between the lines whose equations are $\mathbf{r}_1 = \mathbf{a}_1 + \lambda\mathbf{b}_1$ and $\mathbf{r}_2 = \mathbf{a}_2 + \mu\mathbf{b}_2$ is

(a) $\arccos \dfrac{\mathbf{b}_1 \cdot \mathbf{b}_2}{b_1 b_2}$ (b) $\mathbf{b}_1 \cdot \mathbf{b}_2$ (c) $\arccos \dfrac{\mathbf{a}_1 \cdot \mathbf{a}_2}{a_1 a_2}$

(d) $\arcsin \mathbf{b}_1 \times \mathbf{b}_2$ (e) $\arccos \lambda \cdot \mu$.

9) The equation of the plane normal to $4\mathbf{i}+3\mathbf{j}$ and 5 units from 0 is
(a) $\mathbf{r}.(4\mathbf{i}+3\mathbf{j})=5$ (b) $\mathbf{r}.5\mathbf{k}=5$ (c) $\mathbf{r}.(4\mathbf{i}+3\mathbf{j})=1$
(d) $\mathbf{r}.(4\mathbf{i}+3\mathbf{j})=25$ (e) $\mathbf{r}.5\mathbf{k}=0$.

10) The points A, B and C are collinear and $\overrightarrow{OA}=\mathbf{i}+\mathbf{j}$, $\overrightarrow{OB}=2\mathbf{i}-\mathbf{j}+\mathbf{k}$, $\overrightarrow{OC}=3\mathbf{i}+a\mathbf{j}+b\mathbf{k}$
(a) $a=-3,\quad b=2$ (b) $a=3,\quad b=-2$ (c) $a=0,\quad b=1$
(d) $a=-1,\quad b=0$ (e) $a=6,\quad b=-1$.

TYPE II

11) A line has equation $\mathbf{r}=\mathbf{i}+2\mathbf{j}+3\mathbf{k}+\lambda(4\mathbf{i}-\mathbf{j}+7\mathbf{k})$.
(a) The line passes through $(4,-1,7)$.
(b) The length of the line is $\sqrt{14}$.
(c) The line passes through $(1,2,3)$.
(d) The line intersects the line $\mathbf{r}=\lambda\mathbf{i}$.

12) ABCD is a parallelogram and O is the origin.
(a) The area of ABCD is $\frac{1}{2}|\overrightarrow{AB}\times\overrightarrow{AD}|$.
(b) The equation of the line AC is $\mathbf{r}=\lambda\overrightarrow{AC}$.
(c) The equation of the line AB is $\mathbf{r}=\overrightarrow{OA}+\lambda\overrightarrow{AB}$.
(d) $\overrightarrow{AB}+\overrightarrow{BC}+\overrightarrow{CD}+\overrightarrow{DA}=0$.

13) $\mathbf{a}$ and $\mathbf{b}$ are perpendicular vectors.
(a) $\mathbf{a}\times\mathbf{b}=\mathbf{0}$.
(b) $\mathbf{a}.\mathbf{b}=\mathbf{0}$.
(c) $\mathbf{a}\times\mathbf{b}$, $\mathbf{a}$ and $\mathbf{b}$ are mutually perpendicular.
(d) $|\mathbf{a}\times\mathbf{b}|=ab$.

14) $\mathbf{V}=3\mathbf{i}+3\mathbf{j}+3\mathbf{k}$.
(a) $\hat{\mathbf{V}}=\mathbf{i}+\mathbf{j}+\mathbf{k}$.
(b) $\mathbf{V}$ makes equal angles with $\mathbf{i}, \mathbf{j}$ and $\mathbf{k}$.
(c) $\mathbf{V}.(\mathbf{i}+\mathbf{j}+\mathbf{k})=0$.
(d) $\mathbf{V}\times(\mathbf{i}+\mathbf{j}+\mathbf{k})=\mathbf{0}$.

15) A, B and C have position vectors $\mathbf{a}, \mathbf{b}$ and $\mathbf{c}$ respectively and $\mathbf{a}=\lambda\mathbf{b}+\mu\mathbf{c}$.
(a) A, B, C are collinear.
(b) A, B, C are coplanar.
(c) A divides BC in the ratio $\mu:\lambda$.
(d) $\mathbf{a}$ is parallel to $\mathbf{b}+\mathbf{c}$.

TYPE III

16) (a) $\mathbf{a}=\lambda\mathbf{b}$.
(b) $\mathbf{a}\times\mathbf{b}=\mathbf{0}$.

17) (a) $\mathbf{a} \times \mathbf{b} = \mathbf{b} \times \mathbf{a}$.
(b) $\mathbf{a} \cdot \mathbf{b} = \mathbf{b} \cdot \mathbf{a}$.

18) A, B and C are collinear points with position vectors $\mathbf{a}, \mathbf{b}$ and $\mathbf{c}$ respectively.
(a) $2\mathbf{c} = \mathbf{a} + \mathbf{b}$
(b) C is the midpoint of AB.

19) A, B and C are points with position vectors $\mathbf{a}, \mathbf{b}$ and $\mathbf{c}$ respectively.
(a) The position vector of any point P is given by $\mathbf{r} = \lambda\mathbf{a} + \mu\mathbf{b} + \eta\mathbf{c}$.
(b) A, B and C are collinear.

20) (a) $\mathbf{a} \times (\mathbf{b} + \mathbf{c}) = \mathbf{0}$.
(b) $\mathbf{a} + \mathbf{b} + \mathbf{c} = \mathbf{0}$.

TYPE IV

21) Find the angle between the lines L_1 and L_2.
(a) The equation of L_1 is $\mathbf{r} = \mathbf{a} + \lambda\mathbf{b}$.
(b) L_1 and L_2 intersect.
(c) L_2 is parallel to a vector $\mathbf{c}$.
(d) L_2 passes through the point with position vector $\mathbf{d}$.

22) ABC is a triangle. Find the direction cosines of a normal to the plane ABC.
(a) $\overrightarrow{AC}$ is parallel to $2\mathbf{i} + \mathbf{j}$. (b) B is the point $(1, 0, 1)$.
(c) $\overrightarrow{OC}$ is parallel to $\mathbf{j} + 2\mathbf{k}$. (d) A is the point $(3, 2, 1)$.

23) Determine whether two lines, L_1 and L_2, intersect.
(a) L_1 is parallel to $\mathbf{a}$. (b) L_2 is parallel to $\mathbf{b}$.
(c) L_1 passes through O. (d) L_2 is perpendicular to L_1.

24) Does the line $\dfrac{x - x_1}{a} = \dfrac{y - y_1}{b} = \dfrac{z - z_1}{c}$ lie in the plane $Ax + By + Cz + D = 0$?
(a) $Ax_1 + By_1 + Cz_1 + D = 0$.
(b) $aA + bB + cC = 0$.
(c) The line and the plane are both 5 units from O.
(d) $a:b:c = 1:2:3$.

25) Prove that the points A, B, C and D are coplanar.
(a) $\overrightarrow{OA} = \mathbf{i} + \mathbf{j} - 2\mathbf{k}$. (b) $\overrightarrow{AB} \times \overrightarrow{BC} = 4(\overrightarrow{AB} \times \overrightarrow{AD})$.
(c) $\overrightarrow{OD} = \mathbf{i} - \mathbf{j} + 5\mathbf{k}$. (d) $\overrightarrow{OA} = 2\overrightarrow{OB} - 3\overrightarrow{OC}$.

TYPE V

26) If two lines do not intersect they are parallel.

27) ABCD is a parallelogram. The unit vector normal to the plane ABCD is $\overrightarrow{AB} \times \overrightarrow{BC}$.

28) If three non-zero vectors are such that $\mathbf{a}.\mathbf{b} = \mathbf{a}.\mathbf{c}$ then either $\mathbf{b} = \mathbf{c}$ or $|\mathbf{a} \times (\mathbf{b} - \mathbf{c})| = ab - ac$.

29) $\mathbf{a}.\mathbf{b} = 0 \iff \mathbf{a} = \mathbf{0}$ or $\mathbf{b} = \mathbf{0}$.

30) If $a:b:c$ are the direction ratios of a vector then $a^2 + b^2 + c^2 = 1$.

MISCELLANEOUS EXERCISE 5

In Questions 1–8 give proofs based on vector methods.

1) Prove that the line joining the midpoints of two sides of a triangle is parallel to the third side and equal to half of it.

2) Prove that the internal bisectors of the angles of a triangle are concurrent.

3) Prove that the joins of the midpoints of the opposite edges of a tetrahedron bisect each other.

4) Prove that the perpendicular bisectors of the sides of a triangle are concurrent.

5) Prove that the diagonals of a rhombus intersect at right angles.

6) Prove that the lines joining the midpoints of adjacent sides of a skew quadrilateral form a parallelogram.

7) ABCD is a parallelogram and M is the midpoint of AB. Prove that DM and AC cut each other at points of trisection.

8) ABCD is a trapezium and AB and DC are parallel. M is the midpoint of AD. Prove that the area of triangle BMC is half the area of the trapezium.

9) If the position vectors of points P and Q with respect to O as origin are $\mathbf{p}$ and $\mathbf{q}$ respectively, show that the area of the triangle OPQ is $\frac{1}{2}|\mathbf{p} \times \mathbf{q}|$. The position vectors of the vertices A, B, C of a tetrahedron OABC with respect to O as origin are

$$\overrightarrow{OA} = 2\mathbf{i} - \mathbf{j}, \quad \overrightarrow{OB} = \mathbf{j} + \mathbf{k}, \quad \overrightarrow{OC} = \mathbf{i} + 3\mathbf{j} - \mathbf{k}.$$

Find the angle between (a) the edges AB, AC, (b) the faces OAB, OAC. Prove that BC is perpendicular to the plane OAB, and hence prove that the volume of OABC is 3/2. (U of L)

10) The position vectors of the points A, B, C are respectively

$$\mathbf{a} = \mathbf{i} + 2\mathbf{j} + \mathbf{k}$$
$$\mathbf{b} = 2\mathbf{i} + 4\mathbf{j} + 3\mathbf{k}$$
$$\mathbf{c} = 6\mathbf{i} + 6\mathbf{j} + 6\mathbf{k}.$$

Find (a) $\overrightarrow{AB}.\overrightarrow{AC}$,
(b) $\overrightarrow{AB} \times \overrightarrow{AC}$,
(c) the angle BAC to the nearest degree,
(d) the area of the triangle ABC,
(e) either the vector or the Cartesian equation of the plane ABC.
(U of L)

11) Of the vectors

$$\mathbf{a} = \begin{pmatrix} 5 \\ 6 \\ -11 \end{pmatrix} \quad \mathbf{b} = \begin{pmatrix} 2 \\ -2 \\ 5 \end{pmatrix} \quad \mathbf{c} = \begin{pmatrix} 9 \\ 2 \\ -1 \end{pmatrix} \quad \mathbf{d} = \begin{pmatrix} -5 \\ -14 \\ 24 \end{pmatrix}$$

show that **a, b, d** form a set of basis vectors. Express **c** in terms of this basis. If **a, b, c** and **d** are the position vectors of points A, B, C and D respectively, show that the point P(1, −2, 3) lies on AD, find the ratio AP : AD, and show that BP is perpendicular to PC. (U of L)

12) Find a vector equation for the plane Π passing through the points A, B, C with position vectors $(\mathbf{i}-\mathbf{j}+2\mathbf{k})$, $(2\mathbf{i}+\mathbf{j}+\mathbf{k})$, $(3\mathbf{i}-2\mathbf{j}+2\mathbf{k})$ respectively. Find the area of the triangle ABC and the distance from the plane Π of the point D with position vector $(3\mathbf{i}+\mathbf{j}+\mathbf{k})$. (U of L)

13) Explain how the vector equations $\mathbf{r} = \mathbf{a} + t\mathbf{b}$ and $\mathbf{r}.\mathbf{n} = p$ represent a straight line and a plane respectively, interpreting the symbols geometrically and using sketches if you wish. Prove that, provided $\mathbf{b}.\mathbf{n} \neq 0$, the line and plane intersect in a point whose position vector is $\mathbf{a} + (p - \mathbf{a}.\mathbf{n})\mathbf{b}/(\mathbf{b}.\mathbf{n})$.
(JMB)

14) State a relation which exists between the vectors **p** and **q** when these vectors are (a) parallel, (b) perpendicular.
The position vectors of the vertices of a tetrahedron ABCD are

A: $-5\mathbf{i} + 22\mathbf{j} + 5\mathbf{k}$, B: $\mathbf{i} + 2\mathbf{j} + 3\mathbf{k}$,
C: $4\mathbf{i} + 3\mathbf{j} + 2\mathbf{k}$, D: $-\mathbf{i} + 2\mathbf{j} - 3\mathbf{k}$.

Find the angle CBD and show that AB is perpendicular to both BC and BD. Calculate the volume of the tetrahedron.
If ABDE is a parallelogram, find the position vector of E. (U of L)

15) (a) Show that the plane which is at a distance d from the origin O and whose normal is in the direction of the unit vector **n**, which points away from O, has equation $\mathbf{r}.\mathbf{n} = d$.

(b) Find the perpendicular distance between the planes $2x + 2y + z - 6 = 0$, $2x + 2y + z - 10 = 0$. Find also the area of the triangle whose vertices are $(2, 2, 2), (1, 1, 2)$ and $(1, -1, 6)$.
(U of L)

16) (a) In a triangle ABC the altitudes through A and B meet in a point O. Let $\mathbf{a}, \mathbf{b}, \mathbf{c}$ be the position vectors of A, B, C relative to O as origin. Show that $\mathbf{a}.\mathbf{b} = \mathbf{a}.\mathbf{c}$ and $\mathbf{b}.\mathbf{a} = \mathbf{b}.\mathbf{c}$.
Deduce that the altitudes of a triangle are concurrent.

(b) Find the point of intersection of the line through the points $(2, 0, 1)$ and $(-1, 3, 4)$ and the line through the points $(-1, 3, 0)$ and $(4, -2, 5)$. Calculate the acute angle between the two lines. (U of L)

17) Define the scalar product $\mathbf{a}.\mathbf{b}$ and the vector product $\mathbf{a} \wedge \mathbf{b}$ of two vectors $\mathbf{a}$ and $\mathbf{b}$.
The points P, Q, R have coordinates $(1, 1, 1), (1, 3, 2), (2, 1, 3)$ respectively, referred to rectangular axes $Oxyz$.
Calculate the products $\overrightarrow{PQ}.\overrightarrow{PR}$, $\overrightarrow{PQ} \wedge \overrightarrow{PR}$ and deduce the values of the cosine of the angle QPR and the area of the triangle PQR. (JMB)p

18) Define the vector product $\mathbf{a} \wedge \mathbf{b}$ of two vectors $\mathbf{a}$ and $\mathbf{b}$. Three non-collinear points P, Q and R have position vectors $\mathbf{p}$, $\mathbf{q}$ and $\mathbf{r}$, respectively, relative to an origin O, not necessarily in the plane PQR. Prove that the area of the triangle PQR is equal to the magnitude of the vector

$$\tfrac{1}{2}(\mathbf{q} \wedge \mathbf{r} + \mathbf{r} \wedge \mathbf{p} + \mathbf{p} \wedge \mathbf{q}).$$

When O does lie in the plane PQR interpret this result in terms of the areas of the triangles OPQ, OQR and ORP, considering the cases (a) O inside the triangle PQR, (b) O in the region bounded by PR produced and QR produced. (JMB)p

19) (a) The line AB is the common perpendicular to two skew lines AP and BQ, and C and R are the midpoints of AB and PQ respectively. Prove by vector methods, that CR and AB are perpendicular.

(b) Three vectors $\mathbf{a}, \mathbf{b}$ and $\mathbf{c}$ are such that $\mathbf{a} \neq \mathbf{0}$ and

$$\mathbf{a} \wedge \mathbf{b} = 2\mathbf{a} \wedge \mathbf{c}.$$

Show that

$$\mathbf{b} - 2\mathbf{c} = \lambda\mathbf{a},$$

where λ is a scalar.
Given that

$$|\mathbf{a}| = |\mathbf{c}| = 1, \quad |\mathbf{b}| = 4$$

and the angle between $\mathbf{b}$ and $\mathbf{c}$ is $\arccos\frac{1}{4}$, show that

$$\lambda = +4 \quad \text{or} \quad -4.$$

For each of these cases find the cosine of the angle between $\mathbf{a}$ and $\mathbf{c}$.
(JMB)

20) Find the shortest distance between the straight lines with vector equations

$$\mathbf{r} = -3\mathbf{i} + 5s\mathbf{j} + \mathbf{k}, \quad \mathbf{r} = 3\mathbf{i} - 2\mathbf{j} + \mathbf{k} + t(-\mathbf{i} + 2\mathbf{j} - \mathbf{k}).$$ (U of L)

21) Show that the equation of a plane can be expressed in the form $\mathbf{r}.\mathbf{n} = p$. Find the equation of the plane through the origin parallel to the lines

$$\mathbf{r} = 3\mathbf{i} + 3\mathbf{j} - \mathbf{k} + s(\mathbf{i} - \mathbf{j} - 2\mathbf{k}) \quad \text{and} \quad \mathbf{r} = 4\mathbf{i} - 5\mathbf{j} - 8\mathbf{k} + t(3\mathbf{i} + 7\mathbf{j} - 6\mathbf{k}).$$

Show that one of the lines lies in the plane, and find the distance of the other line from the plane. (U of L)

22) Points A, B, C have position vectors $\mathbf{a}, \mathbf{b}, \mathbf{c}$ and λ, μ, ν are variable parameters subject to the condition $\lambda + \mu + \nu = 1$. If the points are not collinear prove that the plane ABC is represented by the equation $\mathbf{r} = \lambda\mathbf{a} + \mu\mathbf{b} + \nu\mathbf{c}$.

Prove that the equation of the line of intersection of the two planes:

$$\mathbf{r} = \lambda_1\mathbf{i} + 2\mu_1\mathbf{j} + 3\nu_1\mathbf{k}, \quad \lambda_1 + \mu_1 + \nu_1 = 1$$

and

$$\mathbf{r} = 2\lambda_2\mathbf{i} + \mu_2\mathbf{j} + 2\nu_2\mathbf{k}, \quad \lambda_2 + \mu_2 + \nu_2 = 1$$

can be written in terms of a single parameter t as

$$6\mathbf{r} = (3 + t)\mathbf{i} + 4t\mathbf{j} + 9(1 - t)\mathbf{k}.$$ (U of L)

23) Prove that the lines with vector equations

$$\mathbf{r} = (-\mathbf{i} + 2\mathbf{j}) + a(-\mathbf{i} + 2\mathbf{j} - \mathbf{k}),$$
$$\mathbf{r} = \mathbf{k} + b(\mathbf{i} + \mathbf{j} - 2\mathbf{k}),$$
$$\mathbf{r} = \mathbf{i} + \mathbf{j} - \mathbf{k} + c(\mathbf{i} + \mathbf{k}),$$
$$\mathbf{r} = 2\mathbf{i} + \mathbf{j} + d(3\mathbf{i} - \mathbf{j}),$$

in the given order, form a (skew) quadrilateral.

Prove that the ratio of the shortest distances between the two pairs of opposite sides of this quadrilateral is $\sqrt{7}:1$. (U of L)

24) Show that $\mathbf{r} = 2\mathbf{i} + \mathbf{k}$ is a solution of the equation

$$\mathbf{r} \times (\mathbf{i} + \mathbf{j}) = -\mathbf{i} + \mathbf{j} + 2\mathbf{k}.$$

Find the general solution of the equation and give a geometrical interpretation. (U of L)p

25) The position vectors of the points A, B, C with respect to the origin O are $\mathbf{a}, \mathbf{b}, \mathbf{c}$ respectively. If OA is perpendicular to BC, and OB is perpendicular to CA, show that OC is perpendicular to AB, and that

$$OA^2 + BC^2 = OB^2 + CA^2 = OC^2 + AB^2.$$

Show that the plane through BC perpendicular to OA meets the plane through AB perpendicular to OC in a line that lies in the plane through OB perpendicular to CA. If this line passes through the centroid of the triangle AOC, show that the angle AOC is $\pi/3$ radians. (U of L)

26) Given that $\mathbf{i}, \mathbf{j}$ are perpendicular unit vectors and that $\mathbf{r}_1 = x_1\mathbf{i} + y_1\mathbf{j}$, $\mathbf{r}_2 = x_2\mathbf{i} + y_2\mathbf{j}$ are any two vectors, the scalar quantity $\mathbf{r}_1 \circ \mathbf{r}_2$ is defined for these two vectors by the equation

$$\mathbf{r}_1 \circ \mathbf{r}_2 = x_1y_2 - x_2y_1.$$

Deduce that $\mathbf{r}_2 \circ \mathbf{r}_1 = -\mathbf{r}_1 \circ \mathbf{r}_2$.
If $\mathbf{r}_3 = x_3\mathbf{i} + y_3\mathbf{j}$ is a further vector, prove that

$$(\mathbf{r}_1 + \mathbf{r}_2) \circ \mathbf{r}_3 = (\mathbf{r}_1 \circ \mathbf{r}_3) + (\mathbf{r}_2 \circ \mathbf{r}_3).$$ (O)p

27) In a triangle ABC the perpendicular from B to the side AC meets the perpendicular from C to the side AB at H. The position vectors of A, B and C relative to H are **a**, **b** and **c** respectively.
Express $\overrightarrow{CA}$ in terms of **a** and **c** and deduce that $\mathbf{a} \cdot \mathbf{b} = \mathbf{b} \cdot \mathbf{c}$.
Prove that AH is perpendicular to BC. (JMB)p

28) In a parallelogram ABCD X is the midpoint of AB and the line DX cuts the diagonal AC at P. Writing $\overrightarrow{AB} = \mathbf{a}$, $\overrightarrow{AD} = \mathbf{b}$, $\overrightarrow{AP} = \lambda\overrightarrow{AC}$ and $\overrightarrow{DP} = \mu\overrightarrow{DX}$, express AP
(a) in terms of λ, **a** and **b**,
(b) in terms of μ, **a** and **b**.
Deduce that P is a point of trisection of both AC and DX. (JMB)p

CHAPTER 6

TRANSFORMATIONS AND LINEAR EQUATIONS

TRANSFORMATIONS OF THREE DIMENSIONAL SPACE

This section extends to transformations of three dimensional space, the work already covered on transformations of the two dimensional xy plane. As before, we are going to consider only those transformations which leave the origin unchanged and which can be expressed by linear equations, i.e. transformations which map the point $P(x, y, z)$ to the point $P'(X, Y, Z)$ where the relationship between the coordinates of P and P′ can be expressed in the form

$$a_1x + b_1y + c_1z = X \qquad [1]$$

$$a_2x + b_2y + c_2z = Y \qquad [2]$$

$$a_3x + b_3y + c_3z = Z \qquad [3]$$

Equation [1] can be written as a product of a row vector and column vector in the form

$$\begin{pmatrix} a_1 & b_1 & c_1 \end{pmatrix} \begin{pmatrix} x \\ y \\ z \end{pmatrix} = X. \qquad [4]$$

Examples of some linear transformations are

$$\begin{pmatrix} a_2 & b_2 & c_2 \end{pmatrix} \begin{pmatrix} x \\ y \\ z \end{pmatrix} = Y \qquad [5]$$

and equation [3] as

$$\begin{pmatrix} a_3 & b_3 & c_3 \end{pmatrix} \begin{pmatrix} x \\ y \\ z \end{pmatrix} = Z \qquad [6]$$

Adjoining the row vectors in equations [4], [5] and [6] gives the single matrix equation

$$\begin{pmatrix} a_1 & b_1 & c_1 \\ a_2 & b_2 & c_2 \\ a_3 & b_3 & c_3 \end{pmatrix} \begin{pmatrix} x \\ y \\ z \end{pmatrix} = \begin{pmatrix} X \\ Y \\ Z \end{pmatrix}$$

Thus any linear transformation of the xyz space may be expressed in the form

$$\mathbf{M} \begin{pmatrix} x \\ y \\ z \end{pmatrix} = \begin{pmatrix} X \\ Y \\ Z \end{pmatrix}$$

where **M** is a 3×3 matrix.

So **M** may be regarded as an operator which maps the point P whose position vector is $\begin{pmatrix} x \\ y \\ z \end{pmatrix}$ to the point P′ whose position vector is $\begin{pmatrix} X \\ Y \\ Z \end{pmatrix}$.

Note that a linear transformation may be visualised by considering a foam rubber cuboid which is fixed at one corner (the origin). If the rubber block is then uniformly stretched or reduced in any direction, rotated about any line (which must obviously pass through its fixed corner), reflected in any plane (which again must obviously pass through its fixed corner) or is subjected to any combination of these distortions, the block is undergoing a linear transformation in which one point (its fixed corner) does not change position. If however the block is bent or *non*-uniformly stretched so that, for example, some planes contained in the block are transformed to curved surfaces, such transformations are *not* linear.

Similarly equation [2] can be expressed as

(a) rotation about any line through the origin,

(b) reflection in any plane through the origin,

(c) enlargement or reduction (the origin is the centre of the enlargement).

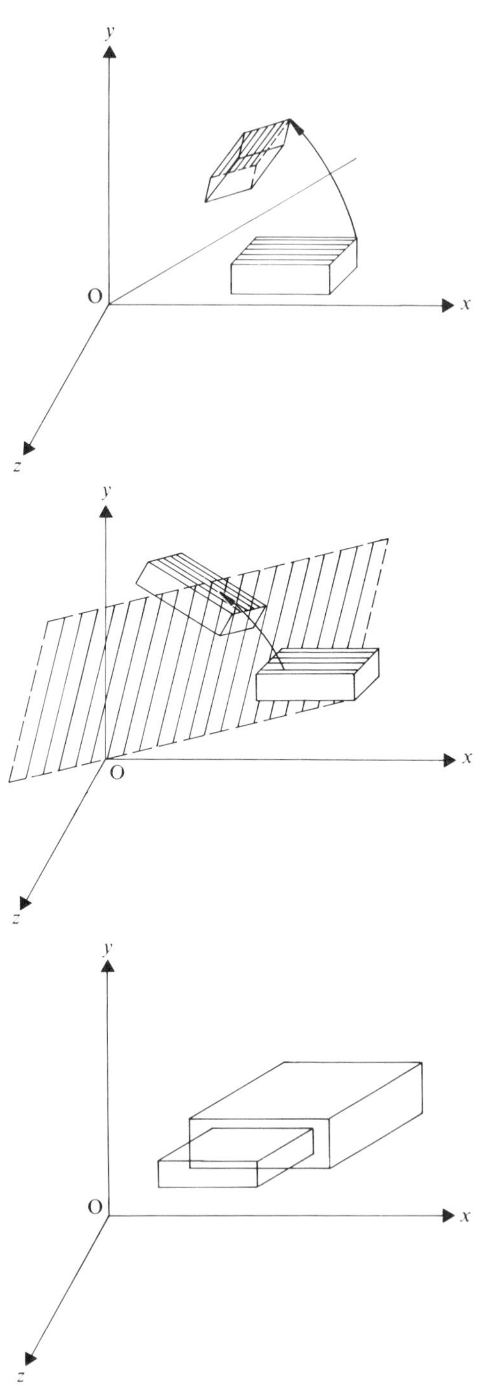

Finding the Matrix Operator to Perform a Given Transformation

Consider the transformation which maps **i** to **a**, **j** to **b**, **k** to **c** where $\mathbf{a} = \begin{pmatrix} a_1 \\ a_2 \\ a_3 \end{pmatrix}$, $\mathbf{b} = \begin{pmatrix} b_1 \\ b_2 \\ b_3 \end{pmatrix}$ and $\mathbf{c} = \begin{pmatrix} c_1 \\ c_2 \\ c_3 \end{pmatrix}$ respectively.

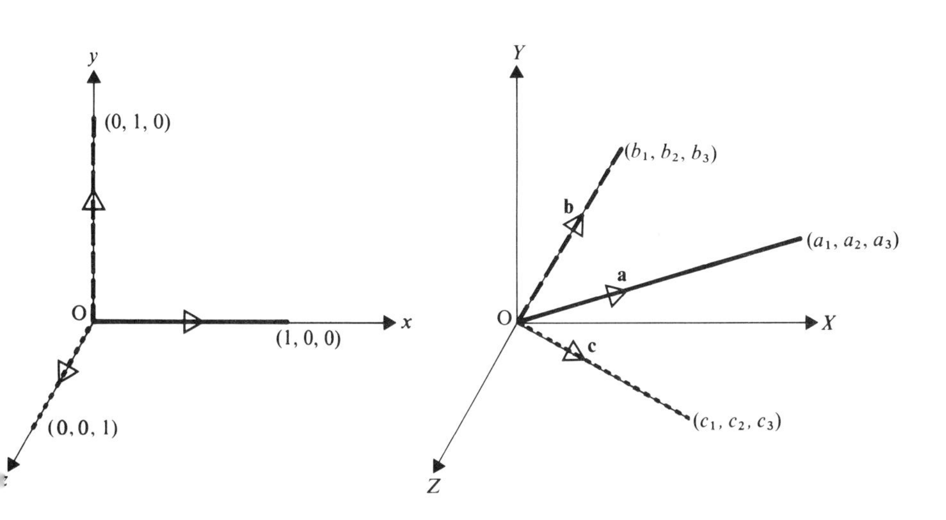

As **i** is mapped to **a**, $x\mathbf{i}$ is mapped to $x\mathbf{a}$,

i.e. $$\begin{pmatrix} x \\ 0 \\ 0 \end{pmatrix} \rightarrow x \begin{pmatrix} a_1 \\ a_2 \\ a_3 \end{pmatrix} = \begin{pmatrix} xa_1 \\ xa_2 \\ xa_3 \end{pmatrix}$$

Similarly $y\mathbf{j}$ maps to $y\mathbf{b}$,

i.e. $$\begin{pmatrix} 0 \\ y \\ 0 \end{pmatrix} \rightarrow y \begin{pmatrix} b_1 \\ b_2 \\ b_3 \end{pmatrix} = \begin{pmatrix} yb_1 \\ yb_2 \\ yb_3 \end{pmatrix}$$

and $z\mathbf{k}$ maps to $z\mathbf{c}$,

i.e. $$\begin{pmatrix} 0 \\ 0 \\ z \end{pmatrix} \rightarrow z \begin{pmatrix} c_1 \\ c_2 \\ c_3 \end{pmatrix} = \begin{pmatrix} zc_1 \\ zc_2 \\ zc_3 \end{pmatrix}$$

Therefore any point whose position vector is $x\mathbf{i} + y\mathbf{j} + z\mathbf{k}$ is mapped to the point whose position vector is $x\mathbf{a} + y\mathbf{b} + z\mathbf{c}$.

Thus $\begin{pmatrix} x \\ y \\ z \end{pmatrix} \to \begin{pmatrix} xa_1 \\ xa_2 \\ xa_3 \end{pmatrix} + \begin{pmatrix} yb_1 \\ yb_2 \\ yb_3 \end{pmatrix} + \begin{pmatrix} zc_1 \\ zc_2 \\ zc_3 \end{pmatrix} = \begin{pmatrix} xa_1 + yb_1 + zc_1 \\ xa_2 + yb_2 + zc_2 \\ xa_3 + yb_3 + zc_3 \end{pmatrix}$

$$= \begin{pmatrix} a_1 & b_1 & c_1 \\ a_2 & b_2 & c_2 \\ a_3 & b_3 & c_3 \end{pmatrix} \begin{pmatrix} x \\ y \\ z \end{pmatrix}$$

Therefore if, under a given transformation,

$$\mathbf{i} \to \begin{pmatrix} a_1 \\ a_2 \\ a_3 \end{pmatrix}, \quad \mathbf{j} \to \begin{pmatrix} b_1 \\ b_2 \\ b_3 \end{pmatrix}, \quad \mathbf{k} \to \begin{pmatrix} c_1 \\ c_2 \\ c_3 \end{pmatrix}$$

the matrix operator for that transformation is

$$\begin{pmatrix} a_1 & b_1 & c_1 \\ a_2 & b_2 & c_2 \\ a_3 & b_3 & c_3 \end{pmatrix}$$

Note that the Cartesian frame of reference, whose base vectors are **i**, **j** and **k** is transformed to the frame of reference whose base vectors are **a**, **b** and **c**.

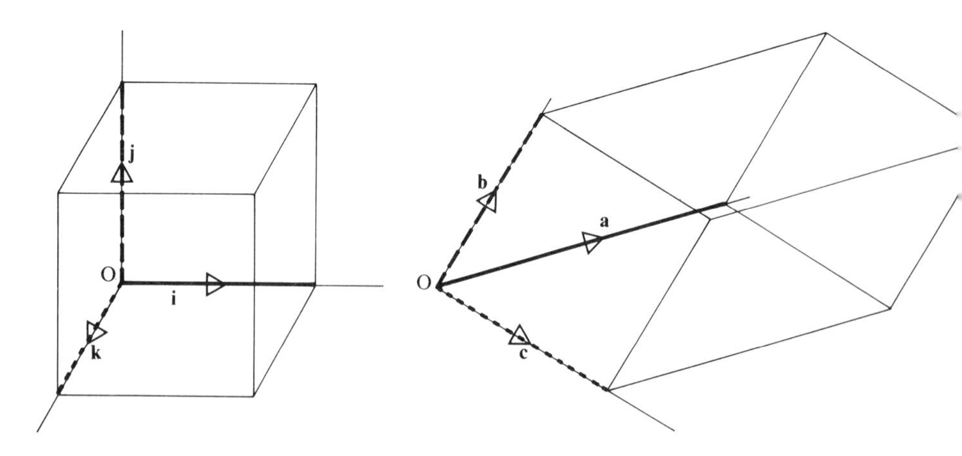

and that a three dimensional grid of unit volume cubes in xyz space is transformed to a three dimensional grid of parallelepipeds in the image space.

EXAMPLES 6a

1) Find the matrix $\mathbf{M}$ such that the point $P(x, y, z)$ is rotated through $90°$ about Oz to the point $P'(X, Y, Z)$ by the transformation

$$\begin{pmatrix} X \\ Y \\ Z \end{pmatrix} = \mathbf{M} \begin{pmatrix} x \\ y \\ z \end{pmatrix}$$

Note that a positive rotation about Oz is in the sense $\mathbf{i} \to \mathbf{j}$, a positive rotation about Ox is in the sense $\mathbf{j} \to \mathbf{k}$ and a positive rotation about Oy is the sense $\mathbf{k} \to \mathbf{i}$.

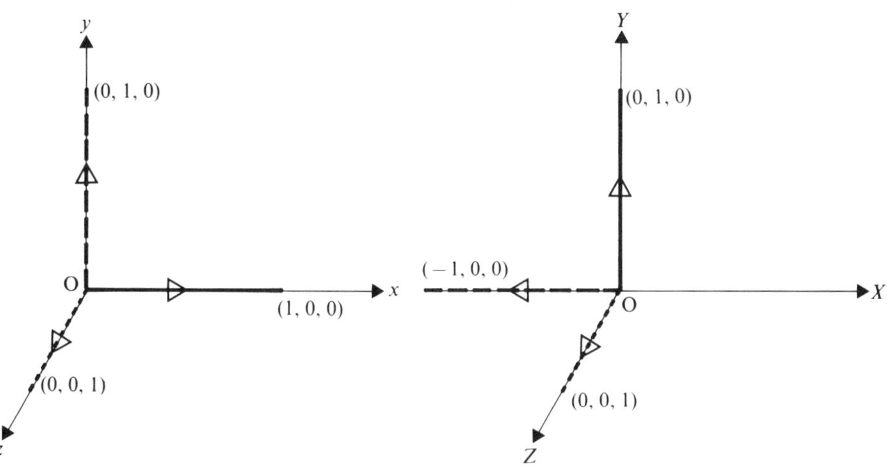

We see from the diagram that

$$\mathbf{M} \text{ maps } \quad \mathbf{i} \to \begin{pmatrix} 0 \\ 1 \\ 0 \end{pmatrix}, \quad \mathbf{j} \to \begin{pmatrix} -1 \\ 0 \\ 0 \end{pmatrix} \quad \text{and} \quad \mathbf{k} \to \begin{pmatrix} 0 \\ 0 \\ 1 \end{pmatrix}$$

Therefore

$$\mathbf{M} = \begin{pmatrix} 0 & -1 & 0 \\ 1 & 0 & 0 \\ 0 & 0 & 1 \end{pmatrix}$$

2) Find the matrix $\mathbf{M}$ which transforms the point $P(x, y, z)$ to the point $P'(X, Y, Z)$ by the equation $\mathbf{M}\,\overrightarrow{OP} = \overrightarrow{OP'}$ where P' is the reflection of P in the plane $x + y + z = 0$.

The direction cosines of the *normal* to the plane Π, $x + y + z = 0$, are

$$\frac{1}{\sqrt{3}}, \frac{1}{\sqrt{3}}, \frac{1}{\sqrt{3}}.$$

Thus the unit vector, $\hat{\mathbf{n}}$, perpendicular to Π is given by

$$\hat{\mathbf{n}} = \frac{1}{\sqrt{3}}\mathbf{i}+\frac{1}{\sqrt{3}}\mathbf{j}+\frac{1}{\sqrt{3}}\mathbf{k}$$

and $\hat{\mathbf{n}}$ is inclined to each coordinate axis at an angle α where $\cos\alpha = \frac{1}{\sqrt{3}}$.

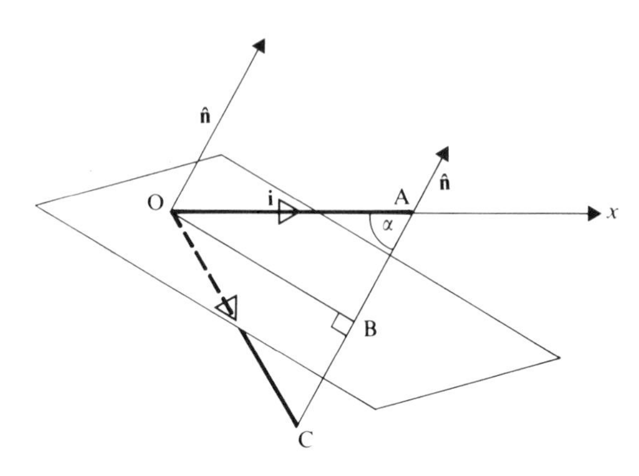

If $\overrightarrow{OA} = \mathbf{i}$, AB is perpendicular to Π and C is the reflection of A in π

then $\quad BA = OA\cos\alpha = \cos\alpha.$

Therefore $\quad \overrightarrow{BA} = (\cos\alpha)\,\hat{\mathbf{n}} = \frac{1}{3}\mathbf{i}+\frac{1}{3}\mathbf{j}+\frac{1}{3}\mathbf{k}.$

Hence the reflection of $\mathbf{i}$ in the plane Π is $\overrightarrow{OC}$, where

$$\begin{aligned}\overrightarrow{OC} &= \mathbf{i}-2\overrightarrow{BA}\\ &= \tfrac{1}{3}\mathbf{i}-\tfrac{2}{3}\mathbf{j}-\tfrac{2}{3}\mathbf{k}.\end{aligned}$$

As $\mathbf{j}$ and $\mathbf{k}$ are also inclined at an angle α to $\hat{\mathbf{n}}$, the reflections of $\mathbf{j}$ and $\mathbf{k}$ in Π are

$$-\tfrac{2}{3}\mathbf{i}+\tfrac{1}{3}\mathbf{j}-\tfrac{2}{3}\mathbf{k} \quad\text{and}\quad -\tfrac{2}{3}\mathbf{i}-\tfrac{2}{3}\mathbf{j}+\tfrac{1}{3}\mathbf{k} \quad\text{respectively.}$$

Therefore $\mathbf{M}$ maps

$$\begin{pmatrix}1\\0\\0\end{pmatrix}\to\begin{pmatrix}\frac{1}{3}\\-\frac{2}{3}\\-\frac{2}{3}\end{pmatrix},\quad \begin{pmatrix}0\\1\\0\end{pmatrix}\to\begin{pmatrix}-\frac{2}{3}\\\frac{1}{3}\\-\frac{2}{3}\end{pmatrix},\quad \begin{pmatrix}0\\0\\1\end{pmatrix}\to\begin{pmatrix}-\frac{2}{3}\\-\frac{2}{3}\\\frac{1}{3}\end{pmatrix}$$

so

$$\mathbf{M} = \begin{pmatrix}\frac{1}{3} & -\frac{2}{3} & -\frac{2}{3}\\ -\frac{2}{3} & \frac{1}{3} & -\frac{2}{3}\\ -\frac{2}{3} & -\frac{2}{3} & \frac{1}{3}\end{pmatrix} = \frac{1}{3}\begin{pmatrix}1 & -2 & -2\\ -2 & 1 & -2\\ -2 & -2 & 1\end{pmatrix}$$

3) A mapping of three dimensional space is defined by the matrix

$$\mathbf{M} = \begin{pmatrix} 0 & 0 & -1 \\ 0 & 1 & 0 \\ 1 & 0 & 0 \end{pmatrix}$$

Interpret geometrically the mappings defined by $\mathbf{M}$ and by $\mathbf{M}^2$.

$$\mathbf{M} = \begin{pmatrix} 0 & 0 & -1 \\ 0 & 1 & 0 \\ 1 & 0 & 0 \end{pmatrix}$$

maps $\quad \mathbf{i} \rightarrow \begin{pmatrix} 0 \\ 0 \\ 1 \end{pmatrix}, \quad \mathbf{j} \rightarrow \begin{pmatrix} 0 \\ 1 \\ 0 \end{pmatrix}, \quad \mathbf{k} \rightarrow \begin{pmatrix} -1 \\ 0 \\ 0 \end{pmatrix},$

i.e.

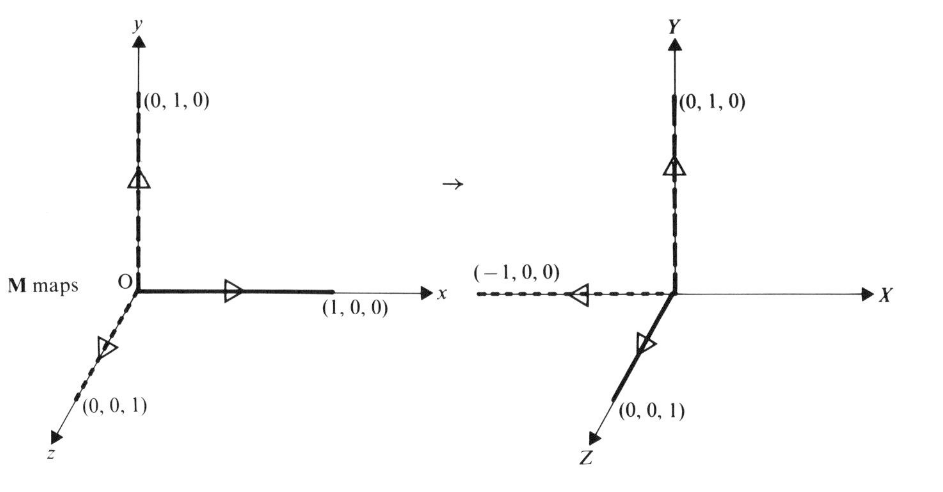

From the diagram we see that $\mathbf{M}$ rotates the xyz space through -90° about Oy.
Now $\mathbf{M}^2 = \mathbf{MM}$ which represents the transformation of a rotation of -90° about Oy followed by a further rotation of -90° about Oy, i.e. $\mathbf{M}^2$ represents a rotation of 180° about Oy.

4) A transformation of three dimensional space is defined as:
a stretch by a factor of 2 in the direction Ox, followed by a rotation of -45° about Oy. Find the matrix which defines this mapping.

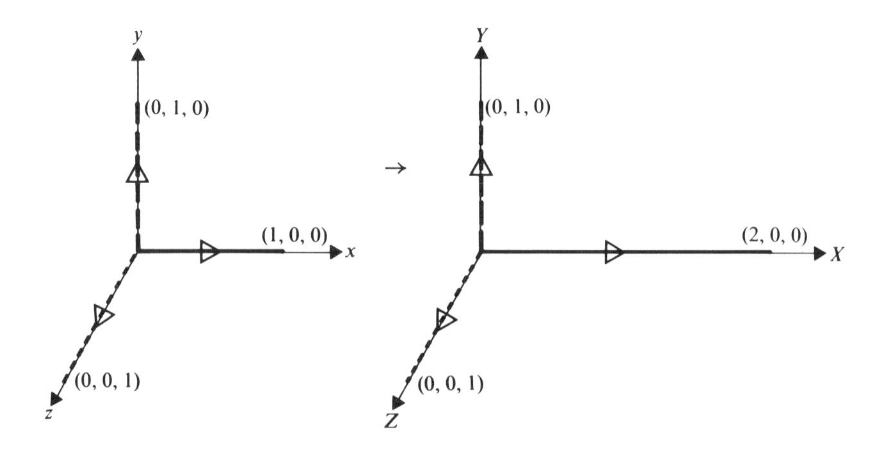

The matrix defining a transformation of a stretch by a factor of 2 in the direction Ox is

$$\mathbf{A} = \begin{pmatrix} 2 & 0 & 0 \\ 0 & 1 & 0 \\ 0 & 0 & 1 \end{pmatrix}$$

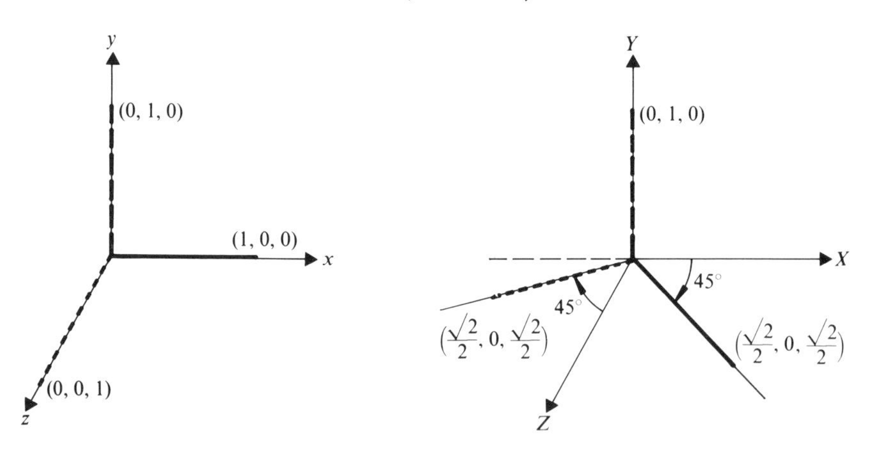

and the matrix defining a rotation of $-45°$ about Oy is

$$\mathbf{B} = \begin{pmatrix} \frac{1}{2}\sqrt{2} & 0 & -\frac{1}{2}\sqrt{2} \\ 0 & 1 & 0 \\ \frac{1}{2}\sqrt{2} & 0 & \frac{1}{2}\sqrt{2} \end{pmatrix} = \frac{1}{2}\sqrt{2}\begin{pmatrix} 1 & 0 & -1 \\ 0 & 2 & 0 \\ 1 & 0 & 1 \end{pmatrix}$$

Therefore a stretch by a factor of 2 in the direction Ox followed by a rotation of $-45°$ about Oy is defined by

$$\mathbf{M} = \mathbf{BA} = \tfrac{1}{2}\begin{pmatrix} \sqrt{2} & 0 & -\sqrt{2} \\ 0 & 2 & 0 \\ \sqrt{2} & 0 & \sqrt{2} \end{pmatrix}\begin{pmatrix} 2 & 0 & 0 \\ 0 & 1 & 0 \\ 0 & 0 & 1 \end{pmatrix}$$

$$= \tfrac{1}{2}\begin{pmatrix} 2\sqrt{2} & 0 & -\sqrt{2} \\ 0 & 2 & 0 \\ 2\sqrt{2} & 0 & \sqrt{2} \end{pmatrix}$$

Note that the *first* named transformation corresponds to the second matrix.

EXERCISE 6a

1) Find the matrices which perform the following transformations of three dimensional space:

(a) a rotation of $\dfrac{\pi}{6}$ about Oy, (b) a rotation of $\dfrac{\pi}{4}$ about Ox,

(c) a reflection in the xy plane, (d) a reflection in the xz plane,

(e) a rotation of π about the line $x = y,\ z = 0$,

(f) a reflection in the plane $x - y - z = 0$,

(g) an enlargement by a factor of 2,

(h) a contraction by a factor of $\frac{1}{3}$,

(i) a stretch by a factor of 2 in the directions Ox and Oz.

2) Describe the effect on three dimensional space of the transformation defined by $\mathbf{M}$ where $\mathbf{M}$ is

(a) $\begin{pmatrix} 0 & 0 & 1 \\ 0 & 1 & 0 \\ 1 & 0 & 0 \end{pmatrix}$ (b) $\begin{pmatrix} -2 & 0 & 0 \\ 0 & -2 & 0 \\ 0 & 0 & 1 \end{pmatrix}$ (c) $\begin{pmatrix} 1 & 0 & 1 \\ 0 & 1 & 0 \\ 0 & 0 & 0 \end{pmatrix}$

3) Using the transformations defined in Question 1, find the matrices which define the compound transformations

(a) (b) followed by (c) (b) (c) followed by (b)

(c) (g) followed by (a) (d) (e) followed by (i)

4) Describe the effects of the transformations defined by **AB**, **AC** and **BA** where

$$\mathbf{A} = \begin{pmatrix} -1 & 0 & 0 \\ 0 & -1 & 0 \\ 0 & 0 & -1 \end{pmatrix}, \quad \mathbf{B} = \begin{pmatrix} 1 & 0 & 0 \\ 0 & 3 & 0 \\ 0 & 0 & 1 \end{pmatrix}, \quad \mathbf{C} = \begin{pmatrix} 2 & 0 & 0 \\ 0 & 2 & 0 \\ 0 & 0 & 2 \end{pmatrix}$$

The Effect on Volume

The transformation defined by $\mathbf{M} = \begin{pmatrix} a_1 & b_1 & c_1 \\ a_2 & b_2 & c_2 \\ a_3 & b_3 & c_3 \end{pmatrix}$ maps a cube of unit volume in xyz space to a parallelepiped in the image space.

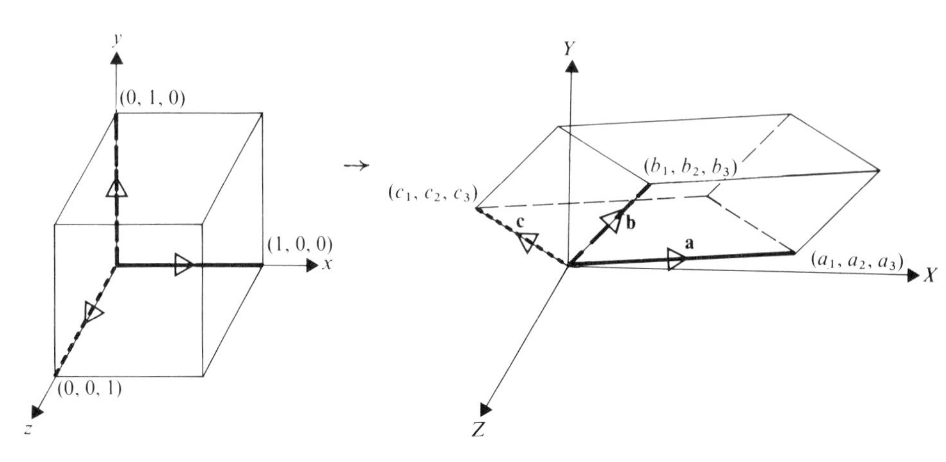

The volume of the parallelepiped is

$$\mathbf{a}.\mathbf{b}\times\mathbf{c} = (a_1\mathbf{i}+a_2\mathbf{j}+a_3\mathbf{k}).\begin{vmatrix} \mathbf{i} & \mathbf{j} & \mathbf{k} \\ b_1 & b_2 & b_3 \\ c_1 & c_2 & c_3 \end{vmatrix}$$

$$= a_1\begin{vmatrix} b_2 & b_3 \\ c_2 & c_3 \end{vmatrix} - a_2\begin{vmatrix} b_1 & b_3 \\ c_1 & c_3 \end{vmatrix} + a_3\begin{vmatrix} b_1 & b_2 \\ c_1 & c_2 \end{vmatrix}$$

$$= \begin{vmatrix} a_1 & a_2 & a_3 \\ b_1 & b_2 & b_3 \\ c_1 & c_2 & c_3 \end{vmatrix} = \begin{vmatrix} a_1 & b_1 & c_1 \\ a_2 & b_2 & c_2 \\ a_3 & b_3 & c_3 \end{vmatrix} = |\mathbf{M}|,$$

i.e. a volume of 1 cubic unit is mapped to a volume $|\mathbf{M}|$ cubic units.
Hence $\mathbf{M}$ alters volume by a factor $|\mathbf{M}|$.

For example, if $\mathbf{A} = \begin{pmatrix} 1 & 0 & 2 \\ -1 & 1 & 4 \\ 0 & 0 & 4 \end{pmatrix}$, $|\mathbf{A}| = 4$

so that under the transformation defined by $\mathbf{A}$, a volume of V cubic units is mapped to a volume of $2V$ cubic units.
If $|\mathbf{M}| = 0$ then volume is destroyed and $\mathbf{M}$ is called singular,

i.e. the volume of the parallelepiped is zero, from which it follows that **a**, **b** and **c** are coplanar.
When a number of vectors are all parallel to a plane, but not necessarily contained in the plane they are described as *paraplanar*. Thus if **a**, **b** and **c** are any three vectors

$|\mathbf{M}| = 0 \iff$ **a**, **b** and **c** are coplanar

and conversely

$|\mathbf{M}| \neq 0 \iff$ **a**, **b** and **c** are not coplanar and so provide a set of basis vectors for three dimensional space analysis

and so provide a set of basis vectors for three dimensional analysis.
So, as **M** maps three dimensional xyz space to the space whose base vectors are **a**, **b** and **c**, when **M** is singular the entire xyz space is mapped to the set of points lying in the plane of **a**, **b** and **c**.

i.e. $|\mathbf{M}| = 0 \iff$ **M** maps three dimensional space to a plane.

If, further, **a**, **b** and **c** are parallel, **M** maps three dimensional space to a line, as in this case $\mathbf{a} = \lambda\mathbf{b} = \mu\mathbf{c}$, so any point (x, y, z) has an image point whose position vector, **r**, is $\mathbf{r} = s\mathbf{a}$.
In the exceptional case when $\mathbf{a} = \mathbf{b} = \mathbf{c} = \mathbf{0}$, **M** maps all points to the origin.
Note that, as the origin is invariant under any transformation of three dimensional space defined by a 3×3 matrix, when three dimensional space is mapped to a plane or a line, the plane or line concerned contains the origin.

EXAMPLES 6b

1) Describe the transformation defined by $\mathbf{M} = \begin{pmatrix} 1 & -1 & 2 \\ 2 & -2 & 4 \\ 1 & -1 & 2 \end{pmatrix}$

From **M** we see that, if $\mathbf{i} \rightarrow \mathbf{a} = \begin{pmatrix} 1 \\ 2 \\ 1 \end{pmatrix}$ then $\mathbf{j} \rightarrow -\mathbf{a}$ and $\mathbf{k} \rightarrow 2\mathbf{a}$.

Hence **M** maps xyz space to a line
and in particular, **M** maps

$$\mathbf{r} = x\mathbf{i} + y\mathbf{j} + z\mathbf{k}$$

to

$$\begin{aligned} \mathbf{r}' &= x\mathbf{a} + y(-\mathbf{a}) + z(2\mathbf{a}) \\ &= \lambda\mathbf{a} \quad \text{where} \quad \lambda = x - y + 2z \\ &= \lambda(\mathbf{i} + 2\mathbf{j} + \mathbf{k}) \end{aligned}$$

So $\mathbf{M}$ maps all points in three dimensional space to points on the line whose vector equation is

$$\mathbf{r}' = \lambda(\mathbf{i} + 2\mathbf{j} + \mathbf{k})$$

and whose Cartesian equations are

$$\frac{x}{1} = \frac{y}{2} = \frac{z}{1}$$

2) Describe the transformation which maps (x, y, z) to (X, Y, Z) by the relation

$$\begin{pmatrix} 1 & 2 & -1 \\ 3 & 4 & -5 \\ 5 & 8 & -7 \end{pmatrix}\begin{pmatrix} x \\ y \\ z \end{pmatrix} = \begin{pmatrix} X \\ Y \\ Z \end{pmatrix}$$

From the matrix, $\mathbf{M}$, of the transformation, we see that $\mathbf{i} \to \mathbf{a}$, $\mathbf{j} \to \mathbf{b}$, $\mathbf{k} \to \mathbf{c}$, where

$$\mathbf{a} = \begin{pmatrix} 1 \\ 3 \\ 5 \end{pmatrix}, \quad \mathbf{b} = \begin{pmatrix} 2 \\ 4 \\ 8 \end{pmatrix}, \quad \mathbf{c} = \begin{pmatrix} -1 \\ -5 \\ -7 \end{pmatrix}$$

and, as $\mathbf{a}$, $\mathbf{b}$ and $\mathbf{c}$ do not have equal direction ratios, they are not parallel.

Now $$|\mathbf{M}| = \begin{vmatrix} 1 & 2 & -1 \\ 3 & 4 & -5 \\ 5 & 8 & -7 \end{vmatrix} = \begin{vmatrix} 1 & 2 & -1 \\ 3 & 4 & -5 \\ 1 & 2 & -1 \end{vmatrix} = 0.$$

So $\mathbf{M}$ maps three dimensional space to a plane.
To find the equation of this plane we can use the facts that
(a) the plane contains the origin,
(b) $\mathbf{a}$ and $\mathbf{b}$ (or $\mathbf{a}$ and $\mathbf{c}$, or $\mathbf{b}$ and $\mathbf{c}$) are the position vectors of two points in the plane.

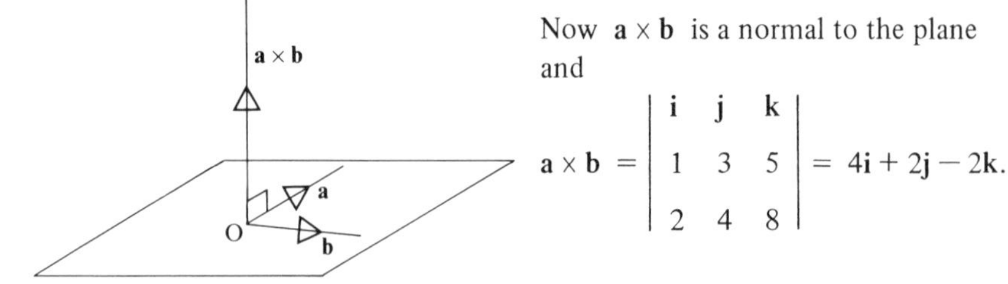

Now $\mathbf{a} \times \mathbf{b}$ is a normal to the plane and

$$\mathbf{a} \times \mathbf{b} = \begin{vmatrix} \mathbf{i} & \mathbf{j} & \mathbf{k} \\ 1 & 3 & 5 \\ 2 & 4 & 8 \end{vmatrix} = 4\mathbf{i} + 2\mathbf{j} - 2\mathbf{k}.$$

So the vector equation of the plane is $\mathbf{r}.(2\mathbf{i} + \mathbf{j} - \mathbf{k}) = 0$
and the Cartesian equation of the plane is $2x + y - z = 0$.

EXERCISE 6b

1) Find the effect on volume of the transformations defined by the following matrices:

(a) $\begin{pmatrix} 2 & -1 & 1 \\ 0 & 1 & 4 \\ 0 & -2 & 0 \end{pmatrix}$ (b) $\begin{pmatrix} -1 & 3 & -1 \\ 0 & 1 & 2 \\ -1 & 0 & 1 \end{pmatrix}$ (c) $\begin{pmatrix} 0 & 0 & 1 \\ 0 & 1 & 0 \\ 1 & 0 & 0 \end{pmatrix}$

(d) $\begin{pmatrix} 2 & -1 & 4 \\ 3 & -2 & 5 \\ 4 & 1 & 2 \end{pmatrix}$

2) Show, in each of the following cases, that the transformation defined by the given matrix maps three dimensional space to a plane and find the Cartesian equation of the plane:

(a) $\begin{pmatrix} 2 & 4 & 6 \\ 1 & -2 & 1 \\ 3 & 2 & 7 \end{pmatrix}$ (b) $\begin{pmatrix} 2 & 0 & 4 \\ 0 & 1 & -1 \\ 1 & 2 & 0 \end{pmatrix}$ (c) $\begin{pmatrix} 3 & -1 & 5 \\ -1 & 2 & 4 \\ 0 & 0 & 0 \end{pmatrix}$

(d) $\begin{pmatrix} 6 & 3 & 4 \\ 2 & 12 & 5 \\ 4 & 9 & 5 \end{pmatrix}$

3) Show that **M** maps three dimensional space to a line and find the Cartesian equations of the line, where **M** is

(a) $\begin{pmatrix} 2 & 1 & 3 \\ 4 & 2 & 6 \\ -2 & -1 & -3 \end{pmatrix}$ (b) $\begin{pmatrix} 5 & -5 & 10 \\ -2 & 2 & -4 \\ 1 & -1 & 2 \end{pmatrix}$ (c) $\begin{pmatrix} 0 & 0 & 0 \\ 1 & 3 & -2 \\ 0 & 0 & 0 \end{pmatrix}$

(d) $\begin{pmatrix} 1 & 2 & -1 \\ 3 & 6 & -3 \\ -1 & -2 & 1 \end{pmatrix}$

4) Describe the effect on three dimensional space of the transformation represented by **M**, where **M** is

(a) $\begin{pmatrix} 3 & -1 & 5 \\ 2 & 3 & -3 \\ 5 & 2 & 2 \end{pmatrix}$ (b) $\begin{pmatrix} 6 & 5 & 1 \\ 2 & -3 & -2 \\ -4 & 8 & 5 \end{pmatrix}$ (c) $\begin{pmatrix} 0 & 1 & 1 \\ 1 & 0 & 1 \\ 1 & 1 & 0 \end{pmatrix}$

(d) $\begin{pmatrix} 1 & -2 & 1 \\ -3 & 6 & -3 \\ 2 & -4 & 2 \end{pmatrix}$ (e) $\begin{pmatrix} 0 & 0 & 0 \\ 0 & 0 & 0 \\ 0 & 0 & 0 \end{pmatrix}$ (f) $\begin{pmatrix} 0 & 0 & 1 \\ 0 & 0 & 0 \\ 0 & 0 & 0 \end{pmatrix}$

5) Find the values of λ for which the vectors $\mathbf{a} = \begin{pmatrix} 1 \\ 2 \\ \lambda \end{pmatrix}$, $\mathbf{b} = \begin{pmatrix} \lambda \\ 2 \\ -1 \end{pmatrix}$, $\mathbf{c} = \begin{pmatrix} 0 \\ 1 \\ \lambda \end{pmatrix}$ are paraplanar.

6) Determine which of the following sets of vectors provide a basis set for three dimensions:

(a) $\begin{pmatrix} 1 \\ -1 \\ 2 \end{pmatrix}, \begin{pmatrix} 3 \\ 5 \\ -2 \end{pmatrix}, \begin{pmatrix} 1 \\ 0 \\ 1 \end{pmatrix}$ (b) $\begin{pmatrix} -1 \\ 4 \\ 6 \end{pmatrix}, \begin{pmatrix} 2 \\ -1 \\ 3 \end{pmatrix}, \begin{pmatrix} 3 \\ -5 \\ -3 \end{pmatrix}$

(c) $\begin{pmatrix} 4 \\ 1 \\ 2 \end{pmatrix}, \begin{pmatrix} -1 \\ 4 \\ 0 \end{pmatrix}, \begin{pmatrix} 2 \\ 1 \\ 1 \end{pmatrix}$

INVERSE TRANSFORMATIONS IN TWO DIMENSIONS

Consider the transformation which rotates the xy plane through $90°$ about the origin.

The matrix for this transformation is $\mathbf{A} = \begin{pmatrix} 0 & -1 \\ 1 & 0 \end{pmatrix}$

Suppose that $\mathbf{A}$ maps a point (x, y) to the point $(2, 3)$,

i.e.
$$\begin{pmatrix} 0 & -1 \\ 1 & 0 \end{pmatrix}\begin{pmatrix} x \\ y \end{pmatrix} = \begin{pmatrix} 2 \\ 3 \end{pmatrix} \qquad [1]$$

If we want to find the point (x, y) from which its image, $(2, 3)$, originates, we must reverse the transformation. In this case we must rotate the plane through $-90°$ about O to map $(2, 3)$ back to (x, y). The matrix for this transformation is $\begin{pmatrix} 0 & 1 \\ -1 & 0 \end{pmatrix}$

so
$$\begin{pmatrix} x \\ y \end{pmatrix} = \begin{pmatrix} 0 & 1 \\ -1 & 0 \end{pmatrix}\begin{pmatrix} 2 \\ 3 \end{pmatrix} \qquad [2]$$

This transformation is called the *inverse* of the original transformation.
Note that the matrices that define the original transformation and its inverse are such that

$$\begin{pmatrix} 0 & 1 \\ -1 & 0 \end{pmatrix}\begin{pmatrix} 0 & -1 \\ 1 & 0 \end{pmatrix} = \begin{pmatrix} 1 & 0 \\ 0 & 1 \end{pmatrix} = \mathbf{I}$$

In general, if a transformation, defined by a matrix $\mathbf{A}$, maps $\begin{pmatrix} x \\ y \end{pmatrix}$ to $\begin{pmatrix} X \\ Y \end{pmatrix}$, the transformation which maps $\begin{pmatrix} X \\ Y \end{pmatrix}$ back to $\begin{pmatrix} x \\ y \end{pmatrix}$ is called the inverse transformation and is defined by the inverse matrix $\mathbf{A}^{-1}$.

Further, if for a given matrix $\mathbf{M}$, we can find a matrix $\mathbf{M}^{-1}$ such that $\mathbf{M}^{-1}\mathbf{M} = \mathbf{I}$ and $\mathbf{M}\mathbf{M}^{-1} = \mathbf{I}$ then $\mathbf{M}^{-1}$ is the inverse of $\mathbf{M}$.

i.e. if $\mathbf{M}\begin{pmatrix} x \\ y \end{pmatrix} = \begin{pmatrix} X \\ Y \end{pmatrix}$ then $\begin{pmatrix} x \\ y \end{pmatrix} = \mathbf{M}^{-1}\begin{pmatrix} X \\ Y \end{pmatrix}$.

THE INVERSE OF A 2 × 2 MATRIX

If $\mathbf{M} = \begin{pmatrix} a & b \\ c & d \end{pmatrix}$ and $\mathbf{M}\begin{pmatrix} x \\ y \end{pmatrix} = \begin{pmatrix} X \\ Y \end{pmatrix}$,

i.e. $$\begin{pmatrix} a & b \\ c & d \end{pmatrix}\begin{pmatrix} x \\ y \end{pmatrix} = \begin{pmatrix} X \\ Y \end{pmatrix},$$

then $$\left.\begin{aligned} ax + by &= X \\ cx + dy &= Y \end{aligned}\right\} \Rightarrow \left\{\begin{aligned} x &= (Xd - Yb)/(ad - bc) \quad [1] \\ y &= (-Xc + Ya)/(ad - bc). \quad [2] \end{aligned}\right.$$

Noting that $ad - bc = |\mathbf{M}|$, we can write equations [1] and [2] in the form

$$x = \frac{1}{|\mathbf{M}|}(d \quad -b)\begin{pmatrix} X \\ Y \end{pmatrix}, \qquad y = \frac{1}{|\mathbf{M}|}(-c \quad a)\begin{pmatrix} X \\ Y \end{pmatrix}$$

which may be adjoined to give

$$\begin{pmatrix} x \\ y \end{pmatrix} = \frac{1}{|\mathbf{M}|}\begin{pmatrix} d & -b \\ -c & a \end{pmatrix}\begin{pmatrix} X \\ Y \end{pmatrix}$$

i.e. if $\mathbf{M} = \begin{pmatrix} a & b \\ c & d \end{pmatrix}$ then $\mathbf{M}^{-1} = \dfrac{1}{|\mathbf{M}|}\begin{pmatrix} d & -b \\ -c & a \end{pmatrix}$

and it can easily be shown that $\mathbf{M}^{-1}\mathbf{M} = \mathbf{M}\mathbf{M}^{-1} = \mathbf{I}$.
For example,

$$\begin{aligned} \mathbf{M}^{-1}\mathbf{M} &= \frac{1}{|\mathbf{M}|}\begin{pmatrix} d & -b \\ -c & a \end{pmatrix}\begin{pmatrix} a & b \\ c & d \end{pmatrix} = \frac{1}{|\mathbf{M}|}\begin{pmatrix} ad - bc & 0 \\ 0 & ad - bc \end{pmatrix} \\ &= \frac{1}{|\mathbf{M}|}\begin{pmatrix} |\mathbf{M}| & 0 \\ 0 & |\mathbf{M}| \end{pmatrix} = \begin{pmatrix} 1 & 0 \\ 0 & 1 \end{pmatrix} = \mathbf{I}. \end{aligned}$$

Note that if $|\mathbf{M}| = 0$, $\mathbf{M}$ has no inverse and the transformation defined by $\mathbf{M}$ has no inverse.

Note also that $\mathbf{M}^{-1}$ is obtained from $\mathbf{M}$ by transposing the elements in the major diagonal, changing the sign of the elements in the minor diagonal and then dividing the resulting matrix by $|\mathbf{M}|$. The steps in this operation are best expressed in a flow diagram:

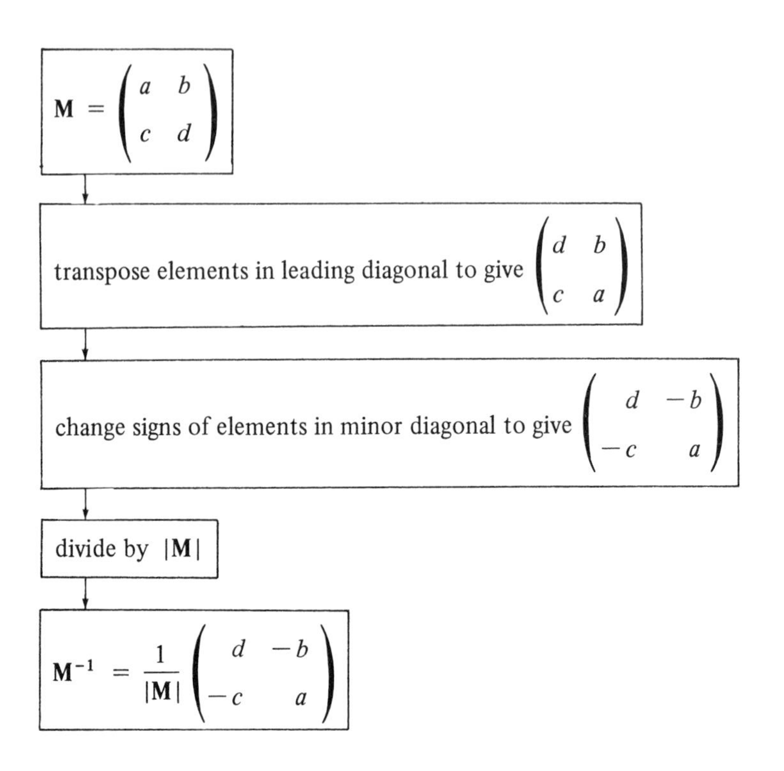

For example, if

$$\mathbf{A} = \begin{pmatrix} 2 & 1 \\ 3 & 2 \end{pmatrix}, \quad |\mathbf{A}| = 1$$

so $$\mathbf{A}^{-1} = \begin{pmatrix} 2 & -1 \\ -3 & 2 \end{pmatrix}$$

and if $$\mathbf{B} = \begin{pmatrix} 2 & 5 \\ -1 & 4 \end{pmatrix}, \quad \mathbf{B}^{-1} = \tfrac{1}{13}\begin{pmatrix} 4 & -5 \\ 1 & 2 \end{pmatrix} = \begin{pmatrix} \frac{4}{13} & -\frac{5}{13} \\ \frac{1}{13} & \frac{2}{13} \end{pmatrix}$$

But if $$\mathbf{C} = \begin{pmatrix} 2 & 1 \\ 4 & 2 \end{pmatrix}, \quad |\mathbf{C}| = 0, \text{ so } \mathbf{C}^{-1} \text{ does not exist.}$$

EXAMPLES 6c

These examples show how inverse transformations can be used to solve problems that have been done by other methods in Chapter 1.

1) Find the image of the line $y = 2x + 1$ under the transformation of the xy plane defined by the relation

$$\begin{pmatrix} -2 & 1 \\ 3 & -1 \end{pmatrix}\begin{pmatrix} x \\ y \end{pmatrix} = \begin{pmatrix} X \\ Y \end{pmatrix}.$$

If $\mathbf{A} = \begin{pmatrix} -2 & 1 \\ 3 & -1 \end{pmatrix}$ then $\mathbf{A}^{-1} = (-1)\begin{pmatrix} -1 & -1 \\ -3 & -2 \end{pmatrix} = \begin{pmatrix} 1 & 1 \\ 3 & 2 \end{pmatrix}$

so $$\begin{pmatrix} x \\ y \end{pmatrix} = \begin{pmatrix} 1 & 1 \\ 3 & 2 \end{pmatrix}\begin{pmatrix} X \\ Y \end{pmatrix} \Rightarrow \begin{cases} x = X + Y \\ y = 3X + 2Y \end{cases}$$

Hence the line $y = 2x + 1$ is mapped by $\mathbf{A}$ to the line

$$(3X + 2Y) = 2(X + Y) + 1,$$

i.e. to $X = 1$.

2) Find the equations of the lines that map to themselves under the transformation defined by $\mathbf{M} = \begin{pmatrix} 2 & 0 \\ 1 & -2 \end{pmatrix}$.

$$\mathbf{M}^{-1} = -\frac{1}{4}\begin{pmatrix} -2 & 0 \\ -1 & 2 \end{pmatrix} = \begin{pmatrix} \frac{1}{2} & 0 \\ \frac{1}{4} & -\frac{1}{2} \end{pmatrix}$$

so (x, y) is mapped to (X, Y) where

$$\begin{pmatrix} x \\ y \end{pmatrix} = \begin{pmatrix} \frac{1}{2} & 0 \\ \frac{1}{4} & -\frac{1}{2} \end{pmatrix}\begin{pmatrix} X \\ Y \end{pmatrix}$$

i.e. $$x = \tfrac{1}{2}X, \qquad y = \tfrac{1}{4}X - \tfrac{1}{2}Y$$

so the line $y = mx + c$ is mapped to the line

$$\tfrac{1}{4}X - \tfrac{1}{2}Y = \frac{m}{2}X + c$$

$$\Rightarrow \qquad Y = \tfrac{1}{2}(1 - 2m)X - 2c.$$

For lines that map to themselves

$$y = mx + c \quad \text{and} \quad Y = \tfrac{1}{2}(1 - 2m)X - 2c$$

represent the same line,

i.e. $$m = \tfrac{1}{2}(1 - 2m) \quad \text{and} \quad c = -2c$$

$$\Rightarrow \qquad m = \tfrac{1}{4} \qquad \text{and} \quad c = 0$$

so $$y = \tfrac{1}{4}x \text{ maps to itself.}$$

EXERCISE 6c

1) Find, where it exists, the inverse of

(a) $\begin{pmatrix} 2 & 0 \\ 0 & 1 \end{pmatrix}$ (b) $\begin{pmatrix} 3 & -1 \\ 4 & 2 \end{pmatrix}$ (c) $\begin{pmatrix} 3 & -1 \\ -6 & 2 \end{pmatrix}$ (d) $\begin{pmatrix} 0 & 1 \\ 0 & 2 \end{pmatrix}$

(e) $\begin{pmatrix} 5 & 1 \\ 2 & 1 \end{pmatrix}$ (f) $\begin{pmatrix} -2 & 4 \\ -1 & -3 \end{pmatrix}$ (g) $\begin{pmatrix} \sin\theta & \cos\theta \\ -\cos\theta & \sin\theta \end{pmatrix}$ (h) $\begin{pmatrix} p & q \\ r & s \end{pmatrix}$.

2) *Write down* the matrix which represents the inverse transformation of
(a) a rotation of θ about the origin,
(b) an enlargement by a factor of 2,
(c) a stretch by a factor of 3 parallel to Ox,
(d) a shear of $45°$ parallel to Oy.

3) Find the equations of the images of the lines
(a) $y = 3x + 2$, (b) $x - 2y + 4 = 0$
under the transformation which maps $\begin{pmatrix} x \\ y \end{pmatrix}$ to $\begin{pmatrix} 2 & -1 \\ 5 & 3 \end{pmatrix}\begin{pmatrix} x \\ y \end{pmatrix}$.

4) Find the equations of the lines which map to themselves under the transformations defined by

(a) $\begin{pmatrix} -1 & 3 \\ 2 & 4 \end{pmatrix}$ (b) $\begin{pmatrix} 3 & -2 \\ 1 & 2 \end{pmatrix}$ (c) $\begin{pmatrix} 0 & 3 \\ 2 & 1 \end{pmatrix}$.

5) Show that the transformation which maps the point (x, y) to the point (X, Y) by the relation

$$\begin{pmatrix} 2 & 1 \\ 4 & 2 \end{pmatrix}\begin{pmatrix} x \\ y \end{pmatrix} = \begin{pmatrix} X \\ Y \end{pmatrix}$$

maps all points in the xy plane to points on a line in the image plane. Show also that all the points on the line $2x + y - 2 = 0$ map to the point $X = 2$, $Y = 4$.

6) A transformation of the xy plane is defined by the matrix $\begin{pmatrix} 2 & -4 \\ -1 & 2 \end{pmatrix}$. Show that this transformation maps all points in the xy-plane to a line and find the equation of this line. Show further that all the points lying on a particular line in the xy plane map to the point $(6, -3)$ and find the equation of that line.

INVERSE TRANSFORMATIONS IN THREE DIMENSIONS

If $\mathbf{M}$ is a 3×3 matrix that maps $\mathbf{r} = \begin{pmatrix} x \\ y \\ z \end{pmatrix}$ to $\mathbf{r}' = \begin{pmatrix} X \\ Y \\ Z \end{pmatrix}$,

i.e. $\mathbf{Mr} = \mathbf{r}'$,

then if the inverse matrix $\mathbf{M}^{-1}$ exists such that $\mathbf{M}^{-1}\mathbf{M} = \mathbf{I}$

the inverse transformation is given by

$$\mathbf{M}^{-1}\mathbf{M}\mathbf{r} = \mathbf{M}^{-1}\mathbf{r}',$$

$$\text{i.e. by} \quad \mathbf{r} = \mathbf{M}^{-1}\mathbf{r}'$$

Starting with a general 3 x 3 matrix $\mathbf{M}$ and finding the images of $\mathbf{i}, \mathbf{j}$ and $\mathbf{k}$ under $\mathbf{M}^{-1}$ gives the steps necessary for the calculation of $\mathbf{M}^{-1}$. However this is a lengthy process so it appears in the appendix to this chapter, and below we give the method for finding $\mathbf{M}^{-1}$ derived from the general process.

Calculation of $\mathbf{M}^{-1}$

Suppose that $\mathbf{M} = \begin{pmatrix} a_1 & a_2 & a_3 \\ b_1 & b_2 & b_3 \\ c_1 & c_2 & c_3 \end{pmatrix}$

and that A_1 is the cofactor of a_1, A_2 is the cofactor of a_2, etc., then the inverse matrix $\mathbf{M}^{-1}$ is given by proceeding as follows.

$$\mathbf{M} = \begin{pmatrix} a_1 & a_2 & a_3 \\ b_1 & b_2 & b_3 \\ c_1 & c_2 & c_3 \end{pmatrix}$$

↓

Transpose rows and columns of $\mathbf{M}$ $\Rightarrow \begin{pmatrix} a_1 & b_1 & c_1 \\ a_2 & b_2 & c_2 \\ a_3 & b_3 & c_3 \end{pmatrix}$ which is called the transpose matrix $\mathbf{M}^T$

↓

Replace elements by their cofactors $\Rightarrow \begin{pmatrix} A_1 & B_1 & C_1 \\ A_2 & B_2 & C_2 \\ A_3 & B_3 & C_3 \end{pmatrix}$ which is called the adjoint or adjugate matrix, $\mathbf{M}^*$

↓

Divide by $|\mathbf{M}|$ $\Rightarrow \mathbf{M}^{-1}$

i.e.

$$\mathbf{M}^{-1} = \frac{1}{|\mathbf{M}|}\mathbf{M}^*$$

Note that, if $|\mathbf{M}| = 0$, $\mathbf{M}^{-1}$ does not exist.
Note also that a matrix of any size can be transposed, e.g.

$$\begin{pmatrix} 2 & -1 & 4 \\ 1 & 3 & 0 \end{pmatrix}^T = \begin{pmatrix} 2 & 1 \\ -1 & 3 \\ 4 & 0 \end{pmatrix}$$

Thus to find the inverse of $\mathbf{A} = \begin{pmatrix} 1 & -2 & 0 \\ 3 & 1 & 5 \\ -1 & 2 & 3 \end{pmatrix}$ we proceed as follows.

$$\mathbf{A}^T = \begin{pmatrix} 1 & 3 & -1 \\ -2 & 1 & 2 \\ 0 & 5 & 3 \end{pmatrix}, \quad \mathbf{A}^* = \begin{pmatrix} -7 & 6 & -10 \\ -14 & 3 & -5 \\ 7 & 0 & 7 \end{pmatrix}, \quad |\mathbf{A}| = 21$$

Hence $$\mathbf{A}^{-1} = \frac{1}{21}\begin{pmatrix} -7 & 6 & -10 \\ -14 & 3 & -5 \\ 7 & 0 & 7 \end{pmatrix}$$

Checking we have

$$\mathbf{A}^{-1}\mathbf{A} = \frac{1}{21}\begin{pmatrix} -7 & 6 & -10 \\ -14 & 3 & -5 \\ 7 & 0 & 7 \end{pmatrix}\begin{pmatrix} 1 & -2 & 0 \\ 3 & 1 & 5 \\ -1 & 2 & 3 \end{pmatrix} = \frac{1}{21}\begin{pmatrix} 21 & 0 & 0 \\ 0 & 21 & 0 \\ 0 & 0 & 21 \end{pmatrix}$$

$$= \mathbf{I}$$

Note that this check on the calculation of $\mathbf{A}^{-1}$ is well worthwhile, because of the high probability of arithmetic mistakes.
Note also that later in this chapter, we give another (shorter) method for calculating an inverse matrix.

PROPERTIES OF INVERSE AND TRANSPOSE MATRICES

1) $$\mathbf{AA}^{-1} = \mathbf{A}^{-1}\mathbf{A} = \mathbf{I}.$$

If $\mathbf{A}^{-1}$ is a matrix such that $\mathbf{A}^{-1}\mathbf{A} = \mathbf{I}$

postmultiplying by $\mathbf{A}^{-1}$ gives

$$(\mathbf{A}^{-1}\mathbf{A})\mathbf{A}^{-1} = \mathbf{A}^{-1}$$

$$\Rightarrow \quad \mathbf{A}^{-1}(\mathbf{AA}^{-1}) = \mathbf{A}^{-1}$$

$$\Rightarrow \quad \mathbf{AA}^{-1} = \mathbf{I},$$

i.e. if $\mathbf{A}^{-1}$ is a left inverse of $\mathbf{A}$ it is also a right inverse of $\mathbf{A}$.

2) $$(\mathbf{A}^{-1})^{-1} = \mathbf{A}.$$

As $\mathbf{A}^{-1}\mathbf{A} = \mathbf{AA}^{-1}$, $\mathbf{A}$ is the inverse of $\mathbf{A}^{-1}$,

i.e. $$(\mathbf{A}^{-1})^{-1} = \mathbf{A}.$$

3) $$(\mathbf{AB})^{-1} = \mathbf{B}^{-1}\mathbf{A}^{-1}$$

Multiplying $\mathbf{B}^{-1}\mathbf{A}^{-1}$ on the left by $\mathbf{AB}$ gives

$$(\mathbf{AB})(\mathbf{B}^{-1}\mathbf{A}^{-1}) = \mathbf{A}(\mathbf{BB}^{-1})\mathbf{A}^{-1} = \mathbf{AIA}^{-1} = \mathbf{AA}^{-1} = \mathbf{I},$$

i.e. $\mathbf{B}^{-1}\mathbf{A}^{-1}$ is the inverse of $\mathbf{AB}$.
hence $(\mathbf{AB})^{-1} = \mathbf{B}^{-1}\mathbf{A}^{-1}$.

4) $$(\mathbf{AB})^T = \mathbf{B}^T\mathbf{A}^T$$

If $\mathbf{A}$ has *rows* $\mathbf{a}_1, \mathbf{a}_2, \mathbf{a}_3$,

i.e. $$\mathbf{A} = \begin{pmatrix} \mathbf{a}_1 \\ \mathbf{a}_2 \\ \mathbf{a}_3 \end{pmatrix}, \quad \text{then} \quad \mathbf{A}^T = (\mathbf{a}_1 \quad \mathbf{a}_2 \quad \mathbf{a}_3),$$

and if $\mathbf{B}$ has *columns* $\mathbf{b}_1, \mathbf{b}_2, \mathbf{b}_3$,

i.e. $$\mathbf{B} = (\mathbf{b}_1 \quad \mathbf{b}_2 \quad \mathbf{b}_3) \quad \text{then} \quad \mathbf{B}^T = \begin{pmatrix} \mathbf{b}_1 \\ \mathbf{b}_2 \\ \mathbf{b}_3 \end{pmatrix}.$$

Now $$\mathbf{AB} = \begin{pmatrix} \mathbf{a}_1 \\ \mathbf{a}_2 \\ \mathbf{a}_3 \end{pmatrix} (\mathbf{b}_1 \quad \mathbf{b}_2 \quad \mathbf{b}_3) = \begin{pmatrix} \mathbf{a}_1 . \mathbf{b}_1 & \mathbf{a}_1 . \mathbf{b}_2 & \mathbf{a}_1 . \mathbf{b}_3 \\ \mathbf{a}_2 . \mathbf{b}_1 & \mathbf{a}_2 . \mathbf{b}_2 & \mathbf{a}_2 . \mathbf{b}_3 \\ \mathbf{a}_3 . \mathbf{b}_1 & \mathbf{a}_3 . \mathbf{b}_2 & \mathbf{a}_3 . \mathbf{b}_3 \end{pmatrix}$$

and

$$\mathbf{B}^T\mathbf{A}^T = \begin{pmatrix} \mathbf{b}_1 \\ \mathbf{b}_2 \\ \mathbf{b}_3 \end{pmatrix} (\mathbf{a}_1 \quad \mathbf{a}_2 \quad \mathbf{a}_3) = \begin{pmatrix} \mathbf{b}_1 . \mathbf{a}_1 & \mathbf{b}_1 . \mathbf{a}_2 & \mathbf{b}_1 . \mathbf{a}_3 \\ \mathbf{b}_2 . \mathbf{a}_1 & \mathbf{b}_2 . \mathbf{a}_2 & \mathbf{b}_2 . \mathbf{a}_3 \\ \mathbf{b}_3 . \mathbf{a}_1 & \mathbf{b}_3 . \mathbf{a}_2 & \mathbf{b}_3 . \mathbf{a}_3 \end{pmatrix} = (\mathbf{AB})^T,$$

i.e. $(\mathbf{AB})^T = \mathbf{B}^T\mathbf{A}^T$.

EXERCISE 6d

For each of the matrices given in Questions 1–5, find (a) the transpose matrix, (b) the adjoint matrix, (c) where it exists, the inverse matrix.

1) $\mathbf{A} = \begin{pmatrix} 1 & 2 & 0 \\ -1 & 1 & 0 \\ 2 & 5 & 1 \end{pmatrix}$ 2) $\mathbf{B} = \begin{pmatrix} 2 & 3 & 1 \\ 5 & 3 & 4 \\ -1 & 2 & 5 \end{pmatrix}$

3) $\mathbf{C} = \begin{pmatrix} 1 & -1 & 3 \\ 2 & 0 & 4 \\ 6 & -2 & 22 \end{pmatrix}$ 4) $\mathbf{D} = \begin{pmatrix} -1 & 0 & 4 \\ 0 & 5 & -2 \\ 1 & 4 & -1 \end{pmatrix}$

5) $\mathbf{E} = \begin{pmatrix} 2 & -1 & 4 \\ 3 & 2 & -1 \\ 5 & -2 & 9 \end{pmatrix}$.

6) For the matrices **A** and **B** given above, confirm that $(\mathbf{AB})^T = \mathbf{B}^T\mathbf{A}^T$.

7) For the matrices **B** and **D** given above, confirm that $(\mathbf{BD})^{-1} = \mathbf{D}^{-1}\mathbf{B}^{-1}$.

8) Using the matrix **A** given above, show that $(\mathbf{A}^{-1})^{-1} = \mathbf{A}$.

9) If $\mathbf{A} = \begin{pmatrix} 2 & 1 & 4 \\ 3 & -1 & 0 \end{pmatrix}$ and $\mathbf{B} = \begin{pmatrix} 4 & 1 \\ 0 & 2 \\ 1 & 0 \end{pmatrix}$

write down the transpose of **A** and of **B**.
Calculate the products **AB** and $\mathbf{B}^T\mathbf{A}^T$ and use your results to confirm that $(\mathbf{AB})^T = \mathbf{B}^T\mathbf{A}^T$.

10) Prove, for three nonsingular matrices, **A**, **B**, **C** that $(\mathbf{ABC})^{-1} = \mathbf{C}^{-1}\mathbf{B}^{-1}\mathbf{A}^{-1}$.

11) Prove that $(\mathbf{A}^T)^{-1} = (\mathbf{A}^{-1})^T$.

12) If $\mathbf{A} = \begin{pmatrix} 1 & 2 \\ 3 & 4 \end{pmatrix}$, find $\mathbf{A}^{-1}$. Find the values of the real constants a and b such that $\mathbf{A} + a\mathbf{A}^{-1} = b\mathbf{I}$. Hence show that **A** satisfies the quadratic equation $\mathbf{A}^2 - b\mathbf{A} + a\mathbf{I} = \mathbf{0}$.

APPLICATIONS OF INVERSE MATRICES

We will now look at some problems where the calculation of an inverse matrix helps in their solution.

EXAMPLES 6e

1) Find the coordinates of the point in the xyz space from which the point $(2, 1, 5)$ in the XYZ space originates under the transformation given by

$$\begin{pmatrix} 2 & -1 & 4 \\ 1 & 0 & 0 \\ 1 & -2 & 0 \end{pmatrix}\begin{pmatrix} x \\ y \\ z \end{pmatrix} = \begin{pmatrix} X \\ Y \\ Z \end{pmatrix}$$

If (x_1, y_1, z_1) is the point from which $(2, 1, 5)$ originates, then

$$\begin{pmatrix} 2 & -1 & 4 \\ 1 & 0 & 0 \\ 1 & -2 & 0 \end{pmatrix}\begin{pmatrix} x_1 \\ y_1 \\ z_1 \end{pmatrix} = \begin{pmatrix} 2 \\ 1 \\ 5 \end{pmatrix} \qquad [1]$$

If $\mathbf{A} = \begin{pmatrix} 2 & -1 & 4 \\ 1 & 0 & 0 \\ 1 & -2 & 0 \end{pmatrix}$, then $\mathbf{A}^{-1} = -\frac{1}{8}\begin{pmatrix} 0 & -8 & 0 \\ 0 & -4 & 4 \\ -2 & 3 & 1 \end{pmatrix}$

so $$\begin{pmatrix} x_1 \\ y_1 \\ z_1 \end{pmatrix} = \frac{1}{8}\begin{pmatrix} 0 & 8 & 0 \\ 0 & 4 & -4 \\ 2 & -3 & -1 \end{pmatrix}\begin{pmatrix} 2 \\ 1 \\ 5 \end{pmatrix} = \frac{1}{8}\begin{pmatrix} 8 \\ -16 \\ -4 \end{pmatrix}$$

$$\Rightarrow \qquad x_1 = 1, \quad y_1 = -2, \quad z_1 = -\tfrac{1}{2}.$$

Note that [1] is a matrix representation of the set of simultaneous equations

$$\begin{aligned} 2x_1 - y_1 + 4z &= 2 \\ x_1 &= 1 \\ x_1 - 2y_1 &= 5 \end{aligned}$$

and that the solution of this set of equations is $x_1 = 1$, $y_1 = -2$, $z_1 = -\frac{1}{2}$.

2) Find the image of the plane $x + 2y - 7z = 2$ under the transformation defined by $\begin{pmatrix} -1 & 2 & 1 \\ -3 & 1 & 4 \\ 0 & 1 & 2 \end{pmatrix}$.

If $\mathbf{M} = \begin{pmatrix} -1 & 2 & 1 \\ -3 & 1 & 4 \\ 0 & 1 & 2 \end{pmatrix}$ maps $\begin{pmatrix} x \\ y \\ z \end{pmatrix}$ to $\begin{pmatrix} X \\ Y \\ Z \end{pmatrix}$

then $\mathbf{M}^{-1}$ maps $\begin{pmatrix} X \\ Y \\ Z \end{pmatrix}$ to $\begin{pmatrix} x \\ y \\ z \end{pmatrix}$

Now $$\mathbf{M}^{-1} = \frac{1}{11}\begin{pmatrix} -2 & -3 & 7 \\ 6 & -2 & 1 \\ -3 & 1 & 5 \end{pmatrix}$$

So $$\begin{pmatrix} x \\ y \\ z \end{pmatrix} = \frac{1}{11}\begin{pmatrix} -2 & -3 & 7 \\ 6 & -2 & 1 \\ -3 & 1 & 5 \end{pmatrix}\begin{pmatrix} X \\ Y \\ Z \end{pmatrix} = \frac{1}{11}\begin{pmatrix} -2X - 3Y + 7Z \\ 6X - 2Y + Z \\ -3X + Y + 5Z \end{pmatrix},$$

i.e. $$\begin{aligned} x &= \tfrac{1}{11}(-2X - 3Y + 7Z) \\ y &= \tfrac{1}{11}(6X - 2Y + Z) \\ z &= \tfrac{1}{11}(-3X + Y + 5Z). \end{aligned}$$

Hence the plane $x + 2y - 7z = 2$ maps to the plane

$$\tfrac{1}{11}(-2X - 3Y + 7Z) + \tfrac{2}{11}(6X - 2Y + Z) - \tfrac{7}{11}(-3X + Y + 5Z) = 2$$

$$\Rightarrow \qquad 31X - 14Y - 26Z = 22$$

Note that a normal to the given plane is $\begin{pmatrix} 1 \\ 2 \\ -7 \end{pmatrix}$

and, under **M**,

$$\begin{pmatrix} 1 \\ 2 \\ -7 \end{pmatrix} \rightarrow \begin{pmatrix} -1 & 2 & 1 \\ -3 & 1 & 4 \\ 0 & 1 & 2 \end{pmatrix}\begin{pmatrix} 1 \\ 2 \\ -7 \end{pmatrix} = \begin{pmatrix} -4 \\ -29 \\ -12 \end{pmatrix}.$$

But a normal to the image plane is $\begin{pmatrix} 31 \\ -14 \\ -26 \end{pmatrix}$,

i.e. in general, the normal to a plane is *not* mapped to the normal of the image plane.
However, any point in a plane *is* mapped to a point in the image plane and an alternative method for finding the image plane uses this fact.
For example, $A(2, 0, 0)$, $B(0, 1, 0)$, $C(0, 0, -\frac{2}{7})$ are points in the given plane.
The images of these points under **M** are $A'(-2, -6, 0)$, $B'(2, 1, 1)$, $C'(-\frac{2}{7}, -\frac{8}{7}, -\frac{4}{7})$ respectively.
So the image of the given plane under **M** is the plane containing the points A', B' and C', and we can find its equation using the method in Chapter 5,
i.e. the equation of the plane $A'B'C'$ is

$$\mathbf{r}.(\overrightarrow{A'B'} \times \overrightarrow{B'C'}) = \overrightarrow{OA'}.(\overrightarrow{A'B'} \times \overrightarrow{B'C'})$$

$$\Rightarrow \qquad \mathbf{r}.\begin{vmatrix} \mathbf{i} & \mathbf{j} & \mathbf{k} \\ -4 & -7 & -1 \\ 16 & 15 & 11 \end{vmatrix} = (-2\mathbf{i} - 6\mathbf{j}).\begin{vmatrix} \mathbf{i} & \mathbf{j} & \mathbf{k} \\ -4 & -7 & -1 \\ 16 & 15 & 11 \end{vmatrix}$$

$$\Rightarrow \qquad \mathbf{r}.(62\mathbf{i} - 28\mathbf{j} - 52\mathbf{k}) = 44$$

$$\Rightarrow \qquad 31x - 14y - 26z = 22$$

There is little to choose between these two methods with regard to the calculation involved. However, if the image of more than one plane is required under a given transformation, the first method is preferable because once the inverse matrix is found, the relationship it provides between (x, y, z) and its image (X, Y, Z) can be used to transform as many planes as required.

3) Find the image of the line L with equations $\dfrac{x-2}{1} = \dfrac{y+1}{2} = \dfrac{z-1}{3}$

under the transformation defined by $\mathbf{A} = \begin{pmatrix} 2 & -1 & 1 \\ 0 & 1 & 0 \\ 1 & 0 & 2 \end{pmatrix}$

Any two points on L are transformed by **A** to two points on the image L_1' of L.

Taking $(2, -1, 1)$ and $(3, 1, 4)$ as two points on L, the images of these two points are

$$\begin{pmatrix} 2 & -1 & 1 \\ 0 & 1 & 0 \\ 1 & 0 & 2 \end{pmatrix}\begin{pmatrix} 2 \\ -1 \\ 1 \end{pmatrix} = \begin{pmatrix} 6 \\ -1 \\ 4 \end{pmatrix} \text{ and } \begin{pmatrix} 2 & -1 & 1 \\ 0 & 1 & 0 \\ 1 & 0 & 2 \end{pmatrix}\begin{pmatrix} 3 \\ 3 \\ 10 \end{pmatrix} = \begin{pmatrix} 9 \\ 1 \\ 11 \end{pmatrix}$$

respectively.

Hence, using $\dfrac{x-x_1}{x_2-x_1} = \dfrac{y-y_1}{y_2-y_1} = \dfrac{z-z_1}{z_2-z_1}$, the equations of L' are

$$\frac{x-6}{3} = \frac{y+1}{2} = \frac{z-4}{7}$$

4) Find the equations of the planes that map to themselves under the transformation defined by $\mathbf{T} = \begin{pmatrix} -1 & 2 & 0 \\ 0 & -3 & 1 \\ 0 & 1 & 0 \end{pmatrix}$.

Under **T**, (x, y, z) is mapped to (X, Y, Z) where

$$\begin{pmatrix} -1 & 2 & 0 \\ 0 & -3 & 1 \\ 0 & 1 & 0 \end{pmatrix}\begin{pmatrix} x \\ y \\ z \end{pmatrix} = \begin{pmatrix} X \\ Y \\ Z \end{pmatrix}$$

Now $$\mathbf{T}^{-1} = \begin{pmatrix} -1 & 0 & 2 \\ 0 & 0 & 1 \\ 0 & 1 & 3 \end{pmatrix}$$

Hence $$\begin{pmatrix} x \\ y \\ z \end{pmatrix} = \begin{pmatrix} -1 & 0 & 2 \\ 0 & 0 & 1 \\ 0 & 1 & 3 \end{pmatrix}\begin{pmatrix} X \\ Y \\ Z \end{pmatrix}$$

$\Rightarrow$
$$x = -X + 2Z$$
$$y = Z$$
$$z = Y + 3Z.$$

Hence any plane $ax + by + cz = D$

maps to $a(-X + 2Z) + b(Z) + c(Y + 3Z) = D$

$\Rightarrow$ $-aX + cY + (2a + b + 3c)Z = D.$

If these equations represent the same plane then comparing

$$x + \frac{b}{a}y + \frac{c}{a}z = \frac{D}{a}$$

with

$$X - \frac{cY}{a} - \left(\frac{2a + b + 3c}{a}\right)Z = -\frac{D}{a}$$

we see that

$$\frac{b}{a} = -\frac{c}{a} \qquad \Rightarrow \quad b = -c \qquad [1]$$

$$\frac{c}{a} = -\left(\frac{2a + b + 3c}{a}\right) \qquad \Rightarrow \quad c = -2a - b - 3c \qquad [2]$$

$$\frac{D}{a} = -\frac{D}{a} \qquad \Rightarrow \quad D = 0. \qquad [3]$$

From [1] and [2] we have $a:b:c = 3:2:-2$.
Hence there is only one plane that maps to itself, and its equation is

$$3x + 2y - 2z = 0$$

Note that, although the plane $3x + 2y - 2z = 0$ maps to the same plane $3X + 2Y - 2Z = 0$ in the image space, a particular point on the plane $3x + 2y - 2z = 0$ will *not*, in general, map to the same point on $3X + 2Y - 2Z = 0$,

e.g. $$\begin{pmatrix} 2 \\ 1 \\ 4 \end{pmatrix} \text{ is on } 3x + 2y - 2z = 0$$

and $$\begin{pmatrix} 2 \\ 1 \\ 4 \end{pmatrix} \rightarrow \begin{pmatrix} -1 & 2 & 0 \\ 0 & -3 & 1 \\ 0 & -1 & 0 \end{pmatrix} \begin{pmatrix} 2 \\ 1 \\ 4 \end{pmatrix} = \begin{pmatrix} 0 \\ 1 \\ -1 \end{pmatrix}.$$

Note also that it may appear, from the examples done so far, that planes and lines that map to themselves must necessarily pass through the origin. However this is not so. Consider, for example, a rotation about the line $x = y = z$.

Any plane that is perpendicular to $x = y = z$ is unchanged. If further, the rotation is $180°$, any line that is perpendicular to, and intersects, $x = y = z$ is unchanged.

EXERCISE 6e

1) Find the coordinates of the point from which the point $(2, 1, 1)$ originates under the transformation defined by $\mathbf{M} = \begin{pmatrix} 2 & -1 & 0 \\ 3 & 1 & 2 \\ -1 & 1 & 0 \end{pmatrix}$

2) Find the inverse of the matrix $\begin{pmatrix} 1 & -2 & 1 \\ 3 & 1 & 2 \\ -1 & 4 & 1 \end{pmatrix}$

Hence solve the equations

$$\begin{aligned} x - 2y + z &= 1 \\ 3x + y + 2z &= 4 \\ -x + 4y + z &= 2. \end{aligned}$$

3) If $\mathbf{A} = \begin{pmatrix} 1 & 2 & 3 \\ -2 & 1 & 2 \\ 4 & -1 & 0 \end{pmatrix}$, find $\mathbf{A}^{-1}$.

Hence solve the equation $\mathbf{A}\begin{pmatrix} x \\ y \\ z \end{pmatrix} = \begin{pmatrix} 2 \\ 1 \\ 3 \end{pmatrix}$.

4) If $\mathbf{M} = \begin{pmatrix} 2 & 1 & 4 \\ 3 & 5 & 1 \\ 1 & 2 & 0 \end{pmatrix}$, find $\mathbf{M}^{-1}$.

Hence find the images, under the transformation $\mathbf{M}\begin{pmatrix} x \\ y \\ z \end{pmatrix} = \begin{pmatrix} X \\ Y \\ Z \end{pmatrix}$, of the planes (a) $x + 2y - z = 4$, (b) $3x + 5y + z = 2$, (c) $-2x + y - 3z = 1$.

5) Find the images of the lines

(a) $\mathbf{r} = \mathbf{i} - \mathbf{j} + \lambda(\mathbf{i} + \mathbf{j} + \mathbf{k})$,

(b) $\mathbf{r} = 2\mathbf{i} - \mathbf{k} + \mu(3\mathbf{i} - \mathbf{j} + \mathbf{k})$,

under the transformation defined in Question 4, giving their equations in vector form.

6) Find the equations of the planes that map to themselves under the transformation defined by $\begin{pmatrix} 1 & 2 & 0 \\ 0 & 1 & -1 \\ 0 & 2 & 1 \end{pmatrix}$

7) Find the equations of the lines that map to themselves under the transformation which maps $\begin{pmatrix} x \\ y \\ z \end{pmatrix}$ to $\begin{pmatrix} X \\ Y \\ Z \end{pmatrix}$ by the relation

$$\begin{pmatrix} 2 & 0 & 0 \\ 0 & 1 & 0 \\ 0 & 0 & 2 \end{pmatrix}\begin{pmatrix} x \\ y \\ z \end{pmatrix} = \begin{pmatrix} X \\ Y \\ Z \end{pmatrix}$$

8) If $\mathbf{A} = \begin{pmatrix} 0 & 1 & 0 \\ 0 & 0 & -1 \\ 1 & 0 & 0 \end{pmatrix}$ interpret the mapping of xyz space represented by $\mathbf{A}^{-1}$ and find $\mathbf{A}^{-1}$. Find the values of the constant λ for which $\mathbf{A}^2 + \lambda\mathbf{A}^{-1} = \mathbf{0}$. Find also the equations of the lines that map to themselves under the transformation defined by $\mathbf{A}^2$.

SYSTEMS OF LINEAR EQUATIONS

Two Variables

Consider the pair of equations

$$a_1x + b_1y + c_1 = 0 \qquad [1]$$

$$a_2x + b_2y + c_2 = 0. \qquad [2]$$

The solution set of this pair of equations is the complete set of values (x, y) that satisfy both equations. The nature of the solution set may be determined by interpreting equations [1] and [2] geometrically as the equations of two straight lines, L_1 and L_2, in the xy plane.

These lines may be such that

(a) L_1 and L_2 intersect, in which case there is only one point whose coordinates satisfy both equations, and the equations have a unique solution.

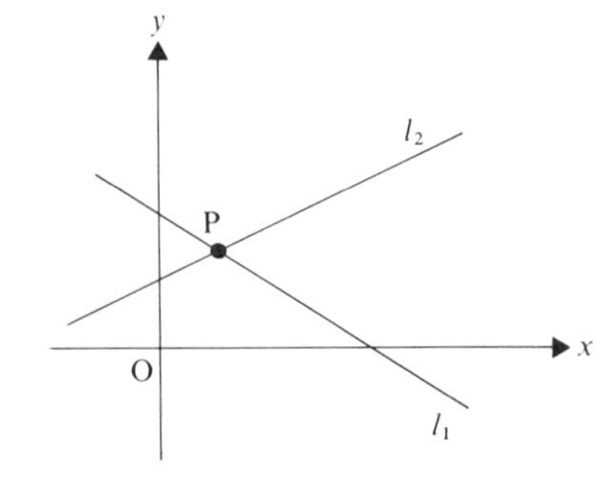

(b) L_1 and L_2 are parallel.
In this case there are no common points so the equations have no solution.
Such a pair of equations are said to be *inconsistent*.

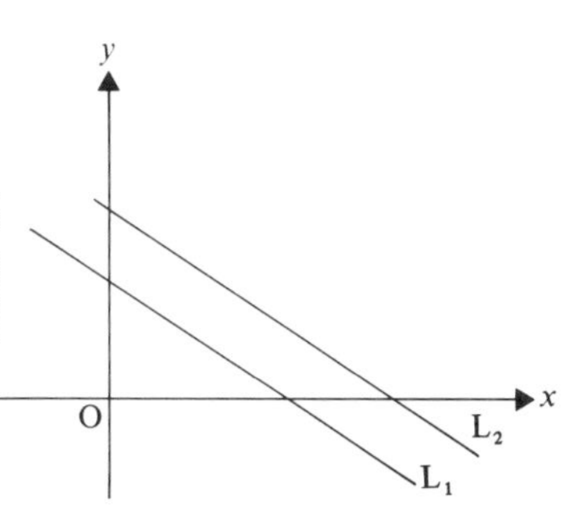

(c) L_1 and L_2 are the same line.
This time, all the points on the line are common to L_1 and L_2 so the equations have an infinite set of solutions. Such equations are said to be *linearly dependent*.

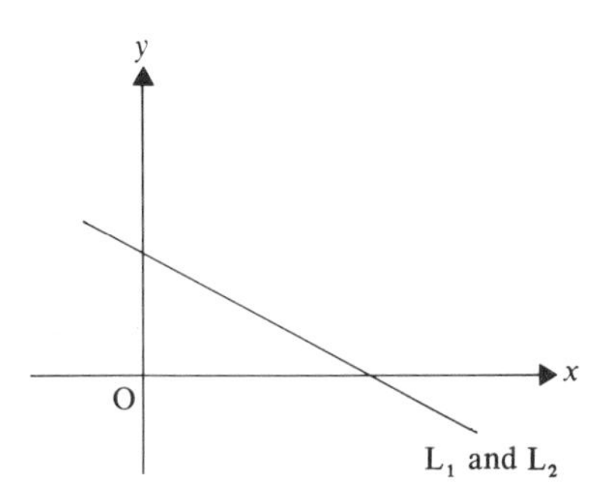

All these properties can easily be observed from the equations. For example

(a)
$$2x - 3y = 1$$
$$x + 4y = 2.$$

The lines represented by these two equations are not parallel, so they intersect in one point, i.e. this pair of equations has a unique solution, which is $x = \frac{10}{11}$, $y = \frac{3}{11}$.

(b)
$$2x - 3y = 1$$
$$4x - 6y = 1.$$

The lines represented by these equations are parallel and distinct. So there is no solution to this pair of equations, i.e. they are inconsistent.

(c)
$$2x - 3y = 1$$
$$4x - 6y = 2.$$

These equations represent the same line. Hence the equations are linearly dependent and their solution set contains the coordinates of all the points on the line. We may express the solution in parametric form, i.e. if $x = \lambda$, then from either equation $y = \frac{1}{3}(2\lambda - 1)$.
Hence the solution set is $x = \lambda$, $y = \frac{1}{3}(2\lambda - 1)$ for all real values of λ.

The Condition for Three Lines to be Concurrent

Consider the equations

$$a_1x + b_1y + c_1 = 0 \qquad [1]$$

$$a_2x + b_2y + c_2 = 0 \qquad [2]$$

$$a_3x + b_3y + c_3 = 0 \qquad [3]$$

representing the lines L_1, L_2 and L_3 in the xy plane.
Assuming that none of the lines are parallel, solving equations [2] and [3] gives

$$x = \frac{b_2c_3 - b_3c_2}{a_2b_3 - a_3b_2}, \qquad y = -\frac{a_2c_3 - a_3c_2}{a_2b_3 - a_3b_2}$$

If the lines are concurrent, the equations have a common solution,
i.e. the values of x and y satisfying equations [2] and [3] also satisfy equation [1],

i.e. $$a_1(b_2c_3 - b_3c_2) - b_1(a_2c_3 - a_3c_2) + c_1(a_2b_3 - a_3b_2) = 0. \qquad [4]$$

But the LHS of [4] is the expansion of the determinant $\begin{vmatrix} a_1 & b_1 & c_1 \\ a_2 & b_2 & c_2 \\ a_3 & b_3 & c_3 \end{vmatrix}$

Hence the lines whose equations are

$$\left.\begin{aligned} a_1x + b_1y + c_1 &= 0 \\ a_2x + b_2y + c_2 &= 0 \\ a_3x + b_3y + c_3 &= 0 \end{aligned}\right\} \text{ are concurrent if } \begin{vmatrix} a_1 & b_1 & c_1 \\ a_2 & b_2 & c_2 \\ a_3 & b_3 & c_3 \end{vmatrix} = 0$$

Three Variables

Now consider the set of equations

$$a_1x + b_1y + c_1z = d_1$$

$$a_2x + b_2y + c_2z = d_2$$

$$a_3x + b_3y + c_3z = d_3$$

which may be written as $\mathbf{A}\begin{pmatrix} x \\ y \\ z \end{pmatrix} = \begin{pmatrix} d_1 \\ d_2 \\ d_3 \end{pmatrix}$ where $\mathbf{A} = \begin{pmatrix} a_1 & b_1 & c_1 \\ a_2 & b_2 & c_2 \\ a_3 & b_3 & c_3 \end{pmatrix}$

Provided that $\mathbf{A}^{-1}$ exists, the solution of this set of equations is

$$\begin{pmatrix} x \\ y \\ z \end{pmatrix} = \mathbf{A}^{-1}\begin{pmatrix} d_1 \\ d_2 \\ d_3 \end{pmatrix}$$

The use of this method for solving a particular set of equations means that $\mathbf{A}^{-1}$ has to be found and so it involves a tedious amount of arithmetic. We now look at another method which considerably reduces the numerical work involved.

Systematic Elimination

Consider the following set of equations and their corresponding matrix representation.

$$\left\{\begin{array}{lr} x+y+\ z = 7 & [1] \\ x-y+2z = 9 & [2] \\ 2x+y-\ z = 1 & [3] \end{array}\right. \quad \text{or} \quad \begin{pmatrix} 1 & 1 & 1 \\ 1 & -1 & 2 \\ 2 & 1 & -1 \end{pmatrix}\begin{pmatrix} x \\ y \\ z \end{pmatrix} = \begin{pmatrix} 7 \\ 9 \\ 1 \end{pmatrix}$$

To solve these equations by elimination we can proceed as follows. Eliminating z from equations [1] and [2], i.e. [1] + [3] and [2] + 2[3] gives

$$\left\{\begin{array}{lr} 3x+2y \qquad = 8 & [4] \\ 5x+\ y \qquad = 11 & [5] \\ 2x+\ y-z = 1 & [6] \end{array}\right. \quad \text{or} \quad \begin{pmatrix} 3 & 2 & 0 \\ 5 & 1 & 0 \\ 2 & 1 & -1 \end{pmatrix}\begin{pmatrix} x \\ y \\ z \end{pmatrix} = \begin{pmatrix} 8 \\ 11 \\ 1 \end{pmatrix}$$

Eliminating y from equation [4], i.e. [4] $-$ 2[5] gives

$$\left\{\begin{array}{l} -7x \qquad\qquad = -14 \\ \quad 5x+y \qquad = 11 \\ \quad 2x+y-z = 1 \end{array}\right. \quad \text{or} \quad \begin{pmatrix} -7 & 0 & 0 \\ 5 & 1 & 0 \\ 2 & 1 & -1 \end{pmatrix}\begin{pmatrix} x \\ y \\ z \end{pmatrix} = \begin{pmatrix} -14 \\ 11 \\ 1 \end{pmatrix}$$

from which we have $x = 2$ from the first equation and by direct substitution into the second and then third equation we have $y = 1$ and $z = 4$. Examining the matrix representation of the equations at each stage of the elimination process shows that the matrices are produced by performing the same operations on their rows as is performed on the equations themselves. So we can work directly on the matrix representation, noting that the row operations must be performed on *both* the matrix on the L.H.S. *and* the vector on the R.H.S. *but not* on the vector $\begin{pmatrix} x \\ y \\ z \end{pmatrix}$.

Thus, for the purpose of elimination, we can represent the equations

$$\left.\begin{array}{r} x+y+\ z = 7 \\ x-y+2z = 9 \\ 2x+y-\ z = 1 \end{array}\right\} \text{ by the } \textit{augmented} \text{ matrix } \left(\begin{array}{ccc:c} 1 & 1 & 1 & 7 \\ 1 & -1 & 2 & 9 \\ 2 & 1 & -1 & 1 \end{array}\right).$$

Denoting the rows by $\mathbf{r}_1, \mathbf{r}_2, \mathbf{r}_3$ and this time choosing to eliminate x from equations [2] and [3], we have

$$\mathbf{r}_3 - 2\mathbf{r}_1 \quad \Rightarrow \quad \left(\begin{array}{ccc|c} 1 & 1 & 1 & 7 \\ 1 & -1 & 2 & 9 \\ 0 & -1 & -3 & -13 \end{array}\right)$$

$$\mathbf{r}_2 - \mathbf{r}_1 \quad \Rightarrow \quad \left(\begin{array}{ccc|c} 1 & 1 & 1 & 7 \\ 0 & -2 & 1 & 2 \\ 0 & -1 & -3 & -13 \end{array}\right)$$

$$2\mathbf{r}_3 - \mathbf{r}_2 \quad \Rightarrow \quad \left(\begin{array}{ccc|c} 1 & 1 & 1 & 7 \\ 0 & -2 & 1 & 2 \\ 0 & 0 & -7 & -28 \end{array}\right) \quad \text{or} \quad \left[\begin{array}{rcl} x + y + z & = & 7 \\ -2y + z & = & 2 \\ -7z & = & -28 \end{array}\right]$$

$$\Rightarrow \quad z = 4, \quad y = 1 \quad \text{and} \quad x = 2.$$

Note that the first solution resulted in a matrix with zeros above the leading diagonal and that in the second solution the matrix has zeros below the leading diagonal.

In both cases, the final form of the matrix is called the reduced or echelon (Greek for ladder) form.

Note that to produce the echelon form, we aim for combinations of rows (*not* columns) giving zeros either above or below the leading diagonal.

This systematic method of elimination (or reduction as it is sometimes called) can be extended to the solution of larger numbers of equations containing more variables. It is this method that is used in computer programmes for the solution of such sets of equations.

However, we are not computers and so should not feel bound to produce an echelon form matrix. We can use the augmented matrix and reduce it to the point where we can easily solve the equations. Also unless a solution by reduction is asked for, a simple algebraic solution is often the best method.

For example, to solve the equations

$$\begin{aligned} x - 2y + z &= 6 \\ x + 5y + z &= -1 \\ 2x - y + 4z &= 15 \end{aligned}$$

we can reduce the augmented matrix $\left(\begin{array}{ccc|c} 1 & -2 & 1 & 6 \\ 1 & 5 & 1 & -1 \\ 2 & -1 & 4 & 15 \end{array}\right)$ as follows.

$$\mathbf{r}_1 - \mathbf{r}_2 \quad \Rightarrow \left(\begin{array}{ccc:c} 0 & -7 & 0 & 7 \\ 1 & 5 & 1 & -1 \\ 2 & -1 & 4 & 15 \end{array}\right)$$

$$2\mathbf{r}_2 - \mathbf{r}_3 \quad \Rightarrow \left(\begin{array}{ccc:c} 0 & -7 & 0 & 7 \\ 0 & 11 & -2 & -17 \\ 2 & -1 & 4 & 15 \end{array}\right) \Rightarrow y = -1, \quad z = 3, \quad x = 1.$$

EXERCISE 6f

Solve for x, y, z the following sets of equations, using systematic reduction to produce an echelon matrix:

1) $2x + y - 3z = 4$
 $x - 2y + z = 1$
 $2x + y - z = 0$

2) $3x - y - z = 2$
 $x + y + z = 4$
 $4x - y + z = 7$

3) $4x - y + 5z = 8$
 $5x + 7y - 3z = 42$
 $3x + 4y + z = 27$

4) $2x - 5y + 2z = 14$
 $9x + 3y - 4z = 13$
 $7x + 3y - 2z = 3.$

Solve for x, y, z the following sets of equations by reducing the augmented matrix:

5) $x - y + 2z = -1$
 $4x + y + z = 13$
 $5x - y + 8z = 5$

6) $x - y - 4z = 1$
 $2x + 5y - z = 2$
 $3x + 2y - 3z = -1$

7) $4x + 2y - z = 24$
 $2x + 3y + 2z = 17$
 $6x - 5y + 7z = -21$

8) $4x - 7y + 6z = -18$
 $5x + y - 4z = -9$
 $3x - 2y + 3z = 12.$

THE SOLUTION SETS FOR LINEAR EQUATIONS IN THREE VARIABLES

The sets of equations in the last section all had unique solutions. However, there are other possibilities.

Consider again the set of equations

$$a_1x + b_1y + c_1z = d_1$$
$$a_2x + b_2y + c_2z = d_2$$
$$a_3x + b_3y + c_3z = d_3.$$

In the xyz coordinate system these equations represent planes Π_1, Π_2 and Π_3. The coordinates of any point common to the three planes is a solution of the equations.

These equations may be such that

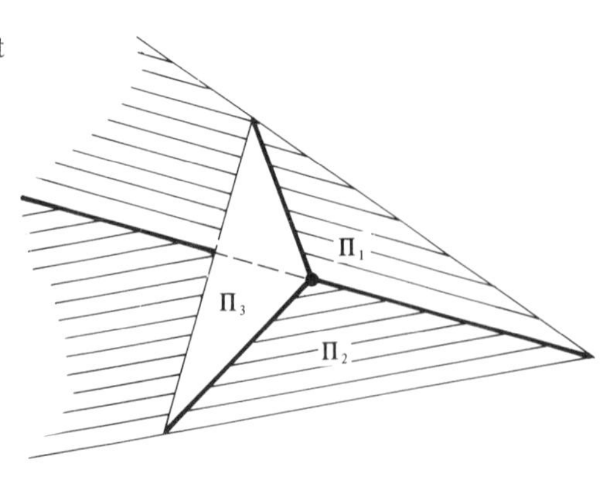

(a) Π_1, Π_2 and Π_3 intersect in one point, in which case the set of equations representing them have a unique solution.

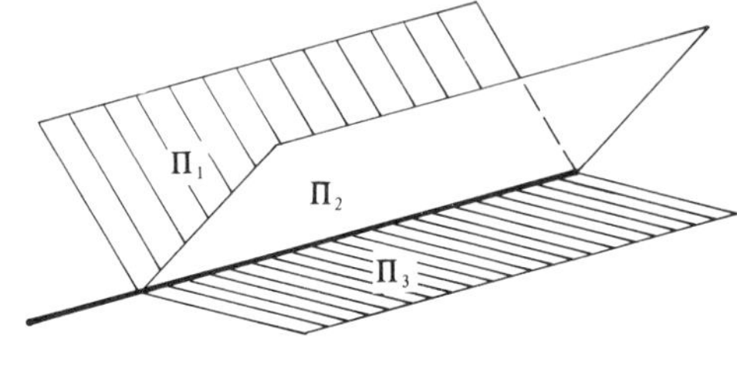

(b) Π_1, Π_2 and Π_3 intersect in a line, (so that the equation of Π_3, say, is a linear combination of the equations of Π_1 and Π_2)

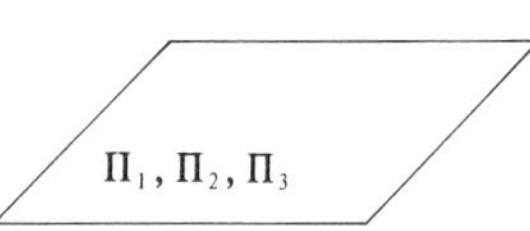

or
Π_1, Π_2, Π_3 are identical.

In either case there is an infinite set of points that are common to all three planes.
planes.
So the set of equations have an infinite set of solutions and the equations are linearly dependent.

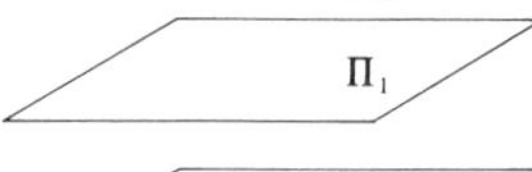

(c) Π_1, Π_2 and Π_3 are all parallel and distinct

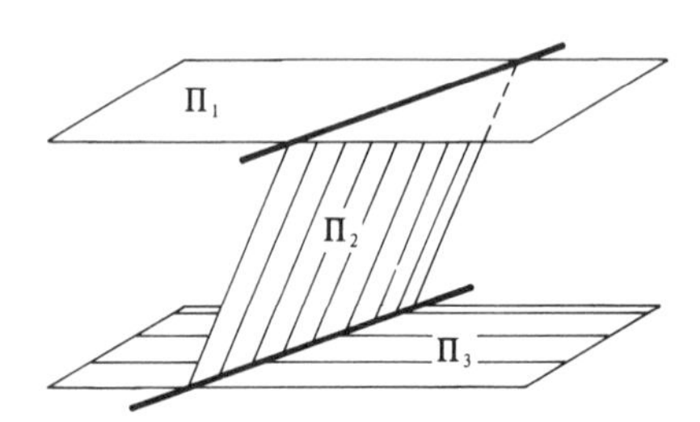

or any two are parallel but distinct

or one plane is parallel to the *line* of intersection of the other two.

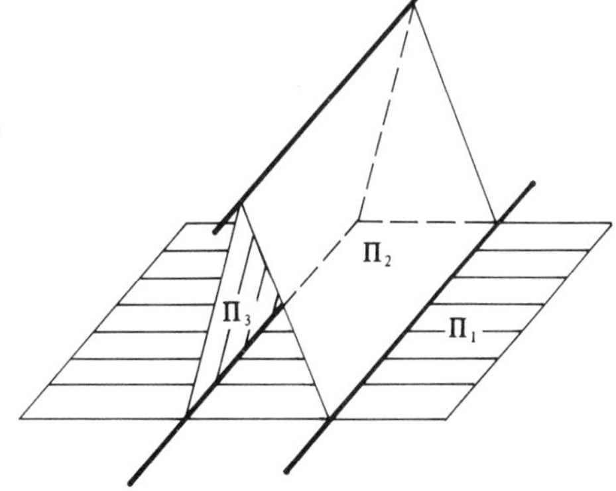

In any of these cases there are no points that are common to all three planes. Hence the set of equations have no solution and are inconsistent.
When the equations represent parallel or identical planes the nature of their solution sets can be determined by observation. (The normals to parallel planes have equal direction ratios.) In the case when none of the planes are parallel, the nature of their solution set quickly becomes apparent when an attempt at solution by reduction is made.

A Unique Solution

In the matrix representation of the set of equations

$$\mathbf{A}\begin{pmatrix} x \\ y \\ z \end{pmatrix} = \begin{pmatrix} d_1 \\ d_2 \\ d_3 \end{pmatrix} \quad \text{where} \quad \mathbf{A} = \begin{pmatrix} a_1 & b_1 & c_1 \\ a_2 & b_2 & c_2 \\ a_3 & b_3 & c_3 \end{pmatrix}$$

we see that the rows of $\mathbf{A}$ represent vectors which are *normal* to the planes Π_1, Π_2 and Π_3.
If the equations do not have a unique solution, then in all five of their possible configurations there is a fourth plane that can be drawn which is perpendicular to Π_1, Π_2 and Π_3.

i.e.

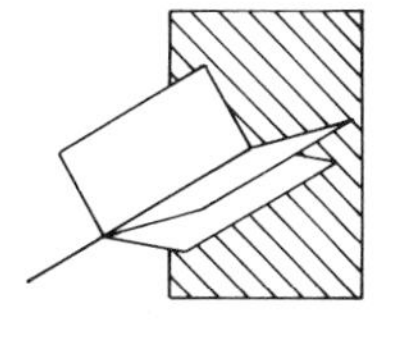

or

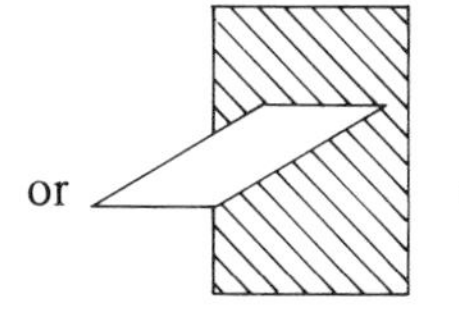

or

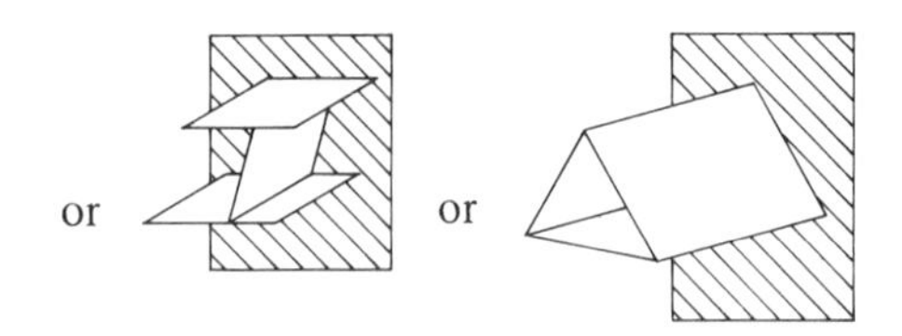

So the normals to Π_1, Π_2 and Π_3 are *paraplanar*, i.e. $|\mathbf{A}| = 0$,

i.e. no unique solution $\Rightarrow$ $|\mathbf{A}| = 0$.

On the other hand, if there is a unique solution, the planes meet in a point and there is *no* fourth plane which is perpendicular to Π_1, Π_2 and Π_3 and so their normals are *not* paraplanar. In this case $|\mathbf{A}| \neq 0$.

Hence the equations have a unique solution $\iff$ $|\mathbf{A}| \neq 0$.

So, by evaluating $|\mathbf{A}|$, we can determine whether or not there is a unique solution. However, in practice it is usually quicker to actually solve the equations by reduction, when the existence, or otherwise, of a unique solution quickly becomes apparent.

An Infinite Set of Solutions

In this case, Π_1, Π_2 and Π_3 have a line in common. So the equation of Π_1 (or Π_2, or Π_3) may be expressed as a linear combination of the equations of the other two planes,
i.e. for some value of k,

$$a_1x + b_1y + c_1z - d_1 = (a_2x + b_2y + c_2z - d_2) + k(a_3x + b_3y + c_3z - d_3)$$

and the equations are *linearly dependent*.
Considering the augmented matrix

$$\begin{pmatrix} a_1 & b_1 & c_1 & \vdots & d_1 \\ a_2 & b_2 & c_2 & \vdots & d_2 \\ a_3 & b_3 & c_3 & \vdots & d_3 \end{pmatrix}$$

we see that, in this case, reduction will lead to *one complete row of zeros*.

Inconsistent Equations

If the equations are inconsistent, the complete rows of the augmented matrix are not linearly dependent, but as $|\mathbf{A}| = 0$, the rows of $\mathbf{A}$ are linearly dependent. So, in this case, reduction of the augmented matrix eventually results in a row of zeros *only* in $\mathbf{A}$. However, it is not always necessary to reduce the augmented matrix to this stage, because evidence of inconsistency often appears earlier.

The following examples show how observation of the given equations for parallel planes followed, if necessary by reduction of the augmented matrix, quickly leads to the determination of the nature of their solution set.

EXAMPLES 6g

1) Solve the equations

$$\begin{aligned} x - 2y + 4z &= 1 \\ 2x - 4y + 8z &= 2 \\ 3x - 6y + 12z &= 3. \end{aligned}$$

By observation we see that the planes Π_1, Π_2 and Π_3 represented by these equations are identical.
(Dividing the second equation by 2 and the third equation by 3 gives the first equation in each case.)
Hence the solution set is the coordinates of all points contained in the plane, i.e. the values of x, y and z such that $x - 2y + 4z = 1$. This solution may be expressed in parametric form, e.g. $x = \lambda, \quad y = \mu \Rightarrow z = \frac{1}{4}(1 - \lambda + \mu)$.

2) Solve the equations

$$\begin{aligned} x - y + z &= 4 \\ 2x + y - 2z &= 1 \\ 5x - 2y + z &= 13. \end{aligned}$$

None of the planes represented by these equations are parallel.
Reducing the augmented matrix

$$\left(\begin{array}{ccc:c} 1 & -1 & 1 & 4 \\ 2 & 1 & -2 & 1 \\ 5 & -2 & 1 & 13 \end{array}\right)$$

gives

$$\mathbf{r}_1 - \mathbf{r}_3 \quad \Rightarrow \quad \left(\begin{array}{ccc:c} -4 & 1 & 0 & -9 \\ 2 & 1 & -2 & 1 \\ 5 & -2 & 1 & 13 \end{array}\right)$$

$$\mathbf{r}_2 + 2\mathbf{r}_3 \quad \Rightarrow \quad \left(\begin{array}{ccc:c} -4 & 1 & 0 & -9 \\ 12 & -3 & 0 & 27 \\ 5 & -2 & 1 & 13 \end{array}\right)$$

Now we see that $\mathbf{r}_2 = -3\mathbf{r}_1$, so the next step would give a complete row of zeros.
Hence the equations are linearly dependent and the planes they represent have a

line in common. So the solution of the equations is the set of coordinates of all the points on this line. The equations of this line are any pair of the three given equations.
So the solution set of the given equations is the set of values of x, y, z such that

$$x - y + z = 4$$
$$2x + y - 2z = 1$$

which may be expressed in parametric form as

$$x = \lambda,\quad y = 4\lambda - 9,\quad z = 3\lambda - 5.$$

3) Solve the equations

$$x + y - z = 1$$
$$x - y + 2z = 5$$
$$2x + y - z = 3.$$

None of the planes represented by these equations are parallel.

The augmented matrix is $\left(\begin{array}{ccc:c} 1 & 1 & -1 & 1 \\ 1 & -1 & 2 & 5 \\ 2 & 1 & -1 & 3 \end{array}\right)$

and $\mathbf{r}_3 - \mathbf{r}_1$ gives $\left(\begin{array}{ccc:c} 1 & 1 & -1 & 1 \\ 1 & -1 & 2 & 5 \\ 1 & 0 & 0 & 2 \end{array}\right)$

From the third row, $x = 2$, but this does not necessarily mean that there is a unique solution. One more step is necessary to establish the nature of the solution set.

$$\mathbf{r}_1 + \mathbf{r}_2 \quad \Rightarrow \quad \left(\begin{array}{ccc:c} 2 & 0 & 1 & 6 \\ 1 & -1 & 2 & 5 \\ 1 & 0 & 0 & 2 \end{array}\right) \quad \Rightarrow \quad x = 2,\quad z = 2,\quad y = 1,$$

i.e. there is a unique solution.

4) Find the solution set, if there is one, of the equations

$$2x + y - z = 1$$
$$x - y + z = 3$$
$$x + 5y - 5z = 2.$$

Again, none of these equations represent parallel planes.

Now

$$\begin{pmatrix} 2 & 1 & -1 & \vdots & 1 \\ 1 & -1 & 1 & \vdots & 3 \\ 1 & 5 & -5 & \vdots & 2 \end{pmatrix} \Rightarrow \begin{pmatrix} 3 & 0 & 0 & \vdots & 4 \\ 1 & -1 & 1 & \vdots & 3 \\ 1 & 5 & -5 & \vdots & 2 \end{pmatrix} \Rightarrow x = \frac{4}{3} \quad \text{(top row)}$$

then

$$(\mathbf{r}_3 + 5\mathbf{r}_2) \Rightarrow \begin{pmatrix} 3 & 0 & 0 & \vdots & 4 \\ 1 & -1 & 1 & \vdots & 3 \\ 6 & 0 & 0 & \vdots & 17 \end{pmatrix} \Rightarrow x = \frac{4}{3} \text{ and } x = \frac{17}{6} \quad \text{(3rd row)}.$$

The values obtained for x from the final matrix are inconsistent. Hence the equations are inconsistent and the planes they represent form a triangular prism.

5) If it is possible, solve the equations

$$\begin{aligned} x+y+z &= 1 \\ 2x-y+z &= 4 \\ x+y+z &= 2. \end{aligned}$$

Two of these equations (the first and last) represent parallel, but distinct, planes. Hence the equations are inconsistent and there is no solution.

EXERCISE 6g

In Questions 1–8 do not solve the equations, but determine whether there is a unique solution, no solution, or an infinite set of solutions, in which case state whether this set is dependent on one or two parameters.

1) $\begin{aligned} x-3y+z &= 4 \\ 2x-y+z &= 2 \\ x+2y &= -2 \end{aligned}$

2) $\begin{aligned} x+y-2z &= 1 \\ 2x-y+z &= 4 \\ 3x+y-z &= 2 \end{aligned}$

3) $\begin{aligned} 2x+y-z &= 1 \\ 6x+3y-3z &= 4 \\ 4x+2y-2z &= 2 \end{aligned}$

4) $\begin{aligned} 2x+y-z &= 1 \\ 6x+3y-3z &= 3 \\ x+y-z &= 2 \end{aligned}$

5) $\begin{aligned} x-y+z &= 0 \\ x+y-2z &= 0 \\ 3x+y-3z &= 0 \end{aligned}$

6) $\begin{aligned} x+y+z &= 1 \\ 2x+2y+2z &= 2 \\ 3x+3y+3z &= 3 \end{aligned}$

7) $\begin{aligned} x-y+z &= 1 \\ 2x+y-z &= 2 \\ 3x-2y+z &= 3 \end{aligned}$

8) $\begin{aligned} x+y-2z &= 1 \\ 3x+y-4z &= 2 \\ 4x+2y-6z &= 5 \end{aligned}$

9) Solve, where possible, the sets of equations in Questions 1–8.

10) By considering the solution to the set of equations

$$\begin{aligned} x+y-z &= 2 \\ 2x-y+z &= 3 \\ x+4y-4z &= 3 \end{aligned}$$

or otherwise, find the set of points which map to the point $(2, 3, 3)$ under the transformation defined by $\mathbf{M} = \begin{pmatrix} 1 & 1 & -1 \\ 2 & -1 & 1 \\ 1 & 4 & -4 \end{pmatrix}$

11) Find the set of points which are mapped to the point $(1, -1, -1)$ by the transformation $\begin{pmatrix} 1 & -2 & 4 \\ 3 & 4 & 6 \\ 1 & 3 & 1 \end{pmatrix}\begin{pmatrix} x \\ y \\ z \end{pmatrix} = \begin{pmatrix} X \\ Y \\ Z \end{pmatrix}$

12) Show that the matrix $\begin{pmatrix} 1 & 0 & 1 \\ -1 & 2 & 1 \\ 1 & 0 & 2 \end{pmatrix}$, which maps $\begin{pmatrix} x \\ y \\ z \end{pmatrix}$ to $\begin{pmatrix} X \\ Y \\ Z \end{pmatrix}$ defines a transformation in which any point in the image space arises from a unique point in xyz space.

13) Find the set of points, which under the transformation defined by $\mathbf{M} = \begin{pmatrix} 1 & -1 & 2 \\ -1 & 1 & -2 \\ 1 & -1 & 2 \end{pmatrix}$ map to the point $(1, -1, 1)$.

14) Show that the matrix $\mathbf{M} = \begin{pmatrix} 1 & 2 & -1 \\ 2 & 4 & -2 \\ 3 & 6 & -3 \end{pmatrix}$ maps all points in xyz space to a line in XYZ space. Find the equations of this line and show that any point on this line arises from a set of coplanar points in xyz space.

15) Find the values of a for which the lines

$$\begin{aligned} 2x - y + a &= 0 \\ ax + y + 1 &= 0 \\ x - y - a &= 0 \end{aligned}$$

are concurrent.

16) What is the condition that the equations

$$\begin{aligned} a_1x + b_1y + c_1z &= 0 \\ a_2x + b_2y + c_2z &= 0 \\ a_3x + b_3y + c_3z &= 0 \end{aligned}$$

should have solutions other than $x = y = z = 0$?
(**Note** that such equations are called homogeneous.)

17) By writing the homogeneous equations

$$\left.\begin{aligned} a_1x + b_1y + c_1z &= 0 \\ a_2x + b_2y + c_2z &= 0 \\ a_3x + b_3y + c_3z &= 0 \end{aligned}\right\} \qquad [1]$$

as

$$\left.\begin{aligned} a_1(x/z) + b_1(y/z) + c_1 &= 0 \\ a_2(x/z) + b_2(y/z) + c_2 &= 0 \\ a_3(x/z) + b_3(y/z) + c_3 &= 0 \end{aligned}\right\} \qquad [2]$$

write down the condition for the set of equations [2] to have a common solution for $\dfrac{x}{z}$ and $\dfrac{y}{z}$.

What is the geometrical significance when this condition is applied to the set of equations [1]?

18) Considering the equations

$$\left.\begin{aligned} a_1x + b_1y + c_1 &= 0 \\ a_2x + b_2y + c_2 &= 0 \\ a_3x + b_3y + c_3 &= 0 \end{aligned}\right\} \qquad [1]$$

as the set of equations

$$\left.\begin{aligned} a_1x + b_1y + c_1z &= 0 \\ a_2x + b_2y + c_2z &= 0 \\ a_3x + b_3y + c_3z &= 0 \end{aligned}\right\} \quad \text{with} \quad z = 1$$

find the condition that the first set should have a unique solution and interpret this solution geometrically.

19) Find the value of a for which the following equations are consistent

$$\begin{aligned} x - 3y + 5z &= 2 \\ x + 4y - z &= 1 \\ 7y - 6z &= a. \end{aligned}$$

20) Find a 3×3 matrix which maps all points of xyz space to points on the plane $x + y + z = 0$.

21) Write down the condition on a_1, b_1 and a_2, b_2 for the equations

$$\left.\begin{aligned} a_1x + b_1y + c_1 &= 0 \\ a_2x + b_2y + c_2 &= 0 \end{aligned}\right\} \text{ to have a unique solution}$$

and show that

$$y = \frac{\begin{vmatrix} a_1 & -c_1 \\ a_2 & -c_2 \end{vmatrix}}{\begin{vmatrix} a_1 & b_1 \\ a_2 & b_2 \end{vmatrix}}$$

Hence show that the quadratic equations

$$\begin{aligned} a_1y^2 + b_1y + c_1 &= 0 \\ a_2y^2 + b_2y + c_2 &= 0 \end{aligned}$$

have a common root if

$$(a_2c_1 - a_1c_2)^2 = (b_1c_2 - b_2c_1)(a_1b_2 - a_2b_1).$$

The Evaluation of an Inverse Matrix by Reduction

The systematic reduction of a set of equations leads to an alternative method for the calculation of an inverse matrix.
Consider the equation

$$\begin{pmatrix} 1 & 1 & 1 \\ 1 & -1 & 2 \\ 2 & 1 & -1 \end{pmatrix}\begin{pmatrix} x \\ y \\ z \end{pmatrix} = \begin{pmatrix} 7 \\ 9 \\ 1 \end{pmatrix}$$

which we can write as

$$\begin{pmatrix} 1 & 1 & 1 \\ 1 & -1 & 2 \\ 2 & 1 & -1 \end{pmatrix}\begin{pmatrix} x \\ y \\ z \end{pmatrix} = \begin{pmatrix} 1 & 0 & 0 \\ 0 & 1 & 0 \\ 0 & 0 & 1 \end{pmatrix}\begin{pmatrix} 7 \\ 9 \\ 1 \end{pmatrix}$$

We find that any row operation performed using

either $\mathbf{I}$ *or* $\begin{pmatrix} 7 \\ 9 \\ 1 \end{pmatrix}$ but *not* both, produces the same result on the R.H.S.,

e.g. $\quad (\mathbf{r}_2 - \mathbf{r}_1)$ on $\mathbf{I} \quad \Rightarrow \quad \begin{pmatrix} 1 & 0 & 0 \\ -1 & 1 & 0 \\ 0 & 0 & 1 \end{pmatrix}\begin{pmatrix} 7 \\ 9 \\ 1 \end{pmatrix} = \begin{pmatrix} 7 \\ 2 \\ 1 \end{pmatrix}$

and $\quad (\mathbf{r}_2 - \mathbf{r}_1) \quad$ on $\quad \begin{pmatrix} 7 \\ 9 \\ 1 \end{pmatrix} \Rightarrow \begin{pmatrix} 1 & 0 & 0 \\ 0 & 1 & 0 \\ 0 & 0 & 1 \end{pmatrix} \begin{pmatrix} 7 \\ 2 \\ 1 \end{pmatrix} = \begin{pmatrix} 7 \\ 2 \\ 1 \end{pmatrix}$

So if we choose to operate on $\mathbf{I}$ on the R.H.S. we work with the matrix

$$\left(\begin{array}{ccc:ccc} 1 & 1 & 1 & 1 & 0 & 0 \\ 1 & -1 & 2 & 0 & 1 & 0 \\ 2 & 1 & -1 & 0 & 0 & 1 \end{array}\right)$$

The elimination process can then be continued until the L.H.S. is reduced to the form $\mathbf{I}\begin{pmatrix} x \\ y \\ z \end{pmatrix}$,

$$(\mathbf{r}_3 - 2\mathbf{r}_1) \quad \Rightarrow \quad \left(\begin{array}{ccc:ccc} 1 & 1 & 1 & 1 & 0 & 0 \\ 1 & -1 & 2 & 0 & 1 & 0 \\ 0 & -1 & -3 & -2 & 0 & 1 \end{array}\right)$$

$$(\mathbf{r}_2 - \mathbf{r}_1) \quad \Rightarrow \quad \left(\begin{array}{ccc:ccc} 1 & 1 & 1 & 1 & 0 & 0 \\ 0 & -2 & 1 & -1 & 1 & 0 \\ 0 & -1 & -3 & -2 & 0 & 1 \end{array}\right)$$

$$(2\mathbf{r}_3 - \mathbf{r}_2) \quad \Rightarrow \quad \left(\begin{array}{ccc:ccc} 1 & 1 & 1 & 1 & 0 & 0 \\ 0 & -2 & 1 & -1 & 1 & 0 \\ 0 & 0 & -7 & -3 & -1 & 2 \end{array}\right)$$

$$(7\mathbf{r}_1 + \mathbf{r}_3) \quad \text{and} \quad (7\mathbf{r}_2 + \mathbf{r}_3) \quad \Rightarrow \quad \left(\begin{array}{ccc:ccc} 7 & 7 & 0 & 4 & -1 & 2 \\ 0 & -14 & 0 & -10 & 6 & 2 \\ 0 & 0 & -7 & -3 & -1 & 2 \end{array}\right)$$

$$2\mathbf{r}_1 + \mathbf{r}_2 \quad \Rightarrow \quad \left(\begin{array}{ccc:ccc} 14 & 0 & 0 & 4 & -2 & 6 \\ 0 & -14 & 0 & -10 & 6 & 2 \\ 0 & 0 & -7 & -3 & -1 & 2 \end{array}\right)$$

$$\tfrac{1}{14}\mathbf{r}_1, \quad -\tfrac{1}{14}\mathbf{r}_2, \quad -\tfrac{1}{7}\mathbf{r}_3 \quad \Rightarrow \quad \left(\begin{array}{ccc:ccc} 1 & 0 & 0 & -\frac{1}{7} & \frac{2}{7} & \frac{3}{7} \\ 0 & 1 & 0 & \frac{5}{7} & -\frac{3}{7} & -\frac{1}{7} \\ 0 & 0 & 1 & \frac{3}{7} & \frac{1}{7} & -\frac{2}{7} \end{array}\right)$$

So we have now reduced the original set of equations to

$$\begin{pmatrix} 1 & 0 & 0 \\ 0 & 1 & 0 \\ 0 & 0 & 1 \end{pmatrix}\begin{pmatrix} x \\ y \\ z \end{pmatrix} = \begin{pmatrix} -\frac{1}{7} & \frac{2}{7} & \frac{3}{7} \\ \frac{5}{7} & -\frac{3}{7} & -\frac{1}{7} \\ \frac{3}{7} & \frac{1}{7} & -\frac{2}{7} \end{pmatrix}\begin{pmatrix} 7 \\ 9 \\ 1 \end{pmatrix},$$

i.e. $\frac{1}{7}\begin{pmatrix} -1 & 2 & 3 \\ 5 & -3 & -1 \\ 3 & 1 & -2 \end{pmatrix}$ is the inverse of $\begin{pmatrix} 1 & 1 & 1 \\ 1 & -1 & 2 \\ 2 & 1 & -1 \end{pmatrix}$

So to find the inverse of $\mathbf{A} = \begin{pmatrix} a_1 & b_1 & c_1 \\ a_2 & b_2 & c_2 \\ a_3 & b_3 & c_3 \end{pmatrix}$

we apply row operations to the matrix

$$\left(\begin{array}{ccc:ccc} a_1 & b_1 & c_1 & 1 & 0 & 0 \\ a_2 & b_2 & c_2 & 0 & 1 & 0 \\ a_3 & b_3 & c_3 & 0 & 0 & 1 \end{array}\right)$$

until the left hand section of this matrix is reduced to $\mathbf{I}$.
The right hand section of this matrix is then $\mathbf{A}^{-1}$.
If $\mathbf{A}$ has no inverse this fact quickly becomes apparent because the attempted reduction of $\mathbf{A}$ produces a row of zeros.
This method has many advantages over the 'cofactor' method introduced earlier in this chapter, because it greatly simplifies the numerical work involved, and so is less likely to give rise to mistakes.
(But mistakes are still possible, so checking the product of $\mathbf{A}$ and the calculated $\mathbf{A}^{-1}$ is still recommended!)
It is interesting to note that computers use this reduction method for calculating inverse matrices.

EXERCISE 6h

Find the inverse, where it exists, of each of the following matrices:

1) $\begin{pmatrix} 1 & 2 & 4 \\ -1 & 2 & 1 \\ 1 & 5 & 3 \end{pmatrix}$ 2) $\begin{pmatrix} 2 & -1 & 1 \\ 1 & -2 & 3 \\ 5 & 1 & 4 \end{pmatrix}$ 3) $\begin{pmatrix} 5 & 7 & 9 \\ -5 & 4 & 6 \\ 0 & 10 & 15 \end{pmatrix}$

4) $\begin{pmatrix} 2 & -1 & 4 \\ 2 & -1 & 5 \\ 1 & 2 & 3 \end{pmatrix}$ 5) $\begin{pmatrix} -1 & 6 & 9 \\ 11 & -12 & -1 \\ 5 & -3 & 4 \end{pmatrix}$ 6) $\begin{pmatrix} 2 & 6 & 3 \\ 5 & -1 & 4 \\ 9 & 11 & 11 \end{pmatrix}$.

SUMMARY

Any transformation of three dimensional space which is linear can be expressed in the form $\mathbf{Mr} = \mathbf{r}'$ where $\mathbf{M} = \begin{pmatrix} a_1 & b_1 & c_1 \\ a_2 & b_2 & c_2 \\ a_3 & b_3 & c_3 \end{pmatrix}$ and $\mathbf{r}'$ is the image of $\mathbf{r}$ under $\mathbf{M}$.

$\mathbf{M}$ maps $\mathbf{i} \to \begin{pmatrix} a_1 \\ a_2 \\ a_3 \end{pmatrix}$, $\mathbf{j} \to \begin{pmatrix} b_1 \\ b_2 \\ b_3 \end{pmatrix}$, $\mathbf{k} \to \begin{pmatrix} c_1 \\ c_2 \\ c_3 \end{pmatrix}$

$\mathbf{M}$ multiplies volume by a factor $|\mathbf{M}|$.

If $|\mathbf{M}| = 0$, volume is destroyed, and $\mathbf{M}$ maps xyz-space to a line or a plane through O.

Three vectors $\mathbf{a} = \begin{pmatrix} a_1 \\ a_2 \\ a_3 \end{pmatrix}$, $\mathbf{b} = \begin{pmatrix} b_1 \\ b_2 \\ b_3 \end{pmatrix}$, $\mathbf{c} = \begin{pmatrix} c_1 \\ c_2 \\ c_3 \end{pmatrix}$ are

paraplanar if

$$\begin{vmatrix} a_1 & b_1 & c_1 \\ a_2 & b_2 & c_2 \\ a_3 & b_3 & c_3 \end{vmatrix} = 0.$$

If $\mathbf{A}^T$ is the transpose of any matrix $\mathbf{A}$, then $\mathbf{A}^T$ is the matrix whose rows are the columns of $\mathbf{A}$.
For any two matrices $\mathbf{A}$ and $\mathbf{B}$ that are compatible for multiplication $(\mathbf{AB})^T = \mathbf{B}^T\mathbf{A}^T$.
For any *square* matrix $\mathbf{A}$, the adjoint matrix, $\mathbf{A}^*$, is the matrix whose elements are the cofactors of $\mathbf{A}^T$.
For any square matrix $\mathbf{A}$, the inverse matrix, $\mathbf{A}^{-1}$, exists if and only if $|\mathbf{A}| \neq 0$ and in this case $\mathbf{A}^{-1} = \dfrac{1}{|\mathbf{A}|}\mathbf{A}^*$.

$\mathbf{A}^{-1}\mathbf{A} = \mathbf{I} = \mathbf{AA}^{-1}$ and $(\mathbf{A}^T)^{-1} = (\mathbf{A}^{-1})^T$.

If $\mathbf{A}$ and $\mathbf{B}$ both have inverses and are of the same size then

$$(\mathbf{AB})^{-1} = \mathbf{B}^{-1}\mathbf{A}^{-1}.$$

If a transformation, defined by $\mathbf{M}$, maps $\mathbf{r}$ to $\mathbf{r}'$
the inverse transformation, defined by $\mathbf{M}^{-1}$, maps $\mathbf{r}'$ back to $\mathbf{r}$.
A set of linear equations has either

a unique solution in which case $|\mathbf{A}| \neq 0$

or an infinite set of solutions in which case $|\mathbf{A}| = 0$ and the equations are linearly dependent

or no solution, in which case $|\mathbf{A}| = 0$ and the equations are inconsistent.

Three lines

$$\left.\begin{aligned} y-m_1x-c_1 &= 0\\ y-m_2x-c_2 &= 0\\ y-m_3x-c_3 &= 0\end{aligned}\right\}\text{ are concurrent if } \begin{vmatrix} 1 & -m_1 & -c_1\\ 1 & -m_2 & -c_2\\ 1 & -m_3 & -c_3\end{vmatrix} = 0.$$

MULTIPLE CHOICE EXERCISE 6

(The instructions for answering the questions are on p. xii)

TYPE I

1) If $\mathbf{M} = \begin{pmatrix} 1 & 0 & 0\\ 1 & 1 & 0\\ 0 & 1 & 0\end{pmatrix}$, $\mathbf{M}^T$ is

(a) $\begin{pmatrix} 1 & 1 & 0\\ 1 & 1 & 0\\ 0 & 1 & 0\end{pmatrix}$ (b) $\begin{pmatrix} 1 & 1 & 0\\ 0 & 1 & 1\\ 0 & 0 & 0\end{pmatrix}$ (c) $\begin{pmatrix} 0 & 0 & 1\\ 0 & 0 & -1\\ 0 & 0 & 1\end{pmatrix}$ (d) $\begin{pmatrix} 0 & 1 & 0\\ 1 & 1 & 0\\ 1 & 0 & 0\end{pmatrix}$

2) The inverse of the matrix $\begin{pmatrix} 1 & -1\\ 2 & 1\end{pmatrix}$ is

(a) $\begin{pmatrix} 1 & 1\\ -2 & 1\end{pmatrix}$ (b) $\frac{1}{3}\begin{pmatrix} 1 & 2\\ -1 & 1\end{pmatrix}$ (c) $\begin{pmatrix} 1 & 0\\ 0 & 1\end{pmatrix}$

(d) $\frac{1}{3}\begin{pmatrix} 1 & 1\\ -2 & 1\end{pmatrix}$ (e) $1\Big/\begin{pmatrix} 1 & -1\\ 2 & 1\end{pmatrix}$

3) If $\mathbf{M} = \begin{pmatrix} 1 & 0 & 0\\ 1 & 1 & 1\\ 0 & 1 & 2\end{pmatrix}$, a reduced echelon form may be

(a) $\begin{pmatrix} 1 & 0 & 0\\ 0 & 0 & -1\\ 0 & 1 & 2\end{pmatrix}$ (b) $\begin{pmatrix} 1 & 0 & 0\\ 1 & \frac{1}{2} & 0\\ 0 & 1 & 2\end{pmatrix}$ (c) $\begin{pmatrix} 0 & 0 & 0\\ 1 & 1 & 1\\ 0 & 1 & 2\end{pmatrix}$ (d) $\begin{pmatrix} 1 & 0 & 0\\ 1 & 1 & 0\\ 0 & 1 & 2\end{pmatrix}$

4) The set of equations

$$\begin{cases} x-2y+4 = 0\\ 2x-4y-3 = 0\end{cases}$$

(a) are inconsistent, (b) have solutions depending on one parameter,
(c) have a unique solution, (d) represent a pair of intersecting lines.

5) If $\mathbf{M} = \begin{pmatrix} 1 & 0 & 0\\ 1 & 1 & 0\\ 0 & 0 & 1\end{pmatrix}$, $\mathbf{M}^*$ is

(a) $\begin{pmatrix} 1 & -1 & 0 \\ 0 & 1 & 0 \\ 0 & 0 & 1 \end{pmatrix}$ (b) $\begin{pmatrix} 1 & 1 & 0 \\ 0 & 1 & 0 \\ 0 & 0 & 1 \end{pmatrix}$ (c) $\begin{pmatrix} 1 & 0 & 0 \\ 0 & 1 & 0 \\ 0 & 0 & 1 \end{pmatrix}$

(d) $\begin{pmatrix} 1 & 0 & 0 \\ -1 & 1 & 0 \\ 0 & 0 & 1 \end{pmatrix}$.

6) The transformation of xyz space defined by $\begin{pmatrix} 2 & 0 & 0 \\ 1 & 2 & 4 \\ 1 & -1 & 3 \end{pmatrix}$

(a) is singular, (b) preserves volume,
(c) maps 3-D space to a 2-D plane, (d) multiplies volume by a factor 20.

TYPE II

7) A transformation of xyz space maps

$$x \text{ to } 2x + y - z$$
$$y \text{ to } x - y + z$$
$$z \text{ to } x + y.$$

(a) The transformation matrix is $\begin{pmatrix} 2 & 1 & -1 \\ 1 & -1 & 1 \\ 1 & 1 & 0 \end{pmatrix}$

(b) Volume is unchanged by the transformation.
(c) The point $(2, 0, 0)$ is mapped to the point $(-3, 0, 0)$.

8) If $\mathbf{M} = \begin{pmatrix} 1 & 0 & 0 \\ 0 & 0 & 1 \\ 0 & 1 & 0 \end{pmatrix}$

(a) $\mathbf{M}$ has an inverse, (b) $\mathbf{M}$ represents a rotation about Ox of $90°$,
(c) $\mathbf{M}$ represents a reflection in the plane $y = z$.

9) $\mathbf{M} = \begin{pmatrix} 2 & -1 & 0 \\ 2 & 1 & 4 \\ 3 & 0 & 2 \end{pmatrix}$.

(a) The equation $\mathbf{M}\begin{pmatrix} x \\ y \\ z \end{pmatrix} = \begin{pmatrix} 0 \\ 0 \\ 0 \end{pmatrix}$ has a unique solution,

(b) $\mathbf{M}$ is singular, (c) $\mathbf{M}$ maps $\begin{pmatrix} 0 \\ 1 \\ 0 \end{pmatrix}$ to $\begin{pmatrix} -1 \\ 1 \\ 0 \end{pmatrix}$.

10) $\mathbf{A} = \begin{pmatrix} 1 & 2 \\ 3 & 4 \end{pmatrix}$.

(a) $\mathbf{A}^T = \begin{pmatrix} 1 & 3 \\ 2 & 4 \end{pmatrix}$, (b) $\mathbf{A}^* = \begin{pmatrix} 4 & -3 \\ -2 & 1 \end{pmatrix}$, (c) $(\mathbf{A}^T)^{-1} = \begin{pmatrix} -2 & \frac{3}{2} \\ 1 & -\frac{1}{2} \end{pmatrix}$.

11) $$\begin{aligned} 2x + y - z &= 4, \\ x + y + z &= 1, \\ 3x - 2y - z &= 2. \end{aligned}$$

(a) The planes represented by these equations are parallel.

(b) The solution is the same as the solution of

$$\begin{pmatrix} 11 & 0 & 0 \\ 4 & -1 & 0 \\ 3 & -2 & -1 \end{pmatrix}\begin{pmatrix} x \\ y \\ z \end{pmatrix} = \begin{pmatrix} 11 \\ 3 \\ 2 \end{pmatrix}.$$

(c) The solution is unique.

TYPE III

12) $a_1x + b_1y + c_1z = d_1$, $a_2x + b_2y + c_2z = d_2$, $a_3x + b_3y + c_3z = d_3$.

(a) $\begin{vmatrix} a_1 & b_1 & c_1 \\ a_2 & b_2 & c_2 \\ a_3 & b_3 & c_3 \end{vmatrix} = 0$

(b) The planes represented by these equations are parallel.

13) $\mathbf{A}$ is a non-singular square matrix.

(a) $|\mathbf{A}^{-1}| = \frac{1}{\mathbf{A}}\Delta$. (b) $|\mathbf{A}| = \frac{1}{\Delta}$.

14) $\mathbf{A}$ is a non-singular square matrix.

(a) $\mathbf{A}$ is of size 3×3.

(b) Under the transformation defined by $\mathbf{A}$, any point in the image space originates from a unique point in xyz space.

15) $\mathbf{a} = \begin{pmatrix} a_1 \\ a_2 \\ a_3 \end{pmatrix}$, $\mathbf{b} = \begin{pmatrix} b_1 \\ b_2 \\ b_3 \end{pmatrix}$, $\mathbf{c} = \begin{pmatrix} c_1 \\ c_2 \\ c_3 \end{pmatrix}$

(a) $\mathbf{a} = \lambda\mathbf{b} + \mu\mathbf{c}$ where λ and μ are scalars.

(b) $\begin{vmatrix} a_1 & a_2 & a_3 \\ b_1 & b_2 & b_3 \\ c_1 & c_2 & c_3 \end{vmatrix} = 0$

16) $2x - 5y + 6 = 0$ and $x + y - 2 = 0$ represent lines L_1 and L_2.

(a) The line $ax + by + c = 0$ is concurrent with L_1 and L_2,

(b) $\begin{vmatrix} 2 & -5 & 6 \\ 1 & 1 & -2 \\ a & b & c \end{vmatrix} = 0$

TYPE V

17) If $\mathbf{A}$ is singular, then $\mathbf{A}$ has no inverse.

18) If three planes form a triangular prism, the equations representing them have no solution.

19) Under a particular transformation, a plane Π is mapped to its image plane Π'. If the normal to Π is mapped to the vector $\mathbf{n}$, then $\mathbf{n}$ is normal to Π'.

MISCELLANEOUS EXERCISE 6

1) The planes

$$\begin{aligned} 2x + y + z &= 4 \\ x + 2y + z &= 2 \\ x + y + 2z &= 6 \end{aligned}$$

meet only in the point $(1, -1, 3)$. The x, y, z coordinate system is transformed by the linear transformation

$$\begin{pmatrix} x \\ y \\ z \end{pmatrix} = \frac{1}{3}\begin{pmatrix} 1 & 2 & 2 \\ 2 & -2 & 1 \\ 2 & 1 & -2 \end{pmatrix}\begin{pmatrix} X \\ Y \\ Z \end{pmatrix}.$$

In the X, Y, Z system, obtain the equations of the planes and the coordinates of the point(s) in which they meet. (U of L)

2) The equations

$$\begin{aligned} 2\lambda x - 3y + \lambda - 3 &= 0 \\ 3x - 2y + 1 &= 0 \\ 4x - \lambda y + 2 &= 0 \end{aligned}$$

represent three straight lines in the xy plane. Find the values of λ for which the lines are concurrent. For each of these values of λ, find the coordinates of the point at which the lines are concurrent. (U of L)p

3) If $\mathbf{M} = \begin{pmatrix} 0 & 1 & 0 \\ -1 & 0 & 0 \\ 0 & 0 & 1 \end{pmatrix}$, find $\mathbf{M}^2$.

Interpret geometrically the transformations of xyz space defined by $\mathbf{M}$ and by $\mathbf{M}^2$. (U of L)

4) (a) Solve for x, y and z the equations

$$\begin{aligned} 2x + 6y + z &= 0 \\ -x + 2y - z &= 10 \\ 4x + 3y + z &= 1. \end{aligned}$$

(b) If $\mathbf{A}$ and $\mathbf{B}$ are non-singular matrices, show that
(a) $(\mathbf{AB})^{-1} = \mathbf{B}^{-1}\mathbf{A}^{-1}$ (b) $(\mathbf{AB})^T = \mathbf{B}^T\mathbf{A}^T$. (U of L)

5) The point $P(x, y, z)$ is transformed to the point $Q(X, Y, Z)$ by the relation

$$\begin{pmatrix} X \\ Y \\ Z \end{pmatrix} = \mathbf{M}\begin{pmatrix} x \\ y \\ z \end{pmatrix}$$

(a) If $\mathbf{M} = \begin{pmatrix} 6 & 8 & 4 \\ 9 & 12 & 6 \\ 4 & -1 & 3 \end{pmatrix}$, show that for all P the corresponding point Q lies on a plane and give an equation of this plane.

(b) If $\mathbf{M} = \begin{pmatrix} 1 & 2 & -1 \\ 3 & 6 & -3 \\ 5 & 10 & -5 \end{pmatrix}$, show that for all P the corresponding point Q lies on a line and give equations for this line.

(c) If $\mathbf{M} = \begin{pmatrix} 0 & -1 & 0 \\ 1 & 0 & 0 \\ 0 & 0 & 1 \end{pmatrix}$, show that for all P the corresponding point Q is in the position P would reach if it were rotated through 90° about Oz and state the inverse matrix in this case. (U of L)

6) Find numbers a, b and c so that the product $\mathbf{BA}$ of the matrices

$$\mathbf{B} = \begin{pmatrix} 1 & 0 & 0 \\ a & 1 & 0 \\ b & c & 1 \end{pmatrix}, \quad \mathbf{A} = \begin{pmatrix} 1 & 2 & 3 \\ -2 & 1 & 4 \\ 2 & 1 & 1 \end{pmatrix}$$

should have only zeros below the leading diagonal.
Hence, or otherwise, solve the equation $\mathbf{Ax} = \mathbf{p}$ where $\mathbf{x}$ is the column vector $\begin{pmatrix} x_1 \\ y_1 \\ z_1 \end{pmatrix}$ and $\mathbf{p}$ is the column vector $\begin{pmatrix} 1 \\ -1 \\ 1 \end{pmatrix}$. (U of L)

7) Find the inverse $\mathbf{A}^{-1}$ of the matrix $\mathbf{A} = \begin{pmatrix} 1 & 0 & 0 \\ -1 & 1 & 0 \\ 3 & 2 & 1 \end{pmatrix}$.

Find also $\mathbf{B}^{-1}$ and $(\mathbf{AB})^{-1}$ where $\mathbf{B} = \begin{pmatrix} 1 & 4 & -2 \\ 0 & 1 & 3 \\ 0 & 0 & 1 \end{pmatrix}$.

Given that $\mathbf{AB}\begin{pmatrix} x_1 \\ x_2 \\ x_3 \end{pmatrix} = \begin{pmatrix} 1 \\ -2 \\ 1 \end{pmatrix}$ find $\begin{pmatrix} x_1 \\ x_2 \\ x_3 \end{pmatrix}$. (U of L)

8) Given that $\mathbf{A} = \begin{pmatrix} 0 & 2 & 3 \\ 2 & 0 & 0 \\ 1 & -1 & 0 \end{pmatrix}$ evaluate $\mathbf{A}^{-1}$ and $\mathbf{A}^2$, and show that $\mathbf{A}^2 + 6\mathbf{A}^{-1} = 7\mathbf{I}$, where $\mathbf{I}$ is the unit matrix of order 3.
Deduce that $\mathbf{A}^3 - 7\mathbf{A} + 6\mathbf{I} = 0$ and use this equation to find $\mathbf{A}^3$. (U of L)

9) Calculate the inverse $\mathbf{A}^{-1}$ of the matrix $\mathbf{A} = \begin{pmatrix} 2 & 2 & 1 \\ 2 & 4 & 1 \\ 3 & 2 & 0 \end{pmatrix}$.

Find the values of λ for which the determinant of the matrix $(\mathbf{A} - \lambda\mathbf{I})$ equals 0, where $\mathbf{I}$ is the unit matrix of order 3.
Show that $\mathbf{A}^2 - 6\mathbf{A} - \mathbf{I}$ is a multiple of $\mathbf{A}^{-1}$. (U of L)

10) If $\mathbf{B}$ is the row vector $(1 \quad 1 \quad 1)$ and $\mathbf{C}$ is the matrix
$$\begin{pmatrix} 2 & 1 & 0 \\ 0 & 3 & 4 \end{pmatrix}$$
find the matrix $\mathbf{A}$ such that $\mathbf{BA}$ is the row vector $(\frac{3}{2} \quad -\frac{1}{4})$ and $\mathbf{CA}$ is a unit matrix. (U of L)

11) If $\mathbf{M} = \begin{pmatrix} 1 & -1 & k \\ 4 & 7 & 3 \\ -1 & 12 & -2 \end{pmatrix}$, evaluate, in terms of k, the determinant of the matrix $\mathbf{M}$.

If $\mathbf{x} = \begin{pmatrix} x \\ y \\ z \end{pmatrix}$, solve the equations

(a) $\mathbf{Mx} = \begin{pmatrix} 1 \\ 11 \\ 21 \end{pmatrix}$ when $k = 2$, (b) $\mathbf{Mx} = \begin{pmatrix} 0 \\ 0 \\ 0 \end{pmatrix}$ when $k = 1$.

Interpret the result (b) geometrically. (U of L)

12) For the matrix equation $\begin{pmatrix} 1 & 1 & 1 \\ 1 & 2 & 3 \\ 1 & 3 & k \end{pmatrix}\begin{pmatrix} x \\ y \\ z \end{pmatrix} = \begin{pmatrix} 3 \\ 6 \\ 4+k \end{pmatrix}$ find the value of k for which the equation does not have a unique solution.
For this value of k, solve the equation and interpret the solution geometrically. (C)p

13) $\mathbf{M}$ denotes the matrix $\begin{pmatrix} 1 & 2 \\ 3 & k \end{pmatrix}$

and $\mathbf{v}$ denotes the vector $\begin{pmatrix} x \\ y \end{pmatrix}$.

Find the solution set of vectors to which $\mathbf{v}$ must belong in each of the following cases:

(a) $k = 5$ and $\mathbf{Mv} = \begin{pmatrix} 3 \\ 1 \end{pmatrix}$ (b) $k = 5$ and $\mathbf{Mv} = \begin{pmatrix} 0 \\ 0 \end{pmatrix}$

(c) $k = 6$ and $\mathbf{Mv} = \begin{pmatrix} 1 \\ 3 \end{pmatrix}$ (d) $k = 6$ and $\mathbf{Mv} = \begin{pmatrix} 1 \\ 1 \end{pmatrix}$.

Comment briefly on the connection between these solution sets and the existence or otherwise of the inverse matrix $\mathbf{M}^{-1}$. (C)

14) Find the matrix $\mathbf{X}$ given that $\mathbf{AXA}^{-1} = \mathbf{B}$ where

$$\mathbf{A} = \begin{pmatrix} 2 & 1 \\ 3 & 2 \end{pmatrix}, \quad \mathbf{B} = \begin{pmatrix} 1 & 0 \\ 0 & 2 \end{pmatrix}$$

(C)p

15) Find the inverse of the matrix

$$\begin{pmatrix} 2 & -1 & 3 \\ 5 & 4 & -3 \\ 3 & -2 & -1 \end{pmatrix}$$

Hence, or otherwise, solve the equations

$$\begin{aligned} 2x - y + 3z &= -25 \\ 5x + 4y - 3z &= -1 \\ 3x - 2y - z &= -17. \end{aligned}$$

(C)

16) Solve the system of equations

$$\begin{aligned} 2x + 3y + z + 1 &= 0, \\ x - 2y - 3z + 4 &= 0, \\ 3x + 4y + 2z - 2 &= 0. \end{aligned}$$

Find the solution set of the system when the last equation is replaced by
(a) $8x + 5y - 3z + 11 = 0$, (b) $4x - y - 5z + 6 = 0$. (O)

17) Use any method to find the inverse of the matrix

$$\begin{pmatrix} 3 & -1 & 5 \\ 4 & 3 & 3 \\ 5 & -4 & -2 \end{pmatrix}$$

Solve the equations

$$\begin{aligned} 3x - \ \ y + 5z &= 4, \\ 4x + 3y + 3z &= 3, \\ 5x - 4y - 2z &= 3. \end{aligned}$$

(C)

18) A transformation in three dimensional space takes the point (x, y, z) to (x_1, y_1, z_1) where

$$\begin{pmatrix} x_1 \\ y_1 \\ z_1 \end{pmatrix} = \begin{pmatrix} 0 & 0 & 1 \\ 1 & 0 & 0 \\ 0 & 1 & 0 \end{pmatrix} \begin{pmatrix} x \\ y \\ z \end{pmatrix}$$

Prove that the transformation leaves unaltered
(a) the distance between two points;
(b) the points of the line $x = y = z$.
Assuming that the transformation is a rotation about a line, find the angle of rotation. (O)

19) Find the complete solutions of the two systems of equations

(a) $3x + 4y + z = 5,$
$2x - y - z = 4,$
$x + 3y + z = 1.$

(b) $3x + 4y + z = 5,$
$2x - y - z = 4,$
$5x + 14y + 5z = 7.$

(O)

20) If the system of equations

$$\begin{aligned} 2x + y + z &= \lambda x, \\ x + 2y + z &= \lambda y, \\ x + y + 2z &= \lambda z \end{aligned}$$

has solutions in which x, y, z are not all zero, find the possible values of λ. For each such value of λ, find the general solution of the system. (O)

21) Show that any real 2×2 matrix $\mathbf{A}\begin{pmatrix} a & b \\ c & d \end{pmatrix}$ satisfies a certain quadratic equation $\mathbf{A}^2 + p\mathbf{A} + q\mathbf{I} = \mathbf{O}$, where $\mathbf{I}$ and $\mathbf{O}$ are the unit and zero 2×2 matrices, and p, q are certain numbers depending on a, b, c and d. If $q \neq 0$, find, in the form $\alpha\mathbf{A} + \beta\mathbf{I}$, the inverse of $\mathbf{A}$ (that is, a matrix $\mathbf{B}$ such that $\mathbf{AB} = \mathbf{I}$).

22) The lines u, v, w are the bisectors of the angles between Oy and Oz, Oz and Ox, Ox and Oy respectively. P is the typical point (x, y, z). Prove that if the reflection P_1 of P in the line u is (x_1, y_1, z_1) then

$$\begin{pmatrix} x_1 \\ y_1 \\ z_1 \end{pmatrix} = \begin{pmatrix} -1 & 0 & 0 \\ 0 & 0 & 1 \\ 0 & 1 & 0 \end{pmatrix} \begin{pmatrix} x \\ y \\ z \end{pmatrix}$$

The reflection of P_1 in y is P_2 and the reflection of P_2 in w is P_3. Find the coordinates of P_3 and show that P_3 coincides with P only if P lies on the y-axis. (O)

APPENDIX: THE DERIVATION OF THE INVERSE 3×3 MATRIX

If $\mathbf{M} = \begin{pmatrix} a_1 & a_2 & a_3 \\ b_1 & b_2 & b_3 \\ c_1 & c_2 & c_3 \end{pmatrix}$ maps $\begin{pmatrix} x \\ y \\ z \end{pmatrix}$ to $\begin{pmatrix} X \\ Y \\ Z \end{pmatrix}$,

i.e.
$$\mathbf{M}\begin{pmatrix} x \\ y \\ z \end{pmatrix} = \begin{pmatrix} X \\ Y \\ Z \end{pmatrix},$$

then the inverse, $\mathbf{M}^{-1}$, of $\mathbf{M}$ maps $\begin{pmatrix} X \\ Y \\ Z \end{pmatrix}$ to $\begin{pmatrix} x \\ y \\ z \end{pmatrix}$,

i.e.
$$\mathbf{M}^{-1}\begin{pmatrix} X \\ Y \\ Z \end{pmatrix} = \begin{pmatrix} x \\ y \\ z \end{pmatrix}$$

If we can find the images of $\mathbf{i}, \mathbf{j}$ and $\mathbf{k}$ under $\mathbf{M}^{-1}$, we can then write down $\mathbf{M}^{-1}$.

If $\mathbf{M}$ maps (x_1, y_1, z_1) to $(1, 0, 0)$,

i.e.
$$\begin{pmatrix} a_1 & b_1 & c_1 \\ a_2 & b_2 & c_2 \\ a_3 & b_3 & c_3 \end{pmatrix} \begin{pmatrix} x_1 \\ y_1 \\ z_1 \end{pmatrix} = \begin{pmatrix} 1 \\ 0 \\ 0 \end{pmatrix}, \qquad [1]$$

then $\mathbf{M}^{-1}\begin{pmatrix} 1 \\ 0 \\ 0 \end{pmatrix} = \begin{pmatrix} x_1 \\ y_1 \\ z_1 \end{pmatrix}$, i.e. $\mathbf{M}^{-1}$ maps $\begin{pmatrix} 1 \\ 0 \\ 0 \end{pmatrix}$ to $\begin{pmatrix} x_1 \\ y_1 \\ z_1 \end{pmatrix}$

Now from [1]

$$a_1x_1 + b_1y_1 + c_1z_1 = 1 \quad [2]$$

$$a_2x_1 + b_2y_1 + c_2z_1 = 0 \quad [3]$$

$$a_3x_1 + b_3y_1 + c_3z_1 = 0. \quad [4]$$

Now the L.H.S. of equation [3] is the scalar product of $(a_2\mathbf{i} + b_2\mathbf{j} + c_2\mathbf{k})$ and $(x_1\mathbf{i} + y_1\mathbf{j} + z_1\mathbf{k})$.
But the R.H.S. of [3] is zero.

Hence $a_2\mathbf{i} + b_2\mathbf{j} + c_2\mathbf{k}$ and $x_1\mathbf{i} + y_1\mathbf{j} + z_1\mathbf{k}$ are perpendicular.

Similarly from equation [4] we can show that $a_3\mathbf{i} + b_3\mathbf{j} + c_3\mathbf{k}$ and $x_1\mathbf{i} + y_1\mathbf{j} + z_1\mathbf{k}$ are perpendicular.

So $x_1\mathbf{i} + y_1\mathbf{j} + z_1\mathbf{k}$ is perpendicular to both $a_2\mathbf{i} + b_2\mathbf{j} + c_2\mathbf{k}$ and $a_3\mathbf{i} + b_3\mathbf{j} + c_3\mathbf{k}$,

i.e. $x_1\mathbf{i} + y_1\mathbf{j} + z_1\mathbf{k}$ is parallel to

$$(a_2\mathbf{i} + b_2\mathbf{j} + c_2\mathbf{k}) \times (a_3\mathbf{i} + b_3\mathbf{j} + c_3\mathbf{k}) = \begin{vmatrix} \mathbf{i} & \mathbf{j} & \mathbf{k} \\ a_2 & b_2 & c_2 \\ a_3 & b_3 & c_3 \end{vmatrix}$$

$$= A\mathbf{i}_1 + B\mathbf{j}_1 + C\mathbf{k}_1$$

where A_1, B_1, C_1 are the cofactors of the elements a_1, b_1, c_1 in $\mathbf{M}$.
Hence $x_1\mathbf{i} + y_1\mathbf{j} + z_1\mathbf{k} = \lambda(A_1\mathbf{i} + B_1\mathbf{j} + C_1\mathbf{k})$,

i.e. $\mathbf{M}^{-1}$ maps $\begin{pmatrix} 1 \\ 0 \\ 0 \end{pmatrix}$ to $\lambda \begin{pmatrix} A_1 \\ B_1 \\ C_1 \end{pmatrix}$.

A similar argument shows that

$$\begin{pmatrix} x_2 \\ y_2 \\ z_2 \end{pmatrix} = \mu \begin{pmatrix} A_2 \\ B_2 \\ C_2 \end{pmatrix} \quad \text{and} \quad \begin{pmatrix} x_3 \\ y_3 \\ z_3 \end{pmatrix} = \eta \begin{pmatrix} A_3 \\ B_3 \\ C_3 \end{pmatrix},$$

i.e. $\mathbf{M}^{-1}$ maps

$$\begin{pmatrix} 1 \\ 0 \\ 0 \end{pmatrix} \to \lambda \begin{pmatrix} A_1 \\ B_1 \\ C_1 \end{pmatrix}, \quad \begin{pmatrix} 0 \\ 1 \\ 0 \end{pmatrix} \to \mu \begin{pmatrix} A_2 \\ B_2 \\ C_2 \end{pmatrix}, \quad \begin{pmatrix} 0 \\ 0 \\ 1 \end{pmatrix} \to \eta \begin{pmatrix} A_3 \\ B_3 \\ C_3 \end{pmatrix}$$

Now consider the matrix $\begin{pmatrix} A_1 & A_2 & A_3 \\ B_1 & B_2 & B_3 \\ C_1 & C_2 & C_3 \end{pmatrix}$ which we denote by $\mathbf{M}^*$ and the product $$\mathbf{M}^*\mathbf{M} = \begin{pmatrix} A_1 & A_2 & A_3 \\ B_1 & B_2 & B_3 \\ C_1 & C_2 & C_3 \end{pmatrix}\begin{pmatrix} a_1 & b_1 & c_1 \\ a_2 & b_2 & c_2 \\ a_3 & b_3 & c_3 \end{pmatrix}.$$

But the product of any column of $\mathbf{M}$ and its cofactors
(e.g. $a_1A_1 + a_2A_2 + a_3A_3$) is equal to $|\mathbf{M}|$
whereas the product of any column of $\mathbf{M}$ with the cofactors of another

column, e.g. $b_1A_1 + b_2A_2 + b_3A_3 = \begin{vmatrix} b_1 & b_2 & b_3 \\ b_1 & b_2 & b_3 \\ c_1 & c_2 & c_3 \end{vmatrix}$, is zero.

Hence $$\mathbf{M}^*\mathbf{M} = \begin{pmatrix} |\mathbf{M}| & 0 & 0 \\ 0 & |\mathbf{M}| & 0 \\ 0 & 0 & |\mathbf{M}| \end{pmatrix} = |\mathbf{M}|\begin{pmatrix} 1 & 0 & 0 \\ 0 & 1 & 0 \\ 0 & 0 & 1 \end{pmatrix},$$

i.e. $$\mathbf{M}^*\mathbf{M} = |\mathbf{M}|\mathbf{I} \quad \text{so} \quad \left(\frac{\mathbf{M}^*}{|\mathbf{M}|}\right)\mathbf{M} = \mathbf{I}$$

$$\Rightarrow \quad \frac{\mathbf{M}^*}{|\mathbf{M}|} = \mathbf{M}^{-1}.$$

So if $\mathbf{M} = \begin{pmatrix} a_1 & b_1 & c_1 \\ a_2 & b_2 & c_2 \\ a_3 & b_3 & c_3 \end{pmatrix}$ then $\mathbf{M}^{-1} = \dfrac{1}{|\mathbf{M}|}\begin{pmatrix} A_1 & A_2 & A_3 \\ B_1 & B_2 & B_3 \\ C_1 & C_2 & C_3 \end{pmatrix}$.

Note that the cofactors in $\mathbf{M}^*$ are not in the same positions as their corresponding elements in $\mathbf{M}$, but they are in the position reached by transposing the rows and columns of $\mathbf{M}$.
Note also that if $|\mathbf{M}| = 0$, $\mathbf{M}^{-1}$ does not exist.

CHAPTER 7

MATHEMATICAL PROOF

FORMAL MATHEMATICS

Anyone who has studied both elementary Euclidean geometry and an experimental science should be aware of the very different ways in which the propositions of these two disciplines are established. In an experimental science, the propositions or 'laws' are accepted because they are confirmed by observation. For example, in Mechanics there is a law that states that an object falling freely has a constant acceleration. This law is accepted because observation of falling bodies verifies its truth.

In Euclidean geometry, the propositions or 'theorems' are accepted because they are deduced by means of a logical proof from previously established, or accepted, truths. For example, the theorem that states that 'the sum of the interior angles of a triangle is two right angles' is accepted because it is deduced logically from previously accepted properties of parallel lines and angles.

It was the ancient Greeks who first used this method to give geometry a formal structure. Later, many other branches of mathematics were subjected to the same formal treatment which is known as the axiomatic method. This consists of accepting *without proof* certain propositions, known as *axioms* (e.g. in Euclidean geometry, the statement that vertically opposite angles are equal, is an axiom). Then all the other statements (called theorems) of the system are derived from the axioms by the principles of logic.

Any game, chess for example, provides a good analogy for this formal structure. If you play the game you accept the rules without question, so the rules are the

'axioms'. The strategies involved in playing the game are then built on these foundations and so form the 'theorems' of the game.
Later in this chapter we look at some of the ways in which Mathematical theorems can be proved, but first we introduce some of the logic concepts that can be used in a proof.

STATEMENTS

A statement, or proposition, is a sentence which is either true or false, but not both.
For example 'It is now 10 o'clock' is a statement
but 'What is the time?' is not a statement.
Statements, or propositions, are denoted by small letters $p, q, r, \ldots$.

Examples are:

p: For all values of x, $\int x \, dx = \frac{1}{2}x^2 + k$. (true)

q: For any two vectors $\mathbf{a}$ and $\mathbf{b}$, $\mathbf{a} \times \mathbf{b} = \mathbf{b} \times \mathbf{a}$. (false)

r: For all values of x, $\frac{d}{dx} \sin x = \cos x$. (true)

s: Any quadratic equation in one variable has two real roots. (false)

Because a statement is either true or false, it must contain enough information for us to be able to decide whether it is correct or not. This information must be either in the sentence itself or in the context in which the sentence appears. This is particularly important in the case of mathematical sentences involving variables.
Consider for example the function $\ln x$. We know that this function is defined for positive values of x but not for negative values of x.

Now consider the sentence: $\int \frac{1}{x} \, dx = \ln x + k$.

If this sentence appears in a context where x has positive values only, it is true. But if it appears in a context where x can take any value it is not true $\left(\text{in this case } \int \frac{1}{x} \, dx = \ln |x| + k \text{ is correct}\right)$.

Taken out of context, $\int \frac{1}{x} \, dx = \ln x + k$ cannot be said to be right or wrong.
Similarly the sentence 'If $x^2 = 4$ then $x = 2$' is true if x is restricted to positive values only, but untrue if x can take any value.

EXERCISE 7a

State whether the following sentences are
(a) true for all real values of x,
(b) true for at least one real value of x, but not all x,
(c) true for no real value of x.

1) $x+1 = 3$.

2) $(x+1)^2 = x^2+2x+1$.

3) $x^2+x+1 = 0$ has two real roots.

4) $(x-1)(x-2) \geqslant 0$.

5) $(x-1)^2 \geqslant 0$.

6) $\begin{pmatrix} x & 1 \\ 0 & 4 \end{pmatrix}$ has an inverse.

7) $\int \tan x \, dx = \ln \sec x + k$.

8) $\cosh x > 0$.

9) $\begin{pmatrix} x & 1 \\ 0 & 4 \end{pmatrix}$ has no inverse.

10) $\int \frac{1}{x+1} dx = -(x+1)^{-2} + k$.

Which of the following statements are true? (x is real.)

11) For all values of x, $\frac{1}{x(x+1)} = \frac{1}{x} - \frac{1}{x+1}$.

12) There is at least one value of x for which $x^2+1=0$.

13) There are no values of x for which $x^2+x-1=0$.

14) For all values of x, $\sin^2 x + \cos^2 x = 1$.

15) For at least one value of x, $\sin^2 x - \cos^2 x = 1$.

Negation

Consider the statement p: 'I have a headache'.
The statement 'I do not have a headache' is called the *negation* of p and is denoted by $\sim p$.
i.e. $\sim p$ (read as 'not p') is the negation of the statement p.
Some examples of statements and their negations are:

a: It is snowing. $\sim a$: It is not snowing.

b: All Englishmen have brown hair. $\sim b$: It is not true to say that all Englishmen have brown hair.

c: 71 is a prime number. $\sim c$: 71 is not a prime number.

d: For all values of x, $x^2+1>0$. $\sim d$: It is not true that for all values of x, $x^2+1>0$.

Looking at the statements and their negations above we see that either the given statement, or its negation, is true. In fact, for any statement p,

if p is true, $\sim p$ is false

if p is false, $\sim p$ is true.

This property is useful when writing down the negation of a given statement. Looking again at the examples given above, we see that some statements are negated by inserting the word 'not' and others by prefacing the given statement with 'It is not true that . . . '. Any statement can be negated by this preface but great care must be taken with any other form of wording. For example, consider again the statement

b: All Englishmen have brown hair. (false)

Now consider the following attempts at the statement $\sim b$.

b_1: All Englishmen do not have brown hair.

b_2: No Englishman has brown hair.

b_3: At least one Englishman does not have brown hair.

b_1 is ambiguous because its meaning depends on the emphasis put on certain words. If the emphasis is on '*All*', b_1 can mean that all Englishmen have a hair colour that is not brown. But if the emphasis is on '*not*', it could mean that some Englishmen have, and some have not, brown hair.
b_2 has a clear meaning and is false (it is equivalent to b_1 with the emphasis on 'all').
b_3 has a clear meaning and is true (it is equivalent to b_1 with the emphasis on 'not').
On closer examination of $\sim b$: 'It is not true to say that all Englishmen have brown hair', we see that $\sim b$ is equivalent to b_3,
i.e. the negation of 'All Englishmen have brown hair' can be written as 'At least one Englishman does not have brown hair'.

Note that there are other correct negations of b, e.g. 'Not all Englishmen have brown hair'.

Now consider a similar mathematical statement, e.g. 'For all values of x, $f(x) \equiv (x-1)(x-2) > 0$ (which is not true).

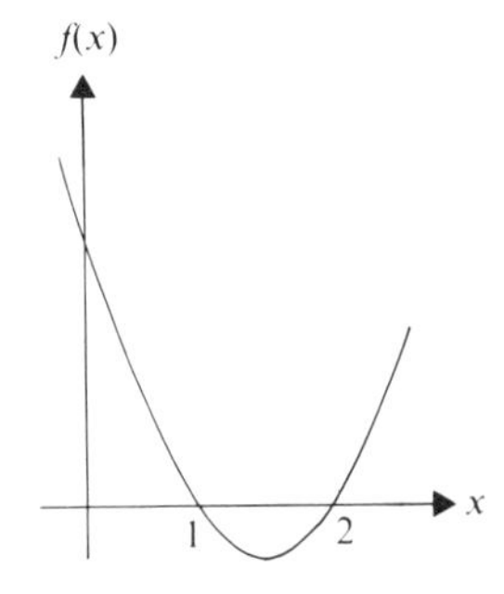

The negation of this statement can be written as

'For at least one value of x, $f(x) \equiv (x-1)(x-2) \leqslant 0$'. (true)

but *not* as

'For all values of x, $f(x) \equiv (x-1)(x-2) \leqslant 0$'. (not true).

EXERCISE 7b

In each of the following questions, state which of the statements b, c, d, e is/are the negation of statement a.

1) a: θ is any positive angle.
   b: θ is any negative angle.
   c: There is at least one value of θ which is not positive.

2) a: 2 is a positive integer.
   b: 2 is not a positive integer.
   c: 2 is a negative integer.

3) a: Marmalade is made from oranges.
   b: Marmalade is not made from oranges.
   c: There is at least one variety of marmalade that is not made from oranges.

4) a: At least one make of car does not have four wheels.
   b: All cars have four wheels.
   c: All cars do not have four wheels.
   d: At least one make of car does have four wheels.

5) a: All quadratic equations in one variable have two real roots.
   b: There is no quadratic equation in one variable that has two real roots.
   c: At least one quadratic equation in one variable does not have real roots.
   d: At least one quadratic equation in one variable has two real roots.

6) a: There is no value of x for which x^2 is negative.
   b: $x^2 \geqslant 0$ for all values of x.
   c: There is no value of x for which x^2 is positive.
   d: x^2 is negative for at least one value of x.

7) a: $f(x) > x$ for all values of $x > 1$.
   b: $f(x) \not> x$ for all values of $x > 1$.
   c: $f(x) \leqslant x$ for at least one value of $x \leqslant 1$.
   d: $f(x) \leqslant x$ for at least one value of $x > 1$.

8) a: For all integral values of n, $f(n) > n$.
   b: $f(n) \leqslant n$ for all integral values of n.
   c: $f(n) > n$ for no integral values of n.
   d: $f(n) < n$ for at least one integral value of n.
   e: $f(n) \leqslant n$ for at least one integral value of n.

Conditional Statements

Working in a context in which x can take any real value, consider the statements

$$a: \quad x = 3$$

$$b: \; x^2 = 9.$$

We do not know whether either of these statements is true or false but we can say that

$$\textit{if} \quad x = 3 \quad \textit{then} \quad x^2 = 9.$$

This is known as an implication, or conditional statement, which we write symbolically as

$$x = 3 \Rightarrow x^2 = 9$$

or
$$a \Rightarrow b.$$

There are several other ways of writing the linguistic equivalent of $a \Rightarrow b$, e.g.

$x = 3$ implies that $x^2 = 9$

$x = 3$ therefore $x^2 = 9$

$x = 3$ is a sufficient condition for $x^2 = 9$

$x = 3$ only if $x^2 = 9$.

In the conditional statement $a \Rightarrow b$, a is called the hypothesis and b is called the conclusion.
The implication $a \Rightarrow b$ will clearly be untrue if a false conclusion is drawn from a true hypothesis.

For example

$$\left.\begin{aligned} x + 1 &= 0 \Rightarrow x = 3 \\ \sin A + \sin B &= 1 \Rightarrow \sin(A+B) = 1 \\ \ln(A+B) &= 2 \Rightarrow \ln A + \ln B = 2 \end{aligned}\right\} \text{are all false,}$$

because our mathematical knowledge tells us that the conclusions do not follow from the hypotheses.
When this is not the case and an implication involves a variable, we must know whether the variable can take *all* real values, or only *some* real values, before we can determine the truth of the implication.
For example, $x^2 = 4 \Rightarrow x = 2$ is not true within the context given at the beginning of this section, but would be true in a context where x can take only positive values.
Similarly, in a context where θ can take any value, $\sin\theta = 0 \Rightarrow \cos\theta = 1$ is not true, as $\cos\theta = \pm 1$ in this case. But if we say that there is at least one value of θ such that $\sin\theta = 0 \Rightarrow \cos\theta = 1$, the implication is correct.

Note that a correct implication does not depend on the truth of the component statements. For example, consider the following solution to the problem 'Solve the equation $\sin\theta + \sin 2\theta = 1$ for values of θ in the range $0 \leqslant \theta \leqslant 90°$'.

$\sin\theta + \sin 2\theta = 1 \Rightarrow \sin 3\theta = 1$ (which is a false implication).

$\sin 3\theta = 1 \Rightarrow 3\theta = 90°$ (which is a true implication but with false component statements).

EXERCISE 7c

All variables in this exercise can take any real value. Determine which of the following implications are true and which are false.

1) $(x+1)(x-2) = 0 \Rightarrow x = -1$ or 2

2) $x^2 = 16 \Rightarrow x = 4$

3) $f(x) \equiv (x+2)^2 \Rightarrow f(x) \geqslant 0$

4) $\sin\theta = 0 \Rightarrow \theta = 0$

5) $ax^2 + bx + c = 0$ has real roots $\Rightarrow b^2 - 4ac \geqslant 0$

6) $\dfrac{dy}{dx} = 2x \Rightarrow y = x^2$

7) $(x+1)(x-2) = 1 \Rightarrow x + 1 = 1$ or $x - 2 = 1$

8) $\cos\alpha\cos\alpha = 1 \Rightarrow \cos 2\alpha = 1$

9) $\sin\alpha\cos\alpha = 1 \Rightarrow \sin 2\alpha = \frac{1}{2}$

10) $\cos\theta = 0 \Rightarrow \sin\theta = \pm 1$

The Converse of a Conditional Statement

We have seen that, when a statement involves a variable, it is important to know what values that variable can take. For the rest of this Chapter it is to be assumed that any variable is free to take all values *unless otherwise stated*.
If in the implication $x = 3 \Rightarrow x + 1 = 4$, we reverse the arrow,

we get $x = 3 \Leftarrow x + 1 = 4$

which might be read as 'If $x + 1 = 4$ then $x = 3$' but this changes the order of the component statements. Keeping to the given order we can say that

'$x = 3 \Leftarrow x + 1 = 4$' means '$x = 3$ is implied by $x + 1 = 4$'

or '$x = 3$ if $x + 1 = 4$'

or '$x = 3$ because $x+1 = 4$'

or '$x = 3$ is a necessary condition for $x+1 = 4$'.

$x = 3 \Leftarrow x+1 = 4$ is called the *converse* of $x = 3 \Rightarrow x+1 = 4$.
In general $p \Leftarrow q$ (or $q \Rightarrow p$) is the *converse* of $p \Rightarrow q$.
In the example given above both the implication and its converse are true.
But the statement $x = 3 \Rightarrow x^2 = 9$ is true whereas its converse, $x = 3 \Leftarrow x^2 = 9$, is false because x can also have the value -3.
Similarly, if a is the statement '$\sin x = 0$' and b is the statement '$\cos x = 1$' then $a \Leftarrow b$ is true but $a \Rightarrow b$ is false because $\sin x = 0 \Rightarrow \cos x = \pm 1$.
Thus a statement and its converse are not necessarily both true, nor both false.
Hence it is unsound to argue from a statement to its converse,
e.g. from

'If this boy is a pupil at school X he wears a green blazer'

to

'If this boy wears a green blazer he is a pupil at school X'.

(This is a common error of reasoning.)

The Negation of a Conditional Statement

The negation of $p \Rightarrow q$ is $\sim(p \Rightarrow q)$.
Note that the *whole statement* is negated, and not p or q separately.
So the negation of 'If it is raining it is wet'
could be 'It is not true to say that if it is raining it is wet'

but not
- 'If it is not raining it is not wet'
- 'If it is raining it is not wet'
- 'If it is not raining it is wet'.

Similarly the negation of '$x^2 = 9 \Rightarrow x = 3$'
can be 'It is not true to say that $x^2 = 9 \Rightarrow x = 3$'.
From these examples we see that the linguistic equivalent of $\sim(p \Rightarrow q)$ is clumsy.
Consider again '$x^2 = 9 \Rightarrow x = 3$'.
We know this is false in a context in which x can take all real values.
So $\sim$'$x^2 = 9 \Rightarrow x = 3$' is true in the same context because it is the negation of a false statement.
But we also know that this is true because there is *one* value of x (-3) for which $x^2 = 9$ and $x \neq 3$.
So, instead of the clumsy statement 'It is not true to say that if x can take any value, then $x^2 = 9$ implies that $x = 3$', we can use 'There is a value of x such that $x^2 = 9$ and $x \neq 3$'.

Now consider the true statement

$$x+1 = 4 \Rightarrow x = 3.$$

The negation of this implication, i.e. the statement

'It is not true to say that if x can take any value then $x+1=4$ implies that $x=3$',

is false.

Now there is no value of x for which $x+1=4$ and $x \neq 3$, the statement

'There is a value of x such that $x+1=4$ and $x \neq 3$'

is also false.

These two examples indicate that

when a variable can take any value, a statement $\sim(p \Rightarrow q)$ involving just that one variable can be replaced by the equivalent statement.

'There is at least one value of the variable such that p and $\sim q$' as either both statements are true or both are false.

When variables are not involved, $\sim(p \Rightarrow q)$ can be replaced by the simpler 'p and $\sim q$'.

For example, if $p \Rightarrow q$ is

'If I am eating then I am hungry'

the negation of this statement, $\sim(p \Rightarrow q)$ can be replaced by the equivalent 'p and $\sim q$', i.e.

'I am eating and I am not hungry'.

But if $p \Rightarrow q$ is

'If all the cars in this showroom are made in Britain then they are all good cars'

$\sim(p \Rightarrow q)$ is

'It is not true that if all the cars in this showroom are made in Britain they are all good cars'

which can be replaced by the equivalent

'There is at least one car in this showroom that is made in Britain and which is not a good car'.

Similarly if $p \Rightarrow q$ is

'When x and y are unrestricted, $\dfrac{dy}{dx} = 2x \Rightarrow y = x^2$'

then $\sim(p \Rightarrow q)$ is

'When x and y are unrestricted, it is not true that $\dfrac{dy}{dx} = 2x \Rightarrow y = x^2$'

which can be replaced by the equivalent

'There are values of x and y such that $\dfrac{dy}{dx} = 2x$ and $y \neq x^2$'

Biconditional Statements

Consider the two statements a: $x = 3$, and b: $x + 1 = 4$. Again, we do not know the truth of either a or b, but we can say

'If $x = 3$ then $x + 1 = 4$ *and* $x = 3$ because $x + 1 = 4$',

i.e. '$a \Rightarrow b$ *and* $a \Leftarrow b$'.
This is known as a *biconditional statement* and may be written symbolically as

$$x = 3 \Longleftrightarrow x + 1 = 4$$

i.e.
$$a \Longleftrightarrow b$$

Some other linguistic equivalents for $a \Longleftrightarrow b$ are

'$x = 3$ implies and is implied by $x + 1 = 4$'

'$x = 3$ if and only if $x + 1 = 4$'

'$x = 3$ is a necessary and sufficient condition for $x + 1 = 4$'.

In general $p \Longleftrightarrow q$ means p if and only if q.

$p \Longleftrightarrow q$ is true only if the arrow in the *true* statement $p \Rightarrow q$ can be *truthfully* reversed, i.e. if $p \Rightarrow q$ and $p \Leftarrow q$ are both true.
For example, for unrestricted values of θ,

$\sin\theta = 0 \Rightarrow \tan\theta = 0$ is true

$\sin\theta = 0 \Leftarrow \tan\theta = 0$ is true

hence $\sin\theta = 0 \Longleftrightarrow \tan\theta = 0$ is true

but $\cos\theta = 1 \Rightarrow \sin\theta = 0$ is true

$\cos\theta = 1 \Leftarrow \sin\theta = 0$ is false

so $\cos\theta = 1 \Longleftrightarrow \sin\theta = 0$ is false

Contrapositive Statements

Consider the statements
$$\begin{cases} a: & x = 3 \\ b: & x^2 = 9 \end{cases}$$

and their negations, i.e.
$$\begin{cases} \sim a: & x \neq 3 \\ \sim b: & x^2 \neq 9. \end{cases}$$

The true conditional relationship between a and b is

$$a \Rightarrow b \quad \text{(i.e. if } x = 3 \text{ then } x^2 = 9)$$

whereas the true conditional relationship between $\sim a$ and $\sim b$ is

$$\sim b \Rightarrow \sim a \quad \text{(i.e. if } x^2 \neq 9 \text{ then } x \neq 3).$$

(**Note** that $b \Rightarrow a$ is false, and so is $\sim a \Rightarrow \sim b$.)

Now consider the statements

c: It is raining

d: It is cloudy

and the following conditional statements

$c \Rightarrow d$: If it is raining then it is cloudy

$d \Rightarrow c$: If it is cloudy it is raining

$\sim d \Rightarrow \sim c$: If it is not cloudy it is not raining

$\sim c \Rightarrow \sim d$: If it is not raining it is not cloudy.

From our observations of the weather we see that

$c \Rightarrow d$ and $\sim d \Rightarrow \sim c$ are both true

but $d \Rightarrow c$ and $\sim c \Rightarrow \sim d$ are both false.

These two examples (the reader can no doubt think of many others) indicate that

if $p \Rightarrow q$ is true, $\sim q \Rightarrow \sim p$ is also true,

if $p \Rightarrow q$ is false, $\sim q \Rightarrow \sim p$ is also false.

The two statements $p \Rightarrow q$ and $\sim q \Rightarrow \sim p$ are each called the *contrapositive* of the other and we have seen that

a statement and its contrapositive are either both true or both false so that $p \Rightarrow q$ and $\sim q \Rightarrow \sim p$ are equivalent statements.

Thus it is sound reasoning to argue from a statement $p \Rightarrow q$ to the contrapositive statement $\sim q \Rightarrow \sim p$,

e.g. from

'If it is snowing it is cold'

to

'If it is not cold it is not snowing'.

Similarly to argue from

'If $f(x)$ has a stationary value when $x = a$ then $f'(a) = 0$'

to

'If $f'(a) \neq 0$ then $f(x)$ does not have a stationary value when $x = a$'

is sound.

Note. The soundness of an argument does not depend on the truth of the statements involved,

e.g. to argue from 'If I am doing A Level Maths then I am mad'

to 'If I am not mad then I am not doing A Level Maths'
is sound, but the implications are (we hope) false.
Note that if we compare the statements $p \Rightarrow q$ and $\sim p \Rightarrow \sim q$ (called the *inverse* of $p \Rightarrow q$) we find that they are not necessarily both true nor both false,
e.g. if $p \Rightarrow q$ is

'If $x = 3$ then $x^2 = 9$' (true)

then the inverse, $\sim p \Rightarrow \sim q$, is

'If $x \neq 3$ then $x^2 \neq 9$' (false because $x = -3 \Rightarrow x^2 = 9$).

So it is not reasonable to argue from an implication to its inverse,
e.g. from 'If it is raining there are clouds in the sky'
to 'If it is not raining there are no clouds in the sky'.

EXERCISE 7d

In Questions 1–16 insert the correct conditional symbol between the given statements p and q (i.e. $p \Rightarrow q$, $p \Leftarrow q$, or $p \Leftrightarrow q$). In any question involving variables it is to be assumed that the variables can take any value.

1) p: $x + 2 = 4$; q: $x = 2$.

2) p: $x^2 = 4$; q: $x = 2$.

3) p: $\theta = \dfrac{\pi}{4}$; q: $\sin \theta = \dfrac{\sqrt{2}}{2}$.

4) p: $\mathrm{P}(x, y)$ is a point on the circle, centre O, radius 2; q: $x^2 + y^2 = 4$.

5) p: $\tan \theta = \pm 1$; q: $\sin \theta = -\dfrac{\sqrt{2}}{2}$.

6) p: $\dfrac{x-1}{(x+2)^2} = 0$; q: $x = 1$.

7) p: x^2 is even; q: x is even.

8) p: x is rational; q: x^2 is rational.

9) α and β are the roots of the equation $x^2 + ax + b = 0$.
p: a and b are real; q: $\alpha\beta$ is real.

10) In triangle ABC p: $\sin \mathrm{C} = \sin 2\mathrm{A}$; q: In triangle ABC, $a = b$.

11) p: α, β are the roots of the equation $z^2 = -1$; q: $\alpha + \beta$ is real.

12) p: $\mathbf{AB} = \mathbf{BA}$; q: $\mathbf{A} = \mathbf{B}$ ($\mathbf{A}$ and $\mathbf{B}$ are square matrices).

13) p: $x > y$; q: $-y > -x$.

14) p: $x^2+y^2<4$; q: (x, y) is the set of points inside the circle $x^2+y^2=4$.

15) p: $f(x) \equiv f(-x)$; q: $f(x) \equiv e^{x^2} \cos x$.

16) p: $S=a+ar+ar^2+ar^3+\ldots$, $|r|<1$; q: $S=\dfrac{a}{1-r}$.

17) Write down a linguistic statement which is (a) the converse, (b) the contrapositive, (c) the inverse, (d) the negation of the following statement: 'If this car is made in Britain then it is a good car'.

18) Repeat Question 17 for the statement:
'If the equation $ax^2+bx+c=0$ has two real roots then $b^2-4ac \geqslant 0$'.

19) p is the statement 'Henry is doing "A" Level Mathematics'.
q is the statement 'Henry is mad'.
Give linguistic equivalents for the following statements:
(a) $p \Rightarrow q$ (b) $p \Leftarrow q$ (c) $\sim p \Rightarrow q$ (d) $p \Leftarrow \sim q$
(e) $\sim(p \Rightarrow q)$ (f) $p \Longleftrightarrow q$ (g) $\sim p \Rightarrow \sim q$ (h) $\sim p \Leftarrow \sim q$.

20) If p is the statement '$x=1$' and q is the statement '$x^2=1$' write down the following statements in symbolic form:

(a) If $x=1$ then $x^2=1$. (b) If $x^2=1$ then $x \neq 1$.

(c) $x^2 \neq 1$ because $x \neq 1$.

(d) It is not true to say that if $x^2=1$ then $x=1$.

(e) $x \neq 1$ because $x^2=1$.

(f) $x=1$ is a necessary and sufficient condition for $x^2=1$.

(g) If $x^2 \neq 1$ then $x \neq 1$.

Which of your statements are true?

21) The conditional statement $p \Rightarrow q$ is 'If $x^2=-1$ then x has no real value'. Write down sentences which are
(a) the converse, (b) the negation, (c) the inverse,
(d) the contrapositive of $p \Rightarrow q$.

22) Determine which of the following statements is equivalent to the statement 'If $f(x) \equiv x^2+5$ then $f(x) \nless 5$'.
(a) There is at least one value of x for which $f(x) \equiv x^2+5$ and $f(x) \geqslant 5$.
(b) If $f(x) \not\equiv x^2+5$ then $f(x)<5$.
(c) If $f(x)<5$ for at least one value of x then $f(x) \not\equiv x^2+5$.

MATHEMATICAL PROOF

In this section we look at some of the ways in which a theorem can be proved. A mathematical theorem is a result that is accepted because it is derived by correct implications and sound arguments from axioms.

(Sound argument is difficult to define, except as an argument that convinces most mathematicians!)
The proof of a theorem is the argument and inferences drawn from the axioms or from previously accepted theorems.
Most mathematical statements which require proof can be expressed in the form of a conditional statement, i.e. $p \Rightarrow q$.
For example:
If, in $\triangle ABC$, $AB = BC$, then $\angle A = \angle B$.
If $x^2 + 2x - 3 = 0$ then $x = -3$ or 1.
If θ is any angle then $\cos^2\theta + \sin^2\theta = 1$.
If n is any positive integer then $\frac{d}{dx}x^n = nx^{n-1}$.

Direct Proof by Deduction

This is probably the form of proof that the reader is most familiar with, and it is illustrated in the following example:
To prove that 'If $x^2 + 2x - 3 = 0$ then $x = -3$ or 1' is true, we proceed as follows.

$$x^2 + 2x - 3 = 0$$

$$\Rightarrow \quad (x + 3)(x - 1) = 0$$

$$\Rightarrow \quad x + 3 = 0 \quad \text{or} \quad x - 1 = 0$$

$$\Rightarrow \quad x = -3 \quad \text{or} \quad x = 1$$

Hence $x^2 + 2x - 3 = 0$ *does* imply $x = -3$ or 1.

The symbolic equivalent of this argument is:
To prove by deduction that $p \Rightarrow q$ is true, start with p then deduce $p \Rightarrow r \Rightarrow s \Rightarrow q$ so $p \Rightarrow q$ is true.
Note that this is a proof of the truth of the *implication* $p \Rightarrow q$.
Whether p is true, or q is true, is another question.

Proof by Induction

Consider the statement 'If n is any positive integer then $\frac{d}{dx}x^n = nx^{n-1}$'
We have not proved that this is true. From first principles we *have* proved that

$$\frac{d}{dx}(x^2) = 2x, \qquad \frac{d}{dx}(x^3) = 3x^2, \qquad \frac{d}{dx}(x^4) = 4x^3.$$

These results *suggest* that $\frac{d}{dx}x^n = nx^{n-1}$, but they certainly *do not prove* that $\frac{d}{dx}x^n = nx^{n-1}$.

Many results are found in a similar manner, that is by extrapolating a general result from a few particular cases. That the general result is correct can often be proved by a method called induction, and this is illustrated in the following example.

Consider the sum of the first n natural numbers and the sum of the cubes of the first n natural numbers.

When $n = 2, 3, 4$ and 5 we have

$$1 + 2 = 3 \qquad 1^3 + 2^3 = 9 \qquad (= 3^2)$$

$$1 + 2 + 3 = 6 \qquad 1^3 + 2^3 + 3^3 = 36 \qquad (= 6^2)$$

$$1 + 2 + 3 + 4 = 10 \qquad 1^3 + 2^3 + 3^3 + 4^3 = 100 \qquad (= 10^2)$$

$$1 + 2 + 3 + 4 + 5 = 15 \qquad 1^3 + 2^3 + 3^3 + 4^3 + 5^3 = 225 \qquad (= 15^2)$$

These results *suggest* that

$$\begin{aligned}(1^3 + 2^3 + 3^3 + \ldots + n^3) &= (1 + 2 + 3 + \ldots + n)^2 \\ &= [\tfrac{1}{2}n(n+1)]^2 \\ &= \tfrac{1}{4}n^2(n+1)^2.\end{aligned}$$

To *prove* this result we proceed as follows.

Let p_n be the statement

'For all positive integral values of n, $1^3 + 2^3 + \ldots + n^3 = \frac{1}{4}n^2(n+1)^2$' [1]

If the formula is valid when $n = k$,

i.e. if $$1^3 + 2^3 + \ldots + k^3 = \tfrac{1}{4}k^2(k+1)^2$$

then, adding $(k+1)^3$ to both sides gives

$$\begin{aligned}1^3 + 2^3 + \ldots + k^3 + (k+1)^3 &= \tfrac{1}{4}k^2(k+1)^2 + (k+1)^3 \\ &= \tfrac{1}{4}(k+1)^2[k^2 + 4(k+1)] \\ &= \tfrac{1}{4}(k+1)^2(k+2)^2\end{aligned}$$

which is the formula when $n = k + 1$.

So we have proved that the *implication* $p_k \Rightarrow p_{k+1}$ is true,

i.e. it is true to say that *if* the formula is valid when $n = k$ *then* the formula is valid when $n = k + 1$.

Now when $n = 1$,

$$\text{LHS of [1] is } 1^3 = 1$$

$$\text{RHS of [1] is } \tfrac{1}{4}(1)^2(2)^2 = 1$$

So the formula *is* valid when $n = 1$, i.e. p_1 is correct.

But, as we have shown that $p_k \Rightarrow p_{k+1}$, it follows that p_2 is correct, p_3 is correct, p_4 is correct, . . . and so on for all positive integral values of n,

i.e. $$p_n \text{ is true}$$

Proof by induction can be used to prove many results obtained by generalising, to any positive integral value of n, a result proved for particular values of n. The example above shows that there are three distinct steps in proof by induction which can be written symbolically as follows.
Let p_n be a statement involving n, where n is any positive integer.
1) Prove directly that the implication $p_k \Rightarrow p_{k+1}$ is correct.
2) Prove directly that p_1 is correct.
3) Combine steps 1 and 2 to showthat $p_2, p_3, \ldots$ are correct.
Note that the third step in this argument is inductive rather than deductive.
Hence p_n is correct.
We now illustrate this method of proof in the following examples.

EXAMPLES 7e

1) Prove by induction that $9^n - 1$ is a multiple of 8 for all positive integral values of n.

Let p_n be the statement 'For all positive integral values of n, $9^n - 1$ is a multiple of 8'.
If the formula is valid when $n = k$
i.e. if $9^k - 1 = 8a$ say

then
$$\begin{aligned} 9^{k+1} - 1 &= 9^k . 9 - 1 \\ &= 9^k . 9 - 9 + 8 \\ &= 9(9^k - 1) + 8 \\ &= 9(8a) + 8 \\ &= 8(9a + 1) \quad \text{which is a multiple of } 8. \end{aligned}$$

i.e. if $9^k - 1$ is a multiple of 8 then so is $9^{k+1} - 1$.
But if $n = 1$, $9^1 - 1 = 8$, so p_1 is true.
So p_2 is true, p_3 is true, . . . and so on for all positive integral values of n.
Hence p_n is true.

2) Prove by induction that $\sum_{r=1}^{n} r^2 = \frac{n}{6}(n+1)(2n+1)$ for all positive integral values of n.

Let p_n be the statement

'$\sum_{r=1}^{n} r^2 = \frac{n}{6}(n+1)(2n+1)$ for all positive integral values of n'.

If the formula is valid when $n = k$

i.e. if $$\sum_{r=1}^{k} r^2 = \frac{k}{6}(k+1)(2k+1)$$

then adding $(k+1)^2$ to both sides gives

$$\sum_{r=1}^{k+1} r^2 = \frac{k}{6}(k+1)(2k+1) + (k+1)^2$$

$$= \frac{1}{6}(k+1)[k(2k+1) + 6(k+1)]$$

$$= \frac{1}{6}(k+1)(2k^2 + 7k + 6)$$

$$= \frac{1}{6}(k+1)(k+2)(2k+3)$$

which is the formula when $n = k + 1$.
Therefore if p_n is valid when $n = k$, p_n is also valid when $n = k + 1$.
When $n = 1$,

L.H.S. of p_n is $1^2 = 1$

R.H.S. of p_n is $\frac{1}{6}(1+1)(2+1) = 1$

So p_n is valid when $n = 1$, i.e. p_1 is true.
Therefore p_2 is true, p_3 is true, . . . and so on. Therefore p_n is true.

3) Use proof by induction to show that $\dfrac{d}{dx}x^n = nx^{n-1}$.

Let p_n be the statement 'If n is any positive integer then $\dfrac{d}{dx}(x^n) = nx^{n-1}$'

If the formula is valid when $n = k$

i.e. if $$\frac{d}{dx}(x^k) = kx^{k-1}$$

then $$\frac{d}{dx}(x^{k+1}) = \frac{d}{dx}(x^k . x)$$

$$= (kx^{k-1})(x) + (x^k)(1) \quad \text{(using the product rule)}$$

$$= (k+1)x^k$$

which is the formula when $n = k + 1$.
So if the formula is valid when $n = k$ it is valid when $n = k + 1$.
Now if $n = 1$, from first principles

$$\frac{d}{dx}(x^1) = \lim_{\delta x \to 0} \left[\frac{(x + \delta x) - x}{\delta x} \right]$$

$$= 1$$

$$= 1 . x^0$$

Hence p_1 is true.
So p_2 is true, p_3 is true, . . . and so on.
Hence p_n is true.

4) Prove by induction that, for all positive integral values of $n > 1$,

$$|z_1 + z_2 + \ldots + z_n| \leqslant |z_1| + |z_2| + \ldots + |z_n|.$$

Let p_n be the statement

$$|z_1 + \ldots + z_n| \leqslant |z_1| + \ldots + |z_n|$$

for all integral values of $n > 1$.

If the inequality is valid when $n = k$

i.e. if $$|z_1 + \ldots + z_k| \leqslant |z_1| + \ldots + |z_k| \qquad [1]$$

then from the diagram,

$$\overrightarrow{OB} = z_1 + \ldots + z_k + z_{k+1}$$

$$\overrightarrow{OA} = z_1 + \ldots + z_k$$

$$\overrightarrow{AB} = z_{k+1}$$

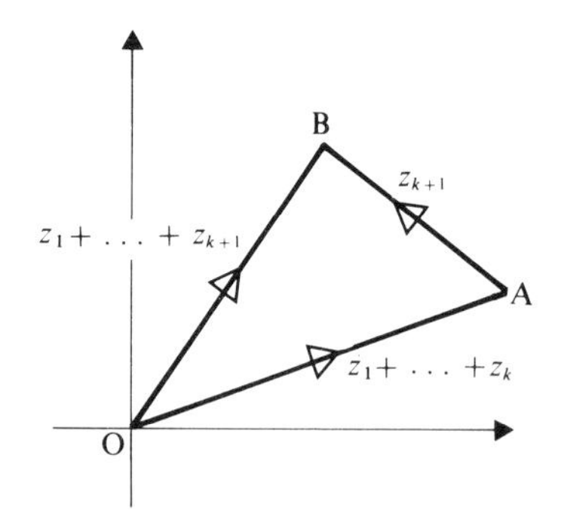

The points OAB form either a triangle or a line.

So $$|\overrightarrow{OB}| \leqslant |\overrightarrow{OA}| + |\overrightarrow{AB}|$$

i.e. $$|z_1 + \ldots + z_k + z_{k+1}| \leqslant |z_1 + z_2 + \ldots + z_k| + |z_{k+1}|$$

Using [1] we have

$$|z_1 + \ldots + z_{k+1}| \leqslant |z_1| + |z_2| + \ldots + |z_k| + |z_{k+1}|$$

which is the inequality when $n = k + 1$.
So if the inequality is valid when $n = k$ then it is valid when $n = k + 1$.

When $n = 2$,

L.H.S. of p_n is $|z_1 + z_2|$

R.H.S. of p_n is $|z_1| + |z_2|$

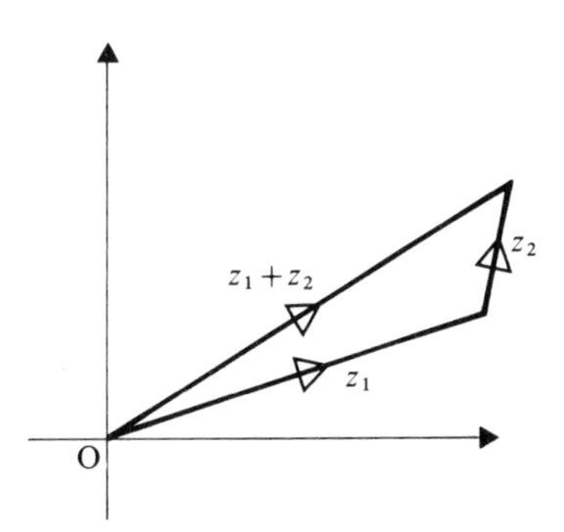

From the diagram, using the property of the lengths of the sides of a triangle again, it follows that

$$|z_1 + z_2| \leqslant |z_1| + |z_2|$$

So p_n is valid when $n = 2$, therefore p_n is valid when $n = 3, \ldots$ and so on.
Hence p_n is valid.

5) Prove by induction that $(1 + x)^n \equiv 1 + {}^nC_1x + {}^nC_2x^2 + \ldots + {}^nC_nx^n$ for all positive integral values of n.

Let p_n be the statement

'If n is any positive integer then

$$(1 + x)^n \equiv 1 + {}^nC_1x + {}^nC_2x^2 + \ldots + {}^nC_nx^n\text{'}$$

If the formula is valid when $n = k$

i.e. if $\quad (1 + x)^k \equiv 1 + {}^kC_1x + {}^kC_2x^2 + \ldots + {}^kC_kx^k$

then, multiplying both sides by $(1 + x)$,

$$(1 + x)^{k+1}$$

$$\equiv (1 + {}^kC_1x + {}^kC_2x^2 + \ldots + {}^kC_kx^k)(1 + x)$$

$$\equiv 1 + ({}^kC_1 + 1)x + ({}^kC_2 + {}^kC_1)x^2 + \ldots + ({}^kC_r + {}^kC_{r-1})x^r + \ldots + {}^kC_kx^{k+1}$$

but $\quad {}^kC_r + {}^kC_{r-1} = {}^{k+1}C_r \quad$ and $\quad {}^kC_k = {}^{k+1}C_{k+1} = 1$

$$\equiv 1 + {}^{k+1}C_1x + {}^{k+1}C_2x^2 + \ldots + {}^{k+1}C_rx^r + \ldots + {}^{k+1}C_{k+1}x^{k+1}$$

which is the formula when $n = k + 1$.
So if the formula is valid when $n = k$ then it is valid when $n = k + 1$.
Now if $n = 1$,

$$(1 + x)^1 \equiv 1 + x \equiv 1 + {}^1C_1x^1$$

So p_1 is true.
Hence p_2 is true, p_3 is true, p_4 is true, . . . and so on.
Hence p_n is true.

EXERCISE 7e

In Questions 1–6 rephrase the statements in the form $p \Rightarrow q$ and give a direct proof to show that they are true. Also state, without proof, if the converse statement (i.e. $p \Leftarrow q$) is true.

1) When $x = \sqrt{2}$, $\dfrac{2+x}{2-x} = 3 + 2\sqrt{2}$.

2) Using the usual notation for a triangle:

in any triangle ABC, $\dfrac{a}{\sin A} = \dfrac{b}{\sin B} = \dfrac{c}{\sin C}$

3) For any angle θ, $\sin\theta + \sin 2\theta + \sin 3\theta = \sin 2\theta(2\cos\theta + 1)$.

4) In a triangle ABC in which $AB = BC$, $\angle A = \angle C$.

5) If $y = f(x)$ has a maximum point at $x = x_1$, then $\dfrac{dy}{dx} = 0$ when $x = x_1$.

6) For a parabolic mirror:
rays of light parallel to its axis are reflected through its focus.

In Questions 7–12, use proof by induction to verify the statements.

7) $\sum_{r=1}^{n} r = \dfrac{n}{2}(n+1)$.

8) $n^3 - n$ is a multiple of 6 for all positive integral values of n.

9) $\sum_{r=1}^{n} \dfrac{1}{r(r+1)} = \dfrac{n}{n+1}$.

10) For all integral values of n, $2^{n+2} + 3^{2n+1}$ is exactly divisible by 7.

11) $\dfrac{d^n}{d\theta^n}(\sin a\theta) = a^n \sin\left(a\theta + \dfrac{n\pi}{2}\right)$

12) $2^n > 2n$ for all integral values of n greater than 2.

13) In the sequence $u_1, u_2, u_3, \ldots, u_n$, $u_1 = 1$ and $u_{r+1} = \dfrac{2u_r - 1}{3}$
Write down the values of u_2, u_3 and u_4 and prove by induction that $u_n = 3(\frac{2}{3})^n - 1$.

14) Prove by induction that $\int x^n \, dx = \dfrac{x^{n+1}}{n+1} + k$ for all positive integral values of n. (*Hint*: use integration by parts.)

15) Prove (a) by direct deduction, (b) by induction, that

$$\sum_{r=1}^{n} ap^r = \frac{a(1-p^n)}{1-p}$$ for all positive integral values of n.

16) Prove by induction that for all positive integral values of n

$$\frac{d}{dx}(x^{-n}) = -nx^{-n-1}.$$

Indirect Proof

Sometimes, when it is difficult to prove a statement in the form $p \Rightarrow q$, a direct proof of the contrapositive statement $\sim q \Rightarrow \sim p$ can be used. This provides a sound proof of $p \Rightarrow q$ because we know that
$(p \Rightarrow q \text{ true}) \iff (\sim q \Rightarrow \sim p \text{ true})$.
Statements whose truth is very nearly obvious can often be proved by this form of indirect proof as illustrated in the following examples.

EXAMPLES 7f

1) Prove that if n is a natural number such that n^2 is even then n is even.

The contrapositive of the statement
'n is a natural number such that n^2 is even $\Rightarrow$ n is even'

is 'n is an odd integer $\Rightarrow$ n^2 is an odd integer'

Now n is an odd integer $\Rightarrow$ $n = 2k+1$ where k is an integer

$$\Rightarrow n^2 = (2k+1)^2$$

$$\Rightarrow n^2 = 4k^2 + 4k + 1$$

$$\Rightarrow n^2 = 2(2k^2 + 2k) + 1$$

$\Rightarrow$ n^2 is an odd integer.

Hence if n is an odd integer then n^2 is an odd integer.
Hence if n is a natural number such that n^2 is an even integer then n is an even integer.

2) In $\triangle ABC$, prove that if $\angle A = \angle C$ then $AB = BC$.

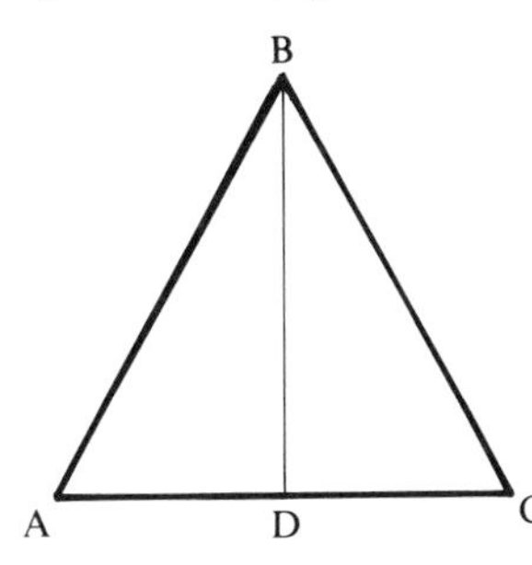

The contrapositive statement is
if $AB \neq BC$ then $\angle A \neq \angle C$.
Now

$$AB \neq BC \Rightarrow \frac{BD}{AB} \neq \frac{BD}{BC}$$

$$\Rightarrow \sin A \neq \sin C$$

$$\Rightarrow \angle A \neq \angle C.$$

Hence if $AB \neq BC$ then $\angle A \neq \angle C$.
Hence if $\angle A = \angle C$ then $AB = BC$.

Proof by Contradiction (*Reductio Ad Absurdum*)

A contradiction is a statement that is always false, regardless of the truth of its component statements.
For example, '$x = 2$ and $x \neq 2$' is false regardless of whether x is 2 or not, as x cannot be both 2 and not 2. So any statement of the form 'q and $\sim q$' is false.
To verify a statement p by contradiction we start with $\sim p$ and deduce a statement of the form 'q and $\sim q$'. As this is false we can argue that $\sim p$ is false, hence that p is true. This is illustrated in the following examples.
To prove that there is an infinite number of prime numbers we start with the negated statement

the number of primes is finite

$\Rightarrow$ there exists an integer p, such that p is the largest prime.

Now $p! + 1 > p$, and $(p! + 1)$ is not divisible by p or by any number less than p†

$\Rightarrow$ either $(p! + 1)$ is not divisible by an integer other than 1 or $(p! + 1)$, in which case $(p! + 1)$ is prime,

or $(p! + 1)$ is divisible by a number between p and $(p! + 1)$

$\Rightarrow$ there is a prime number larger than p.

We have now arrived at the implication

(the number of primes in finite) $\Rightarrow$ (p is the largest prime and there is a prime larger than p)

which contains a contradiction.
Hence 'the number of primes is finite' is false
so 'the number of primes is infinite' is true.
To prove that $\sqrt{2}$ is irrational
The negation of '$\sqrt{2}$ is irrational' is '$\sqrt{2}$ is rational'.
Now $\sqrt{2}$ is rational

$\Rightarrow$ $\sqrt{2} = \dfrac{p}{q}$ where p and q are integers with no common factor

$\Rightarrow$ $2 = \dfrac{p^2}{q^2}$

† If 2 is a factor of a number n, 2 is not a factor of $n + 1$. Similarly if 3 is a factor of n, 3 is not a factor of $n + 1$. Now $2, 3, \ldots, p$ are all factors of $p!$ so none of these are factors of $p! + 1$.

$$\begin{aligned}
&\Rightarrow \quad 2q^2 = p^2\\
&\Rightarrow \quad p^2 \text{ is even}\\
&\Rightarrow \quad p \text{ is even}\\
&\Rightarrow \quad p = 2k \quad \text{where } k \text{ is an integer.}\\
&\Rightarrow \quad 2q^2 = 4k^2\\
&\Rightarrow \quad q^2 = 2k^2\\
&\Rightarrow \quad q^2 \text{ is even}\\
&\Rightarrow \quad q \text{ is even}\\
&\Rightarrow \quad q = 2m.
\end{aligned}$$

Hence we have

$$(\sqrt{2} \text{ is rational}) \Rightarrow (\sqrt{2} = p/q \quad \text{where } p \text{ and } q \text{ are integers with no common factors and} \quad p = 2k, \quad q = 2m)$$

which contains a contradiction.
As the conclusion of this implication is false, the hypothesis that $\sqrt{2}$ is rational is also false.
i.e. $\sqrt{2}$ is irrational.

Both the examples of proof by contradiction given above were proofs of an unconditional statement. We will now look at how the same basic logic can be applied to the proof of a conditional statement.

For example, if $f(x) \equiv x^2 + bx + c$, we will prove by contradiction that

$$b^2 - 4c < 0 \Rightarrow f(x) > 0 \quad \text{for all values of } x. \qquad [1]$$

Starting with the negation of this statement, i.e.
it is not true that $b^2 - 4c < 0 \Rightarrow f(x) > 0$ for all values of x,
we replace this by the equivalent

$$b^2 - 4c < 0 \quad \text{and} \quad f(x) \leqslant 0 \quad \text{for at least one value of } x. \qquad [2]$$

We now take each component statement and make deductions from it.

So '$b^2 - 4ac < 0 \Rightarrow f(x) = 0$ has complex roots'
and '$f(x) \leqslant 0$ for at least one value of x
$\Rightarrow f(x) = 0$ for at least one value of x
(as $f(x)$ is continuous and $f(x) \to +\infty$ as $x \to \infty$)
$\Rightarrow f(x)$ has real roots'.

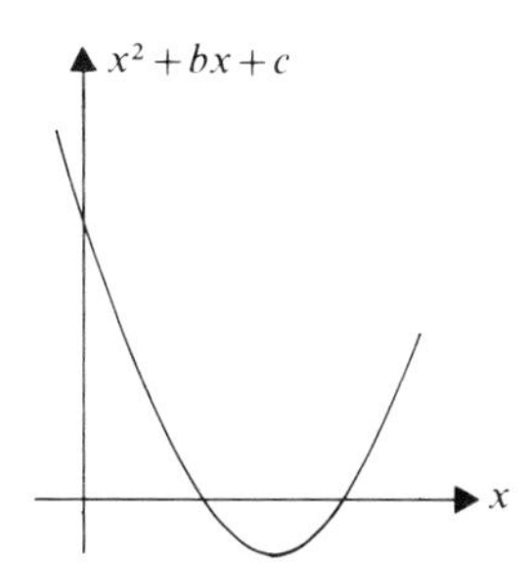

Hence

$$\begin{bmatrix} b^2 - 4c < 0 \quad \text{and} \quad f(x) \leqslant 0 \\ \text{for at least one value of } x \end{bmatrix} \Rightarrow \begin{bmatrix} f(x) = 0 \quad \text{has complex roots} \\ \text{and} \quad f(x) = 0 \quad \text{has real roots} \end{bmatrix}$$

which contains a contradiction.
So [2] is false, therefore [1] is true.
This argument can be written symbolically as follows.

To prove $p \Rightarrow q$ by contradiction.
start with the negation $\sim(p \Rightarrow q)$
and replace by the equivalent p and $\sim q$.
Deduce independently that

$$p \Rightarrow r \quad \text{and} \quad \sim q \Rightarrow \sim r.$$

$$\text{Thus} \qquad (p \text{ and } \sim q) \Rightarrow (r \text{ and } \sim r).$$

As r and $\sim r$ is a contradiction, p and $\sim q$ is false, so $\sim(p \Rightarrow q)$ is false.
Hence $p \Rightarrow q$ is true.

EXAMPLES 7f (continued)

3) Prove by contradiction that, if any series $u_1 + u_2 + \ldots$ is convergent then $\lim_{n \to \infty} u_n = 0$.

Starting with the negation of the given statement, i.e.
'There is at least one series such that $u_1 + u_2 + \ldots$ converges, and $\lim_{n \to \infty} u_n \neq 0$'

and using the notation $\sum_{r=1}^{n} u_r = S_n$ and $\sum_{r=1}^{\infty} u_r = S$,

then

$$u_1 + u_2 + \ldots \text{ converges} \Rightarrow \lim_{n \to \infty} S_n = S$$
$$\Rightarrow \lim_{n \to \infty} (S - S_n) = 0$$

and

$$\lim_{n \to \infty} u_n \neq 0 \Rightarrow \lim_{n \to \infty} u_{n+1} + \lim_{n \to \infty} u_{n+2} + \ldots \neq 0$$
$$\Rightarrow \lim_{n \to \infty} (u_{n+1} + u_{n+2} + \ldots) \neq 0$$
$$\Rightarrow \lim_{n \to \infty} (S - S_n) \neq 0.$$

So $[u_1 + u_2 + \ldots \text{ converges, and } \lim_{n \to \infty} u_n \neq 0]$

$$\Rightarrow [\lim_{n \to \infty} (S - S_n) = 0 \quad \text{and} \quad \lim_{n \to \infty} (S - S_n) \neq 0]$$

This contains a contradiction, therefore no such series exists.
Hence if a series converges then its nth term tends to zero as $n \to \infty$.

The Use of a Counter Example

We have already seen that if a statement $p \Rightarrow q$ is true then its converse $p \Leftarrow q$ is not necessarily true.

In the case of mathematical theorems, it is important to know whether the converse is true or false.
For example, in $\triangle ABC$,

'if $AB = BC$ then $\angle A = \angle C$' is true

and the converse

'if $\angle A = \angle C$ then $AB = BC$' is also true.

However 'If $u_1 + u_2 + \ldots$ converges then $\lim_{n \to \infty} u_n = 0$' is true

but its converse,

'If $\lim_{n \to \infty} u_n = 0$ then $u_1 + u_2 + \ldots$ converges' is false.

To prove that a general implication is false is usually very much easier than proving that it is true. A general statement is always in a context where any variables can take all values,

e.g. All Englishmen have brown hair.

For all values of x and y, $x^2 = y^2 \Rightarrow x = y$.

To prove that such a statement is false, all that we have to do is to produce *just one case* to show that the statement is indeed false. This is called a *counter example*.
So to prove that the statement 'All Englishmen have brown hair' is false, all we have to do is to produce just one Englishman with (say) blond hair.
To prove that 'For all values of x and y, $x^2 = y^2 \Rightarrow x = y$' is false consider the counter example $x = 2$ and $y = -2$. With these values of x and y, $x^2 = y^2$ and $x \neq y$.
(**Note** that a counter example proves that the negation of the given statement is true, i.e. for at least one value of the variable, $\sim p$ is true, so showing p to be false.)

EXAMPLES 7f (continued)

4) Find a counter example to show that the following statement is false.

If, for any series, $\lim_{n \to \infty} u_n = 0$ then $u_1 + u_2 + \ldots$ converges.

Consider the series

$$\tfrac{1}{2} + \tfrac{1}{3} + \tfrac{1}{4} + \tfrac{1}{5} + \tfrac{1}{6} + \ldots$$

$$= \tfrac{1}{2} + (\tfrac{1}{3} + \tfrac{1}{4}) + (\tfrac{1}{5} + \tfrac{1}{6} + \tfrac{1}{7} + \tfrac{1}{8}) + (\tfrac{1}{9} + \tfrac{1}{10} + \tfrac{1}{11} + \tfrac{1}{12} + \tfrac{1}{13} + \tfrac{1}{14} + \tfrac{1}{15} + \tfrac{1}{16}) + \ldots$$

This sum is greater than the following sum:

$$\tfrac{1}{2} + (\tfrac{1}{4} + \tfrac{1}{4}) + (\tfrac{1}{8} + \tfrac{1}{8} + \tfrac{1}{8} + \tfrac{1}{8}) + (\tfrac{1}{16} + \tfrac{1}{16} + \tfrac{1}{16} + \tfrac{1}{16} + \tfrac{1}{16} + \tfrac{1}{16} + \tfrac{1}{16} + \tfrac{1}{16}) + \ldots$$

$$= \tfrac{1}{2} + \tfrac{1}{2} + \tfrac{1}{2} + \tfrac{1}{2} + \ldots$$

which clearly diverges.

Hence $\frac{1}{2}+\frac{1}{3}+\frac{1}{4}+\frac{1}{5}+\ldots$ also diverges *and* $\lim_{n\to\infty}\frac{1}{n}=0$

so proving that the given statement is false.

EXERCISE 7f

In Questions 1–5 prove the given statements by a direct proof of the contrapositive statement.

1) If n is a natural number such that n^2 is odd, then n is odd.

2)

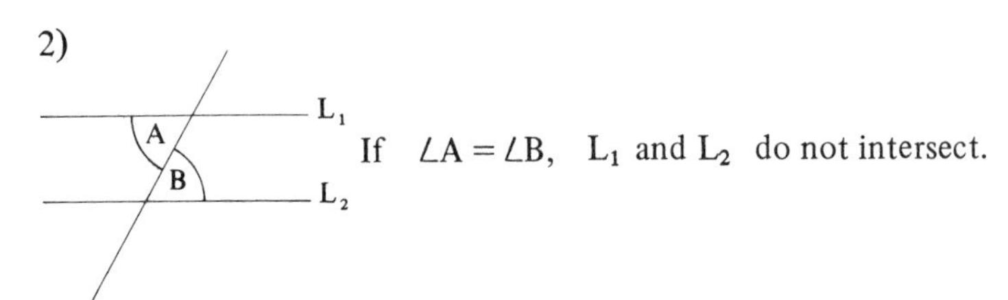

If $\angle A=\angle B$, L_1 and L_2 do not intersect.

3) If n is a perfect number, then n is not a prime number.
(A perfect number is equal to the sum of its prime factors, e.g. $6=3+2+1$.)

4) If two lines are perpendicular, the product of their gradients is -1.

5) If $(x-2)(1-x)>0$ then $1<x<2$.

In Questions 6–12 use proof by contradiction to verify the given statements.

6) If n is a natural number then n^2-n is even.
(Consider two cases, n even, n odd.)

7) $\sqrt{3}$ is irrational.

8) If a and b are real then $a^2+b^2>2ab$.

9) If $ax^2+bx+c=0$ has roots α and β then $cx^2+bx+a=0$ has roots $\dfrac{1}{\alpha}$ and $\dfrac{1}{\beta}$.

10) If $f(x)\equiv\dfrac{1}{x}$ then $f(x)$ has a discontinuity.

11) If $x^2-3x+2<0$ then $1<x<2$.

12) For all $x>0$, $x+\dfrac{1}{x}\geqslant 2$.

In Questions 13–18, find a counter example to show that the given statements are false.

13) For any two vectors **a** and **b**, $\mathbf{a}\times\mathbf{b}=\mathbf{b}\times\mathbf{a}$.

14) For all real values of a and b, $a-b>0 \Rightarrow a^2-b^2>0$.

15) If, in the equation $ax^2 + bx + c = 0$, a, b and c are real then the roots of the equation are real.

16) For any two vectors $\mathbf{a}$ and $\mathbf{b}$, $\mathbf{a} \times \mathbf{b} = \mathbf{0} \Rightarrow \mathbf{a} = \mathbf{0}$ or $\mathbf{b} = \mathbf{0}$.

17) If $f''(x) = 0$ when $x = a$ then $f(x)$ has a point of inflexion when $x = a$.

18) For any three matrices $\mathbf{A}$, $\mathbf{B}$ and $\mathbf{C}$ that are compatible for the products $\mathbf{AB}$ and $\mathbf{AC}$, $\mathbf{AB} = \mathbf{AC} \Rightarrow \mathbf{A} = \mathbf{0}$ or $\mathbf{B} - \mathbf{C} = \mathbf{0}$.

CHAPTER 8

SUMMATION OF SERIES

NUMBER SERIES

A series, each of whose terms has a fixed numerical value, is called a number series. For example, $2 + 4 + 8 + 16 + \ldots$ is a number series, whereas $x + x^2 + x^3 + \ldots$ is called a power series as the terms involve powers of a variable quantity.

In this Chapter we are going to investigate some of the methods available for finding the sum of a number series. The method adopted depends to some extent on the form of the general term of the series. One of the most straightforward ways of summing a given series uses *recognition* of a standard series.

For instance it might be recognised as an arithmetic progression (A.P.) or a geometric progression (G.P.) or as the standard expansion of a familiar function, such as $\ln(1 + x)$, with a numerical value assigned to the variable.

Consider, for example, the series expansion of e^x,

i.e. $$e^x = 1 + x + \frac{x^2}{2!} + \frac{x^3}{3!} + \frac{x^4}{4!} + \ldots + \frac{x^r}{r!} + \ldots \qquad [1]$$

Substituting 1 for x in the R.H.S. gives the number series

$$1 + 1 + \frac{1}{2!} + \frac{1}{3!} + \frac{1}{4!} + \ldots + \frac{1}{r!} + \ldots = \sum_{r=0}^{\infty} \frac{1}{r!}$$

Substituting 1 for x in the L.H.S. of [1] gives e, so we deduce that

$$\sum_{r=0}^{\infty} \frac{1}{r!} = e$$

At this point we recommend that the reader revise the work on standard series in Chapter 16 of Volume 1.

EXAMPLES 8a

1) Evaluate $\sum_{r=1}^{n} 2^{2r-1}$

$$\sum_{r=1}^{n} 2^{2r-1} = 2 + 2^3 + 2^5 + \ldots + 2^{2n-1}$$

which we *recognise* as a G.P. whose first term is 2, and whose common ratio is 2^2.

Using the formula for the sum of the first n terms of a G.P. gives

$$\sum_{r=1}^{n} 2^{2r-1} = \frac{2[(2^2)^n - 1]}{2^2 - 1} = \tfrac{2}{3}(4^n - 1).$$

2) Find the sum to infinity of the series

$$1 - 2 + \frac{4}{2!} - \frac{8}{3!} + \frac{16}{4!} - \ldots$$

In each term of this series, the numerator is a power of 2, so the general term can be expressed in the form $\dfrac{(-1)^r\,2^r}{r!}$.

When we rewrite the series as

$$1 - 2 + \frac{2^2}{2!} - \frac{2^3}{3!} + \frac{2^4}{4!} - \ldots + \frac{(-1)^r\,2^r}{r!} + \ldots$$

we recognise that it is similar to the series expansion of e^{-x},

i.e. $$e^{-x} = 1 - x + \frac{x^2}{2!} - \frac{x^3}{3!} + \ldots + (-1)^r\frac{x^r}{r!} + \ldots$$

Replacing x by 2 gives

$$1 - 2 + \frac{2^2}{2!} - \frac{2^3}{3!} + \frac{2^4}{4!} + \ldots = e^{-2}$$

Note that care must be taken when assigning a value to x in any standard expansion. The number chosen must lie within the range of values of x for which the expansion is valid.

Sometimes a given series, although recognised basically as a known series, has slight deviations from the standard series. The sum of the given series can still be deduced from the known series allowing first for any adjustments that may be necessary. These can usually be identified by comparing general terms, noting particularly any variations in sign and power. It is also necessary to compare the limits of the summation, making sure that the first few terms correspond and that the same set of terms is being summed.
These points are illustrated in the following examples.

3) Find the sum to infinity of the series

$$\frac{1}{2}\left(\frac{2}{3}\right)^2 - \frac{1}{3}\left(\frac{2}{3}\right)^3 + \frac{1}{4}\left(\frac{2}{3}\right)^4 - \dots$$

The general term of the given series is $(-1)^r \frac{1}{r}\left(\frac{2}{3}\right)^r$

which is recognised as basically similar to the general term in the expansion of $\ln(1+x)$, i.e. $(-1)^{(r+1)}\frac{1}{r}x^r$

Now $$\ln(1+x) = x - \frac{x^2}{2} + \frac{x^3}{3} - \frac{x^4}{4} + \dots$$

Substituting $\frac{2}{3}$ for x (which is in the range $-1 < x \leqslant 1$) gives

$$\ln\left(1+\frac{2}{3}\right) = \frac{2}{3} - \frac{1}{2}\left(\frac{2}{3}\right)^2 + \frac{1}{3}\left(\frac{2}{3}\right)^3 - \frac{1}{4}\left(\frac{2}{3}\right)^4 + \dots$$

$$= \frac{2}{3} - \left[\frac{1}{2}\left(\frac{2}{3}\right)^2 - \frac{1}{3}\left(\frac{2}{3}\right)^3 + \frac{1}{4}\left(\frac{2}{3}\right)^4 - \dots\right]$$

$$= \frac{2}{3} - S$$

where S is the sum of the given series.
Hence $S = \frac{2}{3} - \ln\frac{5}{3}$.

4) Evaluate $\sum_{r=n}^{\infty}\left(\frac{1}{3}\right)^r\left(\frac{1}{2}\right)^{r-2}$

$$\sum_{r=n}^{\infty}\left(\frac{1}{3}\right)^r\left(\frac{1}{2}\right)^{r-2} = \left(\frac{1}{3}\right)^n\left(\frac{1}{2}\right)^{n-2} + \left(\frac{1}{3}\right)^{n+1}\left(\frac{1}{2}\right)^{n-1} + \dots$$

Each term of this series has a factor $\left(\frac{1}{3}\right)^n\left(\frac{1}{2}\right)^{n-2}$

so we may write

$$\sum_{r=n}^{\infty}\left(\frac{1}{3}\right)^r\left(\frac{1}{2}\right)^{r-2} = \left(\frac{1}{3}\right)^n\left(\frac{1}{2}\right)^{n-2}\left[1+\left(\frac{1}{3}\right)\left(\frac{1}{2}\right)+\left(\frac{1}{3}\right)^2\left(\frac{1}{2}\right)^2+\dots\right]$$

$$= \left(\frac{1}{3}\right)^n\left(\frac{1}{2}\right)^{n-2}\left[1+\frac{1}{6}+\left(\frac{1}{6}\right)^2+\dots\right]$$

The series in the square bracket is recognised as an infinite G.P. whose sum to infinity is $\dfrac{1}{1-\frac{1}{6}} = \dfrac{6}{5}$.

Hence
$$\sum_{r=n}^{\infty}\left(\frac{1}{3}\right)^r\left(\frac{1}{2}\right)^{r-2} = \frac{6}{5}\left(\frac{1}{3}\right)^n\left(\frac{1}{2}\right)^{n-2}$$

$$= \frac{4}{5}\left(\frac{1}{6}\right)^{n-1}$$

5) Find the sum to infinity of the series

$$\frac{p^3}{2!}+\frac{p^5}{3!}+\frac{p^7}{4!}+\dots$$

Writing the general term of this series in the form

$$\frac{p^{2r-1}}{r!} \equiv \frac{1}{p}\left(\frac{p^{2r}}{r!}\right)$$

we see that it is similar to the general term in the expansion of e^x, i.e. $\dfrac{x^r}{r!}$

Starting with

$$e^x = 1+x+\frac{x^2}{2!}+\frac{x^3}{3!}+\dots+\frac{x^r}{r!}+\dots$$

and replacing x by p^2 we have

$$e^{p^2} = 1+p^2+\frac{p^4}{2!}+\frac{p^6}{3!}+\dots+\frac{p^{2r}}{r!}+\dots$$

Dividing by p gives

$$\frac{1}{p}e^{p^2} = \frac{1}{p}+p+\frac{p^3}{2!}+\frac{p^5}{3!}+\dots+\frac{p^{2r-1}}{r!}+\dots$$

$$= \frac{1}{p}+p+\left[\frac{p^3}{2!}+\frac{p^5}{3!}+\frac{p^7}{4!}+\dots\right]$$

$$= \frac{1}{p}+p+S.$$

Therefore the sum to infinity of the given series is given by

$$S = \frac{1}{p}(e^{p^2} - 1) - p$$

6) Find the sum to infinity of the series

$$\frac{2}{1!} + \frac{3}{2!} + \frac{4}{3!} + \frac{5}{4!} + \dots$$

The general term of this series is $\frac{r+1}{r!}$, which is not immediately recognisable as the general term of a known series.

However, if we write $\frac{r+1}{r!}$ as *two separate fractions*, i.e. $\frac{r}{r!} + \frac{1}{r!}$, then the given series may be expressed as the sum of two series as follows.

$$\frac{2}{1!} + \frac{3}{2!} + \frac{4}{3!} + \dots + \left(\frac{r}{r!} + \frac{1}{r!}\right) + \dots = \sum_{r=1}^{\infty} \left(\frac{r}{r!} + \frac{1}{r!}\right)$$

$$= \sum_{r=1}^{\infty} \frac{r}{r!} + \sum_{r=1}^{\infty} \frac{1}{r!}$$

$$= \left(\frac{1}{1!} + \frac{2}{2!} + \frac{3}{3!} + \dots\right) + \left(\frac{1}{1!} + \frac{1}{2!} + \frac{1}{3!} + \frac{1}{4!} + \dots\right)$$

$$= \left(1 + 1 + \frac{1}{2!} + \frac{1}{3!} + \dots\right) + \left(1 + \frac{1}{2!} + \frac{1}{3!} + \frac{1}{4!} + \dots\right)$$

The first bracket is now recognised as the series expansion of e^x when $x = 1$ and the second bracket is the same series except for the absent first term.

Hence $$\left(1 + 1 + \frac{1}{2!} + \frac{1}{3!} + \dots\right) + \left(1 + \frac{1}{2!} + \frac{1}{3!} + \frac{1}{4!} + \dots\right)$$

$$= (e) + (e - 1)$$

Therefore the sum to infinity of $\frac{2}{1!} + \frac{3}{2!} + \frac{4}{3!} + \dots$ is $2e - 1$

7) Find the sum to infinity of the series

$$1 - \frac{2}{4} + \frac{2.5}{4.8} - \frac{2.5.8}{4.8.12} + \dots$$

$$1 - \frac{2}{4} + \frac{2.5}{4.8} - \frac{2.5.8}{4.8.12} + \dots$$

$$= 1 - \frac{2}{1}\left(\frac{1}{4}\right) + \frac{2.5}{1.2}\left(\frac{1}{4}\right)^2 - \frac{2.5.8}{1.2.3}\left(\frac{1}{4}\right)^3 + \dots$$

This series bears some resemblance to the binominal expansion of $(1+x)^n$.

The general term, u_r, of the given series is $\dfrac{(2)(5)(8)\dots(3r-1)}{r!}\left(-\dfrac{1}{4}\right)^r$ and

dividing each term of the numerator by -3 gives

$$u_r = \frac{(-\frac{2}{3})(-\frac{5}{3})(-\frac{8}{3})\dots(\frac{1}{3}-r)}{r!}(-3)^r\left(-\frac{1}{4}\right)^r$$

$$= \frac{(-\frac{2}{3})(-\frac{5}{3})(-\frac{8}{3})\dots(\frac{1}{3}-r)}{r!}\left(\frac{3}{4}\right)^r.$$

As the general term in the expansion of $(1+x)^n$ is

$$\frac{n(n-1)(n-2)\dots(n-r+1)}{r!}x^r$$

we recognise the given series as $(1+x)^n$ with $x=\frac{3}{4}$ and $n=-\frac{2}{3}$
Hence the sum to infinity of the given series

$$\left(1+\frac{3}{4}\right)^{-\frac{2}{3}} = \left(\frac{4}{7}\right)^{\frac{2}{3}} = \sqrt[3]{\frac{16}{49}}$$

EXERCISE 8a

1) Find the sums, to the number of terms indicated, of the following series.

(a) $1+\dfrac{1}{2!}+\dfrac{1}{4!}+\dfrac{1}{6!}+\dots$

(b) $2-2.3+2.3^2-2.3^3+\dots+2.3^{10}$

(c) $1+\dfrac{1}{3}+\dfrac{1}{3^2 2!}+\dfrac{1}{3^3 3!}+\dots$

(d) $1+10\left(\dfrac{1}{2}\right)+\dfrac{10.9}{2!}\left(\dfrac{1}{2}\right)^2+\dfrac{10.9.8}{3!}\left(\dfrac{1}{2}\right)^3+\dots+\left(\dfrac{1}{2}\right)^{10}$

(e) $-\dfrac{1}{2}-\dfrac{1}{2}\left(\dfrac{1}{2}\right)^2-\dfrac{1}{3}\left(\dfrac{1}{2}\right)^3-\dfrac{1}{4}\left(\dfrac{1}{2}\right)^4-\dots$

(f) $3-\dfrac{27}{3!}+\dfrac{243}{5!}-\dots$

(g) $1-2+\dfrac{4}{2!}-\dfrac{8}{3!}+\dfrac{16}{4!}-\dots$

(h) $1-\dfrac{1}{2}+\dfrac{1}{4}-\dfrac{1}{8}+\dfrac{1}{16}-\dots$

2) Find the value of

(a) $\sum_{r=2}^{n} ab^{2r}$ (b) $\sum_{r=1}^{\infty} \frac{2^{r-1}}{r!}$ (c) $\sum_{r=n+1}^{2n} \frac{2^{r+1}}{3^{r-1}}$

(d) $1-\frac{1}{2!}+\frac{1}{3!}-\frac{1}{4!}+\dots$ (e) $\sum_{r=1}^{\infty} \frac{1}{r \times 2^r}$

(f) $\frac{1}{2^2 2!}-\frac{1}{2^4 4!}+\frac{1}{2^6 6!}-\dots$

(g) $\frac{1}{3!}-\frac{1}{5!}+\frac{1}{7!}-\dots$ (h) $\sum_{r=3}^{\infty} \frac{1}{e^r}$

3) Find the sums of the following series.

(a) $1+\frac{3}{1!}+\frac{5}{2!}+\frac{7}{3!}+\dots+\frac{(2r+1)}{r!}+\dots$

(b) $\sum_{r=0}^{\infty} \left(\frac{r!+1}{r!}\right)\frac{1}{2^r}$

(c) $\frac{2}{1}\left(\frac{1}{3}\right)+\frac{2}{3}\left(\frac{1}{3}\right)^3+\frac{2}{5}\left(\frac{1}{3}\right)^5+\dots+\frac{2}{2r-1}\left(\frac{1}{3}\right)^{2r-1}+\dots$

(d) $\frac{1}{0!}-\frac{5}{2!}+\frac{17}{4!}-\frac{65}{6!}+\dots+\frac{(-1)^r(1+2^{2r})}{(2r)!}+\dots$

(e) $1-\frac{2}{4}+\frac{2.4}{4.8}-\frac{2.4.6}{4.8.12}+\dots$

(f) $1+\frac{1}{4}-\frac{1}{4.8}+\frac{1.3}{4.8.12}-\frac{1.3.5}{4.8.12.16}+\dots$

DIFFERENTIATION OR INTEGRATION OF STANDARD SERIES

When a power of x is differentiated, the power becomes a factor in the numerator,

i.e. $$\frac{d}{dx}(x^r) = rx^{r-1}$$

Similarly when a power of x is integrated, the power becomes a factor in the denominator,

i.e. $$\int_0^x x^r\,dx = \frac{x^{r+1}}{r+1}$$

So if the general term of a number series has such a factor in either numerator or denominator, the sum of the series may sometimes be derived from the differential or integral of a standard power series. The examples that follow are based on the assumption that the differential of a function is equal to the sum to infinity of differentials of the terms of the series expansion of that function.

That is, if $$f(x) = a_0 + a_1x + a_2x^2 + \ldots + a_rx^r + \ldots$$

then $$\frac{d}{dx}f(x) = \sum_{r=0}^{\infty} \frac{d}{dx}(a_rx^r)$$

Similarly, we assume that

$$\int_0^x f(x)\,dx = \sum_{r=0}^{\infty}\left[\int_0^x a_rx^r\,dx\right]$$

EXAMPLES 8b

1) Find the sum to infinity of the series

$$1 + 2(\tfrac{1}{2}) + 3(\tfrac{1}{2})^2 + 4(\tfrac{1}{2})^3 + \ldots$$

The general term of the series is $r(\frac{1}{2})^{r-1}$.

Now $$rx^{r-1} = \frac{d}{dx}(x^r)$$

Also x^r is a general term of the G.P. $1 + x + x^2 + \ldots$ and the sum to infinity of this G.P. is $(1-x)^{-1}$.
Starting with

$$(1-x)^{-1} = 1 + x + x^2 + x^3 + \ldots$$

and differentiating, we have

$$(1-x)^{-2} = 1 + 2x + 3x^2 + \ldots + rx^{r-1} + \ldots$$

Substituting $\frac{1}{2}$ for x (which is a value for which the expansion of $(1-x)^{-1}$ is valid) gives

$$(1-\tfrac{1}{2})^{-2} = 1 + 2(\tfrac{1}{2}) + 3(\tfrac{1}{2})^2 + \ldots + r(\tfrac{1}{2})^{r-1} + \ldots$$

So the sum to infinity of the given series is 4.

2) Evaluate $\sum_{r=1}^{n} r.3^r$

$$\sum_{r=1}^{n} r.3^r = 1(3) + 2(3)^2 + 3(3)^3 + 4(3)^4 + \ldots + n(3^n)$$

Comparing the general term, $r.3^r$, with rx^r we see that rx^r is *not* the differential of x^r.

However, $$rx^r = x(rx^{r-1}) = x\left(\frac{d}{dx}x^r\right)$$

So, as x^r is the general term of a G.P., we could sum the given series using $x\frac{d}{dx}[\Sigma x^r]$. However this involves a lot of work and a better method for this series, as it bears some relation to a G.P., is a method similar to that used for summing a G.P.

If $$S = \sum_{r=1}^{n} r.3^r$$

i.e. $$S = 1.3 + 2.3^2 + 3.3^3 + \ldots + (n-1).3^{n-1} + n.3^n$$

then $$3S = 1.3^2 + 2.3^3 + \ldots + (n-2).3^{n-1} + (n-1).3^n + n.3^{n+1}$$

$\Rightarrow$ $$2S = -3 - 3^2 - 3^3 - \ldots - 3^{n-1} - 3^n + n.3^{n+1}$$

$$= -(3 + 3^2 + \ldots + 3^n) + n.3^{n+1}$$

$$= n.3^{n+1} - \frac{3(3^n - 1)}{3 - 1}$$

$$= \frac{(2n-1).3^{n+1} + 3}{2}$$

Hence $$S = \frac{(2n-1).3^{n+1} + 3}{4}$$

3) Evaluate $\displaystyle\sum_{r=2}^{n} \frac{2^r}{r}$

Comparing $\frac{2^r}{r}$ with $\frac{x^r}{r}$ we see that $\frac{x^r}{r} = \int_0^x x^{r-1}\,dx$

Note that although $\frac{x^r}{r}$ is a general term in the expansion of $\ln(1-x)$ we cannot use this series for comparison because
(a) it is infinite whereas the given series is finite
(b) it is valid only within the range $-1 \leqslant x < 1$, so x cannot be given the value 2.
Now x^{r-1} is a general term of the G.P.

$$1 + x + x^2 + \ldots + x^{r-1} + \ldots + x^{n-1}$$

whose sum is $\dfrac{x^n - 1}{x - 1}$

Hence $$\int_0^x \left(\frac{x^n - 1}{x - 1}\right) dx = \int_0^x (1 + x + x^2 + \ldots + x^{n-1})\, dx$$

$$= x + \frac{x^2}{2} + \frac{x^3}{3} + \ldots + \frac{x^n}{n}$$

Substituting 2 for x gives

$$\int_0^2 \left(\frac{x^n - 1}{x - 1}\right) dx = 2 + \frac{2^2}{2} + \frac{2^3}{3} + \ldots + \frac{2^n}{n} \qquad [1]$$

Noting that the given series starts with the term $\frac{2^2}{2}$, we find that

$$\sum_{r=2}^{n} \frac{2^r}{r} = \left[\int_0^2 \left(\frac{x^n - 1}{x - 1}\right) dx\right] - 2$$

(Integration is not easily done at this stage, so the answer is left in integral form.)

EXERCISE 8b

1) Evaluate $\sum_{r=2}^{\infty} r\left(\frac{1}{2}\right)^{r+1}$

2) Find the sum of the first n terms of the series

$$1 - 2(2) + 3(2)^2 - 4(2)^3 + \ldots$$

3) Find the sum to infinity of the series

$$\frac{x^2}{(2)(1)} - \frac{x^3}{(3)(2)} + \frac{x^4}{(4)(3)} - \ldots$$

4) Evaluate $\sum_{r=1}^{10} r(^{10}C_r)(-\tfrac{1}{2})^{r-1}$

5) Evaluate $\sum_{r=n}^{2n} r2^{r-1}$

The summation techniques adopted so far in this Chapter have relied on the recognition of standard expansions. We will now look at some methods for summing series which are unrelated to known series.

USE OF PARTIAL FRACTIONS

Consider the series $\frac{1}{1.2} + \frac{1}{2.3} + \frac{1}{3.4} + \frac{1}{4.5} + \ldots + \frac{1}{(n-1)n}$

A general term of this series is $\dfrac{1}{r(r+1)}$ which can be expressed in partial fractions,

i.e. $$\frac{1}{r(r+1)} \equiv \frac{1}{r} - \frac{1}{r+1}.$$

Hence $$\frac{1}{1.2} + \frac{1}{2.3} + \frac{1}{3.4} + \ldots + \frac{1}{(n-1)n} = \sum_{r=1}^{n-1} \frac{1}{r(r+1)}$$

$$= \sum_{r=1}^{n-1} \left(\frac{1}{r} - \frac{1}{r+1}\right) \qquad [1]$$

$$= \left(\frac{1}{1} - \frac{1}{2}\right) + \left(\frac{1}{2} - \frac{1}{3}\right) + \left(\frac{1}{3} - \frac{1}{4}\right) + \ldots + \left(\frac{1}{n-1} - \frac{1}{n}\right)$$

Regrouping the terms gives

$$\sum_{r=1}^{n-1} \frac{1}{r(r+1)} = 1 + \left(-\frac{1}{2} + \frac{1}{2}\right) + \left(-\frac{1}{3} + \frac{1}{3}\right) + \ldots + \left(-\frac{1}{n-1} + \frac{1}{n-1}\right) - \frac{1}{n}$$

from which we see that all the terms cancel except 1 and $-\dfrac{1}{n}$,

i.e. $$\sum_{r=1}^{n-1} \frac{1}{r(r+1)} = 1 - \frac{1}{n}$$

This cancellation is easier to see if the terms of [1] are listed vertically:

$$\begin{aligned}
\sum_{r=1}^{n-1} \frac{1}{r(r+1)} = \ & \frac{1}{1} - \frac{1}{2} \\
& + \frac{1}{2} - \frac{1}{3} \\
& + \frac{1}{3} - \frac{1}{4} \\
& + \frac{1}{4} - \frac{1}{5} \\
& \ldots\ldots\ldots \\
& + \frac{1}{n-2} - \frac{1}{n-1} \\
& + \frac{1}{n-1} - \frac{1}{n} \qquad = 1 - \frac{1}{n}.
\end{aligned}$$

Many series whose general terms can be expressed in partial fractions may be summed in a similar way.

EXAMPLE 8c

Find the sum

$$S_n = \sum_{r=1}^{n} \frac{4r}{(2r-1)(2r+1)(2r+3)}$$

Expressing the general term, u_r, in partial fractions gives

$$u_r = \frac{4r}{(2r-1)(2r+1)(2r+3)} \equiv \frac{1}{4(2r-1)} + \frac{1}{2(2r+1)} - \frac{3}{4(2r+3)}.$$

Hence

$$\begin{aligned}
u_1 &= \frac{1}{4} + \frac{1}{6} - \frac{3}{20} \\
u_2 &= \frac{1}{12} + \frac{1}{10} - \frac{3}{28} \\
u_3 &= \frac{1}{20} + \frac{1}{14} - \frac{3}{36} \\
u_4 &= \frac{1}{28} + \frac{1}{18} - \frac{3}{44} \\
&\dots \\
u_{n-2} &= \frac{1}{4(2n-5)} + \frac{1}{2(2n-3)} - \frac{3}{4(2n-1)} \\
u_{n-1} &= \frac{1}{4(2n-3)} + \frac{1}{2(2n-1)} - \frac{3}{4(2n+1)} \\
u_n &= \frac{1}{4(2n-1)} + \frac{1}{2(2n+1)} - \frac{3}{4(2n+3)}
\end{aligned}$$

$$\Rightarrow \quad \sum_{r=1}^{n} u_r = \frac{1}{4} + \frac{1}{6} + \frac{1}{12} + \frac{1}{2(2n+1)} - \frac{3}{4(2n+1)} - \frac{3}{4(2n+3)}$$

$$= \frac{1}{2} - \frac{4n+3}{2(2n+1)(2n+3)}$$

Note that sufficient terms should be tabulated at the beginning *and* at the end of the series to give a clear picture of the pattern of cancelling.

EXERCISE 8c

1) Find the value of

(a) $\displaystyle\sum_{r=1}^{n} \frac{1}{r(r+2)}$ (b) $\displaystyle\sum_{r=3}^{n} \frac{1}{(r+1)(r+2)}$

(c) $\displaystyle\sum_{r=n}^{2n} \frac{1}{r(r+1)}$ (d) $\displaystyle\sum_{r=1}^{n} \frac{r}{(2r-1)(2r+1)(2r+3)}$

2) Find the sums of the following series:

(a) $\dfrac{1}{2.4}+\dfrac{1}{3.5}+\dfrac{1}{4.6}+\ldots+\dfrac{1}{(n+1)(n+3)}$

(b) $\dfrac{1}{2.3.4}+\dfrac{2}{4.5.6}+\dfrac{3}{6.7.8}+\ldots+\dfrac{n}{(n+1)(n+2)(n+3)}$

(c) $\dfrac{1}{4.8}+\dfrac{1}{6.10}+\ldots+\dfrac{1}{2n(2n+4)}$

(d) $\dfrac{1}{3.5}+\dfrac{1}{4.12}+\dfrac{1}{5.21}+\ldots+\dfrac{1}{n(n^2-4)}$

3) Find the sums of the first n terms of the following series:

(a) $\dfrac{1}{1.3}+\dfrac{1}{2.4}+\dfrac{1}{3.5}+\ldots$

(b) $\dfrac{1}{3.5}+\dfrac{1}{5.7}+\dfrac{1}{7.9}+\ldots$

(c) $\dfrac{1}{1.5}+\dfrac{1}{3.7}+\dfrac{1}{5.9}+\ldots$

METHOD OF DIFFERENCES

If the general term of a series, u_r, can be expressed as $f(r+1)-f(r)$, then

$$\begin{aligned}\sum_{r=1}^{n} u_r &= \sum_{r=1}^{n} [f(r+1)-f(r)] \\ &= f(2) - f(1) \\ &\quad + f(3) - f(2) \\ &\quad + f(4) - f(3) \\ &\quad \ldots\ldots\ldots\ldots\ldots\ldots \\ &\quad \ldots\ldots\ldots\ldots\ldots\ldots \\ &\quad + f(n) - f(n-1) \\ &\quad + f(n+1) - f(n) = f(n+1)-f(1)\end{aligned}$$

This technique for finding the sum of a series is known as the method of differences. The use of partial fractions in the previous section is a difference method but the general technique is not limited to series which can be summed using partial fractions.

For example, consider the series

$$1.2+2.3+3.4+\ldots+n(n+1)$$

A general term of the series, u_r, is $r(r+1)$.

Now $\qquad u_r = r(r+1) \equiv \tfrac{1}{3}r(r+1)(r+2)-\tfrac{1}{3}(r-1)r(r+1)$

So we have expressed u_r in the form $f(r+1)-f(r)$

where $\qquad f(r) \equiv \tfrac{1}{3}(r-1)(r)(r+1)$

Therefore

$$\sum_{r=1}^{n} r(r+1) = \sum_{r=1}^{n} \tfrac{1}{3}(r)(r+1)(r+2)-\tfrac{1}{3}(r-1)(r)(r+1)$$

$$\begin{aligned}
= &\quad \tfrac{1}{3}.1.2.3 \quad - \quad 0\\
+ &\quad \tfrac{1}{3}.2.3.4 \quad - \quad \tfrac{1}{3}.1.2.3\\
+ &\quad \tfrac{1}{3}.3.4.5 \quad - \quad \tfrac{1}{3}.2.3.4\\
&\ldots\ldots\ldots\ldots\ldots\ldots\ldots\ldots\\
+ &\quad \tfrac{1}{3}n(n+1)(n+2) \quad - \quad \tfrac{1}{3}(n-1)(n)(n+1) \quad = \tfrac{1}{3}n(n+1)(n+2)
\end{aligned}$$

i.e. $\qquad 1.2+2.3+\ldots+n(n+1) = \tfrac{1}{3}n(n+1)(n+2)$

At this stage the reader is not generally expected to derive the difference identity from the general term. Suitable help will be given in problems of this type.

EXAMPLES 8d

1) If $f(r)\equiv r(r+1)!$ simplify $f(r)-f(r-1)$ and hence sum the series

$$5.2!+10.3!+17.4!+\ldots+(n^2+1)n!$$

$$\begin{aligned}
f(r)-f(r-1) &\equiv r(r+1)!-(r-1)r!\\
&\equiv r![r(r+1)-(r-1)]\\
&\equiv r!\,(r^2+1)
\end{aligned}$$

Hence

$$\sum_{r=2}^{n} (r^2+1)r! = \sum_{r=2}^{n} [f(r)-f(r-1)]$$

$$= \sum_{r=2}^{n} [r(r+1)! - (r-1)r!]$$

$$\begin{aligned} & 2.3! - 1.2! \\ + & 3.4! - 2.3! \\ + & 4.5! - 3.4! \\ & \dots\dots\dots\dots \\ + & (n-1)n! - (n-2)(n-1)! \\ + & n(n+1)! - (n-1)n! \qquad = n(n+1)! - 2! \end{aligned}$$

i.e. $5.2! + 10.3! + \ldots + (n^2+1)n! = n(n+1)! - 2.$

Note that it is not necessary to evaluate each term on the R.H.S. The next example shows the alternative form.

2) If $f(r) \equiv (r - \frac{1}{2})^3$, simplify $f(r+1) - f(r)$. Hence find $\sum_{r=1}^{n} r^2$.

$$\begin{aligned} f(r+1) - f(r) &\equiv (r + 1 - \tfrac{1}{2})^3 - (r - \tfrac{1}{2})^3 \\ &\equiv (r + \tfrac{1}{2})^3 - (r - \tfrac{1}{2})^3 \\ &\equiv 3r^2 + \tfrac{1}{4} \end{aligned}$$

Hence $$\sum_{r=1}^{n} (3r^2 + \tfrac{1}{4}) = \sum_{r=1}^{n} [f(r+1) - f(r)]$$

$$\begin{aligned} = \quad & f(2) - f(1) \\ + & f(3) - f(2) \\ + & f(4) - f(3) \\ & \dots\dots\dots\dots \\ + & f(n) - f(n-1) \\ + & f(n+1) - f(n) \qquad = f(n+1) - f(1) \end{aligned}$$

$$= (n + \tfrac{1}{2})^3 - (\tfrac{1}{2})^3 \qquad [1]$$

Now $$\sum_{r=1}^{n} (3r^2 + \tfrac{1}{4}) = \sum_{r=1}^{n} 3r^2 + \sum_{r=1}^{n} \tfrac{1}{4}$$

$$= 3\sum_{r=1}^{n} r^2 + (\tfrac{1}{4} + \tfrac{1}{4} + \ldots + \tfrac{1}{4})$$

$$= 3\sum_{r=1}^{n} r^2 + \tfrac{1}{4}n \qquad [2]$$

Therefore from [1] and [2]

$$3\sum_{r=1}^{n} r^2 + \tfrac{1}{4}n = (n+\tfrac{1}{2})^3 - \tfrac{1}{8}$$

$$\Rightarrow \qquad 3\sum_{r=1}^{n} r^2 = (n+\tfrac{1}{2})^3 - \tfrac{1}{8} - \tfrac{1}{4}n$$

$$= \tfrac{1}{2}n(n+1)(2n+1)$$

$$\Rightarrow \qquad \sum_{r=1}^{n} r^2 = \tfrac{1}{6}n(n+1)(2n+1)$$

This is a standard result which may be quoted, unless proof is asked for.

STANDARD RESULTS

Some natural number series have sums that are quotable.
These are

(a) the sum of the first n natural numbers,

$$\sum_{r=1}^{n} r = \frac{n}{2}(n+1)$$

(b) the sum of the squares of the first n natural numbers,

$$\sum_{r=1}^{n} r^2 = \frac{n}{6}(n+1)(2n+1)$$

(c) the sum of the cubes of the first n natural numbers,

$$\sum_{r=1}^{n} r^3 = \frac{n^2}{4}(n+1)^2$$

The proof for (a), as it is an A.P., is in Volume I, Chapter 16.
The proof for (b) is given in Question 2 of Examples 8d and the proof of (c) is left to the reader as a problem in Exercise 8d.

EXAMPLES 8d (continued)

3) Find $\sum_{r=1}^{n} r(r+1)(r+2)$ using the standard results above.

$$\sum_{r=1}^{n} r(r+1)(r+2) = \sum_{r=1}^{n} (r^3+3r^2+2r)$$

$$= \sum_{r=1}^{n} r^3 + 3\sum_{r=1}^{n} r^2 + 2\sum_{r=1}^{n} r$$

$$= \frac{n^2}{4}(n+1)^2 + 3\left[\frac{n}{6}(n+1)(2n+1)\right] + 2\left[\frac{n}{2}(n+1)\right]$$

$$= \frac{n}{4}(n+1)(n+2)(n+3).$$

Note. This series can also be summed by the method of differences using the identity

$$4r(r+1)(r+2) \equiv r(r+1)(r+2)(r+3) - (r-1)r(r+1)(r+2).$$

4) Find the sum of the squares of the first n odd numbers.

The odd numbers can be represented by $2r-1$ where $r = 1, 2, 3, \ldots$

Hence we require $\sum_{r=1}^{n} (2r-1)^2$.

Now $$\sum_{r=1}^{n} (2r-1)^2 = \sum_{r=1}^{n} (4r^2-4r+1)$$

$$= 4\sum_{r=1}^{n} r^2 - 4\sum_{r=1}^{n} r + \sum_{r=1}^{n} 1$$

$$= 4\left[\frac{n}{6}(n+1)(2n+1)\right] - 4\left[\frac{n}{2}(n+1)\right] + n$$

$$= \frac{n(4n^2-1)}{3}$$

EXERCISE 8d

1) If $f(r) \equiv \dfrac{1}{r(r+1)}$, simplify $f(r+1) - f(r)$

Hence find $\sum_{r=1}^{n} \dfrac{1}{r(r+1)(r+2)}$

2) If $f(r) \equiv \dfrac{1}{r^2}$, simplify $f(r) - f(r+1)$

Hence find the sum of the first n terms of the series

$$\frac{3}{1^2.2^2} + \frac{5}{2^2.3^2} + \frac{7}{3^2.4^2} + \ldots$$

3) If $f(r) \equiv r(r+1)(r+2)$, simplify $f(r+1) - f(r)$

Hence find $\sum_{r=1}^{n} 3(r^2+3r+2)$ and deduce that $\sum_{r=1}^{n} r^2 = \frac{1}{6}n(n+1)(2n+1)$

4) Verify that $4r^3 + r \equiv (r+\frac{1}{2})^4 - (r-\frac{1}{2})^4$

and hence find $\sum_{r=1}^{n} (4r^3 + r)$

Deduce that $\sum_{r=1}^{n} r^3 = \frac{n^2}{4}(n+1)^2$

5) If $f(r) \equiv r(r+1)(r+2)(r+3)$, simplify $f(r+1) - f(r)$

and use your result to find $\sum_{r=1}^{n} r^3$

6) If $f(r) \equiv \frac{r}{(r+1)(r+2)}$, simplify $f(r) - f(r+1)$

and hence find $\sum_{r=n}^{2n} \frac{r-1}{(r+1)(r+2)(r+3)}$

7) If $f(r) \equiv \frac{1}{r!}$, simplify $f(r) - f(r+1)$

and hence find $\sum_{r=1}^{n} \frac{r}{(r+1)!}$

8) Given $f(r) \equiv r!$, find $f(r+1) - f(r)$ and use your result to find the sum of the first $2n$ terms of the series

$$1.1! + 2.2! + 3.3! + 4.4! + \ldots$$

9) If $f(r) \equiv \cos 2r\theta$, simplify $f(r) - f(r+1)$.
Use your result to find the sum of the first n terms of the series

$$\sin 3\theta + \sin 5\theta + \sin 7\theta + \ldots$$

10) Use standard results for $\sum_{r=1}^{n} r$, $\sum_{r=1}^{n} r^2$, $\sum_{r=1}^{n} r^3$ to find

(a) $\sum_{r=1}^{n} r(r+1)$ (b) $\sum_{r=1}^{n} r(r+1)(r+2)$

(c) The sum of the squares of the first n even numbers

(d) $\sum_{r=n}^{2n} r^2(1+r)$ (e) $1.3+2.4+3.5+\ldots+(n-1)(n+1)$

(f) $1^2-2^2+3^2-4^2+5^2-6^2+\ldots-(2n)^2$

Hint. Consider two series, the sum of the squares of even numbers and sum of the squares of odd numbers.

(g) $\sum_{r=10}^{20} r^3$

PROOF BY INDUCTION

As we have seen in Chapter 7, proof by induction is a powerful method for proving the validity (or otherwise) of a formula obtained by generalising from a few particular cases. Hence, as far as the sum of a series is concerned, induction can be used to *prove* that a formula given for S_n is correct, but it cannot be used to *find* S_n.

EXAMPLES 8e

1) Prove that $\sum_{r=1}^{n} r = \frac{n}{2}(n+1)$

(There is no specific instruction to use induction and so any suitable method can be used, and we illustrate three such methods of proof.)

(a) *Proof by induction*

Let p_n be $\sum_{r=1}^{n} r = \frac{n}{2}(n+1)$

When $n=1$, L.H.S. $= \sum_{r=1}^{n} r = 1$

R.H.S. $= \frac{n}{2}(n+1) = \frac{1}{2}(2) = 1$

i.e. p_1 is valid

If the formula is valid when $n=k$

i.e. *if* $\sum_{r=1}^{k} r = \frac{k}{2}(k+1)$

then, when $n=k+1$

$$\sum_{r=1}^{k+1} r = \sum_{r=1}^{k} r + (k+1) = \frac{k}{2}(k+1) + (k+1)$$

$$= \frac{(k+1)}{2}(k+2)$$

$$\Rightarrow \qquad \sum_{r=1}^{n} r = \frac{n}{2}(n+1) \quad \text{when} \quad n = k+1$$

so we see that

(if the formula is valid for $n = k$ then it is also valid for $n = k + 1$)

But we have already established that the formula *is* valid when $n = 1$. Therefore $p_{n=2}$ is valid, $p_{n=3}$ is valid, . . . and so on for all positive integral values of n.

(b) *Difference method*

Let $\qquad f(r) \equiv (r-1)r$

then $\qquad f(r+1) - f(r) \equiv (r)(r+1) - (r-1)r$

$$\equiv 2r$$

$$\Rightarrow \qquad \sum_{r=1}^{n} (2r) = \sum_{r=1}^{n} [f(r+1) - f(r)]$$

$$= \sum_{r=1}^{n} r(r+1) - (r-1)r$$

$$\begin{aligned} = \; & 1.2 - 0 \\ & + 2.3 - 1.2 \\ & + 3.4 - 2.3 \\ & \cdots\cdots\cdots\cdots \\ & + n(n+1) - (n-1)n \quad = n(n+1), \end{aligned}$$

i.e. $\qquad \displaystyle\sum_{r=1}^{n} 2r = n(n+1)$

$$\Rightarrow \qquad \sum_{r=1}^{n} r = \frac{n}{2}(n+1)$$

(c) *Summation of an A.P. from first principles*

If $$S_n = \sum_{r=1}^{n} r$$

then $$S_n = 1 + 2 + 3 + \ldots + n$$

and $$S_n = n + (n-1) + (n-2) + \ldots + 1$$

$$\Rightarrow \quad 2S_n = (n+1) + (n+1) + \ldots + (n+1)$$

$$= n(n+1)$$

$$\Rightarrow \quad S_n = \frac{n}{2}(n+1)$$

From this example we see that, when proof of a given result is asked for, any suitable method can be used, provided that a particular method is not specified. The common methods available for the summation of a number series are

(a) partial fractions,
(b) induction,
(c) difference method,
(d) standard summation for an A.P. or a G.P.

Note that when *proof* is asked for, standard results cannot be quoted.

EXERCISE 8e

1) Prove by induction that

(a) $\sum_{r=1}^{n} r^3 = \frac{n^2}{4}(n+1)^2$ (b) $\sum_{r=1}^{n} \frac{1}{r(r+1)} = \frac{n}{n+1}$

(c) $\sum_{r=1}^{n} r(3^r) = \frac{3}{4}[1 + 3^n(2n-1)]$

2) Prove, by any suitable method, that

(a) $\sum_{r=1}^{n} r^2 = \frac{1}{6}n(n+1)(2n+1)$ (b) $\sum_{r=1}^{n} 4^r = \frac{4}{3}(4^n - 1)$

(c) $\sum_{r=1}^{n} (2+3r) = \frac{n}{2}(3n+7)$

(d) $\sum_{r=1}^{n} \frac{1}{r(r+1)(r+2)} = \frac{n(n+3)}{4(n+1)(n+2)}$

(e) $\sum_{r=1}^{n} \frac{r}{2^r} = 2 - (\frac{1}{2})^n(2+n)$

SEQUENCES

Consider the sequence

$$1.1, 1.01, 1.001, 1.0001, \ldots, 1 + (\tfrac{1}{10})^n, \ldots$$

The nth term of this sequence, u_n, is $1 + (\frac{1}{10})^n$
and as $n \to \infty$, $1 + (\frac{1}{10})^n \to 1$,

i.e. $$\lim_{n \to \infty} \{u_n\} = \lim_{n \to \infty} \{1 + (\tfrac{1}{10})^n\} = 1$$

and we say that this sequence converges to the value unity.

In general, if u_n is the nth term of a sequence and $\lim_{n \to \infty} \{u_n\}$ exists, the sequence is said to converge and the value of $\lim_{n \to \infty} \{u_n\}$ is called the limiting value, or limit of the sequence.

A Formal Definition of the Limit of $f(n)$ as $n \to \infty$

If $f(n)$ is a function of positive integral values of n, and if a number k exists such that, as n *increases*, $|f(n) - k|$ becomes and *remains* less than an arbitrarily chosen positive quantity ϵ, however small ϵ is, then

$$\lim_{n \to \infty} \{f(n)\} \quad \text{exists and is equal to } k.$$

For example, if $f(n) \equiv 1 + (\frac{1}{10})^n$ for $n = 1, 2, \ldots$ and if $k = 1$

then $$|f(n) - k| = |(\tfrac{1}{10})^n|$$

If $\epsilon = 10^{-6}$ then $|f(n) - 1| < 10^{-6}$ for *all* values of $n > 6$.
If $\epsilon = 10^{-20}$ then $|f(n) - 1| < 10^{-20}$ for *all* values of $n > 20$

Similarly for any given value of ϵ we can find a value of n, N say, such that

$$|f(n) - 1| < \epsilon \quad \text{for all} \quad n > N,$$

i.e. $$\lim_{n \to \infty} \{f(n)\} = 1$$

From this example we can see that the formal definition above implies that if $\lim_{n \to \infty} \{f(n)\} = k$ then, given a value for ϵ, a value of n, N say, exists such that for all $n > N$, $|f(n) - k| < \epsilon$,

i.e. $$\lim_{n\to\infty} \{f(n)\} = k \iff \begin{cases} \text{Given } \epsilon, \text{ there exists a number } N \text{ such} \\ \text{that } |f(n) - k| < \epsilon \quad \text{for all} \quad n > N \end{cases}$$

Note that this definition does not enable the value of a limit to be *found*. All it does is to formalise an intuitive concept of a limit.

The Evaluation of the Limit of $f(n)$ as $n \to \infty$

When evaluating $\lim_{n\to\infty}[f(n)]$, the following assumptions may be made

$$\text{as} \quad n \to \infty \quad \begin{cases} \dfrac{1}{n} \to 0, \quad \dfrac{1}{n!} \to 0 \\ n \to \infty \quad \text{if} \quad a^n > 1, \quad a^n \to 0 \quad \text{if} \quad 0 < a < 1 \\ \dfrac{a^n}{n!} \to 0 \end{cases}$$

If, as $n \to \infty$, $f(n) \to \dfrac{\infty}{\infty}$ or $\dfrac{0}{0}$, both of which are indeterminate, it may be possible to evaluate $\lim_{n\to\infty}[f(n)]$ by one of the following methods.

(a) Express $f(n)$ as a proper fraction

e.g. if $f(n) \equiv \dfrac{n-1}{n+1}$, then $\lim_{n\to\infty} \dfrac{n-1}{n+1}$ is indeterminate.

But $$f(n) \equiv \frac{n-1}{n+1} \equiv 1 - \frac{2}{n+1}$$

so $$\lim_{n\to\infty} \frac{n-1}{n+1} \equiv \lim_{n\to\infty} 1 - \frac{2}{n+1} = 1$$

(b) Replace n by $\dfrac{1}{m}$,

e.g. if $f(n) \equiv \dfrac{n^2 - 5n + 6}{2n^2 + 7n - 2}$

replacing n by $\dfrac{1}{m}$ gives

$$f\left(\frac{1}{m}\right) \equiv \frac{\dfrac{1}{m^2} - \dfrac{5}{m} + 6}{\dfrac{2}{m^2} - \dfrac{7}{m} - 2} \equiv \frac{1 - 5m + 6m^2}{2 + 7m - 2m^2}$$

Now as $n \to \infty$, i.e. as $\dfrac{1}{m} \to \infty$, $m \to 0$.

So $\lim_{n\to\infty}[f(n)]$ becomes $\lim_{m\to 0}\left[f\left(\dfrac{1}{m}\right)\right]$,

i.e. $$\lim_{n\to\infty} \frac{n^2 - 5n + 6}{2n^2 + 7n - 2} \equiv \lim_{m\to 0} \frac{1 - 5m + 6m^2}{2 + 7m - 2m^2} = \frac{1}{2}$$

EXAMPLES 8f

1) The nth term of a sequence, u_n, is given by

$$u_n = \frac{2n}{n+1}$$

Find $\lim_{n\to\infty} \{u_n\}$ and a value N, such that for all values of $n > N$, u_n differs from $\lim_{n\to\infty} \{u_n\}$ by less than 10^{-4}.

$$u_n = \frac{2n}{n+1} \equiv 2 - \frac{2}{n+1}$$

Hence

$$\lim_{n\to\infty} \{u_n\} \equiv \lim_{n\to\infty} 2 - \frac{2}{n+1} = 2$$

Now $|u_n - 2| = \dfrac{2}{n+1}$, hence $|u_n - 2| < 10^{-4}$ when

$$\frac{2}{n+1} < 10^{-4}$$

i.e. $\quad 2 < 10^{-4}(n+1) \qquad (n+1 > 0 \text{ as } n > 0)$

i.e. $\quad n+1 > 2 \times 10^4$

i.e. $\quad n > 2 \times 10^4 - 1,$

i.e. $\quad n > 19\,999$

so $\quad N = 19\,999$

SUM TO INFINITY OF A NUMBER SERIES

Consider the series

$$\frac{1}{1.2} + \frac{1}{2.3} + \frac{1}{3.4} + \frac{1}{4.5} + \dots$$

If S_n is the sum of the first n terms of this series then

$$S_n = \sum_{r=1}^{n} \frac{1}{r(r+1)} = \sum_{r=1}^{n} \left(\frac{1}{r} - \frac{1}{r+1}\right) = 1 - \frac{1}{n+1} = \frac{n}{n+1}$$

As $n \to \infty$, $S_n \to 1$,
so we say that the series is *convergent* and that its sum to infinity, S, is unity.

In general, if S_n is the sum of the first n terms of a series, and if $\lim_{n\to\infty} [S_n]$ exists, then the series is said to be convergent with a sum to infinity, S, where

$$S = \lim_{n\to\infty} [S_n]$$

By using the more formal definition of a limit, the sum to infinity of a series may be defined as follows.
A series is convergent with a sum to infinity S if, given ϵ, there exists a finite number N such that $|S_n - S| < \epsilon$ for all $n > N$.
Conversely, if $\lim_{n\to\infty} [S_n]$ does not exist, the series is not convergent.

So, to sum an infinite number series which is not recognized as a standard expansion, we first find S_n and then evaluate $\lim_{n\to\infty} S_n$.

EXAMPLES 8f (continued)

2) Find the sum to infinity, S, of the series

$$\frac{1}{2.3.4}+\frac{2}{3.4.5}+\frac{3}{4.5.6}+\frac{4}{5.6.7}+\dots$$

and find the smallest value of n for which $|S_n - S| < 10^{-2}$

$$\frac{1}{2.3.4}+\frac{2}{3.4.5}+\frac{3}{4.5.6}+\dots$$

has a general term $u_r = \dfrac{r}{(r+1)(r+2)(r+3)}$

where $$\frac{r}{(r+1)(r+2)(r+3)} \equiv -\frac{1}{2(r+1)}+\frac{2}{(r+2)}-\frac{3}{2(r+3)}$$

Therefore $$S_n = \sum_{r=1}^{n} u_r = \sum_{r=1}^{n}\left(-\frac{1}{2(r+1)}+\frac{2}{r+2}-\frac{3}{2(r+3)}\right)$$

$$u_1 = -\frac{1}{4}+\frac{2}{3}-\frac{3}{8}$$

$$u_2 = -\frac{1}{6}+\frac{2}{4}-\frac{3}{10}$$

$$u_3 = -\frac{1}{8}+\frac{2}{5}-\frac{3}{12}$$

$$u_4 = -\frac{1}{10}+\frac{2}{6}-\frac{3}{14}$$

....

$$u_{n-1} = -\frac{1}{2n}+\frac{2}{n+1}-\frac{3}{2(n+2)}$$

$$u_n = -\frac{1}{2(n+1)}+\frac{2}{n+2}-\frac{3}{2(n+3)}$$

Therefore

$$S_n = \sum_{r=1}^{n} u_r = -\frac{1}{4}+\frac{2}{3}-\frac{1}{6}+\frac{2}{n+2}-\frac{3}{2(n+2)}-\frac{3}{2(n+3)}$$

$$= \frac{1}{4}-\frac{2n+3}{2(n+2)(n+3)}$$

Hence the sum to infinity, S, is given by

$$S = \lim_{n\to\infty}[S_n] = \lim_{n\to\infty}\left[\frac{1}{4}-\frac{2n+3}{2(n+2)(n+3)}\right]$$

$$= \frac{1}{4}.$$

For $|S_n - S| < 10^{-2}$

$$\left|\frac{1}{4}-\frac{2n+3}{2(n+2)(n+3)}-\frac{1}{4}\right| < 10^{-2}$$

$$\Rightarrow \qquad \frac{2n+3}{2(n+2)(n+3)} < 10^{-2} \quad (n>0)$$

$$\Rightarrow \qquad n^2-95n-144 > 0.$$

Solving the equation

$$n^2-95n-144 = 0$$

gives

$$n = 96.49 \quad (n>0).$$

Hence the smallest value of n for which $|S_n - S| < \frac{1}{100}$ is 97 (n takes positive integral values only).

EXERCISE 8f

1) Evaluate the following limits:

(a) $\lim_{n\to\infty}\frac{n+1}{n^2}$ (b) $\lim_{n\to\infty}\frac{n}{2^n}$ (c) $\lim_{n\to\infty}\frac{n^2+2}{n^2+1}$

2) Find the limit of the sequence, whose rth term, u_r is given by

(a) $u_r = \frac{r^2+r}{r^2+r+1}$ (b) $u_r = \frac{2^r}{r!}$ (c) $u_r = \frac{r!}{r^r}$

3) Find the sums to infinity of the following series:

(a) $\frac{1}{3.4}+\frac{1}{4.5}+\frac{1}{5.6}+\dots$

(b) $\frac{2}{1.3.5}+\frac{4}{3.5.7}+\frac{6}{5.7.9}+\dots$

(c) $\dfrac{1}{1.3}+\dfrac{1}{3.5}+\dfrac{1}{5.7}+\ldots$ (d) $\dfrac{1}{1.4}+\dfrac{1}{2.5}+\dfrac{1}{3.6}+\ldots$

4) Find the least value of n for which $|S_n - S| < 10^{-4}$ where S_n is the sum of the first n terms of a series and S is its sum to infinity and S_n is given by

(a) $\dfrac{n}{n-2}$ (b) $\dfrac{2n+1}{n-1}$ (c) $1+\dfrac{3}{n^2}$

MISCELLANEOUS EXERCISE 8

1) Assuming that $|x| < 1$, expand

$$\frac{1-x}{(1+x)^2}+\frac{1+x}{(1-x)^2}$$

in ascending powers of x as far as and including the term in x^6.

Find the coefficient of x^{2n} in the expansion.
By giving a suitable value to x, find

$$\sum_{n=0}^{\infty} \frac{4n+1}{2^{2n}}$$

(U of L)

2) If $|r| \neq 1$ find the sum S_n of the series

$$r + r^3 + r^5 + \ldots + r^{2n-1}.$$

Show that if $r = \frac{1}{2}$, $S_n \to \frac{2}{3}$ as $n \to \infty$. (U of L)p

3) Show that the sum of the series

$$1 + 3(\tfrac{1}{2}) + 5(\tfrac{1}{2})^2 + \ldots + (2n-1)(\tfrac{1}{2})^{n-1}$$

is $6 - R_n$, where $R_n = \dfrac{(2n+3)}{2^{n-1}}$.

Show that for $n \geqslant 4$, $\dfrac{R_{n+1}}{R_n} \leqslant \dfrac{13}{22}$

and deduce that for $n \geqslant 4$, $R_n \leqslant \dfrac{11}{8}\left(\dfrac{13}{22}\right)^{n-4}$ (JMB)

4) For each of the series

$$\sum_{r=1}^{\infty} \frac{(-1)^{r-1}}{r}\frac{x^r}{2^r} \quad \text{and} \quad \sum_{r=1}^{\infty} \frac{1}{r}\frac{x^r}{3^r}$$

state the set of real values of x for which the series converges and find the sum of the series when it does converge. If the series have the same sum, find the two possible values of x. (U of L)p

5) Express $f(x)$, where $f(x) \equiv \dfrac{x}{(x+2)(x+3)(x+4)}$, as the sum of three partial fractions.

Hence find $\sum_{r=1}^{n} \dfrac{r}{(r+2)(r+3)(r+4)}$ (U of L)p

6) By expressing the rth term of the series $\sum_{r=1}^{n} \dfrac{(-1)^{r+1}(2r+1)}{r(r+1)}$ in partial fractions, or otherwise, find the sum of the series.
If this sum is denoted by S_n prove that, for all odd values of n, $S_n S_{n+1} = 1$.
(JMB)

7) The rth term of a series (for $r = 1, 2, \dots$) is $4/(4r^2 - 1)$
Prove that the sum S of the first n terms of the series is given by

$$S = \frac{4n}{2n+1}$$

Show that
$$0 < \frac{4}{4r^2-1} - \frac{1}{r^2} \leqslant \frac{1}{4r^2-1}$$

Hence prove that the error in the approximate formula

$$1^{-2} + 2^{-2} + \dots + n^{-2} \simeq S$$

cannot exceed $\dfrac{S}{4}$ (JMB)p

8) (a) The rth term of a series is $\dfrac{1}{r(r+2)}$. Find the sum of the first n terms of the series and deduce the sum to infinity.

(b) Prove by induction that $\sum_{r=1}^{n} r^2 = \dfrac{n}{6}(n+1)(2n+1)$

Hence, or otherwise, evaluate $\sum_{r=1}^{51} (98+2r)^2$
(AEB)'74

9) (a) Simplify $r(r+1)(r+2) - (r-1)r(r+1)$ and use your result to prove that

$$\sum_{r=1}^{n} r(r+1) = \tfrac{1}{3}n(n+1)(n+2).$$

Deduce that

$$\sum_{r=1}^{n} r^2 = \frac{n}{6}(n+1)(2n+1)$$

(b) Find the sum of the series

$$1.2.3 + 3.4.5 + 5.6.7 + \dots + (2n-1)(2n)(2n+1).$$

(U of L)

10) Show that if $f(r) \equiv r(r+1)(r+2)$ then

$$f(r) - f(r-1) \equiv 3r(r+1)$$

Hence find the sum of the series

$$2+6+12+\ldots+r(r+1)+\ldots+n(n+1)$$ (U of L)p

11) If $f(n)$ denotes $\dfrac{1}{n(n+1)}$, simplify $f(n)-f(n+1)$ and hence find the sum

$$S_n = \sum_{r=1}^{n} \frac{1}{r(r+1)(r+2)}$$

Find the smallest integer n for which S_n differs from $\frac{1}{4}$ by less than 10^{-4}. (U of L)p

12) Show that $n(n+1)\equiv(n-1)(n-2)+4(n-1)+2$ and hence that, for $n\geqslant 3$,

$$\frac{n(n+1)}{(n-1)!} = \frac{1}{(n-3)!}+\frac{4}{(n-2)!}+\frac{2}{(n-1)!}$$

Deduce, or find otherwise, the sum to infinity of the series

$$1.2+\frac{2.3}{1!}+\frac{3.4}{2!}+\ldots+\frac{n(n+1)}{(n-1)!}+\ldots$$ (U of L)p

13) If $f(r)\equiv r(r+1)(r+2)(r+3)$, simplify $f(r)-f(r-1)$, and hence find the sum of the first n terms of the series in which the rth term is $r(r+1)(r+2)$.
Hence, or otherwise, show that $1^3+2^3+3^3+\ldots+n^3=\frac{1}{4}n^2(n+1)^2$ (U of L)p

14) Show that the rth term of the infinite series

$$\frac{1.2^3}{3!}+\frac{3.2^5}{5!}+\ldots+\frac{(2r-1)2^{2r+1}}{(2r+1)!}+\ldots$$

can be expressed in the form

$$\frac{2^{2r+1}}{(2r)!}-\frac{2^{2r+2}}{(2r+1)!}$$

and hence find the sum of the series in terms of e. (U of L)p

15) (a) Assuming the formulae for $\sum_{r=1}^{n} r$ and $\sum_{r=1}^{n} r^2$, or otherwise, find in terms of m the value of

$$\sum_{r=m}^{2m} 3(r-2)(r+1)$$

(b) By expressing $2\sin\theta\cos 2r\theta$ as the difference between two sines, show that

$$\sin\theta\sum_{r=1}^{n}\cos 2r\theta = \sin n\theta\cos(n+1)\theta$$

Evaluate $\sum_{r=1}^{100} \cos^2 \left\{\frac{r\pi}{100}\right\}$ (U of L)

16) Prove by induction, or otherwise, that

$$1.1! + 2.2! + 3.3! + \ldots + n.n! = (n+1)! - 1. \quad \text{(U of L)p}$$

17) If $3u_{n+1} = 2u_n - 1$ for all positive integral values of n and $u_1 = 1$, prove by induction that $u_n = 3(\frac{2}{3})^n - 1$.
Find the sum of the first n terms of the series whose nth term is u_n.

(AEB'75)p

18) Prove, by induction or otherwise, that

$$2.1! + 5.2! + 10.3! + \ldots + (n^2+1)n! = n(n+1)!$$

(U of L)p

19) Prove, by induction or otherwise, that the sum of the cubes of the first n positive integers is $\frac{1}{4}n^2(n+1)^2$. Hence, or otherwise, obtain a formula for the sum of the cubes of the first n odd positive integers. (JMB)p

20) Prove, by induction or otherwise that

$$\sum_{r=2}^{n} (r-1)r(r+2) = \tfrac{1}{12}(n-1)(n)(n+1)(3n+10) \quad \text{(U of L)p}$$

21) Prove, by induction or otherwise, that

$$\sum_{r=1}^{n} r2^{r-1} = 1 + (n-1)2^n \quad \text{(U of L)p}$$

22) Define a prime number and prove that the number of primes is infinite.

Show that no member of the sequence $1, 3, 6, \ldots, \sum_{r=1}^{n} r, \ldots$ after the first two, is a prime number. (JMB)

23) If $-1 < r < 1$ and $S_n = \sum_{p=1}^{n} r^p$ find

(a) $\lim_{n\to\infty} S_n$ (b) $\sum_{n=1}^{\infty} nr^n$ (c) $\sum_{q=1}^{n} S_q$ (U of L)

24) Find the values of A, B and C if the nth term of the sequence 4, 13, 26, 43 ... is $An^2 + Bn + C$. Find the sum of the infinite series

$$\frac{4}{1!} + \frac{13}{2!} + \frac{26}{3!} + \frac{43}{4!} + \ldots + \frac{An^2 + Bn + C}{n!} + \ldots \quad \text{(AEB'72)p}$$

25) If $f(r) \equiv \log\left(1 + \frac{1}{r}\right)$, show that

$$f(1)+f(2)+\dots+f(n) = f\left(\frac{1}{n}\right)$$ (U of L)p

26) (a) Assuming the formula for $\sum_{r=1}^{n} r^2$, write down

(i) the sum of the squares of the first $2n$ positive integers,
(ii) the sum of the squares of the first n even integers.
Hence find $1^2+3^2+5^2+\dots+(2n-1)^2$.

(b) If $S_n = \sum_{r=0}^{n} a^r(1+a+a^2+\dots+a^r)$, $(|a| \neq 1)$ by considering $(1-a)S_n$, show that

$$S_n = \frac{1-a^{2n+2}}{(1-a^2)(1-a)} - \frac{a^{n+1}(1+a^{n+1})}{(1-a)^2}$$

State the set of values of a for which S_n approaches a limit as $n \to \infty$ and find the sum to infinity of the series for these values of a.
(U of L)

27) The positive integers are bracketed as follows:

$$(1), (2, 3), (4, 5, 6), \dots$$

where there are r integers in the rth bracket. Find expressions for the first and last integers in the rth bracket.
Find the sum of all the integers in the first 20 brackets. Prove that the sum of the integers in the rth bracket is $\frac{1}{2}(r^2+1)$. (C)

28) (a) Prove that $\sum_{r=1}^{n} \frac{1}{r(r+1)} = \frac{n}{n+1}$

(b) Sum the series $1+x+x^2+\dots+x^n$ for $x \neq 1$.
By differentiation with respect to x, or otherwise, find the value of

$$1+2x+3x^2+\dots+nx^{n-1}$$

and deduce the value of

$$1.2+2.2^2+3.2^3+\dots+n.2^n$$ (U of L)

29) Prove that

$$1^2+2^2+3^2+\dots+n^2 = \tfrac{1}{6}n(n+1)(2n+1)$$

Show that

$$a^2+(a+d)^2+(a+2d)^2+\dots+(a+nd)^2$$
$$= \tfrac{1}{6}(n+1)[6a(a+nd)+d^2n(2n+1)]$$

Hence, or otherwise, prove that

$$2^2+4^2+6^2+\dots+l^2 = \tfrac{1}{6}l(l+1)(l+2) \qquad (l \text{ even})$$

and $$1^2 + 3^2 + 5^2 + \ldots + l^2 = \tfrac{1}{6}l(l+1)(l+2). \qquad (l \text{ odd})$$ (JMB)

30) Find the sum of the series

$$\frac{4}{3} + \frac{9}{8} + \frac{16}{15} + \ldots + \frac{n^2}{n^2-1}.$$ (U of L)p

31) Sum to infinity the series

$$\frac{3}{2!} + \frac{7}{3!} + \ldots + \frac{n^2-n+1}{n!} + \ldots$$ (U of L)p

32) (a) Find, in terms of n, the sum of the finite series

$$1 + (1+2) + (1+2+3) + \ldots + (1+2+\ldots+n)$$

expressing your answer in its simplest form.

(b) Express, in terms of e, the sum to infinity of the series

$$3 + \frac{5}{2!} + \frac{7}{3!} + \ldots + \frac{2n+1}{n!} + \ldots$$ (U of L)

CHAPTER 9

SOME FUNCTIONS AND THEIR PROPERTIES

EVEN FUNCTIONS

A function $f(x)$ is said to be *even* if $f(x) = f(-x)$ for all values of x. The graphs of all even functions are therefore symmetrical about the vertical axis. Some common even functions are:

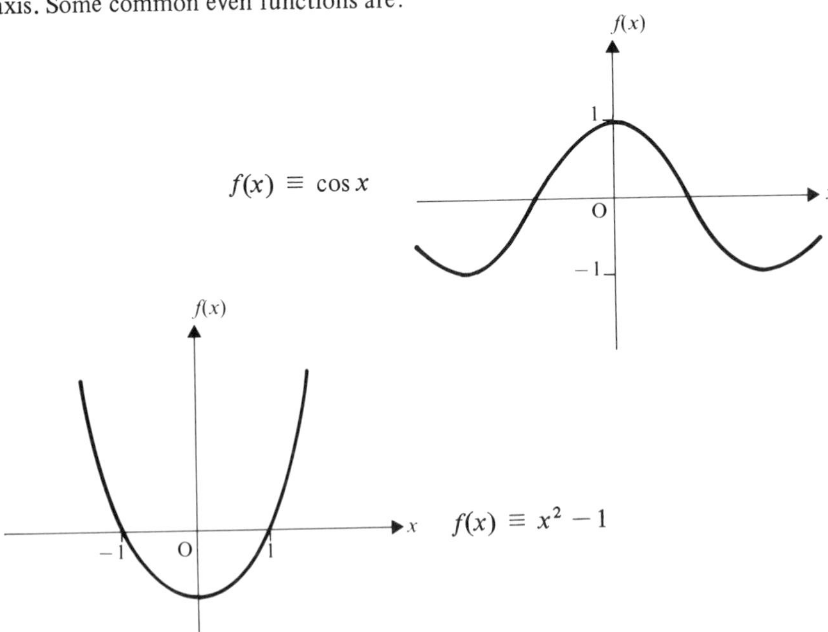

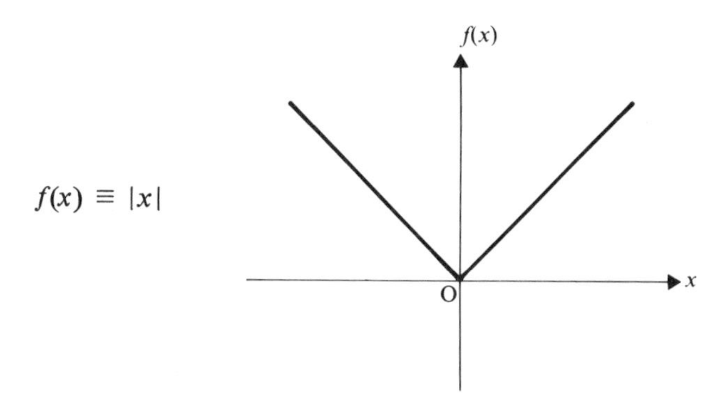

ODD FUNCTIONS

A function $f(x)$ is said to be *odd* if $f(x) = -f(-x)$ for all values of x. The graphs of odd functions are 'symmetrical about the origin', i.e. the origin, O, divides the graph into two sections each of which can be rotated about O, through an angle π, to give the other section.
Some common odd functions are:

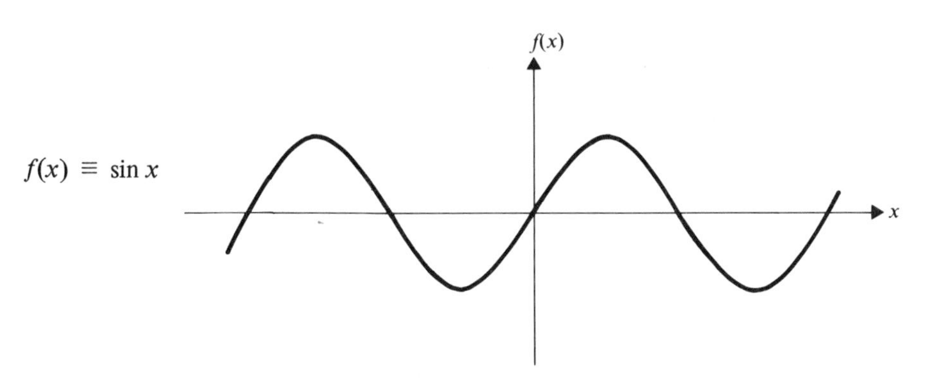

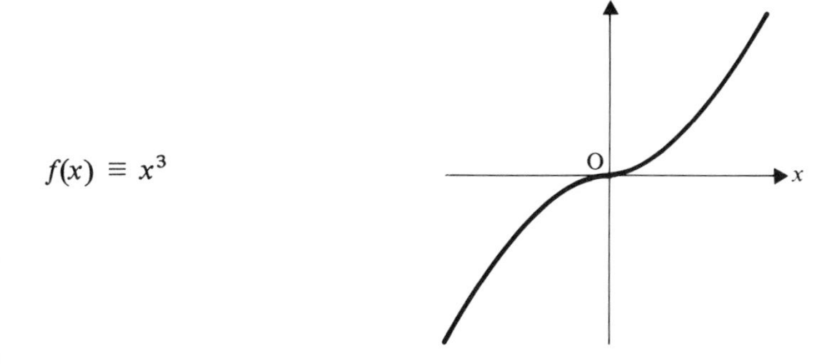

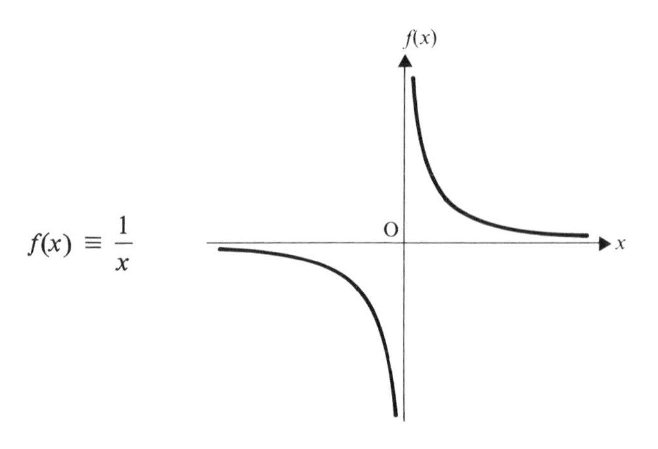

CONTINUOUS FUNCTIONS

A function $f(x)$ is *continuous* when $x = a$ if:

$$\lim_{x \to a,} f(x) \text{ is defined and is equal to } f(a)$$

i.e. $f(x) \to f(a)$ as $x \to a$ from above and from below.
A continuous function satisfies this condition for all values of a, so the graph of a continuous function is unbroken.
The graphs of some common continuous functions are shown below.

(a) $f(x) \equiv \sin x$.

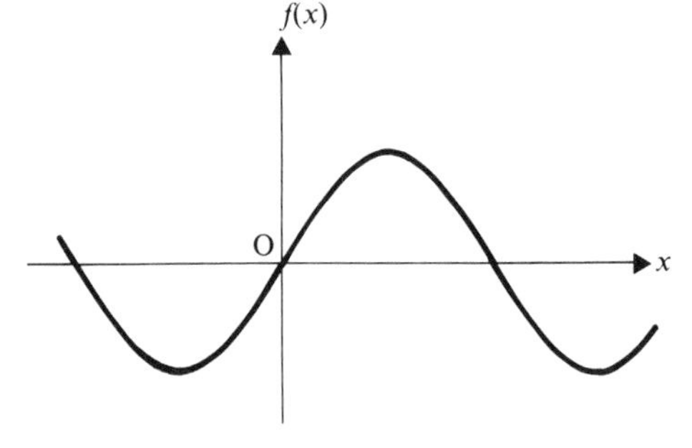

(b) $f(x) \equiv x(x-1)(x-2)$.

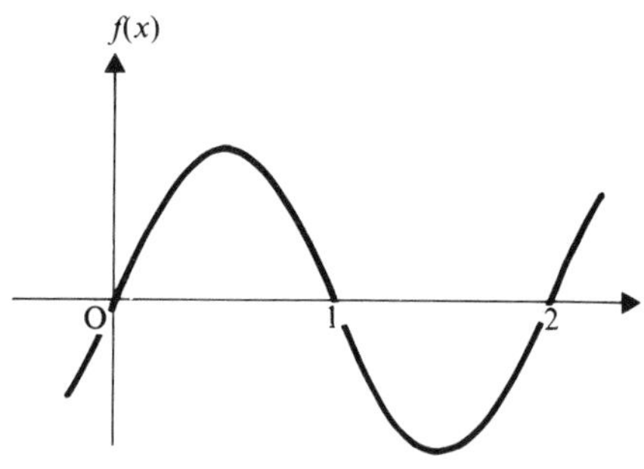

(c) $f(x) \equiv e^x$.

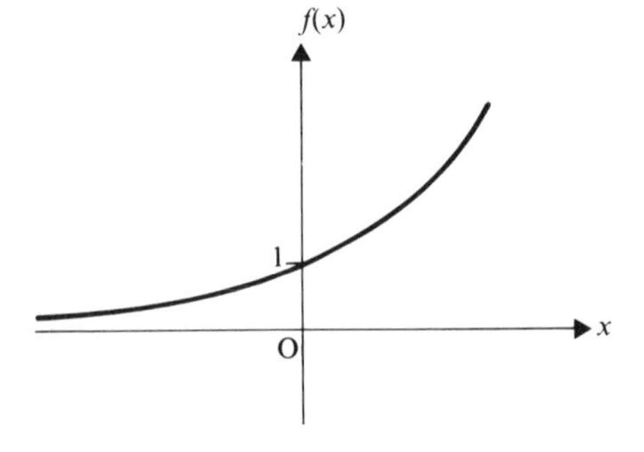

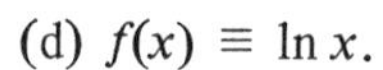
(d) $f(x) \equiv \ln x$.

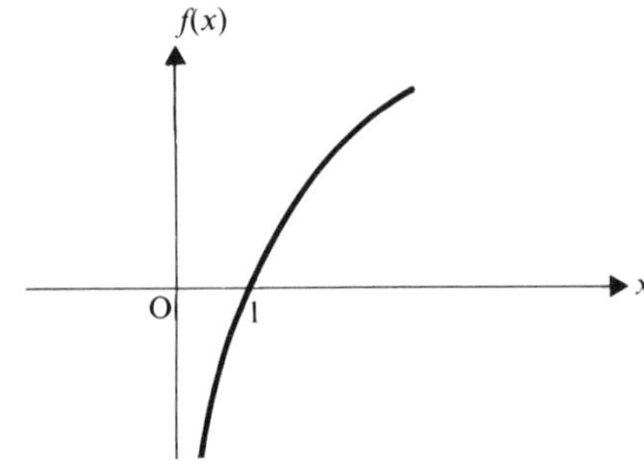

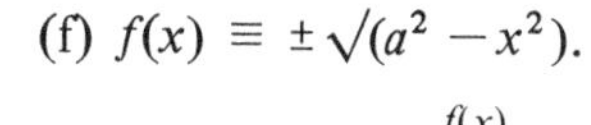

(e) $f(x) \equiv |x|$.

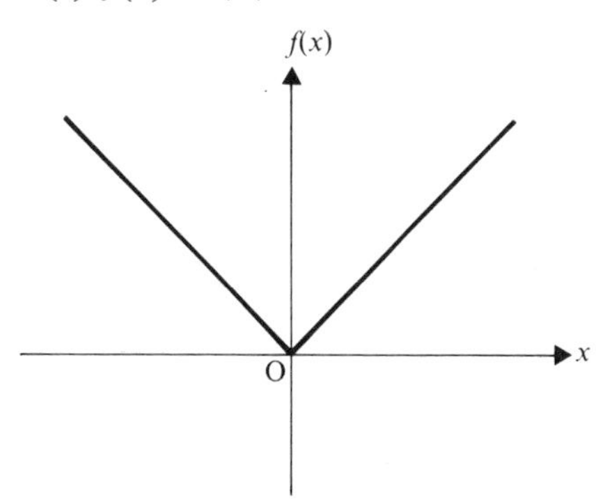

(f) $f(x) \equiv \pm\sqrt{(a^2 - x^2)}$.

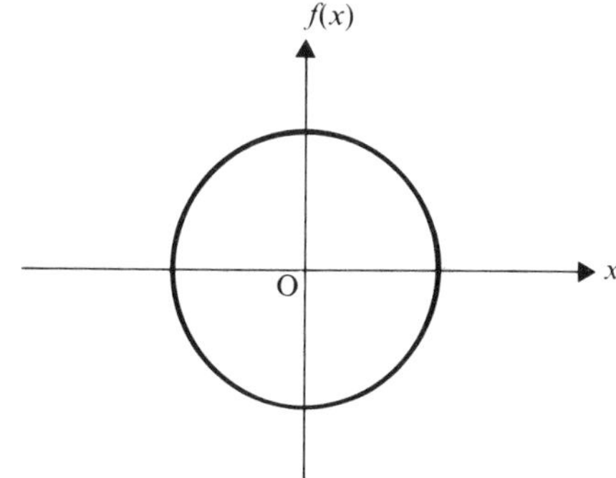

A sudden change in the direction of the graph of a function does not necessarily affect the continuity. For instance, in example (e) above, the gradient of the graph changes suddenly from -1 to $+1$ as the graph passes through O, but $f(x) \equiv |x|$ is continuous at $x = 0$,

as $\begin{cases} f(x) \to 0 & \text{as} \quad x \to 0 \quad \text{from below zero} \\ f(x) \to 0 & \text{as} \quad x \to 0 \quad \text{from above zero,} \end{cases}$

i.e. $\lim_{x \to 0} f(x)$ is defined and is equal to 0.

Note that the graph of any continuous function can be sketched without removing pencil from paper.

DISCONTINUITY

If, when passing through $x = a$, there is a sudden change in the value of $f(x)$, so that there is a break in the graph of $f(x)$, the function is said to have a discontinuity at $x = a$ because the condition for continuity is not satisfied in these circumstances. For example,

(a)

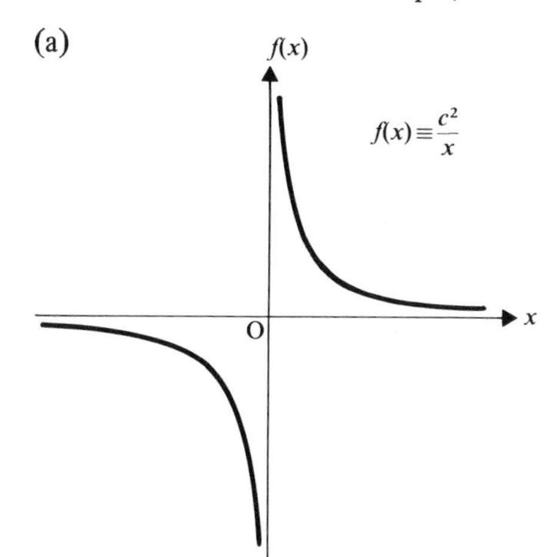

as $x \to 0$ from below zero, $f(x) \to -\infty$.
As $x \to 0$ from above zero, $f(x) \to \infty$.
So $\lim_{x \to 0} f(x)$ is not defined and there is a discontinuity at $x = 0$.

(b)

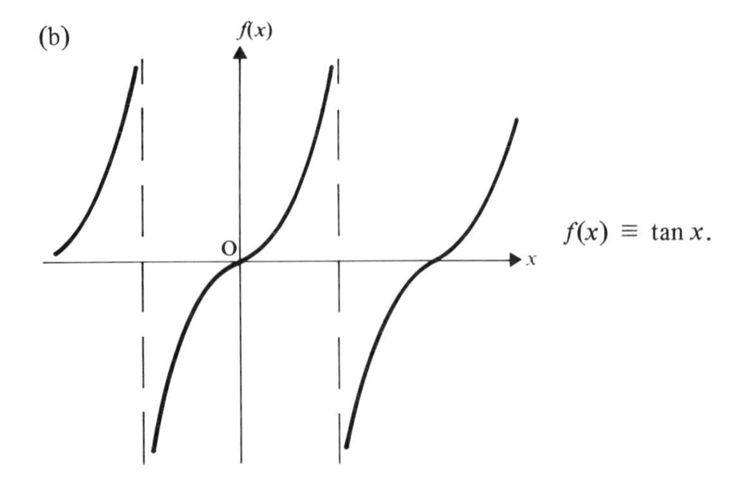

$f(x) \equiv \tan x.$

As $x \to (2n+1)\frac{\pi}{2}$ from below, $f(x) \to \infty$

as $x \to (2n+1)\frac{\pi}{2}$ from above, $f(x) \to -\infty$.

So again $\lim f(x)$ is undefined when $x \to (2n+1)\frac{\pi}{2}$ and there is a discontinuity whenever $x = (2n+1)\frac{\pi}{2}$.

On each of these graphs, the value of $f(x)$ undergoes an infinitely large change at each break in the graph. So the functions have *infinite discontinuities*.
At each infinite discontinuity there is a vertical asymptote.
Not all discontinuities are of this type however.
Consider, for instance, the function $f(x) \equiv [x]$,
where $[x]$ is used to denote 'the greatest integer $\leqslant x$',

e.g. if $x = 2.4$, $[x] = 2$

if $x = -0.8$, $[x] = -1$

if $x = 3$, $[x] = 3$.

The graph of $f(x) \equiv [x]$ is shown opposite and we see that there is a *finite discontinuity* at every integral value of x.

Note that at each discontinuity a different Cartesian equation can be introduced for the graph of the function

i.e. for $0 \leqslant x < 1$, $y = 0$

for $1 \leqslant x < 2$, $y = 1$ etc.

$f(x) \equiv [x]$

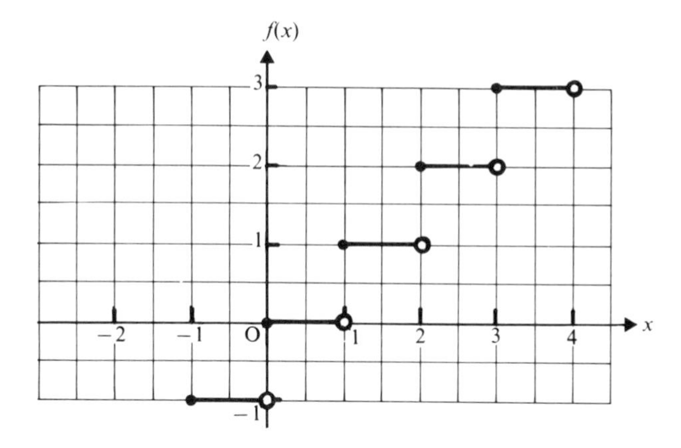

DISCONTINUITY AND DIFFERENTIATION

A function $f(x)$ is said to be differentiable at $x = a$ if its derivative $f'(x)$ is defined and is equal to $f'(a)$,
i.e. if $f'(x)$ has the same value as $x \to a$ from above and from below.
A function whose graph undergoes no sudden change in direction satisfies this condition at every point and so is differentiable for every value of x.
But there are some continuous functions that are not differentiable for all values of x. Consider, for example, the function $f(x) \equiv |x|$.

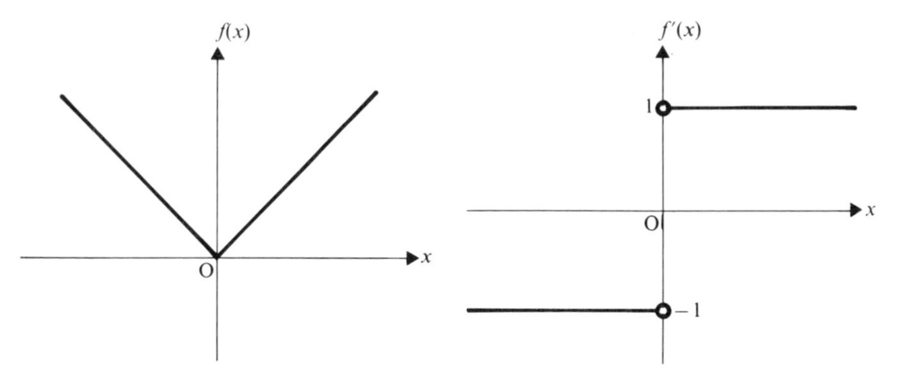

$f(x) \equiv |x|$ is continuous for all values of x, including $x = 0$.
But for $x < 0$, $f'(x) = -1$,
i.e. $f'(x) = -1$ as $x \to 0$ from below.
Whereas for $x > 0$, $f'(x) = 1$,
i.e. $f'(x) = 1$ as $x \to 0$ from above.
Thus $f'(x)$ is not defined when $x = 0$ and so $f(x) \equiv |x|$ is *not* differentiable at $x = 0$ although $f(x)$ is continuous at this point.

i.e. **Continuous $\not\Rightarrow$ Differentiable**

Now consider a function $f(x)$ which has a discontinuity when $x = a$, that causes the value of $f(x)$ to undergo a sudden change of magnitude h, say.

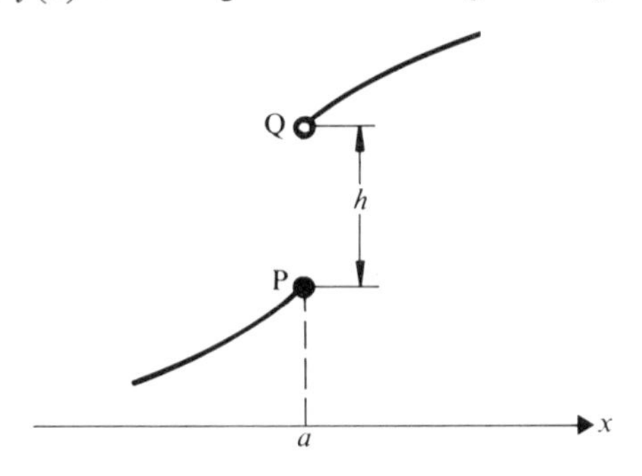

The gradient of the curve as $x \to a$ from below is given by

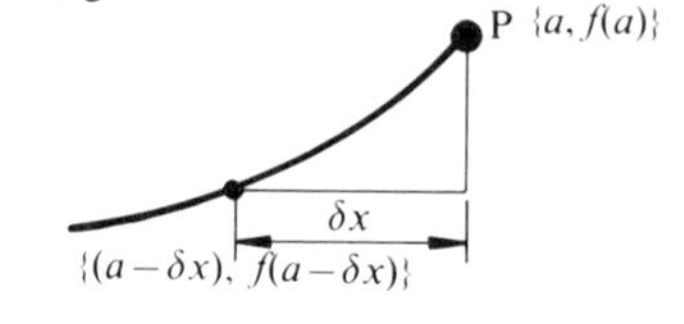

$$\lim_{\delta x \to 0} \frac{f(a) - f(a - \delta x)}{\delta x}$$

and $f(a) - f(a - \delta x) \to 0$ as $\delta x \to 0$.

The gradient of the curve as $x \to a$ from above is given by

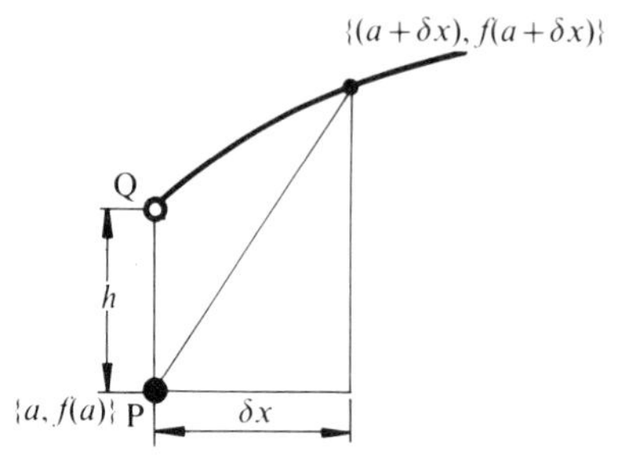

$$\lim_{\delta x \to 0} \frac{f(a + \delta x) - f(a)}{\delta x}$$

and $f(a + \delta x) - f(a) \to h$ as $\delta x \to 0$.

These limiting values are clearly not equal so $f'(x)$ does not exist and $f(x)$ is therefore *not* differentiable when $x = a$,

i.e. **Discontinuity $\Rightarrow$ not differentiable**

THE EFFECT OF DISCONTINUITY ON INTEGRATION

The definite integral $\int_a^b f(x)\,dx$ can be calculated using

$$\left[\int f(x)\,dx\right]_{x=a} - \left[\int f(x)\,dx\right]_{x=b} \qquad \text{unless}$$

(a) $f(x)$ is undefined for any value of x in the range $a \leqslant x \leqslant b$.
(b) $f(x)$ undergoes a finite discontinuity at $x = c$ where $a < c < b$.

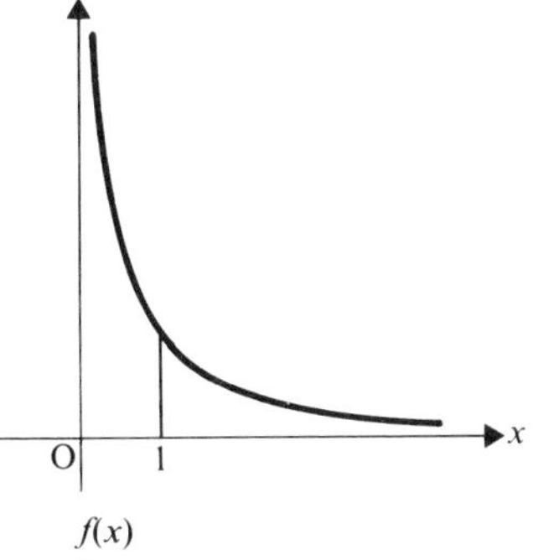

In the first case the integration cannot (at this stage) be carried out at all,

e.g. $$\int_0^1 \frac{1}{x}\,dx$$

cannot be found because $\frac{1}{x}$ is not defined when $x = 0$.

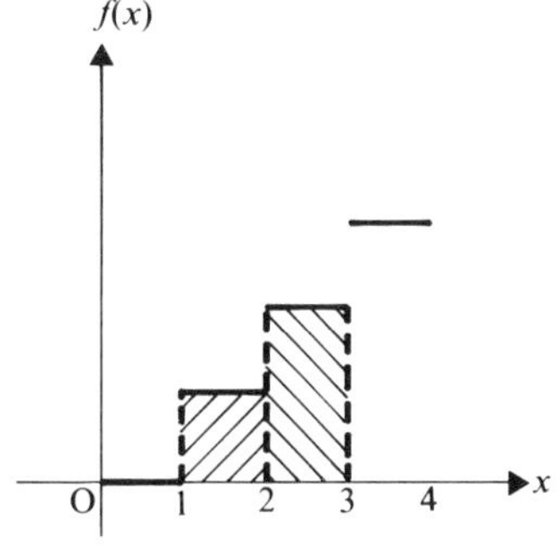

In the second case the integration can be done in sections separated by the value of x at the discontinuity,

e.g. $$\int_1^3 [x]\,dx$$

can be calculated using

$$\int_1^2 [x]\,dx + \int_2^3 [x]\,dx = \int_1^2 1\,dx + \int_2^3 2\,dx$$

because there is a finite discontinuity at $x = 2$.

A function whose graph contains a sudden change in direction also requires sectional integration (although a change in direction does not necessarily imply discontinuity),

e.g. $\int_{-1}^2 |x|\,dx$ is found using

$$\int_{-1}^0 |x|\,dx + \int_0^2 |x|\,dx.$$

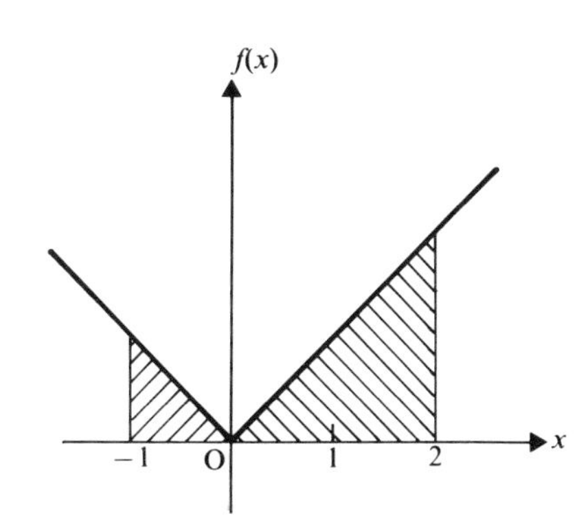

Note that the equation of the graph of $f(x) \equiv |x|$ can be expressed as the pair of equations
for $x > 0$, $y = x$
for $x < 0$, $y = -x$

Thus $$\int_{-1}^{2} |x|\,dx = \int_{-1}^{0} (-x)\,dx + \int_{0}^{2} x\,dx$$

EXERCISE 9a

1) Sketch each of the following functions and state which are continuous.

(a) $\cos x$ (b) $\arctan x$ (c) e^{-x} (d) $\ln(1+x)$
(e) $\cot x$ (f) $(x-1)^2$ (g) $|x-1|$ (h) $\ln|x|$
(i) $\pm\sqrt{(a^2-x^2)}$ (j) $\pm\sqrt{x}$
(k) $x-[x]$ (l) $\dfrac{1}{x}$ (m) $\dfrac{1}{x^2+1}$

2) For each non-continuous function in Question 1, give the value(s) of x at which there is a discontinuity. State whether the discontinuity is finite or infinite.

3) State the value(s) of x, if any, at which each of the functions in Question 1 is not differentiable.

4) State which, if any, of the functions in Question 1 is
(a) even (b) odd.

5) Which of the following graphs represent functions that are
(a) continuous (b) even (c) odd.

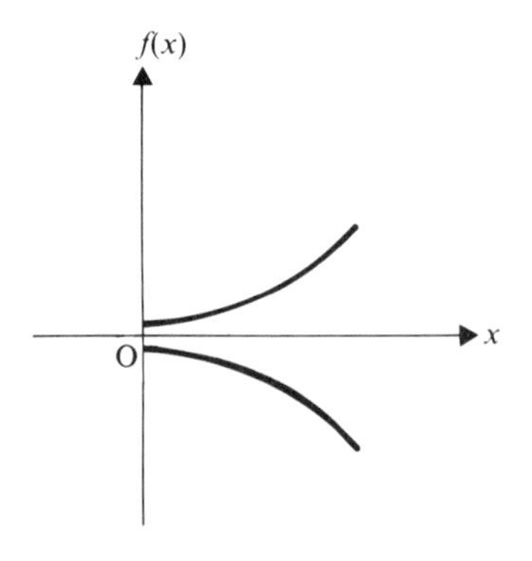

(i)

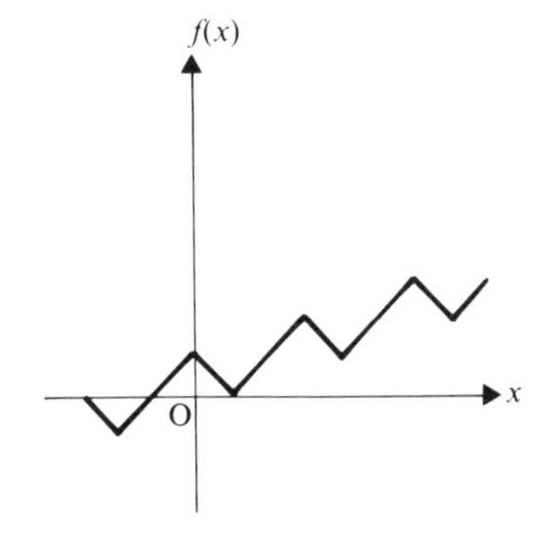

(ii)

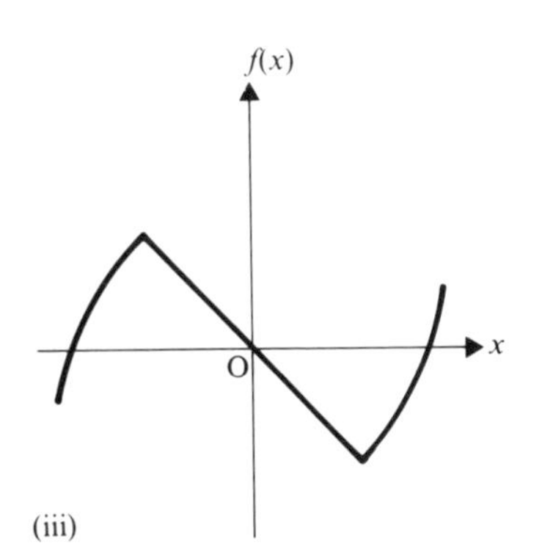

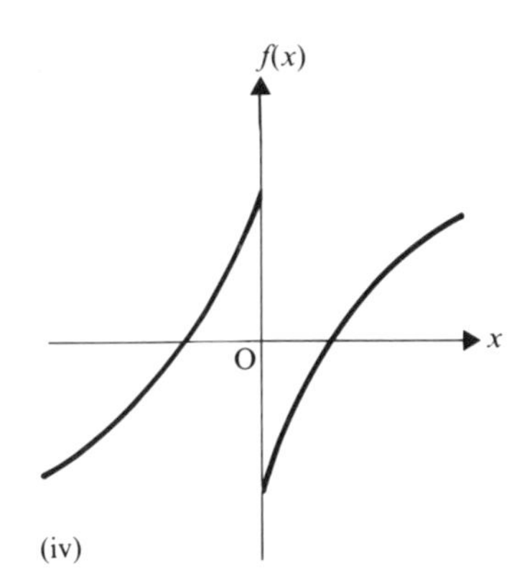

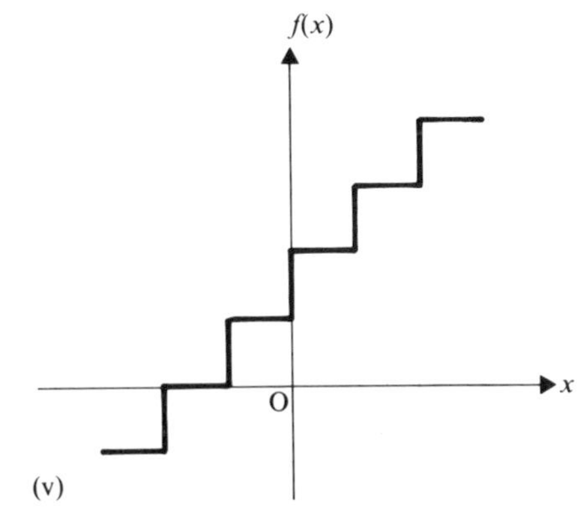

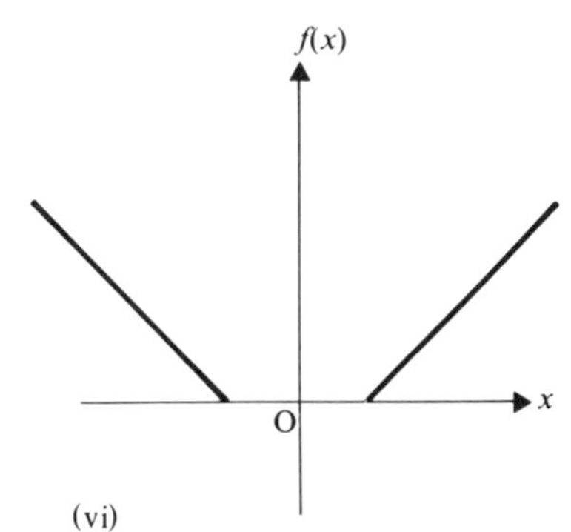

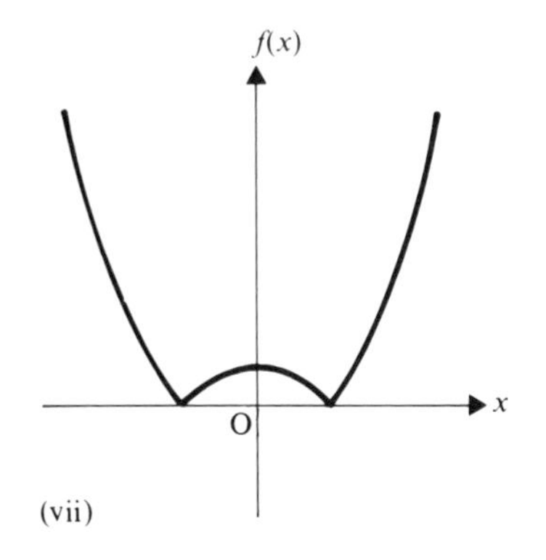

6) Which of the following definite integrals can be evaluated? (Do not carry out the integration.)

(a) $\int_0^{\frac{\pi}{2}} \tan x \, dx$ (b) $\int_1^3 \frac{1}{x-1} \, dx$ (c) $\int_{-1}^1 \frac{1}{x} \, dx$

(d) $\int_{\frac{\pi}{6}}^{\frac{\pi}{3}} \operatorname{cosec} x \, dx$ (e) $\int_0^4 |x-2| \, dx.$

Evaluate the following definite integrals.

7) $\int_0^2 |x-1|\,dx$ (8) $\int_1^3 \{x-[x]\}\,dx$

9) $\int_{-2}^2 |x^2-1|\,dx$.

Give the values of x at which the following functions have discontinuities

10) $\dfrac{x}{(x-1)(x-2)}$ 11) $\dfrac{(x-3)(x-4)}{x}$ 12) $\dfrac{1}{x}+\dfrac{1}{x-1}$

13) Discuss the continuity or otherwise of $f(x)$ and $f'(x)$ where $f(x)$ is

(a) $\dfrac{1}{|x|}$ (b) $|\cos x|$ (c) $e^{|x|}$.

14) Prove that:
(a) if $f(x)$ is an even function,

$$\int_{-a}^{a} f(x)\,dx = 2\int_0^a f(x)\,dx$$

(b) if $f(x)$ is an odd function,

$$\int_{-a}^{a} f(x)\,dx = 0.$$

PERIODIC FUNCTIONS

A function whose graph consists of a basic pattern which repeats at regular intervals is said to be *periodic*. The width of the basic pattern is the *period* of the function,
e.g. $\sin x$ is periodic and its period is 2π

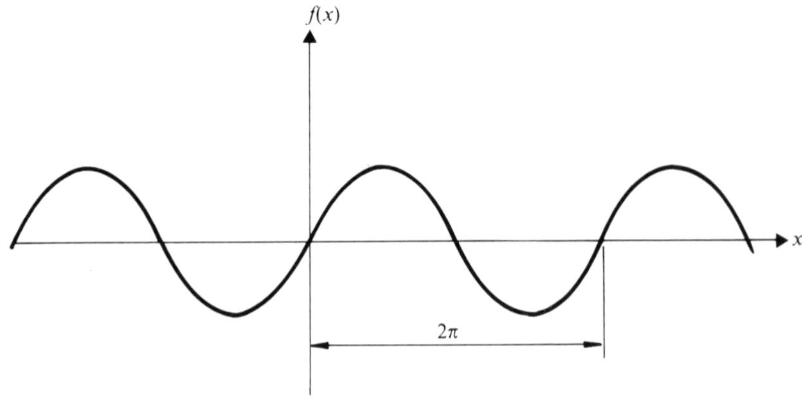

and $\tan x$ is periodic with a period π

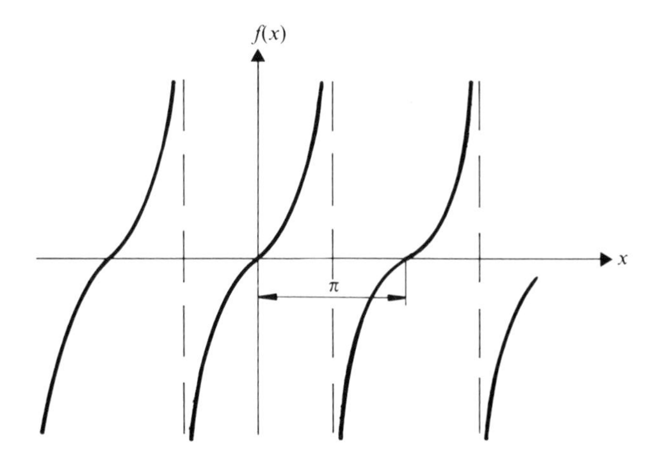

For a periodic function with period a,

$$f(x) = f(x+a) = f(x+2a) = \dots$$
$$= f(x+ka) \quad \text{where } k \text{ is an integer}$$

Similarly $f(x) = f(x-a) = f(x-2a) \dots$.
So a periodic function is defined by the condition $f(x+ka) = f(x)$, or $f(x \pm a) = f(x)$ for all values of x, where a is the period of the function. The definition of the function within one period, i.e. the interval $0 < x \leqslant a$, then defines the whole function.

e.g. if $f(x) \equiv 2x - 1$ for $0 < x \leqslant 1$
and $f(x+1) = f(x)$ for all values of x
then we know that the function is periodic with a period of 1.
So if we draw $f(x) \equiv 2x - 1$ for the range $0 < x \leqslant 1$ this pattern then repeats at unit intervals,
i.e.

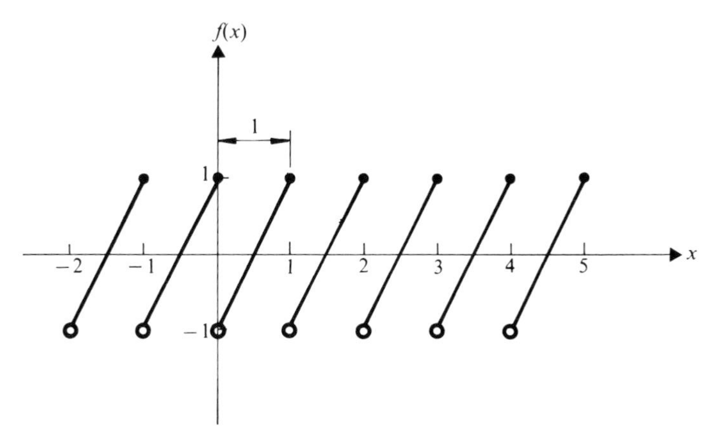

Note that this function has finite discontinuities at each integral value of x.

Compound Periodic Functions

A function may have a compound definition within the basic range. For instance, if

$$f(x) \equiv x^2 \quad \text{for} \quad 0 < x \leqslant 1$$

$$f(x) \equiv 3 - 2x \quad \text{for} \quad 1 < x \leqslant 3$$

and $f(x + 3) = f(x)$ for all values of x

then $f(x)$ is periodic with a period of 3 and the portion of the graph that forms the basis of the repetitive pattern is drawn

from $x = 0$ to $x = 1$ using $f(x) \equiv x^2$

and from $x = 1$ to $x = 3$ using $f(x) \equiv 3 - 2x$.

Thus the graph of $f(x)$ is

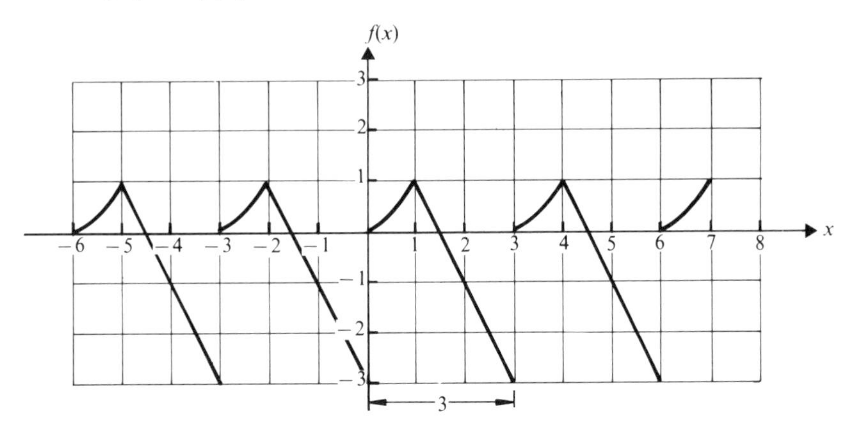

Certain calculations involving a periodic function can be obtained by using its periodic nature, as illustrated in the following example.

EXAMPLE 9b

Given that

$$f(x) \equiv 9 - x^2 \quad \text{for} \quad 0 < x \leqslant 2$$

$$f(x) \equiv 3x - 1 \quad \text{for} \quad 2 < x \leqslant 4$$

and

$$f(x + 4) = f(x) \quad \text{for all values of } x$$

sketch $f(x)$ for the range $-5 < x \leqslant 13$. Evaluate

(a) $f(23)$ (b) $\int_{-2}^{8} f(x)\,dx$.

Because $f(x + 4) = f(x)$, we know that the function is periodic with a period of 4.

For $0 < x \leqslant 2$

$f(x) \equiv 9 - x^2 \Rightarrow$

For $2 < x \leqslant 4$

$f(x) \equiv 3x - 1 \Rightarrow$

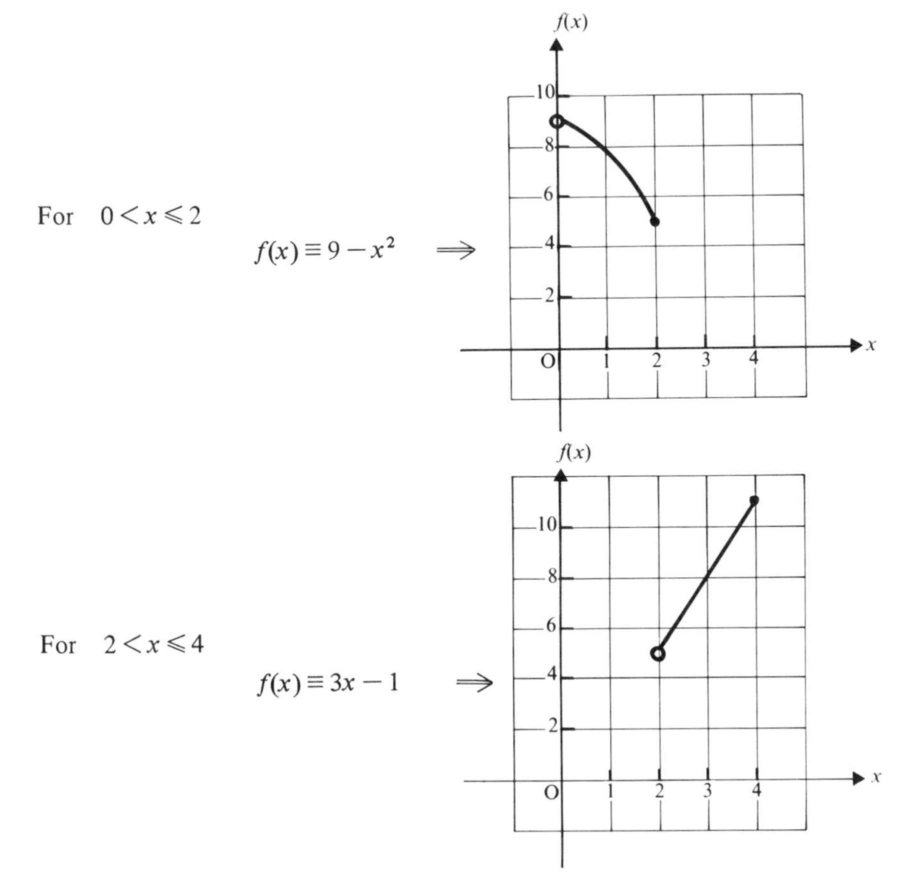

Combining these sections and repeating them at intervals of 4 units gives:

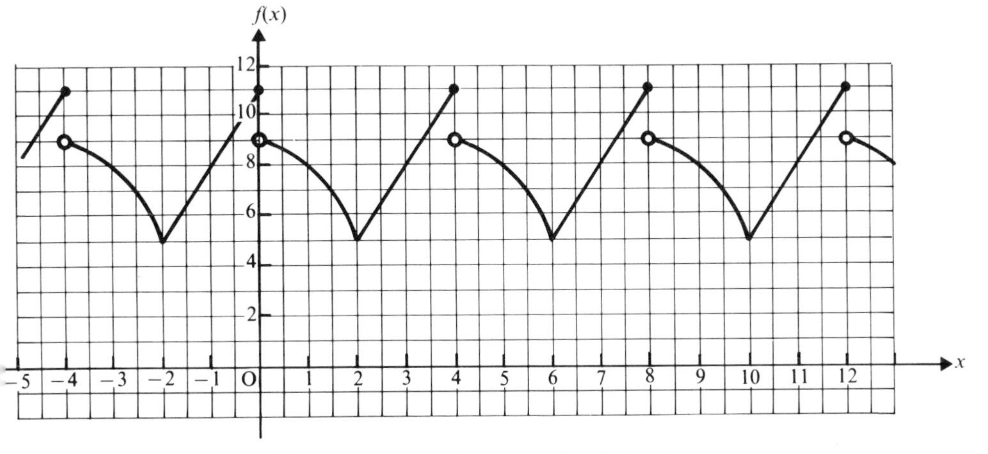

(a) As any value of $f(x)$ occurs at regular intervals of 4 units we can say

$$f(x) = f(x+4n) \quad \text{where } n \text{ is an integer}$$

So $$f(23) = f(3+4\times 5) = f(3) = 8$$

(b)

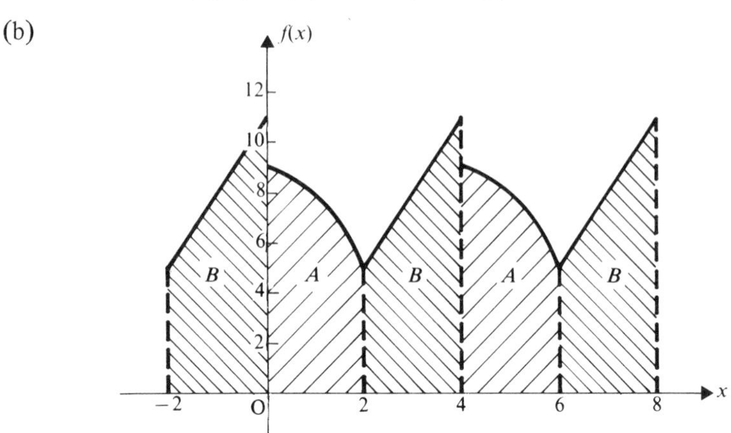

$$\int_{-2}^{8} f(x)\,dx = 2A+3B$$

where $$A = \int_0^2 (9-x^2)\,dx = \left[9x-\frac{x^3}{3}\right]_0^2 = \frac{46}{3}$$

and $$B = \int_2^4 (3x-1)\,dx = \left[\frac{3x^2}{2}-x\right]_2^4 = 16$$

Hence $$\int_{-2}^{8} f(x)\,dx = 2\left(\frac{46}{3}\right)+3(16) = 78\tfrac{2}{3}$$

EXERCISE 9b

1) Sketch the graph of $f(x)$ within the range $-4<x\leqslant 6$ if

$$f(x) \equiv 4-x^2 \quad \text{for} \quad 0<x\leqslant 2$$

and $$f(x) = f(x-2) \quad \text{for all values of } x.$$

2) If $$f(\theta) \equiv \sin\theta \quad \text{for} \quad 0<\theta\leqslant\frac{\pi}{2}$$

$$f(\theta) \equiv \cos\theta \quad \text{for} \quad \frac{\pi}{2}<\theta\leqslant\pi$$

and $$f(\theta+\pi) = f(\theta) \quad \text{for all values of } \theta,$$

sketch the function $f(\theta)$ for the range $-2<\theta\leqslant 2\pi$.

3) A function $f(x)$ is periodic with a period of 4. Sketch the graph of the function from -6 to $+6$, given that

$$f(x) \equiv -x \quad \text{for} \quad 0 < x \leqslant 3$$

$$f(x) \equiv 3x \quad \text{for} \quad 3 < x \leqslant 4.$$

4) If $f(x) \equiv 2x + 1$ for $0 < x \leqslant 2$
and $f(x + 2) = f(x)$ for all values of x, calculate

(a) $f(-11)$ (b) $f(17)$ (c) $\int_{-4}^{4} f(x)\,dx$.

5) Give full definitions of the periodic functions represented by the following graphs.

(a)

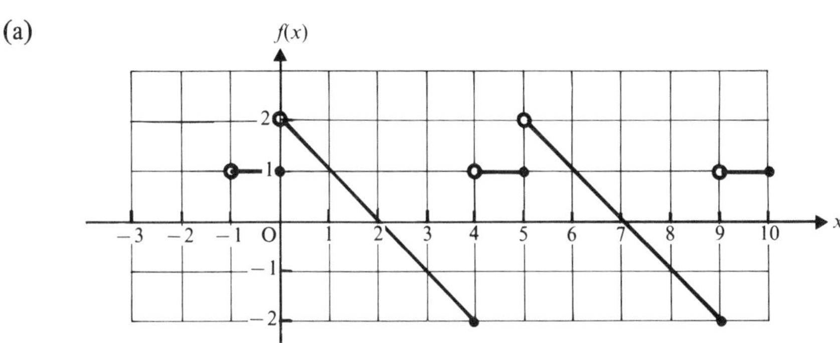

(b)

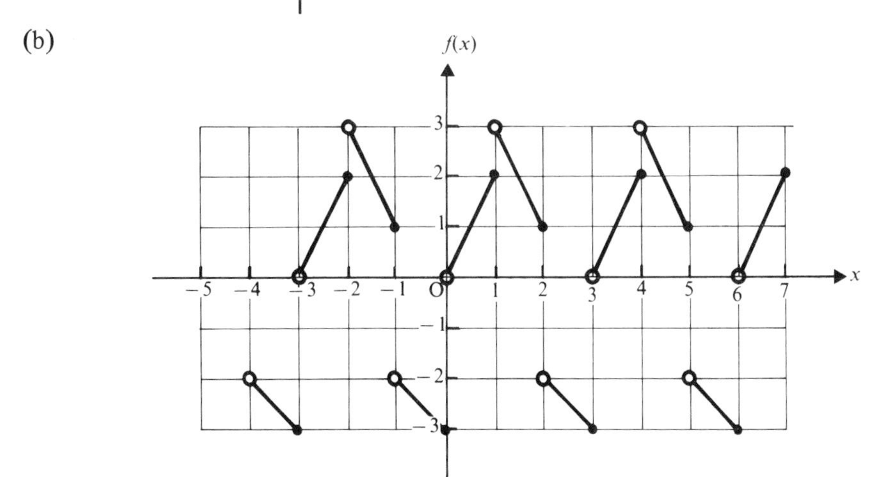

A NEW APPROACH TO THE LOGARITHMIC FUNCTION

The logarithmic function has, up to now, been regarded as the inverse of the exponential function. The logarithmic laws were derived from the laws of indices, the derivative of the log function was derived from the derivative of the

exponential function; even the property $\int \frac{1}{x}\,dx = \ln|kx|$ was obtained indirectly from the exponential function. In fact the log function has not yet been defined as an independent function.
We are now going to take a new look at this function, assuming none of the properties previously obtained from the exponential function, but *defining* the function $\ln x$ in the following way,

$$\ln x = \int_1^x \frac{1}{t}\,dt \quad \text{for} \quad x > 0,$$

i.e. $\ln x$ is represented by the area bounded by the curve $f(t) \equiv \frac{1}{t}$, the t axis and the lines $t = 1$ and $t = x$ where $x > 0$.

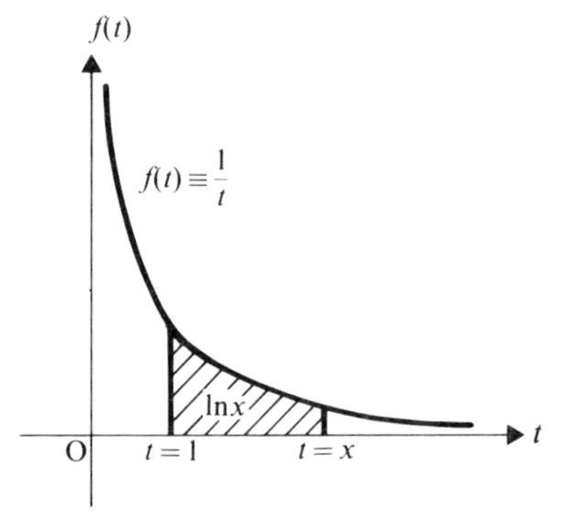

From this definition, it is now possible to derive properties of the log function.

To Prove that $\ln a + \ln b = \ln ab$

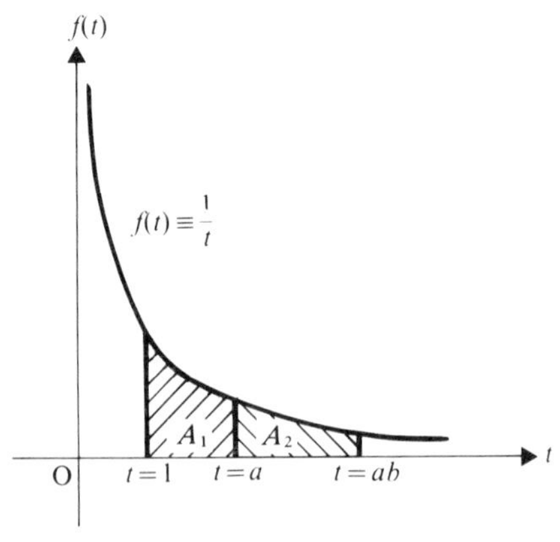

Consider the two areas A_1 and A_2 in the above diagram.
Using the new definition of the log function,

$$A_1 = \int_1^a \frac{1}{t}\,\mathrm{d}t = \ln a \qquad [1]$$

and
$$A_1 + A_2 = \int_1^{ab} \frac{1}{t}\,\mathrm{d}t = \ln ab \qquad [2]$$

But
$$A_2 = \int_a^{ab} \frac{1}{t}\,\mathrm{d}t$$

which, in this form, is not a log function (the lower limit is a instead of 1). However if we make the substitution

$$t \equiv au$$

so that
$$\ldots \mathrm{d}t \equiv \ldots a\,\mathrm{d}u$$

and

t	a	ab
u	1	b

then
$$A_2 = \int_a^{ab} \frac{1}{t}\,\mathrm{d}t \equiv \int_1^b \frac{1}{au}a\,\mathrm{d}u \equiv \int_1^b \frac{1}{u}\,\mathrm{d}u.$$

Now, by definition,

$$A_2 = \ln b. \qquad [3]$$

So, combining [1], [2] and [3],

$$A_1 + A_2 = \ln a + \ln b = \ln ab.$$

It is important to appreciate that, when a function is defined, no properties must be assumed unless proved from that definition. The reader is given the opportunity, in Exercise 9c, to prove some more of the familiar laws of logarithms from the definition $\ln x = \int_1^x \frac{1}{t}\,\mathrm{d}t$.

EXAMPLES 9c

1) Use Simpson's Rule with 7 ordinates to find an approximate value for $\int_1^4 \frac{1}{t}\,\mathrm{d}t$. Hence estimate $\ln 4$ to 3 decimal places.

t	1	1.5	2	2.5	3	3.5	4
$\frac{1}{t}$	1	0.6667	0.5000	0.4000	0.3333	0.2857	0.2500

Using Simpson's Rule to find the area, A, bounded by the curve $y = \frac{1}{t}$, the t axis and the lines $t = 1, t = 4$ we have:

$$A \simeq \frac{0.5}{3}\{1 + 0.2500 + 4(0.6667 + 0.4000 + 0.2857) + 2(0.5000 + 0.3333)\}$$

$$= 1.3877$$

Now, by definition,

$$\ln x = \int_1^x \frac{1}{t}\,dt$$

so $$\ln 4 = \int_1^4 \frac{1}{t}\,dt \simeq 1.3877.$$

i.e. $$\ln 4 \simeq 1.388 \qquad \text{(to 3 d.p.)}.$$

2) By considering the areas bounded by the x axis, the lines $t = 1$, $t = x$ and

(a) the line $y = \dfrac{1}{x}$, (b) the curve $y = \dfrac{1}{t}$,

(c) the line joining the points $(1, 1)$ and $\left(x, \dfrac{1}{x}\right)$ on the curve $y = \dfrac{1}{t}$,

prove that, for $x \geqslant 1$, $\left(\dfrac{x}{x-1}\right)\ln x \to 1$ as $x \to 1$.

(a)

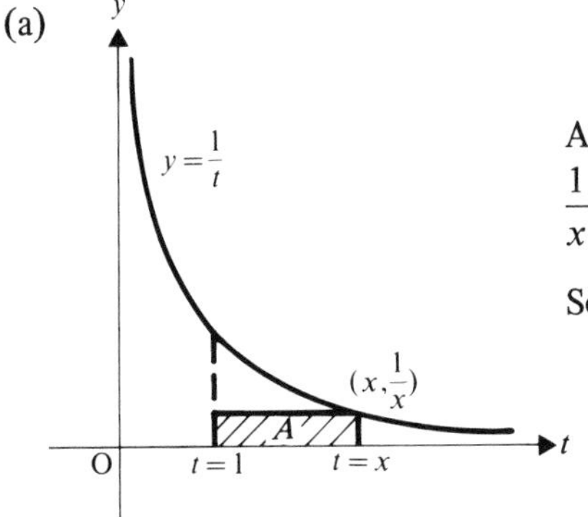

Area A is the area of a rectangle of height $\dfrac{1}{x}$ and base $(x - 1)$.

So $$\text{Area } A = \frac{1}{x}(x-1)$$

(b)

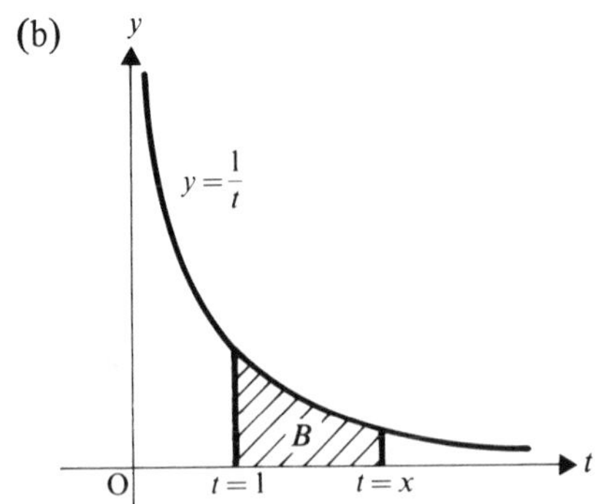

$$\text{Area } B = \int_1^x \frac{1}{t}\,dt$$

$$= \ln x \quad \text{by definition}$$

So $$\text{Area } B = \ln x$$

(c)

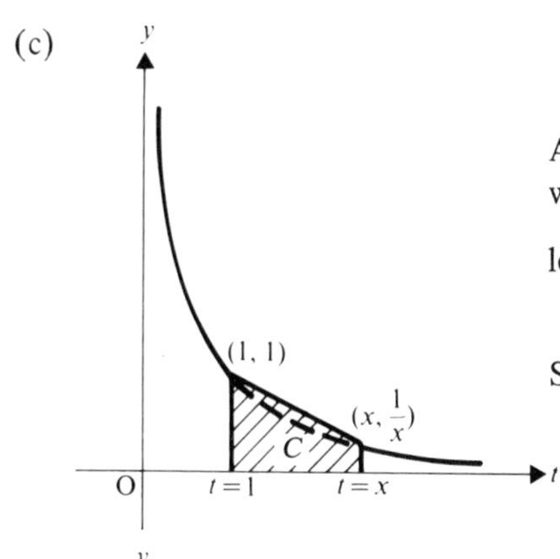

Area C is the area of a trapezium of width $(x-1)$ and with parallel sides of lengths 1 and $\frac{1}{x}$.

So Area $C = \left(\frac{x-1}{2}\right)\left(1+\frac{1}{x}\right)$

$$= \frac{x^2-1}{2x}$$

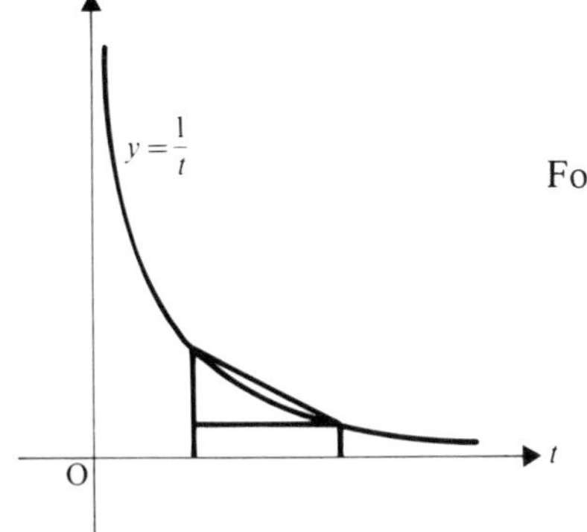

For all $x > 1$,

Area $A <$ Area $B <$ Area C

So
$$\frac{x-1}{x} < \ln x < \frac{x^2-1}{2x}$$

or
$$1 < \frac{x}{x-1}\ln x < \frac{x+1}{2}$$

$\left(\text{Multiplication throughout by } \frac{x}{x-1} \text{ is valid because } x > 1.\right)$

Now if $x \to 1$, $\frac{x+1}{2} \to 1$.

Hence $\frac{x}{x-1}\ln x$ lies between 1 and a quantity that approaches 1.

So $\frac{x}{x-1}\ln x \to 1$ as $x \to 1$

Note. Other forms of restriction on the range of possible values of $\ln x$ can be derived by choosing other areas related to the graph of $\frac{1}{t}$. For instance a cruder relationship is based on the same areas A and B, and a third area bounded by the t axis and the lines $t = 1$, $t = x$ and $y = 1$.

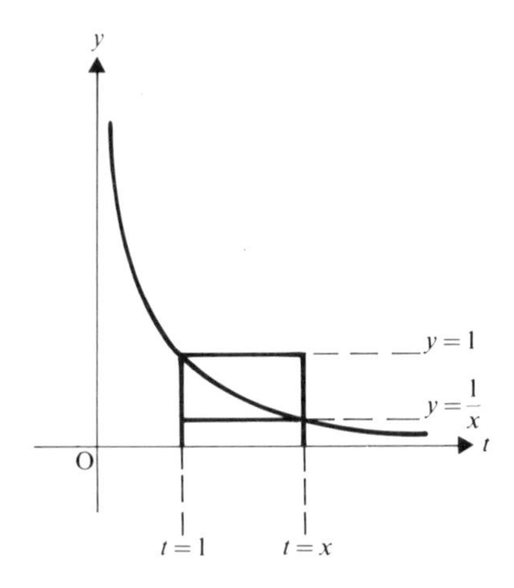

EXERCISE 9c

1) Prove that $\ln a - \ln b = \ln \frac{a}{b}$. $\left(\text{Use the substitution } t \equiv \frac{a}{u} \text{ where appropriate.}\right)$

2) Use the definition $\ln a^n = \int_1^{a^n} \frac{1}{t}\,dt$ and the substitution $t \equiv u^n$ to prove that $\ln a^n = n \ln a$.

3) By using the substitution $t \equiv \frac{1}{u}$ prove from the definition $\ln x = \int_1^x \frac{1}{t}\,dt$ that $\ln x = -\ln \frac{1}{x}$. Hence show that the areas bounded by the curve $y = \frac{1}{t}$, the t axis and

(a) the lines $t = \frac{1}{2}$ and $t = 1$,
(b) the lines $t = 1$ and $t = 2$,
are equal.

4) Using the definition $\ln(1+x) = \int_1^{1+x} \frac{1}{t}\,dt$ and considering the area under the graph $y = \frac{1}{t}$, prove that, for $x > 0$,

$$x(1+x) > (1+x)\ln(1+x) > x.$$

Hence show that $\ln(1+x) > x - x^2 + x^3 - x^4 \ldots$ and that, as $x \to 0$, $\frac{1}{x}\ln(1+x) \to 1$.

5) Use Simpson's Rule with 5 ordinates to evaluate $\int_1^2 \frac{1}{t}\,dt$ correct to 4 decimal places.

Hence find an approximate value for $\ln 2$.

6) Show that $\displaystyle\int_1^{1.1} \frac{1}{t}\,dt = \int_0^{0.1} \frac{1}{1+u}\,du.$

By using the binomial expansion of $(1+u)^{-1}$ find $\displaystyle\int_0^{0.1} \frac{1}{1+u}\,du$ correct to five decimal places. Hence find $\ln 1.1$ correct to 5 decimal places.

7) Use a method similar to that indicated in Question 6 to find, without using tables or calculator, the value of $\ln \frac{6}{5}$.

8) Show that $\displaystyle\int_2^{3} \frac{1}{t}\,dt = \ln \frac{3}{2}.$

9) Show that $\ln 1 = 0$.

10) By using the definition $\ln(1+x) = \displaystyle\int_1^{1+x} \frac{1}{t}\,dt$, the substitution $t \equiv 1+u$ and the binomial series, show that, for a certain range of values of x

$$\ln(1+x) = x - \frac{x^2}{2} + \frac{x^3}{3} - \frac{x^4}{4} + \dots .$$

HYPERBOLIC FUNCTIONS

Consider the two exponential functions e^x and e^{-x}.

$$e^x = 1 + x + \frac{x^2}{2!} + \frac{x^3}{3!} + \frac{x^4}{4!} + \dots + \frac{x^{2n}}{(2n)!} + \frac{x^{2n+1}}{(2n+1)!} + \dots$$

$$e^{-x} = 1 - x + \frac{x^2}{2!} - \frac{x^3}{3!} + \frac{x^4}{4!} - \dots + \frac{x^{2n}}{(2n)!} - \frac{x^{2n+1}}{(2n+1)!} + \dots$$

So $$e^x + e^{-x} = 2 + 2\left(\frac{x^2}{2!}\right) + 2\left(\frac{x^4}{4!}\right) + \dots + 2\left(\frac{x^{2n}}{(2n)!}\right) + \dots$$

$$\Rightarrow \quad \tfrac{1}{2}(e^x + e^{-x}) = 1 + \frac{x^2}{2!} + \frac{x^4}{4!} + \dots + \frac{x^{2n}}{(2n)!} + \dots$$

Also

$$e^x - e^{-x} = 2x + 2\left(\frac{x^3}{3!}\right) + \dots + 2\left(\frac{x^{2n+1}}{(2n+1)!}\right) + \dots$$

$$\Rightarrow \quad \tfrac{1}{2}(e^x - e^{-x}) = x + \frac{x^3}{3!} + \frac{x^5}{5!} + \dots + \frac{x^{2n+1}}{(2n+1)!} + \dots$$

The function $\frac{1}{2}(e^x + e^{-x})$ is called the *hyperbolic cosine* function which is written *cosh x*

The function $\frac{1}{2}(e^x - e^{-x})$ is called the *hyperbolic sine* function which is written *sinh x*,

i.e.
$$\cosh x \equiv \tfrac{1}{2}(e^x + e^{-x}) = \sum_{r=0}^{\infty} \frac{x^{2r}}{(2r)!}$$

and
$$\sinh x \equiv \tfrac{1}{2}(e^x - e^{-x}) = \sum_{r=0}^{\infty} \frac{x^{2r+1}}{(2r+1)!}$$

The series expansions of e^x and e^{-x} are convergent for all values of x so it follows that the series expansions of $\cosh x$ and $\sinh x$ are also universally convergent.

HYPERBOLIC RELATIONSHIPS

The names of the two hyperbolic functions we have so far met, i.e. the hyperbolic sine and cosine, suggest that these functions have certain properties that are similar to those of trig functions.
Consider, for instance, the relationship between $\cosh^2 x$ and $\sinh^2 x$

$$\cosh^2 x \equiv \tfrac{1}{4}(e^{2x} + 2 + e^{-2x})$$
$$\sinh^2 x \equiv \tfrac{1}{4}(e^{2x} - 2 + e^{-2x}).$$

Hence $\quad \cosh^2 x - \sinh^2 x \equiv 1$

and $\quad \cosh^2 x + \sinh^2 x \equiv \tfrac{1}{2}(e^{2x} + e^{-2x}) \equiv \cosh 2x.$

These identities can be compared with

$$\cos^2 x + \sin^2 x \equiv 1$$

and
$$\cos^2 x - \sin^2 x \equiv \cos 2x.$$

In each case it can be seen that the term $\sin^2 x$ becomes $-\sinh^2 x$ in the corresponding hyperbolic identity, but that otherwise there is a direct analogy. This is an example of Osborn's Rule which can be used to convert many trig identities into analogous hyperbolic identities.

The rule is:
Change each trig ratio into the comparative hyperbolic function. Whenever a *product of two sines* occurs, change the sign of that term.
Osborn's Rule should be used with care and restraint as there are a number of ways in which a product of two sine ratios can be disguised, e.g. in $\tan^2 x$.
Osborn's Rule is justified in Chapter 12, where specific relationships between trig and hyperbolic functions are derived.
Because of these close links, trig terminology is used to define further hyperbolic functions. So we have:

$$\tanh x \equiv \frac{\sinh x}{\cosh x} \equiv \frac{e^x - e^{-x}}{e^x + e^{-x}}$$

$$\operatorname{cosech} x \equiv \frac{1}{\sinh x} \equiv \frac{2}{e^x - e^{-x}}$$

$$\operatorname{sech} x \equiv \frac{1}{\cosh x} \equiv \frac{2}{e^x + e^{-x}}$$

$$\coth x \equiv \frac{1}{\tanh x} \equiv \frac{e^x + e^{-x}}{e^x - e^{-x}}.$$

The Graphs of Hyperbolic Functions

The graphs of $\sinh x$ and $\cosh x$ can be obtained by combining the graphs of e^x and e^{-x} as shown below.

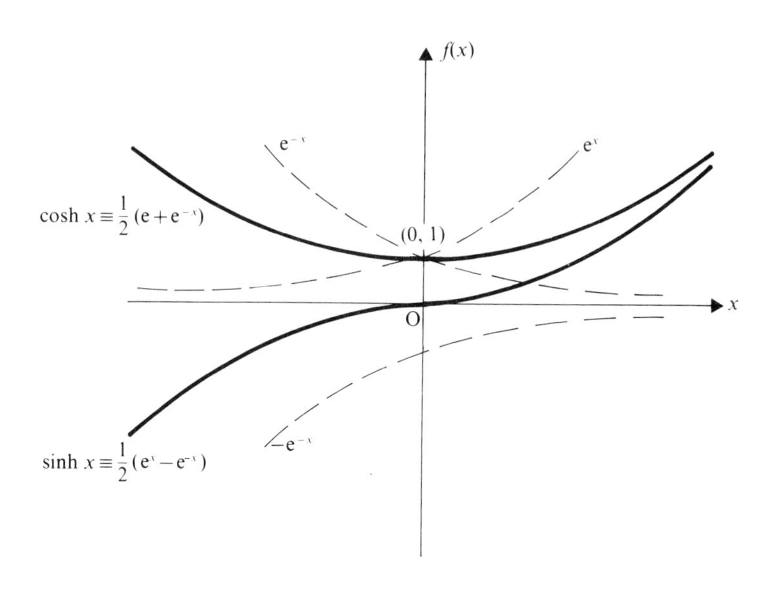

Note that

$\cosh x \geqslant 1$ for all values of x

$\sinh x \to \cosh x$ as $x \to \infty$

$\sinh x \to -\cosh x$ as $x \to -\infty$

the graph of $\cosh x$ has one turning point (minimum)
the graph of $\sinh x$ has no turning point.

The graph of $\tanh x$ is given by dividing $\sinh x$ by $\cosh x$, i.e.

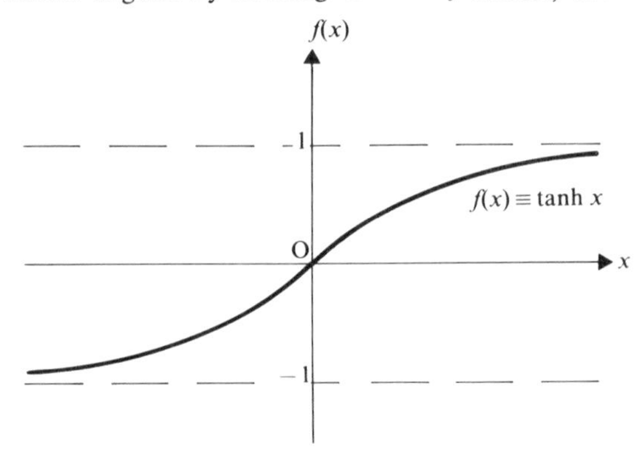

Note that the similarity between hyperbolic and trig functions is limited to their occurrence in identities. Their graphs have no corresponding relationship because, whilst all the trig functions are periodic, none of the hyperbolic functions is periodic.

EXAMPLES 9d

1) Prove that $\sinh 2x \equiv 2 \sinh x \cosh x$.

Starting on the R.H.S.,

$$\begin{aligned} 2\sinh x \cosh x &\equiv 2 \times \tfrac{1}{2}(e^x - e^{-x})\tfrac{1}{2}(e^x + e^{-x}) \\ &\equiv \tfrac{1}{2}(e^{2x} - e^{-2x}) \\ &\equiv \sinh 2x, \end{aligned}$$

i.e. $$\sinh 2x \equiv 2\sinh x \cosh x$$

(compare with $\sin 2x \equiv 2 \sin x \cos x$).
Note that, while Osborn's Rule can be used to *find* a hyperbolic identity, it cannot be used to *prove* that identity. *Proof* is usually based, as in this example, on the *definitions* of the hyperbolic functions.

2) From the trig identity $\tan^2 x + 1 \equiv \sec^2 x$ deduce the corresponding hyperbolic identity and then prove it.
In the trig identity the term $\tan^2 x$ contains $\sin^2 x$ but the term $\sec^2 x$ does not. So we deduce the corresponding hyperbolic identity by converting $\tan^2 x$ to $-\tanh^2 x$ and $\sec^2 x$ to $\text{sech}^2 x$, giving

$$1 - \tanh^2 x \equiv \text{sech}^2 x$$

Considering the L.H.S.

$$1 - \tanh^2 x \equiv 1 - \left(\frac{e^x - e^{-x}}{e^x + e^{-x}}\right)^2$$

$$\equiv \frac{(e^x + e^{-x})^2 - (e^x - e^{-x})^2}{(e^x + e^{-x})^2}$$

$$\equiv \frac{4}{(e^x + e^{-x})^2}$$

$$\equiv \left(\frac{2}{e^x + e^{-x}}\right)^2$$

$$\equiv \operatorname{sech}^2 x$$

This confirms that our deduction was correct.

EXERCISE 9d

Write down the first three terms of the series expansion of each of the following functions:

1) $\cosh 2x$ 2) $\sinh \frac{x}{2}$ 3) $\cosh x + \sinh x$

4) $\cosh^2 x$ 5) $\sinh^2 2x$ 6) $e^x - \sinh x$

7) $\ln(\cosh x)$ 8) $e^{\sinh x}$

Prove the following identities from the basic definitions of the hyperbolic functions:

9) $\coth^2 x - 1 \equiv \operatorname{cosech}^2 x$

10) $\sinh(x + y) \equiv \sinh x \cosh y + \cosh x \sinh y$

11) $\cosh 3x \equiv 4 \cosh^3 x - 3 \cosh x$

12) $\tanh 2x \equiv \dfrac{2 \tanh x}{1 + \tanh^2 x}$

Use Osborn's Rule to deduce the hyperbolic identities corresponding to the trig identities given in Questions 13–18. In each case prove that your result is correct.

13) $\sin(x + y) \equiv \sin x \cos y + \cos x \sin y$

14) $\cos(x - y) \equiv \cos x \cos y + \sin x \sin y$

15) $\sin 3x \equiv 3 \sin x - 4 \sin^3 x$

16) $\cos x + \cos y \equiv 2 \cos \dfrac{x + y}{2} \cos \dfrac{x - y}{2}$

17) $\sin 2x \equiv \dfrac{2 \tan x}{1 + \tan^2 x}$

18) $\cos 2x \equiv \dfrac{1-\tan^2 x}{1+\tan^2 x}$

19) If $\sinh x = \frac{3}{4}$ calculate $\operatorname{sech} x$, $\tanh x$, $\cosh 2x$ and $\tanh 2x$.

20) If $\tanh x = \frac{5}{13}$ calculate $\operatorname{cosech} x$, $\cosh x$, $\sinh 2x$ and $\tanh 2x$.

21) Sketch the graphs of $\coth x$, $\operatorname{cosech} x$, $\operatorname{sech} x$, $\cosh x + \sinh x$, $\sinh^2 x$, $\tanh(-x)$ and $\sinh(-x)$.
State which, if any, of these functions are
(a) even (b) odd (c) continuous.

Prove that:

22) $\tanh x \equiv \sqrt{\left(\dfrac{\cosh 2x - 1}{\cosh 2x + 1}\right)}$ $(x > 0)$.

23) $\cosh 3x \cosh^3 x + \sinh 3x \sinh^3 x \equiv \cosh^3 2x$.

24) Show that the point with coordinates $(a \cosh x,\ b \sinh x)$ lies on the hyperbola $\dfrac{x^2}{a^2} - \dfrac{y^2}{b^2} = 1$.

HYPERBOLIC EQUATIONS

To solve equations involving hyperbolic functions, use can be made of the identities proved in the previous section, as well as of the basic definitions of the functions.

EXAMPLES 9e

1) Solve the equation $3 \sinh x - \cosh x = 1$.

As no convenient identity relates $\sinh x$ to $\cosh x$ we will use the definitions of these functions, giving

$$\tfrac{3}{2}(e^x - e^{-x}) - \tfrac{1}{2}(e^x + e^{-x}) = 1$$

$$\Rightarrow \quad 2e^x - 4e^{-x} = 2$$

$$\Rightarrow \quad e^x - 1 - 2e^{-x} = 0$$

$$\Rightarrow \quad (e^x)^2 - e^x - 2 = 0$$

$$\Rightarrow \quad (e^x - 2)(e^x + 1) = 0$$

So either $e^x - 2 = 0$ or $e^x + 1 = 0$.
But $e^x = -1$ has no real solution.
So $e^x = 2 \Rightarrow x = \ln 2$ as the only solution.

2) Find the values of x for which $12\cosh^2 x + 7\sinh x = 24$.

Using the identity $\cosh^2 x - \sinh^2 x \equiv 1$ gives

$$12(1+\sinh^2 x) + 7\sinh x - 24 = 0$$

$$\Rightarrow \quad 12\sinh^2 x + 7\sinh x - 12 = 0$$

$$\Rightarrow \quad (3\sinh x + 4)(4\sinh x - 3) = 0$$

Hence $\sinh x = -\frac{4}{3}$ or $\frac{3}{4}$.

Now using the definition of $\sinh x$, $\frac{1}{2}(e^x - e^{-x}) = -\frac{4}{3}$ or $\frac{3}{4}$,

i.e. $\quad 3e^x + 8 - 3e^{-x} = 0 \quad$ or $\quad 2e^x - 3 - 2e^{-x} = 0$

$\Rightarrow \quad 3e^{2x} + 8e^x - 3 = 0 \quad$ or $\quad 2e^{2x} - 3e^x - 2 = 0$

$\Rightarrow \quad (3e^x - 1)(e^x + 3) = 0 \quad$ or $\quad (2e^x + 1)(e^x - 2) = 0$

$$\Rightarrow \quad e^x = \tfrac{1}{3},\ -3,\ -\tfrac{1}{2},\ 2.$$

But negative values of e^x do not give real values of x, so real solutions are given only by $e^x = \frac{1}{3}, 2$.

Hence $x = \ln\frac{1}{3}$ or $\ln 2$.

EXERCISE 9e

Solve, for real values of x, the equations given in Questions 1–10.

1) $\sinh x + 4 = 4\cosh x$

2) $7 + 2\cosh x = 6\sinh x$

3) $2\sinh x + 6\cosh x = 9$

4) $5\cosh x + \sinh x = 7$

5) $\cosh 2x - 7\cosh x + 7 = 0$

6) $4\tanh^2 x - \operatorname{sech} x = 1$

7) $4\cosh x - e^{-x} = 3$

8) $\sinh^2 x - 5\cosh x + 5 = 0$

9) $4\sinh x + 3e^x + 3 = 0$

10) $20\cosh 2x - 21\sinh x = 200$.

11) Express $4\cosh x + 5\sinh x$ in the form $r\sinh(x+y)$ giving the values of r and $\tanh y$.

12) By expressing $13\cosh x + 5\sinh y$ in the form $r\cosh(x+y)$, find its minimum value.

THE CALCULUS OF HYPERBOLIC FUNCTIONS

If $\quad f(x) \equiv \cosh x \equiv \frac{1}{2}(e^x + e^{-x})$

then $\quad f'(x) = \frac{1}{2}(e^x - e^{-x}) \equiv \sinh x$,

i.e. $$\frac{d}{dx}(\cosh x) = \sinh x$$

Also if $$f(x) \equiv \sinh x \equiv \tfrac{1}{2}(e^x - e^{-x})$$

then $$f'(x) = \tfrac{1}{2}(e^x + e^{-x}) \equiv \cosh x$$

i.e. $$\frac{d}{dx}(\sinh x) = \cosh x.$$

From these results it follows that

$$\int \sinh x \, dx = \cosh x + K$$

and $$\int \cosh x \, dx = \sinh x + K.$$

Other hyperbolic functions can now be differentiated and integrated, using the results proved above.
Note that Osborn's Rule does *not* apply to calculus operations.

EXAMPLES 9f

1) Differentiate (a) $\operatorname{cosech} x$ (b) $\tanh^2 x$.

(a) $$f(x) \equiv \operatorname{cosech} x \equiv \frac{1}{\sinh x} \equiv (\sinh x)^{-1}$$

so $$f'(x) = -(\sinh x)^{-2} \frac{d}{dx}(\sinh x)$$

$$= -\frac{\cosh x}{\sinh^2 x}.$$

(b) $$f(x) \equiv \tanh^2 x$$

so $$f'(x) = 2 \tanh x \frac{d}{dx}(\tanh x)$$

but $$\tanh x \equiv \frac{\sinh x}{\cosh x}$$

so $$\frac{d}{dx}(\tanh x) = \frac{\cosh x(\cosh x) - \sinh x(\sinh x)}{\cosh^2 x}$$

$$= \frac{1}{\cosh^2 x}$$

$$= \operatorname{sech}^2 x.$$

Hence $$\frac{d}{dx}(\tanh^2 x) = 2 \tanh x \operatorname{sech}^2 x.$$

2) Find (a) $\int \tanh x \, dx$ (b) $\int \sinh^3 x \, dx$.

(a)
$$\int \tanh x \, dx \equiv \int \frac{\sinh x}{\cosh x} \, dx$$
$$\equiv \int \frac{f'(x)}{f(x)} \, dx \quad \text{where} \quad f(x) \equiv \cosh x$$
$$= \ln [f(x)] + K$$

So
$$\int \tanh x \, dx = \ln (\cosh x) + K$$

(b)
$$\int \sinh^3 x \, dx \equiv \int \sinh x \sinh^2 x \, dx$$
$$\equiv \int \sinh x (\cosh^2 x - 1) \, dx$$
$$\equiv \int \sinh x \cosh^2 x \, dx - \int \sinh x \, dx$$
$$= \tfrac{1}{3} \cosh^3 x - \cosh x + K$$

EXERCISE 9f

Differentiate the following functions w.r.t x.

1) $\operatorname{sech} x$ 2) $\tanh 2x$ 3) $\coth x$ 4) $\operatorname{sech}^2 x$ 5) $\sinh 4x$

6) $\cosh^3 2x$ 7) $x \sinh x$ 8) $\sinh x \tanh x$ 9) $e^x \sinh x$

10) $\sqrt{(\cosh 5x)}$ 11) $x^2 \tanh^2 3x$ 12) $\dfrac{e^x}{\sinh 2x}$ 13) $e^{\cosh x}$

14) $\ln \sinh x$ 15) $\sqrt{(\tanh x)}$ 16) $e^{\tanh^2 x}$.

Find the following integrals.

17) $\int \sinh 5x \, dx$ 18) $\int \cosh 4x \, dx$ 19) $\int \tanh 2x \, dx$

20) $\int \coth x \, dx$ 21) $\int \sinh^2 x \, dx$ 22) $\int \cosh^3 x \, dx$

23) $\int x \sinh 2x \, dx$ 24) $\int \dfrac{\sinh x}{\cosh^2 x} \, dx$ 25) $\int e^x \cosh x \, dx$

26) $\int \cosh 3x \cosh x \, dx$ 27) $\int \sinh 5x \cosh 3x \, dx$.

Evaluate the integrals in Questions 28–31.

28) $\int_0^4 \cosh 3x \, dx$ 29) $\int_1^2 e^x \sinh x \, dx$

30) $\int_0^1 \tanh 4x \, dx$ 31) $\int_1^4 \operatorname{sech}^2 5x \, dx$.

32) Find the equation of the tangent and normal to the hyperbola $\frac{x^2}{a^2} - \frac{y^2}{b^2} = 1$ at the point $(a \cosh u, b \sinh u)$.

33) Investigate the stationary value(s) of $25 \cosh x - 7 \sinh x$.

34) If $y = A \cosh kx + B \sinh kx$, prove that $\frac{d^2y}{dx^2} = k^2 y$. Hence find y as a function of x given that $\frac{d^2y}{dx^2} = 4y$, and that, when $x = 0$, $y = 2$ and $\frac{dy}{dx} = 2$.

INVERSE HYPERBOLIC FUNCTIONS

If $p = \sinh q$, then $q = \operatorname{arsinh} p$, i.e. q is the variable whose sinh is p, and $\operatorname{arsinh} p$ is called the *inverse hyperbolic sine function*.
Other inverse hyperbolic functions are $\operatorname{arcosh} u$, $\operatorname{artanh} z$, $\operatorname{arcosech} x$, etc.
An alternative notation, $\sinh^{-1} p$, $\cosh^{-1} u$, $\tanh^{-1} z$, $\operatorname{cosech}^{-1} x$, etc., is sometimes used.

The Graphs of Inverse Hyperbolic Functions

As $y = \operatorname{arcosh} x \iff x = \cosh y$ it is clear that these two equations express the same relationship between x and y.
Hence the graphs representing the two equations are the same curve.
The reader will recall that, in the case of inverse trig functions, a particular value of x corresponds to an infinite set of values of $\arcsin x$, $\arccos x$ or $\arctan x$ (because the trig functions are periodic). It is necessary, therefore, to distinguish between the principal value, $\arcsin x$, and the general set of values, $\operatorname{Arcsin} x$.
In the case of the inverse hyperbolic functions, however, one suitable value of x corresponds to only one value of $\operatorname{arsinh} x$ or $\operatorname{artanh} x$ and only two values of $\operatorname{arcosh} x$. So the need for the notation $\operatorname{Arsinh} x$, etc., does not arise.

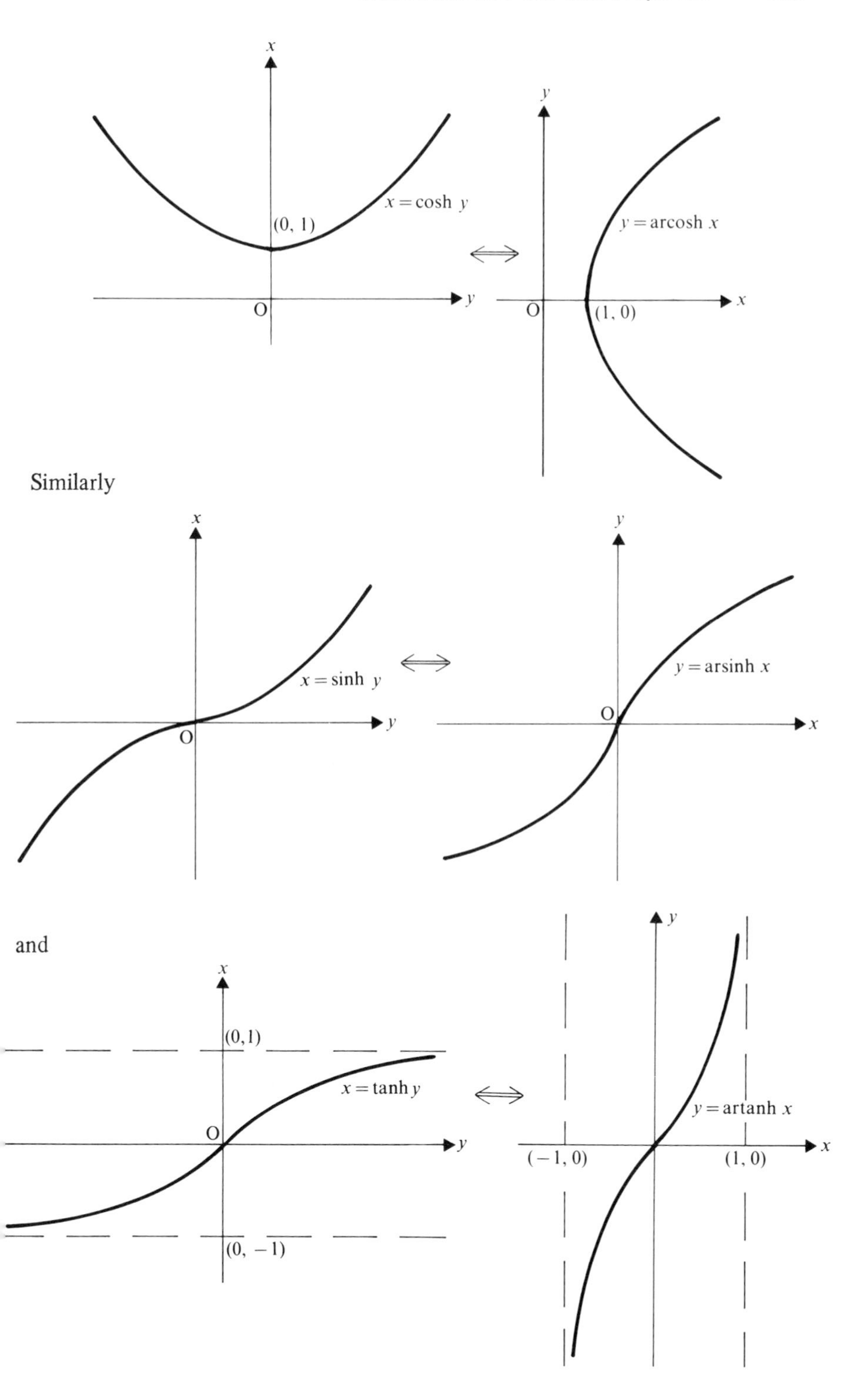
x
(0, 1)
x = cosh y
O
y
y
y = arcosh x
O
(1, 0)
x
Similarly
x
x = sinh y
O
y
y
y = arsinh x
O
x
and
x
(0,1)
x = tanh y
O
y
(0, −1)
y
y = artanh x
(−1, 0)
(1, 0)
x

THE LOGARITHMIC FORM

Using $y = \text{arcosh}\, x \iff x = \cosh y$

and $\cosh y \equiv \frac{1}{2}(e^y + e^{-y})$

we have $x = \frac{1}{2}(e^y + e^{-y})$

$\Rightarrow \quad e^y - 2x + e^{-y} = 0$

$\Rightarrow \quad e^{2y} - 2xe^y + 1 = 0.$

Hence $e^y = x \pm \sqrt{(x^2 - 1)},$

i.e. $y = \ln\,[x \pm \sqrt{(x^2 - 1)}].$

But
$$x - \sqrt{(x^2 - 1)} \equiv \frac{[x - \sqrt{(x^2 - 1)}]\,[x + \sqrt{(x^2 - 1)}]}{[x + \sqrt{(x^2 - 1)}]}$$
$$\equiv \frac{1}{x + \sqrt{(x^2 - 1)}}.$$

So $y = \ln\,[x + \sqrt{(x^2 - 1)}]^{\pm 1}$

i.e. $\text{arcosh}\, x \equiv \pm \ln\,[x + \sqrt{(x^2 - 1)}].$

This form of the inverse hyperbolic cosine function verifies that the graph is symmetrical about the x-axis. (It also verifies that $\text{arcosh}\, x$ is real only when $x \geqslant 1$.)

Similarly $y = \text{arsinh}\, x \iff x = \sinh y$

and $\sinh y \equiv \frac{1}{2}(e^y - e^{-y})$

$\Rightarrow \quad y = \ln\,[x \pm \sqrt{(x^2 + 1)}].$

But $x - \sqrt{(x^2 + 1)} < 0$

So $\text{arsinh}\, x \equiv \ln\,[x + \sqrt{(x^2 + 1)}].$

Note that the logarithmic form of an inverse hyperbolic function is useful when solving some hyperbolic equations.
For instance, the latter part of the solution of the equation

$$12\cosh^2 x + 7\sinh x = 24,$$

given on p. 391, can now be abbreviated as follows:

$$12\cosh^2 x + 7\sinh x = 24$$

$\Rightarrow \quad 12\sinh^2 x + 7\sinh x - 12 = 0$

$\Rightarrow \quad (3\sinh x + 4)(4\sinh x - 3) = 0$

$\Rightarrow \quad \sinh x = -\frac{4}{3} \quad \text{or} \quad \frac{3}{4}$

Hence $\qquad x = \operatorname{arsinh}(-\tfrac{4}{3}) = \ln[-\tfrac{4}{3} + \sqrt{(\tfrac{16}{9} + 1)}]$

or $\qquad x = \operatorname{arsinh}\tfrac{3}{4} = \ln[\tfrac{3}{4} + \sqrt{(\tfrac{9}{16} + 1)}]$

Thus $\qquad x = \ln\tfrac{1}{3} \quad \text{or} \quad \ln 2$

EXAMPLES 9g

1) Prove that $\operatorname{artanh} x - \operatorname{artanh} y \equiv \operatorname{artanh}\dfrac{x-y}{1-xy}$.

Let $\qquad \operatorname{artanh} x \equiv u \Rightarrow \tanh u \equiv x$

and $\qquad \operatorname{artanh} y \equiv v \Rightarrow \tanh v \equiv y.$

Then the L.H.S. of the required identity is $u - v$.

$$\tanh(u-v) \equiv \frac{\tanh u - \tanh v}{1 - \tanh u \tanh v}$$

$$\left(\text{using Osborn's Rule on } \tan(u-v) \equiv \frac{\tan u - \tan v}{1 + \tan u \tan v}\right)$$

$$\equiv \frac{x-y}{1-xy}$$

Hence $\qquad u - v \equiv \operatorname{artanh}\dfrac{x-y}{1-xy}$

i.e. $\qquad \operatorname{artanh} x - \operatorname{artanh} y \equiv \operatorname{artanh}\dfrac{x-y}{1-xy}$.

2) Find x if $\operatorname{arsinh} x = \ln 3$.

If $\qquad \operatorname{arsinh} x = \ln 3$

then $\qquad x = \sinh(\ln 3)$

$$= \tfrac{1}{2}\{e^{\ln 3} - e^{-\ln 3}\}$$

$$= \tfrac{1}{2}\{3 - \tfrac{1}{3}\}$$

Hence $\qquad x = \tfrac{4}{3}.$

EXERCISE 9g

1) Evaluate $\operatorname{arcosh} 3$, $\operatorname{arsinh}(-1)$, $\operatorname{artanh}\tfrac{1}{2}$.

2) Sketch the graph with equation $y = \operatorname{arcosh} 3x$ and state the range of values of x for which $\operatorname{arcosh} 3x$ is real.

3) Sketch the graph with equation $y = \operatorname{artanh}\dfrac{x}{2}$ and give the range of values of x for which y is real.

4) If $y = \operatorname{artanh} x$, express y as a logarithmic function of x.

5) Express $\operatorname{arcosh} 2$ as a logarithm.

6) Prove that

(a) $\operatorname{artanh} x + \operatorname{artanh} y \equiv \operatorname{artanh}\dfrac{x+y}{1+xy}$,

(b) $\operatorname{artanh}\dfrac{2x}{1+x^2} \equiv 2\operatorname{artanh} x$, (c) $2\operatorname{artanh} x \equiv \ln\dfrac{1+x}{1-x}$.

7) Express each of the following as logarithms:

(a) $\operatorname{arsech} x$ (b) $\operatorname{arcosh}\dfrac{1}{x}$ (c) $\operatorname{arsinh}(x^2-1)$.

8) Solve the equation $2\operatorname{artanh} x = \ln 3$.

9) Find x if $\operatorname{arcosh} 5x = \operatorname{arsinh} 4x$.

10) Solve the equation $\operatorname{artanh}\left(\dfrac{x^2-1}{x^2+1}\right) = \ln 2$.

DIFFERENTIATION

If $y = \operatorname{arcosh} x$

$$x = \cosh y$$

$$\Rightarrow \quad \frac{dx}{dy} = \sinh y$$

$$= \pm\sqrt{(\cosh^2 y - 1)}$$

$$= \pm\sqrt{(x^2-1)}.$$

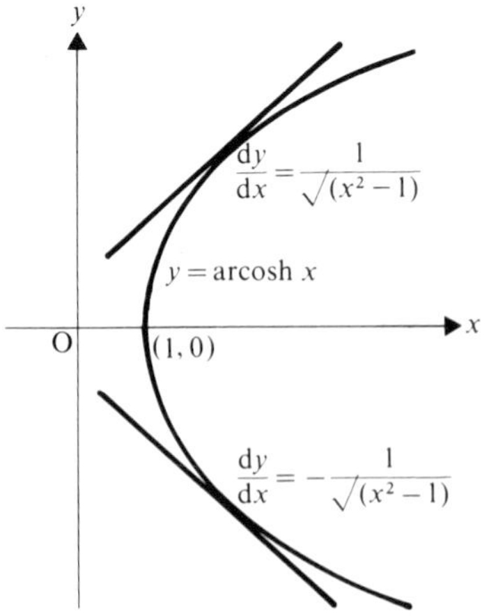

Hence $$\frac{d}{dx}(\operatorname{arcosh} x) = \frac{\pm 1}{\sqrt{(x^2-1)}}.$$

So $$\frac{d}{dx}(\operatorname{arcosh} x) = \frac{1}{\sqrt{(x^2-1)}}$$

when $\operatorname{arcosh} x > 0$

and $$\frac{d}{dx}(\operatorname{arcosh} x) = \frac{-1}{\sqrt{(x^2-1)}}$$

when $\operatorname{arcosh} x < 0$.

It can be shown in a similar way that

$$\frac{d}{dx}(\text{arsinh}\, x) = \frac{1}{\sqrt{(1+x^2)}}$$

and

$$\frac{d}{dx}(\text{artanh}\, x) = \frac{1}{1-x^2}$$

Note that there is no ambiguity of sign in the derivatives of arsinh x and artanh x as these functions have positive gradients for all values of x.

THE USE OF HYPERBOLIC FUNCTIONS IN INTEGRATION

The reader may recall that trig substitutions played a useful part in integrating certain expressions of the type $\sqrt{(1-x^2)}$, $\dfrac{1}{\sqrt{(4-x^2)}}$. The trig identity $\cos^2 x \equiv 1 - \sin^2 x$ suggested using the substitutions $x \equiv \sin u$ and $x \equiv 2 \sin u$ in these two cases.

Similarly, expressions such as $\sqrt{(x^2-1)}$, $\dfrac{1}{\sqrt{(4+x^2)}}$ suggest using the substitutions $x \equiv \cosh u$ and $x \equiv 2 \sinh u$, based on the identity $\cosh^2 x - \sinh^2 x \equiv 1$.

Almost all expressions which are the sum or difference of two squares can be integrated by using a trig or hyperbolic substitution based on the similarity of the given expression to a standard identity.

EXAMPLES 9h

1) Find $\displaystyle\int_0^1 \frac{1}{\sqrt{(1+9x^2)}}\,dx$.

Let $3x \equiv \sinh u$ (using $\cosh^2 u \equiv 1 + \sinh^2 u$)
so that $\ldots 3dx \equiv \ldots \cosh u\, du$

and

x	0	1
u	0	arsinh 3

Then

$$\int_0^1 \frac{1}{\sqrt{(1+9x^2)}}\,dx \equiv \int_0^{\text{arsinh}\,3} \frac{1}{\sqrt{(1+\sinh^2 u)}}\left(\frac{1}{3}\cosh u\right) du$$

$$\equiv \int_0^{\text{arsinh}\,3} \frac{1}{\cosh u}\left(\frac{1}{3}\cosh u\right) du$$

$$= \left[\frac{u}{3}\right]_0^{\text{arsinh}\,3}.$$

So

$$\int_0^1 \frac{1}{\sqrt{(1+9x^2)}}\,dx = \frac{1}{3}\,\text{arsinh}\, 3.$$

2) Integrate $\sqrt{(x^2+2x-1)}$ w.r.t. x.

First we convert x^2+2x-1 into a difference of two squares

i.e. $$x^2+2x-1 \equiv (x+1)^2-2$$

Then $$\int \sqrt{(x^2+2x-1)}\,dx \equiv \int \sqrt{[(x+1)^2-(\sqrt{2})^2]}\,dx.$$

Let $x+1 \equiv \sqrt{2}\cosh u$ (using $\cosh^2 u - 1 \equiv \sinh^2 u$)
so that $\ldots dx \equiv \ldots \sqrt{2}\sinh u\,du$.

Then $$\int \sqrt{(x^2+2x-1)}\,dx \equiv \sqrt{2}\int \sqrt{(\cosh^2 u-1)}\sqrt{2}\sinh u\,du$$

$$\equiv 2\int \sinh^2 u\,du$$

$$\equiv \int (\cosh 2u - 1)\,du$$

$$= \tfrac{1}{2}\sinh 2u - u + K.$$

But $$\sinh 2u \equiv 2\sinh u\cosh u$$

$$\equiv 2\cosh u\sqrt{(\cosh^2 u - 1)}$$

$$\equiv 2\left(\frac{x+1}{\sqrt{2}}\right)\sqrt{\left(\frac{(x+1)^2}{2}-1\right)}$$

$$\equiv (x+1)\sqrt{(x^2+2x-1)}$$

and $$u \equiv \operatorname{arcosh}\frac{x+1}{\sqrt{2}}.$$

So

$$\int \sqrt{(x^2+2x-1)}\,dx = \tfrac{1}{2}(x+1)\sqrt{(x^2+2x-1)} - \operatorname{arcosh}\left(\frac{x+1}{\sqrt{2}}\right) + K.$$

Certain integrals can be obtained by recognition, thus avoiding the need to make a substitution. The standard results that can be quoted are:

$$\int \frac{1}{\sqrt{(x^2+a^2)}}\,dx = \operatorname{arsinh}\frac{x}{a} + K$$

$$\int \frac{1}{\sqrt{(x^2-a^2)}}\,dx = \operatorname{arcosh}\frac{x}{a} + K$$

It is left to the reader to derive these results by substitution and thereafter to quote them where needed.

EXERCISE 9h

1) Prove, by using a suitable substitution, that:

(a) $\displaystyle\int \frac{1}{\sqrt{(x^2+a^2)}}\,dx = \text{arsinh}\,\frac{x}{a} + K$

(b) $\displaystyle\int \frac{1}{\sqrt{(x^2-a^2)}}\,dx = \text{arcosh}\,\frac{x}{a} + K.$

2) *Write down* the integral w.r.t. x of:

(a) $\dfrac{1}{\sqrt{(x^2+4)}}$ (b) $\dfrac{1}{\sqrt{(x^2-9)}}$ (c) $\dfrac{1}{\sqrt{(4x^2-16)}}$ (d) $\dfrac{1}{\sqrt{(4x^2+9)}}$.

[*Hint*. In parts (c) and (d) take out a factor 4 from under the square root.]

3) Differentiate:

(a) $\text{arcosh}\,(x+1)$ (b) $\text{arsinh}\,2x$ (c) $\text{artanh}\,x$

(d) $x\,\text{arcosh}\,x$ (e) $\text{arsinh}\,\dfrac{1}{x}$ (f) $\text{arcosh}\,\sqrt{x}$

(g) $\text{arsech}\,x$ (h) $\text{arcosech}\,(x-1)$ (i) $\text{arcosh}\,(\sinh 2x)$

(j) $e^{\text{arcosh}\,x}$

By making a hyperbolic substitution, find the following integrals.

4) $\displaystyle\int \sqrt{(9x^2+4)}\,dx$ 5) $\displaystyle\int (4x^2-1)^{\frac{3}{2}}\,dx$ 6) $\displaystyle\int \sqrt{(x^2+4x+5)}\,dx$

7) $\displaystyle\int \sqrt{(x^2+4x+3)}\,dx$ 8) $\displaystyle\int \frac{1}{\sqrt{(x^2+2)}}\,dx$ 9) $\displaystyle\int \frac{1}{\sqrt{(2x^2-1)}}\,dx$

10) $\displaystyle\int \frac{1}{\sqrt{(x^2+2x+2)}}\,dx$ 11) $\displaystyle\int \frac{1}{\sqrt{(x^2+2x)}}\,dx$ 12) $\displaystyle\int \sqrt{(4x+x^2)}\,dx$

13) $\displaystyle\int \frac{1}{\sqrt{(4x^2-x)}}\,dx$ 14) $\displaystyle\int \frac{x+1}{\sqrt{(x^2+1)}}\,dx$ 15) $\displaystyle\int \frac{x-1}{\sqrt{(x^2-1)}}\,dx.$

Evaluate the following integrals.

16) $\displaystyle\int_1^2 \frac{1}{\sqrt{(9x^2-1)}}\,dx$ 17) $\displaystyle\int_0^1 \frac{1}{\sqrt{(x^2+4)}}\,dx$ 18) $\displaystyle\int_0^1 \frac{1}{\sqrt{(x^2+6x+5)}}\,dx$

19) $\displaystyle\int_2^3 \frac{x+1}{\sqrt{(x^2-4)}}\,dx$ 20) $\displaystyle\int_0^2 \sqrt{(x^2+4)}\,dx$ 21) $\displaystyle\int_{\frac{4}{3}}^2 \sqrt{(9x^2-16)}\,dx.$

The following integrals can be found by making a hyperbolic substitution, by making a trig substitution, by using partial functions or by recognition. Choose the appropriate method and hence find each integral.

22) $\int \sqrt{(1-4x^2)}\,\mathrm{d}x$ 23) $\int \frac{1}{\sqrt{(4-x^2)}}\,\mathrm{d}x$ 24) $\int \frac{1}{\sqrt{(x^2-9)}}\,\mathrm{d}x$

25) $\int \frac{1}{9x^2-1}\,\mathrm{d}x$ 26) $\int \frac{1}{9x^2+1}\,\mathrm{d}x$ 27) $\int \frac{1}{\sqrt{(9x^2+1)}}\,\mathrm{d}x$

28) $\int \frac{x+2}{\sqrt{(x^2-4)}}\,\mathrm{d}x$ 29) $\int \frac{x+2}{x^2-4}\,\mathrm{d}x$ 30) $\int \frac{x}{x^2-4}\,\mathrm{d}x$

31) $\int \frac{x}{\sqrt{(x^2-4)}}\,\mathrm{d}x$ 32) $\int \sqrt{\left(\frac{x+1}{x^2-1}\right)}\,\mathrm{d}x$

33) $\int \frac{\sqrt{(x^2+1)}}{x}\,\mathrm{d}x.$

By sketching the appropriate graphs, state the number of real roots of the following equations. (Do *not* solve the equations.)

34) $\sin x = \sinh x$ 35) $\operatorname{arcosh} x = \mathrm{e}^{-x}$ 36) $\cosh x = \tan x$

37) $\ln x = \operatorname{arsinh} x$.

GENERAL INVERSE FUNCTIONS

If a relationship between x and y can be expressed in the two forms

$$y = f(x) \quad \text{and} \quad x = g(y)$$

then the function g is the inverse of the function f and conversely. The usual notation for the inverse of a function f is f^{-1}

i.e. $$\text{if} \quad y = f(x) \quad \text{then} \quad x = f^{-1}(y)$$

A number of functions and their inverses have already arisen in the text, e.g.

(a) $y = \mathrm{e}^x \iff x = \ln y$.
So if f is the exponential function, f^{-1} is the logarithmic function.

(b) $y = \sin x \iff x = \arcsin y$.
In this case f is the sine function and f^{-1} is the inverse sine function.

(c) $y = \cosh x \iff x = \operatorname{arcosh} y$.
Here f is a hyperbolic function and f^{-1} is an inverse hyperbolic function.

Note that f and f^{-1} refer to the *function* and not to the variable.
So both f and f^{-1} can be written as functions of x. For instance
if $f(x) \equiv \mathrm{e}^x$ then the inverse function, $f^{-1}(x)$, is $\ln x$.
When a function and its inverse are both expressed in terms of x, the graph of the inverse function is the reflection, in the line $y = x$, of the graph of the function, and conversely,

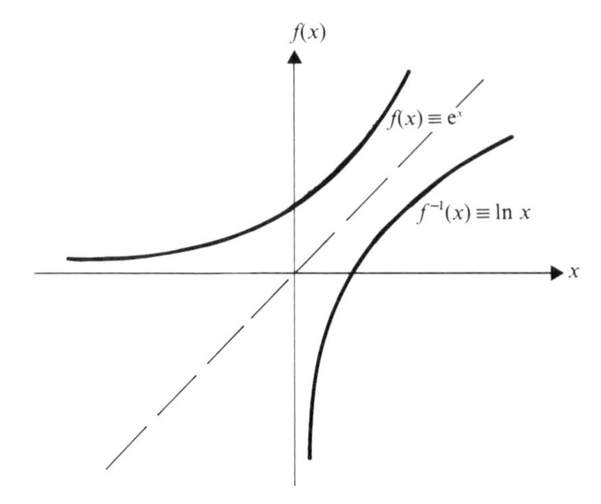

The inverse function can be found only when the relationship $y = f(x)$ can be rearranged in the form $x = g(y)$. If this is not possible then the function f has no inverse,
e.g. the cubic function $f(x) \equiv x^3 - x^2 + 1$ has no inverse, as $y = x^3 - x^2 + 1$ cannot be expressed in the form $x = g(y)$.
But the cubic function $f(x) \equiv x^3 - 1$ does have an inverse which is $f^{-1}(x) \equiv \sqrt[3]{(x+1)}$.
Consider the set of points $P(x, y)$ which lie on the curve $y = f(x)$ and the set of points $Q(X, Y)$ which lie on the curve $Y = f^{-1}(X)$.
The set of points Q is transformed into the set of points P by reflection in the line $y = x$, and this transformation is produced by the matrix $\begin{pmatrix} 0 & 1 \\ 1 & 0 \end{pmatrix}$,

i.e. $$\begin{pmatrix} 0 & 1 \\ 1 & 0 \end{pmatrix}\begin{pmatrix} X \\ f^{-1}(X) \end{pmatrix} = \begin{pmatrix} x \\ f(x) \end{pmatrix}$$

This equation can be used to find $f^{-1}(X)$.
For example, if $f(x) \equiv x^2$, then

$$\begin{pmatrix} x \\ x^2 \end{pmatrix} = \begin{pmatrix} 0 & 1 \\ 1 & 0 \end{pmatrix}\begin{pmatrix} X \\ f^{-1}(X) \end{pmatrix}$$

$$\Rightarrow \quad \begin{cases} x = f^{-1}(X) \\ x^2 = X. \end{cases}$$

Hence $$f^{-1}(X) = \pm\sqrt{X}$$

SUMMARY

If $f(x) = f(-x)$ for all values of x, then $f(x)$ is an even function.
If $f(x) = -f(-x)$ for all values of x, then $f(x)$ is an odd function.

If a function $f(x)$ is such that $f(x+a)=f(x)$ for all values of x, then $f(x)$ is periodic with a period a.
A continuous function $f(x)$ satisfies the conditions

$$\begin{cases} f(a) \text{ is defined} \\ \lim_{x\to a} f(x) \text{ is } f(a) \end{cases}$$

for all values of x, and its graph is unbroken.

The function $\ln x$ can be defined as $\int_1^x \frac{1}{t}\,\mathrm{d}t$.

If $y=f(x) \iff x=g(y)$ then each of the functions f and g is the inverse of the other.

$$\cosh x \equiv \frac{1}{2}(\mathrm{e}^x+\mathrm{e}^{-x}) = 1+\frac{x^2}{2!}+\frac{x^4}{4!}+\ldots+\frac{x^{2n}}{(2n)!}+\ldots$$

$$\sinh x \equiv \frac{1}{2}(\mathrm{e}^x-\mathrm{e}^{-x}) = x+\frac{x^3}{3!}+\frac{x^5}{5!}+\ldots+\frac{x^{2n-1}}{(2n-1)!}+\ldots$$

Relationships between hyperbolic functions can be found from trig identities by changing the sign of any term that contains the product of two sines (Osborn's Rule).

$$\frac{\mathrm{d}}{\mathrm{d}x}(\cosh x) = \sinh x \text{ and } \frac{\mathrm{d}}{\mathrm{d}x}(\sinh x) = \cosh x$$

$$\int \cosh x\,\mathrm{d}x = \sinh x+K \quad\text{and}\quad \int \sinh x\,\mathrm{d}x = \cosh x+K$$

$$\operatorname{arcosh} x \equiv \pm\ln[x+\sqrt{(x^2-1)}]$$

$$\operatorname{arsinh} x \equiv \ln[x+\sqrt{(x^2+1)}]$$

$$\frac{\mathrm{d}}{\mathrm{d}x}(\operatorname{arcosh} x) = \frac{\pm 1}{\sqrt{(x^2-1)}} \quad\text{and}\quad \int \frac{1}{\sqrt{(x^2-1)}}\mathrm{d}x = \operatorname{arcosh} x+K$$

$$\frac{\mathrm{d}}{\mathrm{d}x}(\operatorname{arsinh} x) = \frac{1}{\sqrt{(x^2+1)}} \quad\text{and}\quad \int \frac{1}{\sqrt{(x^2+1)}}\mathrm{d}x = \operatorname{arsinh} x+K$$

$$\frac{\mathrm{d}}{\mathrm{d}x}(\operatorname{artanh} x) = \frac{1}{1-x^2} \quad\text{and}\quad \int \frac{1}{1-x^2}\mathrm{d}x = \operatorname{artanh} x+K.$$

Multiple Choice Questions on this work are given at the end of Chapter 10.

MISCELLANEOUS EXERCISE 9

1) The function f is periodic with period π and

$$f(x) = \sin x \quad\text{for}\quad 0\leqslant x\leqslant \pi/2,$$

$$f(x) = 4(\pi^2-x^2)/(3\pi^2) \quad\text{for}\quad \pi/2<x<\pi.$$

Sketch the graph of $f(x)$ in the range $-\pi\leqslant x\leqslant 2\pi$. (U of L)

2) The function $f(x)$ is defined for $0 \leqslant x \leqslant 2$ by

$$f(x) = x \quad \text{for} \quad 0 \leqslant x \leqslant 1,$$
$$f(x) = (2-x)^2 \quad \text{for} \quad 1 < x \leqslant 2.$$

Sketch the graph of this function for $0 \leqslant x \leqslant 2$ and find

$$\int_0^2 f(x)\,dx.$$ (U of L)p

3) The function f is defined by

$$f(x) = \sin x \quad \text{for} \quad x \leqslant 0,$$
$$f(x) = x \quad \text{for} \quad x > 0.$$

Sketch the graphs of $f(x)$ and its derivative $f'(x)$ for $-\pi/2 < x < \pi/2$ and decide whether the functions f and f' are continuous at $x = 0$ or not.

(U of L)

4) The function f is periodic with period 3 and

$$f(x) = \sqrt{(9-4x^2)} \quad \text{for} \quad 0 < x \leqslant 1.5,$$
$$f(x) = 2x - 3 \quad \text{for} \quad 1.5 < x \leqslant 3.$$

Sketch the graph of $f(x)$ in the range $-3 \leqslant x \leqslant 6$.

(U of L)

5) Defining $\ln x^n$ as $\int_1^{x^n} \frac{1}{t}\,dt$, for $x > 0$, use the substitution $t = u^n$ to prove that $\ln x^n = n \ln x$.

By considering the area under the graph of $y = \frac{1}{t}$ from $t = 1$ to $t = 1 + x$, or otherwise, show that, for $x > 0$,

$$\frac{x}{1+x} < \ln(1+x) < x$$

and deduce that, as x decreases to zero, $\frac{1}{x}\ln(1+x)$ tends to 1.

A periodic function is defined by

$$\begin{cases} f(x) = \dfrac{1}{x}\ln(1+x) & \text{for} \quad 0 < x \leqslant 1, \\ f(x+1) = f(x) & \text{for all } x. \end{cases}$$

Sketch the graph $y = f(x)$ for values of x from -2 to 2.

(U of L)

6) (a) By making the substitution $t = 1/u$ in the integral $\int_1^x \frac{dt}{t}$, prove that, for $x > 0$, $\ln(1/x) = -\ln x$.

(b) The function f is such that $f(x + \pi) = f(x)$ for all values of x. In the interval $0 \leqslant x < \pi$, $f(x) = x - \sin x$. Sketch the curve $y = f(x)$ for $-2\pi \leqslant x \leqslant 2\pi$, and state all the values of x for which the function f is discontinuous. Evaluate the integrals

(i) $\displaystyle\int_{-\pi/2}^{\pi/2} f(x)\,dx$, (ii) $\displaystyle\int_{0}^{3\pi/2} f(x)\,dx$. (U of L)

7) Defining $\ln t$ for $t > 1$ as $\displaystyle\int_1^t \frac{dx}{x}$, prove that

$$1 - \frac{1}{t} < \ln t < t - 1. \qquad \text{(U of L)p}$$

8) (a) By the substitution $x = u^n$, show that

$$\int_1^{a^n} \frac{1}{x}\,dx = n\int_1^a \frac{1}{x}\,dx.$$

(b) Defining e by the equation

$$\int_1^e \frac{1}{x}\,dx = 1,$$

show that

$$\int_1^t \frac{1}{x}\,dx = \log_e t$$

and

$$\int_{t_2}^{t_1} \frac{1}{x}\,dx = \log_e\left(\frac{t_2}{t_1}\right).$$

9) Define $\ln x$ and deduce from the definition that

$$\ln\frac{xy}{z^2} = \ln x + \ln y - 2\ln z. \qquad \text{(AEB'71)p}$$

10) Define $\operatorname{cosech} x$ and $\coth x$ in terms of exponential functions and from your definitions prove that

$$\coth^2 x \equiv 1 + \operatorname{cosech}^2 x. \qquad \text{(AEB'73)p}$$

11) Find the minimum value of $(5\cosh x + 3\sinh x)$. (U of L)

12) Solve, for real x, the equation

$$5\cosh x - 3\sinh x = 5. \qquad \text{(U of L)}$$

13) Find

(a) $\displaystyle\int \frac{dx}{\sqrt{(1+9x^2)}}$, (b) $\displaystyle\int \sinh^2 3x\,dx$. (U of L)

14) If $y = (\operatorname{arcosh} x)^2$ prove that $(x^2-1)\left(\dfrac{dy}{dx}\right)^2 = 4y$. (U of L)

15) Find the possible values of $\sinh x$ if

$$\begin{vmatrix} \cosh x & -\sinh x \\ \sinh x & \cosh x \end{vmatrix} = 2. \qquad \text{(U of L)}$$

16) Solve the equation $3\operatorname{sech}^2 x + 4\tanh x + 1 = 0$. (AEB '71)p

17) If $x = \cosh\theta$ and $y = \sinh\theta$, obtain $\dfrac{dy}{dx}$ in terms of the parameter θ. Sketch the graphs of y and $\dfrac{dy}{dx}$ regarded as functions of θ. (U of L)

18) Sketch the graphs of $y = \text{arsinh}\, x$ and $y = \text{arcosh}\,(x+2)$
Find the coordinates of the point of intersection of the two curves.

19) (a) State the value of $\tanh x$ in terms of e^x and e^{-x}.

Prove that (i) $\dfrac{2\tanh x}{1+\tanh^2 x} \equiv \tanh 2x$

(ii) $\text{artanh}\, x \equiv \dfrac{1}{2}\ln\dfrac{1+x}{1-x}$

If $\tanh 2y = -\frac{4}{5}$, show that $y = -\frac{1}{2}\ln 3$ and find the value of $\tanh y$.

(b) Evaluate $\displaystyle\int_3^5 \frac{1}{\sqrt{(x^2-9)}}\,dx$ and $\displaystyle\int_0^1 \text{arsinh}\, x\, dx$ (JMB)

20) State the expansions of $\cosh(x+y)$ and $\sinh(x+y)$ in terms of hyperbolic functions of x and of y. Hence, or otherwise, express $\cosh 2x$ and $\cosh 3x$ in terms of $\cosh x$.
The three values of $\cosh x$ given by the equation

$$a\cosh 3x + b\cosh 2x = 0$$

where a and b are non-zero, are y_1, y_2 and y_3.
Show that $y_1y_2 + y_2y_3 + y_3y_1$ is independent of a and b. (JMB)

21) Show that the curve $y = \cosh 2x - 4\sinh x$ has just one stationary point and find its coordinates, giving the x-coordinate in logarithmic form. Determine the nature of the stationary point. (JMB)

22) Prove that

$$16\sinh^2 x\cosh^3 x \equiv \cosh 5x + \cosh 3x - 2\cosh x.$$

Hence or otherwise evaluate:

$$\int_0^1 16\sinh^2 x\cosh^3 x\, dx$$

giving your answer in terms of e. (JMB)

23) Show that $\dfrac{1+\tanh^2 x}{1-\tanh^2 x} \equiv \cosh x$.

By means of the substitution $t \equiv \tanh x$, or otherwise, find the indefinite integral $\int \text{sech}\, 2x\, dx$. (JMB)

24) Prove that $\text{arsinh}\, x \equiv \ln[x + \sqrt{(1+x^2)}]$, and draw a rough graph of the function.
Prove that $\text{arsinh}\,\frac{3}{4} + \text{arsinh}\,\frac{5}{12} = \text{arsinh}\,\frac{4}{3}$. (O)

25) Express arcosech x in logarithmic form.
Solve the equation arcosech $x + \ln x = \ln 3$. (JMB)

26) By using the power series expansions for $\cosh x$ and $\cos x$ show that, if powers of x higher than the fifth are negligible
(a) $\cosh x - \cos x \simeq x^2$
(b) $\operatorname{sech} x - \sec x \simeq -x^2$. (JMB)

27) Solve the equations

$$\cosh x - 3 \sinh y = 0,$$
$$2 \sinh x + 6 \cosh y = 5,$$

giving answers in logarithmic form. (AEB '73)p

28) (a) Find $\displaystyle\int \frac{dx}{\sqrt{(x^2 - 2x + 10)}}$.

(b) Find $\displaystyle\int \frac{dx}{x^2 - 2x + 10}$. (U of L)

29) Prove that arsinh $x = \ln \{x + \sqrt{(x^2 + 1)}\}$.

Show that $\dfrac{d}{dx}(\text{arsinh } x) = \dfrac{1}{\sqrt{(x^2 + 1)}}$.

Evaluate $\displaystyle\int_1^8 \frac{1}{\sqrt{(x^2 - 2x + 2)}}\,dx$ expressing your answer as a natural logarithm.

Show that $\displaystyle\int_1^2 \frac{3}{\sqrt{(x^2 - 2x + 2)}}\,dx = \int_1^8 \frac{1}{\sqrt{(x^2 - 2x + 2)}}\,dx$. (JMB)

30) Define $\cosh x$ and $\sinh x$ and hence prove that

$$\frac{1}{\cosh 2x + \sinh 2x} \equiv \cosh 2x - \sinh 2x.$$

Hence, or otherwise, show that

$$\int_0^1 \frac{dx}{\cosh 2x + \sinh 2x} = \tfrac{1}{2}(1 - e^{-2}).$$ (AEB'77)p

31) Find the area enclosed by the curve $y = \cosh x$, the ordinate at $x = \ln 2$ the x axis and the y axis. Find also the volume obtained when this area is rotated completely about the x axis. (AEB'74)p

32) Prove that $\sinh^{-1}x \equiv \ln [x + \sqrt{(1 + x^2)}]$ and write down a similar expression for $\cosh^{-1}x$.
If $2 \cosh y - 7 \sinh x = 3$ and $\cosh y - 3 \sinh^2 x = 2$, find the real values of x and y in logarithmic form. (AEB'75)p

33) (a) Show graphically that the equation

$$\operatorname{arsinh} x = \operatorname{arsech} x$$

has only one real root. Prove that this root is $\{(\sqrt{5}-1)/2\}^{\frac{1}{2}}$.

(b) Prove that

$$\int_{\frac{4}{5}}^{1} \operatorname{arsech} x \, dx = 2 \arctan 2 - \frac{\pi}{2} - \frac{4}{5} \ln 2.$$

CHAPTER 10

SOME CURVES AND THEIR PROPERTIES

THE CYCLOID

Some curves arise naturally in physical situations. Consider, for instance, the curve known as a cycloid. This is the path traced out by a marked point on the rim of a disc when the disc is rolled along a plane. The parametric equations of a cycloid can be found as follows.
Take P as the marked point on the rim of a disc of radius a and let O be the fixed point on the plane which is initially in contact with P.

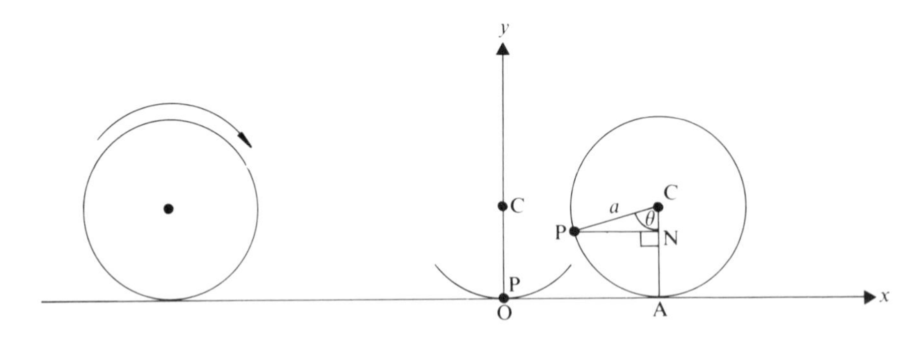

When the disc has rotated through an angle θ, the point of contact is A and the distance OA is equal to the length of the arc AP, i.e. $a\theta$.
Now $\quad PN = a \sin \theta \quad$ and $\quad CN = a \cos \theta$

so the coordinates of P relative to axes along and perpendicular to the plane as shown are

$$x = \text{OA} - \text{PN} = a\theta - a \sin \theta$$

$$y = \text{AC} - \text{CN} = a - a \cos \theta.$$

Thus the parametric equations of the cycloid are

$$x = a(\theta - \sin \theta), \qquad y = a(1 - \cos \theta).$$

It is not easy to eliminate θ from these equations so a cycloid cannot be represented simply in Cartesian form. Consequently its graph is drawn by taking a set of values of θ and plotting the corresponding values of x and y. The resulting curve is shown below and it is seen to be periodic with a period $2\pi a$.

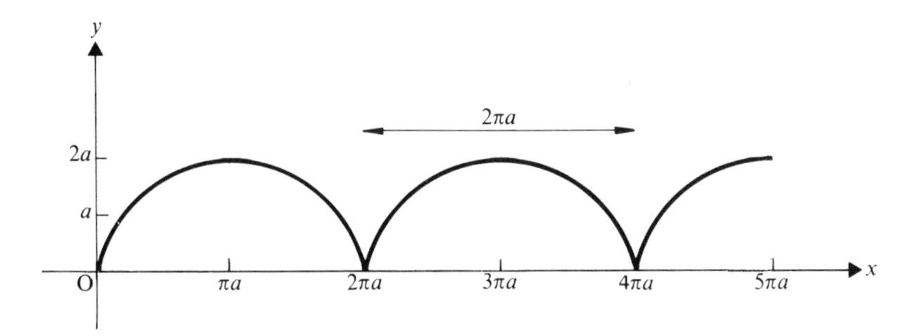

Other curves whose sketches the reader may find useful to remember are,
(a) the semi-cubical parabola,

$$y^2 = x^3;$$

(b) the astroid,

$$x = a \cos^3 \theta, \quad y = a \sin^3 \theta.$$

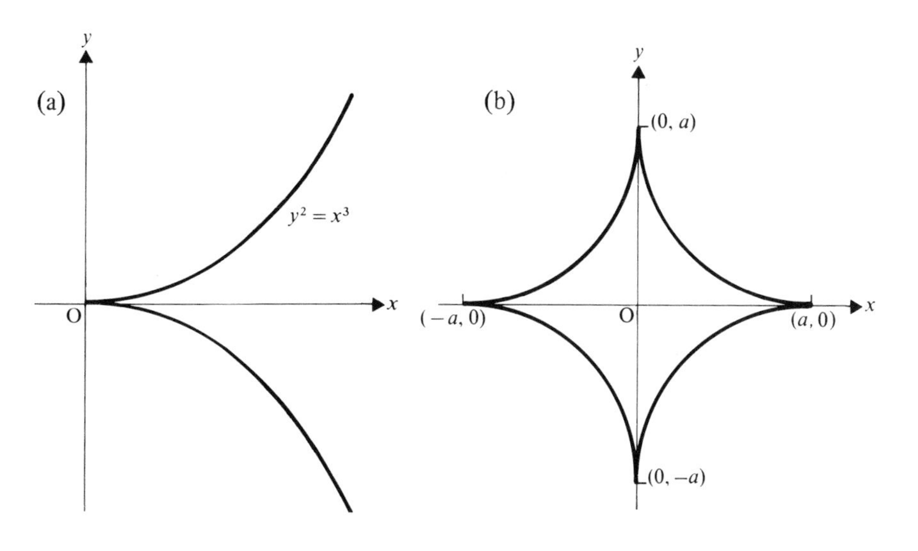

Note that the part of the curve in the first quadrant corresponds to values of θ in the range $0 < \theta \leqslant \frac{\pi}{2}$.

These curves will feature in further work both in this chapter and in Chapter 13.

THE TANGENT TO A CURVE AT THE ORIGIN

Consider first a curve that passes through the origin and which has a second order Cartesian equation. The general equation of such a curve is

$$ax^2 + 2hxy + by^2 + 2gx + 2fy = 0.$$

The gradient of the curve at any point can be found by differentiating this equation with respect to x,

i.e. $$2ax + 2h\left(y + x\frac{dy}{dx}\right) + 2by\frac{dy}{dx} + 2g + 2f\frac{dy}{dx} = 0$$

so $$\frac{dy}{dx} = -\frac{ax + hy + g}{hx + by + f}.$$

The gradient at the origin, where x and y are both zero, is therefore $-\frac{g}{f}$, and the equation of the tangent to the given curve at the origin

$$y = -\frac{g}{f}x \quad \Rightarrow \quad gx + fy = 0.$$

This result can be obtained more simply by using the linear approximation to the given equation when x and y are both very small. In these circumstances, x^2, y^2 and xy are negligible compared with x and y, so the equation of the curve approximates to

$$2gx + 2fy = 0$$

for points very near to the origin.

Thus the equation of the tangent at the origin is

$$gx + fy = 0.$$

For example, the equation of the parabola $y^2 = 4ax$ approximates to $4ax = 0$ near to the origin (as y^2 is insignificant compared with $4ax$ when x and y are both very small).

So the tangent at the origin has equation

$$4ax = 0$$

i.e. $$x = 0$$

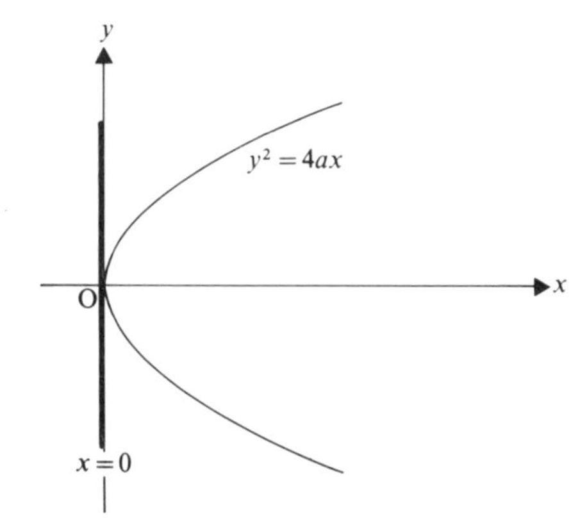

Similarly, to find the equation of the tangent at the origin to the circle

$$x^2 + y^2 - 2x - 4y = 0$$

we delete x^2 and y^2 giving

$$x + 2y = 0.$$

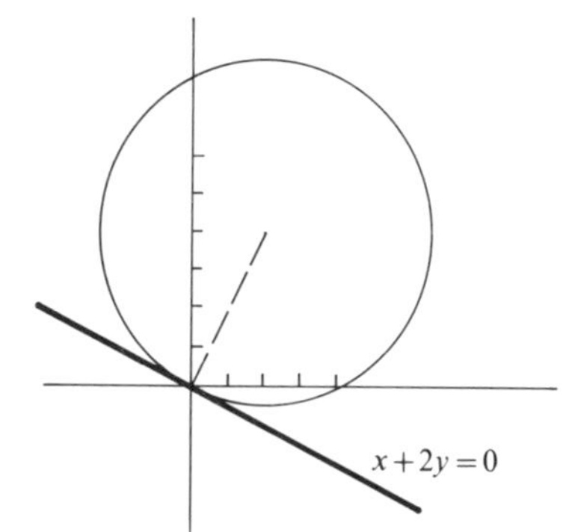

Curves whose equations are of higher degree in x and/or y can be treated in the same way:
e.g., the tangent at O to the curve $x^2 + y^3 - 2y^2 + 3y - 2x = 0$ is $3y - 2x = 0$ (deleting all non-linear terms).
If the equation of a particular curve contains no linear terms, the equation(s) of the tangent(s) at the origin is/are found by deleting all terms except those whose power is least.
For example, the curve with equation $2x^3 + y^3 + 4x^2 = 9y^2$ has two tangents at the origin, whose equations are

$$4x^2 = 9y^2 \Rightarrow 3y = \pm 2x.$$

Note. The equation of the tangent at the origin, to a curve whose equation is not algebraic, can sometimes be found by using a series expansion,

e.g. $$y = x + \sin x \Rightarrow y = x + \left(x - \frac{x^3}{3!} + \dots\right)$$

which approximates to $y = 2x$ when x is small, so $y = 2x$ is the equation of the tangent at the origin. Otherwise the gradient of the tangent at O can be found from the value of $\frac{dy}{dx}$ when $x = 0$ and $y = 0$.

INFLEXION

A point on a curve where the sense of the curvature changes, is called a *point of inflexion*,

e.g.

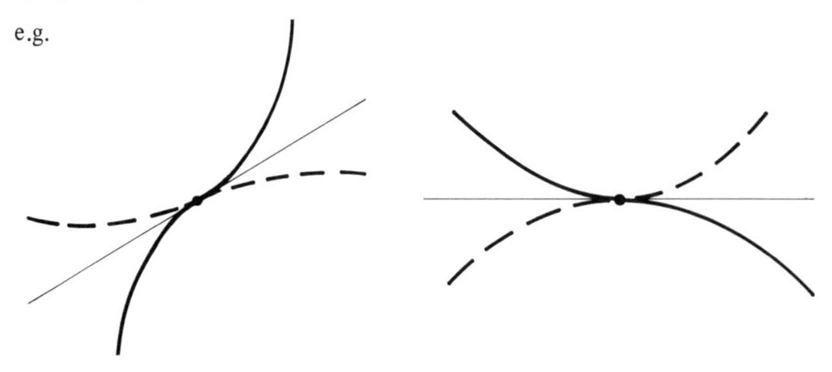

Note that, at any point of inflexion, the curve crosses the tangent drawn at that point.
On one side of a point of inflexion, P, a curve is concave and on the other side it is convex. So on one side of P the gradient is increasing as x increases, while on the other side of P the gradient decreases as x increases,

e.g.

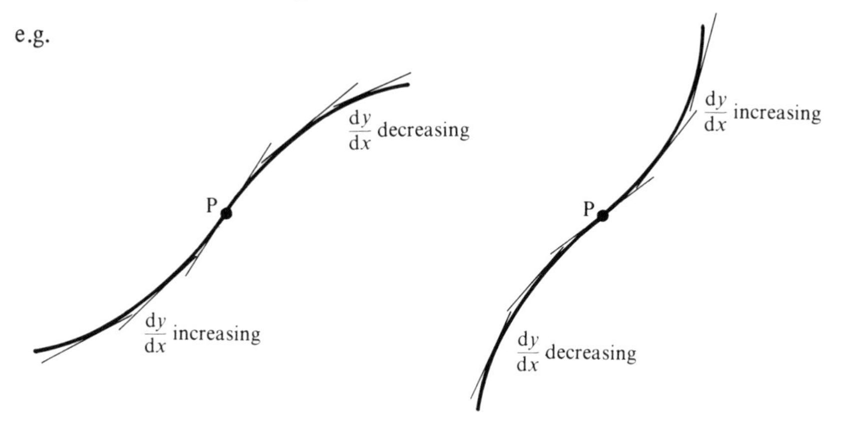

Now if the gradient is increasing, $\dfrac{d}{dx}\left(\dfrac{dy}{dx}\right) > 0$

and if the gradient is decreasing, $\dfrac{d}{dx}\left(\dfrac{dy}{dx}\right) < 0.$

Therefore, as a curve passes through a point of inflexion, $\dfrac{d}{dx}\left(\dfrac{dy}{dx}\right)$ changes sign,

i.e. P is a point of inflexion $\Rightarrow \dfrac{d^2y}{dx^2} = 0$ at P.

The converse of this statement is not always true however so we *cannot* assume that we have an inflexion whenever $\dfrac{d^2y}{dx^2} = 0.$

For instance, if $y = x^4$, $\dfrac{d^2y}{dx^2} = 12x^2$ so $\dfrac{d^2y}{dx^2} = 0$ when $x = 0.$

But we already know (Vol. 1, Chapter 11) that this curve has a minimum turning point when $x = 0$ and not a point of inflexion.

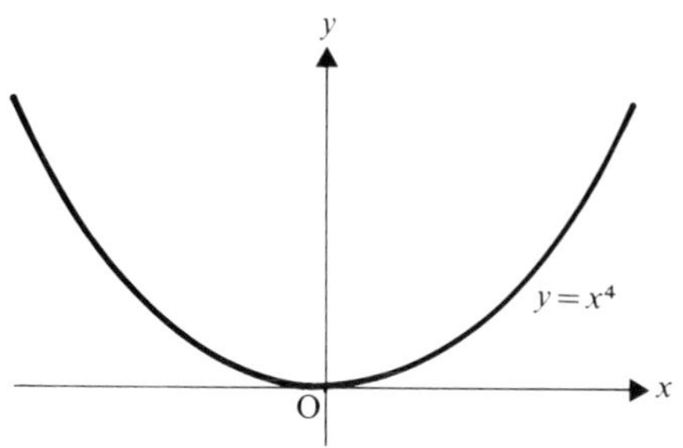

Thus the condition $\dfrac{d^2y}{dx^2} = 0$, although *necessary* is not *sufficient* to define a point of inflexion. The further condition we need is obtained by noting that, because a curve crosses the tangent at a point of inflexion, the *sign* of the gradient *does not change* as the curve passes through this point (whereas at a turning point $\dfrac{dy}{dx}$ *does* change sign) and that $\dfrac{d^2y}{dx^2}$ changes sign.

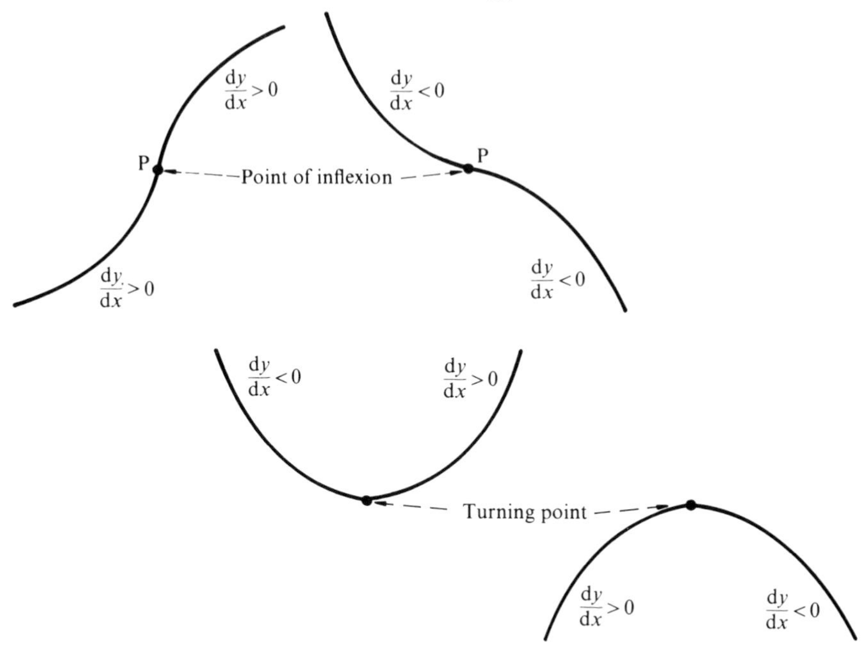

So the *necessary and sufficient* conditions for a point P to be a point of inflexion are that, as the curve passes through P,

$$\frac{d^2y}{dx^2} = 0 \text{ and changes sign}$$

$$\frac{dy}{dx} \text{ does not change sign.}$$

EXAMPLES 10a

1) Prove that the cycloid $x = a(\theta - \sin\theta)$, $y = a(1 - \cos\theta)$
(a) is periodic with period p where $p = 2\pi a$,
(b) is such that y is an even function of x.
Prove that the cycloid has no point of inflexion.

(a) A curve $y = f(x)$ is periodic if $f(x) = f(x + np)$ where p is the period and n is an integer, i.e. if values of x separated by regular intervals give the same value of y.
In this example the value suggested for p is $2\pi a$.
When the parameter is θ,

$$x = a(\theta - \sin\theta)$$

When the parameter is $\theta + 2n\pi$,

$$\begin{aligned} x &= a\{(\theta + 2n\pi) - \sin(\theta + 2n\pi)\} \\ &= a(\theta - \sin\theta) + 2\pi na. \end{aligned}$$

So when θ increases by regular integral multiples of 2π, x increases by regular intervals of $2\pi a$.
When the parameter is $\theta + 2n\pi$,

$$\begin{aligned} y &= a\{1 - \cos(\theta + 2n\pi)\} \\ &= a(1 - \cos\theta) \end{aligned}$$

Hence y takes the value $a(1 - \cos\theta)$ for values of x at regular intervals of $2\pi a$ showing that the cycloid is periodic with a period $2\pi a$.

(b) y is an even function of x if $y = f(x)$ and $f(x) = f(-x)$ for all values of x.

For the cycloid, $x = a(\theta - \sin\theta) \Rightarrow -x = a(-\theta + \sin\theta)$

$$= a\{(-\theta) - \sin(-\theta)\}$$

So for parameters θ and $-\theta$ the values of x are equal in value and opposite in sign.

At the point with parameter θ, $y = a(1 - \cos\theta)$.
At the point with parameter $-\theta$, $y = a\{1 - \cos(-\theta)\}$

$$= a(1 - \cos\theta),$$

i.e. the values of y are equal when x is replaced by $-x$, showing that, for the cycloid, y is an even function of x.

If there is a point of inflexion, then at that point $\dfrac{d^2y}{dx^2} = 0$

$$\frac{dy}{dx} = \frac{a\sin\theta}{a(1-\cos\theta)} = \cot\frac{\theta}{2}$$

$$\frac{d^2y}{dx^2} = \frac{d}{dx}\left(\cot\frac{\theta}{2}\right) = \left(-\frac{1}{2}\operatorname{cosec}^2\frac{\theta}{2}\right)\frac{d\theta}{dx}$$

$$= -\frac{1}{2}\operatorname{cosec}^2\frac{\theta}{2}\Big/ a(1-\cos\theta)$$

$$= -\frac{1}{4a}\operatorname{cosec}^4\frac{\theta}{2}$$

Now $\operatorname{cosec}\dfrac{\theta}{2}$ is never zero so $\dfrac{d^2y}{dx^2}$ cannot be zero
Thus there is no point of inflexion on the cycloid.

2) Show that the cubic curve $y = \dfrac{x^3}{6} - \dfrac{x^2}{2} + x$ has one point of inflexion and give the coordinates of this point.
Write down the equation of the tangent to this curve at the origin and sketch the curve.

At any point of inflexion, $\dfrac{d^2y}{dx^2} = 0$.

For the given curve $\dfrac{dy}{dx} = \dfrac{x^2}{2} - x + 1$

and $\dfrac{d^2y}{dx^2} = x - 1$.

As $\dfrac{d^2y}{dx^2} = 0$ when $x = 1$, there *may* be an inflexion at the point $P(1, \frac{2}{3})$.
Taking values of x close to, and on either side of, this point we can check whether $\dfrac{dy}{dx}$ changes sign at P.

When $x = 0.9$, $\dfrac{dy}{dx} > 0$ and when $x = 1.1$, $\dfrac{dy}{dx} > 0$.

So $\dfrac{dy}{dx}$ does not change sign as the curve passes through $P(1, \frac{2}{3})$ and P is therefore a point of inflexion and at this point $\dfrac{dy}{dx} = \frac{1}{2}$.

Near to the origin, x^3 and x^2 are insignificant compared with x and y, so the equation of the tangent at the origin is

$$y = x$$

Noting also that $y = 0$ for only one real value of x, i.e. $x = 0$, the curve can now be sketched.

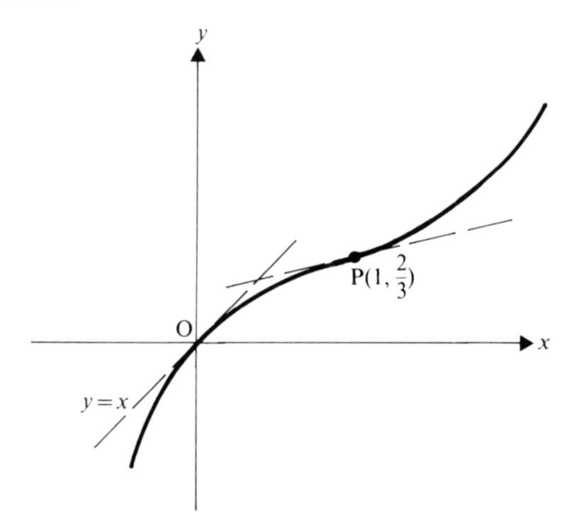

EXERCISE 10a

Write down the equation of the tangent(s) at the origin to each of the following curves, illustrating your result on a diagram.

1) $\dfrac{(x-2)^2}{4} + y^2 = 1$

2) $y = x(x-3)$

3) $(y+1)^2 + (x-1)^2 = 2$

4) $(x-3)^2 - \dfrac{(y-4)^2}{2} = 1$

5) $y^2 = 8x$

6) $x^2 + 2y = 0$

7) $y^3 = x^2 - y^2$

8) $2x^4 + y^4 - 3x^3 = 4y^2 - x^2$.

Determine which of the following six curves contain a point of inflexion, finding the coordinates of this point when it exists.

9) $y = x(x-1)(x-2)$

10) $y = x^4 - 1$

11) $y = x^4 + x^2 + 1$

12) $y = x^3 - 1$

13) $y = x^3$ 14) $x = at^2, \quad y = a(t-1)^3$

15) Determine whether either of the curves $y^3 = x$ and $y^3 = x^2$ is such that
(a) y is an odd function of x,
(b) y is an even function of x,
(c) it has a point of inflexion.
Find the equation of the tangent at the origin to each curve.
Sketch both curves.

16) A cubic curve, $y = f(x)$, has a point of inflexion at the point $(2, -22)$.
The equation of the tangent to this curve at the origin is $y + 3x = 0$.
Determine the equation of the curve and sketch it.

17) The equation of a curve is $y = f(x)$, where $f(x)$ is a quadratic function of x.
Find the equation of the curve if it touches the line $4y = 5x$ at the origin and passes through the point $(4, 21)$.

18) A curve with equation $y = f(x)$ passes through the origin and has a point of inflexion at $(1, \frac{2}{3})$. If $f(x) \equiv x^3 + ax^2 + bx + c$, find the equation of the tangent to the curve at the origin.

INEQUALITIES

For any real number, n, we know that

$$n^2 \geqslant 0 \quad \text{and} \quad -n^2 \leqslant 0.$$

These simple properties can be used in proving the validity of many inequalities, some of which are standard.

The Sum of a Positive Real Number and its Reciprocal Cannot be Less than Two

i.e. $$p + \frac{1}{p} \geqslant 2 \quad \text{if} \quad p > 0.$$

The proof of this quotable property begins by considering the expression

$$p + \frac{1}{p} - 2 \equiv \frac{p^2 + 1 - 2p}{p}$$

$$\equiv \frac{(p-1)^2}{p}$$

But $p > 0$ and $(p-1)^2 \geqslant 0$.

So $$\frac{(p-1)^2}{p} \geqslant 0 \quad \Rightarrow \quad p + \frac{1}{p} - 2 \geqslant 0,$$

i.e. $$p+\frac{1}{p} \geqslant 2 \quad \text{if} \quad p>0.$$

The Geometric Mean of Two Positive Real Numbers Cannot Exceed their Arithmetic Mean

If p and q are positive real numbers we can say

$$p \equiv m^2 \quad \text{and} \quad q \equiv n^2.$$

The geometric mean of p and q is $\sqrt{(pq)} \equiv mn$.
The arithmetic mean of p and q is $\frac{1}{2}(p+q) \equiv \frac{1}{2}(m^2+n^2)$.
We are required to prove that

$$\sqrt{(pq)} \leqslant \tfrac{1}{2}(p+q),$$

i.e. $$mn \leqslant \tfrac{1}{2}(m^2+n^2).$$

To do this we begin by considering

$$\begin{aligned} \tfrac{1}{2}(m^2+n^2)-mn &\equiv \tfrac{1}{2}(m^2+n^2-2mn) \\ &\equiv \tfrac{1}{2}(m-n)^2 \end{aligned}$$

But $$(m-n)^2 \geqslant 0.$$

So $$\tfrac{1}{2}(m^2+n^2)-mn \geqslant 0,$$

i.e. $$\sqrt{(pq)} \leqslant \tfrac{1}{2}(p+q).$$

Note that each of the above proofs began with consideration of an expression obtained by rearranging the required inequality in a form with zero on one side. This practice should be adopted generally when attempting to prove any inequality.

EXAMPLES 10b

1) If x, y and z are unequal real numbers prove that

$$x^2+y^2+z^2 > xy+yz+zx.$$

Consider $$\begin{aligned} &x^2+y^2+z^2-xy-yz-zx \\ &\equiv \tfrac{1}{2}[2x^2+2y^2+2z^2-2xy-2yz-2zx]. \end{aligned}$$

This step is introduced because the aim is to convert our expression into perfect squares, and the product term (e.g. xy) in a perfect square contains a factor of 2.

Then $$\begin{aligned} &\tfrac{1}{2}[2x^2+2y^2+2z^2-2xy-2yz-2zx] \\ &\equiv \tfrac{1}{2}[x^2-2xy+y^2+y^2-2yz+z^2+z^2-2zx+x^2] \\ &\equiv \tfrac{1}{2}[(x-y)^2+(y-z)^2+(z-x)^2]. \end{aligned}$$

But, as x, y and z are unequal, $(x-y)^2>0$, $(y-z)^2>0$ and $(z-x)^2>0$.
So $x^2+y^2+z^2-xy-yz-zx>0$.

RANGE OF VALUES OF A QUADRATIC RATIONAL FUNCTION

A function such as $\dfrac{x^2-x+2}{x^2-x+1}$ can be evaluated for any chosen value of x but it does not follow that the function itself can take all possible values.
Suppose that the function has a value N, so that

$$N = \frac{x^2-x+2}{x^2-x+1}$$

$$\Rightarrow \qquad (N-1)x^2-(N-1)x+(N-2) = 0.$$

For a particular value of N, x can now be calculated provided that

$$(N-1)^2-4(N-1)(N-2)\geqslant 0,$$

i.e. $$(N-1)(7-3N)\geqslant 0,$$

i.e. $$1\leqslant N\leqslant \tfrac{7}{3}.$$

So we see that, regardless of the value of x, the function can never take a value less than 1 or greater than $\frac{7}{3}$. The reader may recall that this method was applied when sketching the graphs of rational functions (Volume 1, Chapter 12).

EXAMPLES 10b (continued)

2) Prove that $\dfrac{2}{27}\leqslant\dfrac{x^2-2x+2}{x^2+3x+9}\leqslant 2.$

This can be proved in two quite different ways.
The first method involves finding the range of possible values of the function $\dfrac{x^2-2x+2}{x^2+3x+9}=N$, say, as illustrated above.
In the second method, we prove separately

that $$\frac{x^2-2x+2}{x^2+3x+9}\geqslant\frac{2}{27} \qquad [1]$$

and $$\frac{x^2-2x+2}{x^2+3x+9}\leqslant 2. \qquad [2]$$

For [1] we consider

$$\frac{x^2-2x+2}{x^2+3x+9}-\frac{2}{27} \equiv \frac{25x^2-60x+36}{27(x^2+3x+9)}$$

$$\equiv \frac{(5x-6)^2}{27[(x+\frac{3}{2})^2+\frac{27}{4}]}$$

But $(5x-6)^2 \geqslant 0$
and $(x+\frac{3}{2})^2 \geqslant 0$ so $(x+\frac{3}{2})^2+\frac{27}{4}>0$.

Hence $$\frac{x^2-2x+2}{x^2+3x+9}-\frac{2}{27} \geqslant 0.$$

Now for [2] we consider

$$\frac{x^2-2x+2}{x^2+3x+9}-2 \equiv -\frac{x^2+8x+16}{x^2+3x+9}$$

$$\equiv -\frac{(x+4)^2}{(x+\frac{3}{2})^2+\frac{27}{4}}$$

But $(x+4)^2 \geqslant 0$ and $(x+\frac{3}{2})^2+\frac{27}{4}>0$.

So $$\frac{x^2-2x+2}{x^2+3x+9}-2 \leqslant 0.$$

Hence $$\frac{x^2-2x+2}{x^2+3x+9} \leqslant 2.$$

GRAPHICAL REPRESENTATION OF INEQUALITIES

A Cartesian equation relating x and y corresponds to a line or a curve in the xy plane such that the coordinates of every point on this locus satisfy the given equation.
We saw in Volume 1 Chapter 4 that a Cartesian inequality relating x and y corresponds to an area in the xy plane such that the coordinates of every point in this area satisfy the inequality. The boundary of the area is the line or curve whose equation is given by replacing the inequality sign by an equality sign.
For example, $y^2>4x$ corresponds to an area in the xy plane bounded by the parabola $y^2=4x$.

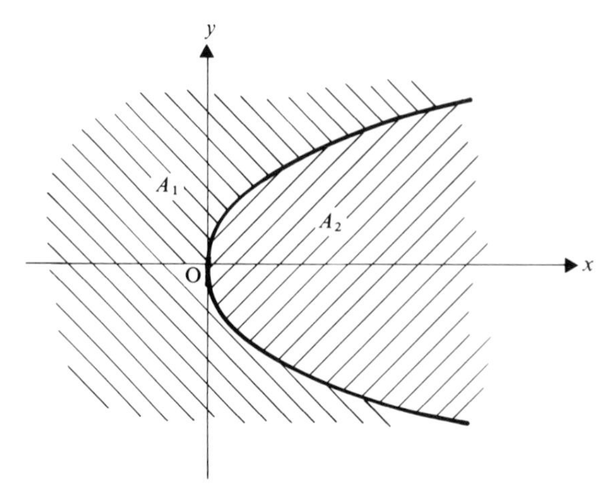

Now the parabola divides the plane into two areas A_1 and A_2, *one* of which represents $y^2 > 4x$. A quick way to decide which is the required area is to take the coordinates of some simple point within one of the two areas and check whether it satisfies the given inequality. In this example, for instance, we could choose the point $(1, 0)$, which is obviously in A_2, and for which

$$y^2 = 0$$

$$4x = 4.$$

So for the point $(1, 0)$, $y^2 \ngtr 4x$, and we deduce that A_2 is *not* the correct area and therefore that A_1 *is* the required area.
Shading the area that is *not* required we have

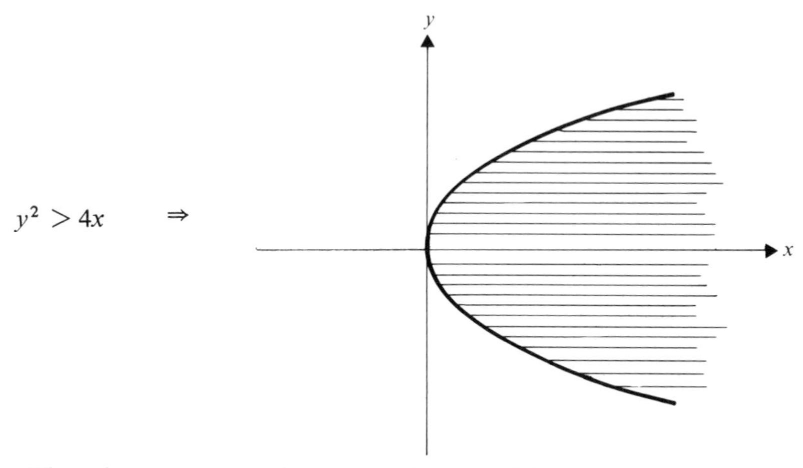

Note. When choosing a particular point within one of the possible areas bounded by a line or curve, the origin is the best choice unless it lies *on* the line or curve.

A different situation arises when the curve forming the boundary of an area has a vertical asymptote.
In this case separate investigation must be carried out, on either side of the asymptote, to determine the appropriate area.

Consider, for example, the area representing the inequality $y < \frac{1}{x}$

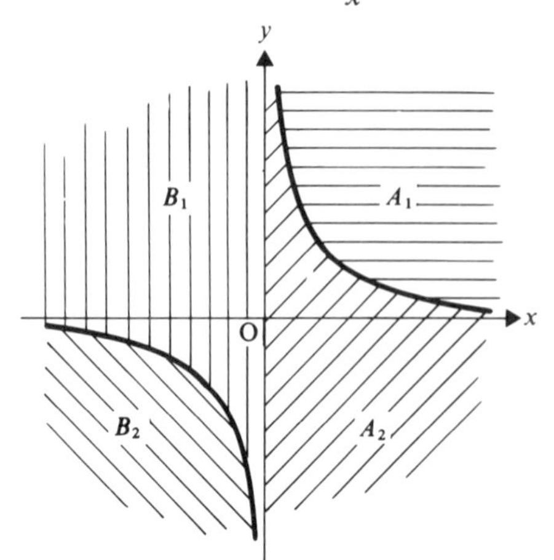

The curve $y = \frac{1}{x}$ has a vertical asymptote $x = 0$, so we check separately the areas bounded by this curve,

where $x > 0$ (i.e. A_1 and A_2) and where $x < 0$ (i.e. B_1 and B_2).

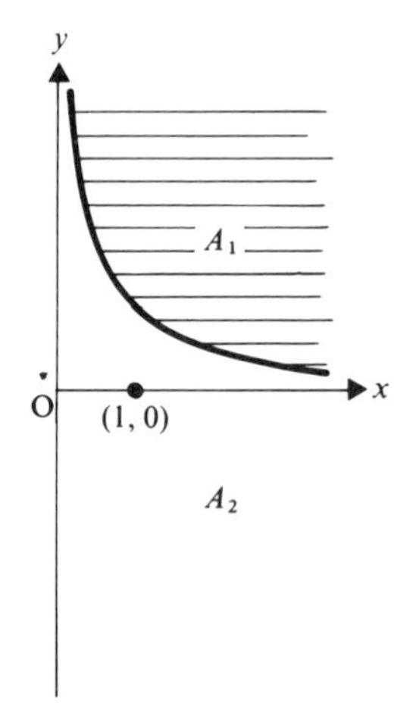

In the region where $x > 0$, the point $(1, 0)$ satisfies the inequality $y < \frac{1}{x}$, showing that A_2 (and not A_1) is the correct area.

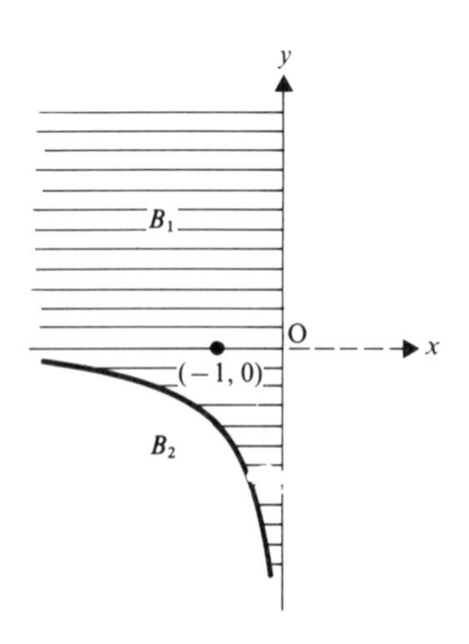

In the region where $x < 0$, the point $(-1, 0)$ does not satisfy the inequality, showing that B_2 (and not B_1) is the correct area.

So, shading the areas *not* required, we have

$$y < \frac{1}{x} \quad \Rightarrow$$

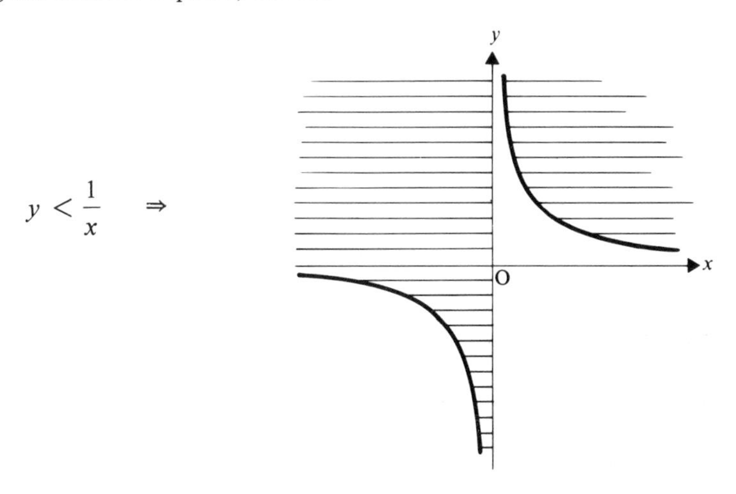

Note that $xy < 1$ is *not* the same inequality as $y < \frac{1}{x}$. This is because, when x is negative, dividing $xy < 1$ by x must be accompanied by a reversal of the inequality sign,

i.e. if $x > 0$, $xy < 1 \Rightarrow y < \frac{1}{x}$

but if $x < 0$, $xy < 1 \Rightarrow y > \frac{1}{x}$

THE SIGN OF $f(xy)$

Considering again the inequality $y^2 > 4x$ we see that it can be written in the form

$$y^2 - 4x > 0$$

or $f(xy) > 0$ where $f(xy) \equiv y^2 - 4x$.

But $y^2 > 4x$ is represented in the xy plane by an area bounded by the parabola $y^2 = 4x$.
So we see that an inequality of the form

$f(xy) > 0$ (i.e. $f(xy)$ has a positive sign)

or $f(xy) < 0$ (i.e. $f(xy)$ has a negative sign)

is represented by an area bounded by the curve $f(xy) = 0$.

Simultaneous Inequalities

The solution of a number of simultaneous inequalities is the set of coordinates (x, y) that satisfy all of the inequalities. The graphical representation of the solution set is therefore obtained by sketching separately the areas that represent each of the individual inequalities and then locating the common area.
This method was used in Volume 1 Chapter 4 for simultaneous linear inequalities and we shall now apply it to those of higher degree.

EXAMPLES 10b (continued)

3) Indicate in a diagram, the area represented by $\dfrac{x^2}{4} + y^2 - 1 < 0.$

If, also, $x^3 - y^2 > 0$, indicate the appropriate area.

The equation $\dfrac{x^2}{4} + y^2 - 1 = 0$ represents an ellipse, so the inequality

$\dfrac{x^2}{4} + y^2 - 1 < 0$ represents an area bounded by the ellipse.

As the origin $(0, 0)$ satisfies the inequality we see that the required area is *inside* the ellipse as shown.

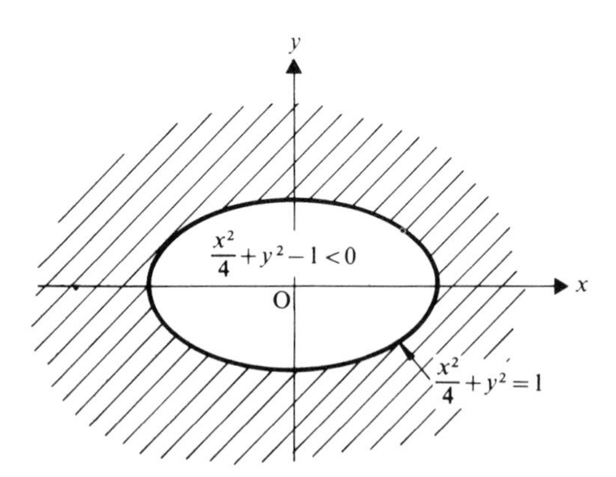

The points that satisfy $x^3 - y^2 > 0$ lie in an area which is bounded by the semi-cubical parabola $y^2 = x^3$. As $(1, 0)$ satisfies $x^3 - y^2 > 0$, the area that represents this inequality is on the right of the curve $y^2 = x^3$ as shown.

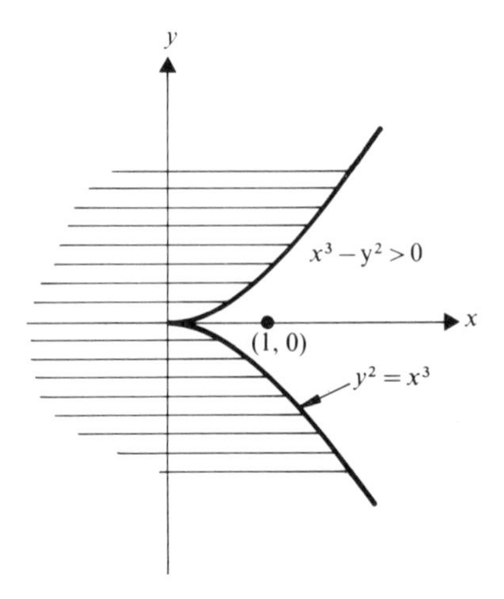

Hence the area common to the two given inequalities is the unshaded region in the diagram below.

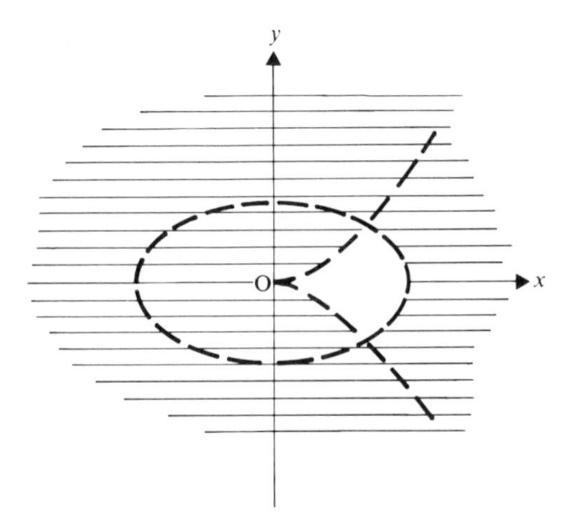

4) Interpret the solution of the inequality $(x^2 + y^2 - 4)(xy - 1) \geqslant 0$ and illustrate your explanation with a diagram.

The product $(x^2 + y^2 - 4)(xy - 1)$ is positive if $x^2 + y^2 - 4$ and $xy - 1$ have the same sign,

i.e. if $x^2 + y^2 - 4 > 0$ *and* $xy - 1 > 0$

or if $x^2 + y^2 - 4 < 0$ *and* $xy - 1 < 0$.

Also the given product is zero if

$$\text{either} \quad x^2 + y^2 - 4 = 0 \quad \text{or} \quad xy - 1 = 0.$$

So the solution of the given inequality is the set of points for which

$$x^2 + y^2 - 4 \geqslant 0 \quad and \quad xy - 1 \geqslant 0$$

and $$x^2 + y^2 - 4 \leqslant 0 \quad and \quad xy - 1 \leqslant 0.$$

The relevant domains in the xy plane are bounded by

the circle $\qquad x^2 + y^2 = 4$

and the rectangular hyperbola $\qquad xy = 1$

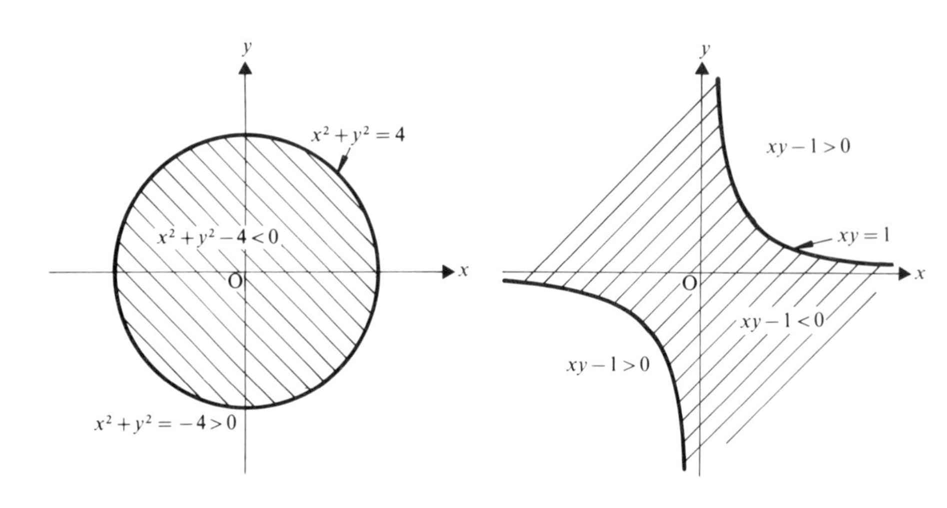

From these separate diagrams it can be seen that the unshaded area in the diagram below represents $\begin{cases} x^2 + y^2 - 4 \geqslant 0 \\ xy - 1 \geqslant 0 \end{cases}$

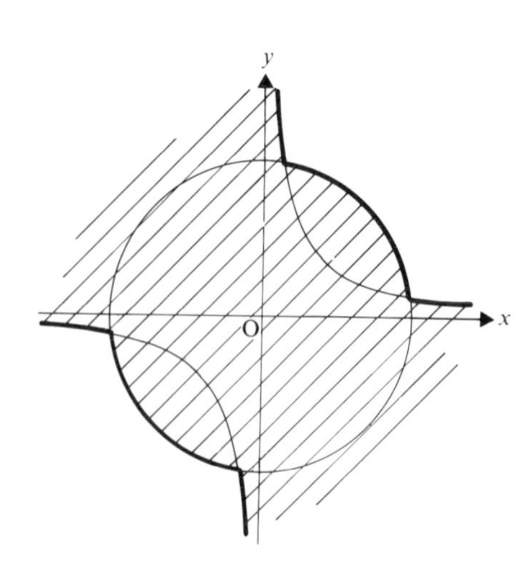

similarly $\begin{cases} x^2 + y^2 - 4 \leqslant 0 \\ xy - 1 \leqslant 0 \end{cases} \Rightarrow$

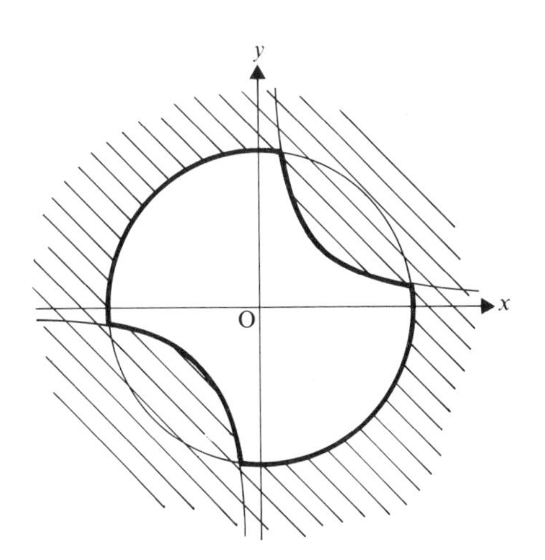

So the solution of the inequality

$$(x^2 + y^2 - 4)(xy - 1) \geqslant 0$$

is represented by the unshaded area shown below.

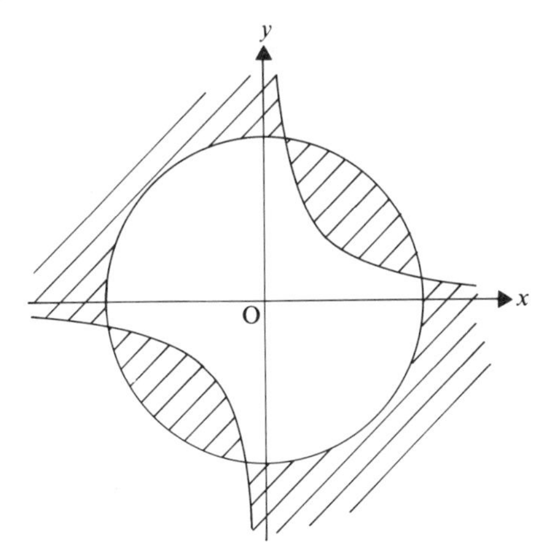

5) If $f(xy) \equiv (x^2 - y^2 - 1)(y^2 - 4x)$ indicate the region in the xy plane where the sign of $f(xy)$ is negative.

If the sign of $f(xy)$ is negative, then $(x^2 - y^2 - 1)(y^2 - 4x) < 0$.

In this case

either $\qquad x^2 - y^2 - 1 < 0$ *and* $y^2 - 4x > 0$

or $\qquad x^2 - y^2 - 1 > 0$ *and* $y^2 - 4x < 0$.

Considering $x^2 - y^2 - 1$ and shading the region where $x^2 - y^2 - 1 < 0$ we have

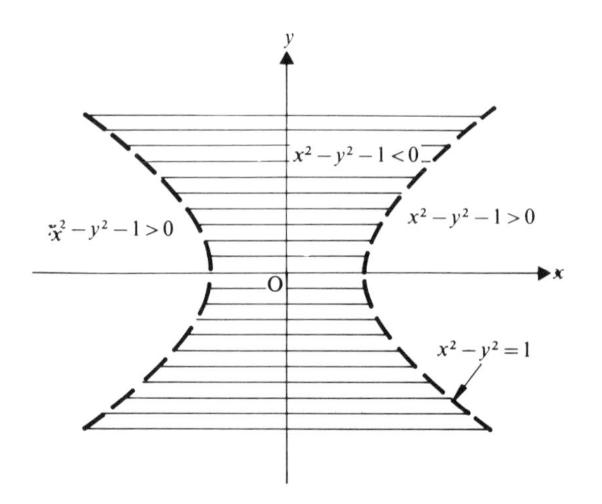

Similarly considering $y^2 - 4x$, we have

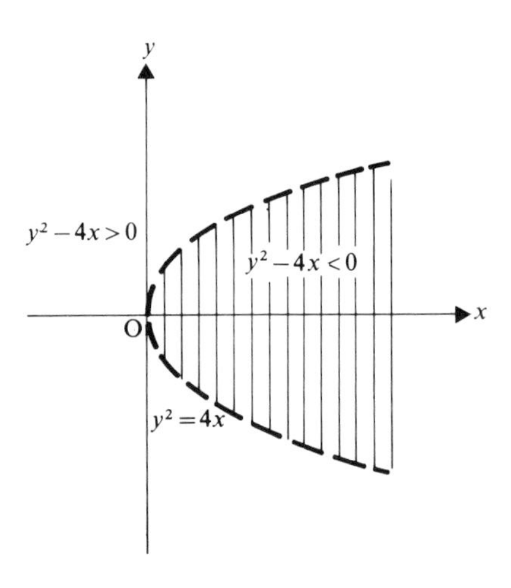

Combining these diagrams shows that

$$\left.\begin{array}{r} x^2 - y^2 - 1 < 0 \\ \text{and} \quad y^2 - 4x > 0 \end{array}\right\} \text{ corresponds to the unshaded area in diagram (i).}$$

$$\left.\begin{array}{r} x^2 - y^2 - 1 > 0 \\ \text{and} \quad y^2 - 4x < 0 \end{array}\right\} \text{ corresponds to the unshaded area in diagram (ii).}$$

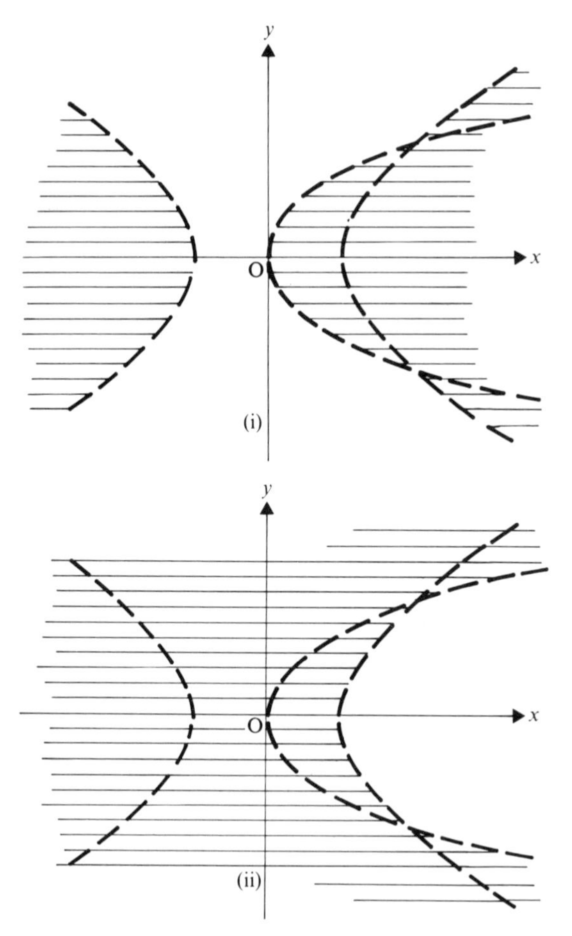

So the domain where $f(xy) \equiv (x^2 - y^2 - 1)(y^2 - 4x)$ has a negative sign corresponding to the unshaded area shown below.

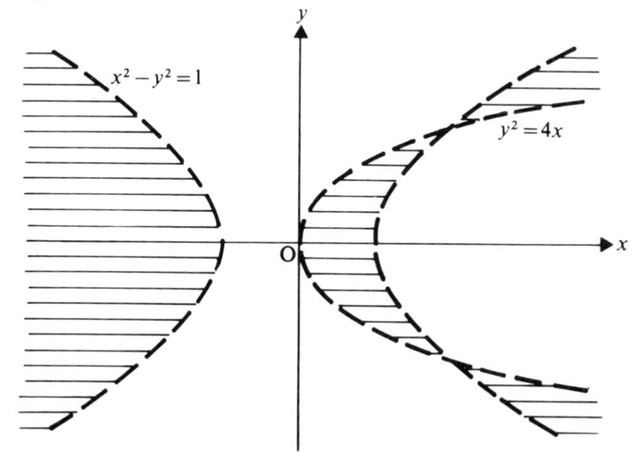

EXERCISE 10b

Prove the following inequalities, for all real values of the variables unless otherwise stated.

1) $x^2 + y^2 - 6y + 9 \geqslant 0$

2) $x^2 + 10x + y^2 + 26 > 0$

3) $2x^3 + 2y^3 \geqslant (x^2 + y^2)(x + y)$ given that $x > 0$ and $y > 0$.

4) $\dfrac{1}{x^2} + \dfrac{1}{y^2} \geqslant \dfrac{8}{(x+y)^2}$ given that $x > 0$ and $y > 0$.

5) $a^2b^2 + b^2c^2 + c^2a^2 \leqslant a^4 + b^4 + c^4$

6) $3x^2 + 3y^2 + 3z^2 \geqslant (x + y + z)^2$

7) $x^2 + y^2 \geqslant \frac{1}{2}(x + y)^2$

8) $-1 \leqslant \dfrac{4x}{x^2 + 4} \leqslant 1$

9) $(a + b)^2 \geqslant 4ab$

10) $a^4 + b^4 + c^4 + d^4 \geqslant 4abcd.$

Interpret in the xy plane the meaning of $f(xy) > 0$ if $f(xy)$ is:

11) $y - x$ 12) $y - |x|$ 13) $|y| - x$ 14) $1 - x^2 + y^2$

15) $y^2 + 2x$

Illustrate the following inequalities graphically:

16) $x^2 + y^2 \leqslant 9$ *and* $y^2 > 4x$

17) $\dfrac{(x-1)^2}{9} + \dfrac{y^2}{4} < 1$ *and* $xy > 1$

18) $(x^2 + y^2 - 4)(x + y) \geqslant 0$

19) $(y - x)(y^2 - x^3) \geqslant 0$

20) $x(y^2 - x) < 0$

21) $(x^2 - 4y^2 - 4)(4x^2 + 9y^2 - 36) \leqslant 0$

22) $(y^2 - 4x)(xy - 1)(y - 1) \leqslant 0.$

Write down the inequalities that define the following shaded areas.

23) 24)

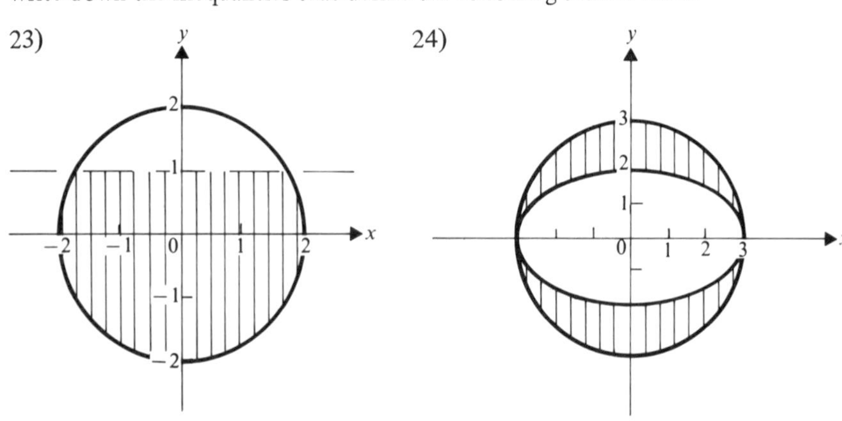

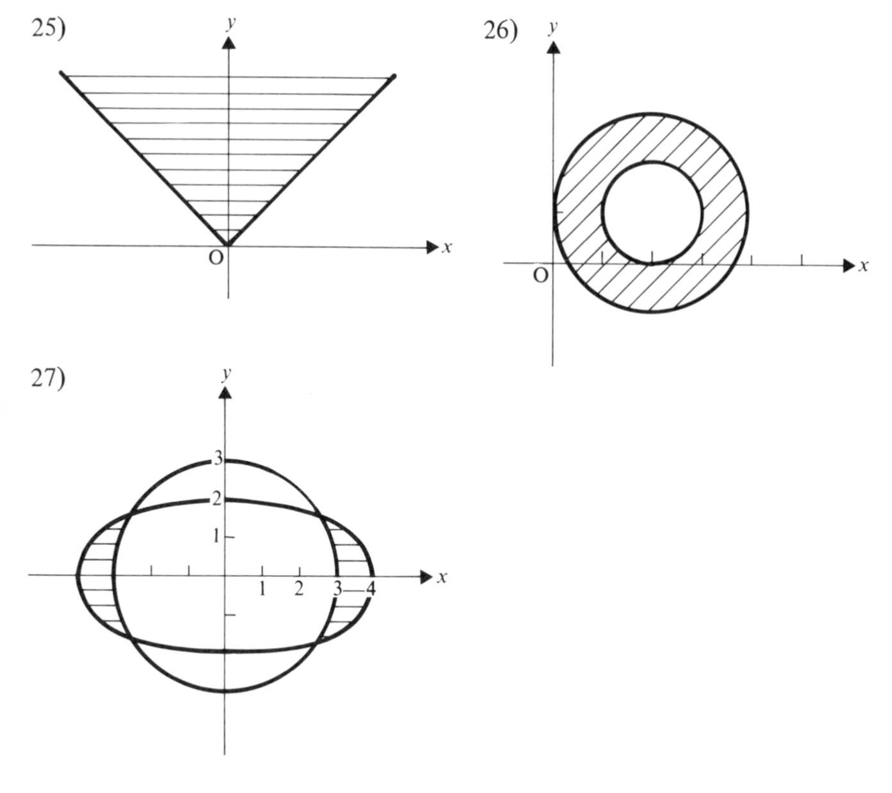

THE USE OF SPECIALISED GRAPH PAPER

In Volume 1 we saw that certain non-linear relationships between two variables could be represented by a linear graph.

For instance, if $y = ab^x$

then $\log y = \log a + x \log b$.

So plotting $\log y$ against x produces a straight line with gradient $\log b$ and whose intercept on the vertical axis is $\log a$.

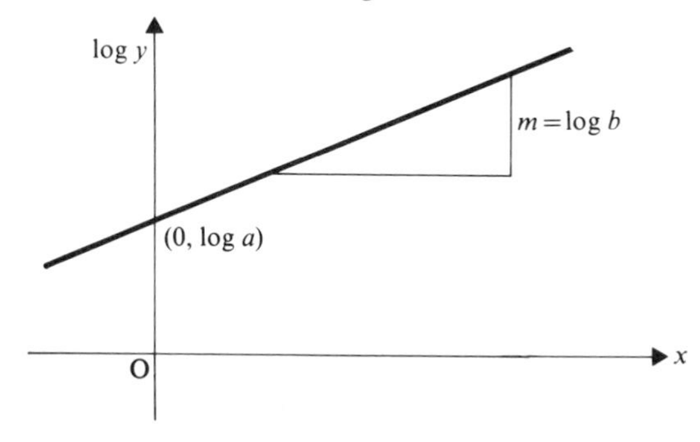

If a set of corresponding values of x and y is given then, in order to produce this graph, it is necessary to find the logarithm of each of the given values of y to give the ordinate of each point. This is a tedious process and can be eliminated by using specially designed graph paper.

Logarithmic Graph Paper

Consider the following values of x and of $\lg x$.

Point	A	B	C	D	E	F	G
x	1	10	50	100	300	600	1000
$\lg x$	0	1	1.699	2	2.477	2.778	3

A line graduated in values of $\lg x$ would display these points as shown below on scale (i)

values of $\lg x$

A B C D E F G

0 1 2 3

scale (i)

On this scale the relative positions of the points depend upon the values of their logarithms. Without moving the points, the graduations can be altered so that each point is *labelled* with the original value of x, as follows,

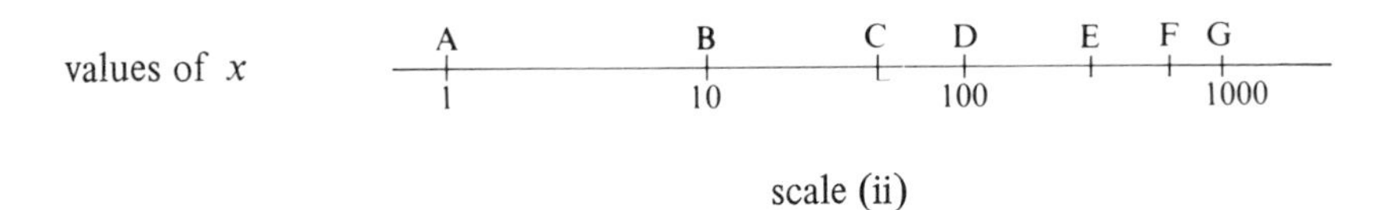

scale (ii)

If scale (ii) were fully graduated in values of x, then a point *marked* on this scale for, say, $x = 20$, would be in the *position* corresponding to $\lg 20$ on scale (i)
i.e. the line graduated in logarithmic divisions, but marked up with ordinary numbers, automatically 'finds the logarithm' of any number plotted on the line. Now the graduations on scale (ii) are not uniform and would be difficult to produce for an individual graph, so graph paper already divided into these logarithmic divisions has been produced, examples of which are shown on page 447.
(A reader familiar with a slide rule will recognise the similarity between its graduations and those of the logarithmic graph paper.)
If the graph paper has logarithmic divisions in one direction and uniform divisions in the other direction it is called log–linear or semi-log, whereas if both sets of divisions are logarithmic it is called log–log graph paper.

Note that the logarithmic scale is marked in *cycles* of 1 to 9, i.e. 1, 2, . . . , 8, 9, 1, 2 . . . . Thus the characteristic of the logarithm is not included in the scale.
In this way the scale can be used for numbers of any size, e.g. a division marked 4 on the log scale can be used for 4, 40, 400, etc., depending upon the range of values to be plotted.
The use of logarithmic paper is demonstrated in the following examples.

EXAMPLES 10c

1) In an experiment, values of a variable v were measured for selected values of another variable t and the results are tabulated below.

t	2	4	5	7	8
v	13	60	99	380	760

It is thought that v and t are connected by a relationship of the form $v = ab^t$. By plotting a linear graph, verify this relationship and find approximate values for a and b.

Since $v = ab^t \iff \log v = \log a + t \log b$ we see that plotting $\log v$ against t should give a straight line graph.
Values of $\log v$ need not be found however if we use log–linear graph paper, plotting values of v in the direction in which the divisions are logarithmic.
The experimental values plotted in this way (see page 436) are seen to lie on a straight line (slight discrepancies being accounted for by experimental error), thus verifying the suggested relationship.
Now from the graph we can find $\log b$ (the gradient) and $\log a$ (the vertical axis intercept, i.e. the value of $\log v$ when $t = 0$).
First let us consider the value of a.
The value of $\log a$ is the ordinate of point $P(0, \log a)$. But, as the vertical scale is graduated in values of a, we can read off directly that $a \simeq 3.8$.
Now to find the gradient, $\log b$, consider the triangle QRS.
The *length* of RS depends upon the *logarithms* of 390 and 14,

i.e. $$RS = \log 390 - \log 14 = \log \frac{390}{14}$$

Hence the gradient of the line is $$\frac{RS}{QR} = \frac{\log 27.86}{5} = 0.29$$

i.e. $$\log b \simeq 0.29 \Rightarrow b \simeq 1.95.$$

Thus, giving values of a and b to the nearest integer, we have $v \simeq 4(2^t)$.

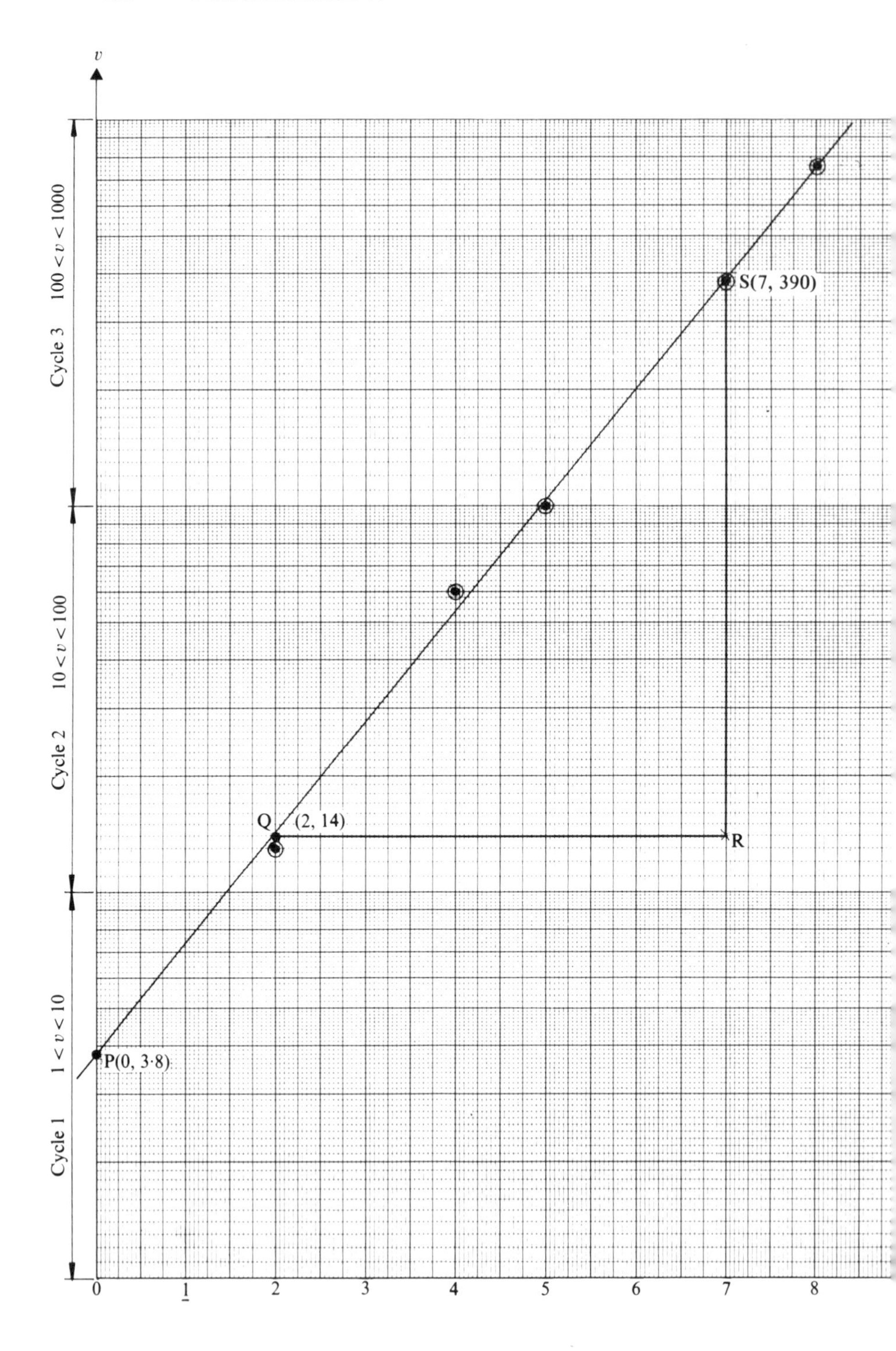

v
Cycle 3 100 < v < 1000
Cycle 2 10 < v < 100
Cycle 1 1 < v < 10
S(7, 390)
Q (2, 14)
R
P(0, 3·8)
0
1
2
3
4
5
6
7
8

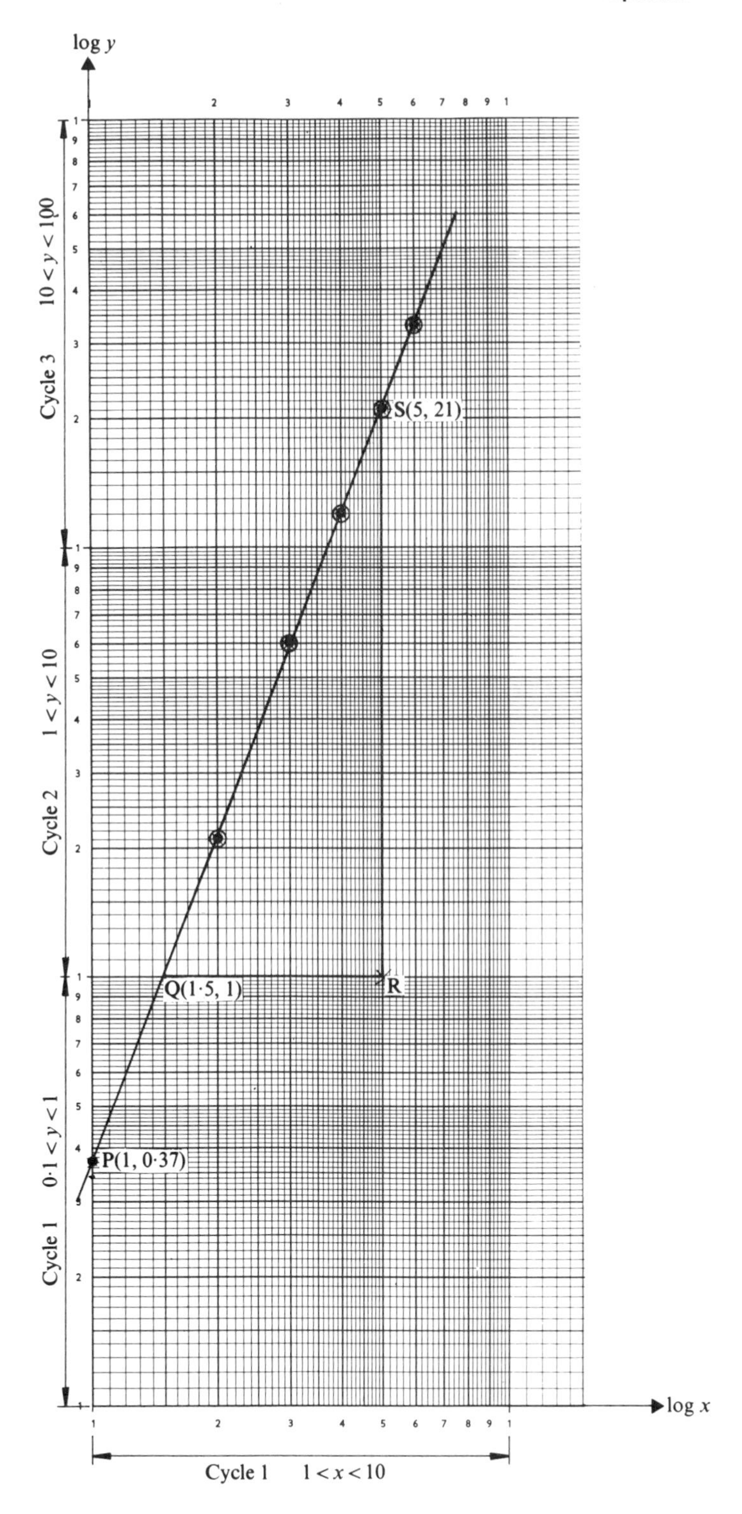
log y
log x
10 < y < 100
Cycle 3
1 < y < 10
Cycle 2
0·1 < y < 1
Cycle 1
S(5, 21)
Q(1·5, 1)
R
P(1, 0·37)
Cycle 1 1 < x < 10

2) Use the following data to show that $x^p = qy^2$ giving values for p and q.

x	2	3	4	5	6
y	2.1	6	12.1	21	33

If $x^p = qy^2$ then

$$p \log x = \log q + 2 \log y$$

or

$$\log y = \frac{p}{2} \log x - \frac{1}{2} \log q.$$

To verify this relationship we would plot $\log y$ against $\log x$ to produce a straight line with a gradient of $\dfrac{p}{2}$
and an intercept of $-\frac{1}{2} \log q$, i.e. $\log q^{-\frac{1}{2}}$.
Using log–log paper this can be done without finding any logarithms.
The straight line graph produced (see page 437) shows that $\log y$ and $\log x$ have a linear relationship.
The gradient can be found by considering any right-angled triangle such as QRS, where

$$\text{RS} = \log 21 - \log 1 = \log 21$$

and

$$\text{QR} = \log 5 - \log 1.5 = \log \tfrac{10}{3}.$$

So the gradient, $\dfrac{p}{2}$, is approximately $\dfrac{\log 21}{\log \frac{10}{3}} = 2.53$

The vertical intercept gives directly the value of $q^{-\frac{1}{2}}$,

i.e.
$$q^{-\frac{1}{2}} \simeq 0.37 \quad \Rightarrow \quad q \simeq 7.3$$

Thus
$$x^{5.06} \simeq 7.3y^2$$

Or, taking integral values for p and q, $x^5 \simeq 7y^2$.

(The value of $\log q^{-\frac{1}{2}}$ is given by the value of $\log y$ when $\log x = 0$ i.e. when $x = 1$. So the vertical intercept is taken on the line $x = 1$. But $\text{P}(1, 0.37)$ gives the value $q^{-\frac{1}{2}}$ directly.)

MULTIPLE CHOICE EXERCISE 10

(Instructions for answering these questions are given on p. xii.)

TYPE I

1) $\int \frac{1}{\sqrt{(x^2-9)}}\,dx$ is:

(a) $\arcsin\frac{x}{3}+K$ (b) $\text{arsinh}\frac{x}{3}+K$ (c) $\arccos\frac{x}{3}+K$

(d) $\text{arcosh}\frac{x}{3}+K$ (e) none of these.

2) $\frac{d}{dx}(\ln\cosh 2x)$ is:

(a) $\frac{1}{\cosh 2x}$ (b) $\frac{2}{\sinh 2x}$ (c) $\frac{1}{2\sinh 2x}$

(d) $2\tanh 2x$ (e) $\frac{1}{2}\tanh 2x$.

3) $f(x) \equiv \sinh x$. The inverse function, $f^{-1}(x)$, is

(a) $\text{cosech}\, x$ (b) $\text{arsinh}\, x$ (c) $\frac{2}{e^x+e^{-x}}$

(d) $\sinh\frac{1}{x}$ (e) none of these.

4) When x is small, $\cosh\frac{x}{2}$ is approximately equal to:

(a) $1+\frac{x^2}{4}$ (b) $\frac{x}{2}+\frac{x^3}{48}$ (c) $1-\frac{x^2}{8}$

(d) $1+\frac{x}{2}$ (e) none of these.

5) The function represented by the graph on the right could be:

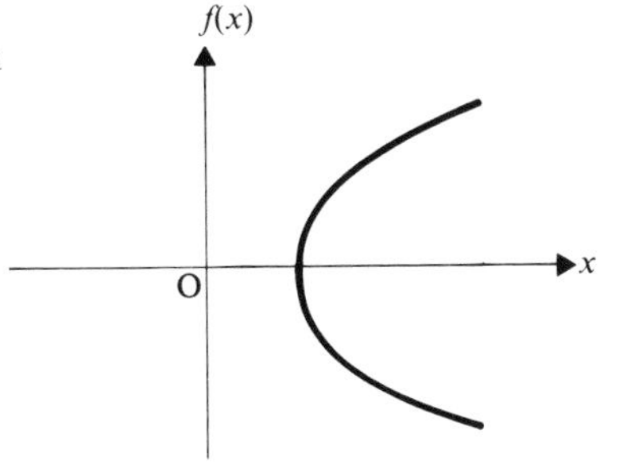

(a) $f(x) \equiv x^2+1$ (b) the inverse of the function x^2+1

(c) $\cosh x$ (d) $\text{arsinh}\, x$ (e) the inverse of the function $\frac{1}{x}$.

6) If x is a real number, $|x+1|<2$ and $2|x-2|>3$ are together equivalent to:

(a) $-3<x<\frac{1}{2}$ (b) $-3<x<-1$ (c) $-1<x<\frac{1}{2}$
(d) $-1<x<1$ (e) $x>3\frac{1}{2}$ or $x<-3$. (U of L)

7) State the line in which an error is first made in the following proof that $a^2+b^2>ab$ where a and b are real.

(a) $(a-b)^2 \equiv a^2+b^2-2ab$

(b) $(a-b)^2 \geqslant 0$

(c) $a^2+b^2 \geqslant 2ab$

(d) $2ab>ab$

(e) $a^2+b^2>ab$.

8)

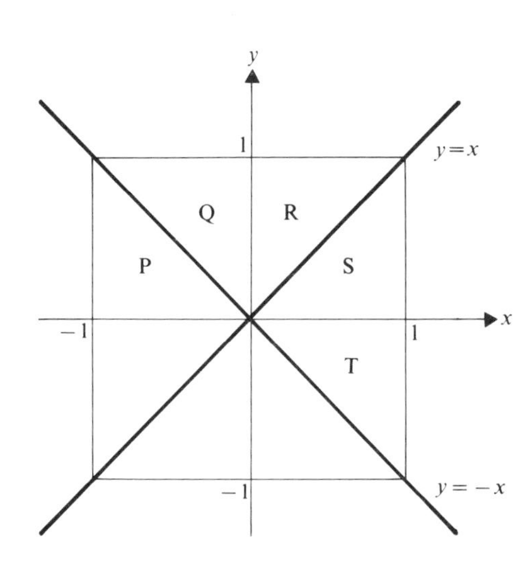

All points (x, y) satisfying both $y>|x|$ and $|y|<1$ lie in the region:
(a) P and S (b) Q and R (c) R (d) S (e) S and T. (U of L)

9) The tangent at the origin to the curve $x^2-y^2+4x+2y=0$ has equation
(a) $y+2x=0$ (b) $2y+x=0$ (c) $y=2x$
(d) $x+y=0$ (e) $x-y=0$. (U of L)

10) If $f(x) \equiv \sin(x-\alpha)\cos(x-\alpha)$, the period of $f(x)$ is
(a) $\pi-\alpha$ (b) 2π (c) $2\pi-2\alpha$ (d) π (e) $\frac{1}{2}(\pi-\alpha)$.

11)

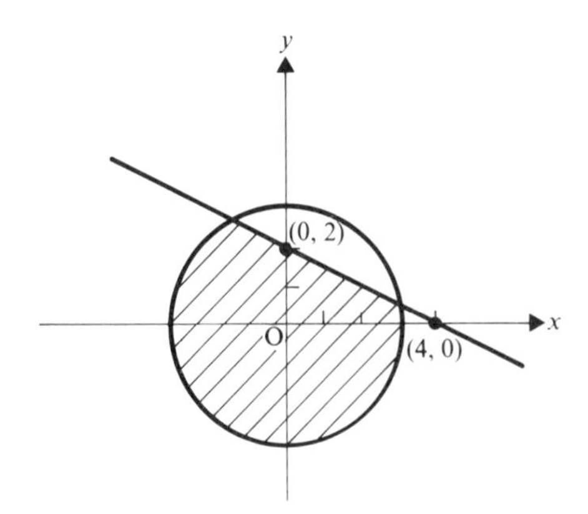

The diagram shows the circle $x^2 + y^2 = 9$ and the straight line $2y + x = 4$. The shaded region consists of those points (x, y) for which:

(a) $x^2 + y^2 - 9 < 2y + x - 4$

(b) $\begin{cases} x^2 + y^2 < 9 \\ 2y + x > 4 \end{cases}$ (c) $\begin{cases} x^2 + y^2 > 9 \\ 2y + x < 4 \end{cases}$

(d) $\begin{cases} x^2 + y^2 > 9 \\ 2y + x > 4 \end{cases}$ (e) $\begin{cases} x^2 + y^2 < 9 \\ 2y + x < 4 \end{cases}$ (U of L)

12) The necessary and sufficient conditions for a point of inflexion are:

(a) $\dfrac{dy}{dx} = 0$ and $\dfrac{d^2y}{dx^2} = 0$

(b) $\dfrac{d^2y}{dx^2} = 0$ and $\dfrac{dy}{dx} \neq 0$

(c) $\dfrac{dy}{dx} = 0$ and $\dfrac{d^2y}{dx^2}$ does not change sign

(d) $\dfrac{d^2y}{dx^2} = 0$ and $\dfrac{dy}{dx}$ does not change sign.

TYPE II

13) The function which is represented by this graph

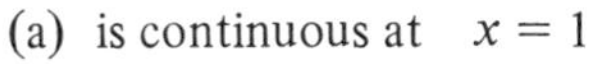

(a) is continuous at $x = 1$
(b) has a discontinuity at $x = 2$
(c) is an even function
(d) is periodic.

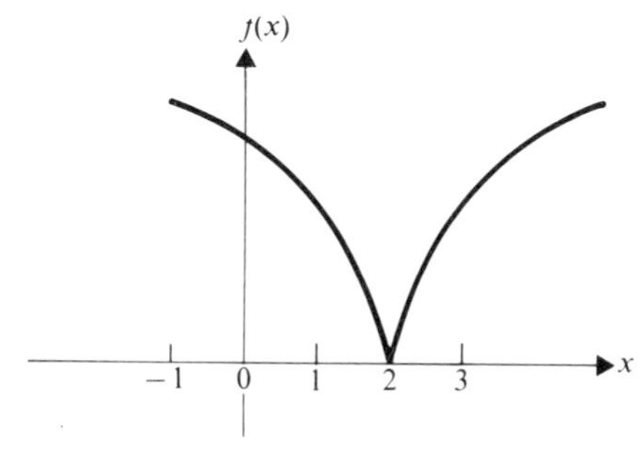

14) If $f(x) \equiv \sinh x \cosh x$

(a) $f(x) = x + \dfrac{4x^3}{3!} + \dfrac{16x^5}{5!} + \dots$ (b) $f'(x) = \cosh 2x$

(c) the graph of $f(x)$ is continuous (d) $f(x) \geqslant 0$.

15) Which of the following identities is/are correct?

(a) $\cosh^2 x + \sinh^2 x \equiv 1$ (b) $1 - \tanh^2 x \equiv \text{sech}^2 x$

(c) $\cosh(x + y) \equiv \cosh x \cosh y - \sinh x \sinh y$

(d) $\sinh(x - y) \equiv \sinh x \cosh y - \cosh x \sinh y$.

16) Which of the following functions is/are even?

(a) $\cosh x$ (b) $\tanh x$ (c) $\text{arcosh}\, x$ (d) $\text{arsinh}\, x$.

17) In which of the following statements is the sign $\Leftrightarrow$ correctly used?

(a) $f(x) = f(-x) \Leftrightarrow f(x)$ is an even function

(b) $y = \cosh x \Leftrightarrow y \geqslant 1$

(c) $f(x) \equiv \sin x \Leftrightarrow f^{-1}(x) \equiv \arcsin x$

(d) $\displaystyle\int f(x)\, dx = \cosh x \Leftrightarrow f(x) \equiv \sinh x$.

18) $\cosh 2x$ is equal to:

(a) $\cosh^2 x - \sinh^2 x$ (b) $\frac{1}{2}(e^{2x} + e^{-2x})$

(c) $\displaystyle\int_0^x \sinh 2x\, dx$ (d) $\dfrac{d}{dx}(\sinh 2x)$.

19) $f(x) \equiv |2x - 1|$

(a) $f(x)$ is continuous

(b) $f'(x)$ is continuous

(c) $y = f(x)$ undergoes a sudden change in gradient at $x = \frac{1}{2}$

(d) f^{-1} exists.

20) $f(x) \equiv \tanh x$

(a) $f(x)$ is periodic (b) $f(x)$ is continuous

(c) $f'(x)$ is discontinuous (d) $f(x)$ is odd.

21) The value of $\begin{vmatrix} \sinh x & \cosh x \\ \cosh x & \sinh x \end{vmatrix}$ is:

(a) 1 (b) $\sinh^2 x - \cosh^2 x$ (c) -1 (d) $\cosh 2x$.

22) The equation of the tangent at the origin to a curve $f(xy) = 0$ can be found if:

(a) $f(xy) \equiv x^2 + y^2 - x + 2y$ (b) $f(xy) \equiv 2x^2 + y^2 + 4x$

(c) $f(xy) \equiv x^2 - y^2 + 4x + 1$ (d) $f(xy) \equiv 2x^2 + 3y^2$.

23) In which of the following statements is the sign $\Leftrightarrow$ correctly used?
(a) $(|x|>|y|) \Leftrightarrow (x^2>y^2)$
(b) $(x^2+y^2>2xy) \Leftrightarrow (x>y)$
(c) $(x^2=4) \Leftrightarrow (x=2)$. (U of L)

TYPE III

24) (a) $f(x)$ is an even function of x.
(b) The graph of $f(x)$ is symmetrical about the x-axis.

25) (a) $y=\cosh x$.
(b) $y \geqslant 1$.

26) (a) $f(4)=f(2)$.
(b) $f(x)$ is a periodic function with period 2.

27) A function $f(x)$ is continuous for $-3 \leqslant x \leqslant 3$.
(a) $\int_{-3}^{3} f(x)\,\mathrm{d}x=0$.
(b) $f(x)$ is an odd function.

28) (a) $\mathrm{e}^p=\mathrm{e}^q$.
(b) $\cosh p=\cosh q$.

29) a, b, c are real numbers.
(a) $ab<ac$.
(b) $b<c$. (U of L)

30) (a) $0<x<1$.
(b) $|2-x|<2$.

31) (a) P is a point of inflexion on a curve $y=f(x)$.
(b) At P, $\frac{\mathrm{d}^2 y}{\mathrm{d}x^2}=0$.

TYPE V

32) The graph of a continuous function contains no sudden change in direction.

33) The graph of $f(x)$ is the reflection of the graph of $f^{-1}(x)$ in the line $y=x$ when f^{-1} is the inverse of f.

34) $f(x) \equiv \sinh x$ is an odd function.

35) If $f(x)$ is continuous in the range $0 \leqslant x \leqslant 1$, then $f'(x)$ is also continuous in the same range.

36) If $f(x)$ is periodic with period $2a$, then $f(x) = f(x+2a)$ for all values of x.

37) If $y = f(x)$ is continuous for all values of x then so is $y^2 = f(x)$.

38) The point $(a \sinh t, b \cosh t)$ lies on the hyperbola $\dfrac{x^2}{a^2} - \dfrac{y^2}{b^2} = 1$ for all values of t.

39) If $f(x)$ is periodic with period a, then $f'(x)$ is also periodic with period a.

40) $n + \dfrac{1}{n} \geqslant 2$ for all integral values of n.

MISCELLANEOUS EXERCISE 10

1) Sketch the cycloid $x = t - \sin t$, $y = 1 - \cos t$ for $0 \leqslant t \leqslant 2\pi$, showing that it is symmetrical about the line $x = \pi$. Find the centroid of the area contained between this curve and the x-axis. (AEB'73)p

2) Write down the equation of the tangent at the origin to the curve $y = 4x - x^3$. Sketch the curve and this tangent for $-2 < x < 2$. (U of L)

3) Find the points of inflexion of the curve $y = \dfrac{x}{x^2+1}$, and show that they lie on a straight line. (O)p

4) Find the (finite) points of inflexion of the curve $y = \dfrac{1}{1+x+x^2}$.
Find also the point of intersection of the tangents at these points. (O)

5) Find the coordinates of all the points of inflexion of the graph of the function

$$y = x - \sin x$$

in the range $0 \leqslant x \leqslant 4\pi$. Sketch the graph in this range. (JMB)

6) Find the points of inflexion of the curve

$$y = \frac{4(x+3)}{x^2+6x+12}$$

and show that they lie on a straight line. (O)

7) Find the values of x for which the function

$$f(x) = e^x(2x^2 - 3x + 2)$$

has (a) a maximum, (b) a minimum, (c) an inflexion.
Draw a rough sketch of the graph of the function. (O)

8) If $y^2 = x^2(x-2)$, obtain an expression for dy/dx in terms of x, and hence show that on the graph of y against x there are no turning points. Show that when $x = 2\frac{2}{3}$, $dy/dx = \pm\sqrt{6}$ and $d^2y/dx^2 = 0$.
Sketch the form of the graph of y against x, paying special attention to the point $(2, 0)$ and to the points where $d^2y/dx^2 = 0$. (C)

9) A region R of the plane is defined by $y^2 - 4ax \leqslant 0$, $x^2 - 4ay \leqslant 0$, $x + y - 3a \leqslant 0$. Find the area of R. (O)p

10) Sketch the region in the xy plane within which all three of the following inequalities are satisfied:
(a) $y < x + 1$, (b) $y > (x-1)^2$, (c) $xy < 2$.
Determine the area of this region. (JMB)

11) Show that, for real x,

$$-1 \leqslant \frac{2x}{x^2+1} \leqslant 1.$$ (U of L)

12) Sketch the three lines whose equation are

$$y - x - 6 = 0, \quad 2y + x - 18 = 0, \quad 2y - x = 0.$$

Shade on your diagram the domain defined by

$$y - x - 6 < 0, \quad 2y + x - 18 < 0, \quad 2y - x > 0.$$ (U of L)

13) Shade the region or regions of the xy plane within which $(y^2 - 8x)(x^2 + y^2 - 9)$ is negative. (U of L)

14) Show, by shading on a sketch of the xy plane, the region for which $x^2 + y^2 \leqslant 1$, $y \geqslant x$ and $y \leqslant x + 1$.
Hence find
(a) the greatest value of y,
(b) the least value of $x + y$
for which these inequalities hold.

15) The equation of a curve is $x^2y^2 - x^2 + y^2 = 0$.
(a) Find the equations of the tangents at the origin.
(b) Find the equations of the real asymptotes.
(c) Show that the numerical value of y is never greater than the corresponding value of x.
(d) Show that the numerical value of y is always less than unity.
(e) Sketch the curve. (AEB '71)

16) The corresponding values of two variables x and y found by experiment are

x	16.7	42	110	298	802
y	1	2	3	4	5

They are believed to be connected by a law of the form $x-2=2b^{y+x}$ where a and b are constants. By using semi-logarithmic graph paper show that this is so and determine probable values of a and b. (AEB '73)p

17) The vapour pressure P mm of carbon tetrachloride at a temperature T kelvin is given in the following table.

T	273	283	293	303	313	323	333
P	33.1	56.0	89.5	138.6	210.9	305.6	439.0

Verify graphically that P and T are related by an equation of the form $P=Ae^{n/T}$. Find probable values of A and n. (AEB '66)

18)

x	0	1	2	3	4	5
y	2.28	3.45	5.17	7.76	11.65	17.46

The table shows approximate values of a variable y corresponding to certain values of another variable x. By drawing a suitable linear graph, verify that the values of x and y satisfy approximately a relationship of the form $y=ab^x$. Use your graph to estimate values of a and b, giving answers to one decimal place. (U of L)

19) Two variables x and y have corresponding values as shown in the table. Plot the points on semi-logarithmic graph paper and determine a probable relationship between the variables.

x	1	2	3	4	5
y	0.2800	0.1960	0.1372	0.0960	0.0672

(AEB'72)p

Log-linear Graph Paper

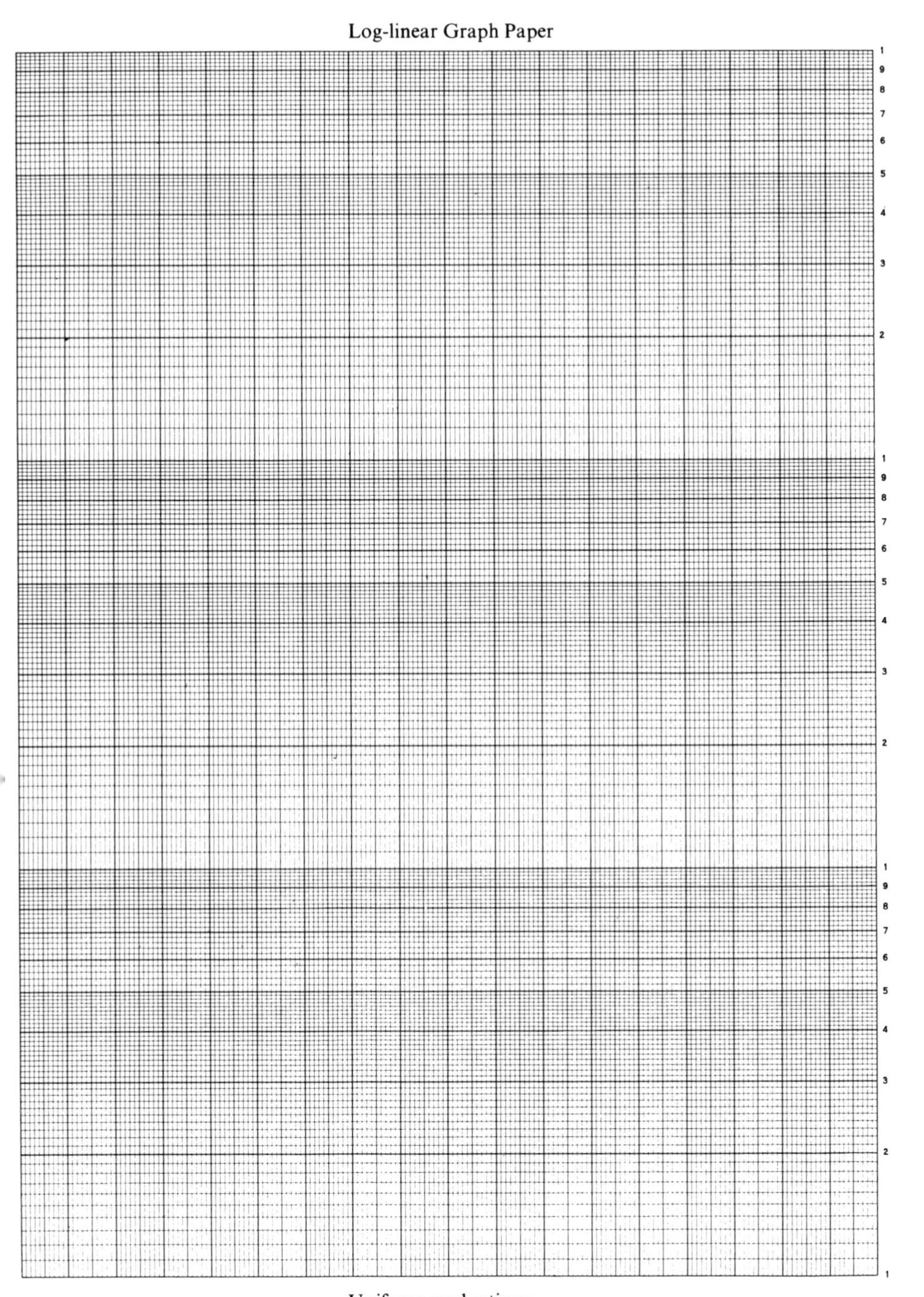

Uniform graduations

Log-log Graph Paper

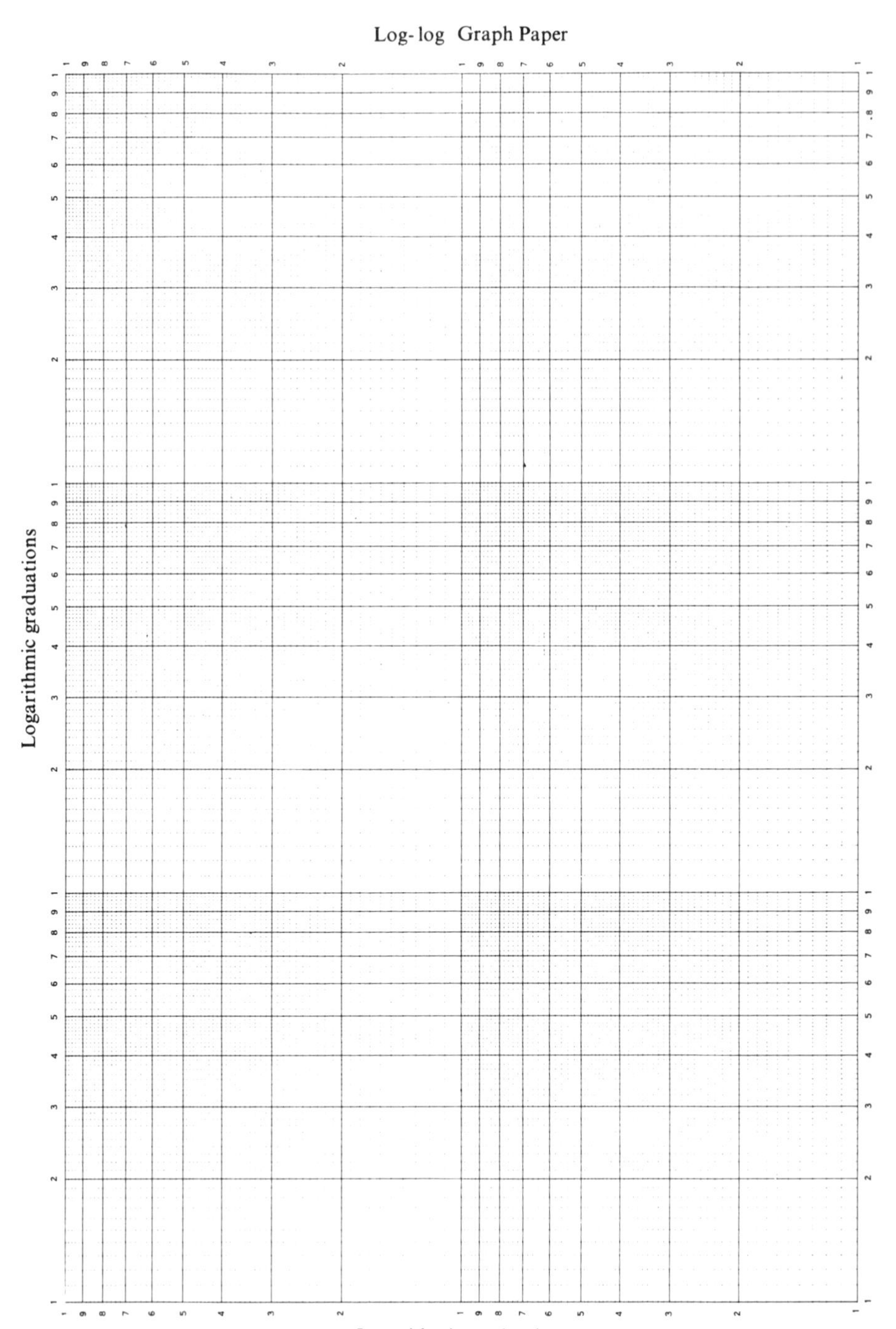

CHAPTER 11

POLYNOMIAL FUNCTIONS AND EQUATIONS

POLYNOMIAL FUNCTIONS OF ONE VARIABLE

Remainders and Factors

In Volume 1 we saw that a polynomial of degree n has the form

$$f(x) \equiv p_n x^n + p_{n-1} x^{n-1} + \ldots + p_0$$

where n is a positive integer and $p_n, p_{n-1}, \ldots$ are constants, of which at least p_n is non-zero.
If $f(x)$ is divided by the linear function $(x - a)$, the quotient, $Q(x)$, is a polynomial of degree $n - 1$ and the remainder, R, is a constant,

i.e. $$f(x) \equiv (x - a)Q(x) + R.$$

Substituting a for x gives $f(a) = R$. This result, which is known as the Remainder Theorem, is introduced in Volume 1 and is restated below.

When a polynomial $f(x)$ is divided by $(x - a)$ the remainder is $f(a)$.

Division of a Polynomial by a Quadratic Function

Consider the polynomial $f(x) \equiv 4x^6 - 2x^4 + x^2 - 2$.
Dividing $f(x)$ by $x^2 - x - 2$ by long division gives

$$
\begin{array}{r|l}
 & x^4 + x^3 + x^2 + 3x + 6 \\
\hline
x^2 - x - 2 & x^6 \quad - 2x^4 \quad + x^2 \quad - 2 \\
 & x^6 - x^5 - 2x^4 \\
\cline{2-2}
 & \quad x^5 \quad + x^2 \quad - 2 \\
 & \quad x^5 - x^4 - 2x^3 \\
\cline{2-2}
 & \qquad x^4 + 2x^3 + x^2 \quad - 2 \\
 & \qquad x^4 - x^3 - 2x^2 \\
\cline{2-2}
 & \qquad\quad 3x^3 + 3x^2 \quad - 2 \\
 & \qquad\quad 3x^3 - 3x^2 - 6x \\
\cline{2-2}
 & \qquad\qquad 6x^2 + 6x - 2 \\
 & \qquad\qquad 6x^2 - 6x - 12 \\
\cline{2-2}
 & \qquad\qquad\quad 12x + 10
\end{array}
$$

So $x^6 - 2x^4 + x^2 - 2$, when divided by $x^2 - x - 2$, gives a quotient $x^4 + x^3 + x^2 + 3x + 6$ and a remainder $12x + 10$.
In general, when a polynomial $f(x)$ of degree n is divided by a quadratic function, the quotient, $Q(x)$, is a polynomial of degree $n - 2$ and the remainder is a linear function $Ax + B$,

i.e. $$f(x) \equiv (ax^2 + bx + c)Q(x) + Ax + B$$

Similarly we can deduce that, when a polynomial is divided by a cubic function, the remainder is quadratic.
If the divisor factorizes, the remainder can be found by adapting the remainder theorem as follows:
Consider again $f(x) \equiv x^6 - 2x^4 + x^2 - 2$ when divided by $x^2 - x - 2$.

As $$x^2 - x - 2 \equiv (x-2)(x+1)$$

we have $$f(x) \equiv (x-2)(x+1)Q(x) + Ax + B$$

$$\Rightarrow \quad \begin{cases} f(2) = 2A + B \\ f(-1) = -A + B \end{cases}$$

$$\Rightarrow \quad \begin{cases} 34 = 2A + B \\ -2 = -A + B \end{cases} \Rightarrow A = 12, \quad B = 10$$

In general, if dividing $f(x)$ by $(x - \alpha)(x - \beta)$ gives a remainder $Ax + B$ then

$$f(x) \equiv (x - \alpha)(x - \beta)Q(x) + Ax + B$$

$$\Rightarrow \quad f(\alpha) = A\alpha + B \quad \text{and} \quad f(\beta) = A\beta + B.$$

This method for finding the remainder may be extended to division of a polynomial by a cubic function.

For example, if $f(x) \equiv 2x^6 - x^5 - 2x^3 - 2$ is divided by $(x-1)(x+1)(2x-1)$, the form of the remainder is $Ax^2 + Bx + C$,

i.e. $$f(x) \equiv (x-1)(x+1)(2x-1)Q(x) + Ax^2 + Bx + C$$

$$\left.\begin{aligned} f(1) &= A+B+C \\ f(-1) &= A-B+C \\ f(\tfrac{1}{2}) &= \frac{A}{4}+\frac{B}{2}+C \end{aligned}\right\} \Rightarrow \left\{\begin{aligned} -3 &= A+B+C \\ 3 &= A-B+C \\ -\frac{9}{4} &= \frac{A}{4}+\frac{B}{2}+C \end{aligned}\right.$$

$\Rightarrow$ $$A = 1, \quad C = -1, \quad B = -3.$$

Note that if the divisor does *not* factorize, and/or the quotient is required, then long division must be used.

Repeated Factors

The factor theorem, which follows from the remainder theorem, is introduced in Volume 1 and states that

if, for a polynomial $f(x)$, $f(a) = 0$, then $(x-a)$ is a factor of $f(x)$.

If $f(x)$ has a repeated factor $(x-a)$,

i.e. $$f(x) \equiv (x-a)^2 g(x)$$

then $$\begin{aligned} f'(x) &= \frac{d}{dx}[(x-a)^2 g(x)] \\ &= (x-a)^2 g'(x) + 2(x-a)g(x) \qquad \left[\text{using } \frac{d}{dx}(uv)\right] \\ &= (x-a)[(x-a)g'(x) + 2g(x)]. \end{aligned}$$

i.e. if $f(x)$ has a repeated factor $(x-a)$, then $f'(x)$ has a factor $(x-a)$.

Also
$f(a) = 0$ *and* $f'(a) = 0$
$\Rightarrow f(x)$ has a stationary value of zero when $x = a$

Note. A linear factor of $f'(x)$ is *not necessarily* a repeated factor of $f(x)$. For example, in the diagram above

$$f'(b) = 0 \Rightarrow (x-b) \text{ is a factor of } f'(x)$$

But $\quad f(b) \neq 0$ so $(x-b)$ is *not* a factor of $f(x)$

So the necessary and sufficient condition for $f(x)$ to have a repeated factor $(x-a)$ is that *both* $f(a)=0$ *and* $f'(a)=0$.

EXAMPLES 11a

1) Determine whether $f(x) \equiv 3x^4 - 8x^3 - 6x^2 + 24x - 13$ has any repeated factors, and, if so, find them.

$$\begin{aligned} f'(x) &\equiv 12x^3 - 24x^2 - 12x + 24 \\ &\equiv 12(x^3 - 2x^2 - x + 2) \\ &\equiv 12(x-1)(x^2 - x - 2) \\ &\equiv 12(x-1)(x+1)(x-2). \end{aligned}$$

Now $f'(x) = 0$ when $x = 1, -1$ or 2.
Checking the value of $f(x)$ for these values of x we have

$$\begin{aligned} f(1) &= 3 - 8 - 6 + 24 - 13 = 0 \\ f(-1) &= 3 + 8 - 6 - 24 - 13 \neq 0 \\ f(2) &= 48 - 64 - 24 + 48 - 13 \neq 0 \end{aligned}$$

So $(x+1)$ and $(x-2)$ are not factors of $f(x)$.
Hence $(x-1)$ is the only repeated factor of $f(x)$

2) If the equation $3x^4 + 2x^3 - 6x^2 - 6x + p = 0$ has two equal roots, find the possible values of p.

If $f(x) \equiv 3x^4 + 2x^3 - 6x^2 - 6x + p$, the equation $f(x) = 0$ has two equal roots if $f(x)$ has two equal factors, i.e. a repeated factor.
Any linear factor of $f'(x)$ is a possible repeated factor of $f(x)$.
Now

$$\begin{aligned} f'(x) \equiv 12x^3 + 6x^2 - 12x - 6 &\equiv 6(2x^3 + x^2 - 2x - 1) \\ &\equiv 6(x-1)(x+1)(2x+1) \end{aligned}$$

If $(x-1)$ is a repeated factor of $f(x)$ then $f(1) = 0$

$$\Rightarrow \qquad 3 + 2 - 6 - 6 + p = 0$$

$$\Rightarrow \qquad p = 7.$$

Similarly $(x+1)$ is a repeated factor of $f(x)$ if $f(-1) = 0$.

$$\Rightarrow \qquad 3 - 2 - 6 + 6 + p = 0 \Rightarrow p = -1$$

and $(2x+1)$ is a repeated factor of $f(x)$ if $f(-\tfrac{1}{2}) = 0$

$$\Rightarrow \qquad \tfrac{3}{16} - \tfrac{2}{8} - \tfrac{6}{4} + \tfrac{6}{2} + p = 0 \quad \Rightarrow \quad p = -\tfrac{23}{16}$$

So the possible values of p are $7, -1, -\frac{23}{16}$.

3) Without performing long division, find the remainder when $x^3 - 5x^2 + 6x - 2$ is divided by $(x-2)^2$.

$f(x) \equiv x^3 - 5x^2 + 6x - 2$ has a remainder $Ax + B$ when divided by $(x-2)^2$,

i.e. $$f(x) \equiv (x-2)^2 Q(x) + Ax + B \qquad [1]$$

Now $$f(2) = 2A + B$$

$$\Rightarrow \qquad 2A + B = -2$$

No other equation linking A and B can be obtained from [1] without involving $Q(x)$. But differentiating [1] w.r.t. x gives

$$f'(x) = \frac{\mathrm{d}}{\mathrm{d}x}[(x-2)^2 Q(x)] + A$$

$$\Rightarrow \qquad 3x^2 - 10x + 6 \equiv (x-2)^2 Q'(x) + 2(x-2)Q(x) + A$$

$$\Rightarrow \qquad f'(2) = 12 - 20 + 6 = A \quad \Rightarrow \quad A = -2$$

So $$B = 2$$

Therefore the remainder is $-2x + 2$.

4) Find the constant p such that $x^2 + 2$ is a factor of $x^4 - 6x^2 + p$. Hence factorize $x^4 - 6x^2 + p$.

If $x^2 + 2$ is a factor of $x^4 - 6x^2 + p$

then $$x^4 - 6x^2 + p \equiv (x^2+2)f(x)$$

where $f(x)$ is of degree 2, i.e. $f(x) \equiv ax^2 + bx + c$.

So $$x^4 - 6x^2 + p \equiv (x^2+2)(ax^2 + bx + c).$$

Comparing coefficients of x^4 gives $1 = a$.
Comparing coefficients of x^3 gives $0 = b$.
Comparing coefficients of x^2 gives $-6 = 2a + c \Rightarrow c = -8$.
Comparing constants gives $p = 2c = -16$.

Therefore $$x^4 - 6x^2 + p \equiv x^4 - 6x^2 - 16 \equiv (x^2+2)(x^2-8)$$
$$\equiv (x^2+2)(x - 2\sqrt{2})(x + 2\sqrt{2}).$$

Common Factors

If two polynomials $f(x)$ and $g(x)$ have a common factor $(x-a)$ then

$$f(x) \equiv (x-a)h(x) \qquad [1]$$

$$g(x) \equiv (x-a)j(x). \qquad [2]$$

For any constant K

$$f(x)+Kg(x) \equiv (x-a)h(x)+K(x-a)j(x)$$

$$\equiv (x-a)[h(x)+Kj(x)]$$

i.e. if $f(x)$ and $g(x)$ have a common factor $(x-a)$ then, for any constant K, $(x-a)$ is a factor of $f(x)+Kg(x)$.

This property is sometimes useful for solving problems concerning two polynomials with a common factor.

Also, from identities [1] and [2], it follows that

$$f(a) = 0 \quad \text{and} \quad g(a) = 0$$

giving a pair of simultaneous equations which provides another approach to problems involving a common factor. Both of these approaches are illustrated in the following examples.

EXAMPLES 11a (continued)

5) Find the constants p and q such that $x-2$ is a common factor of

$$x^3-x^2-2px+3q \quad \text{and} \quad qx^3-px^2+x+2.$$

If $(x-2)$ is a factor of $f(x) \equiv x^3-x^2-2px+3q$, then

$$f(2) \equiv 0 \quad \Rightarrow \quad 4-4p+3q = 0 \qquad [1]$$

If, also $(x-2)$ is a factor of $g(x) \equiv qx^3-px^2+x+2$ then

$$g(2) = 0 \Rightarrow 8q-4p+4 = 0. \qquad [2]$$

Solving equations [1] and [2] simultaneously gives

$$p = 1, \qquad q = 0.$$

6) Show that if $(x+1)$ is a common factor of x^3-ax^2+b and $x^4-ax^3+bx^2+c$ then $2a = 2b-2 = -2-c$.

Let $f(x) \equiv x^3-ax^2+b$ and $g(x) \equiv x^4-ax^3+bx^2+c$.

For *any* value of K, $(x+1)$ is factor of $f(x)+Kg(x)$

therefore $$f(-1)+Kg(-1) = 0$$

i.e. for any value of K.

$$(-1-a+b)+K(1+a+b+c) = 0.$$

When $K=1$, $2b+c=0$
When $K=0$, $a-b=-1$

Hence $$a = b-1 = -\frac{c}{2}-1$$

$\Rightarrow$ $$2a = 2b-2 = -2-c.$$

7) Find the relationship between a, b and c such that the equations $x^2-ax+b=0$ and $ax^2+x-c=0$ have a common root.
If the common root is α, then α satisfies both given equations,

i.e. $$\begin{cases} \alpha^2-a\alpha+b = 0 & [1] \\ a\alpha^2+\alpha-c = 0 & [2] \end{cases}$$

To find a relationship between a, b and c, we must eliminate α and α^2 from equations [1] and [2].

[1] + a[2] $\Rightarrow$ $$(1+a^2)\alpha^2+(b-ac) = 0$$

$\Rightarrow$ $$\alpha^2 = \frac{ac-b}{1+a^2} \quad [3]$$

[2] − a[1] $\Rightarrow$ $$(1+a^2)\alpha-(c+ab) = 0$$

$\Rightarrow$ $$\alpha = \frac{c+ab}{1+a^2} \quad [4]$$

From equations [3] and [4] we have

$$\frac{ac-b}{1+a^2} = \left(\frac{c+ab}{1+a^2}\right)^2$$

$\Rightarrow$ $$(ac-b)(1+a^2) = (c+ab)^2.$$

EXERCISE 11a

1) Find the remainder when $x^4-5x^3+6x^2-7$ is divided by $(x-1)(x-3)$.

2) Find the remainder when x^4+x^2-7 is divided by x^2-4.

3) Find the constants p and q such that when x^3-px+q is divided by x^2-3x+2, the remainder is $4x-1$.

4) Find the constants a, b and c such that when x^5-7x^3+4x-2 is divided by $(x-1)(x+1)(x-3)$ the remainder is ax^2+bx+c.

5) Find the remainder when x^3-5x^2+7 is divided by $(x-1)^2$.

6) Find the constants m and n such that when x^4-mx^2+n is divided by $(x+1)^2$ the remainder is $5x-2$.

7) Determine whether the given functions have any repeated factors and, if they have, find them.

(a) x^4-16 (b) x^4-18x^2+81 (c) $2x^3-3x^2+1$ (d) $x(x^2-4)$.

8) If the equation $2x^3-9x^2+12x+p=0$ has two equal roots, find the possible values of p.

9) Find the value of a for which the function $2x^3-ax^2-12x-7$ has a repeated factor.

10) Find the constant m for which x^2+1 is a factor of mx^4+x^2-1.

11) Show that x^2+3 is a factor of x^3-x^2+3x-3.

12) Show that if $(x-a)$ is a factor of $p_nx^n+p_{n-1}x^{n-1}+\ldots+p_0$ then $|a|$ is a factor of $|p_0|$.

13) Find the constant a for which the functions $f(x)\equiv ax^2+2x-1$ and $g(x)\equiv x^2+4x+a$ have a common factor.

14) Find the constants p and q such that $(x-1)$ is a common factor of x^4-2px^2+2 and x^4+x^2-q.

15) Show that if the cubic functions x^3+ax^2+b and ax^3+bx^2+x-a have a common factor, it is also a factor of the quadratic function $(b-a^2)x^2+x-a(1+b)$.

16) Determine the condition that the equations

$$px^2+qx+1=0 \quad \text{and} \quad x^2+px+q=0$$

have a common root.

17) Determine the value of m such that the equations

$$x^3+mx-1=0 \quad \text{and} \quad x^3-3x+m=0$$

have a common root.

POLYNOMIAL FUNCTIONS OF MORE THAN ONE VARIABLE

A polynomial in three variables x, y, z consists of the sum of terms such as $px^ly^mz^n$ where p is a constant and l, m, n are positive integers. This definition can be extended to any number of variables. The *degree* of such a term is the *sum* of the powers of the variables. For example $2x^2y^3$ is a term of degree five, $5xy^3z^5$ is of degree nine, $4a^3b^3$ is of degree six. The term of highest degree determines the degree of the polynomial.

Homogeneous Functions

A homogeneous function is a polynomial, each of whose terms is of the same degree. For example

$x^2 + xy - y^2$ is a homogeneous polynomial in x and y of degree 2.

$\alpha^2\beta + \beta^2\gamma + \gamma^2\alpha$ is a homogeneous polynomial in α, β, γ of degree 3.

Cyclic Functions

Consider the polynomial $(x-y)(y-z)(z-x)$.
If x and y (or any other *pair* of x, y, z) are interchanged, the polynomial changes.
But if x is replaced by y, y is replaced by z and z is replaced by x (i.e. the variables are interchanged in the cyclic order $x \to y \to z \to x$) the polynomial becomes $(y-z)(z-x)(x-y)$ which is identical to the original function.
Any function that remains the same when *all* the variables are interchanged in *cyclic order* is said to be a *cyclic function*.

For example

$$\left.\begin{array}{l} \alpha\beta + \beta\gamma + \gamma\alpha \\ xy^2 + yz^2 + zx^2 \\ (a^2-b^2)(b^2-c^2)(c^2-d^2)(d^2-a^2) \end{array}\right\} \text{are cyclic functions}$$

but

$$\left.\begin{array}{l} x^2 + xy - y^2 \\ (a^2+b^2)(b^2-c^2)(c^2+a^2) \end{array}\right\} \text{are not cyclic.}$$

The Sigma Notation for Cyclic Functions

The sigma notation provides a useful shorthand form for a cyclic function which is a *sum* of terms.
For example $f(xyz) \equiv xy + yz + zx$ can be written as $f(xyz) \equiv \sum xy$ where $\sum xy$ means the *sum* of all *different* terms found by interchanging x, y and z in cyclic order. Similarly $f(xyz) \equiv \sum x(y+z)$ means the sum of all different terms found by interchanging x, y and z in cyclic order,

i.e. $$\sum x(y+z) \equiv x(y+z) + y(z+x) + z(x+y).$$

Note that $\sum x(y+z)$ is *not* the same as $\sum xy$, in fact $\sum x(y+z) = 2\sum xy$.

EXERCISE 11b

State the degree of the following functions and state also whether they are homogeneous and/or cyclic.

1) $x+y+z$

2) $x^2-y^2+z^2$

3) $ab+a^2+b^2$

4) $a(b-c)+b(a-c)$

5) $\alpha^2(\beta+\gamma)+\beta^2(\gamma+\alpha)+\gamma^2(\beta-\alpha)$

6) $\alpha^2(\beta+1)+\beta^2(\gamma+1)+\gamma^2(\alpha+1)$

7) $(p-q)(q-r)(r-p)$

8) $a^2-b^2-c^2$

9) $ab+bc+ca+a^2+b^2+c^2$

10) $2(x^2+y^2+z^2)-3(xy+yz+zx)$

11) Write out the following cyclic functions in full.

(a) $f(\alpha\beta\gamma) \equiv \sum \alpha(\alpha^2-\beta^2)$

(b) $f(xyz) \equiv \sum xy^2$

(c) $f(\alpha\beta\gamma) \equiv \sum \alpha(\beta^2-\gamma^2)$

(d) $f(abcd) \equiv \sum ab^2$

12) Write the following cyclic functions in the sigma notation.

(a) $x^2(y^2+z^2)+y^2(z^2+x^2)+z^2(x^2+y^2)$

(b) $\alpha^2(\beta+\gamma)+\beta^2(\gamma+\alpha)+\gamma^2(\alpha+\beta)$

(c) $\alpha+\beta+\gamma$

(d) $a^2b+b^2c+c^2d+d^2a$

FACTORIZATION OF HOMOGENEOUS AND CYCLIC POLYNOMIALS

When attempting to factorize a polynomial in several variables, the following considerations should be noted.

(a) If the polynomial is homogeneous, its factors also are homogeneous.

(b) If the polynomial is cyclic, the product of its factors is a cyclic function.

These points help in selecting likely linear factors of a particular polynomial. The factor theorem can then be used to determine whether they are, or are not, factors.

EXAMPLES 11c

1) Factorize a^3-b^3.

$f(ab) \equiv a^3-b^3$ is homogeneous, so any linear factors will be of the form $a \pm b$.

When $a=-b$, $\qquad f(-bb) = -b^3-b^3 \neq 0$

so $a+b$ is not a factor.

When $a=b$, $\qquad f(bb) = b^3-b^3 = 0$

so $a-b$ is a factor.

Therefore $\qquad a^3-b^3 \equiv (a-b)g(ab).$

Now $\quad g(ab)$ is homogeneous, and of degree 2.

So the general form of $g(ab)$ is $Aa^2 + Bab + Cb^2$ where A, B, and C are constants,

i.e. $$a^3 - b^3 \equiv (a-b)(Aa^2 + Bab + Cb^2)$$

Comparing coefficients of a^3, b^3, a^2b gives $A = B = C = 1$

Therefore $$a^3 - b^3 \equiv (a-b)(a^2 + ab + b^2).$$

As $a^2 + ab + b^2$ has no real linear factors, $a^3 - b^3$ cannot be factorized further.

Note that in this and the following examples, what would normally be 'thought processes' are written down to clarify the working. In practice, the factorization of $a^3 - b^3$ would be written down simply as:

When $a = b$, $$a^3 - b^3 = 0.$$

Therefore $$a^3 - b^3 \equiv (a-b)(a^2 + kab + b^2).$$

Comparing coefficients of a^2b gives $k = 1$.

Hence $$a^3 - b^3 \equiv (a-b)(a^2 + ab + b^2).$$

The result in the example above is a special case of the following more general results.
Consider the function $x^n - a^n$ where n is any positive integer.

When $x = a$, $$x^n - a^n = a^n - a^n = 0.$$

So $(x-a)$ is a factor of $x^n - a^n$.
Now consider the function $x^n + a^n$.
It is clear that $(x-a)$ is not a factor of this polynomial.
But if n is odd, i.e. $n = 2m+1$
then when $x = -a$,

$$x^n + a^n = (-a)^{2m+1} + a^{2m+1} = 0.$$

So $(x+a)$ is a factor of $x^n + a^n$ when n is odd.
These results are quotable and are summarized below:

> If n is a positive integer
> $(x-a)$ is a factor of $x^n - a^n$ for all values of n
> $(x+a)$ is a factorof $x^n + a^n$ for odd values of n.

In particular

$$x^3 - y^3 \equiv (x-y)(x^2 + xy + y^2)$$
$$x^3 + y^3 \equiv (x+y)(x^2 - xy + y^2).$$

EXAMPLES 11c (continued)

2) Factorize $a(b^2-c^2)+b(c^2-a^2)+c(a^2-b^2)$.

Let $$f(abc) \equiv a(b^2-c^2)+b(c^2-a^2)+c(a^2-b^2)$$

$f(abc)$ is homogeneous so any linear factors are of the form
$$a \pm b, \; b \pm c, \; c \pm a.$$

When $a=b$, $f(abc) = b(b^2-c^2)+(c^2-b^2)+0 = 0$

so $(a-b)$ is a factor.
As $f(abc)$ is cyclic, it follows that $(a-b)$ is one of a set of factors that, as a whole, is cyclic, so $(b-c)$ and $(c-a)$ also are factors.
Now $f(abc)$ is of degree 3 and we have found three linear factors. Hence the only other possible factor is a constant, k,

i.e. $$f(abc) \equiv k(a-b)(b-c)(c-a)$$

Comparing coefficients of ab^2 gives $k=1$.
Therefore
$$a(b^2-c^2)+b(c^2-a^2)+c(a^2-b^2) \equiv (a-b)(b-c)(c-a).$$

3) Factorize $(x+y)^3(x-y)+(y+z)^3(y-z)+(z+x)^3(z-x)$.

Let $$f(xyz) \equiv (x+y)^3(x-y)+(y+z)^3(y-z)+(z+x)^3(z-x).$$

$f(xyz)$ is homogeneous so likely linear factors are of the form
$$x \pm y, \; x \pm z, \; y \pm z.$$

When $x=y$,
$$f(yyz) = 0+(y+z)^3(y-z)+(z+y)^3(z-y) = 0.$$

Hence $(x-y)$ is a factor.
Similarly $(y-z)$ and $(z-x)$ are factors.
Therefore $(x-y)(y-z)(z-x)$ is a factor of $f(xyz)$.
Now $f(xyz)$ is cyclic and of degree 4 and the product of the factors that we have found so far is also cyclic and of degree 3. Hence the remaining factor is linear and must be cyclic in x, y *and* z. So it can only be of the form $x+y+z$. The only other possible factor is a constant k.

Therefore $$f(xyz) \equiv k(x+y+z)(x-y)(y-z)(z-x).$$

Comparing coefficients of x^3y gives $k=1$.

So $$f(xyz) \equiv (x+y+z)(x-y)(y-z)(z-x).$$

4) Factorize $a^3+b^3+c^3-3abc$.

Let $$f(abc) \equiv a^3+b^3+c^3-3abc.$$

$f(abc)$ is homogeneous, so any linear factors are of the form

$$a \pm b, \quad b \pm c, \quad c \pm a, \quad a \pm b \pm c$$

when $a = \pm b$, $f(\pm bbc) \neq 0$ so $(a+b)$ and $(a-b)$ are not factors.
Similarly $b \pm c$, $c \pm a$ are not factors.
When $a = -(b+c)$

$$\begin{aligned} f(-\{b+c\}bc) &= -(b+c)^3 + b^3 + c^3 + 3(b+c)bc \\ &= -b^3 - 3b^2c - 3bc^2 - c^3 + b^3 + c^3 + 3b^2c + 3bc^2 \\ &= 0. \end{aligned}$$

Hence $(a+b+c)$ is a factor.
Note that $f(abc)$ is cyclic, so its factors *as a whole* must make a cyclic function. The one factor found so far, $a+b+c$, is cyclic so the remaining factor(s) must form a cyclic group. Hence $a-b-c$ and $a+b-c$ cannot be factors.
Now $f(abc)$ is cyclic and of degree 3, and the factor that we have found is also cyclic. So the other factor must be cyclic and of degree 2. The most general form for such an expression is $k_1(a^2+b^2+c^2) + k_2(ab+bc+ca)$ where k_1 and k_2 are constants.
Therefore

$$a^3 + b^3 + c^3 - 3abc \equiv (a+b+c)[k_1(a^2+b^2+c^2) + k_2(ab+bc+ca)]$$

Comparing coefficients of a^3 and of abc gives $k_1 = 1$ and $k_2 = -1$.
Hence

$$a^3 + b^3 + c^3 - 3abc \equiv (a+b+c)(a^2+b^2+c^2-ab-bc-ca).$$

EXERCISE 11c

Factorize the following functions.

1) $(a-b)^3 + (a+b)^3$

2) $x^2 + y^2 + z^2 + 2xy + 2yz + 2zx$

3) $x^2(y-z) + y^2(z-x) + z^2(x-y)$

4) $a(b^2-c^2) + b(c^2-a^2) + c(a^2-b^2)$

5) $(a-b)^3 + (b-c)^3 + (c-a)^3$

6) $x^4(y-z) + y^4(z-x) + z^4(x-y)$

7) $a^6 - b^6$

8) $x^6 - 64$

9) $pq(p-q) + qr(q-r) + rp(r-p)$

10) $a^3 + b^3 + c^3 + 3bc(b+c)$

11) Find the sum of the n terms of the geometric progression

$$x^{n-1} + ax^{n-2} + a^2x^{n-3} + \ldots + a^{n-1}.$$

Hence show that

$$x^n - a^n \equiv (x-a)(x^{n-1} + ax^{n-2} + \ldots + a^{n-1}).$$

Use this result to write down the factors of

$$x^5 - 32 \quad \text{and} \quad a^5 - b^5.$$

12) If m and n are integers, show that $(x-y)$ is a factor of

$$x^n(y^m - z^m) + y^n(z^m - x^m) + z^n(x^m - y^m).$$

13) Show that $(x-a)^2$ is a factor of $x^3 - ax^2 - a^2x + a^3$.
Hence factorize $p^3 - p^2q - pq^2 + q^3$.

14) Factorize $a^3 + 8b^3 + 27c^3 - 18abc$.

POLYNOMIAL EQUATIONS

The Nature of the Roots

It was seen in Volume 1, Chapter 14, that a quadratic equation has either two real roots (distinct or equal) or two conjugate complex roots.
Now consider a cubic equation $f(x) = 0$ which, as we already know, *must* have at least one real root. Therefore the cubic function $f(x)$ has at least one linear factor, $x - \alpha$.

So $$f(x) \equiv (x-\alpha)(ax^2 + bx + c)$$

Now $$f(x) = 0 \Rightarrow x = \alpha \quad \text{or} \quad ax^2 + bx + c = 0.$$

But $ax^2 + bx + c = 0$ has either two real roots $(b^2 - 4ac \geqslant 0)$ or two conjugate complex roots $(b^2 - 4ac < 0)$.
Thus a cubic equation has

either three real roots (not necessarily distinct)

or one real root and a pair of conjugate complex roots.

Polynomial equations of degree higher than three have a similar property,

i.e. if a polynomial equation has any complex roots, they occur in conjugate pairs.

(A proof of this property is given in the appendix to this chapter.)
It therefore follows that no polynomial equation can have an odd number of complex roots.
Assuming that a polynomial equation of degree n has n roots, it also follows that such an equation has

at least one real root if n is odd, but *may* have no real roots if n is even.

Solution of Polynomial Equations

The real roots of a quadratic equation can be found, either by factorising, or by using the standard formula. So, even when these roots are irrational, they can always be found.
The roots of an equation of higher degree also, can sometimes be found by using the factor theorem, but only when the roots are integers or simple rational fractions. If this approach fails, there is no formula that can be used instead, and *exact* solutions of such an equation can be found only in certain special cases, some of which are demonstrated below.

EXAMPLES 11d

1) Solve the equation

$$3x^4 - 4x^3 - 14x^2 - 4x + 3 = 0.$$

This equation has *symmetrical coefficients*, and can be expressed as follows:

$$3x^2 - 4x - 14 - \frac{4}{x} + \frac{3}{x^2} = 0$$

$$\Rightarrow \quad 3\left(x^2 + \frac{1}{x^2}\right) - 4\left(x + \frac{1}{x}\right) - 14 = 0 \qquad [1]$$

Now we use the substitution

$$y \equiv x + \frac{1}{x}$$

$$\Rightarrow \quad y^2 \equiv x^2 + 2 + \frac{1}{x^2}.$$

So [1] becomes

$$3(y^2 - 2) - 4y - 14 = 0$$

$$\Rightarrow \quad 3y^2 - 4y - 20 = 0$$

$$\Rightarrow \quad (3y - 10)(y + 2) = 0$$

$$\Rightarrow \quad y = -2 \quad \text{or} \quad \tfrac{10}{3}$$

If $y = -2$, $\quad x + \frac{1}{x} = -2 \quad \Rightarrow \quad x^2 + 2x + 1 = 0$

$$\Rightarrow \quad x = -1, -1$$

If $y = \frac{10}{3}$, $\quad x + \frac{1}{x} = \frac{10}{3} \quad \Rightarrow \quad 3x^2 - 10x + 3 = 0$

$$\Rightarrow \quad x = 3, \tfrac{1}{3}$$

So the roots of the given equation are $-1, -1, \frac{1}{3}, 3$.

Note that the substitution $y \equiv x + \dfrac{1}{x}$ reduces any quartic equation with symmetrical coefficients, to a quadratic equation.

2) Solve the equation

$$\sqrt{(x+8)} - \sqrt{(x+3)} = \sqrt{(2x-1)}$$

To solve an equation of this type, we must eliminate all square roots.
But whenever both sides of an equation are squared, an extra equation, and hence possible extra solutions, are included,
e.g. if we square both sides of the equation $x = 2$, we get $x^2 = 4$.
But $x^2 = 4$ includes *both* $x = 2$ and $x = -2$.
Thus, whenever a solution involves squaring both sides of an equation, all roots must be checked back in the original equation.
Considering the given equation, and squaring both sides, we have

$$x + 8 - 2\sqrt{(x+8)}\sqrt{(x+3)} + x + 3 = 2x - 1$$

$$\Rightarrow \qquad 12 = 2\sqrt{(x+8)}\sqrt{(x+3)}$$

Squaring both sides again gives

$$36 = (x+8)(x+3)$$

$$\Rightarrow \qquad x^2 + 11x - 12 = 0$$

$$\Rightarrow \qquad x = 1 \quad \text{or} \quad -12.$$

Returning to the given equation,

when $x = 1$, $\qquad$ L.H.S. $= \sqrt{9} - \sqrt{4} = 1$

R.H.S. $= \sqrt{1} = 1$.

So $x = 1$ satisfies the given equation.
When $x = -12$, $\sqrt{(-12+8)}$ is not real,
so $x = -12$ is *not* a solution of the given equation.
Hence the only root is 1.

EXERCISE 11d

Solve the following symmetrical equations.

1) $5x^4 - 16x^3 - 42x^2 - 16x + 5 = 0$

2) $6x^4 + 5x^3 - 38x^2 + 5x + 6 = 0$

3) $63x^4 - 1024x^3 + 4226x^2 - 1024x + 63 = 0$

4) $4x^4 + 17x^3 + 8x^2 + 17x + 4 = 0$.

Solve the following equations using any suitable method.

5) $1+\sqrt{x}=\sqrt{(3x-3)}$ 6) $\sqrt{(2x-5)}-\sqrt{(x-2)}=1$

7) $\sqrt{(3x+1)}+\sqrt{(x-1)}=\sqrt{(7x+1)}$ 8) $x^{\frac{4}{3}}-5x^{\frac{2}{3}}+4=0$

9) $\dfrac{x^2}{4}+y^2=1$ and $xy=1$

10) $x^2+y^2+4x-6y=3$ and $y=x+1$

11) $x^2+y^2+8x-4y+15=0$ and $x^2+y^2+6x+2y-15=0$

12) $x^4-x^3-12x^2-4x+16=0$ use $y\equiv x+\dfrac{4}{x}$

RELATIONSHIPS BETWEEN ROOTS AND COEFFICIENTS

It has already been established that if a quadratic equation $ax^2+bx+c=0$ has roots α and β, then

$$\alpha+\beta = -\frac{b}{a} \quad \text{and} \quad \alpha\beta = \frac{c}{a}$$

Similar relationships between the roots and the coefficients of polynomial equations of higher degree can be found as follows.

Cubic Equations

The general cubic equation can be written

$$ax^3+bx^2+cx+d = 0 \qquad [1]$$

and if its roots are α, β, γ, then the equation can also be written in the form

$$(x-\alpha)(x-\beta)(x-\gamma) = 0. \qquad [2]$$

Dividing equation [1] by a
and comparing it with the expansion of equation [2], we have

$$\begin{cases} x^3+\dfrac{b}{a}x^2+\dfrac{c}{a}x+\dfrac{d}{a} = 0 \\ x^3-(\alpha+\beta+\gamma)x^2+(\alpha\beta+\beta\gamma+\gamma\alpha)x-\alpha\beta\gamma = 0. \end{cases}$$

As these two forms represent the same equation *and* the terms in x^3 are identical, the terms in x^2, x and the constant, must also be identical,

i.e.

$$\frac{b}{a} = -(\alpha+\beta+\gamma) \equiv -\sum\alpha$$

$$\frac{c}{a} = \alpha\beta+\beta\gamma+\gamma\alpha \equiv \sum\alpha\beta$$

$$\frac{d}{a} = -\alpha\beta\gamma$$

or

$$\sum\alpha = -\frac{b}{a}$$

$$\sum\alpha\beta = \frac{c}{a}$$

$$\alpha\beta\gamma = -\frac{d}{a}$$

Quartic Equations

Carrying out a similar investigation of the general quartic equation

$$ax^4 + bx^3 + cx^2 + dx + e = 0$$

with roots $\alpha, \beta, \gamma, \delta$, leads to comparing

$$x^4 + \frac{b}{a}x^3 + \frac{c}{a}x^2 + \frac{d}{a}x + \frac{e}{a} = 0$$

with

$$x^4 - \left(\sum\alpha\right)x^3 + \left(\sum\alpha\beta\right)x^2 - \left(\sum\alpha\beta\gamma\right)x + \alpha\beta\gamma\delta = 0.$$

Equating coefficients of corresponding terms gives

$$\sum\alpha = -\frac{b}{a}$$

$$\sum\alpha\beta = \frac{c}{a}$$

$$\sum\alpha\beta\gamma = -\frac{d}{a}$$

$$\alpha\beta\gamma\delta = \frac{e}{a}.$$

The results obtained for the relationships between the roots and the coefficients of quadratic, cubic and quartic equations, establish a pattern which suggests further relationships for higher degree equations.

Quadratic	Cubic	Quartic	Quintic
$\sum\alpha = -\frac{b}{a}$	$\sum\alpha = -\frac{b}{a}$	$\sum\alpha = -\frac{b}{a}$	$\sum\alpha = -\frac{b}{a}$
$\alpha\beta = \frac{c}{a}$	$\sum\alpha\beta = \frac{c}{a}$	$\sum\alpha\beta = \frac{c}{a}$	$\sum\alpha\beta = \frac{c}{a}$
	$\alpha\beta\gamma = -\frac{d}{a}$	$\sum\alpha\beta\gamma = -\frac{d}{a}$	$\sum\alpha\beta\gamma = -\frac{d}{a}$
		$\alpha\beta\gamma\delta = \frac{e}{a}$	$\sum\alpha\beta\gamma\delta = \frac{e}{a}$
			$\alpha\beta\gamma\delta\epsilon = -\frac{f}{a}$

The reader can verify the truth of the results quoted above, by extrapolation, for a fifth degree (quintic) equation, by adopting the method already used for the lower powers.
Note that even if some of the roots are complex, the sum and the product of the roots are both real, verifying that complex roots occur in conjugate pairs.

APPLICATIONS OF THE RELATIONSHIPS BETWEEN ROOTS AND COEFFICIENTS

Many problems that are based on relationships between roots and coefficients can be approached by using the standard relationships derived in this chapter (but the *solution* of cubic or quartic equations is *not*, in general, assisted by these relationships).
The following examples illustrate some of the methods that can be adopted.

EXAMPLES 11e

1) If the equation $x^3 + px^2 + qx + r = 0$ has roots α, β, γ express in terms of p, q and r,

(a) $\sum\alpha^2$, (b) $\sum\alpha\beta(\alpha+\beta)$.

(a)
$$\sum\alpha^2 \equiv \alpha^2 + \beta^2 + \gamma^2$$
$$\equiv (\alpha+\beta+\gamma)^2 - (2\alpha\beta + 2\beta\gamma + 2\gamma\alpha)$$
$$\equiv \left(\sum\alpha\right)^2 - 2\left(\sum\alpha\beta\right).$$

But $$\sum \alpha = -\frac{b}{a} = -p \quad \text{and} \quad \sum \alpha\beta = \frac{c}{a} = q.$$

So $$\sum \alpha^2 = p^2 - 2q.$$

(b) Terms such as $\alpha^2\beta$ occur, amongst other terms, in the product of $(\alpha+\beta+\gamma)$ and $(\alpha\beta+\beta\gamma+\gamma\alpha)$, so we will consider this product in full.

$$(\alpha+\beta+\gamma)(\alpha\beta+\beta\gamma+\gamma\alpha) \equiv \alpha^2\beta+\beta^2\alpha+\beta^2\gamma+\gamma^2\beta+\gamma^2\alpha+\alpha^2\gamma+3\alpha\beta\gamma$$

$$\equiv \sum \alpha\beta(\alpha+\beta)+3\alpha\beta\gamma$$

So $$\sum \alpha\beta(\alpha+\beta) \equiv \left(\sum \alpha\right)\left(\sum \alpha\beta\right) - 3\alpha\beta\gamma$$

$$= (-p)(q)-3(-r)$$

$$= 3r-pq.$$

Note. The expression $\sum \alpha\beta(\alpha+\beta)$ is an unambiguous representation of the *six* terms

$$\alpha^2\beta+\alpha\beta^2+\beta^2\gamma+\beta\gamma^2+\gamma^2\alpha+\alpha\gamma^2$$

whereas $\sum \alpha^2\beta$ means only the *three* terms in cyclic order

$$\alpha^2\beta+\beta^2\gamma+\gamma^2\alpha.$$

However the reader may well encounter $\sum \alpha^2\beta$ being used to represent the full set of six terms above, and is therefore warned to interpret the meaning of $\sum \alpha^2\beta$ with caution and by taking account of the context.

2) If the equation $x^3+px^2+qx+r=0$ has roots that are in arithmetic progression, show that $2p^3-9pq+27r=0$.

If the roots form an arithmetic progression with common difference λ, and α is the middle root, then the three roots are

$$\alpha-\lambda,\ \alpha,\ \alpha+\lambda.$$

From the given equation we see that

$$a = 1, \quad b = p, \quad c = q, \quad d = r,$$

so $$\sum \alpha = (\alpha-\lambda)+\alpha+(\alpha+\lambda) = -p$$

$\Rightarrow$ $$3\alpha = -p \qquad [1]$$

But α satisfies the given equation so we have

$$\left(-\frac{p}{3}\right)^3 + p\left(-\frac{p}{3}\right)^2 + q\left(-\frac{p}{3}\right) + r = 0$$

$$\Rightarrow \qquad 2p^3 - 9pq + 27r = 0.$$

Equations With Related Roots

Suppose that a cubic equation $ax^3 + bx^2 + cx + d = 0$ has roots α, β, γ and that a second cubic equation $a_1x^3 + b_1x^2 + c_1x + d_1 = 0$ has roots $\alpha_1, \beta_1, \gamma_1$ where α_1, β_1 and γ_1 are functions of α, β and γ, then a_1, b_1, c_1 and d_1, must be related to a, b, c and d. The following examples show how this relationship can be applied in problems.

EXAMPLES 11e (continued)

3) If the roots of the equation $4x^3 + 7x^2 - 5x - 1 = 0$ are α, β and γ, find the equation whose roots are

(a) $\alpha + 1$, $\beta + 1$ and $\gamma + 1$, (b) $\alpha^2, \beta^2, \gamma^2$.

(a) If $4x^3 + 7x^2 - 5x - 1 = 0$
then $x = \alpha, \beta, \gamma$.
If $f(X) = 0$ is the required equation
then $X = \alpha + 1, \beta + 1, \gamma + 1$.
So, for each of the roots,

$$X = x + 1 \quad \Rightarrow \quad x = X - 1$$

But x satisfies the given equation, so

$$4(X-1)^3 + 7(X-1)^2 - 5(X-1) - 1 = 0$$

i.e. the required equation is $$4X^3 - 5X^2 - 7X + 7 = 0$$

(b) This time $x = \alpha, \beta, \gamma$

and $X = \alpha^2, \beta^2, \gamma^2$

So $X = x^2 \quad \Rightarrow \quad x = \pm X^{\frac{1}{2}}$.

The given equation thus becomes

$$\pm 4X^{\frac{3}{2}} + 7X \mp 5X^{\frac{1}{2}} - 1 = 0$$

This is not in a satisfactory form so, first, we rearrange it to isolate $X^{\frac{1}{2}}$, then square both sides, as follows:

$$\pm(4X^{\frac{3}{2}} - 5X^{\frac{1}{2}}) = 1 - 7X$$

$$\Rightarrow \quad [\pm X^{\frac{1}{2}}(4X-5)]^2 = (1-7X)^2$$

$$\Rightarrow \quad X(16X^2 - 40X + 25) = 1 - 14X + 49X^2$$

Thus the required equation is

$$16X^3 - 89X^2 + 39X - 1 = 0.$$

Note. A relationship between X and x can be found in some examples where there does not, initially, appear to be a simple connection.
For example, if the required roots are $\beta\gamma, \gamma\alpha$ and $\alpha\beta$ we can convert them into

$$\frac{\beta\gamma\alpha}{\alpha}, \frac{\gamma\alpha\beta}{\beta} \text{ and } \frac{\alpha\beta\gamma}{\gamma}$$

Then, as $\alpha\beta\gamma = -\dfrac{d}{a}$, the required roots are $-\dfrac{d}{a\alpha}$, $-\dfrac{d}{a\beta}$ and $-\dfrac{d}{a\gamma}$

Thus $$X = -\frac{d}{ax} \Rightarrow x = -\frac{d}{aX}$$

Note also that it is not ***always*** possible to find a simple relationship between X and x. An alternative method for such cases is given in the following example.

4) If α, β and γ are the roots of the equation $x^3 + 7x + 5 = 0$, find the equation whose roots are $\alpha^2 + 1$, $\beta^2 + 1$ and $\gamma^2 + 1$.

From the given equation we have

$$\sum \alpha = -\frac{b}{a} = 0$$

$$\sum \alpha\beta = \frac{c}{a} = 7$$

$$\alpha\beta\gamma = -\frac{d}{a} = -5$$

Now if the required equation is

$$X^3 + \frac{B}{A}X^2 + \frac{C}{A}X + \frac{D}{A} = 0$$

we have

$$-\frac{B}{A} = \sum(\alpha^2 + 1) \equiv \left(\sum \alpha^2\right) + 3$$

$$\equiv \left(\sum \alpha\right)^2 - 2\sum \alpha\beta + 3$$

$$= 0 - 2(7) + 3 = -11$$

$$\Rightarrow \quad \frac{B}{A} = 11$$

$$\frac{C}{A} = \sum (\alpha^2 + 1)(\beta^2 + 1) \equiv \sum \alpha^2\beta^2 + 2\sum \alpha^2 + 3.$$

But $\quad \sum \alpha^2\beta^2 \equiv (\alpha\beta + \beta\gamma + \gamma\alpha)^2 - 2(\alpha\beta^2\gamma + \beta\gamma^2\alpha + \gamma\alpha^2\beta)$

$$\equiv \left(\sum \alpha\beta\right)^2 - 2\alpha\beta\gamma\sum \alpha$$

$$= 49 - 0$$

and $\quad \sum \alpha^2 = \left(\sum \alpha\right)^2 - 2\sum \alpha\beta$

$$= 0 - 14$$

So $\quad \dfrac{C}{A} = 49 - 28 + 3$

$$\Rightarrow \quad \frac{C}{A} = 24$$

$$-\frac{D}{A} = (\alpha^2 + 1)(\beta^2 + 1)(\gamma^2 + 1)$$

$$\equiv \alpha^2\beta^2\gamma^2 + \sum \alpha^2\beta^2 + \sum \alpha^2 + 1$$

$$= (-5)^2 + 49 - 14 + 1 = 61$$

$$\Rightarrow \quad \frac{D}{A} = -61.$$

So the required equation is

$$X^3 + 11X^2 + 24X - 61 = 0.$$

Repeated Roots

If a cubic equation $f(x) = 0$ has two equal roots, α, then $f(x)$ has a repeated factor $(x - \alpha)$.
We saw earlier in the chapter that, in this case, $(x - \alpha)$ is also a factor of $f'(x)$.
So the equation $f'(x) = 0$ has a root, α, i.e. $f'(\alpha) = 0$.
So, if α is a repeated root of a given cubic equation, α is also a root of the quadratic equation given by differentiating the cubic equation.
Note. This property applies also to equations of higher degree.

EXAMPLES 11e (continued)

5) Find the condition that the equation $x^3+px+q=0$ shall have a repeated root.

Let α be the repeated root, so that α satisfies both the given equation and its differential,

i.e. $$\alpha^3+p\alpha+q = 0 \quad [1]$$

and $$3\alpha^2+p = 0 \quad [2]$$

The condition for $x^3+px+q=0$ to have a repeated root must involve p and q so, to find it, we eliminate α from [1] and [2] which can be done as follows:

[2] $\times\alpha$ $\Rightarrow$ $$3\alpha^3+p\alpha = 0$$

[1] $\times 3$ $\Rightarrow$ $$3\alpha^3+3p\alpha+3q = 0.$$

Hence $$2p\alpha+3q = 0$$

$$\Rightarrow \quad \alpha = -\frac{3q}{2p}$$

Then [2] becomes $3\left(-\dfrac{3q}{2p}\right)^2+p = 0$

So the required condition is $27q^2+4p^3=0$.

6) Solve the equation $18x^3-111x^2+224x-147=0$ given that it has two equal roots.

The repeated root of the given equation

$$18x^3-111x^2+224x-147 = 0 \quad [1]$$

also satisfies the equation $$54x^2-222x+224 = 0$$

i.e. $$27x^2-111x+112 = 0$$

i.e. $$(9x-16)(3x-7) = 0$$

So *either* $x=\frac{16}{9}$ *or* $x=\frac{7}{3}$ is a solution of the given equation.
To determine which of these values is the repeated root of equation [1], the following method can be used.
Suppose that $\frac{7}{3}$ is the repeated root (choosing first the simpler of the two possible values), then equation [1] has roots $\frac{7}{3}, \frac{7}{3}, \alpha$, where

$$(\tfrac{7}{3})(\tfrac{7}{3})(\alpha) = -\frac{d}{a} = \tfrac{147}{18} \Rightarrow \alpha = \tfrac{3}{2}.$$

Now $\frac{7}{3}+\frac{7}{3}+\frac{3}{2} = \frac{37}{6} = \frac{111}{18}$ which is the value of $-\dfrac{b}{a}$

So $\frac{7}{3}, \frac{7}{3}, \frac{3}{2}$ are the roots of equation [1].
(If $\frac{16}{9}$ is chosen first, we find that

$$\alpha = \tfrac{47}{18} \quad \text{and that} \quad \tfrac{16}{9} + \tfrac{16}{9} + \tfrac{47}{18} \neq -\frac{b}{a}.$$

So we conclude that $\frac{16}{9}$ is *not* the repeated root of [1].)
Alternatively, the reader may prefer to substitute *each* of the values ($\frac{7}{3}$ and $\frac{16}{9}$) into the L.H.S. of equation [1] to check which is the repeated root. Then the third root can be determined, either by factorising, or by using

$$\sum \alpha = -\frac{b}{a} \quad \text{or} \quad \alpha\beta\gamma = -\frac{d}{a},$$

i.e. if $\qquad f(x) \equiv 18x^3 - 111x^2 + 224x - 147$

then $\qquad f(\tfrac{16}{9}) = \tfrac{8192}{81} - \tfrac{9472}{27} + \tfrac{3584}{9} - 147 \neq 0$

so $\frac{16}{9}$ is *not* a repeated root of equation [1]. It might reasonably be assumed that the repeated root therefore *must* be $\frac{7}{3}$ but it is wiser to check that this is so,

i.e. $\qquad f(\tfrac{7}{3}) = \tfrac{686}{3} - \tfrac{1813}{3} + \tfrac{1568}{3} - 147 = 0$

So $\frac{7}{3}$ *is* the repeated root.
Then, if α is the third root,

$$(\tfrac{7}{3})(\tfrac{7}{3})(\alpha) = -\frac{d}{a} = \tfrac{147}{18} \quad \Rightarrow \quad \alpha = \tfrac{3}{2}$$

So the roots of [1] are $\frac{7}{3}, \frac{7}{3}, \frac{3}{2}$.

EXERCISE 11e

Solve the following equations, given that they each have a repeated root.

1) $80x^3 + 88x^2 - 3x - 18 = 0$ 2) $45x^3 - 69x^2 + 32x - 4 = 0$

3) $20x^3 - 68x^2 + 69x - 18 = 0$ 4) $112x^3 + 152x^2 + 39x - 9 = 0$

5) $x^4 - 12x^3 - 2x^2 + 36x + 2025 = 0$

6) If α, β, γ are the roots of the following equations, write down the values of $\sum \alpha$, $\sum \alpha\beta$ and $\alpha\beta\gamma$.

(a) $4x^3 - x^2 + 2x = 7$ (b) $x^3 - 3x + 1 = 0$ (c) $8x^3 = 1$

(d) $x^3 - x = 0$ (e) $x^3 + 4x^2 = 5$

7) Find the values of $\sum \alpha^2$, $\sum \alpha^2\beta^2$, $\sum \alpha\beta(\alpha + \beta)$ and $\sum \frac{1}{\alpha}$ in each of the following cases.

(a) $x^3 - 3x^2 + x + 5 = 0$ (b) $3x^3 + x^2 - 4x + 1 = 0$
(c) $4x^3 + 3x + 7 = 0$ (d) $x^4 - x^3 + 2x + 3 = 0$
(e) $x^3 + 1 = 0$ (f) $x^4 + x = 1$
(g) $x^4 + x^3 = 0$

8) If the equation $x^3 + 2x^2 - 5x + 1 = 0$ has roots α, β, γ, find the equation with roots

(a) $\alpha - 2,\ \beta - 2,\ \gamma - 2$ (b) $\dfrac{1}{\alpha}, \dfrac{1}{\beta}, \dfrac{1}{\gamma}$
(c) $2\alpha,\ 2\beta,\ 2\gamma$ (d) $\alpha^2,\ \beta^2,\ \gamma^2$
(e) $\beta\gamma,\ \gamma\alpha,\ \alpha\beta$ (f) $\alpha + \beta,\ \beta + \gamma,\ \gamma + \alpha$
(g) $\alpha^2 + \alpha,\ \beta^2 + \beta,\ \gamma^2 + \gamma$

9) The equation $7x^3 - 4x - 11 = 0$ has roots α, β, γ. Prove that

(a) $7\sum\alpha^3 = 33 + 4\sum\alpha$ (b) $7\sum\alpha^4 = 11\sum\alpha + 4\sum\alpha^2$.

Evaluate $\sum\alpha^3$ and $\sum\alpha^4$ and write down a similar expression for $\sum\alpha^5$.

10) Find the relationship between a, b, c and d if the roots of the equation $ax^3 + bx^2 + cx + d = 0$ are:

(a) in geometric progression,
(b) such that one root is equal to the sum of the other two,
(c) all equal.

11) If $\alpha + \beta + \gamma = 2$, $\alpha\beta + \beta\gamma + \gamma\alpha = -5$ and $\alpha\beta\gamma = -6$, write down the equation with roots α, β, γ and hence evaluate α, β and γ.

12) If $\alpha, \beta, \gamma, \delta$ are the roots of the equation

$$3x^4 + 4x^3 - 7x^2 + 5x - 3 = 0$$

write down the values of $\sum\alpha$, $\sum\alpha\beta$, $\sum\alpha\beta\gamma$ and $\alpha\beta\gamma\delta$.

Find the values of $\sum\alpha^2$ and $\sum\dfrac{1}{\alpha}$

Find, also, the equation whose roots are $\dfrac{\alpha}{2}, \dfrac{\beta}{2}, \dfrac{\gamma}{2}$ and $\dfrac{\delta}{2}$

13) If $f(x)$ is a quartic function of x with a repeated factor, $(x - a)$, prove that $(x - a)$ is also a factor of $f'(x)$.
Hence solve the equation

$$3x^4 - 8x^3 - 6x^2 + 24x = 8.$$

14) If the roots of the equation

$$x^4 - px^3 + qx^2 - pqx + 1 = 0$$

are α, β, γ and δ, show that

$$(\alpha+\beta+\gamma)(\alpha+\beta+\delta)(\alpha+\gamma+\delta)(\beta+\gamma+\delta) = 1.$$

15) Find the equation whose roots are given by adding 2 to the roots of the equation

$$x^4 + 3x^3 - 13x^2 - 51x - 36 = 0.$$

Hence solve the given equation.

THE SOLUTION OF EQUATIONS

Some polynomial equations of degree higher than two cannot be solved either by using the Remainder Theorem or by one of the special methods previously demonstrated. In such cases we can find only approximate values for the roots, using a graphical or a numerical approach.
Consider, first, a curve with a polynomial equation $y = f(x)$, (which is known to be continuous),

e.g.

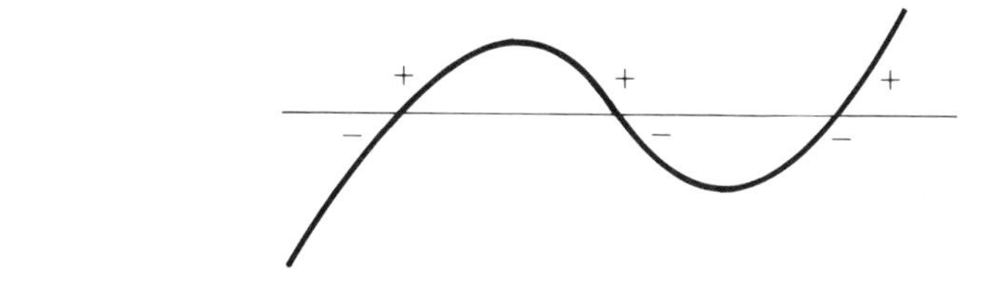

or

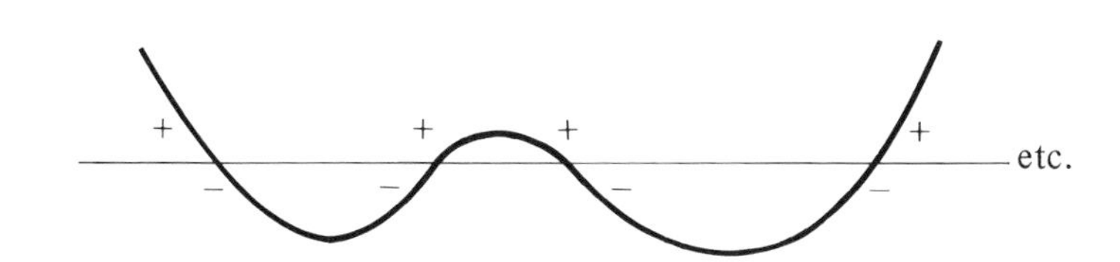

On each occasion that the curve crosses the x axis, the sign of y changes.
If, therefore, one only of these crossing points lies between two values of x, $x = x_1$ and $x = x_2$, then $f(x_1)$ and $f(x_2)$ are of opposite sign.
Conversely, if we can find two values for x such that $f(x_1)$ and $f(x_2)$ have opposite signs, then we know that the curve $y = f(x)$ has crossed the x axis between x_1 and x_2.
We cannot assume that there is only one crossing point, however, as is shown by the following diagrams.

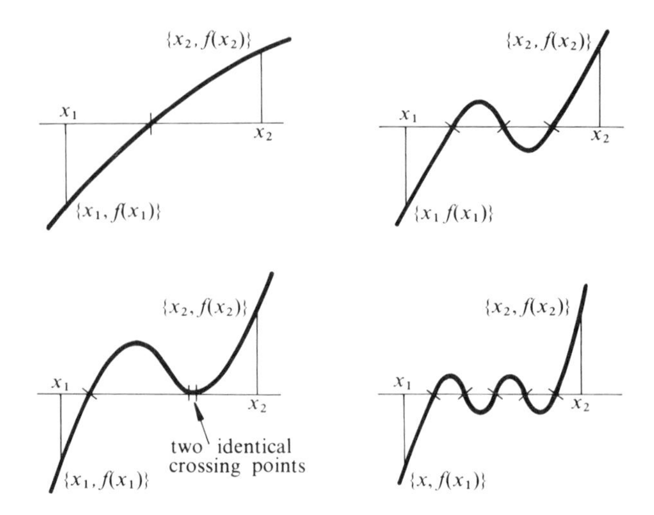

These examples illustrate that

if $f(x_1)f(x_2) < 0$, then the equation $f(x) = 0$ has an odd number of real roots between x_1 and x_2 (some of these roots may be equal)

Now if $f(x_1)$ and $f(x_2)$ have the same sign then the curve $y = f(x)$ has either not crossed the x-axis at all, or has met it an even number of times between x_1 and x_2, as the following diagrams illustrate.

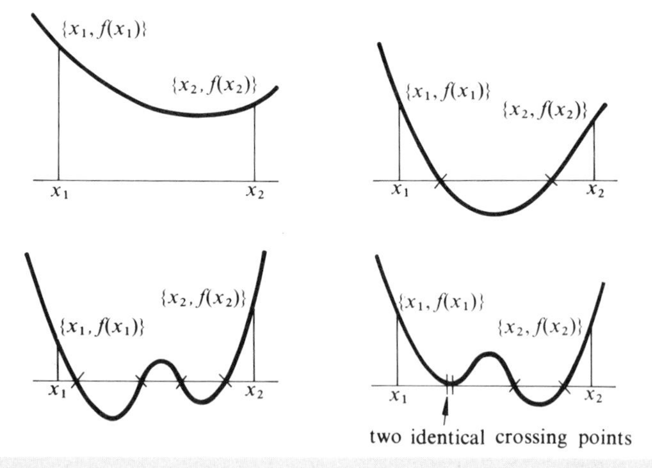

So, if $f(x_1)f(x_2) > 0$, then the equation $f(x) = 0$ has an even number of real roots (distinct or equal) between x_1 and x_2.

EXAMPLES 11f

1) Given that the roots of the equation $12x^3 - 112x^2 + 267x - 77 = 0$ all lie in the range $0 < x < 10$, find the integral values of x between which each of these roots lies.

If $$f(x) \equiv 12x^3 - 112x^2 + 267x - 77 \qquad [1]$$

then
$$\left.\begin{aligned} f(0) &= -77 \\ f(1) &= 90 \end{aligned}\right\}$$
$$f(2) = 105$$
$$\left.\begin{aligned} f(3) &= 40 \\ f(4) &= -33 \end{aligned}\right\}$$
$$\left.\begin{aligned} f(5) &= -42 \\ f(6) &= 85 \end{aligned}\right\}$$

Because $f(x)$ changes sign between $x = 0$ and 1, $x = 3$ and 4, $x = 5$ and 6, we know that, between each of these pairs of integers, there is an odd number of roots of equation [1]. But a cubic equation cannot have more than three roots, so one root lies in each of these intervals, and no further sign changes can occur.
Hence the roots of the given equation lie between
$x = 0$ and 1, 3 and 4, 5 and 6.

THE NUMBER OF REAL ROOTS OF A POLYNOMIAL EQUATION

Although a polynomial equation of degree n may have, at most, n real roots, the actual number of real roots may be less than n. In this case an attempt to find n intervals within which $f(x)$ changes sign would be a fruitless task. So it is useful to be able to assess the *number* of real roots of a given equation before trying to locate them. It is not always possible to do this, but it is worth trying one or both of the methods given below, each of which can, in some cases, be very helpful.

(1) Consider a cubic equation $f(x) = 0$.
The curve $y = f(x)$ has either two turning points or a point of inflexion (which is not necessarily parallel to the x axis).
When there are two turning points, the signs of y_{max} and y_{min} can be used to determine the number of real roots of the equation $f(x) = 0$, as follows:

(a)

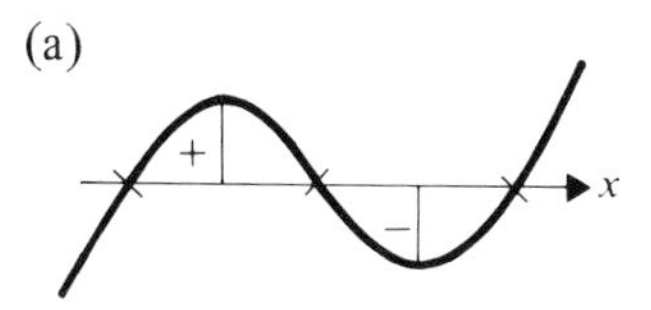

if $(y_{max})(y_{min}) < 0$ there are three real distinct roots.

(b)

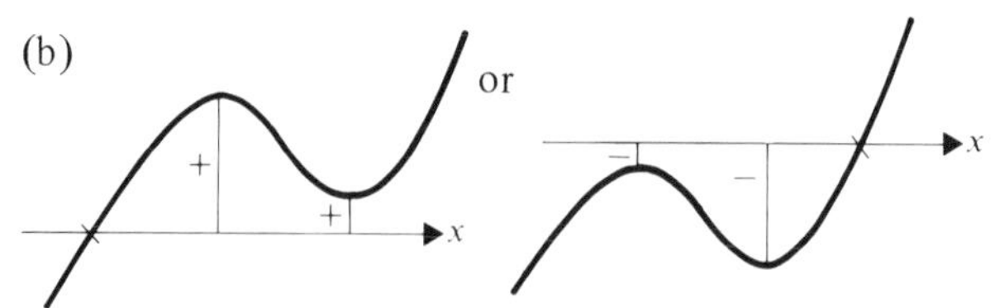

if $(y_{max})(y_{min}) > 0$ there is only one real root.

(c)

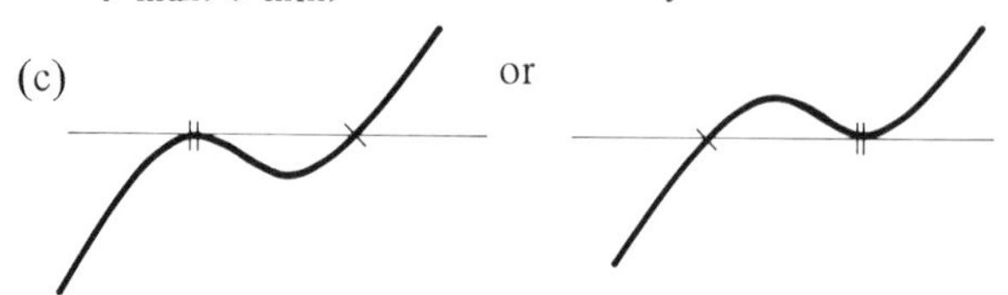

if $(y_{max})(y_{min}) = 0$ there are three real roots one of which is repeated.

If there is a point of inflexion then

(d)

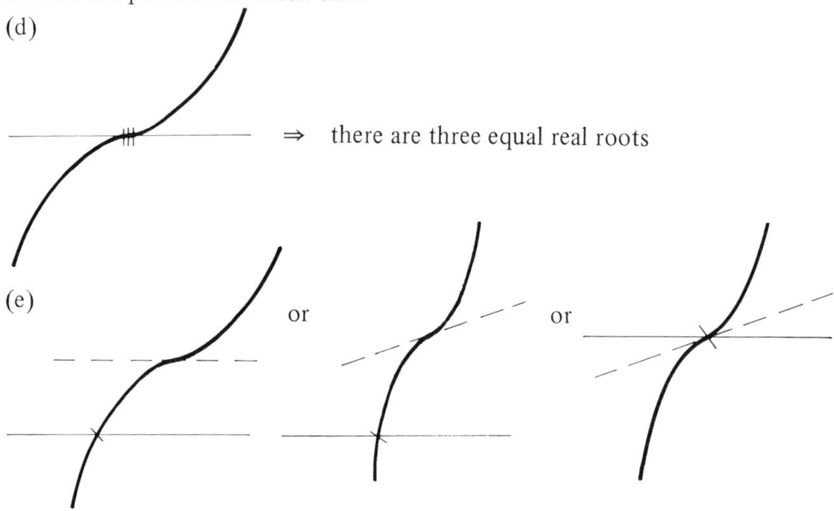

⇒ there is only one real root.

Consider, for example, the equation $2x^3 - 9x^2 + 12x - 1 = 0$.

If $\quad y = 2x^3 - 9x^2 + 12x - 1,$

then $\quad \dfrac{dy}{dx} = 6x^2 - 18x + 12$

At turning points $\quad \dfrac{dy}{dx} = 0$

$\Rightarrow \quad x = 1$ or 2.

When $x = 1, \quad y = 4$

When $x = 2, \quad y = 3$

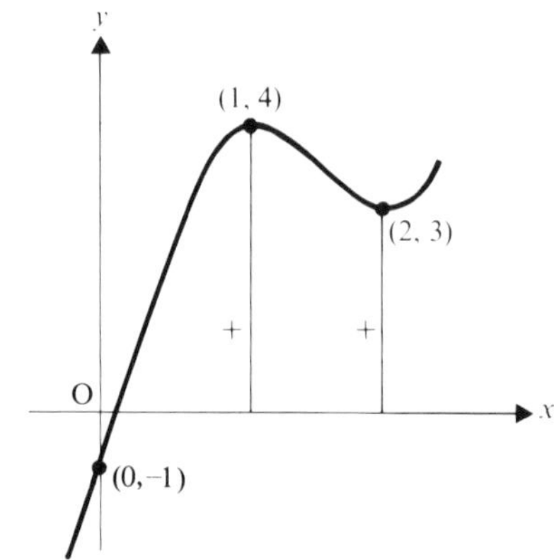

The maximum and minimum values of y have the *same* sign, so the given equation has only one real root.
The coordinates of the turning points, together with a rough sketch of the curve, indicate that this real root is between 1 and zero.
Note. If this method is attempted for equations of degree higher than 3, there is no guarantee that the derived equation, $f'(x) = 0$, can be solved.

(2) Another way of trying to find the number of real roots of a polynomial equation, $f(x) = 0$, is given below. It is offered, without any attempt at proof of its validity, as a useful aid.

First Step. Count the number of times that consecutive coefficients of the terms in $f(x)$ are of opposite sign. If this happens p times then we can say that the equation $f(x) = 0$ has *not more than* p *positive real roots.*

Second Step. Consider the coefficients of $f(-x)$ in the same way. If the signs of consecutive coefficients change q times we can say that the original equation $f(x) = 0$ *has not more than* q *negative real roots.*

This process is very quickly completed, and, if $p + q < n$, reduces the amount of work subsequently carried out in locating the real roots.
Consider, for example, the equation

$$x^3 - 2x^2 - 7 = 0 \qquad [1]$$

which, being cubic, may have three real roots.
However, the coefficients $+1, -2, -7$ include only one sign change,
so there is not more than one real positive root
Now examining the equation given by replacing x by $-x$,

i.e. $$-x^3 - 2x^2 - 7 = 0$$

we see that the coefficients have no sign changes,
so there are no real negative roots.
Thus we see that there is only one real root of [1] and that it is positive.
The single root can now be located approximately as before.
The technique described above does not always help however.
For example, if $x^3 + 2x^2 - 7 = 0$
we see from $+1, +2, -7$ that there is not more than one real positive root.
Then, replacing x by $-x$, we see from $-1, +2, -7$ that there are not more than two real negative roots.
So this time the cubic equation can have three real roots (but is not *certain* to have three) and the only useful information we have gained is that only one of the roots can be positive.
Note that the information obtained by this technique is often of value only when used in conjunction with basic knowledge of the graph of the polynomial function $f(x)$.

EXAMPLES 11f (continued)

2) Find the number of real roots of the equation $x^4 + 2x^3 + 8x + 15 = 0$ and find the integral values of x between which each of these roots lies.

If $f(x) \equiv x^4 + 2x^3 + 8x + 15$, the coefficients of $f(x)$, i.e. 1, 2, 8, 15, include no sign change.
So $f(x) = 0$ has *no* real positive roots.
The coefficients of $f(-x) \equiv x^4 - 2x^3 - 8x + 15$ are $1, -2, -8, 15$ and include two sign changes.
So $f(x) = 0$ has not more than two real negative roots.
This result is not conclusive as the equation $f(x) = 0$ can have either, two real negative roots (distinct or equal)

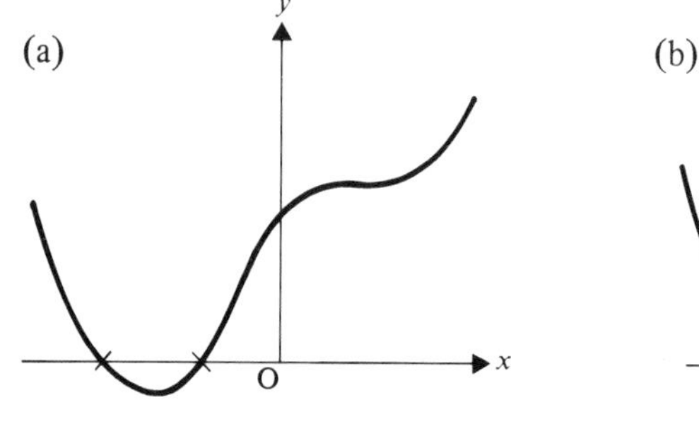

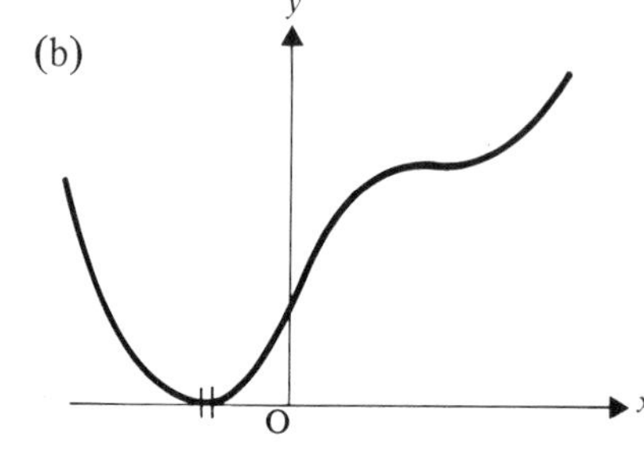

or, no real roots

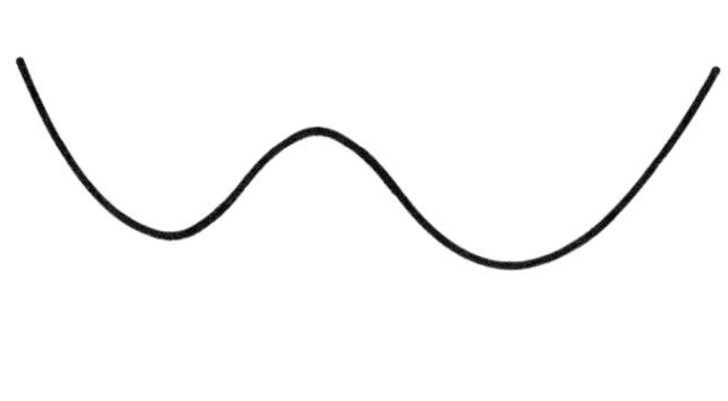

($f(x) = 0$ *cannot* have only one real root as its complex roots must occur in conjugate *pairs*.)
To determine which is the case we can now look for turning points on the curve $y = f(x)$

$$\frac{dy}{dx} = 4x^3 + 6x^2 + 8 \equiv 2(x+2)(2x^2 - x + 2)$$

$$\Rightarrow \qquad \frac{dy}{dx} = 0 \quad \text{only when} \quad x = -2 \quad \text{and} \quad y = -1.$$

So there *is* a turning point below the x axis, showing that the curve crosses the x axis twice as in diagram (a) above ($f(x) \to \infty$ when $x \to \pm\infty$).
Thus the equation $f(x) = 0$ has exactly two real negative roots, which lie on opposite sides of $x = -2$.

Then,
$$f(0) = 15$$
$$f(-1) = 6$$
$$f(-2) = -1$$
$$f(-3) = 18.$$

$f(x)$ changes sign between $x = -1$ and -2
and between $x = -2$ and -3.
So one real root lies in each of these intervals.

EXERCISE 11f

Find the maximum number of real roots of each of the following equations.

1) $x^3 + x + 2 = 0$

2) $3x^4 - 4x^3 + x^2 + 7 = 0$

3) $2x^5 + x^3 = 1$

4) $x^4 + 4x - 9 = 0$

Determine the *exact* number of real roots of each of the following equations.

5) $x^3 - 3x^2 - 4 = 0$

6) $3x^4 - 4x^3 + 6x^2 - 12x + 5 = 0$

7) $3x^4 + 8x^3 + 24x^2 + 96x - 10 = 0$

8) $x^3 - 3x^2 + 1 = 0$

9) In Questions 5–8, find the consecutive integers between which each real root lies.

10) If $x^3 + \lambda x^2 + 2 = 0$ explain why, if $\lambda > 0$, the equation must have two complex roots.

11) Find the range of values of λ for which the equation $2x^3 + 9x^2 + 12x + \lambda = 0$ has three real distinct roots.

APPROXIMATE SOLUTION OF EQUATIONS

The graphical method for finding an approximate value for a root of an equation was explained in Volume 1. In this book we shall examine a numerical approach to this problem. There is a variety of numerical methods, but the three given below are probably those in commonest use.

Iterative Methods

METHOD 1

Suppose that we wish to find the smallest positive root of the equation
$$x^3 + 2x^2 + 5x - 1 = 0.$$

First we locate the consecutive integral values of x between which the required root lies.

We find that, if $f(x) \equiv x^3 + 2x^2 + 5x - 1$, then $f(0) = -1$ and $f(1) = 7$.
So there is a root between 0 and 1 (which is likely to be closer to 0 than to 1).
A more accurate value for this root can be found as follows:
Let the root be $0 + h_1$, so that $f(x) = 0$ becomes

$$h_1{}^3 + 2h_1{}^2 + 5h_1 - 1 = 0.$$

But h_1 is small, so the terms $h_1{}^2$ and $h_1{}^3$ can be neglected, giving $5h_1 - 1 \simeq 0$.

i.e. $$h_1 \simeq 0.2$$

Now, for a better approximation, let the root be $(0.2 + h_2)$, giving

$$(0.2 + h_2)^3 + 2(0.2 + h_2)^2 + 5(0.2 + h_2) - 1 = 0$$

Neglecting terms in $h_2{}^2$ and $h_2{}^3$ gives

$$(0.2)^3 + 3(0.2)^2 h_2 + 2(0.2)^2 + 4(0.2)h_2 + 1 + 5h_2 - 1 \simeq 0$$

$\Rightarrow$ $$h_2 \simeq -0.015$$

So a better approximation to the root between 0 and 1 is $0.2 - 0.015$
i.e. 0.185
This process can be repeated as often as required, the next step being to let the root be $(0.185 + h_3)$, giving

$$(0.185)^3 + 3(0.185)^2 h_3 + 2(0.185)^2 + 4(0.185)h_3 + 5(0.185) + 5h_3 - 1 \simeq 0$$

(again neglecting terms containing $h_3{}^2$ and $h_3{}^3$).
Hence h_3 can be found, giving $(0.185 + h_3)$ as a better approximation for the value of the root, and so on.
Note. This method can be used to find *any* real root of a polynomial equation. It is of no significance that, in the example above, the required root is small.

METHOD 2 $x = g(x)$

This method can often be used to find a root of an equation $f(x) = 0$ which can be written in the form $x = g(x)$.
The roots of the equation $x = g(x)$ are the values of x at the points of intersection of the line $y = x$ and the curve $y = g(x)$.
Taking x_1 as a first approximation to a root α then

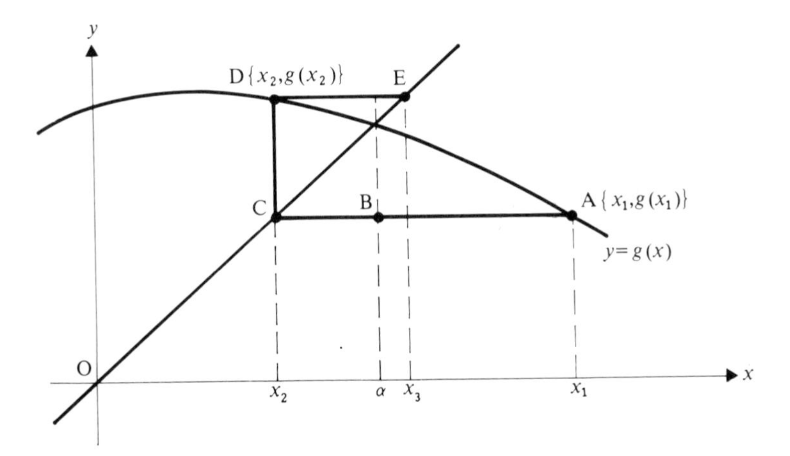

in the diagram,

A is the point on the curve where $x = x_1$, $y = g(x_1)$
B is the point where $x = \alpha$, $y = g(x_1)$
C is the point on the line where $x = x_2$, $y = g(x_1)$

If in the region of α the slope of $y = g(x)$ is less steep than that of the line

i.e. provided $|g'(x)| < 1$

then CB < BA

so x_2 is closer to α than is x_1
i.e. x_2 is a better approximation to α.
But C is on the line $y = x$

therefore $x_2 = g(x_1)$

Now taking the point D on the curve where $x = x_2$, $y = g(x_2)$ and repeating the argument above we find that x_3 is a better approximation to α than is x_2

where $x_3 = g(x_2)$

This process can be repeated as often as necessary to achieve the required degree of accuracy.
The rate at which these approximations converge to α depends on the value of $|g'(x)|$ near α. The smaller $|g'(x)|$ is, the more rapid is the convergence.
It should be noted that this method fails if $|g'(x)| > 1$ near α.

The following diagrams illustrate some of the factors which determine the success, or otherwise, of this method.

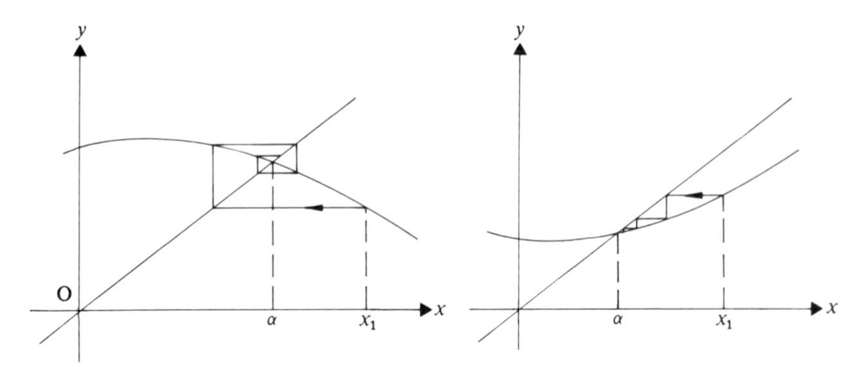

Rapid rate of convergence ($|g'(x)|$ small).

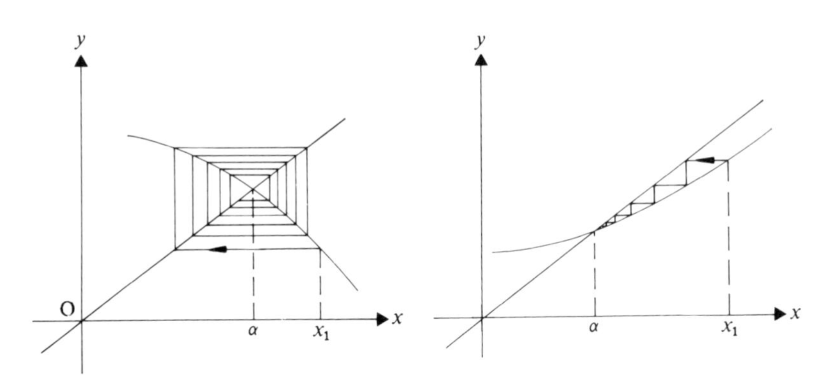

Slow rate of convergence ($|g'(x)| < 1$ but close to 1)

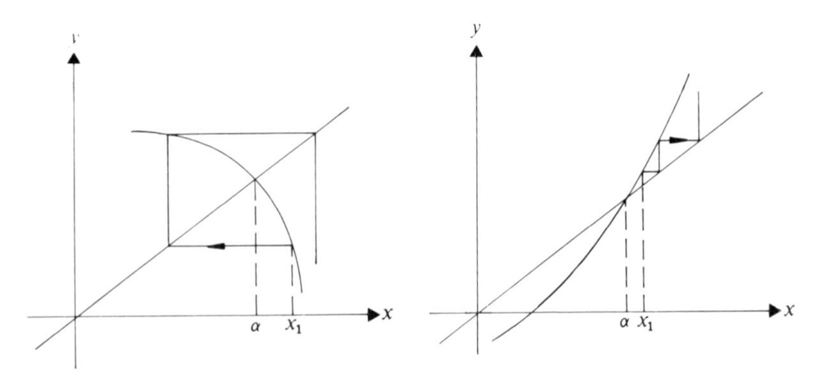

Divergence, i.e. failure ($|g'(x)| > 1$)

The reader is recommended to vary the shape of the curve and the position of x_1 to investigate the effect of these changes on the convergence or otherwise of approximations.
As an example consider again the equation

$$x^3 + 2x^2 + 5x - 1 = 0$$

which we saw in method 1 above has a root α for which $x = 0.2$ is a first approximation.
This equation can be written in the form

$$x = g(x)$$

where $$g(x) \equiv -\tfrac{1}{5}(x^3 + 2x^2 - 1)$$

A better approximation, x_2, is found from

$$x_2 = g(x_1) = -\tfrac{1}{5}\{(0.2)^3 + 2(0.2)^2 - 1\}$$
$$= 0.1824$$

Further improvements are obtained by repeating this step

i.e. $$x_3 = g(x_2) = -\tfrac{1}{5}\{(0.1824)^3 + 2(0.1824)^2 - 1\}$$
$$= 0.1855 \qquad \text{(to 4 d.p.)}$$
$$x_4 = g(x_3) = -\tfrac{1}{5}\{(0.1855)^3 + 2(0.1855)^2 - 1\}$$
$$= 0.1850 \qquad \text{(to 4 d.p.)}$$

and so on.

Both of the methods described above involve using successive approximations step-by-step to find better approximations. Such a method is called an *iterative method.* The degree of accuracy at any stage can be checked by determining the sign of $f(x)$ on either side of the value so far obtained for the root, e.g. taking $x \simeq 0.1850$ we find that $f(0.1846)$ is negative and $f(0.1854)$ is positive, so $x = 0.185$ correct to 3 d.p.
Note. This does *not* show that $x = 0.1850$ to 4 d.p.

Linear Interpolation

This method, which can be applied to equations other than polynomials, is based on the use of linear proportion in successive steps.
Consider a function $f(x)$ which changes sign between $x = a$ and $x = b$, so that the equation $f(x) = 0$ has a root, α, in this range.

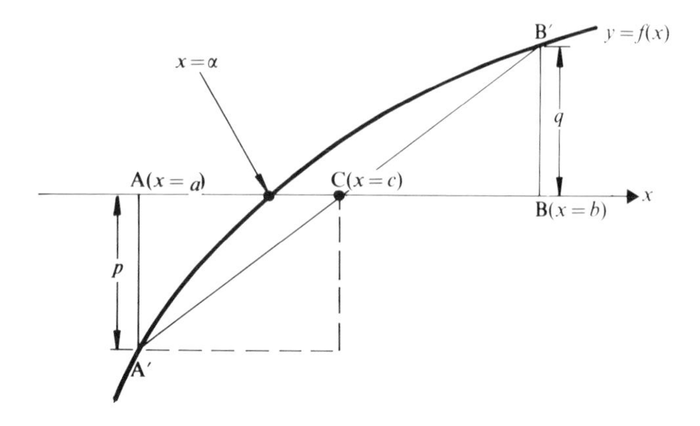

In the diagram we see that the line joining $A'(a, f(a))$ and $B'(b, f(b))$ crosses the x-axis at C where C divides AB in the ratio $p:q$ (from the similar triangles $A'AC$ and $B'BC$).

So, at C, $\quad x = \dfrac{aq + bp}{p + q} \quad$ where $\quad p = |f(a)| \quad$ and $\quad q = |f(b)|$.

Now if $b - a$ is small, $c - \alpha$ is very small, so the value of x at C is an approximate root of $f(x) = 0$.
For a better approximation we can now take two points C and D on either side of the root, α, and use linear interpolation between C and D, to find the value of x at E,

i.e. $$x = \frac{dr + cs}{r + s}.$$

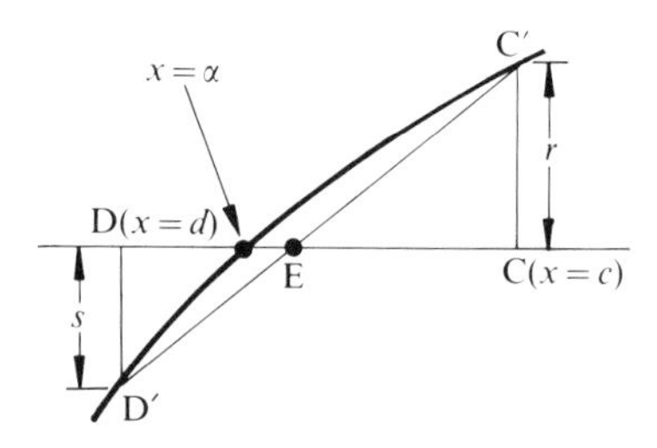

Each successive step gives a more accurate estimate of the root, α.

EXAMPLES 11g

1) Use linear interpolation to find the larger root of the equation $\ln x = x - 2$ correct to 3 decimal places, checking the accuracy of your result.

Before we can begin using linear interpolation, we must find the approximate location of the specified root by plotting the graphs $y = \ln x$ and $y = x - 2$ on the same axes,

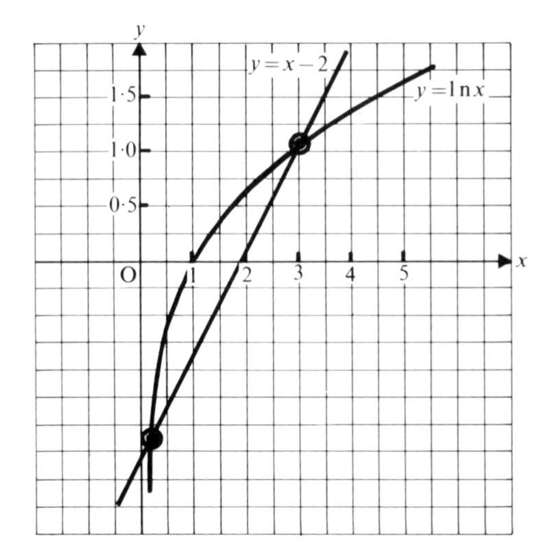

From the graph we see that there are two real roots of the equation $\ln x = x - 2$ and that the larger root is slightly larger than 3. So, considering $f(x) \equiv \ln x - x + 2$, the graph of $y = f(x)$ crosses the x axis between $x = 3$ and 3.5

When $x = 3$, $f(x) = 0.099$
and when $x = 3.5$, $f(x) = -0.247$
Therefore, at C,

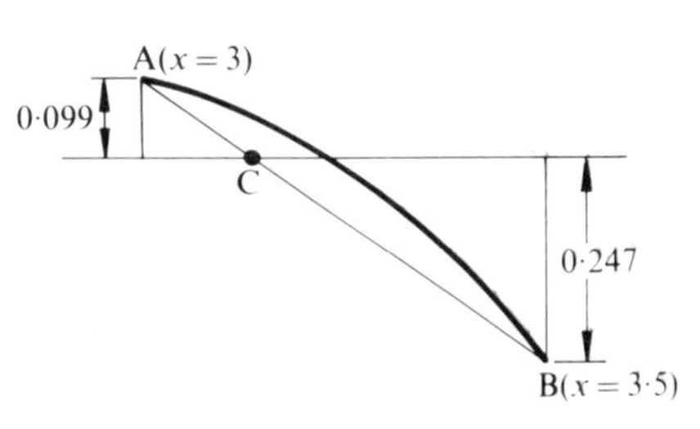

$$x = \frac{0.099 \times 3.5 + 0.247 \times 3}{0.099 + 0.247}$$

$$\simeq 3.14$$

Now we take the points where x is 3.14 and 3.15 for the next linear interpolation (checking that $f(3.14)$ and $f(3.15)$ have opposite signs so that the required root lies between these values of x).

When $x = 3.14$, $f(x) = 0.0042$
and when $x = 3.15$, $f(x) = -0.0026$
Therefore, at E,

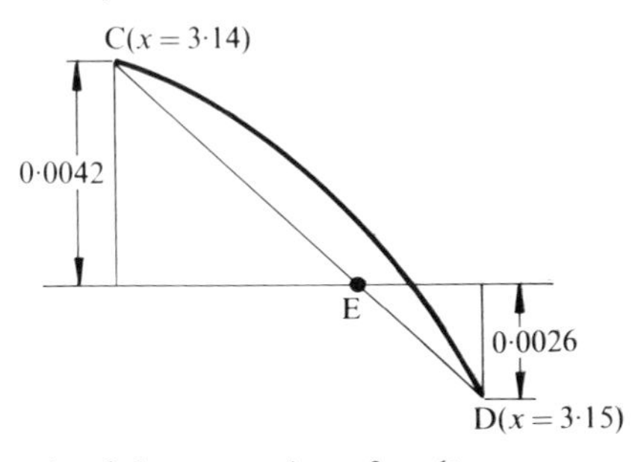

$$x = \frac{0.0026 \times 3.14 + 0.0042 \times 3.15}{0.0026 + 0.0042}$$

$$\simeq 3.146$$

This value is correct to 3 decimal places only if the true value of x lies between 3.1455 and 3.1464, so to check the accuracy of the third decimal place we can evaluate $f(3.1455)$ and $f(3.1464)$, and we find that $f(3.1455) > 0$ and $f(3.1464) < 0$.

So the root lies between these values and is therefore equal to 3.146 correct to 3 d.p.

Newton's Method (or the Newton–Raphson Method)

This method is based on determining, first, a better linear approximation for the equation of a curve than was used in Method 1.

In the diagram, P and Q are two points on a curve $y = f(x)$, separated by a small horizontal distance h.
So P is $(a, f(a))$
and Q is $\{(a+h), f(a+h)\}$.
The gradient of PR, the tangent at P,

is $\dfrac{\text{RS}}{h} \Rightarrow f'(a) = \dfrac{\text{RS}}{h}$

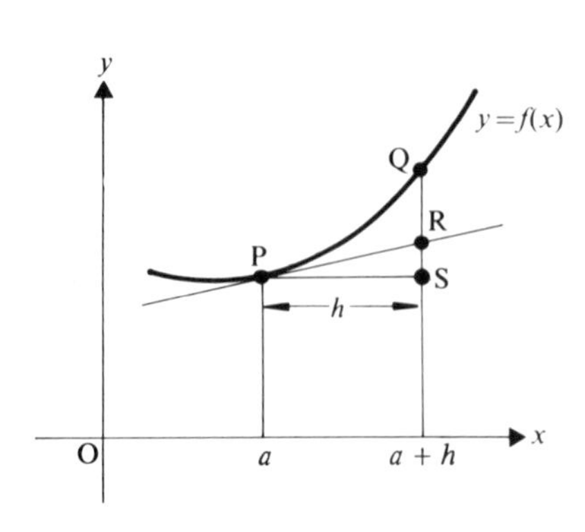

Now, if h is small, $\text{QS} \simeq \text{RS}$.

So $$f'(a) \simeq \frac{\text{QS}}{h} = \frac{f(a+h) - f(a)}{h}$$

i.e. $$f(a+h) \simeq f(a) + hf'(a)$$

Alternatively, if $x \simeq a + h$, this relationship can be expressed in the form

$$f(x) \simeq f(a) + (x-a)f'(a)$$

The RHS of this equation is a linear approximation for $f(x)$ in the region of $x = a$.
(**Note** also that it is the first two terms of Taylor's series.)

EXAMPLES 11g (continued)

2) Given that $1^\circ = 0.0175^c$, estimate the value of $\tan 46^\circ$.

If $f(x) \equiv \tan\theta$, then $f'(x) \equiv \sec^2\theta$.

Using $a = \dfrac{\pi}{4}$ and $h = 0.0175^\circ$ in

$$f(a+h) \simeq f(a) + hf'(a)$$

gives $$\tan\left(\frac{\pi}{4} + 0.0175^c\right) \simeq \tan\frac{\pi}{4} + 0.0175\sec^2\left(\frac{\pi}{4}\right)$$

$$\simeq 1 + 0.035$$

Hence $$\tan 46^\circ \simeq 1.035$$

APPROXIMATE SOLUTION OF EQUATIONS USING NEWTON'S METHOD

Consider a function $f(x)$ which has a zero value between two integers a and $a+1$ and which is continuous in this region. Then the equation $f(x)=0$ has a root $a+h$ where $0<h<1$.

Now
$$f(a+h) \simeq f(a)+hf'(a).$$

But $f(a+h)=0$ since $a+h$ is a root of the equation $f(x)=0$.

So
$$h \simeq -\frac{f(a)}{f'(a)}.$$

Thus, if $x = a$ is first approximation to a solution of the given equation,
$x = a-\dfrac{f(a)}{f'(a)}$ is a second (and better) approximation.

The result can be illustrated geometrically as follows:

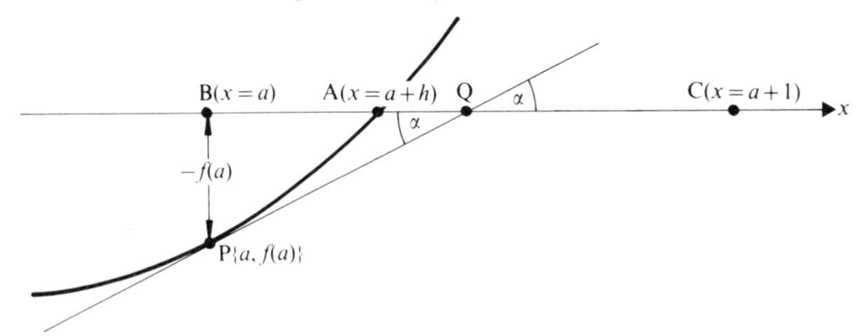

In the diagram,
$$\text{BQ} = \text{BP} \cot \alpha$$

But the gradient of PQ is $f'(a)=\tan \alpha$.

So
$$\text{BQ} = \frac{-f(a)}{f'(a)}$$

Taking $\text{BA} \simeq \text{BQ}$ when h is small, gives
$$\text{BA} \simeq -\frac{f(a)}{f'(a)}$$

So, at A,
$$x \simeq a-\frac{f(a)}{f'(a)}$$

i.e. at B we have the first approximation to the root at A, and at Q the second approximation to this root.
Further approximations can be obtained in the same way using successively more accurate values for a.
Note. There are certain cases where this method fails, because the approximations made above are too crude. These can be understood by referring to the graphs when

(a) h is too large

BA $\not\approx$ BQ.

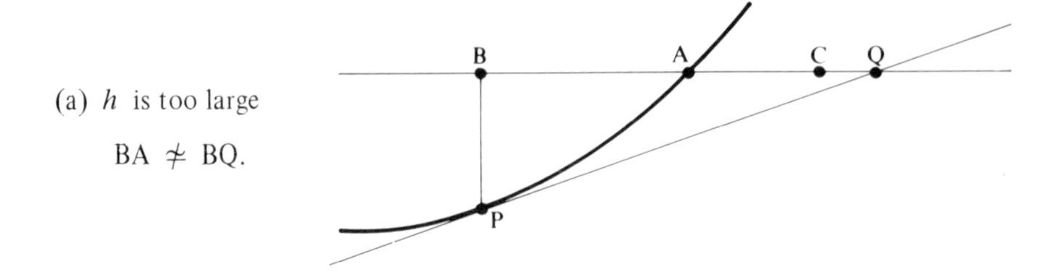

(b) $f'(a)$ is too small

BA $\not\approx$ BQ.

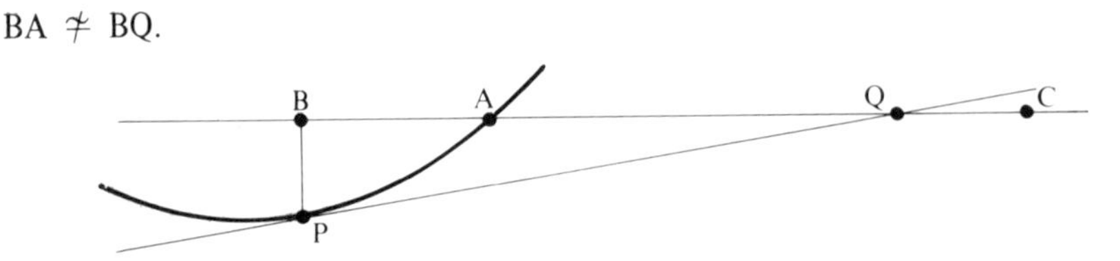

(c) $f''(a)$ is too large

BA $\not\approx$ BQ.

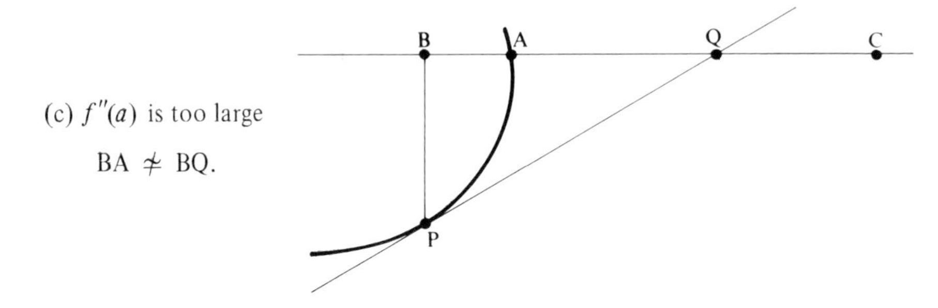

Clearly Newton's Method can be used only when none of these extreme cases is being considered.

Note also, that when valid, Newton's Method can be applied to finding approximate solutions of equations other than polynomial equations. In these cases, a first approximation for the root can be found by a graphical method.

EXAMPLES 11g (continued)

3) Find the root of $xe^x = 3$ correct to 3 decimal places, using Newton's Method.

A first approximation to the root is found by drawing the graphs of $y = \dfrac{3}{x}$ and $y = e^x$.

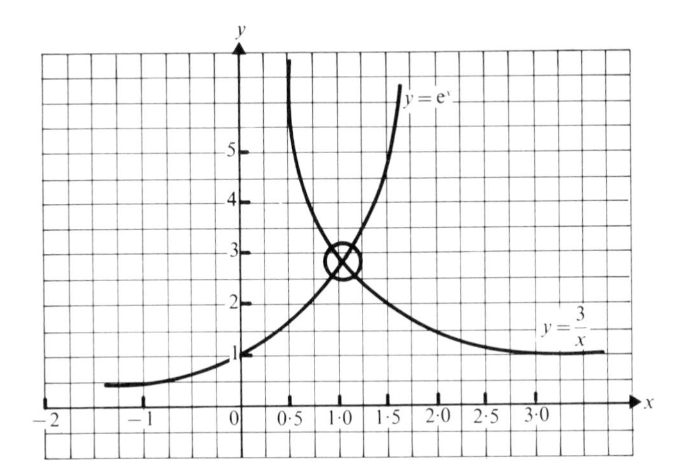

From these graphs we see that $\frac{3}{x} = e^x$ when $x \simeq 1$.

Taking this value as a first approximation for the root of the equation $xe^x = 3$, we will use Newton's Method to find a better approximation as follows:

If $$f(x) \equiv xe^x - 3$$

then $$f'(x) \equiv (x+1)e^x$$

Now, for an approximate root, a, a better approximation is $a + h$, where

$$h \simeq -\frac{f(a)}{f'(a)}$$

In this case $a = 1$, so

$$h \simeq -\frac{(e-3)}{2e} \simeq 0.05 \quad \Rightarrow \quad a + h \simeq 1.05$$

Now using the better approximation $a_1 = 1.05$

$$h_1 \simeq -\frac{(1.05e^{1.05} - 3)}{2.05e^{1.05}} \simeq -0.0001 \quad \Rightarrow \quad a_1 + h_1 \simeq 1.0499$$

So, to three decimal places, the root is likely to be 1.050 and we can check that this is so by calculating $f(1.0495)$ and $f(1.0504)$. Now $f(1.0495)$ is negative and $f(1.0504)$ is positive. Thus the root lies between these values and, correct to 3 d.p., is 1.050.

EXERCISE 11g

In Questions 1–3, use the relationship $f(a+h) \simeq f(a) + hf'(a)$.

1) Given that $e^{0.5} = 1.6487$, find an approximate value for $e^{0.501}$.

2) Given that arcsin (0.6) is $36°52'$ and that $52' \simeq 0.051^c$, find an approximation for $\sin 36°$.

3) Given that $1° = 0.0175^c$, find an approximate value for
(a) $\cos 61°$, (b) $\tan 29°$, (c) $\cos 89°$ (take $\sqrt{3} = 1.7321$).

Show that each of the following equations has a root between $x = 0$ and $x = 1$ and find this root correct to 2 decimal places.

4) $x^3 - x^2 + 10x - 2 = 0$

5) $3x^3 - 2x^2 - 9x + 2 = 0$

6) $2x^3 + x^2 + 6x - 1 = 0$

Find, correct to 3 significant figures, the smallest positive root of each of the following equations.

7) $x^3 + x - 11 = 0$

8) $x^4 - 4x^3 - x^2 + 4x - 10 = 0$

9) $4 + 5x^2 - x^3 = 0$

Use a graphical method to find a first approximation to the root(s) of the following six equations. Then apply two stages of Newton's Method to give a better approximation. State the accuracy of each of your results.

10) $\tan x = 2x$ (the positive root)

11) $x^2 = \ln(x+1)$

12) $e^x = 2x + 1$

13) $\sin x = 1 - x$

14) $x^3 - 6x + 3 = 0$ (the negative root)

15) $e^x(1+x) = 2$

Use linear interpolation to find the root(s) of the following equations, correct to 3 decimal places.

16) $e^x = 3x + 1$

17) $x = 1 + \ln x$

18) $3 + x - 2x^2 = e^x$

19) $e^x = 2\cos x$ $\left(\text{the roots between } -\frac{\pi}{2} \text{ and } \frac{\pi}{2}\right)$

20) $x^3 - 3x^2 - 1 = 0$

SUMMARY

When a polynomial function $f(x)$ is divided by $(x-a)(x-b)$ the remainder is of the form $\mathrm{P}x + \mathrm{Q}$, and

$$\begin{cases} f(a) = \mathrm{P}a + \mathrm{Q} \\ f(b) = \mathrm{P}b + \mathrm{Q}. \end{cases}$$

When a polynomial function $f(x)$ has a repeated factor $(x-a)$, then $(x-a)$ is also a factor of $f'(x)$,

i.e. $$f(a) = 0 \quad \textit{and} \quad f'(a) = 0.$$

If $\alpha, \beta, \gamma, \delta \ldots$ are the roots of a polynomial equation

$$ax^n + bx^{n-1} + cx^{n-2} + \ldots = 0$$

then

$$\sum\alpha = -\frac{b}{a}, \quad \sum\alpha\beta = \frac{c}{a}, \quad \sum\alpha\beta\gamma = -\frac{d}{a}, \quad \sum\alpha\beta\gamma\delta = \frac{e}{a}\ldots$$

A polynomial equation of degree n has
(a) n roots (real and/or complex),
(b) complex roots, if any, in conjugate pairs,
(c) at least one real root if n is odd.

If there are p occasions when there is a change in the sign of consecutive coefficients of the terms in a polynomial equation $f(x)=0$, then the equation has

not more than p real positive roots

and if there are q sign changes in the coefficients of $f(-x)$, then the equation $f(x)=0$ has

not more than q real negative roots.

If a polynomial function $f(x)$ is such that $f(a)f(a+1)<0$ then there is at least one real root of the equation $f(x)=0$ between $x=a$ and $x=a+1$.

If $x=a$ is a first approximation to a solution of the equation $f(x)=0$, then

$$x = a - \frac{f(a)}{f'(a)}$$

is a better approximation (the Newton–Raphson Method).

APPENDIX. PROOF THAT THE COMPLEX ROOTS, IF ANY, OF A POLYNOMIAL EQUATION, OCCUR IN CONJUGATE PAIRS.

(De Moivre's Theorem is used in this proof so the reader is recommended to refer to Chapter 12 at this stage.)

Consider the polynomial equation

$$f(x) \equiv a_n x^n + a_{n-1}x^{n-1} + \ldots + a_0 = 0$$

If $z = r(\cos\theta + i\sin\theta)$ is a root, then

$$\begin{aligned} f(z) &= a_n r^n(\cos\theta + i\sin\theta)^n \\ &\quad + a_{n-1}r^{n-1}(\cos\theta + i\sin\theta)^{n-1} + \ldots + a_0 = 0 \\ &= a_n r^n(\cos n\theta + i\sin n\theta) \\ &\quad + a_{n-1}r^{n-1}\{\cos(n-1)\theta + i\sin(n-1)\theta\} + \ldots + a_0 = 0 \end{aligned}$$

Therefore $\quad \operatorname{Re}\{f(z)\} = 0 \quad$ and $\quad \operatorname{Im}\{f(z)\} = 0,$

i.e. $$\begin{cases} [a_n r^n \cos n\theta + a_{n-1}r^{n-1}\cos(n-1)\theta + \ldots + a_0] = 0 \\ i[a_n r^n \sin n\theta + a_{n-1}r^{n-1}\sin(n-1)\theta + \ldots + a_1\sin\theta] = 0. \end{cases}$$

Now consider $f(\bar{z})$ where $\bar{z} = r(\cos\theta - i\sin\theta)$.

$$\begin{aligned} f(\bar{z}) &= a_n r^n(\cos n\theta - i\sin n\theta) \\ &\quad + a_{n-1}r^{n-1}\{\cos(n-1)\theta - i\sin(n-1)\theta\} + \ldots + a_0 \\ &= [a_n r^n \cos n\theta + a_{n-1}r^{n-1}\cos(n-1)\theta + \ldots + a_0] \\ &\quad - i[a_n r^n \sin n\theta + a_{n-1}r^{n-1}\sin(n-1)\theta + \ldots + a_1\sin\theta] \\ &= [0] - i[0], \end{aligned}$$

i.e. $f(\bar{z}) = 0$.
Thus, if z is a root of $f(x) = 0$, so is $\bar{z}$,
i.e. complex roots always occur in conjugate pairs.

MULTIPLE CHOICE EXERCISE 11

(The instructions for answering these questions are on p. xii.)

TYPE I

1) When $x^3 + 5x - 2$ is divided by $(x-1)(x-2)$ the remainder is:
(a) 4 (b) $12x - 8$ (c) $12x$ (d) 16 (e) 0.

2) The function $p^3 + q^3 + r^3 - 3pqr$ has a factor:
(a) $p - q$ (b) $p + q + r$ (c) $p + r$ (d) $p - q - r$ (e) $q - p$.

3) The function $x^4 - 8x^3 + 22x^2 - 24x + 9$ has a repeated factor:
(a) $x - 1$ (b) $x^2 + 1$ (c) $x - 2$ (d) $x^2 - 1$ (e) $x + 1$.

4) The value of a for which $x^2 - ax - 2$ and $x^3 + x^2 - x - 1$ have a common factor is:
(a) -2 (b) -1 (c) $\frac{1}{2}$ (d) 2 (e) 0.

5) The sum of the squares of the roots of the equation $4x^3 + 3x^2 - 2x + 1 = 0$ is:
(a) $\frac{9}{16}$ (b) $-\frac{9}{16}$ (c) $-\frac{7}{16}$ (d) $\frac{25}{16}$ (e) none of these.

6) If α, β, γ are the roots of the equation $x^3 - 2x^2 + 3x - 4 = 0$, the equation whose roots are $\frac{1}{\alpha}, \frac{1}{\beta}, \frac{1}{\gamma}$ is:

(a) $x^3 - \frac{1}{2}x^2 + \frac{1}{3}x - \frac{1}{4} = 0$ (b) $4x^3 - 3x^2 + 2x - 1 = 0$

(c) $\frac{x^3}{4} - \frac{x^2}{3} + \frac{x}{2} - 1 = 0$ (d) $\frac{1}{x^3 - 2x^2 + 3x - 4} = 0$

(e) none of these.

TYPE II

7) The equation $2x^3 - 5x + 1 = 0$ has a root between the following values of x:
(a) -1 and 0 (b) 1 and 2 (c) -1 and -2 (d) -3 and -2.

8) $f(ab) \equiv a^n - b^n$.
(a) $f(ab)$ is cyclic.
(b) $f(ab)$ is homogeneous.
(c) $a - b$ is a factor of $f(ab)$.

9) $f(x) \equiv x^2 - 2x + 1$ and $g(x) \equiv x^3 - x^2 - x + 1$.
(a) $f(x)$ and $g(x)$ have a common factor.
(b) $g(x)$ has a repeated factor $x - 1$.
(c) $f(x) + Kg(x)$ has a real linear factor for all real values of K.

10) The equation $5x^3 - 9x^2 + 12x + 4 = 0$:
(a) has a root between 0 and -1,
(b) has three real roots,
(c) has roots α, β, γ where $\alpha\beta\gamma = \frac{4}{5}$,
(d) has a repeated root.

11) If $x^3 + px + q = 0$ has a repeated root, α, then

(a) $3\alpha^2 = p$,

(b) $\alpha^3 + p\alpha + q = 0$,

(c) the third root is -2α.

TYPE III

12) (a) $f(abc)$ is a cyclic polynomial of degree 3 with 3 real linear factors.
(b) $f(abc) \equiv (a+b)(b+c)(c+a)$.

13) (a) $f(abc)$ is a cyclic polynomial of degree 2.
(b) $f(abc) \equiv \pm(a+b+c)^2$.

14) (a) $f(xy)$ is a homogeneous polynomial.
(b) $f(xy)$ is a cyclic polynomial.

15) (a) $f'(x)$ has a factor $(x-\alpha)$.
(b) $f(x)$ has a repeated factor $(x-\alpha)$.

16) (a) $f(x) \equiv x^n - a^n$.
(b) $(x-a)$ is a factor of $f(x)$.

17) (a) $f(x) + Kg(x)$ has a factor $x-1$ for all real values of K.
(b) $x-1$ is a common factor of $f(x)$ and $g(x)$.

18) $f(x)$ is a polynomial function.
(a) $f(x)$ is never zero between $x = a$ and $x = b$.
(b) $[f(a)]\,[f(b)] > 0$.

19) The roots of the equation $x^3 + qx^2 + rx + s = 0$ are α, β, γ.
(a) α, β and γ are all real.
(b) q, r and s are all real. (U of L)

20) $f(x)$ is a polynomial function.
(a) The equation $f'(x) = 0$ has two equal roots.
(b) The equation $f(x) = 0$ has three equal roots.

21) $f(x) = 0$ is a polynomial equation.
(a) $f(x) = 0$ has exactly one real root.
(b) $f'(x) = 0$ has no real roots.

TYPE IV

22) Evaluate the sum of the squares of the roots of the equation $x^4 + px^3 + qx^2 + rx + s = 0$.

(a) $p = 1$.

(b) $q = -3$.

(c) $r = 2$.

(d) $s = 5$.

23) Determine the number of real roots of the equation $x^3 + bx^2 + cx + d = 0$.
(a) $f'(x) = 0$ when $x = p$ and q.
(b) $d > 0$.
(c) $f(r) = 0$ where $p < r < q$.

24) Use the Newton–Raphson method to find an approximate value of a root of the equation $f(x) = 0$.
(a) $[f(a)][f(a+1)] < 0$.
(b) $f'(x) > 0$ for $a \leqslant x \leqslant a+1$.
(c) $f(x)$ is continuous for $a \leqslant x \leqslant a+1$.
(d) $f'(a)$ is not small.

TYPE V

25) $f(x)$ is a polynomial function of x of degree n. When $f(x)$ is divided by x^m $(m < n)$, the remainder is a polynomial of degree $n - m$.

26) If $f(x)$ has a stationary value of zero when $x = a$, then $x - a$ is a repeated factor of $f(x)$.

27) If $f(xyz)$ is cyclic then each of its linear factors (if it has any) is also cyclic.

28) If $f(xyz)$ is homogeneous, any factor of $f(xyz)$ will also be homogeneous.

29) If $f(x)$ is a polynomial of degree n, $f(x) = 0$ has n roots.

30) The number of real roots of a polynomial equation $f(x) = 0$ is given by counting the number of sign changes in consecutive coefficients of $f(x)$ and $f(-x)$.

31) One root of the equation $x^3 + 3x - 5 = 0$ is approximately equal to the solution of the equation $3x - 5 = 0$.

MISCELLANEOUS EXERCISE 11

1) Let $f(x) \equiv 2x^4 + ax^2 + bx - 60$. The remainder when $f(x)$ is divided by $(x-1)$ is -94. One factor of $f(x)$ is $(x-3)$. Determine the constants a and b. (U of L)

2) Find integers m and n such that $(x+1)^2$ is a factor of $x^5 + 2x^2 + mx + n$. (U of L)p

3) Determine the quadratic function $f(x)$ which is exactly divisible by $(2x+1)$ and has remainders -6 and -5 when divided by $(x-1)$ and $(x-2)$ respectively. Determine $g(x) \equiv (px+q)^2 f(x)$, where p, q are constants, given that, on division by $(x-2)^2$, the remainder is $-39 - 3x$. (AEB '72)

4) When a polynomial in x is divided by $(x-a)$ the remainder is R_1 and when it is divided by $(x-b)$ the remainder is R_2. Find the remainder when the polynomial is divided by $(x-a)(x-b)$. (U of L)p

5) Given that x^2+1 is a factor of x^4+px^3+3x+q, find the values of p and q. Hence find the real roots of the equation

$$x^4+px^3+x^2+3x+q+1 = 0.$$ (U of L)p

6) When the polynomial $f(x)$ is divided by

$$(x-1)(x-2)(x-3)$$

the remainder equals

$$a(x-2)(x-3)+b(x-3)(x-1)+c(x-1)(x-2).$$

Express the constants a, b, c in terms of $f(1), f(2)$ and $f(3)$.
Without performing the division, find the value of the constant k for which the remainder when (x^5+kx^2) is divided by $(x-1)(x-2)(x-3)$ contains no term in x^2. (U of L)p

7) Show that the remainder when the polynomial $f(x)$ is divided by $(x-a)$ is $f(a)$. Show further that, if $f(x)$ is divided by $(x-a)(x-b)$, where $a \neq b$, then the remainder is

$$\left[\frac{f(a)-f(b)}{a-b}\right]x+\frac{af(b)-bf(a)}{a-b}$$ (C)p

8) A polynomial $f(x)$ is divided by x^2-a^2, where $a \neq 0$, and the remainder is $px+q$. Prove that

$$p = \frac{1}{2a}[f(a)-f(-a)]$$

$$q = \tfrac{1}{2}[f(a)+f(-a)].$$

Find the remainder when x^n-a^n is divided by x^2-a^2 for the cases when
(a) n is even, (b) n is odd. (JMB)

9) Show that if $(x+t)$ is a common factor of x^3+px^2+q and ax^3+bx+c, then it is also a factor of $apx^2-bx+aq-c$. Show that $x^3+\sqrt{7}x^2-14\sqrt{7}$ and $2x^3-13x-\sqrt{7}$ have a common factor and hence find all the roots of the equation $2x^3-13x-\sqrt{7}=0$. (U of L)

10) Show that $(x-y)$ is a factor of

$$x(y-z)^3+y(z-x)^3+z(x-y)^3$$

and hence factorize the expression completely. (U of L)p

11) Show, by putting $a=b+c$ or otherwise, that $a-b-c$ is a factor of

$$a^4+b^4+c^4-2b^2c^2-2c^2a^2-2a^2b^2.$$

Factorize this expression completely. (C)

12) Given that m and n are positive integers, prove that

$$x^m(b^n - c^n) + b^m(c^n - x^n) + c^m(x^n - b^n)$$

is divisible by $x^2 - x(b+c) + bc$. (U of L)

13) Express $4b^2c^2 - (b^2 + c^2 - a^2)^2$ as the product of four factors and hence determine the sign of the expression when a, b and c denote the lengths of the sides of a triangle. (U of L)p

14) Factorize $$a^2(b-c) + b^2(c-a) + c^2(a-b)$$

and $$a^4(b-c) + b^4(c-a) + c^4(a-b)$$

and show that, if a, b, c are real quantities, no two of which are equal, $a^4(b-c) + b^4(c-a) + c^4(a-b)$ cannot be zero. (U of L)

15) Prove that, if the roots of the equation

$$ax^3 + bx^2 + cx + d = 0, \quad (a \neq 0)$$

are α, β, γ, then $\alpha + \beta + \gamma = -b/a$, $\alpha\beta\gamma = -d/a$.
Solve the equation

$$32x^3 - 14x + 3 = 0$$

given that one root is twice another. (U of L)p

16) The roots of the equation $x^3 + 3x + 2 = 0$ are α, β and γ. Find the equation whose roots are $\alpha + \dfrac{1}{\alpha}$, $\beta + \dfrac{1}{\beta}$, $\gamma + \dfrac{1}{\gamma}$. (AEB '73)

17) If the roots of the equation

$$ax^3 + bx^2 + cx + d = 0$$

are in arithmetic progression (that is, one root is half the sum of the other two), prove that

$$2b^3 - 9abc + 27a^2d = 0.$$

Solve the equation

$$18x^3 + 27x^2 + x - 4 = 0$$ (O)

18) If α, β, γ are the roots of the equation

$$x^3 - x^2 - 4x + 5 = 0$$

find cubic equations whose roots are
(a) 2α, 2β and 2γ,
(b) $1/\alpha$, $1/\beta$ and $1/\gamma$,
(c) $\alpha + \beta$, $\beta + \gamma$ and $\gamma + \alpha$.

Evaluate $\sum(\alpha + \beta)^2$. (U of L)

19) If α, β, γ are the three roots of the cubic equation $x^3 + ax^2 + bx + c = 0$, show that $\alpha^2 + \beta^2 + \gamma^2 = a^2 - 2b$ and $\alpha^3 + \beta^3 + \gamma^3 = -a^3 + 3ab - 3c$.
If $\alpha + \beta + \gamma = 2$, $\alpha^2 + \beta^2 + \gamma^2 = 14$ and $\alpha^3 + \beta^3 + \gamma^3 = 20$, find the value of $\alpha^4 + \beta^4 + \gamma^4$. (O)

20) The roots of the equation $x^3 - 26x^2 + 156x + p = 0$ are in geometric progression. Find p. (U of L)

21) The roots of the equation $x^3 + px^2 + qx + 30 = 0$ are in the ratios $2:3:5$. Find the values of p and q. (U of L)

22) If $\alpha, \beta, \gamma, \delta$ are the roots of the equation

$$x^4 + ax^3 + bx^2 + cx + d = 0$$

and $\alpha + \beta = \gamma + \delta$, show that the roots of the equation

$$ay^2 - 2cy + ad = 0$$

are $\alpha\beta$ and $\gamma\delta$. Prove, also, that

$$a^3 - 4ab + 8c = 0.$$

Solve the equation

$$4x^4 - 8x^3 - 23x^2 + 27x + 18 = 0.$$

(O)

23) If the roots of the equation $x^3 - 9x^2 + 3x - 39 = 0$ are α, β, γ, show that an equation whose roots are $\alpha - 3$, $\beta - 3$, and $\gamma - 3$ is

$$x^3 - 24x - 84 = 0.$$

Show also that the equation $x^3 - 24x - 84 = 0$ has only one real root, and show that this root lies between 6 and 7.
Sketch the two curves $y = x^3 - 9x^2 + 3x - 39$ and $y = x^3 - 24x - 84$ on the same diagram. (U of L)

24) By investigating the turning values of

$$f(x) = x^3 + 3x^2 + 6x - 38$$

or otherwise, show that the equation $f(x) = 0$ has only one real root.
Find two consecutive integers, n and $n + 1$, which enclose the root.
Describe a method by which successive approximations to the root can be obtained. Starting with the value of n as a first approximation, calculate two further successive approximations to the root.
Give your answers correct to 3 significant figures. (C)

25) If α, β, γ are the roots of the equation $x^3 - px^2 - q = 0$, prove that $\alpha^2 + \beta^2 + \gamma^2 = p^2$ and express

$$\beta^2\gamma^2 + \gamma^2\alpha^2 + \alpha^2\beta^2$$

in terms of p and q.
Use the Remainder Theorem to find the remainder when $x^3 - 7x^2 + 36$ is divided by $x + 2$. Hence solve the equation $x^3 - 7x^2 + 36 = 0$ and verify

that your expression for

$$\beta^2\gamma^2 + \gamma^2\alpha^2 + \alpha^2\beta^2$$

is correct in this case. (U of L)

26) Use tables to show that when $y = \ln x$ the error in the approximation

$$y(a+h) \simeq y(a) + hy'(a)$$

with $a \geqslant 2$ and $h = 0.1$, is less than 1%.
Calculate the values of $\ln(2.1)$, $\ln(2.2)$ and $\ln(2.3)$ by applying this approximation to the differential equation

$$dy/dx = 1/x,$$

where $y(2) = 0.6931$.
Explain with the help of a diagram why the calculated values are larger than the true values. (U of L)

27) Show that the equation $x^3 - 5x - 3 = 0$ has a root between 2 and 3.
Use linear interpolation to find an approximation to this root.
Determine the root correct to two places of decimals. (C)

28) Show that the equation $x = \frac{1}{5}(x^4 + 2)$ has two real roots, both of which are positive.
Evaluate the smaller root correct to 3 places of decimals, showing that your answer has the required degree of accuracy. (O)

29) Find the positive root of the equation $x = \ln(1 + x + x^2)$, correct to three significant figures, showing that your answer has the required degree of accuracy. (O)

30) (a) Show that the equation

$$x^4 - 2x - 1 = 0$$

has exactly two real roots. Find integers a and b such that one of these roots lies between a and $a+1$ and another between b and $b+1$.

(b) Using the substitution $y \equiv x + x^{-1}$, obtain the four roots of the equation

$$6x^4 - 25x^3 + 37x^2 - 25x + 6 = 0.$$

(U of L)

31) By drawing appropriate graphs show that the equation

$$x(\pi - x) = 4 \sin 2x$$

has four real roots. Show that the root between $x = 0$ and $x = \pi$ occurs at approximately $x = 1.2$ and by means of a single application of Newton's Method determine a more accurate estimate of its value. (AEB '72)

32) Prove that, if $x = a$ is an approximation to one root of the equation $f(x) = 0$, then $x = a - \dfrac{f(a)}{f'(a)}$ is a closer approximation.

Hence

(a) establish the formula $x_{r+1} = x_r(2 - Nx_r)$ as a method of successive approximation to the reciprocal of N,

(b) show that $x^4 + x^2 - 80 = 0$ has a root near $x = 3$.

Taking $x = 3$ as the first approximation use Newton's Method twice to obtain a better approximation. (AEB '74)

33) Sketch the curve $y = x^4 + 3x^3 + x^2$, giving the coordinates and nature of its turning points.

For the equation $x^4 + 3x^3 + x^2 - k = 0$

(a) find the complex roots in the form $a + ib$, a and b being real, when $k = -4$,

(b) find the integer root when $k = 9$, and find two consecutive integers between which the other real root must lie in this case,

(c) calculate the sum of the squares of the roots, showing that it is independent of k. (JMB)

34) Obtain an approximation to the smaller positive root of the equation $3 \log_e (1 + x) = (x - 1)(x - 2)$ by

(a) expanding $\log_e (1 + x)$ and neglecting terms in x^n $(n > 2)$,

(b) drawing the graphs of $y = 3 \log_e (1 + x)$ and $y = (x - 1)(x - 2)$ for values of x between $x = 0$ and $x = 2$.

Taking $x = 0.4$ as a first approximation, obtain an improved value of the root by a single application of Newton's Method. (AEB '74)

35) Given that $p + iq$, where p and q are real and $q \neq 0$, is a root of the equation

$$a_0 z^n + a_1 z^{n-1} + \ldots + a_n = 0$$

where $a_0, a_1, \ldots, a_n$ are all real, prove that $p - iq$ is also a root.

Given that $1 + 3i$ is a root of the equation

$$z^4 - 6z^3 + az^2 + bz + 70 = 0$$

where a and b are real, find a, b and the other three roots of the equation. (JMB)

36) A student, asked to find a root of the equation

$$f(x) \equiv x^3 - 14x^2 + 49x - 8 = 0$$

did not notice the solution $x = 8$ but chose, instead, to use Newton's Method, taking $x = 7.2$ as the first approximation. He then calculated, correctly, $f(7.2) = -7.712$, $f'(7.2) = 2.92$ and deduced (again correctly) the second approximation 9.84. By means of a graph, or otherwise, explain why Newton's Method failed to give a better approximation in this case. Prove that using Newton's Method, a first approximation α, for a value of α in the interval $7.2 < \alpha < 8$, would give a second approximation which is closer to the root $x = 8$ provided that

$$2(8-\alpha)f'(\alpha) + f(\alpha) > 0$$

and deduce that any value of α in the above range exceeding $5 + \sqrt{(5.6)}$ would in fact give improvement.
[You may, if you wish, assume without proof that $f'(x)$ is positive and increasing for $x > 7.2$; also that

$$2(8-\alpha)f'(\alpha) + f(\alpha) = (8-\alpha)(5\alpha^2 - 50\alpha + 97).]$$

(C)

37) An equation can be written in the form $x = F(x)$ and it is known that the equation has only one real root and that this root is near $x = x_1$. Explain, with the aid of a diagram, how, if $|F'(x)| < 1$, the iterative formula

$$x_{r+1} = F(x_r), \quad (r = 1, 2, \ldots)$$

will give the root to whatever degree of accuracy is required.
Show that the cubic equation $x^3 + 3x - 15 = 0$ has only one real root and that this root is near $x = 2$.
This cubic equation can be written in any one of the forms

(a) $x = \frac{1}{3}(15 - x^3)$

(b) $x = 15/(x^2 + 3)$

(c) $x = (15 - 3x)^{\frac{1}{3}}$

Determine which of these forms would be suitable for the use of the previous iterative formula. (U of L)

CHAPTER 12

COMPLEX NUMBER

Any complex number $x + yi$ can be expressed in the form $r(\cos\theta + i\sin\theta)$ where r is the modulus and θ is the argument of $x + yi$.
If $z = x + yi$ has unit modulus (in which case z behaves as a unit vector in the xy plane) then

$$z = \cos\theta + i\sin\theta.$$

Now if we consider z^2, we know already (Volume 1, page 505) that

$$|z^2| = |(z)(z)| = |z||z| = 1$$

and $$\arg(z^2) = \arg z + \arg z = 2\arg z = 2\theta.$$

So $$z^2 = \cos 2\theta + i\sin 2\theta.$$

Similarly it can be shown that

$$|z^3| = |(z^2)(z)| = |z^2||z| = 1$$

and $$\arg(z^3) = \arg(z^2) + \arg z = 2\theta + \theta = 3\theta.$$

So $$z^3 = \cos 3\theta + i\sin 3\theta.$$

These two results suggest that, for positive integral values of n,

$$z^n = \cos n\theta + i\sin n\theta.$$

This deduction, which is known as De Moivre's Theorem, can be proved to be valid using the method of induction.

DE MOIVRE'S THEOREM

De Moivre's Theorem states that

$$(\cos\theta + i\sin\theta)^n \equiv \cos n\theta + i\sin n\theta.$$

To prove that it is valid for positive integral values of n we use the method of induction as follows.

If p_n is the statement $(\cos\theta + i\sin\theta)^n \equiv \cos n\theta + i\sin n\theta$

then p_k is $(\cos\theta + i\sin\theta)^k \equiv \cos k\theta + i\sin k\theta$

$\Rightarrow$

$$\begin{aligned} p_{k+1} \text{ is } (\cos\theta + i\sin\theta)^{k+1} &\equiv (\cos k\theta + i\sin k\theta)(\cos\theta + i\sin\theta) \\ &\equiv (\cos k\theta\cos\theta - \sin k\theta\sin\theta) \\ &\quad + i(\sin k\theta\cos\theta + \cos k\theta\sin\theta) \\ &\equiv \cos(k+1)\theta + i\sin(k+1)\theta. \end{aligned}$$

Hence $\qquad p_k \Rightarrow p_{k+1} \qquad$ [1]

Now p_1 is $(\cos\theta + i\sin\theta)^1 \equiv \cos\theta + i\sin\theta$
which is obviously true.
Hence, using [1] above, p_2 is true, p_3 is true, and so on for all positive integral values of n,

e.g. $\qquad z^7 = (\cos\theta + i\sin\theta)^7 \equiv \cos 7\theta + i\sin 7\theta.$

Further, it can be proved that De Moivre's Theorem is valid for any rational value of n.

Proof When n is a Negative Integer

If n is negative, then $n = -m$ where m is a positive integer.

Now

$$(\cos\theta + i\sin\theta)^n \equiv (\cos\theta + i\sin\theta)^{-m} \equiv \frac{1}{(\cos\theta + i\sin\theta)^m}$$

But m is a positive integer,

so $\qquad (\cos\theta + i\sin\theta)^m \equiv \cos m\theta + i\sin m\theta$

and $$\frac{1}{\cos m\theta + i\sin m\theta} \equiv \frac{\cos m\theta - i\sin m\theta}{(\cos m\theta + i\sin m\theta)(\cos m\theta - i\sin m\theta)}$$

$$\equiv \cos m\theta - i\sin m\theta$$

$$\equiv \cos(-m\theta) + i\sin(-m\theta)$$

$$\equiv \cos n\theta + i\sin n\theta.$$

So, when n is a negative integer,

$$(\cos\theta + i\sin\theta)^n \equiv \cos n\theta + i\sin n\theta$$

e.g. $$\frac{1}{z} = z^{-1} = (\cos\theta + i\sin\theta)^{-1}$$

$$= \cos(-\theta) + i\sin(-\theta)$$

$\Rightarrow$ $$\frac{1}{z} = \cos\theta - i\sin\theta$$

Proof When n is a Rational Fraction

If n is a rational fraction, $n = \dfrac{p}{q}$ where p and q are integers.

We wish to prove that

$$(\cos\theta + i\sin\theta)^{p/q} \equiv \cos\frac{p}{q}\theta + i\sin\frac{p}{q}\theta.$$

Raising the R.H.S. to power q we have

$$\left(\cos\frac{p}{q}\theta + i\sin\frac{p}{q}\theta\right)^q \equiv \cos p\theta + i\sin p\theta$$

(as q is an integer).

Then $$(\cos p\theta + i\sin p\theta) \equiv (\cos\theta + i\sin\theta)^p$$ (as p is an integer).

So $$\left(\cos\frac{p}{q}\theta + i\sin\frac{p}{q}\theta\right)^q \equiv (\cos\theta + i\sin\theta)^p$$

$\Rightarrow$ $$\cos\frac{p}{q}\theta + i\sin\frac{p}{q}\theta \equiv (\cos\theta + i\sin\theta)^{p/q}.$$

Hence De Moivre's Theorem applies when n is a rational fraction.

Note that $\cos\frac{p}{q}\theta + i\sin\frac{p}{q}\theta$ is *just one* value of $(\cos\theta + i\sin\theta)^{p/q}$. There are further values, as subsequent work will show.

Note also that De Moivre's Theorem is, in fact, valid also when n is irrational but we will not attempt a proof at this stage.

Applications of De Moivre's Theorem

Certain trig identities can be derived using De Moivre's Theorem. In particular, expressions such as $\cos n\theta$, $\sin n\theta$, $\tan n\theta$ can be expressed in terms of $\cos\theta$, $\sin\theta$ and $\tan\theta$.
For instance we can find an identity for $\cos 5\theta$ as follows:

$$\begin{aligned}\cos 5\theta &\equiv \text{Re}\,(\cos 5\theta + i\sin 5\theta)\\ &\equiv \text{Re}\,(\cos\theta + i\sin\theta)^5 \quad \text{(De Moivre's Theorem)}\\ &\equiv \text{Re}\,(c^5 + 5c^4is + 10c^3i^2s^2 + 10c^2i^3s^3 + 5ci^4s^4 + i^5s^5)\\ &\qquad\qquad (\text{where } c \equiv \cos\theta \text{ and } s \equiv \sin\theta)\\ &\equiv c^5 - 10c^3s^2 + 5cs^4\end{aligned}$$

So $$\cos 5\theta \equiv \cos^5\theta - 10\cos^3\theta\sin^2\theta + 5\cos\theta\sin^4\theta.$$

If required, the R.H.S. can be expressed entirely in terms of $\cos\theta$ by using $\cos^2\theta + \sin^2\theta \equiv 1$.
Note that the method used above to find an identity for $\cos 5\theta$ provides, at the same time, an identity for $\sin 5\theta$,

i.e. $$i\sin 5\theta \equiv \text{Im}\,(\cos 5\theta + i\sin 5\theta)$$

$\Rightarrow$ $$i\sin 5\theta \equiv 5c^4is + 10c^2i^3s^3 + i^5s^5$$

$\Rightarrow$ $$\sin 5\theta \equiv 5\cos^4\theta\sin\theta - 10\cos^2\theta\sin^3\theta + \sin^5\theta.$$

Further, using the identities so derived for $\cos 5\theta$ and $\sin 5\theta$, $\tan 5\theta$ can also be expressed in terms of $\tan\theta$,

i.e. $$\tan 5\theta \equiv \frac{\sin 5\theta}{\cos 5\theta} \equiv \frac{5\cos^4\theta\sin\theta - 10\cos^2\theta\sin^3\theta + \sin^5\theta}{\cos^5\theta - 10\cos^3\theta\sin^2\theta + 5\cos\theta\sin^4\theta}$$

Then dividing every term by $\cos^5\theta$ gives

$$\tan 5\theta \equiv \frac{5\tan\theta - 10\tan^3\theta + \tan^5\theta}{1 - 10\tan^2\theta + 5\tan^4\theta}$$

The technique demonstrated above can clearly be applied to find similar identities for $\cos n\theta$, $\sin n\theta$ and $\tan n\theta$ for any positive integral value of n.

Properties of z and $\frac{1}{z}$

It has been shown that

if $z = \cos\theta + i\sin\theta$ then $\frac{1}{z} = \cos\theta - i\sin\theta$.

Hence $$z + \frac{1}{z} = 2\cos\theta$$

and $$z - \frac{1}{z} = 2i\sin\theta$$

Further, De Moivre's Theorem shows that

$$\frac{1}{z^n} = (\cos\theta + i\sin\theta)^{-n} \equiv \cos(-n\theta) + i\sin(-n\theta).$$

So $$\frac{1}{z^n} = \cos n\theta - i\sin n\theta$$

Also $$z^n = \cos n\theta + i\sin n\theta$$

Hence $$z^n + \frac{1}{z^n} = 2\cos n\theta$$

and $$z^n - \frac{1}{z^n} = 2i\sin n\theta$$

These relationships are useful for expressing $\cos^n\theta$ or $\sin^n\theta$ in terms of the cosines or sines of multiples of θ.
For example, consider $\cos^4\theta$.
Beginning with the relationship containing $\cos\theta$, i.e. $z + \dfrac{1}{z} = 2\cos\theta$,

we have

$$(2\cos\theta)^4 = \left(z + \frac{1}{z}\right)^4$$

$$\Rightarrow \quad 2^4\cos^4\theta = z^4 + 4z^3\left(\frac{1}{z}\right) + 6z^2\left(\frac{1}{z^2}\right) + 4z\left(\frac{1}{z^3}\right) + \frac{1}{z^4}$$

$$\equiv z^4 + 4z^2 + 6 + \frac{4}{z^2} + \frac{1}{z^4}$$

$$\equiv \left(z^4 + \frac{1}{z^4}\right) + 4\left(z^2 + \frac{1}{z^2}\right) + 6$$

Using $z^n + \dfrac{1}{z^n} = 2\cos n\theta$ on the R.H.S. gives

$$2^4\cos^4\theta \equiv (2\cos 4\theta) + 4(2\cos 2\theta) + 6$$

So $$\cos^4\theta \equiv \tfrac{1}{8}[\cos 4\theta + 4\cos 2\theta + 3]$$

Again it is clear that this method can be applied to any positive integral power of $\sin\theta$ or $\cos\theta$. It should be pointed out, however, that when dealing with $\sin^n\theta$ we begin with the relationship $z - \dfrac{1}{z} = 2i\sin\theta$, using

$$(2i \sin \theta)^n = \left(z - \frac{1}{z}\right)^n$$

In this case extra care must be taken with
(a) the power of i on the L.H.S.,
(b) the signs of the terms on the R.H.S.

EXAMPLES 12a

1) Use De Moivre's Theorem to prove that

$$(\cos p\theta + i \sin p\theta)(\cos q\theta + i \sin q\theta) \equiv \cos (p+q)\theta + i \sin (p+q)\theta.$$

Simplify $\dfrac{\cos 3\theta + i \sin 3\theta}{\cos 5\theta - i \sin 5\theta}$

De Moivre's Theorem states that

$$(\cos \theta + i \sin \theta)^n \equiv \cos n\theta + i \sin n\theta$$

Hence $\qquad \cos p\theta + i \sin p\theta \equiv (\cos \theta + i \sin \theta)^p$

and $\qquad \cos q\theta + i \sin q\theta \equiv (\cos \theta + i \sin \theta)^q$

So $\quad (\cos p\theta + i \sin p\theta)(\cos q\theta + i \sin q\theta) \equiv (\cos \theta + i \sin \theta)^{p+q}$

$$\equiv \cos (p+q)\theta + i \sin (p+q)\theta.$$

Now $\quad \dfrac{\cos 3\theta + i \sin 3\theta}{\cos 5\theta - i \sin 5\theta} \equiv (\cos 3\theta + i \sin 3\theta)(\cos 5\theta - i \sin 5\theta)^{-1}$

$$\equiv (\cos 3\theta + i \sin 3\theta)(\cos \{-5\theta\} - i \sin \{-5\theta\})$$

$$\equiv (\cos 3\theta + i \sin 3\theta)(\cos 5\theta + i \sin 5\theta).$$

$$\equiv \cos (3+5)\theta + i \sin (3+5)\theta$$

So $\quad \dfrac{\cos 3\theta + i \sin 3\theta}{\cos 5\theta - i \sin 5\theta} \equiv \cos 8\theta + i \sin 8\theta.$

2) Prove that $\sin^7\theta \equiv \frac{1}{64}(35 \sin \theta - 21 \sin 3\theta + 7 \sin 5\theta - \sin 7\theta)$.

Hence find $\int (35 \sin \theta - 64 \sin^7\theta)\, d\theta$.

Using $z - \dfrac{1}{z} = 2i \sin \theta$ gives

$$(2i \sin \theta)^7 = \left(z - \frac{1}{z}\right)^7$$

$$\Rightarrow \quad 2^7 i^7 \sin^7\theta = z^7 - 7z^5 + 21z^3 - 35z + \frac{35}{z} - \frac{21}{z^3} + \frac{7}{z^5} - \frac{1}{z^7}$$

$$\equiv \left(z^7 - \frac{1}{z^7}\right) - 7\left(z^5 - \frac{1}{z^5}\right) + 21\left(z^3 - \frac{1}{z^3}\right) - 35\left(z - \frac{1}{z}\right)$$

But $$z^n - \frac{1}{z^n} = 2i \sin n\theta$$

so, using $n = 1, 3, 5$ and 7 we have

$$2^7 i^7 \sin^7\theta \equiv (2i \sin 7\theta) - 7(2i \sin 5\theta) + 21(2i \sin 3\theta) - 35(2i \sin \theta)$$

But $i^7 = -i$

So $$-2^7 i \sin^7\theta \equiv 2i[\sin 7\theta - 7 \sin 5\theta + 21 \sin 3\theta - 35 \sin \theta]$$

$\Rightarrow$ $$\sin^7\theta \equiv \tfrac{1}{64}[35 \sin \theta - 21 \sin 3\theta + 7 \sin 5\theta - \sin 7\theta]$$

Hence $$35 \sin \theta - 64 \sin^7\theta \equiv 21 \sin 3\theta - 7 \sin 5\theta + \sin 7\theta.$$

So

$$\int (35 \sin \theta - 64 \sin^7\theta)\, d\theta \equiv \int (21 \sin 3\theta - 7 \sin 5\theta + \sin 7\theta)\, d\theta$$

$$= -7 \cos 3\theta + \tfrac{7}{5} \cos 5\theta - \tfrac{1}{7} \cos 7\theta + K.$$

3) Show that $\tan 4\theta \equiv \dfrac{4t - 4t^3}{1 - 6t^2 + t^4}$ where $t \equiv \tan \theta$.

Use your result to solve the equation

$$t^4 + 4t^3 - 6t^2 - 4t + 1 = 0$$

giving your answers correct to three decimal places.

To express $\tan 4\theta$ in terms of $\tan \theta$ we begin by using the complex number

$$\cos 4\theta + i \sin 4\theta \equiv (c + is)^4 \qquad (c \equiv \cos \theta, \quad s \equiv \sin \theta)$$

$$\equiv c^4 + 4c^3 is + 6c^2 i^2 s^2 + 4c i^3 s^3 + i^4 s^4$$

$$\cos 4\theta \equiv \operatorname{Re}(c + is)^4 \equiv c^4 - 6c^2 s^2 + s^4$$

and $$i \sin 4\theta \equiv \operatorname{Im}(c + is)^4 \equiv i(4c^3 s - 4cs^3).$$

Hence $$\tan 4\theta \equiv \frac{\sin 4\theta}{\cos 4\theta} \equiv \frac{4c^3 s - 4cs^3}{c^4 - 6c^2 s^2 + s^4}.$$

Dividing every term on the R.H.S. by c^4 gives

$$\tan 4\theta \equiv \frac{4t - 4t^3}{1 - 6t^2 + t^4}$$

If $$t^4 + 4t^3 - 6t^2 - 4t + 1 = 0 \qquad [1]$$

then $$\frac{4t - 4t^3}{1 - 6t^2 + t^4} = 1.$$

But $t \equiv \tan \theta$,

so $$\tan 4\theta = 1 \Rightarrow \theta = \frac{1}{4}\left(n\pi + \frac{\pi}{4}\right)$$

The solutions of equation [1] are the tangents of the angles in the set

$$\theta = \frac{1}{4}\left(n\pi + \frac{\pi}{4}\right)$$

Hence, taking $n = 0, 1, 2, 3,$ we get

$$t = \tan\frac{\pi}{16},\ \tan\frac{5\pi}{16},\ \tan\frac{9\pi}{16},\ \tan\frac{13\pi}{16},$$

i.e. $$t = 0.199,\ 1.497,\ -5.027,\ -0.668$$

Note that further values of n, although they give further values of θ, repeat the values of $\tan\theta$ already found.

EXERCISE 12a

1) Use De Moivre's Theorem to express each of the following complex numbers in the form $\cos n\theta + i\sin n\theta$.

(a) $(\cos\theta + i\sin\theta)^7$ (b) $(\cos\theta + i\sin\theta)^{-3}$

(c) $(\cos\theta + i\sin\theta)^{\frac{1}{2}}$ (d) $\left(\cos\frac{\pi}{3} + i\sin\frac{\pi}{3}\right)^3$

(e) $\left(\cos\frac{\pi}{4} + i\sin\frac{\pi}{4}\right)^{-2}$ (f) $(\cos\pi + i\sin\pi)^{\frac{1}{3}}$

2) Express each of the following complex numbers in the form $(\cos\theta + i\sin\theta)^n$.

(a) $\cos 5\theta + i\sin 5\theta$ (b) $\cos 2\theta - i\sin 2\theta$

(c) $\cos\frac{\theta}{3} + i\sin\frac{\theta}{3}$ (d) $\cos\frac{\theta}{2} - i\sin\frac{\theta}{2}$

Use De Moivre's Theorem to simplify the following expressions.

3) $(\cos 2\theta + i\sin 2\theta)(\cos 5\theta + i\sin 5\theta)$

4) $\dfrac{\cos 7\theta + i\sin 7\theta}{\cos 2\theta - i\sin 2\theta}$

5) $\dfrac{(\cos 4\theta + i\sin 4\theta)(\cos 3\theta + i\sin 3\theta)}{\cos 5\theta + i\sin 5\theta}$

6) $(\cos 3\theta + i\sin 3\theta)(\cos 6\theta - i\sin 6\theta)$

7) $(\cos 2\theta + i\sin 2\theta)^2(\cos\theta + i\sin\theta)^3$

8) $\dfrac{\cos\theta - i\sin\theta}{\cos 4\theta - i\sin 4\theta}$

9) $\left(\cos\frac{\pi}{3}+i\sin\frac{\pi}{3}\right)^2\left(\cos\frac{2\pi}{3}+i\sin\frac{2\pi}{3}\right)^4$

10) $\dfrac{\left(\cos\frac{\pi}{4}+i\sin\frac{\pi}{4}\right)^5\left(\cos\frac{3\pi}{4}+i\sin\frac{3\pi}{4}\right)^3}{\left(\cos\frac{\pi}{4}-i\sin\frac{\pi}{4}\right)^2}$

Write down one value of each of the following expressions.

11) $(\cos 2\theta+i\sin 2\theta)^{\frac{1}{2}}$ 12) $\sqrt[3]{\left(\cos\frac{\pi}{2}+i\sin\frac{\pi}{2}\right)}$

13) $\sqrt[5]{(\cos\pi+i\sin\pi)}$ 14) $\sqrt[n]{(\cos\theta+i\sin\theta)}$

Prove the following trig identities using methods based on De Moivre's Theorem.

15) $\cos 4\theta\equiv 8\cos^4\theta-8\cos^2\theta+1$

16) $\sin^5\theta\equiv\frac{1}{16}(\sin 5\theta-5\sin 3\theta+10\sin\theta)$

17) $\tan 6\theta\equiv 2t\left(\dfrac{3-10t^2+3t^4}{1-15t^2+15t^4-t^6}\right)$ where $t\equiv\tan\theta$

18) $\cos^4\theta\equiv\frac{1}{8}(\cos 4\theta+4\cos 2\theta+3)$

19) $\sin^6\theta\equiv\frac{1}{32}(10-15\cos 2\theta+6\cos 4\theta-\cos 6\theta)$

20) $\cos^4\theta+\sin^4\theta\equiv\frac{1}{4}(\cos 4\theta+3)$

21) Prove that $\tan 3\theta\equiv\dfrac{3\tan\theta-\tan^3\theta}{1-3\tan^2\theta}$ and *hence* solve the equation $1-3t^2=3t-t^3$. Give your answers correct to 3 significant figures and verify these results by an algebraic method.

22) Use De Moivre's Theorem to find the following integrals.

(a) $\int 8\cos^4\theta\,d\theta$ (b) $\int(32\cos^6\theta-\cos 6\theta)\,d\theta$

(c) $\int(\cos 4\theta-8\sin^4\theta)\,d\theta$

CUBE ROOTS OF COMPLEX NUMBERS

If z is a cube root of a complex number w then $z^3=w$ or $z^3-w=0$. As $z^3-w=0$ is a cubic equation we expect it to have three roots, z_1, z_2 and z_3, and, because w is complex, it is likely that some or all of these roots will be complex.

De Moivre's Theorem can be used to find all three cube roots of the complex number w if w is first expressed in modulus and *general argument* form, i.e. if θ is the principal argument of w, the general argument is $(\theta + 2n\pi)$.

If $$w = r\{\cos(\theta + 2n\pi) + i\sin(\theta + 2n\pi)\}$$

then $$z = \sqrt[3]{r}\{\cos(\theta + 2n\pi) + i\sin(\theta + 2n\pi)\}^{\frac{1}{3}}$$

$$= \sqrt[3]{r}\left\{\cos\left(\frac{\theta + 2n\pi}{3}\right) + i\sin\left(\frac{\theta + 2n\pi}{3}\right)\right\}$$

Taking three consecutive integral values of n then gives three values, z_1, z_2 and z_3, which are the three cube roots of w. (Further values of n repeat the values of z already found.)
The argument of *one* of the cube roots is $\theta/3$ (i.e. $\frac{1}{3}$ of the principal argument of w) and the arguments of the other cube roots are seen to differ by $\frac{2}{3}\pi$.
The modulus of all three cube roots is $\sqrt[3]{r}$ so we see that the points P_1, P_2, P_3, representing the cube roots on an Argand diagram are such that

> P_1, P_2 and P_3 lie on a circle, centre O, with radius $\sqrt[3]{r}$, where
> $\angle P_1OP_2 = \angle P_2OP_3 = \angle P_3OP_1 = \frac{2}{3}\pi$

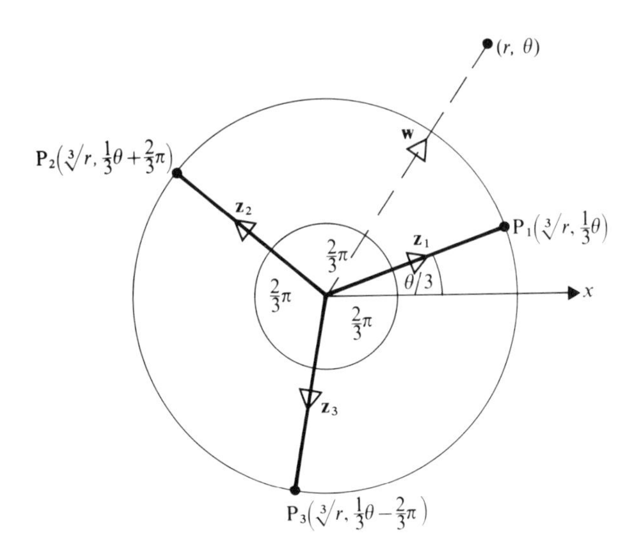

From this diagram, certain properties of the vectors $\mathbf{z}_1, \mathbf{z}_2$ and $\mathbf{z}_3$ can be deduced.

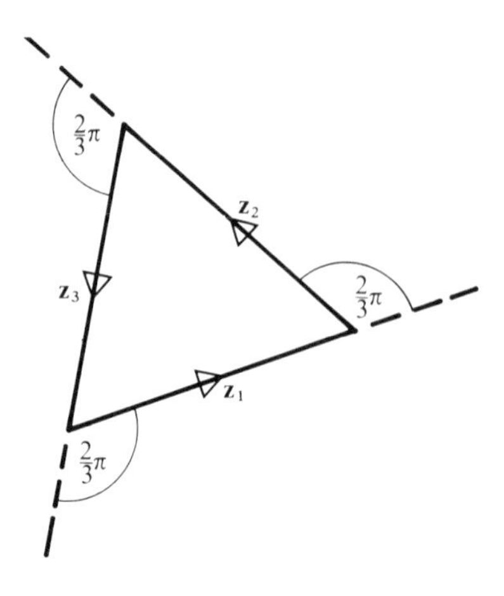

For example, lines representing $\mathbf{z}_1$, $\mathbf{z}_2$ and $\mathbf{z}_3$ form an equilateral triangle

$\Rightarrow$ $$\mathbf{z}_1 + \mathbf{z}_2 + \mathbf{z}_3 = \mathbf{0}.$$

Also, if we join P_1P_2, P_2P_3 and P_3P_1 then $P_1P_2P_3$ is also an equilateral triangle

i.e. $$|\overrightarrow{P_1P_2}| = |\overrightarrow{P_2P_3}| = |\overrightarrow{P_3P_1}|.$$

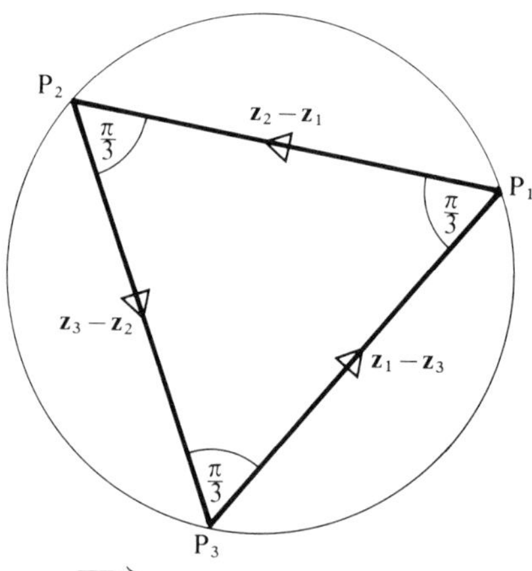

But $$\overrightarrow{P_1P_2} = \mathbf{z}_2 - \mathbf{z}_1 \quad \text{etc.}$$

So $$|\mathbf{z}_2 - \mathbf{z}_1| = |\mathbf{z}_3 - \mathbf{z}_2| = |\mathbf{z}_1 - \mathbf{z}_3|$$

Further, the angles $P_1P_2P_3$, $P_2P_3P_1$ and $P_3P_1P_2$ are all $\frac{\pi}{3}$

But $$\angle P_3P_1P_2 = \arg(z_2 - z_1) - \arg(z_1 - z_3)$$

So $$\begin{aligned} \arg(z_2 - z_1) - \arg(z_1 - z_3) &= \arg(z_3 - z_2) - \arg(z_2 - z_1) \\ &= \arg(z_1 - z_3) - \arg(z_3 - z_2) \\ &= \frac{\pi}{3} \end{aligned}$$

All the analysis carried out above on the cube roots of a general complex number w, is simplified if w has unit modulus. A particularly important example of this type involves finding the cube roots of unity.

CUBE ROOTS OF UNITY

The modulus of unity is 1 and its principal argument is 0.
The modulus of each cube root is therefore 1. The argument of *one* cube root is 0 and the other two therefore have arguments $-\frac{2}{3}\pi$ and $\frac{2}{3}\pi$.
So the cube roots of 1 are

$$\left.\begin{aligned} \mathbf{z}_1 &= \cos\left(-\tfrac{2}{3}\pi\right) + i\sin\left(-\tfrac{2}{3}\pi\right) \\ &= -\frac{1}{2} - i\,\frac{\sqrt{3}}{2} \\ \mathbf{z}_2 &= \cos 0 + i\sin 0 = 1 \\ \mathbf{z}_3 &= \cos\tfrac{2}{3}\pi + i\sin\tfrac{2}{3}\pi \\ &= -\frac{1}{2} + i\,\frac{\sqrt{3}}{2} \end{aligned}\right\} \Rightarrow$$

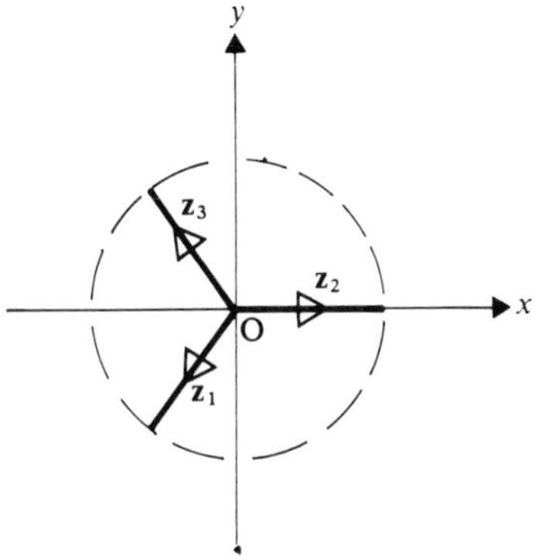

Note that these cube roots were found algebraically in Volume 1, Chapter 14, by solving the equation $z^3 - 1 = 0$.
In the same chapter it was also shown algebraically that each of the complex cube roots of 1 is the square of the other, and that the sum of the three cube roots is zero.
An alternative proof of these facts uses the properties of a cubic equation.
Referring to the three cube roots of unity as $1, \omega, \Omega$ we regard them as the solutions of the cubic equation

$$z^3 - 1 = 0. \qquad [1]$$

Using the standard notation for this equation,

$$a = 1, \quad b = c = 0, \quad d = -1,$$

hence, $\qquad$ product of roots $= -\dfrac{d}{a} \Rightarrow \omega\Omega = 1$

But $\qquad \omega^3 = 1 \Rightarrow \omega\Omega = \omega^3$

$$\Rightarrow \quad \Omega = \omega^2$$

So the cube roots of unity can be denoted by $1, \omega, \omega^2$.

Further, in equation [1],

$$\text{sum of roots} = -\frac{b}{a} \Rightarrow 1 + \omega + \omega^2 = 0.$$

Note that this property can also be deduced from the Argand diagram using vector consideration (as we saw in the general case).
The techniques used in the above work can be extended to finding the pth roots of any complex number (where p is a positive integer).

The pth Roots of $r(\cos\theta + i\sin\theta)$

Writing the given complex number with a general argument gives

$$r(\cos\{\theta + 2n\pi\} + i\sin\{\theta + 2n\pi\})$$

If z is one of its pth roots, then

$$z = [r(\cos\{\theta + 2n\pi\} + i\sin\{\theta + 2n\pi\})]^{1/p}$$

$$= r^{1/p}\left(\cos\left\{\frac{\theta}{p} + \frac{2n}{p}\pi\right\} + i\sin\left\{\frac{\theta}{p} + \frac{2n}{p}\pi\right\}\right)$$

Taking p consecutive integral values of n gives all possible values of z and we see that

each pth root has a modulus $\sqrt[p]{r}$

one pth root has an argument $\dfrac{\theta}{p}$

successive arguments differ by $\dfrac{2\pi}{p}$

So, on an Argand diagram, the points representing the pth roots of $r(\cos\theta + i\sin\theta)$ lie on a circle of radius $\sqrt[p]{r}$ and are separated by angles $\dfrac{2\pi}{p}$

The general approach is demonstrated in the examples that follow.

EXAMPLES 12b

1) Find the cube roots of $2 + 2i$ and mark this number, and its cube roots, on an Argand diagram. Deduce that the sum of the cube roots is zero.

$$|2 + 2i| = \sqrt{8} \quad \text{and} \quad \text{Arg}(2 + 2i) = \frac{\pi}{4} + 2n\pi.$$

If z is any cube root of $2 + 2i$, then

$$z = \left[\sqrt{8}\left(\cos\left\{\frac{\pi}{4} + 2n\pi\right\} + i\sin\left\{\frac{\pi}{4} + 2n\pi\right\}\right)\right]^{\frac{1}{3}}$$

$$= \sqrt{2}\left(\cos\left\{\frac{\pi}{12} + \frac{2}{3}n\pi\right\} + i\sin\left\{\frac{\pi}{12} + \frac{2}{3}n\pi\right\}\right)$$

So, taking $n = -1, 0, 1$, we have

$$z_1 = \sqrt{2}(\cos\{-\tfrac{7}{12}\pi\} + i\sin\{-\tfrac{7}{12}\pi\})$$
$$z_2 = \sqrt{2}(\cos\tfrac{1}{12}\pi + i\sin\tfrac{1}{12}\pi)$$
$$z_3 = \sqrt{2}(\cos\tfrac{3}{4}\pi + i\sin\tfrac{3}{4}\pi)$$

The modulus of each cube root is $\sqrt{2}$, so, on an Argand diagram, the points P_1, P_2, P_3 representing the cube roots z_1, z_2, z_3 lie on a circle, centre O, with radius $\sqrt{2}$.

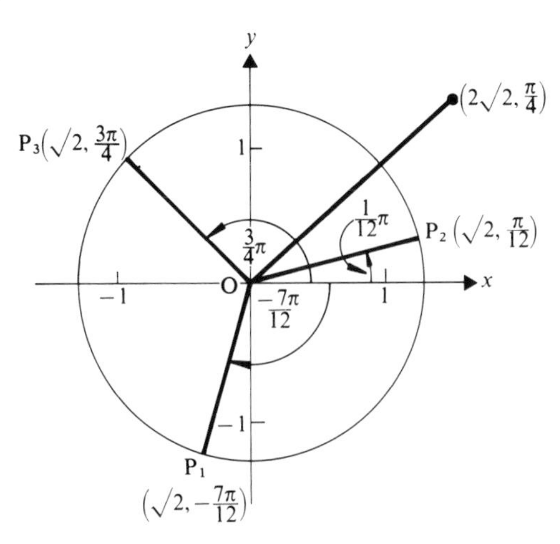

Again $\angle P_1OP_2 = \angle P_2OP_3 = \angle P_3OP_1 = \tfrac{2}{3}\pi$.
So the vectors represented by $\overrightarrow{OP}_1$, $\overrightarrow{OP}_2$, $\overrightarrow{OP}_3$ form an equilateral triangle and therefore their vector sum is zero

so $$z_1 + z_2 + z_3 = 0.$$

2) Find the fifth roots of -1 and hence, by considering the solutions of the equation $z^5 + 1 = 0$, prove that $\cos\tfrac{3}{5}\pi + \cos\tfrac{1}{5}\pi = \tfrac{1}{2}$.

The modulus and argument of -1, are 1 and π,

i.e. $$-1 = \cos(\pi + 2n\pi) + i\sin(\pi + 2n\pi).$$

If z is any of the five fifth roots of -1, then

$$z = (\cos\{\pi + 2n\pi\} + i\sin\{\pi + 2n\pi\})^{\frac{1}{5}}$$
$$= \cos\left\{\frac{\pi}{5} + \frac{2}{5}n\pi\right\} + i\sin\left\{\frac{\pi}{5} + \frac{2}{5}n\pi\right\}$$

Taking $n = -2, -1, 0, 1, 2$ gives

$$z_1 = \cos(-\tfrac{3}{5}\pi) + i\sin(-\tfrac{3}{5}\pi)$$

$$z_2 = \cos(-\tfrac{1}{5}\pi) + i\sin(-\tfrac{1}{5}\pi)$$
$$z_3 = \cos(\tfrac{1}{5}\pi) + i\sin(\tfrac{1}{5}\pi)$$
$$z_4 = \cos(\tfrac{3}{5}\pi) + i\sin(\tfrac{3}{5}\pi)$$
$$z_5 = \cos\pi + i\sin\pi = -1.$$

Now z_1, z_2, z_3, z_4 and z_5 are the solutions of the equation $z^5+1=0$. In this equation $a=1$, $b=c=d=e=0$, $f=1$. So the sum of its solutions, $-\dfrac{b}{a}$, is zero,

i.e. $$z_1+z_2+z_3+z_4+z_5 = 0.$$

Hence $$\text{Re}\,(z_1+z_2+z_3+z_4+z_5) = 0.$$

[Also $\text{Im}\,(z_1+z_2+z_3+z_4+z_5)=0$, but is not relevant to this problem.]

So $$\cos(-\tfrac{3}{5}\pi)+\cos(-\tfrac{1}{5}\pi)+\cos(\tfrac{1}{5}\pi)+\cos(\tfrac{3}{5}\pi)-1 = 0$$

But $$\cos(-\tfrac{3}{5}\pi) = \cos(\tfrac{3}{5}\pi) \quad \text{and} \quad \cos(-\tfrac{1}{5}\pi) = \cos(\tfrac{1}{5}\pi)$$

$\Rightarrow$ $$2\cos(\tfrac{3}{5}\pi)+2\cos(\tfrac{1}{5}\pi)-1 = 0$$

i.e. $$\cos\tfrac{3}{5}\pi + \cos\tfrac{1}{5}\pi = \tfrac{1}{2}.$$

3) Solve the equation $x^6-2x^3+4=0$.

First regarding the equation as quadratic in x^3, i.e. $(x^3)^2-2(x^3)+4=0$,

gives $$x^3 = 1 \pm i\sqrt{3}.$$

Considering $$x^3 = 1+i\sqrt{3} = 2\left(\cos\left\{\frac{\pi}{3}+2n\pi\right\}+i\sin\left\{\frac{\pi}{3}+2n\pi\right\}\right)$$

$\Rightarrow$ $$x = \sqrt[3]{2}\left(\cos\left\{\frac{\pi}{9}+\frac{2}{3}n\pi\right\}+i\sin\left\{\frac{\pi}{9}+\frac{2}{3}n\pi\right\}\right) \qquad [1]$$

Similarly $$x^3 = 1-i\sqrt{3} = 2\left(\cos\left\{-\frac{\pi}{3}+2n\pi\right\}+i\sin\left\{-\frac{\pi}{3}+2n\pi\right\}\right)$$

$\Rightarrow$ $$x = \sqrt[3]{2}\left(\cos\left\{-\frac{\pi}{9}+\frac{2}{3}n\pi\right\}+i\sin\left\{-\frac{\pi}{9}+\frac{2}{3}n\pi\right\}\right) \qquad [2]$$

Taking $n=-1,0,1$ in [1] and [2] gives the six values of x that satisfy the given equation, i.e.

from [1] $$x_1 = \sqrt[3]{2}(\cos\{-\tfrac{5}{9}\pi\}+i\sin\{-\tfrac{5}{9}\pi\})$$
$$x_2 = \sqrt[3]{2}(\cos\{\tfrac{1}{9}\pi\}+i\sin\{\tfrac{1}{9}\pi\})$$
$$x_3 = \sqrt[3]{2}(\cos\{\tfrac{7}{9}\pi\}+i\sin\{\tfrac{7}{9}\pi\})$$

from [2] $\quad x_4 = \sqrt[3]{2}(\cos\{-\tfrac{7}{9}\pi\} + i\sin\{-\tfrac{7}{9}\pi\})$

$x_5 = \sqrt[3]{2}(\cos\{-\tfrac{1}{9}\pi\} + i\sin\{-\tfrac{1}{9}\pi\})$

$x_6 = \sqrt[3]{2}(\cos\{\tfrac{5}{9}\pi\} + i\sin\{\tfrac{5}{9}\pi\})$.

Hence $\quad x = \sqrt[3]{2}(\cos k\pi \pm i\sin k\pi)$

where $k = \tfrac{1}{9}, \tfrac{5}{9}$ or $\tfrac{7}{9}$.

EXERCISE 12b

1) Use De Moivre's Theorem to find the square roots of:
(a) $1-i$ (b) $3+4i$ (c) $-5+12i$ (d) $\sqrt{3}+i$
(e) $-2-2i$ (f) i (g) -1

2) Find the cube roots of each complex number given in Question 1.

3) Without first calculating them, illustrate the fifth roots of z on an Argand diagram, where z is
(a) 32 (b) $-4-4i$ (c) $9-9\sqrt{2}i$ (d) $-1+\sqrt{3}i$.

4) Find the cube roots of -1. Show that they can be denoted by $-1, \lambda, -\lambda^2$ and prove that $\lambda^2 - \lambda + 1 = 0$.

Solve the following equations, giving any complex roots in the form $r(\cos\theta + i\sin\theta)$.

5) $x^6 - 4x^3 + 8 = 0$

6) $x^4 - 6x^2 + 25 = 0$

7) $x^6 - 1 = 0$

8) $(x-2)^3 = 1$ (*Hint*. Find the three values of $x-2$.)

9) Prove that the fifth roots of 1 can be denoted by $1, \alpha, \alpha^2, \alpha^3, \alpha^4$ and show that $1 + \alpha + \alpha^2 + \alpha^3 + \alpha^4 = 0$.

10) Without using any series expansions, prove that

$$(\sqrt{3}+i)^n + (\sqrt{3}-i)^n \text{ is real.}$$

Find the value of this expression when $n = 12$.

11) By considering the seventh roots of -1 prove that

$$\cos\frac{\pi}{7} + \cos\frac{3\pi}{7} + \cos\frac{5\pi}{7} = \frac{1}{2}.$$

12) 1 and α are two of the fifth roots of unity, and α has a positive acute argument. If $u = \alpha + \alpha^4$ and $v = \alpha^2 + \alpha^3$ prove that

$$u+v = uv = -1$$

and $$u-v = \sqrt{5}.$$

Deduce that $\cos 72° = \frac{1}{4}(\sqrt{5}-1)$.

THE EXPONENTIAL FORM FOR A COMPLEX NUMBER

Because any complex number can be given in the form $r(\cos\theta + i\sin\theta)$, other ways of writing a complex number can be found by expressing $\cos\theta$ and/or $\sin\theta$ in alternative forms. A particularly interesting form is found from the series expansions of $\sin\theta$ and $\cos\theta$.

$$\cos\theta = 1 - \frac{\theta^2}{2!} + \frac{\theta^4}{4!} - \frac{\theta^6}{6!} + \dots$$

$$\sin\theta = \theta - \frac{\theta^3}{3!} + \frac{\theta^5}{5!} - \dots$$

Adding these series together gives

$$\cos\theta + \sin\theta = 1 + \theta - \frac{\theta^2}{2!} - \frac{\theta^3}{3!} + \frac{\theta^4}{4!} + \frac{\theta^5}{5!} - \dots$$

This series is very similar to the expansion of e^θ,

i.e. $$e^\theta = 1 + \theta + \frac{\theta^2}{2!} + \frac{\theta^3}{3!} + \frac{\theta^4}{4!} + \frac{\theta^5}{5!} + \dots$$

The discrepancy is that the signs of *pairs of consecutive terms* alternate. As this type of sign change is a property of powers of i,

i.e. $$i, i^2, i^3, i^4, i^5, i^6, i^7 \dots = i, -1, -i, 1, i, -1, -i \dots$$

it suggests examining the expansion of $e^{i\theta}$.

$$\begin{aligned} e^{i\theta} &= 1 + i\theta + \frac{(i\theta)^2}{2!} + \frac{(i\theta)^3}{3!} + \frac{(i\theta)^4}{4!} + \frac{(i\theta)^5}{5!} + \dots \\ &= 1 + i\theta - \frac{\theta^2}{2!} - \frac{i\theta^3}{3!} + \frac{\theta^4}{4!} + \frac{i\theta^5}{5!} - \dots \\ &= 1 - \frac{\theta^2}{2!} + \frac{\theta^4}{4!} - \dots \\ &\quad + i\left(\theta - \frac{\theta^3}{3!} + \frac{\theta^5}{5!} - \dots\right) \end{aligned}$$

Comparing this series with those for $\cos\theta$ and $\sin\theta$ we see that

$$e^{i\theta} \equiv \cos\theta + i\sin\theta$$

Thus any complex number can be written in any of the forms

$$z = x + yi = r(\cos\theta + i\sin\theta) = re^{i\theta}$$

Most of the work carried out earlier in this chapter using $r(\cos\theta + i\sin\theta)$ can now be done using $re^{i\theta}$, and some readers may prefer the brevity that this form offers.
For instance, to find the cube roots of $4 + 3i$, whose modulus is 5 and whose general argument is $(\alpha + 2n\pi)$ (where $\tan\alpha = \frac{3}{4}$), we can use

$$4 + 3i = 5e^{i(\alpha + 2n\pi)}$$

Then, if z is any cube root of $4 + 3i$,

$$z = \{5\,e^{i(\alpha+2n\pi)}\}^{\frac{1}{3}} = \sqrt[3]{5}\,e^{i(\frac{\alpha}{3}+\frac{2}{3}n\pi)}$$

Taking $n = -1, 0, 1$ gives

$$\begin{cases} z_1 = \sqrt[3]{5}\,e^{i(\frac{\alpha}{3}-\frac{2}{3}\pi)} \\ z_2 = \sqrt[3]{5}\,e^{i\frac{\alpha}{3}} \\ z_3 = \sqrt[3]{5}\,e^{i(\frac{\alpha}{3}+\frac{2}{3}\pi)} \end{cases}$$

Note that when using $re^{i\theta}$, θ *must* be given in radians because the series expansions for $\sin\theta$ and $\cos\theta$ are valid only for an angle measured in radians.

EXAMPLES 12c

1) Express in the form $re^{i\theta}$, each of the complex numbers

$$z_1 = 1 + i,\quad z_2 = \sqrt{3} - i,\quad z_3 = \frac{1+i}{\sqrt{3}-i},\quad z_4 = (1+i)(\sqrt{3}-i)$$

and represent them on an Argand diagram.

$$|z_1| = \sqrt{2} \quad \text{and} \quad \arg z_1 = \frac{\pi}{4} \quad \Rightarrow \quad z_1 = \sqrt{2}e^{i\frac{\pi}{4}}$$

$$|z_2| = 2 \quad \text{and} \quad \arg z_2 = -\frac{\pi}{6} \quad \Rightarrow \quad z_2 = 2e^{-i\frac{\pi}{6}}$$

$$\left.\begin{aligned} |z_3| &= \left|\frac{z_1}{z_2}\right| = \frac{\sqrt{2}}{2} \\ \arg z_3 &= \arg\left(\frac{z_1}{z_2}\right) = \arg z_1 - \arg z_2 = \frac{5\pi}{12} \end{aligned}\right\} \Rightarrow \quad z_3 = \frac{\sqrt{2}}{2}\,e^{i\frac{5}{12}\pi}$$

$$\left.\begin{aligned} |z_4| &= |z_1|\,|z_2| = 2\sqrt{2} \\ \arg z_4 &= \arg z_1 + \arg z_2 = \frac{\pi}{12} \end{aligned}\right\} \Rightarrow \quad z_4 = 2\sqrt{2}e^{i\frac{\pi}{12}}$$

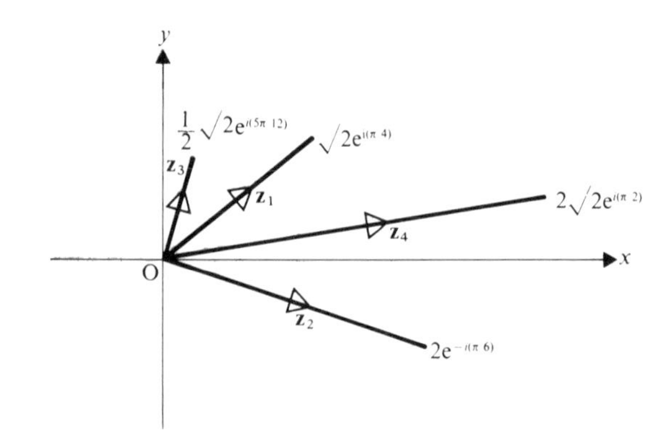

2) Without integrating by parts, find $\int_0^{\frac{\pi}{2}} e^{\theta} \cos\theta \, d\theta$.

$$\cos\theta \equiv \text{Re}(\cos\theta + i\sin\theta) \equiv \text{Re}(e^{i\theta})$$

Then
$$I = \text{Re}\int_0^{\frac{\pi}{2}} e^{\theta}e^{i\theta}\,d\theta = \text{Re}\int_0^{\frac{\pi}{2}} e^{(1+i)\theta}\,d\theta$$

$$= \text{Re}\left[\frac{1}{1+i}e^{(1+i)\theta}\right]_0^{\frac{\pi}{2}}$$

$$= \text{Re}\left[\frac{e^{\theta}}{1+i}(\cos\theta + i\sin\theta)\right]_0^{\frac{\pi}{2}}$$

$$= \text{Re}\left[\frac{(1-i)}{2}e^{\theta}(\cos\theta + i\sin\theta)\right]_0^{\frac{\pi}{2}}$$

$$= \left[\tfrac{1}{2}e^{\theta}\cos\theta + \tfrac{1}{2}e^{\theta}\sin\theta\right]_0^{\frac{\pi}{2}}$$

$$= \tfrac{1}{2}e^{\frac{\pi}{2}} - \tfrac{1}{2},$$

i.e.
$$I = \tfrac{1}{2}(e^{\frac{\pi}{2}} - 1).$$

Note. Assumptions are made in this solution that integration of complex functions is carried out in the same way as for real functions.

EXERCISE 12c

1) Express in the form $re^{i\theta}$.

(a) $1+i$ (b) i (c) $2-2\sqrt{3}i$ (d) $-1+i$

(e) 4 (f) $3+4i$ (g) $\dfrac{1+\sqrt{3}i}{1-\sqrt{3}i}$

2) Express in the form $a + bi$

(a) $e^{-i\frac{\pi}{3}}$ (b) $2e^{i\frac{5\pi}{6}}$ (c) $5e^{i\pi}$ (d) $e^{-i\frac{\pi}{2}}$ (e) $4e^{-i\pi}$

Find, in the form $re^{i\theta}$,

3) the cube roots of -1, 4) the fifth roots of $-4-4i$,

5) the seventh roots of 1, 6) the fourth roots of $3-\sqrt{7}i$.

Use the relationship $e^{i\theta} = \cos\theta + i\sin\theta$ to evaluate the following integrals.

7) $\int_0^{\frac{\pi}{2}} e^{-\theta}\sin\theta\, d\theta$ 8) $\int_0^{\frac{\pi}{4}} e^{\theta}\cos 2\theta\, d\theta$

9) Express $\cos\theta$ and $\sin\theta$ each in terms of $e^{i\theta}$ and $e^{-i\theta}$.
Use your results to prove that

$$16\cos^3\theta\sin^2\theta \equiv 2\cos\theta - \cos 3\theta - \cos 5\theta.$$

RELATIONSHIPS BETWEEN HYPERBOLIC AND TRIGONOMETRIC FUNCTIONS

Examination of the series expansions of $\cos x$, $\sin x$, $\cosh x$ and $\sinh x$ discloses a number of interesting relationships connecting these functions and the exponential function.

$$\cos x = 1 - \frac{x^2}{2!} + \frac{x^4}{4!} - \frac{x^6}{6!} + \ldots \quad [1]$$

$$\cosh x = 1 + \frac{x^2}{2!} + \frac{x^4}{4!} + \frac{x^6}{6!} + \ldots \quad [2]$$

$$\sin x = x - \frac{x^3}{3!} + \frac{x^5}{5!} - \frac{x^7}{7!} + \ldots \quad [3]$$

$$\sinh x = x + \frac{x^3}{3!} + \frac{x^5}{5!} + \frac{x^7}{7!} + \ldots . \quad [4]$$

Remembering that $i^{4n} = 1$ and $i^{(4n-2)} = -1$
it is clear that replacing x by ix in [1] gives

$$\cos ix = 1 + \frac{x^2}{2!} + \frac{x^4}{4!} + \frac{x^6}{6!} + \ldots$$

$$\Rightarrow \qquad \cos ix = \cosh x$$

Replacing x by ix in [3] gives

$$\sin ix = i\left\{x + \frac{x^3}{3!} + \frac{x^5}{5!} + \frac{x^7}{7!} + \ldots\right\}$$

$\Rightarrow$ $$\sin ix = i \sinh x.$$

If, however, x is replaced by ix in equations [2] and [4] we find that

$$\cosh ix = 1 - \frac{x^2}{2!} + \frac{x^4}{4!} - \frac{x^6}{6!} + \ldots$$

$$= \cos x$$

$\Rightarrow$ $$\cosh ix = \cos x.$$

$$\sinh ix = i\left\{x - \frac{x^3}{3!} + \frac{x^5}{5!} - \frac{x^7}{7!} + \ldots\right\}$$

$\Rightarrow$ $$\sinh ix = i \sin x.$$

The two latter relationships present yet another way of denoting a complex number,

i.e. $$\cos\theta + i\sin\theta = \cosh i\theta + \sinh i\theta.$$

OSBORN'S RULE

The reader will recall that, when applying Osborn's Rule to convert a trig identity into a hyperbolic identity, every term containing $\sin^2\theta$ becomes $-\sinh^2 x$.

The reason for this change of sign can now be appreciated by considering the relationships

$$\begin{cases} \cos ix = \cosh x \\ \sin ix = i\sinh x \end{cases}$$

Replacing ix by θ gives

$$\begin{cases} \cos\theta = \cosh x \\ \sin\theta = i\sinh x \end{cases}$$

Now any term containing $\sin^2\theta$ becomes $(i\sinh x)^2$
and any term containing $\cos^2\theta$ becomes $(\cosh x)^2$,

i.e. $$\begin{cases} \sin^2\theta = -\sinh^2 x \\ \cos^2\theta = \cosh^2 x, \end{cases}$$

e.g. $$\cos^2\theta + \sin^2\theta \equiv 1 \quad \Rightarrow \quad \cosh^2 x - \sinh^2 x \equiv 1.$$

Exponential Form for Trigonometric Functions

Using the relationships

$$\cosh ix = \cos x$$
$$\sinh ix = i \sin x$$

and the definitions

$$\cosh ix \equiv \tfrac{1}{2}(e^{ix} + e^{-ix})$$
$$\sinh ix \equiv \tfrac{1}{2}(e^{ix} - e^{-ix}),$$

we have

$$\cos x = \tfrac{1}{2}(e^{ix} + e^{-ix})$$
$$\sin x = \frac{1}{2i}(e^{ix} - e^{-ix})$$

All the relationships derived in this section, and others that the reader may produce, are based on the assumption that the expansions of trig, hyperbolic and exponential functions have the same form for an imaginary variable as for a real variable. In fact at this stage we are actually *defining* trig, hyperbolic and exponential functions of an imaginary variable.

TRANSFORMATIONS IN THE ARGAND DIAGRAM

When the coordinates of a point $P(x, y)$ represent a complex number $z = x + yi$, and the coordinates of a point $Q(u, v)$ represent a complex number $w = u + vi$, then a relationship between z and w maps P to Q in an Argand diagram.
If, also, the locus of P is known, this locus will be mapped to the locus of Q. For instance, if $w = z - 3$ and the locus of $P(x, y)$ is $x - 2y = 5$ the locus of $Q(u, v)$ can be found as follows:

$$w = z - 3 \quad \Rightarrow \quad u + vi = x + yi - 3$$
$$\Rightarrow \quad u + 3 = x \quad \text{and} \quad v = y,$$

i.e. $$w = z - 3 \quad \text{maps} \quad \begin{cases} x \text{ to } u + 3 \\ y \text{ to } v \end{cases}$$

or $$x - 2y = 5 \quad \text{maps to} \quad u + 3 - 2v = 5$$

So the locus of $Q(u, v)$ is $u - 2v = 2$.

Alternatively, the equation of the locus of $P(x, y)$ can be expressed in the form

$$\arg(z-5) = \alpha$$

where $\tan\alpha = \frac{1}{2}$

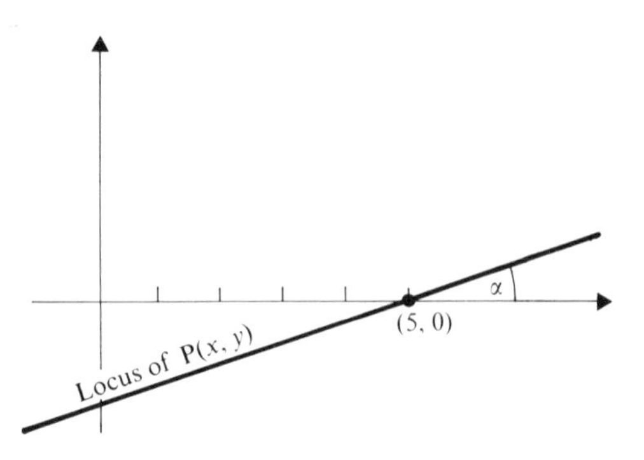

Then, using $w = z - 3$, we have

$$\arg(w-2) = \alpha$$

So the equation of the locus of $Q(u, v)$ is

$$v = \tfrac{1}{2}(u-2)$$

i.e. $\quad 2v = u - 2$

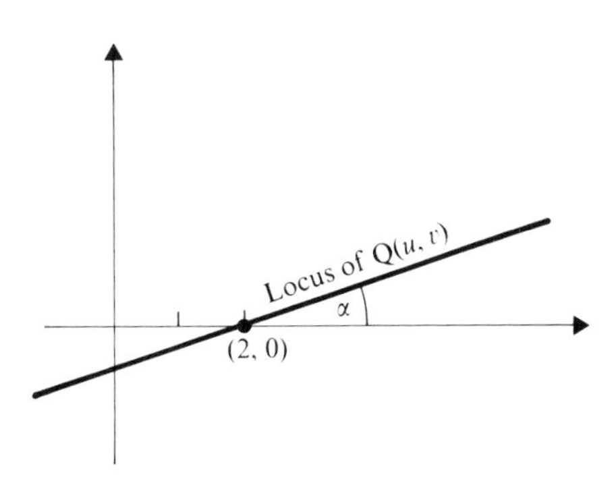

EXAMPLES 12d

1) The relationship $w = z^2$, where $z = x + yi$ and $w = u + vi$, maps the point $P(x, y)$ to the point $Q(u, v)$ in an Argand diagram. If P describes

(a) the circle $x^2 + y^2 = 25$,

(b) the curve $x^2 - y^2 = 4$,

find the equation of the locus of Q in each case.

(a) The equation $x^2 + y^2 = 25$ can be written in the form $|z| = 5$.

Now $\quad w = z^2$

Hence $\quad |w| = |z^2| = |z||z| = 25$

So the equation of the locus of Q is $|w| = 25$
i.e. Q describes a circle, centre O, and radius 25.

(b) As $x^2 - y^2 = 4$ cannot easily be expressed in complex form we use the relationships between x and y, u and v.

$$w = z^2 \Rightarrow u + vi = x^2 - y^2 + 2xyi$$

$$\Rightarrow \begin{cases} u = x^2 - y^2 \\ v = 2xy \end{cases}$$

(not needed)

So the equation $x^2 - y^2 = 4$
maps to the equation $u = 4$
and we see that the locus of $Q(u, v)$ is the vertical line $u = 4$.

EXERCISE 12d

In this exercise x, y, u, v are real numbers and z, w are complex numbers such that $z = x + yi$, $w = u + vi$. Describe the locus of the point (u, v) in the Argand diagram, in Questions 1–4.

1) $w = z + 4$ and
(a) $|z| = 3$ (b) $\arg(z) = \dfrac{\pi}{3}$ (c) $|z + 4| = 5$
(d) $x^2 + y^2 = 1$ (e) $y = 2x$ (f) $y^2 = 4x$.

2) $w = -2z$ and
(a) $x^2 - y^2 = 1$ (b) $\arg(z) = \dfrac{\pi}{2}$ (c) $|z| = 4$
(d) $y = 2x + 3$.

3) $w = \dfrac{1}{z}$ and
(a) $|z| = 6$ (b) $\arg z = -\dfrac{\pi}{4}$ (c) $\dfrac{x^2}{4} + y^2 = 1$
(d) $xy = 4$ (e) $y = 3x + 4$.

4) $w^2 = z$ and P describes
(a) the circle $x^2 + (y-1)^2 = 1$,
(b) the positive x-axis,
(c) the negative y-axis.

5) Given that $z = w^2$, prove that
(a) if u varies but v is constant, then the locus of $P(x, y)$ is a parabola,
(b) if v varies but u is constant, the locus of P is again a parabola.
Show that these two parabolas have the same axis and focus.

SUMMARY

$(\cos\theta + i\sin\theta)^n \equiv \cos n\theta + i\sin n\theta$ (De Moivre's Theorem)

$$\left.\begin{aligned} z^n + \frac{1}{z^n} &= 2\cos n\theta \\ z^n - \frac{1}{z^n} &= 2i\sin n\theta \end{aligned}\right\} \text{ where } z = \cos\theta + i\sin\theta.$$

If $z = r(\cos\theta + i\sin\theta) = re^{i\theta}$, the pth roots of z are given by

$$\sqrt[p]{r}\, e^{i\left(\frac{\theta}{p} + \frac{2n}{p}\pi\right)}$$

The cube roots of 1 can be expressed as $1, \omega, \omega^2$, and $1 + \omega + \omega^2 = 0$.

MISCELLANEOUS EXERCISE 12

1) Simplify, without the use of tables,

$$\frac{(\cos \frac{1}{7}\pi - i \sin \frac{1}{7}\pi)^3}{(\cos \frac{1}{7}\pi + i \sin \frac{1}{7}\pi)^4}$$ (U of L)p

2) Prove that, when n is a positive integer,

$$(\cos \theta + i \sin \theta)^n = \cos n\theta + i \sin n\theta.$$

Find the modulus and argument of

$$\frac{[(\sqrt{3})(\cos \theta + i \sin \theta)]^4}{\cos 2\theta - i \sin 2\theta}$$ (JMB)

3) If $z = \cos \theta + i \sin \theta$, show that

(a) $z + \dfrac{1}{z} = 2 \cos \theta$, (b) $z^n + \dfrac{1}{z^n} = 2 \cos n\theta$.

Hence, or otherwise, show that

$$\cos^4\theta = \tfrac{1}{8}(\cos 4\theta + 4 \cos 2\theta + 3).$$ (U of L)p

4) Use De Moivre's theorem to prove that, if θ is not a multiple of π,

$$\frac{\sin 5\theta}{\sin \theta} = 16 \cos^4\theta - 12 \cos^2\theta + 1.$$ (U of L)p

5) Prove by induction that if n is a positive integer

$$(\cos \theta + i \sin \theta)^n = \cos n\theta + i \sin n\theta.$$

Evaluate $(1 + i)^n + (1 - i)^n$ when $n = 20$. (U of L)p

6) Assuming De Moivre's theorem for an index which is a positive integer, prove that one value of $(\cos \theta + i \sin \theta)^{p/q}$ is

$$\cos (p\theta/q) + i \sin (p\theta/q)$$

when p and q are positive integers and p/q is in its lowest terms.
What are the other values?
Find, in the form $a + ib$, the six roots of the equation $z^6 + 1 = 0$, and represent them on the Argand diagram. (O)

7) Expand $\left(z + \dfrac{1}{z}\right)^4$ and $\left(z - \dfrac{1}{z}\right)^4$

By putting $z = \cos \theta + i \sin \theta$, deduce that

$$\cos^4\theta + \sin^4\theta = \tfrac{1}{4}(\cos 4\theta + 3).$$ (U of L)p

8) By using De Moivre's theorem, or otherwise, express $\sin 4\theta$ and $\cos 4\theta$ in terms of powers of $\sin \theta$ and $\cos \theta$ and show that

$$\tan 4\theta = \frac{4 \tan \theta(1 - \tan^2\theta)}{1 - 6 \tan^2\theta + \tan^4\theta}$$ (U of L)p

9) By comparing the expressions for $(\cos\theta + i\sin\theta)^5$ given by De Moivre's theorem and by the binomial theorem, prove that

$$\cos 5\theta = 16\cos^5\theta - 20\cos^3\theta + 5\cos\theta.$$

By considering the equation $\cos 5\theta = 0$, prove that

$$\cos(\pi/10).\cos(3\pi/10) = \tfrac{1}{4}\sqrt{5}.$$ (U of L)p

10) Assuming De Moivre's theorem for an integral index, prove that one value of $(\cos\theta + i\sin\theta)^{p/q}$ (where p and q are integers having no common factor) is $\cos(p\theta/q) + i\sin(p\theta/q)$. What are the other values?
Solve the equation

$$z^6 + z^3 + 1 = 0.$$

[Give your answers in the form $\cos\theta + i\sin\theta$, but do not work out the values of the cosines and sines.] (O)

11) Show that the solutions of $z^6 + z^3 + 1 = 0$ are among the solutions of $z^9 - 1 = 0$. Hence find the solutions of $z^6 + z^3 + 1 = 0$ in the form $\cos\phi + i\sin\phi$. Deduce the values of θ between 0 and 2π which satisfy both the equations

$$\cos 6\theta + \cos 3\theta + 1 = 0,$$

and

$$\sin 6\theta + \sin 3\theta = 0.$$ (JMB)

12) Draw the locus $|z| = 1$ on an Argand diagram. Mark also a point $z = a + bi$, for which $a > b > 0$ and $a^2 + b^2 > 1$.
On the same diagram join the points representing

$$z^2, \ \frac{1}{z} \text{ and } z + \frac{1}{z},$$

to the origin indicating any equal angles.
(a) If $u = x + yi$, find the complete set of values of u such that $u + \dfrac{1}{u}$ is real.
(b) If $\omega = \frac{1}{2}(-1 + i\sqrt{3})$, express in the form $p + qi$ the complex numbers ω^4 and ω^5. (U of L)

13) Find the modulus and the argument of each of the roots of the equation

$$z^5 + 32 = 0.$$

Hence express

$$z^4 - 2z^3 + 4z^2 - 8z + 16$$

as the product of two quadratic factors of the form $z^2 - az\cos\theta + b$, where a, b and θ are real. (JMB)

14) Find the two complex numbers whose squares are equal to i.
Hence, or otherwise, find the four (complex) roots of the equation $z^4 + 4 = 0$, giving the modulus and argument of each.
Hence express $z^4 + 4$ as the product of two real quadratic polynomials. (O)

15) (a) If $\omega = \cos 120° + i \sin 120°$, represent the numbers ω, ω^2 and ω^3 in an Argand diagram, and find all the possible values of $(\omega^n + \omega^{2n} + \omega^{3n})$, where n is an integer.

(b) If a^2 is greater than $(b^2 + c^2)$ in the triangle ABC, show that the value of the expression

$$\frac{(1 + \cos 2B + i \sin 2B)(1 + \cos 2C + i \sin 2C)}{(1 + \cos 2A - i \sin 2A)}$$

is real and positive. (U of L)

16) (a) Express the complex number $1 + i$ in the form $r(\cos\theta + i \sin\theta)$, where $r > 0$. Hence, or otherwise, express $(1 + i)^{11}$ in the form $a + ib$, where a and b are real.

(b) Solve the equation $z^4 + 14z^2 + 625 = 0$, giving the four roots in the form $x + iy$, where x and y are real. (O)

17) Obtain the three cube roots of unity in the form $a + ib$ where a and b are real.
These three roots are denoted by $\omega_1, \omega_2, \omega_3$. Show that the real parts of $\dfrac{1}{1+\omega_1}, \dfrac{1}{1+\omega_2}$ and $\dfrac{1}{1+\omega_3}$ are all equal to $\frac{1}{2}$, and interpret this result geometrically. (JMB)

18) (a) If $z = 3 - 4i$, express z^2 and $1/z$ in the form $a + bi$ where a and b are real and represent them in an Argand diagram.
If $w^2 = z$, express the two values of w in the form $a + bi$.

(b) State de Moivre's theorem for a positive integral power and use it to express $\sin 3\theta$ and $\cos 3\theta$ in terms of $\sin\theta$ and $\cos\theta$ respectively. Hence show that, if $\sin 3\theta + \cos 3\theta = 0$, either $\tan\theta = 1$ or $\sin 2\theta = -\frac{1}{2}$. (U of L)

19) Prove De Moivre's Theorem,

$$(\cos\theta + i \sin\theta)^n = \cos n\theta + i \sin n\theta$$

in the case where n is a positive integer.
Express $\cos 6\theta$ as a polynomial in $\cos\theta$.

If $z = \cos\theta + i \sin\theta$, show by expanding $\left(z + \dfrac{1}{z}\right)^5 \left(z - \dfrac{1}{z}\right)^5$, or otherwise, that

$$\sin^5\theta \cos^5\theta = \frac{1}{2^9}(\sin 10\theta - 5 \sin 6\theta + 10 \sin 2\theta).$$

Evaluate

$$\int_0^{\frac{\pi}{2}} \sin^5\theta \cos^5\theta \, d\theta.$$

20) (a) Find the cube roots of -1, either by using De Moivre's theorem or by

factorizing $z^3 + 1$ and hence solving algebraically the equation $z^3 + 1 = 0$.
Show that if either complex root is denoted by λ, the other is $-\lambda^2$, and that

$$(X + \lambda Y - \lambda^2 Z)(X - \lambda^2 Y + \lambda Z) = X^2 + Y^2 + Z^2 - YZ + ZX + XY.$$

(b) Express the complex number $z = 8(1 + i)/\sqrt{2}$ in the form $r(\cos\theta + i\sin\theta)$ and hence show that the three values of $z^{\frac{2}{3}}$ are $-4i$, $2(\sqrt{3} + i)$, $2(-\sqrt{3} + i)$. (U of L)

21) (a) If $1, \omega$ and ω^2 are the three cube roots of unity, find the value of
(i) $1 + \omega + \omega^2$,
(ii) $(1 + 2\omega + 3\omega^2)(1 + 2\omega^2 + 3\omega)$.
Also show that, if the equations $x^3 - 1 = 0$ and $px^5 + qx + r = 0$ have a common root, then

$$(p + q + r)(p\omega^5 + q\omega + r)(p\omega^{10} + q\omega^2 + r) = 0.$$

(b) If $|z - 1| = 3|z + 1|$, prove that the locus of z in an Argand diagram is a circle and find its centre and radius. (U of L)

22) Prove that $\tan\dfrac{\pi}{15}$ is a root of the equation

$$t^4 - 6\sqrt{3}t^3 + 8t^2 + 2\sqrt{3}t - 1 = 0.$$

Give the other roots in the form $\tan\dfrac{r\pi}{15}$. (O)

23) Express in the form $\cos\theta + i\sin\theta$ each of the cube roots of unity. If $\alpha^3 = \beta^3 = 1$ and $\alpha \neq \beta$, use the Argand diagram to find the value of $|\alpha - \beta|$. On the same diagram plot the points representing the three possible values of $\alpha + \beta$, and evaluate $(\alpha + \beta)^3$. (U of L)p

24) Express $\sqrt{3} - i$ in the form $r\,e^{i\theta}$, where $r > 0$ and $-\pi < \theta \leqslant \pi$. Hence show that, when n is a positive integer,

$$(\sqrt{3} - i)^n + (\sqrt{3} + i)^n = 2^{n+1}\cos(n\pi/6).$$ (U of L)p

25) (a) Show on an Argand diagram,
(i) the three roots of the equation $z^3 - 1 = 0$,
(ii) the four roots of the equation $z^4 - 16 = 0$.
(b) If $z = e^{i\theta}$, show that
(i) $z + z^{-1} = 2\cos\theta$,
(ii) $z^n + z^{-n} = 2\cos n\theta$.
Show also that

$$\cos^6\theta = \tfrac{1}{32}(\cos 6\theta + 6\cos 4\theta + 15\cos 2\theta + 10).$$ (U of L)

26) Given that z is one of the three cube roots of unity, find the two possible values of the expression $z^2 + z + 1$.
Given that ω is a complex cube root of unity, simplify each of the expressions

$$(1 + 3\omega + \omega^2)^2 \quad \text{and} \quad (1 + \omega + 3\omega^2)^2,$$

and show that their product is equal to 16 and that their sum is -4. (JMB)

27) Solve the equation

$$z^3 = 8i,$$

giving the roots in the form $re^{i\theta}$ where $r > 0$ and $0 \leqslant \theta < 2\pi$. (U of L)

28) Verify that the complex number $\alpha = e^{2\pi i/5}$ is a root of the equation $z^5 - 1 = 0$, and deduce, or prove otherwise, that $1 + \alpha + \alpha^2 + \alpha^3 + \alpha^4 = 0$. Find a quadratic equation whose roots are $\alpha + \alpha^4$ and $\alpha^2 + \alpha^3$. Hence, or otherwise, show that $\cos\dfrac{2\pi}{5} = \dfrac{\sqrt{5} - 1}{4}$.
The points A, B, C, and D on the Argand diagram represent $\alpha, \alpha^2, \alpha^3$ and α^4 respectively. Find, to three significant figures, the ratio AD/BC. (JMB)

29) When $z = 4\sqrt{3}\, e^{i\pi/3} - 4\, e^{i5\pi/6}$ express z in the form $re^{i\theta}$
Hence
(a) show that $\dfrac{z}{8} + i\left(\dfrac{z}{8}\right)^2 + \left(\dfrac{z}{8}\right)^3 = 2e^{i\pi/2}$,

(b) find the cube roots of z in the (r, θ) form. (AEB '77)p

30) (a) Solve the equation $z^5 = 1$ and represent the roots on an Argand diagram.
If ω denotes any one of the non-real roots of $z^5 = 1$, show that $1 + \omega + \omega^2 + \omega^3 + \omega^4 = 0$.
(b) By expressing $\sin\theta$ and $\cos\theta$ in terms of $e^{i\theta}$ and $e^{-i\theta}$, or otherwise, prove that

$$2^5 \sin^4\theta \cos^2\theta = \cos 6\theta - 2\cos 4\theta - \cos 2\theta + 2. \quad \text{(U of L)}$$

31) (a) Points P and Q represent complex numbers w and z respectively in an Argand diagram. If $w = u + iv$, $z = x + iy$ and $w = \dfrac{1 + zi}{z + i}$, express u and v in terms of x and y.
Prove that when P describes the portion of the imaginary axis between the points representing $-i$ and i, Q describes the whole of the positive half of the imaginary axis.
(b) If $2\cos\theta = z + z^{-1}$, prove that, if n is a positive integer,

$$2\cos n\theta = z^n + z^{-n}.$$

Hence or otherwise solve the equation

$$3z^4 - z^3 + 2z^2 - z + 3 = 0$$

given that no roots are real. (U of L)

32) The coordinates of a point P in one Argand diagram are (x, y) and are expressed in complex form $z = x + iy$. If the coordinates of a point Q in a second Argand diagram are (u, v) express the coordinates of Q in complex form w.
If $z = w^2$, find x and y in terms of u and v and show that, if P lies on the circle $x^2 + y^2 = 16$, then Q lies on the circle $u^2 + v^2 = 4$. (AEB '73)

33) The transformation $w = (z + 1)^2 + 3$ maps the complex number $z = x + iy$ to the complex number $w = u + iv$.
Show that as z moves along the y-axis from the origin to the point $(0, 2)$ in the z-plane, w moves from the point $(4, 0)$ to the point $(0, 4)$ along a curve in the w-plane.
Write down the equation of this curve. (JMB)

CHAPTER 13

FURTHER INTEGRATION AND DIFFERENTIAL EQUATIONS

REDUCTION METHOD OF INTEGRATION

Certain methods of integration which are applicable to functions involving a power n where n is a positive integer, are viable only when n is relatively small. For instance $\int \cos^n x \, dx$ can be found when $n = 4$ by using the identity $\cos^2\theta \equiv \frac{1}{2}(1 - \cos 2\theta)$ as often as necessary, but the same method applied to $\int \cos^{20} x \, dx$ would be extremely unwieldy. In cases like this, a means of systematically reducing the value of n is useful, and is called a reduction method. Usually a reduction method is based on the technique of integration by parts.

EXAMPLES 13a

1) If $I_n \equiv \int \cos^n x \, dx$, show that

$$I_n = \frac{1}{n}\sin x \cos^{n-1} x + \frac{n-1}{n} I_{n-2}$$

Hence find $\int \cos^5 x \, dx$.

To use integration by parts, $\cos^n x$ is written in the product form $\cos x \cos^{n-1} x$, so that

$$I_n = \int \cos x \cos^{n-1} x \, dx.$$

$$\text{If} \quad v = \cos^{n-1} x; \quad \frac{du}{dx} = \cos x$$

$$\text{then} \quad \frac{dv}{dx} = (n-1)\cos^{n-2} x(-\sin x); \quad u = \sin x$$

So $\quad I_n = \sin x \cos^{n-1} x + \int (n-1)\cos^{n-2} x \sin^2 x \, dx$

$$= \sin x \cos^{n-1} x + (n-1)\int \cos^{n-2} x(1-\cos^2 x)\, dx$$

$$= \sin x \cos^{n-1} x + (n-1)\int \cos^{n-2} x \, dx - (n-1)\int \cos^n x \, dx.$$

Hence

$$I_n = \sin x \cos^{n-1} x + (n-1)I_{n-2} - (n-1)I_n$$

or $\quad nI_n = \sin x \cos^{n-1} x + (n-1)I_{n-2}$

So $\quad I_n = \frac{1}{n}\sin x \cos^{n-1} x + \frac{n-1}{n} I_{n-2}$

Now we can apply this reduction formula to find $\int \cos^5 x \, dx$,

first using $n = 5$,

i.e. $\quad I_5 = \frac{1}{5}\sin x \cos^4 x + \frac{4}{5} I_3 \quad [1]$

Further, using $n = 3$ we have,

$$I_3 = \tfrac{1}{3}\sin x \cos^2 x + \tfrac{2}{3} I_1 \quad [2]$$

Now

$$I_1 = \int \cos x \, dx = \sin x + K \quad [3]$$

Combining [1], [2] and [3] gives,

$$I_5 = \tfrac{1}{5}\sin x \cos^4 x + \tfrac{4}{5}\left(\tfrac{1}{3}\sin x \cos^2 x + \tfrac{2}{3}\sin x\right) + K.$$

2) Establish a reduction formula that could be used to find $\int x^n e^x \, dx$ and use it when $n = 4$.

Let $I_n = \int x^n e^x \, dx$.

If $\quad v = x^n \qquad \dfrac{du}{dx} = e^x$

then $\quad \dfrac{dv}{dx} = nx^{n-1} \qquad u = e^x$

Then $\qquad I_n = x^n e^x - \int nx^{n-1} e^x \, dx,$

i.e. $\qquad I_n = x^n e^x - nI_{n-1}$

This is a suitable reduction formula.

Using
$$\begin{cases} n = 4 \text{ gives } I_4 = x^4 e^x - 4I_3 \\ n = 3 \text{ gives } I_3 = x^3 e^x - 3I_2 \\ n = 2 \text{ gives } I_2 = x^2 e^x - 2I_1 \\ n = 1 \text{ gives } I_1 = x e^x - I_0 \end{cases}$$

Now $\qquad I_0 = \int x^0 e^x \, dx = e^x + K$

So $\qquad I_4 = x^4 e^x - 4[x^3 e^x - 3\{x^2 e^x - 2(x e^x - e^x)\}] + K.$

Note that some reduction formulae cause the value of n to fall by 2 in each step while in other cases n falls only by 1 per step.

EXERCISE 13a

Establish a reduction formula for each of the following integrals.

1) $\int \sin^n x \, dx$ 2) $\int \tan^n x \, dx$ (use $\tan^2 x \equiv \sec^2 x - 1$)

3) $\int (x + 1)^n e^{2x} \, dx$ 4) $\int \cos^n 2\theta \, d\theta$

5) $\int x^n e^{ax} \, dx$ 6) $\int x(\ln x)^n \, dx$

7) $\int \cosh^n x \, dx$ 8) $\int \sec^n x \, dx$

Use a reduction method to find each of the following integrals.

9) $\int \cos^6 x \, dx$ 10) $\int \sin^7 x \, dx$

11) $\int (1 - x)^3 e^{4x} \, dx$ 12) $\int \cos^4 (ax + b) \, dx$

13) $\int \sin^4\left(\frac{\pi}{4} - 3\theta\right) d\theta$ 14) $\int \tan^5 x \, dx$

15) $\int x^3 \sin x \, dx$ 16) $\int \sinh^5 x \, dx$

17) $\int \sec^4 x \, dx$ 18) $\int x^6 e^{-x} \, dx$

19) $\int x(\ln x)^3 \, dx$ 20) $\int \cosh^4 3x \, dx$

21) Prove that, if $I_n = \int x^n(1 + x^3)^7 \, dx$ then

$$I_n = \frac{1}{n + 22}\left\{x^{n-2}(1 + x^3)^8 - (n - 2)I_{n-3}\right\}$$

(*Hint*. Use $x^n \equiv x^{n-2}x^2$)

Hence or otherwise determine $\int x^5(1 + x^3)^7 \, dx$

22) If $I_n = \int \frac{\cos^{2n} x}{\sin x} dx$, write down a similar expression for I_{n+1}.

Hence, by using the identity $\cos^2 x + \sin^2 x \equiv 1$, prove that

$$(2n + 1)I_{n+1} = (2n + 1)I_n + \cos^{2n+1} x.$$

Hence or otherwise determine $\int \frac{\cos^6 x}{\sin x} dx.$

DEFINITE INTEGRATION BY REDUCTION METHODS

Again we will consider integrating $\cos^n x$ but this time as a definite integral with boundaries at $x = 0$ and $x = \frac{\pi}{2}$, so that

$$I_n = \int_0^{\frac{\pi}{2}} \cos^n x \, dx$$

Integrating by parts as before gives

$$I_n = \left[\frac{1}{n} \sin x \cos^{n-1} x\right]_0^{\frac{\pi}{2}} + \frac{n-1}{n} I_{n-2}$$

$$= 0 + \frac{n-1}{n} I_{n-2},$$

i.e. $$I_n = \int_0^{\frac{\pi}{2}} \cos^n x \, dx \Rightarrow I_n = \frac{n-1}{n} I_{n-2}$$

If we now consider $\int_0^{\frac{\pi}{2}} \sin^n x \, dx$, it can be shown that

$$I_n = \int_0^{\frac{\pi}{2}} \sin^n x \, dx \Rightarrow I_n = \frac{n-1}{n} I_{n-2}$$

When these reduction formulae are used successively, we find that

$$\begin{aligned} I_n &= \left(\frac{n-1}{n}\right) I_{n-2} \\ &= \left(\frac{n-1}{n}\right)\left(\frac{n-3}{n-2}\right) I_{n-4} \\ &= \left(\frac{n-1}{n}\right)\left(\frac{n-3}{n-2}\right)\left(\frac{n-5}{n-4}\right) I_{n-6}, \quad \text{etc.} \end{aligned}$$

The final term in the formula depends on whether n is odd or even.

(a) If n is even we have

$$I_n = \left(\frac{n-1}{n}\right)\left(\frac{n-3}{n-2}\right) \cdots \left(\frac{3}{4}\right)\left(\frac{1}{2}\right) I_0$$

But $$I_0 = \int_0^{\frac{\pi}{2}} \cos^0 x \, dx \quad \text{or} \quad \int_0^{\frac{\pi}{2}} \sin^0 x \, dx,$$

i.e. $$I_0 = \frac{\pi}{2} \quad \text{in both cases.}$$

So, when n is even,

$$I_n = \left(\frac{n-1}{n}\right)\left(\frac{n-3}{n-2}\right) \cdots \left(\frac{3}{4}\right)\left(\frac{1}{2}\right)\left(\frac{\pi}{2}\right)$$

(b) If n is odd we have

$$I_n = \left(\frac{n-1}{n}\right)\left(\frac{n-3}{n-2}\right) \cdots \left(\frac{4}{5}\right)\left(\frac{2}{3}\right) I_1$$

But $$I_1 = \int_0^{\frac{\pi}{2}} \cos x \, dx \quad \text{or} \quad \int_0^{\frac{\pi}{2}} \sin x \, dx,$$

i.e. $$I_1 = 1 \quad \text{in both cases.}$$

So, when n is odd

$$I_n = \left(\frac{n-1}{n}\right)\left(\frac{n-3}{n-2}\right) \cdots \left(\frac{4}{5}\right)\left(\frac{2}{3}\right)$$

These very simple and convenient reduction formulae can be quoted but their use is, of course, valid only when the limits of integration are 0 and $\dfrac{\pi}{2}$.

e.g. $$\int_0^{\frac{\pi}{2}} \cos^{10} x \, dx = \left(\frac{9}{10}\right)\left(\frac{7}{8}\right)\left(\frac{5}{6}\right)\left(\frac{3}{4}\right)\left(\frac{1}{2}\right)\left(\frac{\pi}{2}\right)$$

and $$\int_0^{\frac{\pi}{2}} \sin^9 x \, dx = \left(\frac{8}{9}\right)\left(\frac{6}{7}\right)\left(\frac{4}{5}\right)\left(\frac{2}{3}\right)$$

but $$\int_0^{\pi} \cos^8 x \, dx \text{ is } \textit{not } \left(\frac{7}{8}\right)\left(\frac{5}{6}\right)\left(\frac{3}{4}\right)\left(\frac{1}{2}\right)\left(\frac{\pi}{2}\right)$$

A Graphical Extension to the use of Reduction Formulae

If a definite integral of $\sin^n x$ or $\cos^n x$ is required between limits 0 and $k\pi/2$ (where k is an integer) a graphical approach makes it possible to use the simplified reduction formula which is valid only for the limits 0 to $\dfrac{\pi}{2}$.
Consider, for instance, the graphs of $\cos x$, $\cos^2 x$ and $\cos^3 x$.

(a)

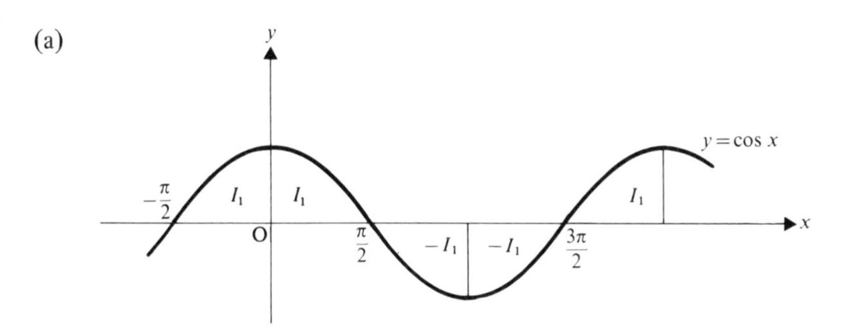

(b)

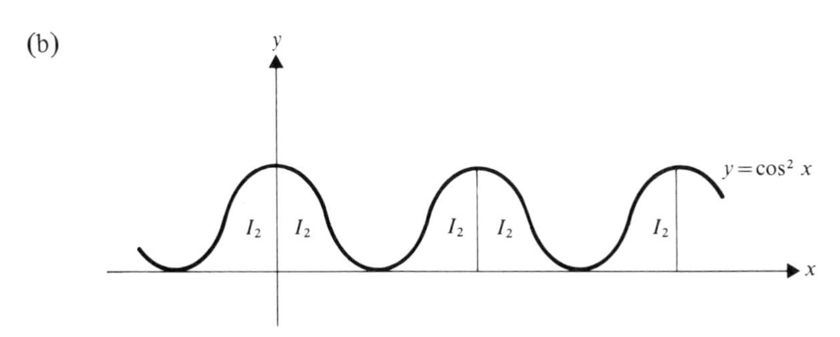

Note in graph (b) that, since $y = (\cos x)^2$, $y \geqslant 0$ for all values of x.

(c)

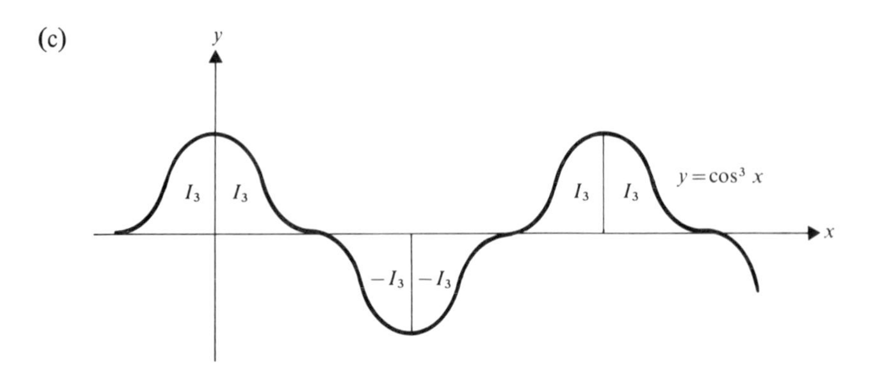

In case (a), if $\int_0^{\frac{\pi}{2}} \cos x \, dx = I_1$

then $\int_{\frac{\pi}{2}}^{\pi} \cos x \, dx = -I_1$ etc.

So $\int_0^{\pi} \cos x \, dx = I_1 - I_1 = 0$

and $\int_0^{\frac{3\pi}{2}} \cos x \, dx = I_1 - I_1 - I_1 = -I_1$

$$\Rightarrow \quad \int_0^{\frac{k\pi}{2}} \cos x \, dx = \begin{cases} 0 & \text{if } k \text{ is even} \\ \pm I_1 & \text{if } k \text{ is odd.} \end{cases}$$

In case (b), if $\int_0^{\frac{\pi}{2}} \cos^2 x \, dx = I_2$

then $\int_0^{\pi} \cos^2 x \, dx = I_2 + I_2 = 2I_2$

and $\int_0^{\frac{3\pi}{2}} \cos^2 x \, dx = I_2 + I_2 + I_2 = 3I_2$ etc.

$$\Rightarrow \quad \int_0^{\frac{k\pi}{2}} \cos^2 x \, dx = kI_2 \quad \text{for any integer } k.$$

Case (c) follows the same pattern as case (a).

In general the graph of $\cos^n x$ is

either

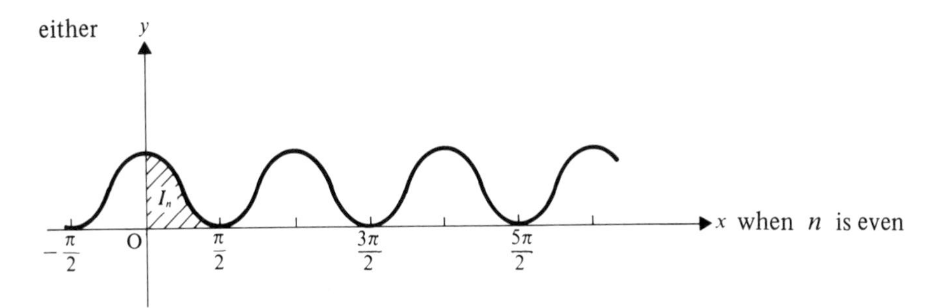

or

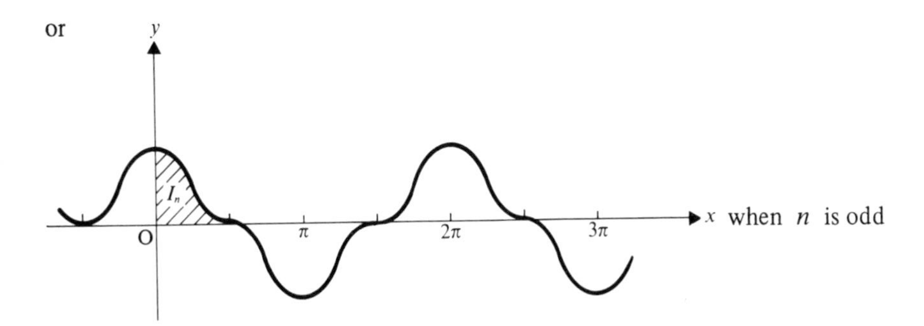

From the graphs it can be seen that

$$\int_0^{\frac{k\pi}{2}} \cos^n x \, dx = kI_n \qquad \text{when } n \text{ is even}$$

and $\int_0^{\frac{k\pi}{2}} \cos^n x \, dx$ is either zero or $\pm I_n$ if n is odd.

These conclusions should not be regarded as quotable but simply as examples of the graphical approach to such integrals. In all cases the required integral is expressed in terms of I_n where $I_n = \int_0^{\frac{\pi}{2}} \cos^n x \, dx$ and which can be calculated using the simple reduction formula, $I_n = \frac{n-1}{n} I_{n-2}$.

Note that the lower limit can also be a multiple of $\frac{\pi}{2}$.

In a similar way, $\int_0^{\frac{k\pi}{2}} \sin^n x \, dx$ can be expressed in terms of $\int_0^{\frac{\pi}{2}} \sin^n x \, dx$,

e.g. to find $\int_{-\frac{\pi}{2}}^{2\pi} \sin^3 x \, dx$, we use the graph of $\sin^3 x$ as follows.

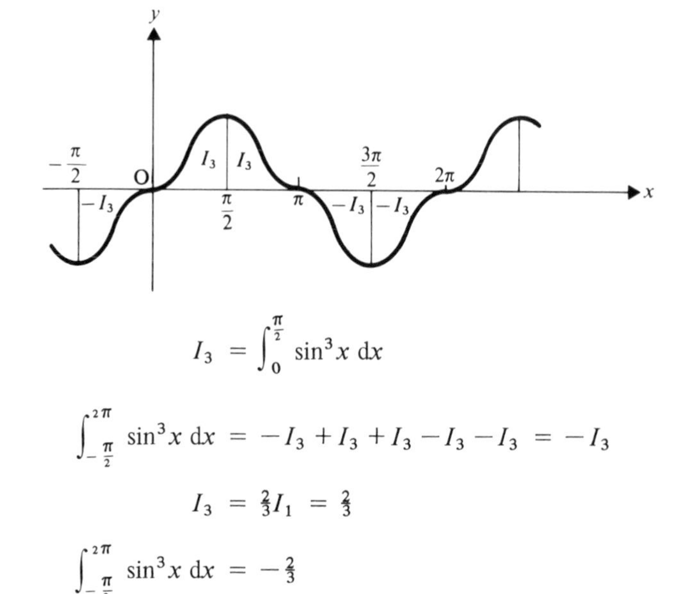

If $$I_3 = \int_0^{\frac{\pi}{2}} \sin^3 x \, dx$$

then $$\int_{-\frac{\pi}{2}}^{2\pi} \sin^3 x \, dx = -I_3 + I_3 + I_3 - I_3 - I_3 = -I_3$$

But $$I_3 = \tfrac{2}{3} I_1 = \tfrac{2}{3}$$

So $$\int_{-\frac{\pi}{2}}^{2\pi} \sin^3 x \, dx = -\tfrac{2}{3}$$

Note that the graph of $\sin^n x$, when n is even, is

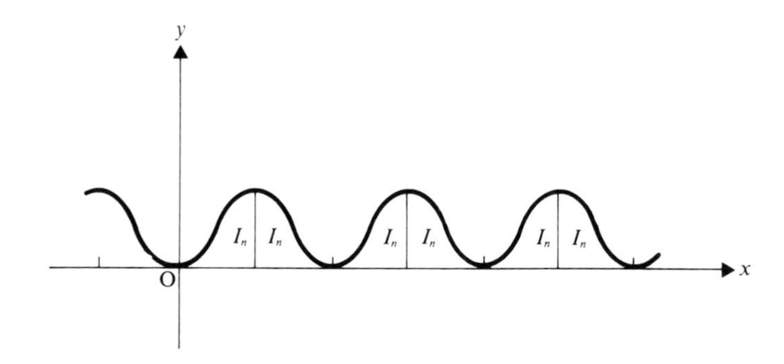

Reduction Formula With Change of Variable

The periodic property of the integrand cannot always be used to carry out the integration and when this is the case a change of variable may *sometimes* help.

For example, to evaluate $\int_0^{\frac{\pi}{3}} \cos^5 3x \, dx$, we can use the substitution

$$u \equiv 3x \quad \Rightarrow \quad \ldots du \equiv \ldots 3 \, dx$$

and

x	0	$\frac{\pi}{3}$
u	0	π

Then $$\int_0^{\frac{\pi}{3}} \cos^5 3x \, dx \equiv \tfrac{1}{3} \int_0^{\pi} \cos^5 u \, du.$$

Now $\int_0^{\pi} \cos^5 u \, du$ can be found using the graph of $\cos^5 u$, and in this way $\int_0^{\frac{\pi}{3}} \cos^5 3x \, dx$ can be evaluated.

EXAMPLES 13b

1) Use a suitable change of variable to evaluate $\int_{-\frac{\pi}{4}}^{\frac{\pi}{4}} \cos^8 2\theta \, d\theta$.

Let $$u \equiv 2\theta \quad \Rightarrow \quad \dots du \equiv \dots 2 \, d\theta$$

and

θ	$-\frac{\pi}{4}$	$\frac{\pi}{4}$
u	$-\frac{\pi}{2}$	$\frac{\pi}{2}$

Then $$\int_{-\frac{\pi}{4}}^{\frac{\pi}{4}} \cos^8 2\theta \, d\theta \equiv \tfrac{1}{2} \int_{-\frac{\pi}{2}}^{\frac{\pi}{2}} \cos^8 u \, du.$$

As the limits are not 0 to $\frac{\pi}{2}$, we now refer to the graph of $\cos^8 u$,

i.e.

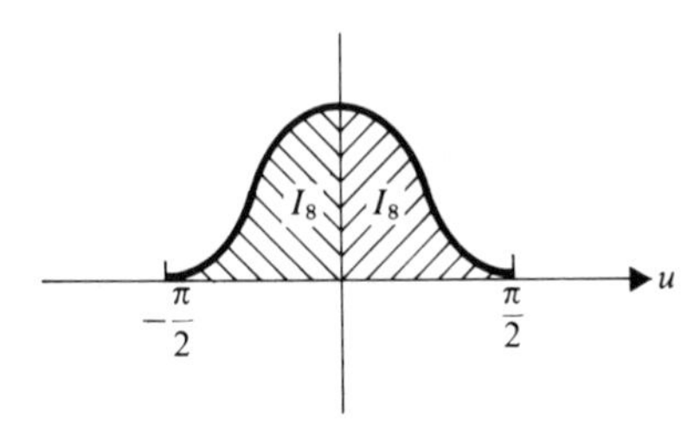

From the graph it is seen that

$$\int_{-\frac{\pi}{2}}^{\frac{\pi}{2}} \cos^8 u \, du = 2I_8$$

$$\Rightarrow \quad \int_{-\frac{\pi}{4}}^{\frac{\pi}{4}} \cos^8 2\theta \, d\theta = I_8$$

$$= \left(\frac{7}{8}\right)\left(\frac{5}{6}\right)\left(\frac{3}{4}\right)\left(\frac{1}{2}\right)\left(\frac{\pi}{2}\right)$$

2) If $I_{p,q} = \int_0^{\frac{\pi}{2}} \sin^p x \cos^q x \, dx$, prove that

$$I_{p,q} = \frac{p-1}{p+q} I_{p-2,q}$$

Use your result to evaluate $\int_0^{\frac{\pi}{2}} \sin^{10}x \cos^2 x \, dx$.

In order to integrate $\sin^p x \cos^q x$ by parts, the integral is arranged in the form

$$I_{p,q} = \int_0^{\frac{\pi}{2}} (\sin^{p-1} x)(\sin x \cos^q x) \, dx.$$

Then we can use $\quad v = \sin^{p-1} x \qquad \dfrac{du}{dx} = \sin x \cos^q x$

$$\frac{dv}{dx} = (p-1)\sin^{p-2} x \cos x \qquad u = -\frac{1}{q+1}\cos^{q+1} x$$

$$\Rightarrow \quad I_{p,q} = \left[-\frac{1}{q+1}\cos^{q+1} x \sin^{p-1} x\right]_0^{\frac{\pi}{2}} + \frac{p-1}{q+1}\int_0^{\frac{\pi}{2}} \sin^{p-2} x \cos^{q+2} x \, dx$$

$$\Rightarrow \quad I_{p,q} = \frac{p-1}{q+1}\int_0^{\frac{\pi}{2}} \sin^{p-2} x \cos^{q+2} x \, dx.$$

The R.H.S. is not yet in the form $I_{p-2,q}$, so we use $\cos^2 x \equiv 1 - \sin^2 x$ to reduce $\cos^{q+2} x$ to $\cos^q x$.

Then

$$\begin{aligned} I_{p,q} &= \frac{p-1}{q+1}\int_0^{\frac{\pi}{2}} \sin^{p-2} x \cos^q x (1 - \sin^2 x) \, dx \\ &= \frac{p-1}{q+1}\int_0^{\frac{\pi}{2}} (\sin^{p-2} x \cos^q x - \sin^p x \cos^q x) \, dx \\ &= \frac{p-1}{q+1}(I_{p-2,q} - I_{p,q}) \end{aligned}$$

$$\Rightarrow \qquad I_{p,q} = \frac{p-1}{p+q} I_{p-2,q}$$

Using $p = 10$, $q = 2$, we have

$$\begin{aligned} \int_0^{\frac{\pi}{2}} \sin^{10} x \cos^2 x \, dx = I_{10,2} &= \tfrac{9}{12} I_{8,2} \\ &= (\tfrac{9}{12})(\tfrac{7}{10}) I_{6,2} \\ &\cdots\cdots\cdots\cdots \\ &= (\tfrac{9}{12})(\tfrac{7}{10})(\tfrac{5}{8})(\tfrac{3}{6})(\tfrac{1}{4}) I_{0,2} \\ &= \frac{9.7.5.3.1}{6.5.4.3.2(2^5)} I_{0,2} \end{aligned}$$

But $$I_{0,2} = \int_0^{\frac{\pi}{2}} \cos^2 x \, dx$$

$$= \left(\frac{1}{2}\right)\left(\frac{\pi}{2}\right) = \frac{\pi}{4}$$

So $$\int_0^{\frac{\pi}{2}} \sin^{10} x \cos^2 x \, dx = \frac{9.7.5.3\pi}{6!(2^7)} = \frac{21\pi}{2^{11}}$$

EXERCISE 13b

Evaluate the following definite integrals:

1) $\int_0^{\frac{\pi}{2}} \sin^{11} x \, dx$

2) $\int_{-\frac{\pi}{2}}^{0} \cos^{10} x \, dx$

3) $\int_0^{\pi} \cos^7 x \, dx$

4) $\int_{-\frac{\pi}{2}}^{\frac{\pi}{2}} \sin^6 \theta \, d\theta$

5) $\int_0^{\frac{3\pi}{2}} \sin^8 x \, dx$

6) $\int_{-\frac{\pi}{2}}^{\pi} \cos^5 \theta \, d\theta$

7) $\int_0^{2\pi} \sin^4 x \, dx$

8) $\int_{-\pi}^{\pi} \cos^{12} x \, dx$

9) $\int_0^{\infty} x^8 \, e^{-x} \, dx$

10) $\int_0^{\frac{\pi}{4}} \tan^6 \theta \, d\theta$

11) $\int_0^{\frac{\pi}{2}} \sin^8 x \cos^2 x \, dx$

12) $\int_0^{\frac{\pi}{4}} \sin^7 2x \, dx$

13) $\int_0^{\frac{\pi}{3}} \cos^5 3\theta \, d\theta$

14) $\int_0^{\frac{\pi}{8}} \sin^4 4x \, dx$.

15) If $I_n = \int_0^1 x^n \, e^x \, dx$, express I_n in terms of I_{n-1}. Find I_5 in terms of e.

16) If $I_n = \int_0^1 x(1 - x^3)^n \, dx$ prove that

$$(3n+2)I_n = 3nI_{n-1}$$

and find I_n in terms of n.

17) Given $I_n = \int_1^0 x^n \cosh x \, dx$, prove that, for $n \geqslant 2$,

$$I_n = \sinh 1 - n \cosh 1 + n(n-1)I_{n-2}.$$

Evaluate (a) $\int_1^0 x^4 \cosh x \, dx$ (b) $\int_1^0 x^3 \cosh x \, dx$.

18) If $I_n = \int_0^1 x^n \sqrt{(1-x)} \, dx$ prove that $(2n+3)I_n = 2nI_{n-1}$.

Hence find I_9.

19) If $I_n = \int_0^{\frac{\pi}{2}} x^n \cos x \, dx$ prove that

$$I_n = \left(\frac{\pi}{2}\right)^n - n(n-1)I_{n-2} \quad \text{for} \quad n \geqslant 2.$$

Hence find I_6.

20) If $I_n = \int_0^1 \frac{\sinh^{2n} x}{\cosh x} \, dx$, prove that

$$I_{n+1} + I_n = \frac{\sinh^{2n+1}(1)}{2n+1}$$

Show that $I_0 = -\frac{\pi}{2} + 2 \arctan \mathrm{e}$ and evaluate I_3.

21) If $I_{m,n} = \int_{-1}^1 (1+x)^m (1-x)^n \, dx$ where m and n are positive integers, prove that $(n+1)I_{m,n} = mI_{m-1,n+1}$.
Hence, or otherwise, evaluate $I_{5,6}$.

FURTHER DIFFERENTIAL EQUATIONS

There are many physical situations in which different variables increase or decrease at certain rates. Whenever there is a relationship between these rates of change, it can be expressed in the form of a differential equation. A direct (i.e. non-differential) relationship between the variables can be found if the differential equation can be solved. So clearly the solution of differential equations is of great importance in many spheres.

There is an endless variety of types of differential equation and the solution of each type has its own special technique.
(There are also some differential equations which cannot be solved to give a direct relationship between the variables.)
Solving a differential equation necessarily involves an integration process and therefore automatically introduces a constant of integration, e.g. a first order differential equation contains $\frac{dy}{dx}$ and therefore is solved by *one* integration operation, thus introducing *one* arbitrary constant.
A second order differential equation contains $\frac{d^2y}{dx^2}$; its solution requires *two* integration processes and therefore contains *two* arbitrary constants.
We saw in Volume 1 that integration is based on *recognising* derivatives. Logically then, the solution of some differential equations will also involve recognition.
There are so many different types of differential equation, each with its own solution technique, that the solution of these equations comprises a vast subject in its own right. In this book, however, we are going to study only two further categories of differential equation, one of first order and one of second order (first order linear equations with variables separable were dealt with in Volume 1).

FIRST ORDER DIFFERENTIAL EQUATIONS OF THE PRODUCT TYPE

Exact Differential Equations

Knowing that $\frac{d}{dx}(uv) = v\frac{du}{dx} + u\frac{dv}{dx}$, we can recognise the R.H.S. of this formula when it occurs in a differential equation, and quote its integral, uv.
For example, consider the differential equation

$$x\frac{dy}{dx} + y = e^x.$$

The L.H.S. can be recognised as the derivative w.r.t. x of the product xy. Thus, integrating both sides of the equation with respect to x gives

$$xy = e^x + A.$$

A differential equation of this type, in which part of it is the exact derivative of a product, is called an *exact differential equation*.

EXAMPLES 13c

1) Find the general solution of the differential equation

$$x^2 \cos y \frac{dy}{dx} + 2x \sin y = \frac{1}{x^2}$$

The L.H.S. is seen to be the derivative w.r.t. x of $x^2 \sin y$ (the identification of this product is made easier by noting that the term containing $\frac{dy}{dx}$ is given by differentiating a term that is a function of y; i.e. $\sin y$ in this case).
Hence, integrating both sides w.r.t. x gives

$$x^2 \sin y = -\frac{1}{x} + A$$

i.e.
$$x^3 \sin y = Ax - 1$$

is the general solution of the given differential equation.

2) Find y as a function of x if

$$2y\, e^x \frac{dy}{dx} + y^2 e^x = e^{2x}$$

and $y = 0$ when $x = 1$.

Rearranging the L.H.S. in the form

$$\left(2y \frac{dy}{dx}\right)(e^x) + (y^2)(e^x)$$

we recognise it as the derivative w.r.t. x of $y^2 e^x$.

Thus
$$y^2 e^x = \int e^{2x}\, dx = \tfrac{1}{2}e^{2x} + A$$

$\Rightarrow$
$$y = \pm\sqrt{\left(\frac{e^x}{2} + \frac{A}{e^x}\right)}$$

But $y = 0$ when $x = 1$, so

$$0 = \frac{e}{2} + \frac{A}{e} \quad \Rightarrow \quad A = -\frac{e^2}{2}$$

Hence
$$y = \pm\sqrt{\left(\frac{e^x}{2} - \frac{e^2}{2e^x}\right)}.$$

The reader may have felt tempted to simplify the differential equation in

Example 2 above by cancelling e^x, giving

$$2y\frac{dy}{dx}+y^2 = e^x. \qquad [1]$$

In this form, however, the L.H.S. is *not* the derivative of a product.
In fact, given equation [1], we would choose to *multiply by* e^x in order to make the L.H.S. exact in form.
The term e^x, which thus makes the L.H.S. of [1] integrable, is called an *integrating factor*. It is often possible to find a suitable integrating factor, either by inspection, or by using the result of the formal method shown below.

The Integrating Factor

Consider a first order differential equation that can be written in the form

$$\frac{dy}{dx}+Fy = G$$

where F and G are both *functions of* x *only*.
The L.H.S. is not yet exact, but suppose that it becomes exact when it is multiplied by I, a function of x,

then $$(I)\left(\frac{dy}{dx}\right)+(y)(FI) = GI$$

Comparing the L.H.S. with $v\frac{du}{dx}+u\frac{dv}{dx}$

we have $$v = I, \quad \frac{du}{dx} = \frac{dy}{dx}$$

$$u = y, \quad \frac{dv}{dx} = FI$$

i.e. $$\frac{dI}{dx} = FI$$

Now this is a first order linear differential equation with variables separable, so

$$\int \frac{1}{I}dI = \int F\,dx$$

$\Rightarrow$ $$\ln I = \int F\,dx$$

$\Rightarrow$ $$I = e^{\int F\,dx}$$

Thus we see that $e^{\int F\,dx}$ is an integrating factor for the expression

$$\frac{dy}{dx}+Fy.$$

So, *provided that* $e^{\int F\,dx}$ *can be found*, we have:

$$\frac{dy}{dx} + Fy = G \Rightarrow I\frac{dy}{dx} + yIF = IG$$

$$\Rightarrow \int\left(I\frac{dy}{dx} + yIF\right)dx = \int IG\,dx$$

$$\Rightarrow Iy = \int IG\,dx$$

Both I and G are functions of x only, so $\int IG\,dx$ can usually be found.

EXAMPLES 13c (continued)

3) Find a suitable integrating factor and hence solve the differential equation

$$x\frac{dy}{dx} + 3y = \frac{e^x}{x^2}$$

First the equation must be written in the standard form $\frac{dy}{dx} + Fy = G$,

i.e. $$\frac{dy}{dx} + \frac{3y}{x} = \frac{e^x}{x^3} \Rightarrow F = \frac{3}{x}$$

The integrating factor is I where $I = e^{\int F\,dx}$

$$\int F\,dx = \int \frac{3}{x}\,dx = 3\ln x = \ln x^3$$

$$\Rightarrow \qquad I = e^{\ln x^3} = x^3$$

Multiplying the *standard form* of the given equation by I, we have

$$x^3\frac{dy}{dx} + 3x^2y = e^x$$

Then integrating both sides w.r.t. x gives

$$x^3y = e^x + A$$

Note. A mistake that is easily made, but which must of course be avoided, is to multiply the *original* differential equation, rather than the standard form, by I.

EXERCISE 13c

Find the general solution of each of the following exact differential equations.

1) $x\frac{dy}{dx} + y = e^x$

2) $\cos x\frac{dy}{dx} - y\sin x = x^2$

3) $\dfrac{x}{y}\dfrac{dy}{dx} + \ln y = x + 1$

4) $\dfrac{1}{x}\dfrac{dy}{dx} - \dfrac{y}{x^2} = \sin x$

5) $e^x y + e^x \dfrac{dy}{dx} = 2$

6) $x\,e^y \dfrac{dy}{dx} + e^y = e^x$

7) $\ln x \dfrac{dy}{dx} + \dfrac{y}{x} = x \ln x$

8) $(1 + x)\dfrac{dy}{dx} + y = x^3$

9) $x \sec^2 y \dfrac{dy}{dx} + \tan y = \tan x$

10) $e^x\, e^y \dfrac{dy}{dx} + e^x\, e^y = e^{2x}$

By using integrating factors, solve the following differential equations.

11) $\dfrac{dy}{dx} + 3y = e^{-3x}$

12) $\dfrac{dy}{dx} + y \cot x = \operatorname{cosec} x$

13) $x^2 \dfrac{dy}{dx} + xy = x + 1$

14) $\dfrac{dy}{dx} - \dfrac{3y}{x + 1} = (x + 1)^4$

15) $\tan x \dfrac{dy}{dx} + y = e^x \tan x$

16) $\dfrac{dv}{dt} = t - 2vt$

17) $x \dfrac{dy}{dx} + 2y = \dfrac{\sin x}{x}$

18) $x \dfrac{dy}{dx} = y - x\,e^{-x}$

19) $\dfrac{dr}{d\theta} + r \cot \theta = \sin \theta$

20) $y + x(x - 1)\dfrac{dy}{dx} = x^3\, e^{-x^2}$

21) $x \dfrac{dy}{dx} = y + x^2(\sin x + x \cos x)$ and $y = 0$ when $x = \dfrac{\pi}{2}$.

22) $\dfrac{dy}{dx} = x - xy$ and $y = 0$ when $x = 0$.

FORMATION OF SECOND ORDER LINEAR DIFFERENTIAL EQUATIONS

If an equation $y = f(x)$ contains two arbitrary constants, A and B, then by differentiating twice two more equations are produced,

i.e.
$$\frac{dy}{dx} = f'(x) \quad \text{and} \quad \frac{d^2y}{dx^2} = f''(x).$$

These two equations, together with the original $y = f(x)$, allow A and B to be eliminated, so forming a second order differential equation. We are now going to consider three types of equation $y = f(x)$, all of which give rise in this way to a linear second order differential equation

Case (a)

Consider $y = A\,e^{2x} + B\,e^{3x}$

$$\Rightarrow \quad \frac{dy}{dx} = 2A\,e^{2x} + 3B\,e^{3x} = 2y + B\,e^{3x}$$

$$\Rightarrow \quad \frac{d^2y}{dx^2} = 2\frac{dy}{dx} + 3B\,e^{3x} = 2\frac{dy}{dx} + 3\left(\frac{dy}{dx} - 2y\right)$$

Thus $\quad y = A\,e^{2x} + B\,e^{3x} \quad \Rightarrow \quad \dfrac{d^2y}{dx^2} - (2+3)\dfrac{dy}{dx} + (2\times 3)y = 0$

There is a marked similarity between the coefficients of this differential equation and those of the quadratic equation $u^2 - 5u + 6 = 0$ whose roots are 2 and 3.

In order to check that this analogy is general, and not merely coincidence, we now consider the more general equation

$$y = A\,e^{\alpha x} + B\,e^{\beta x}$$

$$\Rightarrow \quad \frac{dy}{dx} = A\alpha\,e^{\alpha x} + B\beta\,e^{\beta x}$$

$$\Rightarrow \quad \frac{d^2y}{dx^2} = A\alpha^2\,e^{\alpha x} + B\beta^2\,e^{\beta x}$$

Eliminating A and B from these three equations gives

$$\frac{d^2y}{dx^2} - (\alpha+\beta)\frac{dy}{dx} + \alpha\beta y = 0$$

which compares with the quadratic equation

$$u^2 - (\alpha+\beta)u + \alpha\beta = 0$$

whose roots are α and β.

This is called the *auxiliary quadratic equation* and it can now be used to *recognise* the general solution of a second order linear differential equation with constant coefficients,

i.e. $$a\frac{d^2y}{dx^2} + b\frac{dy}{dx} + cy = 0.$$

If the auxiliary quadratic equation $au^2 + bu + c = 0$ has real distinct roots α and β (i.e. $b^2 - 4ac > 0$) then we can quote, by recognition, the solution

$$y = A\,e^{\alpha x} + B\,e^{\beta x}$$

Case (*b*)

Consider $y = e^{2x}(A + Bx)$

$$\Rightarrow \quad \frac{dy}{dx} = 2y + B\,e^{2x}$$

$$\Rightarrow \quad \frac{d^2y}{dx^2} = 2\frac{dy}{dx} + 2B\,e^{2x} = 2\frac{dy}{dx} + 2\left(\frac{dy}{dx} - 2y\right)$$

Thus $y = e^{2x}(A + Bx) \Rightarrow \dfrac{d^2y}{dx^2} - (2+2)\dfrac{dy}{dx} + (2 \times 2)y = 0.$

This time we see that the auxiliary quadratic equation has equal roots of value 2,

i.e. $u^2 - 4u + 4 = 0$

Again we will check a general example of this type, using

$$y = e^{\alpha x}(A + Bx)$$

$$\Rightarrow \quad \frac{dy}{dx} = \alpha y + B\,e^{\alpha x}$$

$$\Rightarrow \quad \frac{d^2y}{dx^2} = \alpha\frac{dy}{dx} + B\alpha\,e^{\alpha x} = \alpha\frac{dy}{dx} + \alpha\left(\frac{dy}{dx} - \alpha y\right)$$

So $y = e^{\alpha x}(A + Bx) \Rightarrow \dfrac{d^2y}{dx^2} - 2\alpha\dfrac{dy}{dx} + \alpha^2 y = 0$

This provides a second form for the solution, by recognition, of the differential equation

$$a\frac{d^2y}{dx^2} + b\frac{dy}{dx} + cy = 0$$

When the auxiliary quadratic equation $au^2 + bu + c = 0$
has equal roots, α (i.e. $b^2 - 4ac = 0$)

then $y = e^{\alpha x}(A + Bx)$

Case (*c*)

Consider $y = A\,e^{2x}\cos(3x + \epsilon)$

$$\Rightarrow \quad \frac{dy}{dx} = 2A\,e^{2x}\cos(3x + \epsilon) - 3A\,e^{2x}\sin(3x + \epsilon)$$

$$= 2y - 3A\,e^{2x}\sin(3x + \epsilon)$$

$$= 2y - 3A\,e^{2x}\sin(3x+\epsilon)$$

$$\Rightarrow \qquad \frac{d^2y}{dx^2} = 2\frac{dy}{dx} - 6A\,e^{2x}\sin(3x+\epsilon) - 9A\,e^{2x}\cos(3x+\epsilon)$$

$$= 2\frac{dy}{dx} - 2\left(2y - \frac{dy}{dx}\right) - 9y$$

Thus $\quad y = A\,e^{2x}\cos(3x+\epsilon) \Rightarrow \dfrac{d^2y}{dx^2} - 4\dfrac{dy}{dx} + 13y = 0$

In this case the auxiliary quadratic equation is $u^2 - 4u + 13 = 0$ and it has complex roots $2 \pm 3i$.

Applying the procedure above to the general case when the auxiliary quadratic equation has roots $p \pm qi$ we can show that

if the auxiliary quadratic equation $au^2 + bu + c = 0$ has complex roots $p \pm qi$ (i.e. $b^2 - 4ac < 0$) then we can quote, by recognition, the solution $\quad y = A\,e^{px}\cos(qx+\epsilon)$

A reader who is studying mechanics will recognise this form as the equation of damped harmonic motion.

Note that the compound angle identity $A\cos(qx+\epsilon) \equiv B\cos qx + C\sin qx$ provides an alternative form for the solution above,

i.e. $\quad y = e^{px}(B\cos qx + C\sin qx)$

SOLUTION OF SECOND ORDER LINEAR DIFFERENTIAL EQUATIONS WITH CONSTANT COEFFICIENTS

From the work done in the preceding section we know that the differential equation

$$a\frac{d^2y}{dx^2} + b\frac{dy}{dx} + cy = 0$$

has a general solution based on the roots of its auxiliary quadratic equation

$$au^2 + bu + c = 0$$

such that

(a) if $b^2 - 4ac > 0$, $\quad y = A\,e^{\alpha x} + B\,e^{\beta x}$

(b) if $b^2 - 4ac = 0$, $\quad y = e^{\alpha x}(A + Bx)$

(c) if $b^2 - 4ac < 0$, $\quad y = A\,e^{px}\cos(qx+\theta) \equiv e^{px}(B\cos qx + C\sin qx)$

For example,

If $\dfrac{d^2y}{dx^2} + 3\dfrac{dy}{dx} - 4y = 0$, then (using $u^2 + 3u - 4 = 0 \Rightarrow u = -4, 1$)

we have $y = A\,e^{-4x} + B\,e^{x}$.

If $\dfrac{d^2y}{dx^2} + 6\dfrac{dy}{dx} + 9y = 0$, then (using $u^2 + 6u + 9 = 0 \Rightarrow u = -3$)

we have $y = e^{-3x}(A + Bx)$.

If $\dfrac{d^2y}{dx^2} + 2\dfrac{dy}{dx} + 2y = 0$, then (using $u^2 + 2u + 2 = 0 \Rightarrow -1 \pm i$)

we have $y = A\,e^{-x}\cos(x + \epsilon)$.

Note that if the differential equation contains no term in $\dfrac{dy}{dx}$ and the other two terms have the same sign, then the roots of the auxiliary equation are purely imaginary,

e.g. if
$$\frac{d^2y}{dx^2} + k^2y = 0$$

then
$$u^2 + k^2 = 0 \Rightarrow u = \pm ki$$

The general solution then becomes

$$y = A\cos(kx + \epsilon)$$

(from $y = A\,e^{px}\cos(qx + \epsilon)$ when $p = 0$ and $q = k$).
A student of mechanics will notice that this is the equation of simple harmonic motion.

EXERCISE 13d

Write down the general solution for each of the following differential equations.

1) $\dfrac{d^2y}{dx^2} - 3\dfrac{dy}{dx} + 2y = 0$

2) $3\dfrac{d^2y}{dx^2} - 7\dfrac{dy}{dx} + 4y = 0$

3) $\dfrac{d^2y}{dx^2} - 2\dfrac{dy}{dx} + y = 0$

4) $\dfrac{d^2y}{dx^2} + \dfrac{dy}{dx} + y = 0$

5) $\dfrac{d^2y}{dx^2} - 5\dfrac{dy}{dx} + 4y = 0$

6) $\dfrac{d^2y}{dx^2} - 4y = 0$

7) $\dfrac{d^2y}{dx^2} + 4y = 0$

8) $2\dfrac{d^2y}{dx^2} + \dfrac{dy}{dx} + 2y = 0$

9) $\dfrac{d^2y}{dx^2} - 2\dfrac{dy}{dx} = 0$

10) $9\dfrac{d^2y}{dx^2} - 6\dfrac{dy}{dx} + y = 0$

The Particular Integral

The second order linear differential equations that we have so far considered have been of the form

$$a\frac{d^2y}{dx^2}+b\frac{dy}{dx}+cy = 0.$$

Now we will turn our attention to the form in which the R.H.S. is not zero but is a function of x,

i.e. $$a\frac{d^2y}{dx^2}+b\frac{dy}{dx}+cy = f(x) \qquad [1]$$

Suppose, for example, that we want to solve the equation

$$\frac{d^2y}{dx^2}-5\frac{dy}{dx}+6y = e^x$$

The term on the R.H.S. suggests that a solution may be of the form $y=\lambda\,e^x$. Using this as a *trial solution* we have

$$y = \lambda\,e^x \quad\Rightarrow\quad \frac{dy}{dx} = \lambda\,e^x \quad\Rightarrow\quad \frac{d^2y}{dx^2} = \lambda\,e^x$$

Substituting these expressions into the given equation gives

$$\lambda\,e^x - 5\lambda\,e^x + 6\lambda\,e^x = e^x \quad\Rightarrow\quad \lambda = \tfrac{1}{2}$$

So we see that $y=\frac{1}{2}e^x$ does satisfy the given equation.
But $y=\frac{1}{2}e^x$ cannot be the complete solution because it contains no arbitrary constants.
However it must be part of the complete solution,
and is called a *particular integral* (P.I.).
The remainder of the solution can be found by considering the simpler differential equation

$$\frac{d^2y}{dx^2}-5\frac{dy}{dx}+6y = 0$$

whose general solution is $y=A\,e^{2x}+B\,e^{3x}$.
Clearly this solution *alone* does not satisfy the original equation, but, when combining it with the particular integral, we find that

$$\left.\begin{aligned} &\text{if} && y = A\,e^{2x}+B\,e^{3x}+\tfrac{1}{2}e^x \\ &\text{then} && \frac{dy}{dx} = 2A\,e^{2x}+3B\,e^{3x}+\tfrac{1}{2}e^x \\ &\text{and} && \frac{d^2y}{dx^2} = 4A\,e^{2x}+9B\,e^{3x}+\tfrac{1}{2}e^x \end{aligned}\right\} \Rightarrow \frac{d^2y}{dx^2}-5\frac{dy}{dx}+6y = e^x$$

So the general solution of the given differential equation is

$$y = A\,e^{2x} + B\,e^{3x} + \tfrac{1}{2}e^{x}$$

which is obtained by adding the *complementary function* $(A\,e^{2x} + B\,e^{3x})$ and the particular integral.

In fact, for all differential equations of the type

$$a\frac{d^2y}{dx^2} + b\frac{dy}{dx} + cy = f(x)$$

the solution is $\qquad y = \text{C.F.} + \text{P.I.}$

where the complementary function, C.F., is the solution of the equation

$$a\frac{d^2y}{dx^2} + b\frac{dy}{dx} + cy = 0$$

and the particular integral is *another* solution of the complete differential equation.

The Failure Case

If we examine a similar differential equation $\dfrac{d^2y}{dx^2} - 5\dfrac{dy}{dx} + 4y = e^{x}$, we find that the complementary function is $y = A\,e^{4x} + B\,e^{x}$ and the trial solution would appear at first sight to be $y = \lambda\,e^{x}$.

But $\lambda\,e^{x}$ is already included in the term $B\,e^{x}$ in the complementary function and so it is not an *extra* function of x.

This fact emerges if $y = \lambda\,e^{x}$ is used in the given equation, as we find that

$$\lambda\,e^{x} - 5\lambda\,e^{x} + 4\lambda\,e^{x} = e^{x}$$

which is inconsistent.

In this case, when the R.H.S. is $e^{\alpha x}$ (where α is a root of the auxiliary equation)

we use, as a trial solution, $y = \lambda x\,e^{\alpha x}$.

In the example above,

$$y = \lambda x\,e^{x} \quad\Rightarrow\quad \frac{dy}{dx} = \lambda(x\,e^{x} + e^{x})$$

$$\Rightarrow\quad \frac{d^2y}{dx^2} = \lambda(x\,e^{x} + 2\,e^{x})$$

so that $\qquad \lambda(x\,e^{x} + 2\,e^{x}) - 5\lambda(x\,e^{x} + e^{x}) + 4\lambda x\,e^{x} = e^{x}$

i.e. $\qquad -3\lambda\,e^{x} = e^{x} \quad\Rightarrow\quad \lambda = -\tfrac{1}{3}$

Thus the general solution of the given equation is $y = A\,e^{4x} + B\,e^{x} - \tfrac{1}{3}x\,e^{x}$.

The Complex Index Case

If the R.H.S. of a similar differential equation is of the form $k\,\mathrm{e}^{(p+qi)x}$ then a suitable trial solution is, $y=(\lambda+\mu i)\,\mathrm{e}^{(p+qi)x}$.
For example, to solve the equation

$$\frac{d^2y}{dx^2}+2\frac{dy}{dx}-3y = 8\,\mathrm{e}^{(3+2i)x} \qquad [1]$$

we can seek a particular integral by trying

$$y = (\lambda+\mu i)\,\mathrm{e}^{(3+2i)x}$$

$$\Rightarrow \quad \frac{dy}{dx} = (3+2i)(\lambda+\mu i)\,\mathrm{e}^{(3+2i)x}$$

$$\Rightarrow \quad \frac{d^2y}{dx^2} = (3+2i)^2(\lambda+\mu i)\,\mathrm{e}^{(3+2i)x}$$

Substituting these values in [1] gives

$$(3+2i)^2(\lambda+\mu i)+2(3+2i)(\lambda+\mu i)-3(\lambda+\mu i) = 8$$

$$\Rightarrow \quad (\lambda-2\mu)+(2\lambda+\mu)i = 1$$

Hence $\quad \lambda-2\mu = 1 \quad$ and $\quad 2\lambda+\mu = 0$

$$\Rightarrow \quad \lambda = \tfrac{1}{5} \quad \text{and} \quad \mu = -\tfrac{2}{5}.$$

So the P.I. for [1] is $y=\frac{1}{5}(1-2i)\,\mathrm{e}^{(3+2i)x}$.
From the auxiliary quadratic equation we find that the C.F. is $A\,\mathrm{e}^{x}+B\,\mathrm{e}^{-3x}$.
So the complete solution is

$$y = A\,\mathrm{e}^{x}+B\,\mathrm{e}^{-3x}+\tfrac{1}{5}(1-2i)\,\mathrm{e}^{(3+2i)x}.$$

Note that we assume here that exponential functions with complex indices can be differentiated in the same way as when the index is real.

Further Trial Solutions

When $a\dfrac{d^2y}{dx^2}+b\dfrac{dy}{dx}+cy=f(x)$, certain expressions for $f(x)$ suggest standard trial solutions. We have already seen that if $f(x)\equiv \mathrm{e}^{kx}$ a suitable trial solution is $y=\lambda\,\mathrm{e}^{kx}$, unless k is a root of the auxiliary quadratic equation, in which case we use $y=\lambda x\,\mathrm{e}^{kx}$.
Other standard trial solutions are given alongside. Their use is demonstrated in the next set of examples.

$f(x)$	Trial solution
p (a constant)	$y=\lambda$
$px+q$	$y=\lambda x+\mu$
px^2+qx+r	$y=\lambda x^2+\mu x+\eta$
$p\sin x$ or $p\cos x$ or $p\sin x+q\cos x$	$y=\lambda\sin x+\mu\cos x$

Calculation of Arbitrary Constants

Because the solution of a second order differential equation contains two arbitrary constants, their evaluation requires two extra facts or initial conditions, e.g. if $\frac{d^2y}{dx^2}-3\frac{dy}{dx}+2y=0$ and when $x=0$, $y=3$ and $\frac{dy}{dx}=5$, the solution is carried out as follows.

From $u^2-3u+2=0$ we have $u=1, 2$

so that $$y = A\,e^x + B\,e^{2x}$$

$\Rightarrow$ $$\frac{dy}{dx} = A\,e^x + 2B\,e^{2x}$$

When $x=0$, $$\begin{cases} y = 3 \quad \Rightarrow \quad 3 = A+B \\ \frac{dy}{dx} = 5 \quad \Rightarrow \quad 5 = A+2B. \end{cases}$$

Hence $A=1$ and $B=2$.

So $$y = e^x + 2\,e^{2x}$$

EXAMPLES 13e

1) Solve the equation $\frac{d^2y}{dx^2}+3\frac{dy}{dx}+2y=10\cos x$

given that $y=1$ and $\frac{dy}{dx}=0$ when $x=0$.

First we find the Complementary Function using

$$u^2+3u+2 = 0 \quad \Rightarrow \quad u = -1, -2$$

i.e. C.F. is $$y = A\,e^{-x} + B\,e^{-2x}$$

Now for a trial solution we will use

$$y = \lambda\cos x + \mu\sin x$$

$\Rightarrow$ $$\frac{dy}{dx} = -\lambda\sin x + \mu\cos x$$

$\Rightarrow$ $$\frac{d^2y}{dx^2} = -\lambda\cos x - \mu\sin x$$

Then the given equation becomes

$$-\lambda\cos x - \mu\sin x + 3(-\lambda\sin x + \mu\cos x) + 2(\lambda\cos x + \mu\sin x) = 10\cos x.$$

$\Rightarrow$ $$(\lambda+3\mu)\cos x - (3\lambda-\mu)\sin x = 10\cos x.$$

Equating coefficients of $\cos x$ and $\sin x$ gives

$$\left.\begin{aligned}\lambda + 3\mu &= 10\\ 3\lambda - \mu &= 0\end{aligned}\right\} \Rightarrow \lambda = 1, \quad \mu = 3.$$

So the general solution is

$$y = A\,e^{-x} + B\,e^{-2x} + \cos x + 3\sin x$$

$$\Rightarrow \qquad \frac{dy}{dx} = -A\,e^{-x} - 2B\,e^{-2x} - \sin x + 3\cos x.$$

But $y = 1$ and $\dfrac{dy}{dx} = 0$ when $x = 0$, so

$$\left.\begin{aligned}1 &= A + B + 1\\ \text{and} \quad 0 &= -A - 2B + 3\end{aligned}\right\} \Rightarrow A = -3, \quad B = 3$$

Hence $y = 3\,e^{-2x} - 3\,e^{-x} + \cos x + 3\sin x$.

2) If $\dfrac{d^2s}{dt^2} + s = t$, $s = 0$ when $t = 0$ and when $t = \dfrac{\pi}{2}$, find s when $t = \dfrac{\pi}{4}$.

The auxiliary equation is $u^2 + 1 = 0 \Rightarrow u = \pm i$.
So the C.F. is $s = A\cos(t + \epsilon)$.
For a trial solution we will use $s = \lambda t$, so that $\dfrac{ds}{dt} = \lambda$ and $\dfrac{d^2s}{dt^2} = 0$.

Then $$0 + \lambda t = t \Rightarrow \lambda = 1.$$

So $$s = A\cos(t + \epsilon) + t$$

But $$s = 0 \text{ when } t = 0 \Rightarrow A\cos\epsilon = 0$$

and $$s = 0 \text{ when } t = \frac{\pi}{2} \Rightarrow -A\sin\epsilon + \frac{\pi}{2} = 0$$

Hence $$\epsilon = \frac{\pi}{2} \text{ and } A = \frac{\pi}{2}.$$

So $$s = \frac{\pi}{2}\cos\left(t + \frac{\pi}{2}\right) + t.$$

When $t = \dfrac{\pi}{4}$, $$s = \frac{\pi}{2}\cos\left(\frac{3\pi}{4}\right) + \frac{\pi}{4}$$

i.e. $$s = \frac{\pi}{4}(1 - \sqrt{2}).$$

Note. In the differential equation

$$a\frac{d^2y}{dx^2}+b\frac{dy}{dx}+cy = f(x),$$

the R.H.S., $f(x)$, can be any function of x. The methods given above for finding the P.I. deal only with certain special forms of $f(x)$ which occur most frequently. Extra help would be needed to find the P.I. when $f(x)$ is not one of those functions whose trial solutions have already been suggested.

EXERCISE 13e

Find the complete solution of the following differential equations.

1) $\dfrac{d^2y}{dx^2}-6\dfrac{dy}{dx}+5y=3$

2) $\dfrac{d^2y}{dx^2}-2\dfrac{dy}{dx}+y=e^{2x}$

3) $\dfrac{d^2y}{dx^2}-2\dfrac{dy}{dx}+y=e^{x}$

4) $\dfrac{d^2y}{dx^2}+3\dfrac{dy}{dx}+2y=\sin x$

(Take $y=\lambda x^2 e^x$ as a trial solution in Question 3.)

5) $\dfrac{d^2y}{dx^2}+\dfrac{dy}{dx}+y=1+x$

6) $\dfrac{d^2y}{dx^2}-4y=3\,e^{-2x}$

7) $4\dfrac{d^2y}{dx^2}-5\dfrac{dy}{dx}+y=\cos x-\sin x$

8) $9\dfrac{d^2y}{dx^2}+6\dfrac{dy}{dx}+y=x^2+2x+3$

9) $\dfrac{d^2y}{dx^2}+25y=2\,e^{(1+i)x}$

10) $\dfrac{d^2y}{dx^2}-4\dfrac{dy}{dx}+3y=65\sin 2x$

11) $\dfrac{d^2s}{dt^2}-4\dfrac{ds}{dt}+4s=5$

12) $\dfrac{d^2x}{dt^2}+\dfrac{dx}{dt}+2x=e^{(2-3i)t}$

13) $\dfrac{d^2y}{dx^2}+3\dfrac{dy}{dx}+2y=6\,e^x+\sin x$

14) $\dfrac{d^2\theta}{dt^2}+\theta=5\,e^t\sin t$

(Take $\theta=\lambda\,e^t\sin t+\mu\,e^t\cos t$ as a trial solution in Question 14.)

15) If $y=1$ and $\dfrac{dy}{dx}=12$ when $x=0$, solve the differential equation

$$\frac{d^2y}{dx^2}-3\frac{dy}{dx}-10y = 0.$$

16) Find the general solution of the differential equation

$$\frac{d^2y}{dx^2}+8\frac{dy}{dx}+25y = 48\cos x-16\sin x.$$

Find also the particular solution for which $y=8$ and $\dfrac{dy}{dx}=27$ when $x=0$

17) Find a particular integral $f(x)$ of the differential equation

$$\frac{d^2y}{dx^2} - 5\frac{dy}{dx} + 6y = (12x - 7)e^{-x}$$

in the form $f(x) \equiv (Ax + B)e^{-x}$.
Find the solution of the equation that satisfies the conditions $y = 0 = \dfrac{dy}{dx}$ when $x = 0$.

18) Find the complementary function and a particular integral for the differential equation $\dfrac{d^2y}{dx^2} + 4\dfrac{dy}{dx} + 4y = 8\sin 2x$.

If, also, $y = 1$ and $\dfrac{dy}{dx} = 0$ when $x = 0$, find the solution of the equation.

SUMMARY

$$\left.\begin{array}{l}\displaystyle\int_0^{\frac{\pi}{2}} \sin^n\theta\, d\theta \\ \text{and} \\ \displaystyle\int_0^{\frac{\pi}{2}} \cos^n\theta\, d\theta\end{array}\right\} = \begin{cases}\left(\dfrac{n-1}{n}\right)\left(\dfrac{n-3}{n-2}\right)\cdots\left(\dfrac{1}{2}\right)\dfrac{\pi}{2} & \text{when } n \text{ is even} \\ \left(\dfrac{n-1}{n}\right)\left(\dfrac{n-3}{n-2}\right)\cdots\left(\dfrac{2}{3}\right) & \text{when } n \text{ is odd (and } n \geqslant 3)\end{cases}$$

The differential equation $\dfrac{dy}{dx} + Fy = G$ where F and G are functions of x can be solved using an integrating factor I where $I = e^{\int F\,dx}$

The differential equation $a\dfrac{d^2y}{dx^2} + b\dfrac{dy}{dx} + cy = f(x)$ has a complementary function:

(a) $Ae^{\alpha x} + Be^{\beta x}$
if $b^2 - 4ac > 0$ and the auxiliary equation has roots α, β.

(b) $e^{\alpha x}(A + Bx)$
if $b^2 - 4ac = 0$ and the auxiliary equation has a repeated root α.

(c) $Ae^{px}\cos(qx + \epsilon)$ or $e^{px}(B\cos qx + C\sin qx)$
if $b^2 - 4ac < 0$ and $p \pm iq$ are the roots of the auxiliary equation.

MISCELLANEOUS EXERCISE 13

1) Given that $I_n = \int_0^1 x^p(1-x)^n \, dx$, where p and n are positive, show that

$$(n+p+1)I_n = nI_{n-1}.$$

Evaluate I_2 for $p = \frac{1}{2}$. (JMB)

2) If $I_n = \int_0^{\frac{\pi}{2}} e^{2x} \sin^n x \, dx$, $n > 1$, show that

$$(n^2+4)I_n = n(n-1)I_{n-2} + 2e^{\pi}.$$

Hence, or otherwise, find

$$\int_0^{\frac{\pi}{2}} e^{2x} \sin^3 x \, dx.$$

(AEB '71)

3) Given that

$$I_{p,n} = \int_0^1 (1-x)^p x^n \, dx, \quad (p \geqslant 0, \quad n \geqslant 0)$$

prove that, for $p \geqslant 1$,

$$(n+1)I_{p,n} = pI_{p-1,n+1}$$

and also that

$$(p+n+1)I_{p,n} = pI_{p-1,n}$$

Hence prove that, if p and n are positive integers,

$$I_{p,n} = \frac{p!n!}{(p+n+1)!}$$

(JMB)

4) Express $\dfrac{1}{1-x^2}$ in partial fractions and show that

$$\int_0^{\frac{1}{2}} \frac{dx}{1-x^2} = \frac{1}{2}\ln 3$$

Given that

$$I_n = \int_0^{\frac{1}{2}} \frac{dx}{(1-x^2)^n}$$

show that

$$I_{n-1} = I_n - \int_0^{\frac{1}{2}} \frac{x^2}{(1-x^2)^n} \, dx.$$

Hence, or otherwise, prove that, for $n > 1$

$$2(n-1)I_n = (2n-3)I_{n-1} + \tfrac{1}{2}(\tfrac{4}{3})^{n-1}.$$

Evaluate I_2. (JMB)

5) Derive a reduction formula for I_m in terms of I_{m-1} when

$$I_m = \int x^3 (\log_e x)^m \, dx.$$

Hence find $\int x^3 (\log_e x)^3 \, dx$. (AEB'75)p

6) If $I_n = \int \tan^n x \, dx$, obtain a reduction formula for I_n.

Hence, or otherwise, show that $\int_0^{\frac{\pi}{4}} \tan^4 x \, dx = \dfrac{3\pi - 8}{12}$ (AEB'76)p

7) Given that $I_n = \int_0^1 x^n \cosh x \, dx$ prove that, for $n \geqslant 2$,

$$I_n = \sinh 1 - n \cosh 1 + n(n-1)I_{n-2}.$$

Evaluate: (a) $\int_0^1 x^4 \cosh x \, dx$ (b) $\int_0^1 x^3 \cosh x \, dx$

expressing each answer in terms of e.

8) If $I_n = \int_0^1 (1+x^2)^n \, dx$, where n is a positive integer, prove that

$$(2n+1)I_n = 2^n + 2nI_{n-1}.$$

Hence, or otherwise, find a reduction formula for

$$J_m = \int_0^{\frac{1}{4}\pi} \sec^{2m}\theta \, d\theta.$$

Evaluate $\int_0^{\frac{1}{4}\pi} \sec^6\theta \, d\theta$. (O)

9) If $I_n = \int \dfrac{\sin 2n\theta}{\sin \theta} \, d\theta$, prove that

$$I_n = I_{n-1} + \frac{2}{2n-1}\sin(2n-1)\theta.$$

Hence, or otherwise, evaluate

$$\int_0^{\frac{\pi}{2}} \frac{\sin 5\theta}{\sin \theta} \, d\theta.$$ (AEB '76)

10) Given that

$$I_n = \int \frac{\sin nx}{\sin x} \, dx,$$

prove that, for $n > 2$

$$I_n - I_{n-2} = \frac{2\sin(n-1)x}{n-1} + \text{a constant}.$$

Hence find the general solution of the differential equation

$$\frac{dy}{dx} - y \cot x = \sin 5x.$$

Find also the general solution of the differential equation

$$\frac{dy}{dx} + y \cot x = \sin 5x. \qquad \text{(JMB)}$$

11) Show how the differential equation

$$\frac{dy}{dx} + Py = Q,$$

where P and Q are functions of x, can be solved by use of an integrating factor. Find the general solution of the differential equation

$$\frac{dy}{dx} + y \cot x = \cos 3x. \qquad \text{(C)}$$

12) Find the general solution of the differential equation

$$\frac{d^2x}{dt^2} - \frac{dx}{dt} - 2x = 10 \sin t.$$

Determine the solution which remains finite as $t \to \infty$ and for which $x = 4$ at $t = 0$. (JMB)

13) (a) Solve the differential equation

$$(1 + y^2) \sin 2x + (1 + \cos^2 x) \frac{dy}{dx} = 0$$

given that $y = 0$, when $x = \frac{\pi}{2}$.

(b) Solve the differential equation

$$x \frac{dy}{dx} + (1 - x)y = x. \qquad \text{(AEB '67)}$$

14) If $x = e^t$, show that $x \frac{dy}{dx} = \frac{dy}{dt}$ and $x^2 \frac{d^2y}{dx^2} = \frac{d^2y}{dt^2} - \frac{dy}{dt}$.

Use these results to reduce the differential equation

$$x^2 \frac{d^2y}{dx^2} + x \frac{dy}{dx} - 4y = 16$$

to a differential equation in y and t, and hence solve it, given that $y = 0$ and $dy/dx = 0$ when $x = 1$. (O)

15) Solve the differential equation

(a) $\sqrt{(3x - x^2)} \frac{dy}{dx} = 1 + \cos 2y$,

(b) $(1 - x^2) \frac{dy}{dx} = x(1 - 2y + x)$. (AEB '69)

16) The three differential equations

$$\frac{d^2x}{dt^2} + x = 0,$$

$$\frac{d^2x}{dt^2} - 3\frac{dx}{dt} + 2x = 10 \cos t,$$

$$\frac{d^2x}{dt^2} + 3\frac{dx}{dt} = -10 \cos t$$

are all subject to the conditions that $x = 1$ and $\dfrac{dx}{dt} = -3$ when $t = 0$.

Find the solutions. (C)

17) Find the curve $y = f(x)$ which passes through the point $(0, 1)$ and satisfies the differential equation

$$\frac{dy}{dx} + y \tan x = 2 \sin^2 x \cos x. \qquad \text{(JMB)}$$

18) Solve the differential equations

(a) $\dfrac{dy}{dx} + \dfrac{\sqrt{(1+y^2)}}{xy(1+x^2)} = 0,$

(b) $\dfrac{dy}{dx} = y + x$ given that $y = 3$ when $x = 0$. (AEB '66)

19) (a) Find the solution of the differential equation

$$\sin x \frac{dy}{dx} - y \cos x = \sin^2 x \cos x$$

for which $y = 2$ when $x = \frac{1}{2}\pi$.

(b) Find the solution of the differential equation

$$\frac{d^2y}{dx^2} + 2\frac{dy}{dx} + 5y = 2e^{-x}$$

for which $y = 0$ when $x = 0$ and when $x = \frac{1}{4}\pi$. (O)

20) Solve the equation

$$\frac{dy}{dx} + 2y \cot x = \cos x. \qquad \text{(AEB'75)p}$$

21) (a) If $(x^2 + 1)\dfrac{dy}{dx} + 4xy = 12x^3$, and $y = 1$ when $x = 1$, express y in terms of x in as simple a form as possible.

(b) If $\dfrac{d^2y}{dx^2} - 4\dfrac{dy}{dx} + 8y = 2e^{2x}$, and $y = 0$, $dy/dx = 0$ when $x = 0$, find y in terms of x. (O)

22) Solve the differential equations

(a) $(1 + \cos 2x)\dfrac{dy}{dx} - (1 + e^y)\sin 2x = 0$ given that $y = 0$ when $x = \pi/4$;

(b) $(1 + x)\,e^x\dfrac{dy}{dx} + x\,e^x y - 1 = 0.$ (AEB '73)

23) Solve the differential equations

(a) $x^2\dfrac{dy}{dx} - xy = 1,$ $y = 0$ when $x = 1$;

(b) $\dfrac{d^2y}{dx^2} - 2\dfrac{dy}{dx} + 5y = 2e^x,$ $y = 1$ and $\dfrac{dy}{dx} = 1$ when $x = 0.$ (O)

CHAPTER 14

FURTHER APPLICATIONS OF INTEGRATION

THE LENGTH OF AN ARC OF A CURVE

Cartesian Coordinates

There is no simple formula for calculating the length of any portion of a curve other than a circle. So if the length of a particular arc is required, we use the method of summing small elements of arc length.

Suppose that the arc PQ, of length δs, is such an element, then the length, s, of the curve AB is given by $\displaystyle s = \sum_{x=x_1}^{x=x_2} \delta s.$

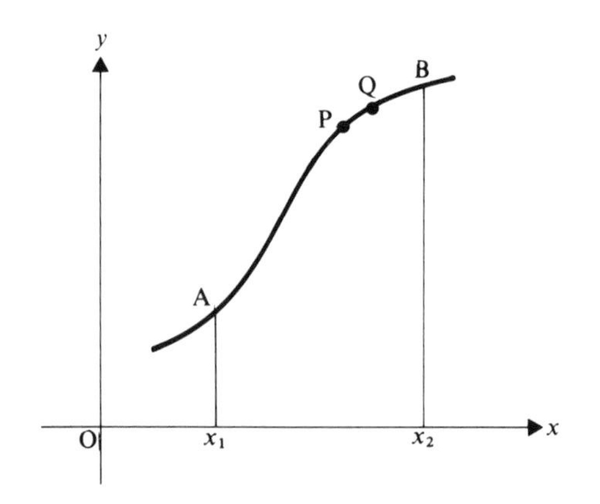

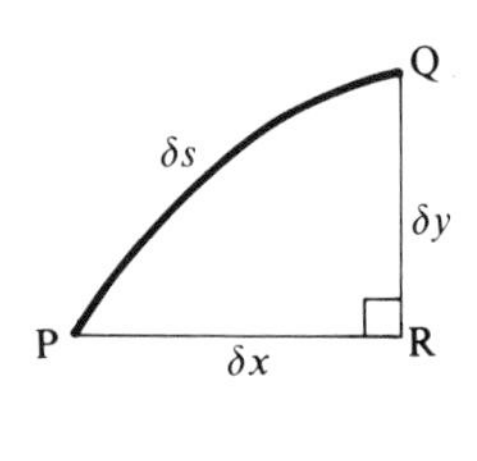

Now between P and Q, x increases by δx, y increases by δy and, as δs is small, δs is approximately equal to the hypotenuse of ΔPQR,

i.e.
$$(\delta s)^2 \simeq (\delta x)^2 + (\delta y)^2$$

$\Rightarrow$
$$\left(\frac{\delta s}{\delta x}\right)^2 \simeq 1 + \left(\frac{\delta y}{\delta x}\right)^2$$

$\Rightarrow$
$$\delta s \simeq \left\{1 + \left(\frac{\delta y}{\delta x}\right)^2\right\}^{\frac{1}{2}} \delta x$$

Now
$$s = \sum_{x=x_1}^{x=x_2} \delta s$$

$$\simeq \sum_{x=x_1}^{x=x_2} \left\{1 + \left(\frac{\delta y}{\delta x}\right)^2\right\}^{\frac{1}{2}} \delta x$$

As
$$\delta x \to 0, \qquad \frac{\delta y}{\delta x} \to \frac{dy}{dx}$$

So
$$s = \lim_{\delta x \to 0} \sum_{x=x_1}^{x=x_2} \left\{1 + \left(\frac{\delta y}{\delta x}\right)^2\right\}^{\frac{1}{2}} \delta x$$

i.e.
$$s = \int_{x_1}^{x_2} \left\{1 + \left(\frac{dy}{dx}\right)^2\right\}^{\frac{1}{2}} dx$$

The length of any particular section of a curve whose Cartesian equation is known can now be evaluated provided that the necessary integration can be performed (otherwise an approximate value for $\Sigma\, \delta s$ can be found by a numerical method, e.g. Simpson's Rule).

EXAMPLES 14a

1) Find the length of the portion of the curve $y = x^2$ between $x = 0$ and $x = 1$.

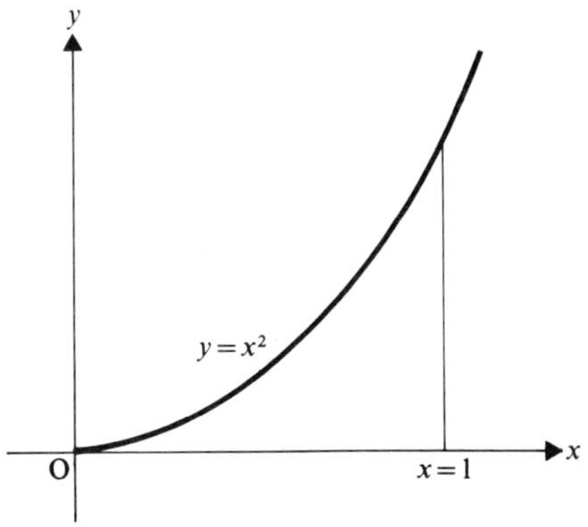

$$y = x^2 \quad \Rightarrow \quad \frac{dy}{dx} = 2x$$

So

$$\text{arc OA} = \int_0^1 \{1 + (2x)^2\}^{\frac{1}{2}}\, dx$$

y
$y = x^2$
O
$x = 1$
x

To carry out the integration we can use the substitution

$$2x \equiv \sinh u \quad \Rightarrow \quad \dots 2\,dx \equiv \dots \cosh u\,du$$

and

x	0	1
u	0	arsinh 2

Then
$$\int (1+4x^2)^{\frac{1}{2}}\,dx \equiv \tfrac{1}{2}\int \cosh^2 u\,du$$

$$\equiv \tfrac{1}{4}\int (1+\cosh 2u)\,du$$

$$= \tfrac{1}{4}(u+\tfrac{1}{2}\sinh 2u).$$

Hence
$$\int_0^1 (1+4x^2)^{\frac{1}{2}}\,dx = \tfrac{1}{4}\Big[u+\sinh u\cosh u\Big]_0^{\operatorname{arsinh} 2}$$

$$= \tfrac{1}{4}\operatorname{arsinh} 2 + \tfrac{1}{2}\sqrt{(1+4)}$$

So the length of the arc OA is $\frac{1}{2}\sqrt{5}+\frac{1}{4}\operatorname{arsinh} 2$
or $\frac{1}{2}\sqrt{5}+\frac{1}{4}\ln(2+\sqrt{5})$

2) Find an approximate value for the length of the portion of the ellipse $\dfrac{x^2}{4}+y^2=1$ that lies in the first quadrant between $x=0$ and $x=1$.

The length, s, of the arc AB is given by

$$s = \int_0^1 \left\{1+\left(\frac{dy}{dx}\right)^2\right\}^{\frac{1}{2}} dx$$

where
$$\frac{dy}{dx} = -\frac{x}{4y}$$

$\Rightarrow$
$$s = \int_0^1 \left\{1+\frac{x^2}{16y^2}\right\}^{\frac{1}{2}} dx$$

As the indefinite integral $\displaystyle\int \left\{1+\frac{x^2}{16y^2}\right\}^{\frac{1}{2}} dx$ cannot easily be found for the given ellipse, we find an approximate value for s as follows.

Using Simpson's Rule with five ordinates, we have

x	0	0.25	0.5	0.75	1.00
$\left\{1+\dfrac{x^2}{16y^2}\right\}^{\frac{1}{2}}$	1.0000	1.0020	1.0083	1.0202	1.0408

Then $s \simeq \dfrac{0.25}{3}[1.0000 + 1.0408 + 4(1.0020 + 1.0202) + 2(1.0083)]$

$\Rightarrow \quad s \simeq 1.0122$

Parametric Coordinates

Let us now consider a curve whose equations are given in terms of a parameter t say.
The approximate relationship

$$(\delta s)^2 \simeq (\delta x)^2 + (\delta y)^2$$

can be used again, but this time we will divide each term by $(\delta t)^2$, giving

$$\left(\frac{\delta s}{\delta t}\right)^2 \simeq \left(\frac{\delta x}{\delta t}\right)^2 + \left(\frac{\delta y}{\delta t}\right)^2$$

or

$$\delta s \simeq \left\{\left(\frac{\delta x}{\delta t}\right)^2 + \left(\frac{\delta y}{\delta t}\right)^2\right\}^{\frac{1}{2}} \delta t$$

Then, as $\delta t \to 0$, $\dfrac{\delta x}{\delta t} \to \dfrac{dx}{dt}$ and $\dfrac{\delta y}{\delta t} \to \dfrac{dy}{dt}$

So

$$s = \lim_{\delta t \to 0} \sum \left\{\left(\frac{\delta x}{\delta t}\right)^2 + \left(\frac{\delta y}{\delta t}\right)^2\right\}^{\frac{1}{2}} \delta t$$

$$\Rightarrow \quad s = \int \left\{\left(\frac{dx}{dt}\right)^2 + \left(\frac{dy}{dt}\right)^2\right\}^{\frac{1}{2}} dt$$

In this way the length of a particular section of a curve bounded by points where $t = t_1$ and $t = t_2$, can be found either by integration or by an approximate method of summation.

EXAMPLES 14a (continued)

3) Find the length of the circumference of the astroid $x = a\cos^3\theta$, $y = a\sin^3\theta$.

As the astroid is symmetrical about both Ox and Oy, the length of the arc AB is one quarter of the circumference.

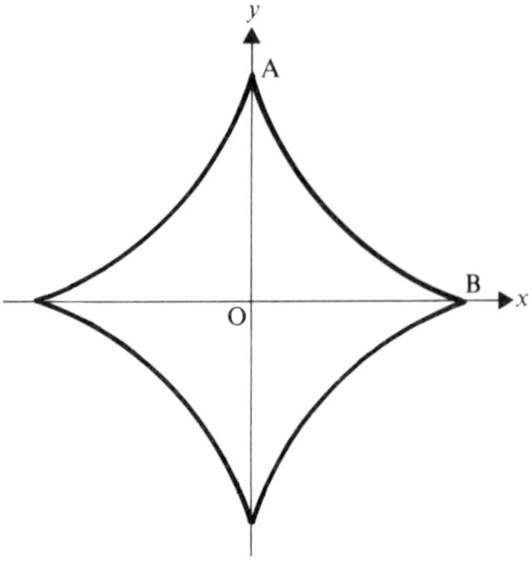

Now at A, $x = 0$ so $\cos\theta = 0$

$$\Rightarrow \qquad \theta = \frac{\pi}{2}$$

and at B, $y = 0$ so $\sin\theta = 0$

$$\Rightarrow \qquad \theta = 0$$

Thus

$$\text{arc AB} = \int_0^{\frac{\pi}{2}} \left\{ \left(\frac{dx}{d\theta}\right)^2 + \left(\frac{dy}{d\theta}\right)^2 \right\}^{\frac{1}{2}} d\theta$$

$$= \int_0^{\frac{\pi}{2}} \{(-3a\cos^2\theta\sin\theta)^2 + (3a\sin^2\theta\cos\theta)^2\}^{\frac{1}{2}} d\theta$$

$$= 3a\int_0^{\frac{\pi}{2}} \cos\theta\sin\theta\{\cos^2\theta + \sin^2\theta\}^{\frac{1}{2}} d\theta$$

$$= \frac{3a}{2}\int_0^{\frac{\pi}{2}} \sin 2\theta \, d\theta$$

$$= \frac{3a}{4}\Big[-\cos 2\theta\Big]_0^{\frac{\pi}{2}} = \frac{3a}{2}$$

So the length of the circumference of the astroid is $6a$.

Note. The reader may find that, occasionally, the integral used to calculate an arc length gives a negative result. If the limits of the integration are reversed, the result is of equal magnitude but is positive. So a negative result is a property only of the sense in which δs is summed and the arc length can therefore be taken as $|\Sigma\, \delta s|$.

Polar Coordinates

In this case we use a right angled triangle with sides along and perpendicular to the radius vector, as shown in the diagram, so that the hypotenuse PQ is approximately equal to δs.

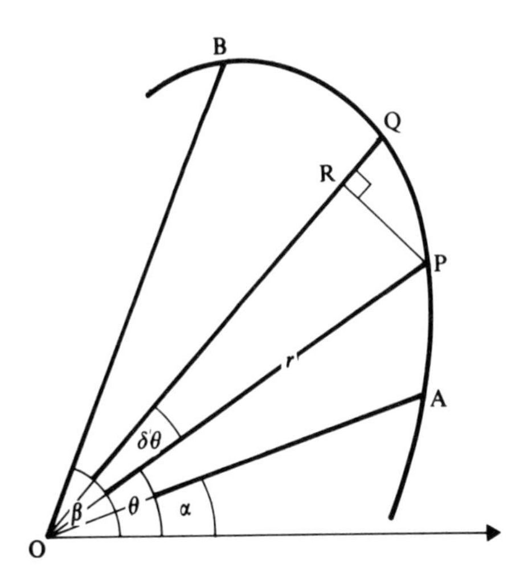

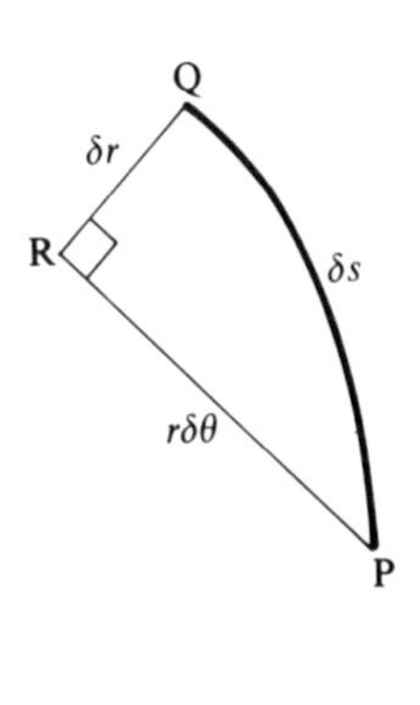

Now $\quad OP = r \quad \text{and} \quad OQ = r + dr$

$\Rightarrow \quad RQ \simeq \delta r$

Also $\quad RP = r \sin \delta\theta$

$\quad \simeq r\,\delta\theta \qquad$ (as $\delta\theta$ is small)

So $\quad (PQ)^2 = (PR)^2 + (RQ)^2$

$\Rightarrow \quad (\delta s)^2 \simeq (r\delta\theta)^2 + (\delta r)^2$

$\Rightarrow \quad \left(\dfrac{\delta s}{\delta\theta}\right)^2 \simeq r^2 + \left(\dfrac{\delta r}{\delta\theta}\right)^2$

i.e. $\quad \delta s \simeq \left\{r^2 + \left(\dfrac{\delta r}{\delta\theta}\right)^2\right\}^{\frac{1}{2}} \delta\theta$

Now the length, s, of the arc AB is given by

$$s = \sum_{\theta=\alpha}^{\theta=\beta} \delta s$$

$$\simeq \sum_{\theta=\alpha}^{\theta=\beta} \left\{r^2 + \left(\frac{\delta r}{\delta\theta}\right)^2\right\}^{\frac{1}{2}} \delta\theta$$

In the limiting case when $\delta\theta \to 0$ and $\dfrac{\delta r}{\delta\theta} \to \dfrac{dr}{d\theta}$,

$$s = \int_{\alpha}^{\beta} \left\{r^2 + \left(\frac{dr}{d\theta}\right)^2\right\}^{\frac{1}{2}} d\theta$$

Note that each of the expressions for finding the length of an arc can easily be derived from the appropriate right angled triangle and so *need* not be memorised. However there is no reason why they should not be quoted unless their proof is required.

Note also that, in each of the preceding cases, the arc length can be found either by integration or by an appropriate numerical method of summation.

EXAMPLES 14a (continued)

4) Find the length of the circumference of the cardioid $r = a(1 + \cos\theta)$

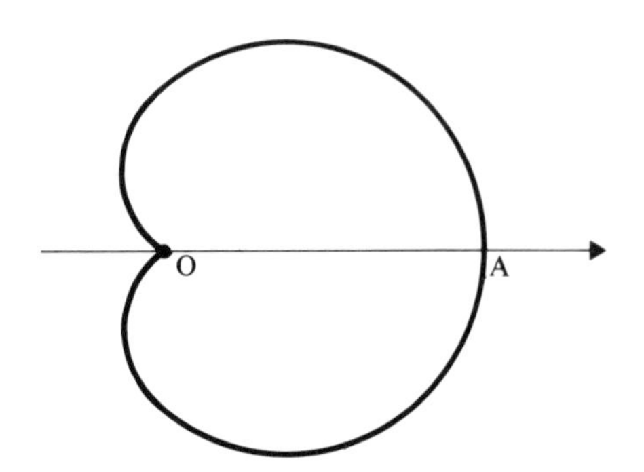

As OA is an axis of symmetry we need calculate only the upper half-circumference

At A, $\theta = 0$ and at O, $\theta = \pi$.

So the length, s, of the curve AO (in the anticlockwise sense) is given by

$$s = \int_0^\pi \left\{r^2 + \left(\frac{dr}{d\theta}\right)^2\right\}^{\frac{1}{2}} d\theta$$

$$= \int_0^\pi \{a^2(1 + \cos\theta)^2 + (-a\sin\theta)^2\}^{\frac{1}{2}} d\theta$$

$$= a\int_0^\pi \{2(1 + \cos\theta)\}^{\frac{1}{2}} d\theta$$

$$= a\int_0^\pi 2\cos\frac{\theta}{2}\, d\theta$$

$$= \left[4a\sin\frac{\theta}{2}\right]_0^\pi$$

$$= 4a.$$

So the length of the circumference of the cardioid $r = a(1 + \cos\theta)$ is $8a$.

EXERCISE 14a

(**Note** that the problems set throughout this chapter are basically exercises in integration. A reader who finds the integration in a particular problem difficult is recommended to omit it and continue with the next one.)

In each of the following questions, the equation(s) of a curve and a specific section of the curve, are given. In each case find the length of the specified portion.

1) The circle $x^2+y^2=1$; between $A(-1,0)$ and $B(1,0)$.

2) $y=c\cosh\dfrac{x}{c}$ (c is a constant); between $A(0,0)$ and $B(x,y)$.

3) The semi-cubical parabola $x^3=y^2$; between $A(0,0)$ and $B(1,1)$.

4) The parabola $2y=x^2$; between $A(0,0)$ and $B(1,\frac{1}{2})$.

5) $y=\ln\sec x$; between $A(x=0)$ and $B\left(x=\dfrac{\pi}{6}\right)$.

6) The cycloid $x=a(\theta-\sin\theta)$, $y=a(1-\cos\theta)$; between $A\left(\theta=\dfrac{\pi}{2}\right)$ and $B(\theta=\pi)$.

7) The parabola $x=at^2$, $y=2at$; between the vertex $A(t=0)$ and $B(at^2,2at)$.

8) The circle $x=a\cos\theta$, $y=a\sin\theta$; *find* the length of the circumference.

9) $x=\tanh t$, $y=\operatorname{sech} t$; between $A(t=0)$ and $B(t=1)$. What is this curve called?

10) The spiral $r=k\theta$ $(k>0)$; between $A(0,0)$ and $B\left(k\dfrac{\pi}{2},\dfrac{\pi}{2}\right)$.

11) The cardioid $r=a(1-\sin\theta)$; find the length of its circumference.

12) The circle $r=a$; between $A(a,0)$ and $B\left(a,\dfrac{\pi}{2}\right)$.

13) The equiangular spiral $r=a\,e^{\theta}$ between $A(\theta=0)$ and $B(\theta=\pi)$.

AREA OF SURFACE OF REVOLUTION

When a section of a curve rotates through one revolution about the x axis, a three dimensional surface is formed. The surface area of such an object can be found by summing the areas of small elements of the total surface. A suitable element is formed by the rotation of an elemental arc of length δs tracing out a 'ring' of radius y and width δs.
The surface area, δA, of this ring element is given by

$$\delta A \simeq 2\pi y\,\delta s.$$

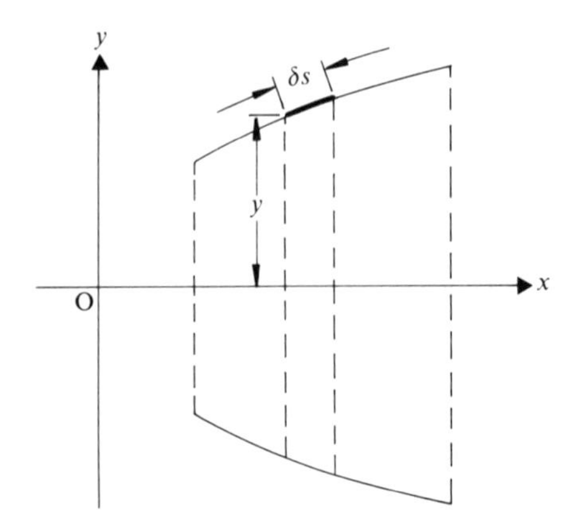

Now the equation of the curve whose rotation has formed the surface may be in Cartesian, parametric or polar form. In each case the total surface area is obtained by evaluating $\Sigma\,\delta A$, but the method of performing this summation depends upon the type of equation.

Cartesian Equation

Using the relationship

$$\delta s \simeq \left\{1+\left(\frac{\delta y}{\delta x}\right)^2\right\}^{\frac{1}{2}}\delta x$$

gives

$$\delta A \simeq 2\pi y\left\{1+\left(\frac{\delta y}{\delta x}\right)^2\right\}^{\frac{1}{2}}\delta x$$

$\Rightarrow$

$$\sum\delta A \simeq \sum 2\pi y\left\{1+\left(\frac{\delta y}{\delta x}\right)^2\right\}^{\frac{1}{2}}\delta x$$

Thus, when $\delta x \to 0$, the total surface area, **A**, is given by

$$A = \lim_{\delta x\to 0}\sum 2\pi y\left\{1+\left(\frac{\delta y}{\delta x}\right)^2\right\}^{\frac{1}{2}}\delta x$$

$\Rightarrow$

$$A = 2\pi\int y\left\{1+\left(\frac{\mathrm{d}y}{\mathrm{d}x}\right)^2\right\}^{\frac{1}{2}}\mathrm{d}x$$

Parametric Equations

This time we use

$$\delta s \simeq \left\{\left(\frac{\delta x}{\delta t}\right)^2+\left(\frac{\delta y}{\delta t}\right)^2\right\}^{\frac{1}{2}}\delta t$$

so that

$$\delta A \simeq 2\pi y\left\{\left(\frac{\delta x}{\delta t}\right)^2+\left(\frac{\delta y}{\delta t}\right)^2\right\}^{\frac{1}{2}}\delta t$$

Again taking the limit, as $\delta x \to 0$, of $\Sigma\, \delta A$, we get

$$A = 2\pi \int y \left\{ \left(\frac{dx}{dt} \right)^2 + \left(\frac{dy}{dt} \right)^2 \right\}^{\frac{1}{2}} dt$$

Polar Equation

Using the third of the forms we derived earlier for δs, we have

$$\delta s \simeq \left\{ r^2 + \left(\frac{\delta r}{\delta \theta} \right)^2 \right\}^{\frac{1}{2}} \delta \theta$$

$$\Rightarrow \qquad \delta A \simeq 2\pi y \left\{ r^2 + \left(\frac{\delta r}{\delta \theta} \right)^2 \right\}^{\frac{1}{2}} \delta \theta$$

But $\qquad y = r \sin \theta,$

so

$$\delta A \simeq 2\pi r \sin \theta \left\{ r^2 + \left(\frac{\delta r}{\delta \theta} \right)^2 \right\}^{\frac{1}{2}} \delta \theta.$$

Taking the limit, as $\delta\theta \to 0$, of $\Sigma\, \delta A$, the total surface area becomes

$$A = 2\pi \int r \sin \theta \left\{ r^2 + \left(\frac{dr}{d\theta} \right)^2 \right\}^{\frac{1}{2}} d\theta$$

Note. The reader is recommended to remember *how these formulae are derived* rather than to memorise all three individual results.

EXAMPLE 14b

Find the area of the surface generated when each of the following curves is rotated through an angle 2π about the x axis:

(a) the arc of the parabola $y^2 = 4x$ between the origin and the point $(4, 4)$,

(b) the cycloid $x = a(\theta - \sin\theta)$, $y = a(1 - \cos\theta)$ between $(0, 0)$ and $(2\pi a, 0)$,

(c) the cardioid $r = a(1 + \cos\theta)$ between $(2a, 0)$ and $(0, \pi)$.

(a)

$$y^2 = 4x \quad \Rightarrow \quad \frac{dy}{dx} = \frac{2}{y}$$

So

$$\left\{ 1 + \left(\frac{dy}{dx} \right)^2 \right\}^{\frac{1}{2}} = \left\{ 1 + \frac{4}{y^2} \right\}^{\frac{1}{2}} = \left\{ \frac{x+1}{x} \right\}^{\frac{1}{2}}$$

Then

$$A = 2\pi \int y \left\{ 1 + \left(\frac{dy}{dx} \right)^2 \right\}^{\frac{1}{2}} dx$$

$$\Rightarrow \qquad A = 2\pi \int_0^4 2x^{\frac{1}{2}} \left\{ \frac{x+1}{x} \right\}^{\frac{1}{2}} dx$$

$$= 4\pi \int_0^4 (x+1)^{\frac{1}{2}}\,dx$$

$$= 4\pi \left[\tfrac{2}{3}(x+1)^{\frac{3}{2}} \right]_0^4$$

$$\Rightarrow \quad A = \frac{8\pi}{3}(5\sqrt{5}-1).$$

(b) $$x = a(\theta - \sin\theta) \quad \Rightarrow \quad \frac{dx}{d\theta} = a(1-\cos\theta)$$

$$y = a(1-\cos\theta) \quad \Rightarrow \quad \frac{dy}{d\theta} = a\sin\theta.$$

So $$\left\{\left(\frac{dx}{d\theta}\right)^2 + \left(\frac{dy}{d\theta}\right)^2\right\}^{\frac{1}{2}} = \{a^2(1-2\cos\theta+\cos^2\theta)+a^2\sin^2\theta\}^{\frac{1}{2}}$$

$$= a\{2-2\cos\theta\}^{\frac{1}{2}}$$

$$= a\left\{4\sin^2\frac{\theta}{2}\right\}^{\frac{1}{2}}$$

$$= 2a\sin\frac{\theta}{2}$$

Then A is given by

$$2\pi \int y \left\{\left(\frac{dx}{d\theta}\right)^2 + \left(\frac{dy}{d\theta}\right)^2\right\}^{\frac{1}{2}} d\theta = 2\pi \int a(1-\cos\theta)2a\sin\frac{\theta}{2}\,d\theta$$

As x goes from 0 to $2\pi a$, θ goes from 0 to 2π.

So $$A = 4\pi a^2 \int_0^{2\pi} 2\sin^3\frac{\theta}{2}\,d\theta$$

$$= 8\pi a^2 \int_0^{2\pi} \sin\frac{\theta}{2}\left(1-\cos^2\frac{\theta}{2}\right)d\theta$$

$$= 8\pi a^2 \left[-2\cos\frac{\theta}{2} + \frac{2}{3}\cos^3\frac{\theta}{2}\right]_0^{2\pi}$$

$$\Rightarrow \quad A = \tfrac{64}{3}\pi a^2.$$

(c) $$r = a(1+\cos\theta) \quad \Rightarrow \quad \frac{dr}{d\theta} = -a\sin\theta$$

So $$\left\{r^2 + \left(\frac{dr}{d\theta}\right)^2\right\}^{\frac{1}{2}} = \{a^2(1+2\cos\theta+\cos^2\theta)+a^2\sin^2\theta\}^{\frac{1}{2}}$$

$$= a\{2 + 2\cos\theta\}^{\frac{1}{2}}$$

$$= 2a\cos\frac{\theta}{2}$$

Then $\quad A = 2\pi\int_0^{\pi} r\sin\theta\left\{r^2 + \left(\frac{dr}{d\theta}\right)^2\right\}^{\frac{1}{2}} d\theta$

$\Rightarrow \quad A = 2\pi\int_0^{\pi} a(1+\cos\theta)\sin\theta\left(2a\cos\frac{\theta}{2}\right) d\theta$

$$= 4\pi a^2\int_0^{\pi}\left(2\cos^2\frac{\theta}{2}\right)\left(2\sin\frac{\theta}{2}\cos\frac{\theta}{2}\right)\left(\cos\frac{\theta}{2}\right) d\theta$$

$$= 16\pi a^2\int_0^{\pi}\cos^4\frac{\theta}{2}\sin\frac{\theta}{2}\, d\theta$$

$$= 16\pi a^2\left[-\frac{2}{5}\cos^5\frac{\theta}{2}\right]_0^{\pi}$$

$\Rightarrow \quad A = \frac{32}{5}\pi a^2$

VOLUME OF REVOLUTION

When an area rotates completely about a fixed line, a solid is formed with circular cross-section. The volume of such a solid of revolution can be found by summing the volumes of small elements of the solid.

The detailed application of this method, when the area is bounded by a curve with a Cartesian equation, was explained in Volume 1, Chapter 18. We shall now adapt the method to deal with the rotation of an area bounded by a curve with parametric equations.

Consider, for example, the volume generated when the area in the first quadrant bounded by the ellipse $x = 4\cos\theta$, $y = 3\sin\theta$ rotates completely about the x axis.

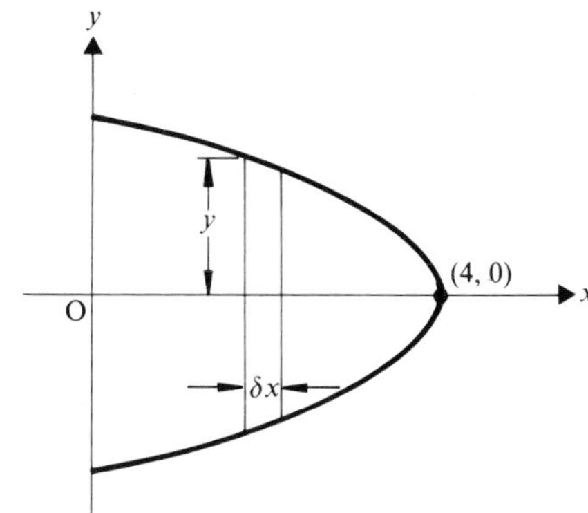

Dividing the volume into thin slices perpendicular to the x axis, we see that each slice is approximately a disc of radius y and thickness δx.

So the volume, δV, of one element is given by

$$\delta V \simeq \pi y^2\,\delta x.$$

The total volume of revolution, V, is found by taking the limit of the sum of the volumes of all the elements,

i.e.
$$V = \lim_{\delta x \to 0} \sum_{x=0}^{x=4} \delta V = \lim_{\delta x \to 0} \sum_{x=0}^{x=4} \pi y^2\ \delta x$$
$$= \int_0^4 \pi y^2\ \mathrm{d}x.$$

But $\qquad x = 4\cos\theta \qquad \Rightarrow \quad \ldots \mathrm{d}x \equiv \ldots -4\sin\theta\ \mathrm{d}\theta$

and $\qquad y = 3\sin\theta$

Also, when $x = 0,\quad \theta = \dfrac{\pi}{2}$

and when $x = 4,\quad \theta = 0.$

So
$$V = \pi\int_{\frac{\pi}{2}}^{0} (3\sin\theta)^2(-4\sin\theta)\ \mathrm{d}\theta$$
$$= 36\pi\int_0^{\frac{\pi}{2}} \sin^3\theta\ \mathrm{d}\theta$$
$$= 36\pi(\tfrac{2}{3}) \qquad \text{(using a reduction formula).}$$

Hence the volume generated by rotating a quadrant of the given ellipse is 24π.

Note that the degree of rotation necessary for an area to generate a complete solid of revolution, is not always a full revolution. For example, a sphere can be generated either when a *circle* rotates about a diameter through an angle π, or when a *semi-circle* rotates about its diameter through an angle 2π.

THE ROOT MEAN SQUARE VALUE OF A FUNCTION

In Volume 1 we saw that the mean value of a function $f(x)$ represented the value of the average ordinate of the graph of $y = f(x)$. Thus the mean value of a function takes into account the sign of the function. It is sometimes desirable to find an average that considers only the numerical value, and not the sign, of the function within the specified range. This can be achieved by *squaring* $f(x)$ and finding the mean of these square values, subsequently taking the square root of the mean found in this way, to give the *root mean square value of the function* (or R.M.S. value).
i.e. the R.M.S. value of a function $f(x)$ within the range $a \leqslant x \leqslant b$ is given by

$$\left\{\frac{1}{b-a}\int_a^b [f(x)]^2\ \mathrm{d}x\right\}^{\frac{1}{2}}$$

If $[f(x)]^2$ is integrable, the R.M.S. value can be determined exactly. Otherwise an approximate value for $\sum_{x=a}^{x=b} [f(x)]^2 \, \delta x$ can be found using Simpson's Rule or some other numerical method.

For example, the R.M.S. value of $\sin x$ between $x = 0$ and $x = \frac{\pi}{2}$ is found from

$$\left\{\frac{1}{(\pi/2-0)}\int_0^{\frac{\pi}{2}} \sin^2 x \, dx\right\}^{\frac{1}{2}} = \left\{\frac{1}{\pi}\int_0^{\frac{\pi}{2}} (1-\cos 2x)\, dx\right\}^{\frac{1}{2}}$$

$$= \left\{\frac{1}{\pi}\left[\frac{\pi}{2}\right]\right\}^{\frac{1}{2}}$$

$$= \tfrac{1}{2}\sqrt{2}.$$

Similarly the R.M.S. value of e^{x^2} between $x = 0$ and $x = 1$ is given by

$$\left\{\frac{1}{1-0}\int_0^1 e^{2x^2}\, dx\right\}^{\frac{1}{2}}$$

but $\int e^{2x^2}\, dx$ cannot be found. So an approximate value for this integral is obtained by using Simpson's Rule with, say, five ordinates,

i.e.

x	0	0.25	0.5	0.75	1.0
e^{2x^2}	1	1.133	1.649	3.080	7.389

$$\Rightarrow \qquad \int_0^1 e^{2x^2}\, dx \simeq \frac{0.25}{3}\{8.389 + 4(4.213) + 2(1.649)\}$$

$$\simeq 2.378$$

Hence the required R.M.S. value is approximately $\left\{\frac{1}{(1-0)}(2.378)\right\}^{\frac{1}{2}} = 1.542.$

EXERCISE 14b

Find the areas of the surfaces generated by the complete rotation about the x axis (or the initial line) of each of the following arcs.

1) The quarter circle in the first quadrant whose equation is $x^2 + y^2 = a^2$.

2) The arc of the astroid $x = a\cos^3 t$, $y = a\sin^3 t$ that lies above the x axis.

3) The line $2y = x$ between the origin and the point $(4, 2)$. (Do not assume any formulae for a cone.)

4) The arc of the curve $r = 2a\cos\theta$ bounded by $\theta = 0$ and $\theta = \frac{\pi}{2}$.

5) The complete curve $r = a$. (Do not assume the formula for surface area of a sphere.)

6) The arc between $t=0$ and $t=4$ of the parabola $x=t^2$, $y=2t$.

7) The section of the spiral $r=e^\theta$ between $\theta=0$ and $\theta=\frac{\pi}{2}$.

8) The part of the rectangular hyperbola $x^2-y^2=a^2$ that lies in the first quadrant between $x=a$ and $x=2a$.

9) The portion of the curve $y=\cosh x$ between $x=0$ and $x=1$.

10) The section between $x=0$ and $x=1$ of the curve $y=e^x$.

11) The arc of the curve $x=t^3$, $y=3t^2$ between the points where $t=0$ and $t=3$.

Find the volume generated by the complete rotation of the following areas about the specified lines.

12) The area bounded by the x axis, the curve $x=2t^2$, $y=8t$ and the line $x=8$, rotated about
(a) the x axis, (b) the y axis.

13) The area in the first quadrant, of the astroid $x=\cos^3 t$, $y=\sin^3 t$, rotated about the x axis.

14) The area bounded by the x axis and the arc of the cycloid $x=a(\theta-\sin\theta)$, $y=a(1-\cos\theta)$ between $\theta=0$ and $\theta=2\pi$, rotated about the x axis.

15) The area between the parabola $x=t^2$, $y=2t$ and the line $x=2$, rotated about the line $x=2$.

16) The area bounded by the lines $x=1$, $x=2$, $y=1$ and the curve $x=4t$, $y=\frac{4}{t}$, rotated about
(a) the line $x=1$ (b) the line $y=1$
(c) the x axis (d) the y axis.

Find the root mean square value of the following functions over the specified ranges.
Where necessary use Simpson's Rule using five ordinates giving these results to three significant figures.

17) $f(x)\equiv\frac{1}{x}$ for $1\leqslant x\leqslant 4$

18) $f(x)\equiv\tan x$ for $0\leqslant x\leqslant\frac{\pi}{4}$

19) $f(x)\equiv 1+\sin x$ for $0\leqslant x\leqslant\frac{\pi}{2}$

20) $f(x)\equiv e^x$ for $1\leqslant x\leqslant 2$

21) $f(x)\equiv\sqrt{(4-x^2)}$ for $0\leqslant x\leqslant 2$

22) $f(x)\equiv\sinh x$ for $1\leqslant x\leqslant 2$

23) $f(x) \equiv x \sin x$ for $0 \leqslant x \leqslant \frac{\pi}{2}$ 24) $f(x) \equiv \ln x$ for $1 \leqslant x \leqslant 2$

25) $f(x) \equiv 1 + x^2$ for $0 \leqslant x \leqslant 4$.

CENTROID AND FIRST MOMENT

In Volume 1, Chapter 18, we saw that the *centroid* of an object is its geometric centre. For a symmetrical object therefore, the centroid lies on the axis of symmetry. For an object made of uniform material, the centroid coincides with the point at which the body can be supported in a perfectly balanced state.

Relative to a specified line, an object has a *first moment* which depends upon
(a) the distance, h, of its centroid from the specified line,
(b) the volume of the object if it is a solid,
the area of the object if it is a surface,
the length of the object if it is an arc of a curve.

In Volume 1 the first moments of a volume and an area were defined,

i.e.
$$\text{first moment of a volume} = \text{volume} \times h$$
$$\text{first moment of an area} = \text{area} \times h.$$

We now define the first moment of an arc in a similar way,

i.e.
$$\text{first moment of an arc} = \text{length} \times h.$$

In cases when h is not known, the first moment of an object can be found by summing the first moments of all the elements into which it can be divided. Thus, if h_e is the distance from the specified line of the centroid of one typical element of a solid object, we have

first moment of volume $= \Sigma V_e h_e$ where V_e is the volume of the element.

But first moment of volume is also given by Vh.

So
$$Vh = \sum V_e h_e \qquad [1]$$

Similarly, for an object with an area of magnitude A we have

$$Ah = \sum A_e h_e \qquad [2]$$

And, for an object that is an arc of a curve of length S and which therefore has a first moment of arc length,

$$Sh = \sum S_e h_e \qquad [3]$$

The application of equations [1] and [2] above to the locating of centroids of volumes and areas is dealt with in Volume 1. The centroid of an arc length, however, was not found at that time, so we will now consider an example of this type.

EXAMPLES 14c

1) Find the first moment, relative to the x axis, of a semi-circular arc of radius a, bounded by the x axis. Hence find the position of the centroid of the arc.

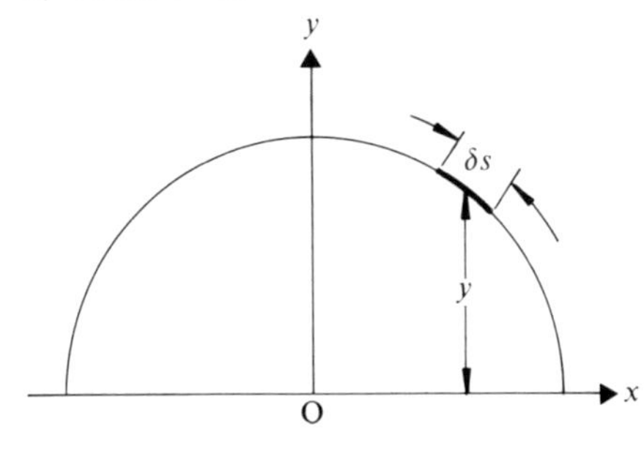

Taking an element of the arc of length δs, whose centroid is at a distance y from Ox we have:

First moment of arc

$$\simeq \sum y\,\delta s$$

But $\quad \delta s = a\delta\theta$

and $\quad y \simeq a \sin\theta$

So $\quad \sum y\,\delta s \simeq \sum a^2 \sin\theta\,\delta\theta$

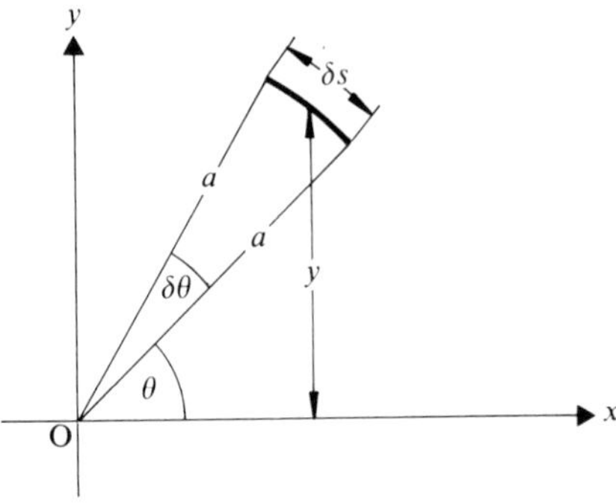

Taking the limit of the summation, and working from $\theta = 0$ to $\theta = \pi$, we find that the first moment, M, of the semi circular arc is given by

$$M = \int_0^{\pi} a^2 \sin\theta\,d\theta$$

$$= a^2\left[-\cos\theta\right]_0^{\pi}$$

i.e. $\quad M = 2a^2$

Now the total length of the arc is πa and, from symmetry, its centroid lies on the y-axis. If the distance of its centroid from Ox is $\bar{y}$, then the first moment of the arc length is also given by

$$M = \pi a\bar{y}$$

$$\Rightarrow \quad \pi a\bar{y} = 2a^2$$

$$\Rightarrow \quad \bar{y} = \frac{2a}{\pi}$$

So the centroid of the arc is the point $\left(0, \dfrac{2a}{\pi}\right)$.

Note that the centroid is *not* on the arc, a result common to most arcs.

Note also that the relationship $(\delta s)^2 \simeq (\delta x)^2 + (\delta y)^2$ can be used where appropriate.

THE THEOREMS OF PAPPUS

Theorem 1

Consider a section of a curve of perimeter s, that does not cross the x axis, and whose centroid is G. If this curve rotates completely about the x axis, then Pappus' first theorem states that the surface area, A, of the surface so formed, is given by

$$A = s \times \text{distance travelled by } G.$$

Proof

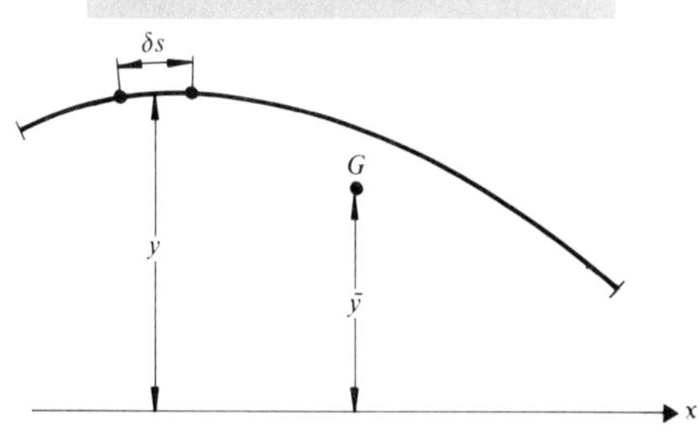

First we find the ordinate, $\bar{y}$, of the centroid, G, using

$$\text{first moment of whole curve} = \sum \text{first moment of element},$$

i.e.
$$s\bar{y} \simeq \sum y\, \delta s$$

$\Rightarrow$
$$\bar{y} \simeq \frac{1}{s} \sum y\, \delta s$$

Now the distance travelled by G when the curve rotates about Ox through one revolution is $2\pi\bar{y}$

where
$$2\pi\bar{y} \simeq \frac{2\pi}{s} \sum y\, \delta s = \frac{1}{s} \sum 2\pi y\, \delta s$$

But $2\pi y\, \delta s$ is the approximate area of an elemental 'ring' formed when δs rotates completely.
So the total surface area of revolution is given approximately by $\Sigma\, 2\pi y\, \delta s$,

i.e.
$$A \simeq \sum 2\pi y\,\delta s$$
$$\simeq s\left[\frac{1}{s}\sum 2\pi y\,\delta s\right]$$

Taking the limit, as $\delta x \to 0$, of this summation we have
$$A = s \times (\text{distance travelled by } G).$$

EXAMPLES 14c (continued)

2) Find the surface area of a lifebelt whose cross-section is a circle of radius 8 cm and whose inner radius is 20 cm.

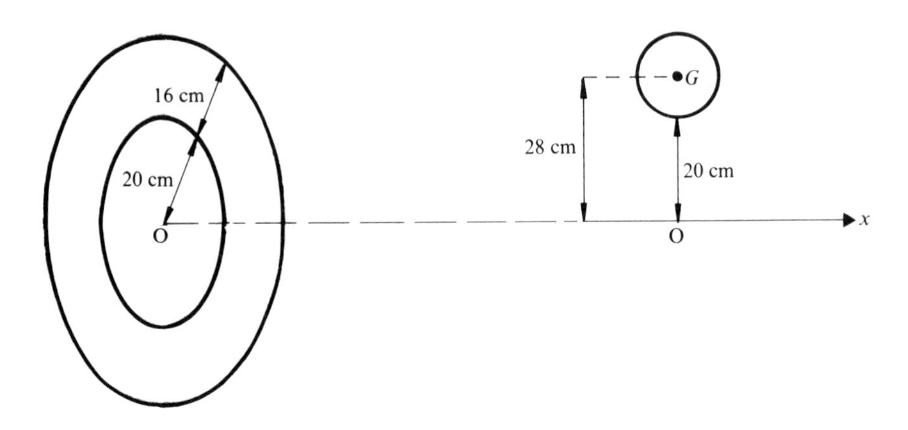

The surface of the lifebelt is formed when a circle of radius 8 cm, with its centre at a distance 28 cm from the x axis, rotates completely about the x axis.
The centroid of the circle is its centre, G, so using Pappus' Theorem, we have

$$\text{Surface Area} = (\text{perimeter of circle}) \times (\text{distance travelled by } G)$$
$$= 2\pi(8) \times 2\pi(28).$$

Hence, taking $\pi^2 \simeq 10$, we find that the surface area of the lifebelt is approximately 8960 square centimetres.

Theorem 2

Pappus second theorem refers to the area, A, enclosed by a curve that does not cross the x axis, and whose centroid (of area) is G. If this area rotates completely about the x axis, then this theorem states that the volume V of the solid of revolution is given by

$$V = A \times \text{distance travelled by } G.$$

Proof

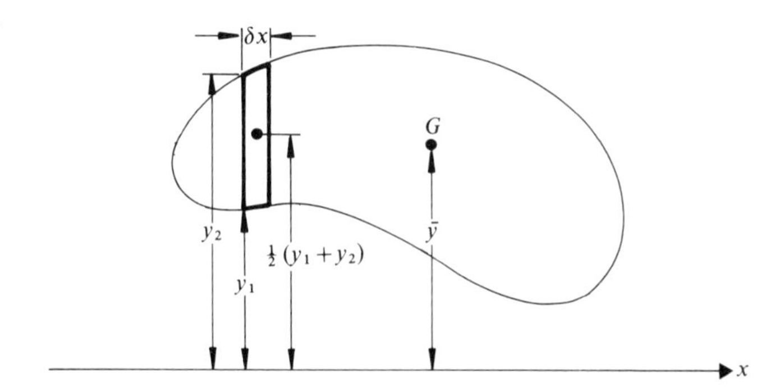

First we find the ordinate, $\bar{y}$, of the centroid G using an element of area as shown in the diagram.
The area of the element $\simeq (y_2 - y_1)\,\delta x$ and the centroid of the element is at an approximate distance $\frac{1}{2}(y_1 + y_2)$ from Ox.
So the first moment about Ox of the element is approximately

$$\tfrac{1}{2}(y_1 + y_2)(y_2 - y_1)\,\delta x$$

But first moment of whole area = Σ first moment of element,

i.e.
$$A\bar{y} \simeq \sum \frac{1}{2}(y_1 + y_2)(y_2 - y_1)\,\delta x$$

⇒
$$\bar{y} \simeq \frac{1}{2A}\sum (y_2^2 - y_1^2)\,\delta x$$

Now the distance travelled by $\bar{y}$ when the area rotates about Ox through one revolution is $2\pi\bar{y}$,

where
$$2\pi\bar{y} \simeq 2\pi\left\{\frac{1}{2A}\sum (y_2^2 - y_1^2)\right\}\delta x$$

$$= \frac{1}{A}\sum (\pi y_2^2 - \pi y_1^2)\,\delta x$$

But $(\pi y_2^2 - \pi y_1^2)\,\delta x$ is the approximate volume of the elemental annulus formed when the element of area rotates about Ox.
So the total volume of revolution, V, is given by

$$V \simeq \sum (\pi y_2^2 - \pi y_1^2)\,\delta x$$

i.e.
$$V \simeq A\left[\frac{1}{A}\sum (\pi y_2^2 - \pi y_1^2)\,\delta x\right]$$

Taking the limit, as $\delta x \to 0$, of this summation, we have

$$V = A \times \text{distance travelled by } G.$$

EXAMPLES 14c (continued)

3) Find the volume of the lifebelt described in Example 2 (page 586).

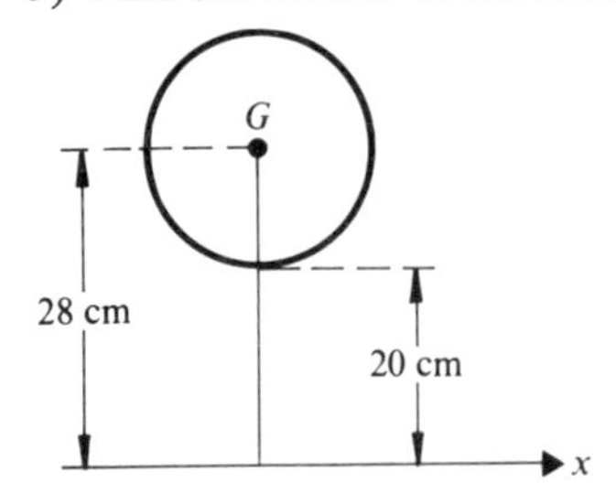

The area enclosed by the circular cross-section is $\pi(8^2) = 64\pi \text{ cm}^2$.

The distance travelled by G when the circle rotates about Ox is $2\pi \times 28$ cm.

Using Pappus' second theorem to find the volume, V, of the lifebelt, we have

$$V = (64\pi)(56\pi) \text{ cm}^3,$$

i.e. $$V \simeq 35\,840 \text{ cm}^3 \qquad (\pi^2 \simeq 10)$$

4) Find the position of the centroid of a semicircle by applying Pappus' second theorem to the volume of a sphere.

Consider a semicircle of radius a, with its diameter on the x axis.
When this semicircle rotates completely about Ox, a sphere is generated whose volume is $\frac{4}{3}\pi a^3$.

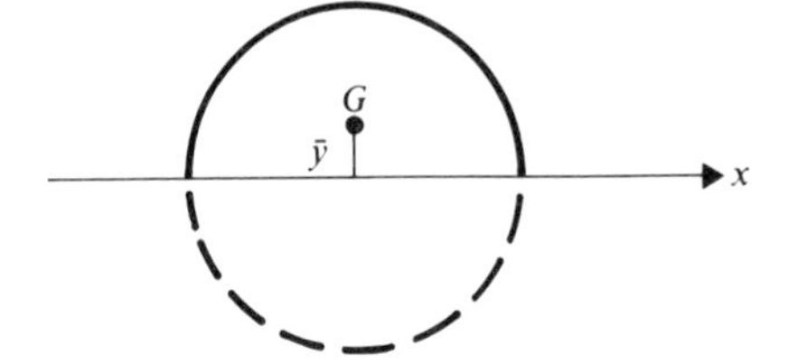

The closed area of the semicircle is $\frac{1}{2}\pi a^2$ and the distance travelled by G during rotation is $2\pi\bar{y}$.
Then Pappus' second theorem states that

$$\text{Volume generated} = \text{Area} \times \text{distance travelled by } G$$

i.e. $$\tfrac{4}{3}\pi a^3 = \tfrac{1}{2}\pi a^2 \times 2\pi\bar{y}$$

$\Rightarrow$ $$\bar{y} = \frac{4a}{3\pi}$$

EXERCISE 14c

1) Find the first moment about Ox of the arc of the curve $y = \sin x$ between $x = 0$ and $x = \pi$.

2) Find the first moment about Oy of the curve $y = x^2$ between $(1, 1)$ and $(2, 4)$.

3) Find the first moment about Ox of the part of the parabola $y = 1 - x^2$ that lies above the x axis.

4) An area is defined by $0 \leqslant x \leqslant 4$, $0 \leqslant y \leqslant 2\sqrt{x}$. Find its first moment (a) about the x axis, (b) about the y axis.

5) Find the first moments about the x axis of

(a) the area under the curve $y = \cos x$ from $x = -\dfrac{\pi}{2}$ to $x = \dfrac{\pi}{2}$.

(b) the area bounded by the curve $y = e^x$ and the lines $x = 0$, $y = 0$, $x = 2$.

(c) the area in the first quadrant enclosed by the astroid $x = a\cos^3\theta$, $y = a\sin^3\theta$.

(d) the area defined by $1 \leqslant x \leqslant 2$ and $\dfrac{1}{x} \leqslant y \leqslant 1$.

6) Evaluate each of the areas defined in Question 5.

7) Use the results of Questions 5 and 6 to find the y coordinate of the centroid of each of the areas described.

8) Each of the areas described in Question 5 rotates completely about the x axis to form a solid of revolution.
Use Pappus' second theorem, together with your answers to Questions 6 and 7 to find the volume of each solid.

9) Find the first moment, about the y axis, of the arc of the parabola $y = (1 - x^2)$ for which $x \geqslant 0$. Find also the length of this arc and hence find the centroid of the arc. Use Pappus' first theorem to find the area of the surface generated when the arc rotates completely about the y axis.

10) Find the volume of the solid generated when the part of the ellipse $x^2 + 4y^2 = 4$ that lies above the x axis, rotates completely about the x axis. Use Pappus' second theorem to find the centroid of this area

$\left(\text{the area of the ellipse } \dfrac{x^2}{a^2} + \dfrac{y^2}{b^2} = 1 \text{ is } \pi ab\right)$.

11) The line $4y = 3x$ between the origin and the point $(4, 3)$ rotates completely about the x axis generating the surface of a cone. Write down the coordinates of the centroid of the line. Use Pappus' first theorem to find the surface area of the cone.

12) Apply Pappus' second theorem to the rotation of a rectangular area about one of its sides, to derive the formula for the volume of a cylinder.

SECOND MOMENT OF AREA

Consider an element in the xy plane, whose area is δA.

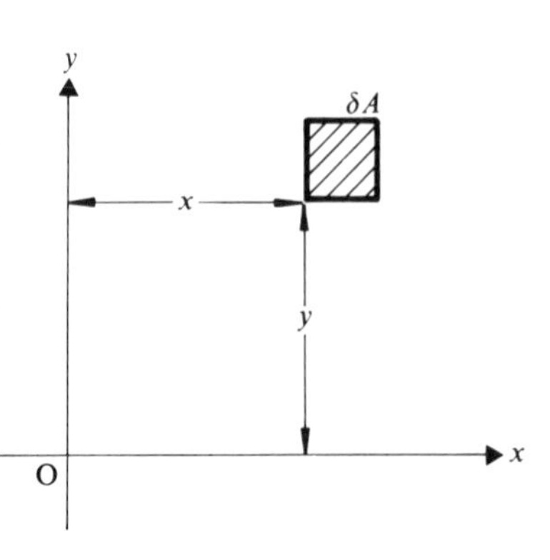

The second moment of the element about the x axis is defined as

$$y^2 \, \delta A.$$

Similarly the second moment of the element about the y axis is defined as

$$x^2 \, \delta A.$$

From these definitions, the second moment of a thin strip parallel to Ox (or to Oy) can be determined.

Second Moment of a Strip

Consider, for example, a thin vertical strip of width d and height h with its lower end on the x axis.

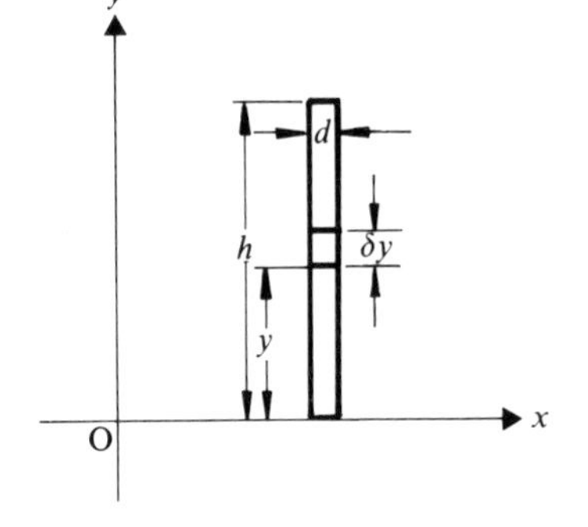

If an element of length δy is at a distance y from the x axis, then the second moment of this element about Ox is approximately

$$y^2 (d \, \delta y)$$

The second moment of the whole strip can then be found by summing the second moments of all the elements in the strip

i.e.
$$\sum_{y=0}^{y=h} y^2 (d \, \delta y) = d \sum_{y=0}^{y=h} y^2 \, \delta y$$

Taking the limit as $\delta y \to 0$ of this summation, we have

$$\text{second moment of strip about O}x = d \int_0^h y^2 \, dy$$

$$= \frac{dh^3}{3}$$

But the area of the strip is hd.

So its second moment becomes $\dfrac{Ah^2}{3}$.

If we now consider a horizontal strip of width d and length h, we find in a similar way that its second moment about Oy is also $\dfrac{Ah^2}{3}$ (the reader is recommended to verify this fact).
Thus, in general,

> the second moment of a thin strip about a perpendicular line through one end of the strip, is
>
> $\frac{1}{3}$(area of strip)(length of strip)^2.

Note that this result is quotable unless proof is requested.

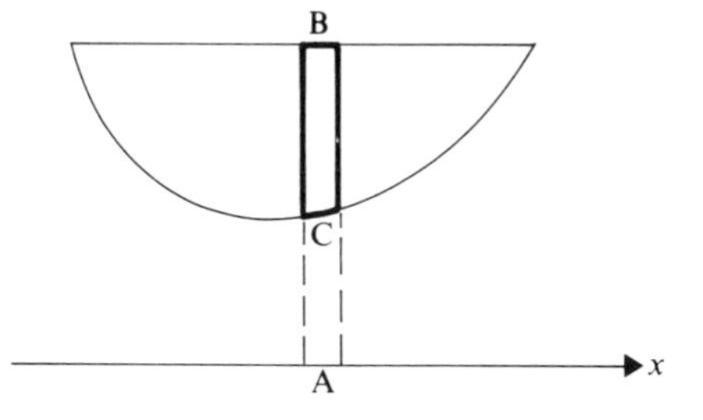

It can also be shown that for a strip whose end is *not* on the x axis (e.g. the strip BC in the diagram) that the second moment of BC about Ox

= the second moment of BA about Ox

− the second moment of AC about Ox.

Second Moment of an Area made up of Strips

Consider an area bounded by perpendicular straight lines and a curve. Such an area can be divided into strips parallel to the two sides, as shown. The second moment of the area about the base AB can therefore be found by summing the second moments of the strips about AB.

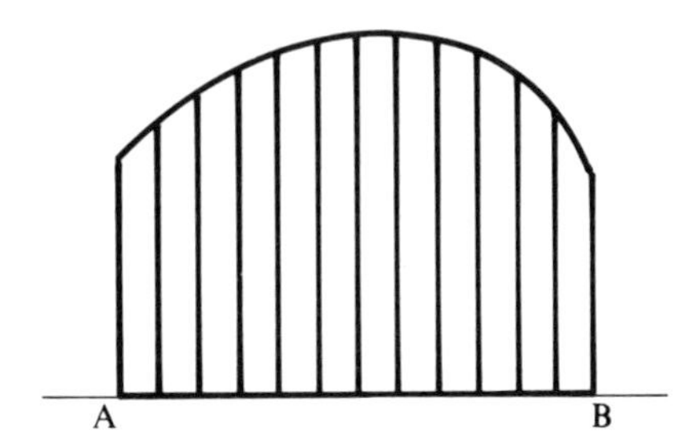

EXAMPLES 14d

1) Find the second moment about the x axis of the area defined by $0 \leqslant x \leqslant 4$ and $0 \leqslant y \leqslant x^2 + 1$.

One thin vertical element of the specified area is approximately a strip of length y and width δx.
So the area δA of the element is given by

$$\delta A \simeq y\,\delta x.$$

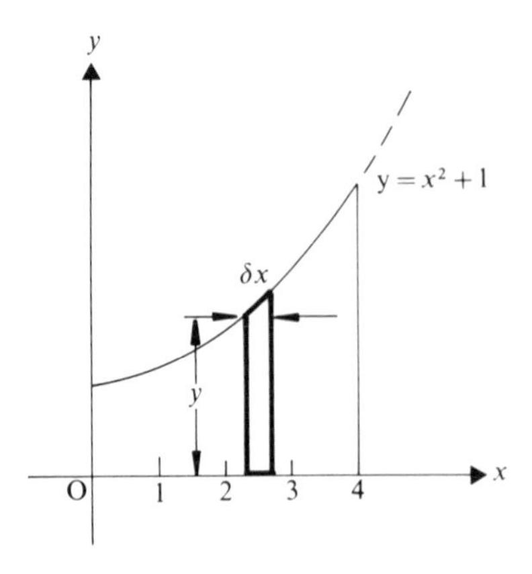

Using the result proved above, the second moment about Ox of the element can be written in the form $\frac{1}{3}(\delta A)y^2 \simeq \frac{1}{3}(y\,\delta x)y^2$.
Taking the limit of the sum of the second moments of all the elements gives,

$$\text{second moment of area about O}x = \lim_{\delta x \to 0} \sum_{x=0}^{x=4} \tfrac{1}{3}y^3\,\delta x$$

$$= \tfrac{1}{3}\int_0^4 y^3\,\mathrm{d}x$$

But $y = x^2 + 1$, so

$$\tfrac{1}{3}\int_0^4 y^3\,\mathrm{d}x = \tfrac{1}{3}\int_0^4 (x^6 + 3x^4 + 3x^2 + 1)\,\mathrm{d}x$$

$$= \tfrac{1}{3}\left[\tfrac{1}{7}x^7 + \tfrac{3}{5}x^5 + x^3 + x\right]_0^4$$

Thus the second moment about Ox of the specified area is 1007.7

2) Find the second moment about the y axis of the area bounded by the parabola $y^2 = x$ and the line $x = 9$.

Dividing the given area into horizontal strips such as BC (see diagram overleaf) and producing this strip to meet the y axis at A, we use,

second moment about Oy of BC

= second moment about Oy of AC

− second moment about Oy of AB.

The width of the strip is δy and the lengths of AB and AC are x and 9 respectively, hence

$$\text{second moment about O}y\text{ of BC} = \tfrac{1}{3}(9\,\delta y)9^2 - \tfrac{1}{3}(x\,\delta y)\,x^2$$

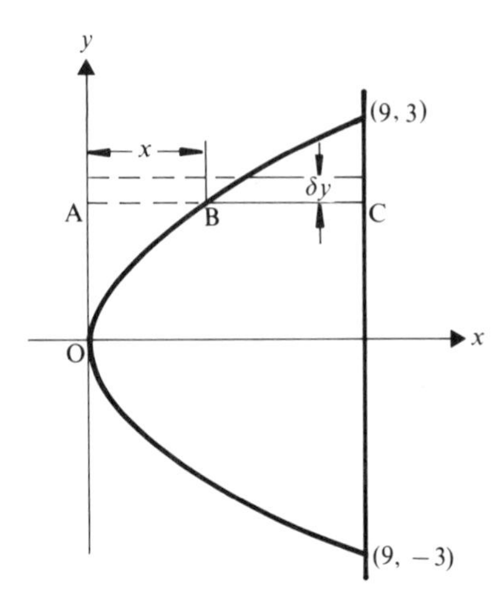

Hence the second moment about Oy of the specified area is given by

$$\lim_{\delta y \to 0} \sum_{y=-3}^{y=3} \tfrac{1}{3}(9^3 - x^3)\,\delta y = \tfrac{1}{3}\int_{-3}^{3} (9^3 - x^3)\,dy$$

$$= \tfrac{1}{3}\int_{-3}^{3} (9^3 - y^6)\,dy$$

$$= \tfrac{1}{3}\left[9^3 y - \tfrac{1}{7}y^7\right]_{-3}^{3}$$

Hence the required second moment is 1249.7

SECOND MOMENT OF VOLUME

If a small element of volume δV is at a distance h from a fixed line, then,

the second moment of volume of the element is defined as $h^2 \delta V$

By summing the second moments of the elements it contains, the second moment of a solid body can be found.

Second Moment of a Ring

Consider, first, a thin, narrow ring of radius r.

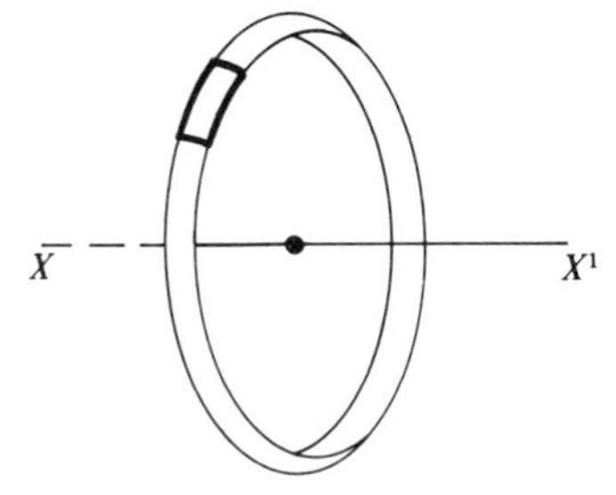

If we take the fixed line through the centre of, and perpendicular to, the ring, then every element of the ring is distant r from this line, XX'. So the second moment of the element about XX' is $r^2 \delta V$.

Hence the second moment of the whole ring is given by $\Sigma r^2 \delta V$.
But r^2 is constant, so

$$\sum r^2 \delta V = r^2 \sum \delta V = Vr^2,$$

i.e. the second moment of a ring about a perpendicular axis through the centre is Vr^2.

We can now use this result to find the second moment of objects made up of ring elements.

Second Moment of a Disc

A disc of radius r and width d, can be divided into ring elements. A typical element has radius x, width d and thickness δx. So its volume, δV, is given by

$$\delta V \simeq (2\pi x)(\delta x)(d).$$

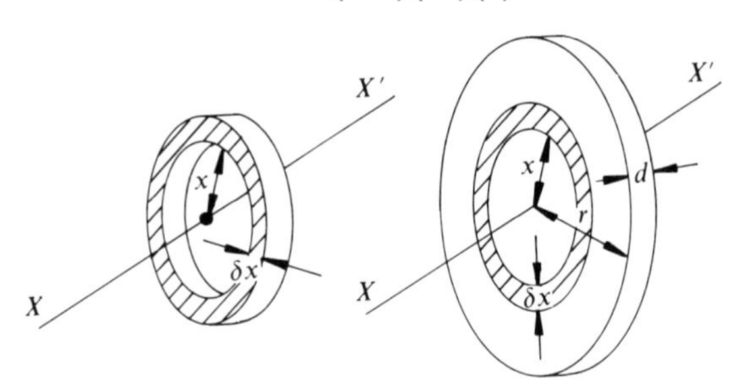

Its second moment about the line XX', through the centre of the disc and perpendicular to it, is therefore

$$x^2 \delta V \simeq 2\pi d x^3 \, \delta x.$$

Hence the second moment of the disc is

$$\lim_{\delta x \to 0} \sum_{x=0}^{x=r} x^2 \delta V \simeq \lim_{\delta x \to 0} \sum_{x=0}^{x=r} 2\pi d x^3 \, \delta x$$

$$= 2\pi d \int_0^r x^3 \, dx$$

$$= \frac{\pi d r^4}{2}$$

But the volume V of the disc is $\pi r^2 d$.

So the second moment of the disc about XX' is $\dfrac{Vr^2}{2}$

Using this result, we can find the second moment of a solid made up of discs.

SECOND MOMENT OF A SOLID OF REVOLUTION

Consider the area bounded by the x axis, the lines $x = a$ and $x = b$, and the curve $y = f(x)$. When this area rotates through a complete revolution about Ox, a solid is formed that can be divided into thin disc-like elements, of width δx, perpendicular to Ox.

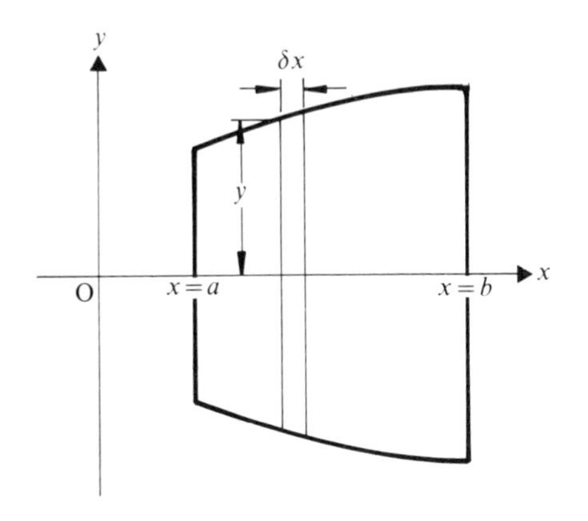

The second moment about Ox of one such element is

$$\frac{Vr^2}{2} \simeq (\pi y^2\,\delta x)\frac{y^2}{2}$$

The second moment about Ox of the whole solid can now be found by summation, and is

$$\lim_{\delta x \to 0} \sum_{x=a}^{x=b} \tfrac{1}{2}\pi y^4\,\delta x = \tfrac{1}{2}\pi \int_a^b y^4\,dx.$$

For a particular solid of revolution, the equation $y = f(x)$ is used to evaluate this integral.

EXAMPLES 14d (continued)

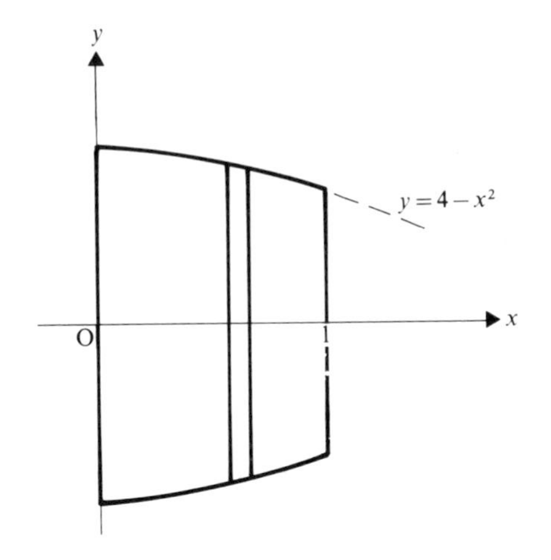

3) Find the second moment about the x axis of the solid generated when the area bounded by the x and y axes, the line $x = 1$ and the curve $y = 4 - x^2$, rotates completely about Ox.

About Ox, the second moment of the solid of revolution generated is

$$\frac{\pi}{2}\int_0^1 y^4\,dx = \frac{\pi}{2}\int_0^1 (4 - x^2)^4\,dx$$

$$= \frac{\pi}{2}\int_0^1 (256 - 256x^2 + 96x^4 - 16x^6 + x^8)\,dx$$

$$= 294.8$$

4) The area enclosed between the curve $y^2 = x$ and the line $x = 4$ rotates completely about the line $x = 4$. Find the second moment about this line of the solid generated.

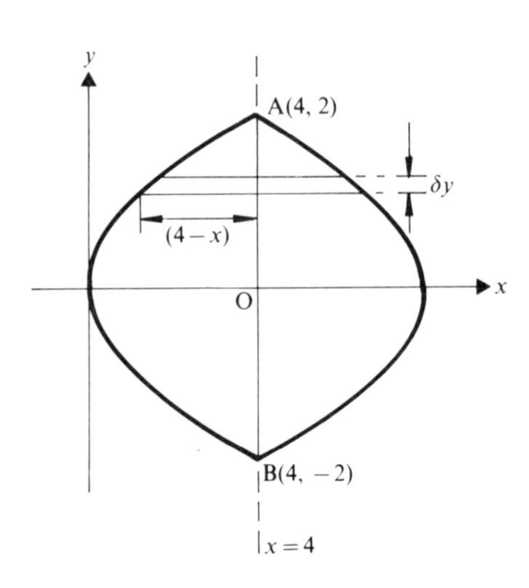

This solid can be divided into horizontal disc elements of radius $r = (4-x)$ and thickness δy.

So the second moment of one element about AB is $\dfrac{Vr^2}{2}$ where V is the volume of the element,

i.e. $$V \simeq \pi(4-x)^2\,\delta y$$

Hence the second moment about AB of the whole solid is given by

$$\lim_{\delta y \to 0} \sum_{y=-2}^{y=2} \{\pi(4-x)^2\,\delta y\}\left\{\frac{(4-x)^2}{2}\right\} = \frac{\pi}{2}\int_{-2}^{2} (4-x)^4\,dy$$

$$= \frac{\pi}{2}\int_{-2}^{2} (4-y^2)^4\,dy$$

$$= 653.6$$

EXERCISE 14d

Find the second moment, about the specified line, of each of the following areas.

1) A rectangle ABCD where $AB = 2a$ and $BC = 2b$,
(a) about AB, (b) about BC.

2) A triangle ABC in which $\angle ABC = 90^\circ$ and $AB = BC = l$, about AB.

3) The area enclosed by the x axis, and the curve $y = \cos x$ from $x = -\pi/2$ to $\pi/2$, about the x axis.

4) The area defined by $0 \leqslant x \leqslant 2$ and $0 \leqslant y \leqslant x^2$, about the x axis.

5) The area between the y axis and the curve $y = \ln x$ from $x = 1$ to $x = \mathrm{e}$, about the y axis.

6) The area bounded by the curve $xy = 4$ and the lines $x = 1$ and $y = 1$, about the line $x = 1$.

7) The area enclosed between the line $x = 3$ and the parabola $y^2 = 3x$
(a) about the line $x = 3$, (b) about the y axis.

Find the second moment, about its axis of symmetry, of each of the following solids of revolution.

8) A cylinder of radius a and length l.

9) A sphere of radius a.

10) A cone of base radius r and perpendicular height h.
(*Hint*. Use the vertex as the origin and the axis of symmetry as the x axis.)

11) The volume generated when the area enclosed between the x axis and the curve $y = 1 - x^2$ rotates about the x axis.

12) The volume generated when the area enclosed between the y axis, the curve $y = \mathrm{e}^x$ and the line $y = \mathrm{e}$ rotates
(a) about the y axis, (b) about the line $y = \mathrm{e}$.

13) The volume generated when the area between the x axis and the curve $y = \sin x$ from $x = 0$ to $x = \dfrac{\pi}{2}$ rotates about the x axis.

14) By dividing the surface into ring elements, find the second moment, about the axis of symmetry, of
(a) a cylindrical surface of radius r and length l,
(b) a conical surface of slant length l and semi-vertical angle 30°.

CURVATURE

The curvature of a curve is a measure of its rate of turning.
Suppose that, at a point P on a curve, the tangent makes an angle ψ with the x axis, and at an adjacent point Q, the angle between the tangent and x axis is $\psi + \delta\psi$.

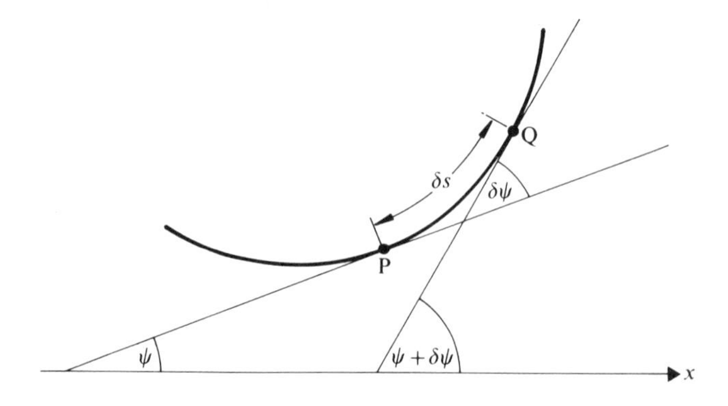

The curvature, κ, at P is defined by

$$\kappa = \lim_{\delta s \to 0} \frac{\delta \psi}{\delta s} = \frac{d\psi}{ds}$$

RADIUS OF CURVATURE

If, at the points P and Q described above, lines are drawn at right angles to the tangents at these points, to meet at C, then C is the centre of a circle passing through P and Q.

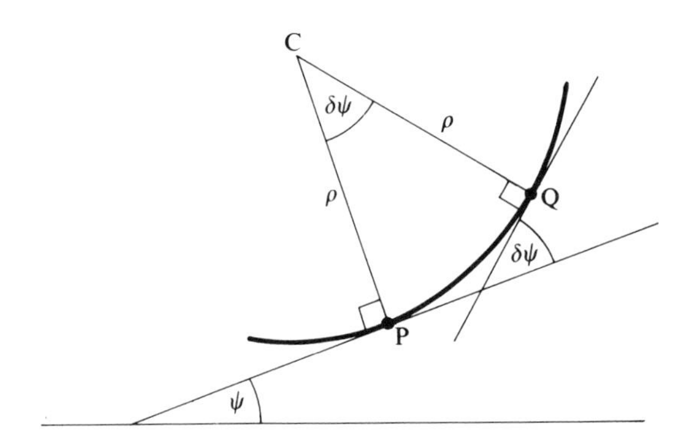

The radius of this circle is denoted by ρ (where $\rho = \text{CP} = \text{CQ}$),
so the length of the *circular* arc between P and Q is $\rho\,\delta\psi$.
Now as Q approaches P, the length of the circular arc PQ approaches δs,

i.e.
$$\rho\,\delta\psi \simeq \delta s$$

$\Rightarrow$
$$\rho \simeq \frac{\delta s}{\delta \psi}$$

In the limiting case, when $\delta s \to 0$, we have

$$\rho = \frac{ds}{d\psi} = \frac{1}{\kappa}$$

and ρ is called the *radius of curvature* at P of the given curve.
Thus the radius of curvature at a point on a curve is the reciprocal of the curvature at that point.

CALCULATION OF RADIUS OF CURVATURE

(a) Cartesian Form

The gradient of the tangent at P to the given curve is $\tan\psi$,

i.e. $$\frac{dy}{dx} = \tan\psi$$

$$\Rightarrow \quad \frac{d^2y}{dx^2} = \frac{d}{dx}(\tan\psi) = \sec^2\psi\,\frac{d\psi}{dx} \qquad [1]$$

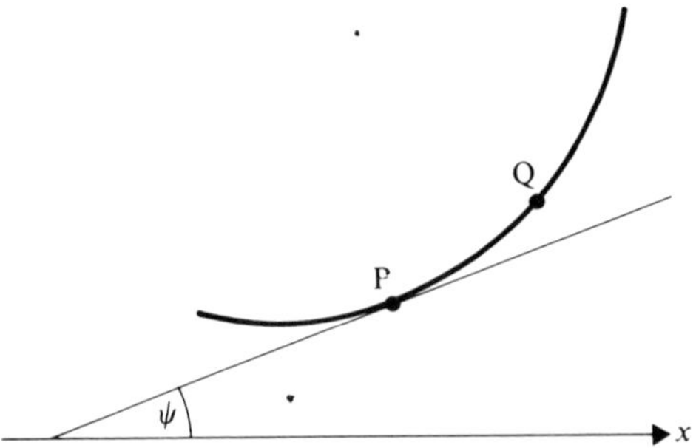

If we consider again the arc PQ of the curve, we see that, when δs is small,

$$\delta x \simeq \delta s \cos\psi$$

or $$\frac{\delta x}{\delta s} \simeq \cos\psi$$

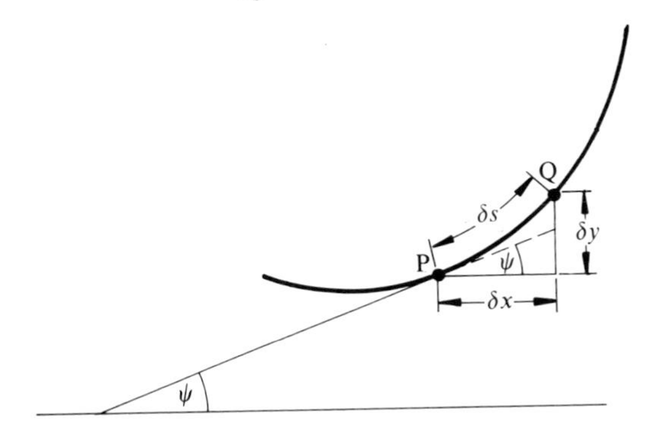

So, when $\delta s \to 0, \quad \dfrac{dx}{ds} = \cos\psi$ [2]

Using results [1] and [2] we can now find an expression for ρ,

i.e. $$\rho = \frac{ds}{d\psi} = \left(\frac{ds}{dx}\right)\left(\frac{dx}{d\psi}\right)$$

$$= (\sec\psi)\left(\sec^2\psi \Big/ \frac{d^2y}{dx^2}\right).$$

But $$\sec^2\psi \equiv 1 + \tan^2\psi = 1 + \left(\frac{dy}{dx}\right)^2$$

So $$\rho = \left[1 + \left(\frac{dy}{dx}\right)^2\right]^{\frac{3}{2}} \Big/ \frac{d^2y}{dx^2}$$

(b) Parametric Form

The Cartesian formula for ρ given above can be adapted for use when the equations of a curve are parametric.
Suppose that $y = f(t)$ and $x = g(t)$,

then $$\frac{dy}{dx} = \frac{f'(t)}{g'(t)}$$

and $$\frac{d^2y}{dx^2} = \frac{1}{f'(t)}\frac{d}{dt}\left(\frac{f'(t)}{g'(t)}\right)$$ (Volume I, p. 280).

Hence, as both $\dfrac{dy}{dx}$ and $\dfrac{d^2y}{dx^2}$ can be found (in terms of t), the Cartesian formula can be used.

(c) Polar Form

If P is a point (r, θ) on a curve with a polar equation, the tangent at P is inclined at ψ to the initial line, and ϕ is the angle between this tangent and the radius vector OP, then $\psi = \theta + \phi$.

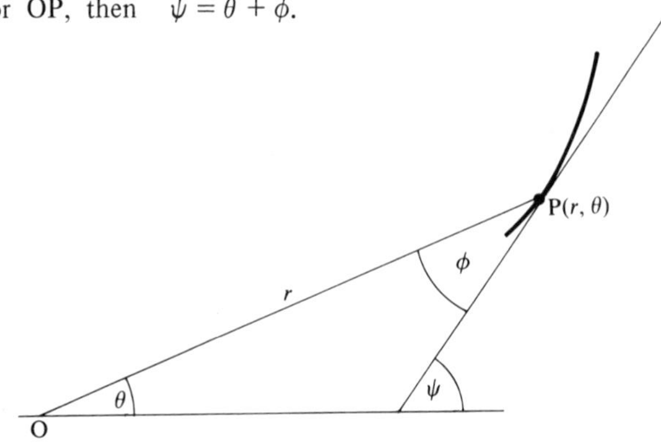

$Q(r+\delta r, \theta+\delta\theta)$ is a point close to P and PR is perpendicular to OQ.

Hence $$RQ \simeq \delta r \quad \text{and} \quad PR \simeq r\,\delta\theta$$

$$\Rightarrow \qquad \tan OQP \simeq \frac{r\,\delta\theta}{\delta r}$$

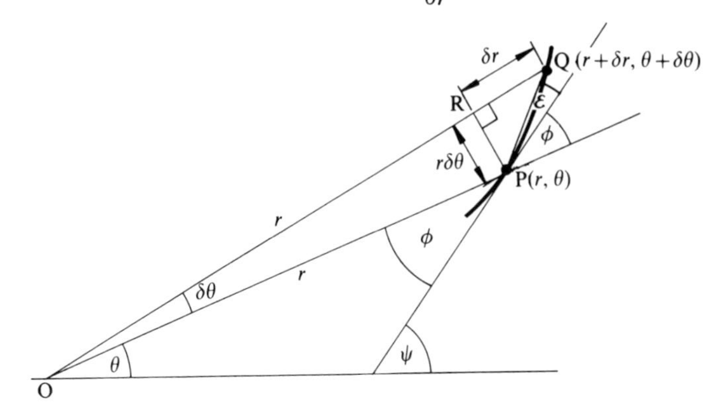

Also, if ϵ is the angle between the tangent at P and the chord PQ, then

$$\phi + \epsilon = OQP + \delta\theta$$

So $$\tan(\phi + \epsilon - \delta\theta) \simeq \frac{r\,\delta\theta}{\delta r}$$

But, as $\delta\theta \to 0$, $\dfrac{\delta\theta}{\delta r} \to \dfrac{d\theta}{dr}$ and $\epsilon \to 0$ (the chord PQ merges with the tangent at P when $\delta\theta \to 0$).

So $$\tan\phi = r\frac{d\theta}{dr}$$

Differentiating w.r.t. r, gives

$$\sec^2\phi \frac{d\phi}{dr} = \frac{d\theta}{dr} + r\frac{d^2\theta}{dr^2} \qquad [1]$$

Now differentiating $\psi = \theta + \phi$ w.r.t. r, we have

$$\frac{d\psi}{dr} = \frac{d\theta}{dr} + \frac{d\phi}{dr} \qquad [2]$$

Also, from 'triangle' PQR above, we see that

$$(\delta s)^2 \simeq (r\,\delta\theta)^2 + (\delta r)^2$$

$$\Rightarrow \qquad \left(\frac{ds}{dr}\right)^2 = \left(r\frac{d\theta}{dr}\right)^2 + 1 \qquad [3]$$

These three expressions can now be used to find ρ for a curve with a polar equation, as

$$\rho = \frac{ds}{d\psi} = \left(\frac{ds}{dr}\right)\left(\frac{dr}{d\psi}\right)$$

and
$$\begin{cases} \dfrac{ds}{dr} & \text{can be obtained from [3],} \\ \dfrac{dr}{d\psi} & \text{can be obtained from [1] and [2].} \end{cases}$$

EXAMPLE 14e

Find the radius of curvature at the point P if

(a) P is the point (x, y) on the curve $y = x^2$,

(b) P is the point $(\sin^3\theta, \cos^3\theta)$ on the astroid $x = \sin^3\theta,\quad y = \cos^3\theta$,

(c) P is the point $\left(3, \dfrac{\pi}{3}\right)$ on the cardioid $r = 2(1 + \cos\theta)$.

(a) Using the quotable Cartesian formula for ρ we have,

$$\rho = \left[1 + \left(\frac{dy}{dx}\right)^2\right]^{\frac{3}{2}} \Big/ \frac{d^2y}{dx^2}$$

where $$y = x^2 \Rightarrow \frac{dy}{dx} = 2x \Rightarrow \frac{d^2y}{dx^2} = 2$$

So $$\rho = \tfrac{1}{2}(1 + 4x^2)^{\frac{3}{2}} \quad \text{at any point on} \quad y = x^2.$$

(b) If $x = \sin^3\theta$ and $y = \cos^3\theta$ then

$$\frac{dy}{dx} = \frac{3\cos^2\theta(-\sin\theta)}{3\sin^2\theta(\cos\theta)} = -\cot\theta$$

and
$$\frac{d^2y}{dx^2} = \frac{d}{dx}(-\cot\theta) = \frac{d}{d\theta}(-\cot\theta)\frac{d\theta}{dx}$$

$$= \frac{\operatorname{cosec}^2\theta}{3\sin^2\theta\cos\theta}$$

$$= \frac{1}{3\sin^4\theta\cos\theta}$$

Then
$$\rho = [1 + (-\cot\theta)^2]^{\frac{3}{2}}(3\sin^4\theta\cos\theta)$$
$$= (\operatorname{cosec}^3\theta)(3\sin^4\theta\cos\theta)$$

So $\rho = 3\sin\theta\cos\theta$ at any point on the given astroid.

(c) For the cardioid $r = 2(1 + \cos\theta)$ we have:

$$\tan\phi = r\frac{d\theta}{dr} = r\Big/\frac{dr}{d\theta}$$

$$= 2(1 + \cos\theta)/(-2\sin\theta)$$

$$= -\cot\frac{\theta}{2}$$

Hence $$\phi = \frac{\pi}{2} + \frac{\theta}{2}$$

$\Rightarrow$ $$\frac{d\phi}{dr} = \frac{1}{2}\frac{d\theta}{dr} = \frac{1}{2(-2\sin\theta)}$$

When $$\theta = \frac{\pi}{3}, \quad \frac{d\theta}{dr} = -\frac{1}{2\sqrt{3}}$$

Now using $\psi = \theta + \phi$, we have

$$\frac{d\psi}{dr} = \frac{d\theta}{dr} + \frac{d\phi}{dr}$$

$$= \frac{-1}{2\sin\theta} + \frac{d\phi}{dr}$$

So, when $\theta = \frac{\pi}{3}$, $$\frac{d\psi}{dr} = -\frac{1}{\sqrt{3}} - \frac{1}{2\sqrt{3}} = -\frac{\sqrt{3}}{2}.$$

Also $$\left(\frac{ds}{dr}\right)^2 = \left(r\frac{d\theta}{dr}\right)^2 + 1$$

$$= \left(\frac{r}{-2\sin\theta}\right)^2 + 1$$

When $\theta = \frac{\pi}{3}$ and $r = 3$, $\left(\frac{ds}{dr}\right)^2 = 4 \Rightarrow \frac{ds}{dr} = 2$

Now we are in a position to find ρ at the point $\left(3, \frac{\pi}{3}\right)$ using

$$\rho = \frac{ds}{d\psi} = \left(\frac{ds}{dr}\right)\left(\frac{dr}{d\psi}\right) = (2)\left(-\frac{2}{\sqrt{3}}\right)$$

The sign obtained for ρ serves only to indicate whether the curve is concave or convex at the given point.

So the radius of curvature of the cardioid at the point $\left(3, \frac{\pi}{3}\right)$ is $\frac{4\sqrt{3}}{3}$

Note. A rigorous explanation of the significance of the sign of ρ is beyond the scope of this book.

EXERCISE 14e

Find the radius of curvature of each of the following curves at the specified point.

1) $y = 2x^2$; $x = 1$

2) $y = \cos x$; $x = \dfrac{\pi}{3}$

3) $y = e^{-x}$; $x = 0$

4) $y = \ln x$; $x = 2$

5) $y^2 = 4ax$; $(a, 2a)$

6) $xy = 1$; $(1, 1)$

7) $x^2 - y^2 = 1$; $(1, 0)$

8) $y = \cosh x$; $x = 1$

9) $x = t^2$, $y = 2t$; $t = 1$

10) $x = a \cos\theta$, $y = b \sin\theta$; $\theta = \dfrac{\pi}{6}$

11) $x = ct$, $y = \dfrac{c}{t}$; $t = 2$

12) $x = a \cos^3\theta$, $y = a \sin^3\theta$; $\theta = \dfrac{\pi}{4}$

13) $x = 2 \cosh u$, $y = \sinh u$, $u = 0$

14) $x = a \cos\theta$, $y = a \sin\theta$; $\theta = \dfrac{\pi}{2}$

Find the curvature of each of the following curves at the specified point.

15) $y^2 = x^3$; $(1, 1)$

16) $x = 2t^2$, $y = 4t$; $(8, 8)$

17) $r = a \cos\theta$; $\theta = \dfrac{\pi}{4}$

18) $r = a \sin 3\theta$; $\theta = \dfrac{\pi}{12}$

19) $r = a(1 + \cos\theta)$; $\theta = 0$

20) $r^2 = a^2 \cos 2\theta$; (r, θ)

Find the angle ϕ between the radius vector OP and the tangent at $P(r, \theta)$ for each of the following curves.

21) $r = a \cos 2\theta$

22) $r = a(1 - \sin\theta)$

23) $r = a\,e^{\theta}$

24) $r = a \sin 3\theta$

25) Prove that the tangents to the cardioid $r = a(1 + \cos\theta)$ at the points $\theta = \alpha$, $\theta = \alpha + \frac{2}{3}\pi$ and $\theta = \alpha + \frac{4}{3}\pi$, are all parallel.

MISCELLANEOUS EXERCISE 14

1) The region bounded by the x axis, the ordinates $x = 2$ and $x = 6$ and the arc of the parabola $y = \sqrt{x}$ between these ordinates, is rotated through 360° about the x axis. Find the area of the curved surface of this solid. (JMB)p

2) The tangent at a point P on the curve whose parametric equations are $x = a\left(t - \dfrac{t^3}{3}\right)$, $y = at^2$ cuts the x axis at T. Prove that the distance of the point T from the origin O is one half of the length of the arc OP.
(AEB '67)

3) Find the area enclosed by the curve $y = \cosh x$, the ordinate at $x = \ln 2$, the x axis and the y axis. Find also
(a) the length of the perimeter enclosing this area,
(b) the volume obtained when this area is rotated completely about the x-axis.
(AEB '74)

4) Show that the length of the arc of the curve

$$y = x^2$$

between the origin and the point $(1, 1)$ is equal to $\frac{1}{4}(2\sqrt{5} + \sinh^{-1}2)$. (JMB)

5) The curve $y = \frac{1}{4}(e^{2x} + e^{-2x})$ meets the y axis at A, and the tangent at a variable point P on the curve is inclined to the x axis at an angle ψ. Show that the length of the curve between the points A and P is $\frac{1}{2}\tan\psi$. (AEB '67)

6) The parametric equations of a curve are

$$x = a(2\cos t - \cos 2t), \qquad y = a(2\sin t - \sin 2t)$$

where a is a constant.
Show that the angle between the tangent at the point whose parameter is t and the positive direction of the x axis is $\dfrac{3t}{2}$. Find the length of the arc of the curve for which $0 \leqslant t \leqslant \pi$. Find also the area of the surface formed when this arc is revolved once about the x axis. (JMB)

7) (a) The curve given parametrically by the equations $x = \sin^3 t$, $y = \cos^3 t$ is rotated through 180° about the x axis. Calculate the area of the surface generated.
(b) Find the radius of curvature of the curve $y = 2\cosh\frac{1}{2}x$ at the point where $x = 2$. Find also the length of the curve from the point where $x = -2$ to the point where $x = 2$. (AEB '78)

8) Find the second moment about the y axis of the area enclosed by the ellipse $4x^2 + 9y^2 = 36$. (AEB'78)p

9) A curve is given by the parametric equations

$$x = a\cos^3 t, \quad y = a\sin^3 t \qquad (-\pi < t \leqslant \pi)$$

Show that near $t = 0$ the equations reduce approximately to

$$a - x \simeq 3at^2/2, \quad y \simeq at^3.$$

Sketch the curve.
Show that the total length of the curve is $6a$.

Find the area of the surface generated when the part of the curve which is in the first quadrant is rotated once about the x axis. (JMB)

10) A surface of revolution is formed by rotating completely about the x axis the arc of the parabola $x = at^2$, $y = 2at$ from $t = 0$ to $t = \sqrt{3}$.
Show that its surface area, S, is $56\pi a^2/3$.
Show also that the x coordinate, $\bar{x}$, of the centroid of this surface is given by

$$S\bar{x} = 8\pi a^3 \int_0^3 t^3\sqrt{(1+t^2)}\,dt.$$

Hence find $\bar{x}$, using the substitution $u^2 = 1 + t^2$ or otherwise. (U of L)

11) Find the second moment about the y axis of the area enclosed by the curve $y^2 = x$ and the straight line $x = 2$.
That portion of the curve $y^2 = x$ on which $0 \leqslant y \leqslant \sqrt{2}$ is rotated completely about the x axis to form a shell of revolution. Show that the curved surface area of the shell is $\dfrac{13\pi}{3}$. (AEB '73)

12) Find the length of the arc of the curve $ay^2 = x^3$ between the origin and the point on the curve where $x = 5a$.
Show that, when this arc is rotated about the axis of x, the volume enclosed by the surface so formed and the plane $x = 5a$ is $625\pi a^3/4$. (JMB)p

13) A curve is given by the parametric equations

$$x = 4\cos t + \cos 2t,$$
$$y = \sin 2t + 4\sin t + 2t.$$

Find the length of the curve between the points $t = 0$ and $t = \pi/4$. Find the smallest positive value of t for which the radius of curvature of the curve has the absolute value 6. (AEB '72)

14) Find the root mean square value of $\sin x$ with respect to x for $0 \leqslant x \leqslant \pi$. (U of L)

15) Use Pappus' theorem for areas to find, for a semi-circular arc of radius r, the distance of the centroid from the bounding diameter. (U of L)

16) Given that the area of the ellipse $\dfrac{x^2}{a^2} + \dfrac{y^2}{b^2} = 1$ is πab and that the volume generated when this area rotates through an angle π about the x axis is $\frac{4}{3}\pi ab^2$, use Pappus' theorem to find
(a) the centroid,
(b) the first moment about Ox of that part of the area for which $y \geqslant 0$.

17) A metal ring is cast in the shape of the solid obtained when a square ABCD of side 1 cm rotates completely about a line parallel to AB and distant 10 cm from AB. Find
(a) the volume,
(b) the total surface area of the ring.

18) Find the coordinates of all the points of inflexion of the graph of the function

$$y = x - \sin x$$

in the range $0 \leqslant x \leqslant 4\pi$. Sketch the graph in this range.
The region bounded by the x axis, the line $x = 4\pi$ and the curve whose equation is

$$y = x - \sin x$$

is rotated once about the x axis. Prove that the volume swept out is $\frac{2}{3}\pi^2(32\pi^2 + 15)$.
Find the mean value of y^2 with respect to x over the range $0 \leqslant x \leqslant 4\pi$.

(JMB)

ANSWERS

Exercise 1a – p. 2

1) $\begin{pmatrix}2\\9\end{pmatrix}$ 2) $\begin{pmatrix}0\\8\end{pmatrix}$ 3) $\begin{pmatrix}1\\7\end{pmatrix}$

4) $\begin{pmatrix}1\\6\end{pmatrix}$ 5) $\begin{pmatrix}\frac{1}{2}\\1\end{pmatrix}$ 6) $\begin{pmatrix}-2\\-1\end{pmatrix}$

7) $\begin{pmatrix}-1\\6\end{pmatrix}$ 8) $\begin{pmatrix}0\\2\end{pmatrix}$ 9) $\begin{pmatrix}\frac{1}{3}\\2\end{pmatrix}$

Exercise 1b – p. 6

1) a) 11 b) -22 c) -3

2) a) $\begin{pmatrix}8\\2\end{pmatrix}$ b) $\begin{pmatrix}-5\\-9\end{pmatrix}$ c) $\begin{pmatrix}-4\\6\end{pmatrix}$

3) $\begin{pmatrix}1\\0\end{pmatrix}, \begin{pmatrix}2\\0\end{pmatrix}, \begin{pmatrix}2\\-3\end{pmatrix}$; reflection in O$x$

4) $\begin{pmatrix}-1\\0\end{pmatrix}, \begin{pmatrix}-2\\0\end{pmatrix}, \begin{pmatrix}-2\\3\end{pmatrix}$;
reflection in Oy

Exercise 1c – p. 6

1) a) $\begin{pmatrix}-1 & 0\\0 & 1\end{pmatrix}$ b) $\begin{pmatrix}\frac{1}{2}\sqrt{2} & -\frac{1}{2}\sqrt{2}\\\frac{1}{2}\sqrt{2} & \frac{1}{2}\sqrt{2}\end{pmatrix}$

c) $\begin{pmatrix}1 & 0\\0 & 2\end{pmatrix}$ d) $\begin{pmatrix}-\frac{3}{5} & \frac{4}{5}\\\frac{4}{5} & \frac{3}{5}\end{pmatrix}$

e) $\begin{pmatrix}1 & 0\\\frac{1}{3}\sqrt{3} & 1\end{pmatrix}$ f) $\begin{pmatrix}3 & 0\\0 & 1\end{pmatrix}$

g) $\begin{pmatrix}3 & 0\\0 & 3\end{pmatrix}$ h) $\begin{pmatrix}2 & 0\\0 & -1\end{pmatrix}$

2) a) reflection in y axis
b) rotation of 180°
c) plane maps to the line $y = x$
d) no change
e) enlargement and shear 11 ∥ to Ox
f) plane maps to x axis
g) plane maps to the origin
h) rotation of $90° - \theta$
i) shear 11 ∥ to negative y axis

3) a) A′(2, 2) B′(4, 2) C′(4, 8)
b) A′(− 3, − 3) B′(− 6, − 3) C′(− 6, − 12)
c) A′(− 1, 3) B′(− 2, 3) C′(− 2, 12)
d) A′(1, 1) B′(2, 2) C′(2, 2)
e) A′(3, 7) B′(4, 10) C′(10, 22)
f) A′(2, − 3) B′(2, − 3) C′(8, − 12)

4) a) $\begin{pmatrix}-\frac{7}{25} & \frac{24}{25}\\\frac{24}{25} & \frac{7}{25}\end{pmatrix}$ b) $\begin{pmatrix}-\frac{3}{5} & -\frac{4}{5}\\-\frac{4}{5} & \frac{3}{5}\end{pmatrix}$

5) a) $\begin{pmatrix}\frac{1}{2} & -\sqrt{3}/2\\\sqrt{3}/2 & \frac{1}{2}\end{pmatrix}$

b) $\begin{pmatrix}\sqrt{2}/2 & \sqrt{2}/2\\-\sqrt{2}/2 & \sqrt{2}/2\end{pmatrix}$

6) a) $y = 2x$ b) $y = -2$
c) $x + y - 2 = 0$

7) a) $y = x, y + 5x = 0$ b) no real lines
c) $x + y = 0, 3y = x$

Exercise 1d – p. 20

2) **A** and **E**, **b** and **h**, **C** and **D**, **F** and **G**

3) $\begin{pmatrix}-2 & 6 & 10\\14 & -8 & 4\end{pmatrix}$, $(-7 \quad 1 \quad 0)$,

$\begin{pmatrix}-4 & 12\\2 & 2\end{pmatrix}$, does not exist, does not exist, $\begin{pmatrix} -2 & 3 \\ -2 & 12 \\ 7 & -6 \end{pmatrix}$

4) a) $\begin{pmatrix} 4 & 24 & 0 \\ 8 & 4 & 20 \end{pmatrix}$

b) $\begin{pmatrix} 6 & 0 & -20 \\ -24 & 18 & 2 \end{pmatrix}$

c) $\begin{pmatrix} -25 & 45 \\ 0 & -15 \end{pmatrix}$ d) $(0 \quad \lambda \quad 3\lambda)$

5) 1, 3, 7, 0, -1, 2, -5, 0

6) a) $\begin{pmatrix} -7 & 12 & 30 \\ 40 & -25 & 7 \end{pmatrix}$

b) $\begin{pmatrix} -11 & 9 \\ 14 & 1 \\ 21 & 32 \end{pmatrix}$ c) $\begin{pmatrix} 6 & -6 \\ 2 & 8 \end{pmatrix}$

Exercise 1e — p. 22

1) -1 2) -11 3) 13

4) $2p^2 - q + 3r^2$ 5) $\begin{pmatrix} -9 \\ 3 \\ -12 \end{pmatrix}$

6) $\begin{pmatrix} 27 \\ -1 \end{pmatrix}$ 7) $\begin{pmatrix} 15 \\ -18 \\ 6 \end{pmatrix}$ 8) $\begin{pmatrix} -10 \\ 15 \\ 6 \\ -32 \end{pmatrix}$

9) $\begin{pmatrix} -8 \\ -2 \\ 1 \\ 0 \end{pmatrix}$ 10) $\begin{pmatrix} 1 \\ 0 \end{pmatrix}$

11) $\begin{pmatrix} 3x^2 + 6xy \\ 6xy + 2y^2 \end{pmatrix}$ 12) $\begin{pmatrix} 0 \\ \sin 2\theta - 2 \end{pmatrix}$

13) $\begin{pmatrix} 0 \\ -3t^2 \\ 3t^2 \end{pmatrix}$ 14) $\begin{pmatrix} i^2 + j^2 \\ i + j \end{pmatrix}$

Exercise 1f — p. 27

1) $\mathbf{AB} = \begin{pmatrix} -4 & 3 \\ 0 & -6 \end{pmatrix}$,

$\mathbf{BA} = \begin{pmatrix} -2 & 1 \\ -8 & -8 \end{pmatrix}$

2) $\mathbf{AB} = \begin{pmatrix} 20 & 0 \\ 22 & -6 \end{pmatrix}$,

$\mathbf{BA} = \begin{pmatrix} -6 & -14 \\ 0 & 20 \end{pmatrix}$

3) $\mathbf{AB} = \begin{pmatrix} -18 & 8 & -25 \\ 24 & -7 & 26 \end{pmatrix}$

4) $\mathbf{AB} = \begin{pmatrix} 13 & -22 \\ -8 & -2 \\ 4 & -13 \end{pmatrix}$

5) $\mathbf{AB} = \begin{pmatrix} -10 & 5 \\ 26 & -13 \\ 56 & -28 \end{pmatrix}$

6) $\mathbf{AB} = \begin{pmatrix} -3 & 1 & 0 \\ 2 & 0 & 4 \\ 6 & -5 & 3 \end{pmatrix}$,

$\mathbf{BA} = \begin{pmatrix} 1 & -2 & 0 \\ 1 & 0 & 8 \\ 3 & -1 & -1 \end{pmatrix}$

7) $\begin{pmatrix} 10 & -12 & 5 & 2 \\ -3 & 4 & -2 & 0 \end{pmatrix}$

8) $\mathbf{AB} = \begin{pmatrix} -2 & 3 & 13 \\ 1 & 4 & 10 \\ -2 & 0 & 4 \end{pmatrix}$,

$\mathbf{BA} = \begin{pmatrix} 4 & 5 \\ -3 & 2 \end{pmatrix}$

9) $\mathbf{AB} = \begin{pmatrix} 3 & 1 & 3 \\ -2 & -1 & 2 \\ 5 & 6 & 1 \end{pmatrix}$,

$\mathbf{BA} = \begin{pmatrix} 6 & -1 & 11 \\ 0 & -8 & 16 \\ 0 & -2 & 5 \end{pmatrix}$

10) $\mathbf{AB} = \mathbf{BA} = \begin{pmatrix} 1 & 2 & 3 \\ 4 & 5 & 6 \\ 7 & 8 & 9 \end{pmatrix}$

11) $\mathbf{AB} = \begin{pmatrix} 0 & 0 \\ 0 & 0 \end{pmatrix}$,

$\mathbf{BA} = \begin{pmatrix} 50 & 20 \\ -125 & -50 \end{pmatrix}$

Exercise 1g — p. 31

1) $\begin{pmatrix} 0 & 1 \\ 1 & 0 \end{pmatrix}$ 2) $\begin{pmatrix} 0 & -1 \\ -1 & 0 \end{pmatrix}$

3) $\begin{pmatrix} 2 & 0 \\ 0 & 6 \end{pmatrix}$ 4) $\begin{pmatrix} -1 & 0 \\ 0 & 1 \end{pmatrix}$

5) $\begin{pmatrix} 2 & 0 \\ 0 & -2 \end{pmatrix}$ 6) $\begin{pmatrix} 0 & -2 \\ 2 & 0 \end{pmatrix}$

7) $\begin{pmatrix} 0 & 2 \\ 2 & 0 \end{pmatrix}$

Exercise 1h – p. 35

1) $\mathbf{A}^2 = \begin{pmatrix} 1 & 0 \\ 0 & 1 \end{pmatrix}, \quad \mathbf{A}^3 = \begin{pmatrix} 1 & 0 \\ 2 & -1 \end{pmatrix}$

2) $\mathbf{B}^2 = \begin{pmatrix} 2 & 0 & -1 \\ -1 & 1 & 1 \\ -1 & 0 & 1 \end{pmatrix},$

$\mathbf{B}^3 = \begin{pmatrix} 3 & 0 & -2 \\ -2 & 1 & 2 \\ -2 & 0 & 1 \end{pmatrix}$

4) $\mathbf{C} = \begin{pmatrix} 0 & -1 \\ 1 & 0 \end{pmatrix}$

5) $\mathbf{A}^2 = \begin{pmatrix} -1 & 0 \\ 0 & -1 \end{pmatrix}, \quad \mathbf{A}^4 = \begin{pmatrix} 1 & 0 \\ 0 & 1 \end{pmatrix}$

6) $\begin{pmatrix} \cos^2\theta & 0 & 0 \\ 0 & \sin^2\theta & 0 \\ 0 & 0 & \cos^2\theta \end{pmatrix}$

7) If $\mathbf{A}$ is $m \times n$ then $\mathbf{B}$ is $n \times m$. They are square.

8) $\mathbf{M}$ maps all points to the line $X = Y$. $\mathbf{M}^2$ maps all points to the line $X = Y$.

9) $\mathbf{AB}$ maps all points to the origin. $\mathbf{BA}$ maps all points to the line $X + Y = 0$.

10) $\begin{pmatrix} 6 & -2 \\ -10 & 6 \end{pmatrix}$

11) $\begin{pmatrix} \pm 3 & 0 \\ 0 & \pm 3 \end{pmatrix}$ or $\begin{pmatrix} a & b \\ \dfrac{9-a^2}{b} & -a \end{pmatrix}$

for all real values of a and b.

Exercise 1i – p. 39

1) 13 2) -5 3) 0 4) -6

5) A parallelogram of area 13 sq. units, a parallelogram of area 5 sq. units (turned over), a line (area destroyed), a parallelogram of area 6 sq. units (turned over).

6) a) $\begin{pmatrix} -1 & 19 \\ -5 & -3 \end{pmatrix}$ b) 7 c) 14
d) 98; yes

7) 72, 72

8) a) true b) true c) true
d) false.

11) a) 1 b) $xy(y-x)$
c) $a^2 - b^2$ d) $a^2 - ab - b^2$
e) $\ln 2 \ln \frac{6}{25}$ f) $\cos\theta$

13) a) $\frac{7}{11}, \frac{5}{11}$ b) $\frac{17}{29}, \frac{9}{29}$ c) $\frac{5}{26}, \frac{29}{26}$

Exercise 1j – p. 43

1) a) $\begin{vmatrix} 6 & -12 & -8 \\ -5 & 11 & 7 \\ -8 & 18 & 12 \end{vmatrix}$

b) $\begin{vmatrix} -2 & -29 & -7 \\ 0 & -12 & 0 \\ -2 & -11 & -1 \end{vmatrix}$

2) a) 25 b) 14 c) -51

3) a) 16 b) 12

4) a) yes b) no c) yes

5) a) $(\cos\theta - \sin\theta)$
$\times(\cos\theta + \sin\theta - \cos\theta\sin\theta - 1)$
b) $(a-b)(b-c)(c-a)abc$
$= abc[bc(c-b) + ac(c-a)$
$+ ab(b-a)]$
c) $1 - \sin 2\theta$ d) 0

Exercise 1k – p. 51

1) 384 2) -320 3) 505×10^3

4) 1190 5) $x^2(1-x)^2(1+2x)$

6) $-4x$

7) $\cos\theta\sin\theta(\cos\theta - \sin\theta)(1-\sin\theta)$
$\times(1-\cos\theta)$

8) $2(a+1)(a^2 - a + 1)$ 9) -2

10) -3

Multiple Choice Exercise 1 – p. 53

1) c 2) c 3) a 4) d
5) a 6) e 7) c 8) c
9) c 10) b 11) a, b 12) c
13) a 14) a, b, c 15) b
16) *B* 17) *C* 18) *A* 19) *D*
20) *D* 21) *E* 22) *A* 23) *D*
24) *a* 25) b, c 26) *I* 27) *F*
28) *F* 29) *T* 30) *T* 31) *F*
32) *T* 33) *F*

Miscellaneous Exercise 1 – p. 57

2) $y = x,\ y = 6x$

4) $\begin{pmatrix} 0 & 1 \\ 1 & 0 \end{pmatrix}$, $\theta = 0,\ 180^\circ,\ 360^\circ$

5) a) $\begin{pmatrix} -\frac{1}{2} & \sqrt{3}/2 \\ \sqrt{3}/2 & \frac{1}{2} \end{pmatrix}$ b) $\begin{pmatrix} 0 & -1 \\ 1 & 0 \end{pmatrix}$

c) $\begin{pmatrix} -\frac{1}{2} & -\sqrt{3}/2 \\ -\sqrt{3}/2 & \frac{1}{2} \end{pmatrix}$

rotation through 30° about O.

7) $-(a-b)(b-c)(c-a)$

8) $2\begin{vmatrix} 1 & 0 & z \\ z & 1 & 0 \\ 0 & z & 1 \end{vmatrix}$

9) $(\sin 4\theta - \sin\theta)(\sin\theta - \frac{1}{2})(\sin 4\theta - \frac{1}{2})$,
$\theta = \dfrac{2n\pi}{3},\ (2n+1)\dfrac{\pi}{5},\ [6n + (-1)^n]\dfrac{\pi}{6},$
$[6n + (-1)^n]\dfrac{\pi}{24}$

12) (1, 3), (2, 1), No

13) $\mathbf{R}_\alpha \mathbf{S} \mathbf{R}_{-\alpha} = \begin{pmatrix} \cos 2\alpha & \sin 2\alpha \\ \sin 2\alpha & -\cos 2\alpha \end{pmatrix}$

14) $\lambda = 1$ or 5, $x + y = 0$, $3y - x = 0$

15) $\mathbf{u} = \begin{pmatrix} a \\ c \end{pmatrix}$, $\mathbf{v} = \begin{pmatrix} b \\ d \end{pmatrix}$

16) False 18) $-9xy^2$ 19) $4n(n^2-1)$

21) a) $\begin{pmatrix} \sqrt{3}/2 & -\frac{1}{2} \\ \frac{1}{2} & \sqrt{3}/2 \end{pmatrix}$

b) $\begin{pmatrix} -\frac{1}{2} & -\sqrt{3}/2 \\ -\sqrt{3}/2 & \frac{1}{2} \end{pmatrix}$

c) $\begin{pmatrix} -\sqrt{3}/2 & \frac{1}{2} \\ -\frac{1}{2} & -\sqrt{3}/2 \end{pmatrix}$

reflection in the line through O at 30° to Ox

22) $p = y/b,\ q = x - ay/b$ 23) -9

25) $\begin{pmatrix} a & b \\ 4b & 3b + a \end{pmatrix}$, $\mathbf{P} = \begin{pmatrix} 0 & 1 \\ 4 & 3 \end{pmatrix}$

26) $3y' = 8x' - 5$

Exercise 2a – p. 67

1) $y^2 = a(x - 3a);\ (3a, 0)$
2) $(at^2, 2at)$
5) $\dfrac{a}{3}\left(p^2 + \dfrac{1}{p^2}\right),\ \dfrac{2a}{3}\left(p - \dfrac{1}{p}\right);$
$9y^2 = 12ax - 8a^2$
9) b) $x = -4a$ 10) $3y^2 = 16x$
11) $y^2 = 2a(x - a)$

Exercise 2b – p. 73

1) a) $4x^2 + 9y^2 = 36$
b) $4x^2 + (y-1)^2 = 16$
c) $3(x-2)^2 + 4y^2 = 12a^2$
d) $16(x-3)^2 + 9(y-4)^2 = 144$

2) a) $(0, 0);\ 6, 8;\ \frac{1}{4}\sqrt{7}$
b) $(0, 1);\ 4, 2;\ \frac{1}{2}\sqrt{3}$
c) $(0, 0);\ 10, 8;\ \frac{3}{5}$
d) $(0, 0);\ 2, \frac{8}{5};\ \frac{3}{5}$
e) $(2, 0);\ 8, 4;\ \frac{1}{2}\sqrt{3}$
f) $(a, b);\ 2a, 2b;\ \frac{1}{a}\sqrt{a^2 - b^2}$
g) $(-3, 4);\ 2\sqrt{3}, 2\sqrt{2};\ \frac{1}{3}\sqrt{3}$

3) a) $(\pm\sqrt{5}, 0);\ x = \pm\dfrac{9\sqrt{5}}{5}$
b) $(0, 1 \pm 2\sqrt{3});\ y = 1 \pm \dfrac{8\sqrt{3}}{3}$
c) $(2 \pm a, 0);\ x = 2 \pm 4a$
d) $(3, 4 \pm \sqrt{7});\ y = 4 \pm \dfrac{16\sqrt{7}}{7}$

4) $\frac{1}{2};\ 6;\ 3x^2 + 4y^2 = 27$
5) a) $7x^2 + 7y^2 - 46x - 62y - 2xy + 199 = 0$
b) $9x^2 + 8y^2 - 36y + 36 = 0$

Exercise 2c – p. 80

1) $\pm\sqrt{37};\ \pm\left(\dfrac{12\sqrt{37}}{37}, -\dfrac{\sqrt{37}}{37}\right)$
3) a) misses
b) meets in two distinct points
c) touches
4) $9y + 8x = 0;\ 2y = x;$
$45y + 40x = \pm 30$
6) $5 \pm \frac{4}{5}y_1$ 7) 7 units
9) $5x^2 + 9y^2 = 180$
11) $(\pm\sqrt{a^2 - b^2}, 0)$ 12) $x^2 + y^2 = 25$

Exercise 2d – p. 88

2) a) $\frac{1}{4}\sqrt{7};\ (0, 0);\ (\pm\sqrt{7}, 0)$
b) $\frac{1}{2}\sqrt{3};\ (0, 0);\ (0, \pm\sqrt{3})$
c) $\frac{1}{4}\sqrt{7};\ (5, 2);\ (5 \pm \sqrt{7}, 2)$
d) $\frac{1}{2}\sqrt{3};\ (-1, 0);\ (-1, \pm\sqrt{3})$

3) a) $9x^2 + 16y^2 = 144$
b) $4x^2 + y^2 = 4$
c) $9(x-5)^2 + 16(y-2)^2 = 144$
d) $4(x+1)^2 + y^2 = 4$

4) a) $x = 1 + 3\cos\theta \quad y = 3 + 2\sin\theta$
b) $x = \sqrt{2}\cos\theta - 2$
$y = \sqrt{3}\sin\theta - 1$
c) $x = 3\cos\theta \quad y = 4\sin\theta$
d) $x = 2\cos\theta - 1 \quad y = \sin\theta + 1$

5) a) $3x + 4y\sqrt{3} = 24$
b) $2x\sqrt{3} - y = 4$
c) $x - 4y + 4\sqrt{2} = 0$
d) $3x + 5\sqrt{3}y + 30 = 0$

6) $x\cos\theta + 2y\sin\theta = 2$;
$y\cos\theta + 2x\sin\theta + 3\sin\theta\cos\theta = 0$

7) $2x(\cos\theta - \sin\theta)$
$+ 3y(\cos\theta + \sin\theta) = 6$

8) a) (i) $(a\sec\theta, 0)$ (ii) $(0, b\operatorname{cosec}\theta)$
b) (i) $\left\{\dfrac{(a^2 - b^2)}{a}\cos\theta, 0\right\}$
(ii) $\left\{0, \dfrac{(b^2 - a^2)}{b}\sin\theta\right\}$

9) b) $x = \cos\theta - 2, \quad y = 2\sin\theta + 5$

10) $4(x + 2\sqrt{5})^2 + 9y^2 = 324$

11) $\tan^2\theta$ 12) a^2/b^2

13) a) $\cos\theta(1 - \cos\theta)$: $\sin\theta(1 - \sin\theta)$
b) $\begin{cases} x = 2(1 - \tan\theta + \sec\theta) \\ y = \frac{3}{2}(1 - \cot\theta + \operatorname{cosec}\theta) \end{cases}$

14) $\pm\left(\dfrac{4}{\sqrt{5}}, \dfrac{4}{\sqrt{5}}\right)$,
$\pm\left(\dfrac{8}{\sqrt{5}}, -\dfrac{2}{\sqrt{5}}\right)$

15) a) $-\dfrac{5\pi}{6}$ b) $3y + \sqrt{3}x = \pm 10$
c) $\sqrt{3}y + x = 0$

Exercise 2e – p. 93

1) a) $3x^2 - y^2 = 27$ b) $8x^2 - y^2 = 288$
c) $72(x - 2)^2 - 9y^2 = 32$
d) $5(y - 1)^2 - 4x^2 = 1$

2) a) $(0, 0); 4$ b) $(0, 0); 8$
c) $(1, 0); 2$ d) $(-1, -2); 4$
e) $(0, 0); 2$ f) $(0, 0); 4$

3) a) $(\pm\sqrt{5}, 0); x = \pm 4/\sqrt{5}$
b) $(\pm\sqrt{41}, 0); x = \pm 16/\sqrt{41}$
c) $(1 \pm \sqrt{10}, 0); x = 1 \pm 1/\sqrt{10}$
d) $(-2, -1 \pm \sqrt{13}); y = -1 \pm 4/\sqrt{13}$
e) $(\pm\sqrt{2}, 0); x = \pm 1/\sqrt{2}$
f) $(0, \pm 2\sqrt{2}); y = \pm\sqrt{2}$

4) $x^2 + y^2 - 4xy + 4x + 2y = 5$

5) $8x^2 - y^2 + 8x - 16 = 0$

Exercise 2f – p.104

1) a) $y + 3 = \sqrt{3}x$ b) $4x - 5y = 9$
c) $2x - y = 1$

2) a) $2y = \pm\sqrt{3}x$ b) $y = \pm x$
c) $y = \pm\sqrt{3}x$

3) $\frac{1}{3}$ 4) $b^2 + c^2 = a^2m^2; ay = \pm bx$

5) $(-4/\sqrt{3}, -1/\sqrt{3})$ 6) $\pm\frac{1}{2}\sqrt{10}$

8) $8y = x$ 12) 2

13) $4(x - a)^2 - 4y^2 = a^2$; $(a, 0)$; $\left(\dfrac{a}{2}, 0\right)$;
$\left(\dfrac{3a}{2}, 0\right)$; ± 1

14) $y = x \pm 1$; $(3, 2), (-3, -2)$; $\sqrt{6}$

Miscellaneous Exercise 2 – p. 105

2) $ty = x + at^2$

3) $(t_1 + t_2)y = 2x + 2at_1t_2$

4) b) $\{a(p^2 - 4p + 8), a(2p - 4)\}$;
$y^2 = 4a(x - 4a)$

5) $y = tx - at - at^2$;
$x + ty = a + 2at + at^3$;
$\left\{a - 2at - \dfrac{4a}{t}, at^2 + 4a + \dfrac{4a}{t^2}\right\}$

6) $x - 4ty + 2t^2 = 0$; $x \pm 4y + 2 = 0$

8) $x^2 + y^2 = a^2 + b^2$

9) $25x^2 + 9y^2 = 64$; $(0, \pm\frac{32}{15})$

10) $(2, 1), (-1, 3)$

11) $x\left(x - \dfrac{a}{\cos\theta}\right)$
$+ y\left(y - \dfrac{b^2 - a^2}{b}\sin\theta\right) = 0$

12) $9a^2x^2 + 9b^2y^2 = (a^2 - b^2)^2$

15) $\left\{0, \dfrac{3}{2}\left(\dfrac{1}{\sin\theta} - \sin\theta\right)\right\}$;
$\dfrac{2x^2}{9} + \dfrac{9}{16y^2} = 1$

18) $y = \pm x$

21) $\dfrac{a^2}{x^2} + \dfrac{b^2}{y^2} = 4$

22) $3y = \pm 4x$

24) $\left(\dfrac{a^2}{p}\cos\alpha, -\dfrac{b^2}{p}\sin\alpha\right)$;
$\pm 3\sqrt{2}x \pm \sqrt{7}y = 15$

25) $x^2 + y^2 = 5$

27) $\left\{\pm a\left(\dfrac{b^2 + c^2}{a^2 + b^2}\right)^{\frac{1}{2}}, \pm b\left(\dfrac{a^2 - c^2}{a^2 + b^2}\right)^{\frac{1}{2}}\right\}$;
$a^2 - b^2 = 2c^2$

28) $bx \sec t - ay \tan t = ab$;
$a^2y^2 = 4x^2y^2 + b^2x^2$
29) $m^2x^2 - y^2 = m^2a^2$
30) $\frac{x}{b}\tan\theta + \frac{y}{a}\sec\theta = \frac{a^2+b^2}{ab}\sec\theta\tan\theta$;
$\left(\frac{2x}{a}\right)^2 - \left(\frac{2yb}{a^2+2b^2}\right)^2 = 1$
31) $a^2y_0(y-y_0) = b^2x_0(x-x_0)$
33) $(\pm 4, 0)$; $9x^2 + 16y^2 = 144$

Exercise 3a – p. 120

1) The answers given are suitable parametric equations but there are other possibilities.
a) $x = 4t, y = 4/t$
b) $x = 5t, y = -5/t$
c) $x = t + 2, y = 1/t$
d) $x = 3t, y = 3/t$
e) $x = 2t + 1, y = 2/t$
f) $x = t/2, y = 1/2t$

2) a) $xy = 4$ b) $xy + 9 = 0$
c) $xy = 1$ d) $y(x-1) = 1$
e) $x(1-y) = 16$ f) $xy + 4 = 0$

3) (i) a) $(0, 0)$; $(4\sqrt{2}, 4\sqrt{2})$, $(-4\sqrt{2}, -4\sqrt{2})$; $(4, 4)$, $(-4, -4)$
b) $(0, 0)$; $(5\sqrt{2}, -5\sqrt{2})$, $(-5\sqrt{2}, 5\sqrt{2})$; $(5, -5)$, $(-5, 5)$
c) $(2, 0)$; $(2+\sqrt{2}, \sqrt{2})$, $(2-\sqrt{2}, -\sqrt{2})$; $(3, 1)$, $(1, -1)$
d) $(0, 0)$; $(3\sqrt{2}, 3\sqrt{2})$, $(-3\sqrt{2}, -3\sqrt{2})$; $(3, 3)$, $(-3, -3)$
e) $(1, 0)$; $(1+2\sqrt{2}, 2\sqrt{2})$, $(1-2\sqrt{2}, -2\sqrt{2})$; $(3, 2)$, $(-1, -2)$
f) $(0, 0)$; $(\sqrt{2}/2, \sqrt{2}/2)$, $(-\sqrt{2}/2, -\sqrt{2}/2)$; $(\frac{1}{2}, \frac{1}{2})$, $(-\frac{1}{2}, -\frac{1}{2})$
(ii) a) $(0, 0)$; $(2\sqrt{2}, 2\sqrt{2})$, $(-2\sqrt{2}, -2\sqrt{2})$; $(2, 2)$, $(-2, -2)$
b) $(0, 0)$; $(3\sqrt{2}, -3\sqrt{2})$, $(-3\sqrt{2}, 3\sqrt{2})$; $(3, -3)$, $(-3, 3)$
c) $(0, 0)$; $(\sqrt{2}, \sqrt{2})$, $(-\sqrt{2}, -\sqrt{2})$; $(1, 1)$, $(-1, -1)$
d) $(1, 0)$; $(1+\sqrt{2}, \sqrt{2})$, $(1-\sqrt{2}, -\sqrt{2})$; $(0, -1)$, $(2, 1)$
e) $(0, 1)$; $(4\sqrt{2}, 1-4\sqrt{2})$, $(-4\sqrt{2}, 1+4\sqrt{2})$; $(4, -3)$, $(-4, 5)$
f) $(0, 0)$; $(2\sqrt{2}, -2\sqrt{2})$, $(-2\sqrt{2}, 2\sqrt{2})$; $(2, -2)$, $(-2, 2)$

4) a) $t^2y + x = 8t$; $t^3x - ty = 4(t^4 - 1)$
b) $4y + x = 4$; $8x - 2y = 15$
c) $t^2y - x = 6t$; $t^3x + ty = 3(1 - t^4)$
d) $x + y = \pm 4$; $y = x$ is the normal in both cases
5) a) $2y + x = \pm 4\sqrt{2}$
b) $y + 4x = 0$; $y = 0$
6) $(-\frac{3}{8}, -24)$; $\frac{51\sqrt{17}}{8}$ 7) $2\sqrt{29}$
8) $\frac{21}{10}\sqrt{29}$ 9) $\left(\frac{c}{m}, cm\right)$ 10) $\frac{17^{\frac{3}{2}}}{4}$
13) $x + 3y = 9$ 14) $n^2 = 4mlc^2$
17) $(x^2 + y^2)^2 = 4c^2xy$

Exercise 3b – p. 126

1) a, c, e, f
2) a) $x + y = 0, x + y = 0$
c) $x + y = 0, x - y = 0$
e) $2x + y = 0, x + y = 0$
f) $x + 3y = 0, x + y = 0$
3) a) $6x^2 - 5xy + y^2 = 0$
b) $2y^2 + 5xy - 12x^2 = 0$
c) $y^2 = 3x^2$
4) a) (i) b) (iii) c) (ii), (v)
(*Note.* (iv) does not represent a line pair through O)
5) a) $\arctan\frac{3}{4}$ b) $\arctan\frac{3}{4}$
c) $\arctan 2$ d) $\arctan\frac{\sqrt{p^2-4q}}{1+q}$
6) a) 1 b) -1 c) $\frac{1}{3}(-5 \pm 2\sqrt{7})$
7) a) $q^2 > 4pr$ b) $p + r = 0$; two coincident lines, $x + y = 0$
9) a) $a + b = 0$ b) $h^2 = ab$
10) $ay = \pm bx$; $\arctan\left|\frac{2ab}{b^2-a^2}\right|$; $a^2 = b^2$
11) $a^2y^2 + b^2x^2 = 0$ does not represent a line pair through O.
12) a) no real asymptote through O
b) $x + 3y = 0, x - y = 0$
c) $x - y = 0, 4x + 5y = 0$
d) no real asymptotes through O
e) $2x + y = 0, 2x - y = 0$
f) no real asymptotes through O

Exercise 3c – p. 134

1) $256x^2 + 800xy + 175y^2 = 0$
2) $35x^2 - 144xy + 143y^2 = 0$
3) $11x^2 + 36xy + 13y^2 = 0$
4) $5y^2 = 18x^2 - 9xy$
5) $(c^2 - m^2a^2)x^2 + 2a^2mxy + (c^2 - a^2)y^2 = 0$
6) $2bx + 2ay = ab$
7) $2x + 3y = 9$
8) $y = 2(x-2)$
9) $x + y = 32$
10) $(2p + 3q - 4)x + (4q + 3p)y = 10 + 4p$
11) crosses 12) crosses
13) misses 14) misses
15) touches

Miscellaneous Exercise 3 – p. 135

1) $\left(-\dfrac{c}{t^3}, -ct^3\right)$; $\left(ct^9, \dfrac{c}{t^9}\right)$; $4x^3y^3 + c^2(x^2-y^2)^2 = 0$
4) $x^2 + y^2 - x\left(3ct - \dfrac{c}{t^3}\right) + 2c^2\left(t^2 - \dfrac{1}{t^2}\right) = 0$
5) $4c\sqrt{\dfrac{m}{1+m^2}}$; $8c^2\left(\dfrac{1-m^2}{1+m^2}\right)$
7) (h, k); $x = h, y = k$; $(\frac{3}{2}, 3), (-\frac{1}{2}, -1)$
8) $p^2y + x = 2cp$; $\left(\dfrac{2cpq}{p+q}, \dfrac{2c}{p+q}\right)$
9) $ty - t^3x = c(1-t^4)$; $\left(-\dfrac{c}{t^3}, -ct^3\right)$
11) $x + t^2y = 2t$; $3 \pm 2\sqrt{2}$
13) $-\dfrac{4c^2}{a^2}$; $\frac{1}{2}$; $2:3$
14) $3x^2 - 8xy - 3y^2 = 0$; 90°
15) $3x + 2y + 4 = 0$
17) $y^2(\lambda^2 - 4\lambda - 1) + 2\lambda xy(2\lambda - 1) + 4\lambda^2x^2 = 0$; 1 or $-\frac{1}{5}$; ∞ or $-\frac{5}{12}$; $12y + 5x + 5 = 0$, $x = -1$.
19) $(5 + 3\sqrt{3})y^2 - (6 + 2\sqrt{3})xy + (7 - 3\sqrt{3})x^2 = 0$
21) $x^2(ar^2 - 2gpr + p^2c) + 2xy(hr^2 - fpr - gqr + pqc) + y^2(br^2 - 2fqr + q^2c) = 0$; $r^2(a+b) - 2r(gp + fq) + c(p^2 + q^2) = 0$

Exercise 4a – p. 145

1) a) $3\mathbf{i} + 6\mathbf{j} + 4\mathbf{k}$, b) $\mathbf{i} - 2\mathbf{j} - 7\mathbf{k}$, c) $\mathbf{i} - 3\mathbf{k}$
2) a) $(5, -7, 2)$, b) $(1, 4, 0)$ c) $(0, 1, -1)$
3) a) $\sqrt{21}$ b) 5 c) 3
4) a) 6 b) 7 c) $\sqrt{206}$
5) a) $3\mathbf{i} + 4\mathbf{k}$, b) $2\mathbf{i} - 2\mathbf{j} + 2\mathbf{k}$, c) $2\mathbf{i} + 3\mathbf{j} + 3\mathbf{k}$ d) $-6\mathbf{i} + 12\mathbf{j} - 8\mathbf{k}$
6) a) $-2\mathbf{i} - 3\mathbf{j} + 7\mathbf{k}$, b) $-\mathbf{j} + 3\mathbf{k}$, c) $-2\mathbf{i} - 2\mathbf{j} + 4\mathbf{k}$
7) $\sqrt{62}, \sqrt{10}, 2\sqrt{6}$ 8) $\sqrt{5}$
9) $\overrightarrow{AB} = -\mathbf{i} + 3\mathbf{j}$, $\overrightarrow{BD} = -\mathbf{j} - 4\mathbf{k}$, $\overrightarrow{CD} = -2\mathbf{i} + 2\mathbf{j} - 6\mathbf{k}$, $\overrightarrow{AD} = -\mathbf{i} + 2\mathbf{j} - 4\mathbf{k}$

Exercise 4b – p. 148

1) a) $\frac{2}{3}, \frac{2}{3}, -\frac{1}{3}$ b) $\frac{6}{7}, -\frac{2}{7}, -\frac{3}{7}$ c) $\frac{3}{5}, 0, \frac{4}{5}$ d) $\frac{1}{9}, \frac{8}{9}, \frac{4}{9}$
2) a) 54.74° b) 90° c) 45° d) 131.81°
3) a) $(\frac{1}{3}, \frac{2}{3}, -\frac{2}{3})$, b) $(-\frac{3}{7}, \frac{2}{7}, \frac{6}{7})$
4) a) $(\frac{12}{7}, -\frac{18}{7}, \frac{36}{7})$ b) $(\frac{16}{9}, \frac{2}{9}, -\frac{8}{9})$ c) $(4/\sqrt{3}, 4/\sqrt{3}, 4/\sqrt{3})$
5) $(\frac{3}{2}, \pm 3/\sqrt{2}, 3/\sqrt{2})$
6) $a = \pm\sqrt{(23/2)}$, $b = 5\sqrt{2}/2$

Exercise 4c – p. 153

1) $\sqrt{14}$; $2/\sqrt{14}, -3/\sqrt{14}, 1/\sqrt{14}$
$4\sqrt{2}$; $-1/\sqrt{2}, -1/\sqrt{2}, 0$
$4\sqrt{2}$; $-1/\sqrt{2}, -1/\sqrt{2}, 0$
2) a) $(3, \frac{4}{3}, \frac{1}{3})$, b) $(2, -\frac{1}{2}, \frac{1}{2})$, c) $(-6, -\frac{7}{2}, -\frac{1}{2})$
3) a) no b) yes c) yes
4) a) no b) yes c) yes d) no
5) $4\mathbf{i} + \frac{3}{2}\mathbf{j} + \frac{1}{2}\mathbf{k}$, $5\mathbf{i} + \mathbf{k}$

Exercise 4d – p. 157

1) a) $\dfrac{x-2}{1} = \dfrac{y+1}{2} = \dfrac{z-4}{3}$
b) $\dfrac{x-3}{4} = \dfrac{y+1}{-1} = \dfrac{z-5}{3}$
c) $\dfrac{x-4}{3} = \dfrac{y+2}{5} = \dfrac{z-1}{7}$
2) a) $12:9:8$; $\frac{12}{17}, \frac{9}{17}, \frac{8}{17}$
b) $3:2:-2$; $3/\sqrt{17}, 2/\sqrt{17}, -2/\sqrt{17}$
c) $\cos\alpha : \cos\beta = 2:1$; $2/\sqrt{5}, 1/\sqrt{5}, 0$
d) $\cos\alpha : \cos\gamma = 1:2$; $1/\sqrt{5}, 0, 2/\sqrt{5}$
e) $2:-3:-7$; $2/\sqrt{62}, -3/\sqrt{62}, -7/\sqrt{62}$

3) a) $x = 3, y + 1 = z - 4$
   b) $\frac{x-3}{1} = \frac{z-7}{-4}, y = 2$
   c) $x = 1, y = 2$
4) a) no b) no c) yes
5) $x = \lambda + 2,\ y = \frac{1}{3}(1 - 2\lambda),\ z = \frac{1}{2}(4 - \lambda)$
   a) $(6, -\frac{7}{3}, 0)$ b) $(\frac{5}{2}, 0, \frac{7}{4})$
   c) $(0, \frac{5}{3}, 3)$ d) $(2, \frac{1}{3}, 2)$

Exercise 4e – p. 163

1) a) parallel   b) intersecting at (1, 2, 0)
   c) skew
2) -3; $(-1, 3, 4)$
3) $x = 1, z = 3$; $\frac{x+2}{3} = \frac{y}{3} = \frac{z-5}{-1}$;
   $\frac{x-4}{-3} = \frac{y-7}{-4} = \frac{z-1}{2}$; (1, 3, 3)
4) a) $\arccos \frac{19}{21}$ b) $\arccos \frac{2}{3}$
   c) $\arccos 8\sqrt{3}/15$ d) $\arccos 3/\sqrt{182}$
6) $(\frac{4}{3}, -\frac{5}{6}, \frac{13}{6})$; $\frac{1}{6}\sqrt{174}$
7) $\frac{x-1}{1} = \frac{y}{10} = \frac{z-1}{23}$
8) $\frac{1}{5}\sqrt{110}$; $\sqrt{11}$
9) $\arccos\left(\frac{a + 2b + 3c}{\sqrt{\{13(a^2 + b^2 + c^2)\}}}\right)$;
   $\arccos\left(\frac{a - b + 3c}{\sqrt{\{11(a^2 + b^2 + c^2)\}}}\right)$
   $a:c = -3:1,\ b = 0$

Exercise 4f – p. 169

1) a) $x + 2y - z = -3$
   b) $x + 3y + 4z = -5$
   c) $2x - y + 2z = 0$
   d) $x + 2y - 3z = 3$
2) a) $23x - 21y - 4z = 51$
   b) $x + y = 1$ c) $2x - y - z = 0$
3) a) $x + 21y - 9z = -56$
   b) $31x + 6y - 11z = 0$
4) $2x - y + 5z = 28$
5) $2x + 2y + z = 4$
9) $5x + 11y + z = -1$
10) $(\frac{3}{4}, \frac{21}{16}, \frac{3}{8})$ 11) $(\frac{3}{4}, -\frac{13}{6}, \frac{7}{12})$

Exercise 4g – p. 175

1) a) $\sqrt{30}/15$ b) 2   c) 2
   d) $2\sqrt{14}/7$
2) a) parallel; $5\sqrt{14}/7$
   b) intersecting; 1:5:2
   c) parallel; $\sqrt{11}/11$
   d) intersecting; $-3:6:7$
3) $\arccos 14/3\sqrt{30}$; $\arccos 29/7\sqrt{30}$
4) a) $9\sqrt{11}/11$ b) $9\sqrt{14}/14$
5) a) $\arcsin 10/\sqrt{418}$
   b) $\arcsin 6/\sqrt{558}$

Miscellaneous Exercise 4 – p. 178

1) a) $(-\frac{48}{13}, -\frac{3}{13}, \frac{12}{13})$
   c) $\frac{1}{3}x = -\frac{1}{2}y = z - 2$
2) (2, 3, 1) a) $7x + 4y - 5z = 21$
   b) $x - 8y - 5z = -27$
   c) $7x - 2y - 8z = 0$
3) a) $\frac{x}{2} = \frac{y-1}{-4} = \frac{z}{1}$ b) $2y + 3x = 0$
   c) $\sqrt{(209/273)}$ d) $1/\sqrt{33}$
4) $\frac{x+1}{2} = \frac{y-2}{-3} = \frac{z-3}{4}$; $(-\frac{53}{29}, \frac{94}{29}, \frac{39}{29})$;
   $a = 5$; $\arccos 6/5\sqrt{29}$
5) a) $\frac{2}{27}$ b) $(0, -5, 19)$
   c) $(-\frac{1}{9}, \frac{1}{9}, -\frac{22}{9})$
6) $x = y = 2z$; $4x - 3y - 2z = 0$; $90°$
7) $\frac{x-14}{2} = \frac{y-5}{2} = \frac{z-2}{-1}$;
   $\cos\theta = \frac{2}{3}$; $x - 2y - 2z = 0$
8) $(x + 1)^2 + (y - 1)^2 = 21,\ z = 0$;
   $z = -3$; $z = 7$
9) B(0, 5, 0); C(0, 0, -4); A(1, 2, -2);
   $\frac{10}{3}$
10) $1/\sqrt{2}, -1/\sqrt{2}, 0$; $4x + 4y + 7z = 0$;
   $8x = 8y = -7z$
11) $x - 2 = \frac{1}{4}(y - 3) = \frac{1}{2}(z - 4)$;
   $2x - 5y + 9z - 25 = 0$;
   55°37′; 58°31′; $80/\sqrt{110}$
12) a) $4x + 3z = 48,\ y = 0$
   b) $\frac{x}{6} = \frac{y}{6\sqrt{3}} = \frac{z-16}{-16}$
   c) $4\sqrt{3}x + 4y + 3\sqrt{3}z = 48\sqrt{3}$; 15;
   $(0, 0, -9)$
13) $3x - 2y - z - 3 = 0$; $10/\sqrt{14}$;
   $\frac{x-1}{1} = \frac{y-1}{4} = \frac{z+2}{-5}$
14) $2x - 8y - 4z + 3 = 0$;
   $\frac{x}{1} = \frac{y-1}{-4} = \frac{z-4}{-2}$; $4x - 3y + 8z = 29$
15) $0, \sqrt{3}/3, -\sqrt{6}/3$;
   $x = 2,\ 2\sqrt{3} - y = z/\sqrt{2}$;
   $y + \frac{\sqrt{2}}{2}z = 2\sqrt{3}$
16) a) $x - y - 2z + 7 = 0$,
   b) $3\sqrt{7}/14$; (0, 5, 7), (4, 1, -1)

17) a) $1-x=\frac{y}{2}=\frac{z}{3}$ b) $1,-2,3$
c) $73°24'$ d) $6x-3y-2z=0$
e) $\frac{6}{7}$
18) $-2x=-2y=5z+9-\sqrt{2}$
19) a) $\frac{x}{4}=\frac{y-1}{-6}=\frac{z+1}{4}$
b) $2x-3y+2z+5=0$; $(1,3,1)$
20) a) $x-y-z=4$
b) $(-\frac{4}{3},\frac{4}{3},\frac{4}{3})$ c) $\frac{2}{3}$

Exercise 5a – p. 188

1) a) $\frac{2}{3},\frac{2}{3},-\frac{1}{3}$ b) $\frac{6}{7},-\frac{2}{7},-\frac{3}{7}$
c) $\frac{3}{5},0,\frac{4}{5}$ d) $\frac{1}{9},\frac{8}{9},\frac{4}{9}$
2) a) $\frac{2}{3}\mathbf{i}+\frac{2}{3}\mathbf{j}-\frac{1}{3}\mathbf{k}$ b) $\frac{6}{7}\mathbf{i}-\frac{2}{7}\mathbf{j}-\frac{3}{7}\mathbf{k}$
c) $\frac{3}{5}\mathbf{i}+\frac{4}{5}\mathbf{k}$ d) $\frac{1}{9}\mathbf{i}+\frac{8}{9}\mathbf{j}+\frac{4}{9}\mathbf{k}$
3) a) $(\frac{1}{3},\frac{2}{3},-\frac{2}{3})$ b) $(\frac{3}{7},\frac{2}{7},\frac{6}{7})$
4) a) 54.74° to Ox, 125.26° to Oy, 54.74° to Oz
b) 63.61° to Ox, 27.27° to Oy, 83.62° to Oz
c) 90° to Ox, 45° to Oy, 135° to Oz
5) a) $4\mathbf{j}+5\mathbf{k}$ b) $16\mathbf{i}-16\mathbf{j}-8\mathbf{k}$
c) $\pm 4\sqrt{2}\mathbf{i}+4\mathbf{j}+4\mathbf{k}$
d) $\pm(\frac{8}{3}\mathbf{i}+\frac{1}{3}\mathbf{j}+\frac{4}{5}\mathbf{k})$
6) a) $5\sqrt{2}$; 64°54′, 55°34′, 45°
b) $\sqrt{3}$; 125°16′, 54°44′, 125°16′
c) $\sqrt{42}$; 39.5°, 81.1°, 51.9°
7) a) $\frac{1}{\sqrt{3}}\mathbf{i}-\frac{1}{\sqrt{3}}\mathbf{j}+\frac{1}{\sqrt{3}}\mathbf{k}$
b) $\frac{3}{7}\mathbf{i}+\frac{2}{7}\mathbf{j}-\frac{6}{7}\mathbf{k}$
8) $2\sqrt{3}(\mathbf{i}+\mathbf{j}+\mathbf{k})$
9) $\sqrt{22}, 3/\sqrt{22}, 2/\sqrt{22}, 3/\sqrt{22}$;
$3\sqrt{2}, \frac{1}{3\sqrt{2}}, \frac{-4}{3\sqrt{2}}, \frac{-1}{3\sqrt{2}}$

Exercise 5b – p. 190

1) $\sqrt{13}$; 3
2) $2/\sqrt{5}, 0, -1/\sqrt{5}; -1/\sqrt{5}, 0, 2/\sqrt{5}$
3) a) $\frac{1}{5}(2\mathbf{i}+\mathbf{j}+7\mathbf{k})$ b) $-2\mathbf{i}+\mathbf{j}+5\mathbf{k}$
4) a) no b) no c) no

Exercise 5c – p. 196

1) a) $x-2=y-3=z+1$
b) $\frac{x}{3}=\frac{z}{5}$ and $y=4$ c) $\frac{x}{2}=\frac{y}{3}=\frac{z}{4}$
2) a) $\mathbf{r}=3\mathbf{i}+\mathbf{j}+7\mathbf{k}+\lambda(2\mathbf{i}+\mathbf{j}+3\mathbf{k})$
$\mathbf{r}=2\mathbf{i}-5\mathbf{j}+\mathbf{k}+\lambda(3\mathbf{i}+\mathbf{j}+4\mathbf{k})$
c) $\mathbf{r}=\mathbf{i}+\lambda(-3\mathbf{i}+5\mathbf{j}+\mathbf{k})$
3) a) $\mathbf{r}=\mathbf{i}-3\mathbf{j}+2\mathbf{k}+\lambda(5\mathbf{i}+4\mathbf{j}-\mathbf{k})$;
$\frac{x-1}{5}=\frac{y+3}{4}=\frac{z-2}{-1}$
b) $\mathbf{r}=2\mathbf{i}+\mathbf{j}+\lambda(3\mathbf{j}-\mathbf{k})$;
$x=2, \frac{y-1}{3}=\frac{z}{-1}$
c) $\mathbf{r}=\lambda(\mathbf{i}-\mathbf{j}-\mathbf{k}); x=-y=-z.$
4) a) no b) yes c) yes d) yes
e) no
5) $\mathbf{r}=4\mathbf{i}+5\mathbf{j}+10\mathbf{k}+\lambda(\mathbf{i}+\mathbf{j}+3\mathbf{k})$,
$4-x=5-y=\frac{10-z}{3}$
$\mathbf{r}=2\mathbf{i}+3\mathbf{j}+4\mathbf{k}+\lambda(\mathbf{i}+\mathbf{j}+5\mathbf{k})$,
$2-x=3-y=\frac{4-z}{5}$; $3\mathbf{i}+4\mathbf{j}+5\mathbf{k}$.
6) a) $\mathbf{r}=3\mathbf{i}+\mathbf{j}-4\mathbf{k}+\lambda(\mathbf{i}-3\mathbf{j}-6\mathbf{k})$;
$(\frac{7}{3},3,0)$; $(0,10,14)$; $(\frac{10}{3},0,-6)$
b) $\mathbf{r}=\mathbf{i}+\mathbf{j}+7\mathbf{k}+\lambda(2\mathbf{i}+3\mathbf{j}-6\mathbf{k})$;
$(\frac{10}{3},\frac{9}{2},0)$; $(0,-\frac{1}{2},10)$; $(\frac{1}{3},0,9)$
7) $\mathbf{r}=5\mathbf{i}-2\mathbf{j}-4\mathbf{k}+\lambda(3\mathbf{i}+4\mathbf{j}+5\mathbf{k})$;
$(\frac{13}{2},0,-\frac{3}{2})$
8) $1:2:3$; $\mathbf{r}=2\mathbf{i}-\mathbf{j}-\mathbf{k}+\lambda(\mathbf{i}+2\mathbf{j}+3\mathbf{k})$

Exercise 5d – p. 199

1) a) parallel b) intersecting; $\mathbf{r}=\mathbf{i}+2\mathbf{j}$
c) skew
2) $a=-3$; $\mathbf{r}=-\mathbf{i}+3\mathbf{j}+4\mathbf{k}$
3) $(\frac{1}{3},-\frac{2}{3},\frac{2}{3})$; $(\frac{2}{7},\frac{3}{7},-\frac{6}{7})$; arccos $-\frac{16}{21}$

Exercise 5e – p. 206

1) a) 30
b) 0, **a** and **b** are perpendicular
c) -1
2) a) $7, \sqrt{7}/3$ b) $14, \sqrt{7}/19$
3) 4
4) a) -4 b) 15 c) 11
d) -12 e) 29
5) $\sqrt{(\frac{7}{34})}$; arccos $-\sqrt{(\frac{7}{10})}$
9) a) arccos $\frac{19}{21}$ b) arccos $\frac{2}{3}$
c) arccos $8\sqrt{3}/15$
10) a) arccos $3/\sqrt{182}$ b) 0
13) a) $(\mathbf{i}-9\mathbf{j}-5\mathbf{k})/\sqrt{107}$
b) $(-\mathbf{i}-2\mathbf{j}+\mathbf{k})/\sqrt{6}$

Exercise 5f – p. 211

1) a) $\mathbf{a}\times\mathbf{b}$ b) 0 c) $2\mathbf{a}\times\mathbf{b}$
d) $\mathbf{a}\times\mathbf{c}.\mathbf{b}$ e) 0 f) 0
2) a) $-3\mathbf{j}-3\mathbf{k}$ b) $-3\mathbf{j}-3\mathbf{k}$
3) $(3\mathbf{i}-\mathbf{j}+\mathbf{k})/\sqrt{11}$; $\sqrt{(\frac{11}{12})}$
4) $-\mathbf{i}-3\mathbf{j}+2\mathbf{k}$; $\mathbf{i}-\mathbf{j}$
5) $\sqrt{(\frac{3}{5})}$; $\pm(\mathbf{i}-\mathbf{j}-2\mathbf{k})/\sqrt{6}$

6) $\pm(4\mathbf{i}-5\mathbf{j}+7\mathbf{k})/3\sqrt{10}$
8) $k\mathbf{a}=3\mathbf{b}-2\mathbf{c}$
10) $-2\mathbf{k}+t(\mathbf{i}-3\mathbf{k})$

Exercise 5g – p. 217

1) $\frac{1}{2}\sqrt{35}$ 2) $\frac{1}{2}\sqrt{5}$ 4) 1
5) $\sqrt{265}$ 6) 3 7) $\frac{1}{3}$
8) $\frac{11}{6}$
9) $\frac{1}{6}|\mathbf{b}\times\mathbf{c}.\mathbf{a}+\mathbf{b}\times\mathbf{a}.\mathbf{d}+\mathbf{c}\times\mathbf{b}.\mathbf{d}+\mathbf{a}\times\mathbf{c}.\mathbf{d}|$
10) a) $\sqrt{2}$ b) $\sqrt{2}/2$ c) $\dfrac{|\mathbf{c}\times\mathbf{b}|}{|\mathbf{c}|}$
d) $\dfrac{5|\mathbf{p}\times\mathbf{r}|}{|\mathbf{r}+2\mathbf{p}|}$
11) $3\sqrt{2}/2$ 12) $8/\sqrt{29}$
13) a) no b) yes c) no

Exercise 5h – p. 224

1) a) $\mathbf{r}=\mathbf{i}+2\mathbf{j}-\mathbf{k}+\lambda(\mathbf{j}+3\mathbf{k})+\mu(\mathbf{i}-2\mathbf{k})$,
$\mathbf{r}.(2\mathbf{i}-3\mathbf{j}+\mathbf{k})=-5$
b) $\mathbf{r}=\mathbf{i}+\mathbf{j}-2\mathbf{k}+\lambda(\mathbf{j}-3\mathbf{k})+\mu(3\mathbf{i}+\mathbf{k})$,
$\mathbf{r}.(-\mathbf{i}+9\mathbf{j}+3\mathbf{k})=2$
2) a) $\mathbf{r}.(5\mathbf{i}-2\mathbf{j}-3\mathbf{k})=7$
b) $\mathbf{r}.(-\mathbf{j}+2\mathbf{k})=3$
c) $\mathbf{r}.(2\mathbf{i}+\mathbf{k})=5$
3) a) $x+y-z=2$
b) $2x+3y-4z=1$
c) $2x-5y-z=-15$
4) $\mathbf{r}.(\mathbf{i}-\mathbf{j}-\mathbf{k})=0$
5) $\mathbf{r}=-3\mathbf{i}-2\mathbf{j}+\lambda(\mathbf{i}-2\mathbf{j}+\mathbf{k})+\mu(2\mathbf{i}-\mathbf{j}+2\mathbf{k})$
6) $\mathbf{r}=\mathbf{i}+\mathbf{j}+\lambda(\mathbf{i}+\mathbf{k})+\mu(\mathbf{i}-\mathbf{j}+\mathbf{k})$; the second line is contained in the plane.
7) $\mathbf{r}.(7\mathbf{i}+2\mathbf{j}-3\mathbf{k})=3$, $3/\sqrt{62}$
8) $3/\sqrt{5}$ 9) 2 10) $\mathbf{r}=\lambda(\mathbf{i}-2\mathbf{j}+\mathbf{k})$
11) $\mathbf{r}=2\mathbf{i}+\mathbf{j}+\mathbf{k}+\lambda(\mathbf{i}+2\mathbf{j}-3\mathbf{k})$
12) $\mathbf{r}.(\mathbf{i}-2\mathbf{j})=-2$
13) $\mathbf{r}.(\mathbf{j}-\mathbf{k})=0$ 14) $(\frac{5}{2},-\frac{1}{2},-\frac{1}{2})$
15) $(1,-1,4)$ 16) $7\sqrt{3}/3$
17) a) intersecting b) intersecting
c) parallel d) contained in the plane.

Exercise 5i – p. 230

1) a) $3/\sqrt{11}$ b) $1/\sqrt{1102}$
c) $20/\sqrt{442}$
2) a) $\sqrt{3}/9$ b) $\sqrt{15}/5$ c) $\frac{8}{21}$
3) a) 2 b) $\frac{22}{9}$ c) $9/\sqrt{5}$
4) a) $\mathbf{r}=-\frac{1}{3}(10\mathbf{j}+11\mathbf{k})+\lambda(3\mathbf{i}+5\mathbf{j}+7\mathbf{k})$
b) $\mathbf{r}=\mathbf{j}+4\mathbf{k}+\lambda(\mathbf{i}-\mathbf{j}-3\mathbf{k})$

Exercise 5j – p. 236

1) 1 2) $\frac{1}{5}(13\mathbf{i}+10\mathbf{j}-7\mathbf{k})$
3) $4/\sqrt{259}$
4) $(4,4,2)$; $\mathbf{r}.(\mathbf{i}+4\mathbf{j}-\mathbf{k})=18$
5) $(0,0,0)$; $(2,0,0)$; $(0,-4,0)$; $(0,0,4)$; $\frac{16}{3}$
6) $\frac{16}{3}$; $\sqrt{50}/6$; $200\pi/27$ 7) $4, 16\pi$
8) $\mathbf{r}=2\mathbf{i}+\mathbf{k}+\lambda(\mathbf{i}-2\mathbf{j}+2\mathbf{k})$; $(\frac{11}{3},-\frac{10}{3},\frac{13}{3})$; $(-\frac{4}{3},-\frac{25}{3},-\frac{17}{3})$; $625\pi/3$
9) $(\mathbf{j}+\mathbf{k})/\sqrt{2}$; $\mathbf{r}=(3\mathbf{i}+2\mathbf{j})+\lambda(\mathbf{j}+\mathbf{k})$; $(3, 2+3\sqrt{2}, 3\sqrt{2})$
10) $\mathbf{r}=\mathbf{i}+\mathbf{j}+\lambda(3\mathbf{i}-\mathbf{j}+5\mathbf{k})+\mu(\mathbf{i}-2\mathbf{j})$
11) $\mathbf{r}=\lambda(2\mathbf{i}+\mathbf{j}-5\mathbf{k})$;
$\mathbf{r}=\mathbf{i}+2\mathbf{j}-\mathbf{k}+\mu(2\mathbf{i}+\mathbf{j}-5\mathbf{k})$;
$\mathbf{r}.(2\mathbf{i}+\mathbf{j}-5\mathbf{k})=0$; $\sqrt{30}/5$

Multiple Choice Exercise 5 – p. 240

1) b	2) c	3) b
4) e	5) c	6) b
7) d	8) a	9) d
10) a	11) c	12) c, d
13) b, c, d	14) b, d	15) b
16) *A*	17) *A*	18) *C*
19) *D*	20) *B*	21) b, d
22) *C*	23) *I*	24) c, d
25) a, c, d	26) *F*	27) *F*
28) *T*	29) *F*	30) *F*

Miscellaneous Exercise 5 – p. 243

9) a) $\arccos\sqrt{2}/2$ b) $\arccos\sqrt{6}/6$
10) a) 23 b) $2\mathbf{i}+5\mathbf{j}-6\mathbf{k}$
c) 19° d) $\frac{1}{2}\sqrt{65}$
e) $2x+5y-6z=6$
11) $\mathbf{c}=\mathbf{a}+2\mathbf{b}$; 2:5
12) $\mathbf{r}.(\mathbf{i}+2\mathbf{j}+5\mathbf{k})=9$; $\frac{1}{2}\sqrt{30}$; $1/\sqrt{30}$
14) a) $\mathbf{p}\times\mathbf{q}=\mathbf{0}$
b) $\mathbf{p}.\mathbf{q}=0$; 90°; $\frac{220}{3}$; $-7\mathbf{i}+22\mathbf{j}-\mathbf{k}$
15) b) $\frac{4}{3}$; 3 16) b) $(1,1,2)$; 70.53°
17) 2; $4\mathbf{i}+\mathbf{j}-2\mathbf{k}$; $\frac{2}{5}$; $\frac{1}{2}\sqrt{21}$
18) $\triangle OPQ+\triangle OQR+\triangle ORP$; $\triangle OPQ-\triangle OQR-\triangle ORP$
19) b) $-\frac{1}{4},\frac{1}{4}$ 20) $3\sqrt{2}$
21) $\mathbf{r}.(2\mathbf{i}+\mathbf{k})=0$; $\sqrt{5}$
24) $\mathbf{r}=\lambda\mathbf{i}+(\lambda-2)\mathbf{j}+\mathbf{k}$ 27) $\mathbf{a}-\mathbf{c}$
28) a) $\lambda(\mathbf{a}+\mathbf{b})$ b) $\mathbf{b}+\mu(\frac{1}{2}\mathbf{a}-\mathbf{b})$

Exercise 6a – p. 257

1) a) $\begin{pmatrix}\sqrt{3}/2 & 0 & \frac{1}{2}\\ 0 & 1 & 0\\ -\frac{1}{2} & 0 & \sqrt{3}/2\end{pmatrix}$

b) $\begin{pmatrix} 1 & 0 & 0 \\ 0 & \sqrt{2}/2 & -\sqrt{2}/2 \\ 0 & \sqrt{2}/2 & \sqrt{2}/2 \end{pmatrix}$

c) $\begin{pmatrix} 1 & 0 & 0 \\ 0 & 1 & 0 \\ 0 & 0 & -1 \end{pmatrix}$ d) $\begin{pmatrix} 1 & 0 & 0 \\ 0 & -1 & 0 \\ 0 & 0 & 1 \end{pmatrix}$

e) $\begin{pmatrix} 0 & 1 & 0 \\ 1 & 0 & 0 \\ 0 & 0 & -1 \end{pmatrix}$ f) $\begin{pmatrix} \frac{1}{3} & \frac{2}{3} & \frac{2}{3} \\ \frac{2}{3} & \frac{1}{3} & -\frac{2}{3} \\ \frac{2}{3} & -\frac{2}{3} & \frac{1}{3} \end{pmatrix}$

g) $\begin{pmatrix} 2 & 0 & 0 \\ 0 & 2 & 0 \\ 0 & 0 & 2 \end{pmatrix}$ h) $\begin{pmatrix} \frac{1}{3} & 0 & 0 \\ 0 & \frac{1}{3} & 0 \\ 0 & 0 & \frac{1}{3} \end{pmatrix}$

i) $\begin{pmatrix} 2 & 0 & 0 \\ 0 & 1 & 0 \\ 0 & 0 & 2 \end{pmatrix}$

2) a) reflection in the plane $x - z = 0$
 b) a rotation of π about Oz together with a stretch by a factor of 2 parallel to Ox and Oy.
 c) all points are mapped to the xy plane.

3) a) $\begin{pmatrix} 1 & 0 & 0 \\ 0 & \sqrt{2}/2 & -\sqrt{2}/2 \\ 0 & -\sqrt{2}/2 & -\sqrt{2}/2 \end{pmatrix}$

 b) $\begin{pmatrix} 1 & 0 & 0 \\ 0 & \sqrt{2}/2 & \sqrt{2}/2 \\ 0 & \sqrt{2}/2 & -\sqrt{2}/2 \end{pmatrix}$

 c) $\begin{pmatrix} \sqrt{3} & 0 & 1 \\ 0 & 2 & 0 \\ -1 & 0 & \sqrt{3} \end{pmatrix}$ d) $\begin{pmatrix} 0 & 2 & 0 \\ 1 & 0 & 0 \\ 0 & 0 & -2 \end{pmatrix}$

4) **AB** is a rotation of 180° about Oz followed by a reflection in the xz plane together with a stretch by a factor of 3 parallel to the y axis
 AC is a rotation of 180° about Oz followed by a reflection in the xz plane together with an enlargement by a factor of 2.
 BA = **AB**

Exercise 6b – p. 261

1) a) enlarges volume by a factor 16
 b) enlarges volume by a factor 8
 c) does not alter volume
 d) enlarges volume by a factor 12

2) a) $x + y - z = 0$ b) $x + 4y - 2z = 0$
 c) $z = 0$ d) $5x + 7y - 11z = 0$

3) a) $x = \frac{y}{2} = \frac{z}{-1}$ b) $\frac{x}{5} = \frac{y}{-2} = z$
 c) $x = z = 0$ d) $x = \frac{y}{3} = \frac{z}{-1}$

4) a) maps all points to the plane $x + y - z = 0$
 b) maps all points to the plane $x - 17y - 7z = 0$
 c) changes a unit cube to a rhomboid of volume 2 cu. units
 d) maps all points to the line $x = \frac{y}{-3} = \frac{z}{2}$
 e) maps all points to the origin
 f) maps all points to the x axis.

5) $1 \pm \sqrt{2}$ 6) (b)

Exercise 6c – p. 266

1) a) $\begin{pmatrix} \frac{1}{2} & 0 \\ 0 & 1 \end{pmatrix}$ b) $\begin{pmatrix} \frac{1}{5} & \frac{1}{10} \\ -\frac{2}{5} & \frac{3}{10} \end{pmatrix}$
 c) – d) – e) $\begin{pmatrix} \frac{1}{3} & -\frac{1}{3} \\ -\frac{2}{3} & \frac{5}{3} \end{pmatrix}$
 f) $\begin{pmatrix} -\frac{3}{10} & -\frac{2}{5} \\ \frac{1}{10} & -\frac{1}{5} \end{pmatrix}$
 g) $\begin{pmatrix} \sin\theta & -\cos\theta \\ \cos\theta & \sin\theta \end{pmatrix}$
 h) $\frac{1}{ps - rq}\begin{pmatrix} s & -q \\ -r & p \end{pmatrix}$

2) a) $\begin{pmatrix} \cos\theta & \sin\theta \\ -\sin\theta & \cos\theta \end{pmatrix}$ b) $\begin{pmatrix} \frac{1}{2} & 0 \\ 0 & \frac{1}{2} \end{pmatrix}$
 c) $\begin{pmatrix} \frac{1}{3} & 0 \\ 0 & 1 \end{pmatrix}$ d) $\begin{pmatrix} 1 & 0 \\ -1 & 1 \end{pmatrix}$

3) a) $14X + Y + 22 = 0$
 b) $13X - 3Y + 44 = 0$

4) a) $y = 2x$ and $3y + x = 0$ b) none
 c) $y = x$ and $3y + 2x = 0$

6) $2Y + X = 0$, $2y = x - 3$

Exercise 6d – p. 269

1) $\mathbf{A}^T = \begin{pmatrix} 1 & -1 & 2 \\ 2 & 1 & 5 \\ 0 & 0 & 1 \end{pmatrix}$,

$$\mathbf{A}^* = \begin{pmatrix} 1 & -2 & 0 \\ 1 & 1 & 0 \\ -7 & -1 & 3 \end{pmatrix}, \quad \mathbf{A}^{-1} = \tfrac{1}{3}\mathbf{A}^*$$

2) $$\mathbf{B}^T = \begin{pmatrix} 2 & 5 & -1 \\ 3 & 3 & 2 \\ 1 & 4 & 5 \end{pmatrix},$$

$$\mathbf{B}^* = \begin{pmatrix} 7 & -13 & 9 \\ -29 & 11 & -3 \\ 13 & -7 & -9 \end{pmatrix},$$

$\mathbf{B}^{-1} = -\frac{1}{60}\mathbf{B}^*$

3) $$\mathbf{C}^T = \begin{pmatrix} 1 & 2 & 6 \\ -1 & 0 & -2 \\ 3 & 4 & 22 \end{pmatrix},$$

$$\mathbf{C}^* = \begin{pmatrix} 8 & 16 & -4 \\ -20 & 4 & 2 \\ -4 & -4 & 2 \end{pmatrix},$$

$\mathbf{C}^{-1} = \frac{1}{16}\mathbf{C}^*$

4) $$\mathbf{D}^T = \begin{pmatrix} -1 & 0 & 1 \\ 0 & 5 & 4 \\ 4 & -2 & -1 \end{pmatrix},$$

$$\mathbf{D}^* = \begin{pmatrix} 3 & 16 & -20 \\ -2 & -3 & -2 \\ -5 & 4 & -5 \end{pmatrix},$$

$\mathbf{D}^{-1} = -\frac{1}{23}\mathbf{D}^*$

5) $$\mathbf{E}^T = \begin{pmatrix} 2 & 3 & 5 \\ -1 & 2 & -2 \\ 4 & -1 & 9 \end{pmatrix},$$

$$\mathbf{E}^* = \begin{pmatrix} 16 & 1 & -7 \\ -32 & -2 & 14 \\ -16 & -1 & 7 \end{pmatrix},$$

$\mathbf{E}^{-1}$ does not exist

9) $$\mathbf{A}^T = \begin{pmatrix} 2 & 3 \\ 1 & -1 \\ 4 & 0 \end{pmatrix},$$

$$\mathbf{B}^T = \begin{pmatrix} 4 & 0 & 1 \\ 1 & 2 & 0 \end{pmatrix}, \quad \mathbf{AB} = \begin{pmatrix} 12 & 4 \\ 12 & 1 \end{pmatrix}$$

12) $$\mathbf{A}^{-1} = \tfrac{1}{2}\begin{pmatrix} -4 & 2 \\ 3 & -1 \end{pmatrix},$$

$a = -2,\ b = 5$

Exercise 6e – p. 275

1) $(3, 4, -6)$

2) $$\frac{1}{16}\begin{pmatrix} -7 & 6 & -5 \\ -5 & 2 & 1 \\ 13 & -2 & 7 \end{pmatrix},$$

$x = \frac{7}{16}, y = \frac{5}{16}, z = \frac{19}{16}$

3) $$\frac{1}{12}\begin{pmatrix} 2 & -3 & 1 \\ 8 & -12 & -8 \\ -2 & 9 & 5 \end{pmatrix},$$

$x = \frac{1}{3}, y = -\frac{5}{3}, z = \frac{5}{3}$

4) $$\begin{pmatrix} -2 & 8 & -19 \\ 1 & -4 & 10 \\ 1 & -3 & 7 \end{pmatrix}$$

a) $X - 3Y + 6Z = -4$ b) $Y = 2$
c) $2X - 11Y + 27Z = 1$

5) a) $\mathbf{r} = \mathbf{i} - 2\mathbf{j} - \mathbf{k} + \lambda(7\mathbf{i} + 9\mathbf{j} + 3\mathbf{k})$
b) $\mathbf{r} = 5\mathbf{j} + 2\mathbf{k} + \mu(9\mathbf{i} + 5\mathbf{j} + \mathbf{k})$

6) $x - z = D$ for all values of D

7) The y axis and any line through O in the xz plane

8) $$\mathbf{A}^{-1} = \begin{pmatrix} 0 & 0 & 1 \\ 1 & 0 & 0 \\ 0 & -1 & 0 \end{pmatrix}$$

and represents a reflection in the plane $x - z = 0$ followed by a rotation of $-90°$ about Ox;
$\lambda = -1,\ x = -y = -z$

Exercise 6f – p. 281

1) $x = -\frac{1}{5}, y = -\frac{8}{5}, z = -2$
2) $x = \frac{3}{2}, y = \frac{3}{4}, z = \frac{7}{4}$
3) $x = 2, y = 5, z = 1$
4) $x = 1, y = -4, z = -4$
5) $x = 3, y = 2, z = -1$
6) $x = -2, y = 1, z = -1$
7) $x = 3, y = 5, z = -2$
8) $x = 3, y = 12, z = 9$

Exercise 6g – p. 287

1) an infinite set of solutions dependent on one parameter
2) unique solution
3) no solution
4) an infinite set of solutions dependent on one parameter
5) an infinite set of solutions dependent on one parameter

6) an infinite set of solutions dependent on two parameters
7) unique solution
8) no solution
9) $x = \lambda, y = -1 - \frac{1}{2}\lambda, z = 1 - \frac{5}{2}\lambda$;
$x = \frac{6}{5}, y = -3, z = -\frac{7}{5}$;
$x = -1, y = \lambda, z = \lambda - 3$;
$x = \lambda, y = \mu, z = 1 - (\lambda + \mu)$;
$x = \lambda, y = 3\lambda, z = 2\lambda$;
$x = 1, y = 0, z = 0$
10) The set of points on the line $x = \frac{5}{3}$, $3y = 3z + 1$
11) The set of points on the line $\dfrac{x + \frac{5}{3}}{-14} = \dfrac{y}{3} = \dfrac{z - \frac{2}{3}}{5}$
13) The set of points in the plane $x - y + 2z = 1$
14) $X = \dfrac{Y}{2} = \dfrac{Z}{3}$; all points on the plane $x + 2y - z = a$ map to a point on the line for all values of a.
15) $\frac{1}{4}(-3 \pm \sqrt{17})$
16) $\begin{vmatrix} a_1 & b_1 & c_1 \\ a_2 & b_2 & c_2 \\ a_3 & b_3 & c_3 \end{vmatrix} = 0$
17) $\begin{vmatrix} a_1 & b_1 & c_1 \\ a_2 & b_2 & c_2 \\ a_3 & b_3 & c_3 \end{vmatrix} = 0$;
the planes represented by [1] have a line in common.
18) $\begin{vmatrix} a_1 & b_1 & c_1 \\ a_2 & b_2 & c_2 \\ a_3 & b_3 & c_3 \end{vmatrix} = 0$;
[1] represent a set of concurrent lines.
19) $a = -1$
20) Any matrix of the form
$$\begin{pmatrix} a_1 & b_1 & c_1 \\ a_2 & b_2 & c_2 \\ -(a_1 + a_2) & -(b_1 + b_2) & -(c_1 + c_2) \end{pmatrix}$$
21) $\begin{vmatrix} a_1 & b_1 \\ a_2 & b_2 \end{vmatrix} \neq 0$

Exercise 6h – p. 292

1) $\dfrac{1}{19}\begin{pmatrix} -1 & -14 & 6 \\ -4 & 1 & 5 \\ 7 & 3 & -4 \end{pmatrix}$

2) $\dfrac{1}{22}\begin{pmatrix} 11 & -5 & 1 \\ -11 & -3 & 5 \\ -11 & 7 & 3 \end{pmatrix}$

3) $\dfrac{1}{75}\begin{pmatrix} 0 & -15 & 6 \\ 75 & 75 & -75 \\ -50 & -50 & 55 \end{pmatrix}$

4) $\dfrac{1}{5}\begin{pmatrix} 13 & -11 & 1 \\ 1 & -2 & 2 \\ -5 & 5 & 0 \end{pmatrix}$

5) no inverse

6) $\dfrac{1}{32}\begin{pmatrix} 55 & 33 & -27 \\ 19 & 5 & -7 \\ -64 & -32 & 32 \end{pmatrix}$

Multiple Choice Exercise 6 – p. 294

1) b	2) d	3) b
4) a	5) d	6) d
7) a	8) a, c	9) c
10) a, c	11) b, c	12) *B*
13) *C*	14) *C*	15) *C*
16) *C*	17) *T*	18) *T*
19) *F*		

Miscellaneous Exercise 6 – p. 297

1) $\left.\begin{aligned} 2X + Y + Z &= 4 \\ 7X + Y + 2Z &= 6 \\ 7X - 2Y - Z &= 18 \end{aligned}\right\}$ meet at $(1\frac{2}{3}, 2\frac{1}{3}, -1\frac{2}{3})$
2) $\lambda = 3$ or 6; $(1, 2)$ or $(-\frac{1}{5}, \frac{1}{5})$
3) $\mathbf{M}^2 = \begin{pmatrix} -1 & 0 & 0 \\ 0 & -1 & 0 \\ 0 & 0 & 1 \end{pmatrix}$

M represents a rotation of $-90°$ about Oz, $\mathbf{M}^2$ a rotation of $180°$ about Oz
4) a) $x = 2, y = 1, z = -10$
5) a) $3X - 2Y = 0$ b) $X = \dfrac{Y}{3} = \dfrac{Z}{5}$
c) $\begin{pmatrix} 0 & 1 & 0 \\ -1 & 0 & 0 \\ 0 & 0 & 1 \end{pmatrix}$
6) $a = 2,\ b = -\frac{4}{5},\ c = \frac{3}{5}$; $\mathbf{x} = \begin{pmatrix} \frac{1}{5} \\ 1 \\ -\frac{2}{5} \end{pmatrix}$
7) $\mathbf{A}^{-1} = \begin{pmatrix} 1 & 0 & 0 \\ 1 & 1 & 0 \\ -5 & -2 & 1 \end{pmatrix}$,

$$\mathbf{B}^{-1} = \begin{pmatrix} 1 & -4 & 14 \\ 0 & 1 & -3 \\ 0 & 0 & 1 \end{pmatrix},$$

$$\begin{pmatrix} x_1 \\ x_2 \\ x_3 \end{pmatrix} = \begin{pmatrix} 5 \\ -1 \\ 0 \end{pmatrix}$$

8) $-\frac{1}{6}\begin{pmatrix} 0 & -3 & 0 \\ 0 & -3 & 6 \\ -2 & 2 & -4 \end{pmatrix},$

$$\begin{pmatrix} 7 & -3 & 0 \\ 0 & 4 & 6 \\ -2 & 2 & 3 \end{pmatrix}, \begin{pmatrix} -6 & 14 & 21 \\ 14 & -6 & 0 \\ 7 & -7 & -6 \end{pmatrix}$$

9) $\mathbf{A}^{-1} = \begin{pmatrix} \frac{1}{3} & -\frac{1}{3} & \frac{1}{3} \\ -\frac{1}{2} & \frac{1}{2} & 0 \\ \frac{4}{3} & -\frac{1}{3} & -\frac{2}{3} \end{pmatrix}$; 1, −1, 6

10) $\begin{pmatrix} \frac{5}{2} & -1 \\ -4 & 2 \\ 3 & -\frac{5}{4} \end{pmatrix}$

11) $55(k-1)$,
a) $x = -3, y = 2, z = 3$,
b) $\frac{x}{10} = -y = -\frac{z}{11}$;
three planes through O with a line in common.

12) 5; $x = \lambda, y = 3 - 2\lambda, z = \lambda$;
three planes intersecting in a line.

13) a) $\begin{pmatrix} -13 \\ 8 \end{pmatrix}$, b) $\begin{pmatrix} 0 \\ 0 \end{pmatrix}$,
c) $\begin{pmatrix} 1-2\lambda \\ \lambda \end{pmatrix}$, d) no solution

$\mathbf{M}^{-1}$ exists in a) and b) but not in c) and d).

14) $\begin{pmatrix} -2 & -2 \\ 6 & 5 \end{pmatrix}$

15) $x = -5, y = 3, z = -4$

16) $x = 2, y = -3, z = 4$;
a) $x = \lambda - 2, y = 1 - \lambda, z = \lambda$;
b) no solution

17) $-\frac{1}{160}\begin{pmatrix} 6 & -22 & -18 \\ 23 & -31 & 11 \\ -31 & 7 & 13 \end{pmatrix}$;

$x = 0.6, y = -0.2, z = 0.4$

18) $\frac{2}{3}\pi$

19) a) $x = 3, y = -2, z = 4$
b) $x = 3\lambda, y = -5\lambda + 3, z = 11\lambda - 7$

20) $\lambda = 1$; $x = \alpha, y = \beta, z = -\alpha - \beta$
$\lambda = 4$; $x = y = z = \gamma$

21) $-(\mathbf{A} + p\mathbf{I})/q$ 22) $(-z, y, x)$

Exercise 7a – p. 307

1) b	2) a	3) c
4) b	5) a	6) b
7) b	8) a	9) b
10) c	11) true	12) false
13) false	14) true	15) true

Exercise 7b – p. 309

1) c	2) b	3) c	4) b
5) c	6) d	7) d	8) e

Exercise 7c – p. 311

1) true	2) false	3) true
4) false	5) true	6) false
7) false	8) true	9) false
10) true		

Exercise 7d – p. 316

1) $p \Leftrightarrow q$	2) $p \Leftarrow q$	3) $p \Rightarrow q$
4) $p \Leftrightarrow q$	5) $p \Leftarrow q$	6) $p \Leftrightarrow q$
7) $p \Leftarrow q$	8) $p \Rightarrow q$	9) $p \Rightarrow q$
10) $p \Leftarrow q$	11) $p \Rightarrow q$	12) $p \Leftarrow q$
13) $p \Leftrightarrow q$	14) $p \Leftrightarrow q$	15) $p \Leftarrow q$
16) $p \Leftrightarrow q$		

17) a) If this car is a good car it is made in Britain
b) If this car is not a good car it is not made in Britain
c) If this car is not made in Britain it is not a good car
d) This car is made in Britain and it is not a good car

18) a) If $b^2 - 4ac \geqslant 0$ then $ax^2 + bx + c = 0$ has two real roots
b) If $b^2 - 4ac < 0$ then $ax^2 + bx + c = 0$ does not have two real roots
c) If $ax^2 + bx + c = 0$ does not have two real roots then $b^2 - 4ac < 0$
d) $ax^2 + bx + c = 0$ has two real roots and $b^2 - 4ac < 0$

19) a) If Henry is doing A level maths he is mad
b) Henry is doing A level maths because he is mad

c) If Henry is not doing A level maths he is mad
d) Henry is doing A level maths because he is not mad
e) It is not true that if Henry is doing A level maths he is mad
f) Henry is doing A level maths if and only if he is mad
g) If Henry is not doing A level maths he is not mad
h) Henry is not doing A level maths because he is not mad

20) a) $p \Rightarrow q$ b) $\sim p \Leftarrow q$
c) $\sim p \Rightarrow \sim q$ d) $\sim (p \Leftarrow q)$
e) $\sim p \Leftarrow q$ f) $p \Longleftrightarrow q$
g) $\sim p \Leftarrow \sim q$
True statements are a), d), g).

21) a) If x has no real value then $x^2 = -1$
b) It if not true that if $x^2 = -1$ then x has no real value
c) If $x^2 \neq -1$ then it is not true that x has no real value
d) If x can have at least one real value then $x^2 \neq -1$

22) (c)

Exercise 7e – p. 324

1) converse true 2) converse true
3) converse true 4) converse true
5) converse false 6) converse false
13) $u_2 = \frac{1}{3}, u_3 = -\frac{1}{9}, u_4 = -\frac{11}{27}$

Exercise 8a – p. 337

1) a) $\frac{1}{2}(e + e^{-1})$ b) $\frac{1}{2}(1 + 3^{11}) = 88\,574$
c) $e^{\frac{1}{3}}$ d) $(\frac{3}{2})^{10}$ e) $\ln \frac{1}{2}$
f) $\sin 3^c = 0.1411$ g) e^{-2} h) $\frac{2}{3}$
2) a) $\dfrac{ab^2(b^2 - b^{2n})}{1 - b^2}$ b) $\frac{1}{2}(e^2 - 1)$
c) $12[(\frac{2}{3})^n - (\frac{2}{3})^{2n}]$ d) $(e - 1)/e$
e) ln 2 f) $1 - \cos \frac{1}{2}^c = 0.1224$
g) $1 - \sin 1^c = 0.1585$
h) $1/e^2(e - 1)$
3) a) 3e b) $2 + e^{\frac{1}{2}}$ c) ln 2
d) $\cos 1^c + \cos \sqrt{2^c}$ e) $\frac{2}{3}$
f) $\sqrt{(\frac{3}{2})}$

Exercise 8b – p. 341

1) $\frac{3}{4}$ 2) $\frac{1}{9}[1 - (3n + 1)(-1)^n 2^n]$
3) $-x + (x + 1)\ln(1 + x)$ for $-1 < x \leqslant 1$
4) $5(\frac{1}{2})^8 = \frac{5}{256}$
5) $(2n - 1)2^{2n} - (n - 2)2^{n-1}$

Exercise 8c – p. 344

1) a) $\dfrac{n(3n + 5)}{4(n + 1)(n + 2)}$ b) $\dfrac{n - 2}{4(n + 2)}$
c) $\dfrac{n + 1}{n(2n + 1)}$ d) $\dfrac{n(n + 1)}{2(2n + 1)(2n + 3)}$
2) a) $\dfrac{n(5n + 13)}{12(n + 2)(n + 3)}$
b) $\dfrac{n(n + 1)}{4(n + 2)(n + 3)}$
c) $\dfrac{5}{48} - \dfrac{2n + 3}{8(n + 1)(n + 2)}$
d) $\dfrac{1}{8}\left[\dfrac{11}{12} - \dfrac{4n^2 + 4n - 2}{n(n + 2)(n^2 - 1)}\right]$
3) a) $\dfrac{n(3n + 5)}{4(n + 2)(n + 1)}$ b) $\dfrac{n}{3(2n + 3)}$
c) $\dfrac{n(4n + 5)}{3(2n + 1)(2n + 3)}$

Exercise 8d – p. 348

1) $\dfrac{n(n + 3)}{4(n + 1)(n + 2)}$ 2) $\dfrac{n(n + 2)}{(n + 1)^2}$
3) $n(n^2 + 6n + 11)$
4) $\dfrac{n}{2}[2n^3 + 4n^2 + 6n + 1]$
5) $4(r + 1)(r + 2)(r + 3)$
6) $\dfrac{n}{(n + 1)(n + 2)} - \dfrac{2n + 1}{(2n + 2)(2n + 3)}$
7) $1 - \dfrac{1}{(n + 1)!}$ 8) $(2n + 1)! - 1$
9) $\dfrac{\cos 2\theta - \cos(2n + 2)\theta}{2 \sin \theta}$
10) a) $\dfrac{n}{3}(n + 1)(n + 2)$
b) $\dfrac{n}{4}(n + 1)(n + 2)(n + 3)$
c) $\dfrac{2n}{3}(n + 1)(2n + 1)$
d) $\dfrac{n}{12}(n + 1)(45n^2 + 37n + 2)$
e) $\dfrac{n}{6}(2n - 1)(n - 1)$
f) $-2n^2 - n$ g) 42 075

Exercise 8f – p. 357

1) a) 0 b) 0 c) 1
2) a) 1 b) 0 c) 0
3) a) $\frac{1}{3}$ b) $\frac{1}{4}$ c) $\frac{1}{2}$ d) $\frac{11}{18}$
4) a) 20 003 b) 30 002 c) 174

Miscellaneous Exercise 8 – p. 358

1) $2+10x^2+18x^4+26x^6$; $2(4n+1)$; $3\frac{1}{9}$
2) $r(1-r^{2n})/(1-r^2)$
4) $-2<x\leqslant 2$, $\ln\left(1+\frac{x}{2}\right)$; $-3\leqslant x<3$, $\ln\left(1-\frac{x}{3}\right)$; $x=0$ or 1
5) $-\frac{1}{x+2}+\frac{3}{x+3}-\frac{2}{x+4}$; $\frac{n(n+1)}{6(n+3)(n+4)}$
6) $1+\frac{(-1)^{n+1}}{n+1}$
8) a) $\frac{1}{2}\left[\frac{3}{2}-\frac{1}{n+1}-\frac{1}{n+2}\right]$; $\frac{3}{4}$
b) 1 191 700
9) b) $n(n+1)(2n^2+2n-1)$
10) $\frac{1}{3}n(n+1)(n+2)$ 11) 70
12) 7e 13) $n(n+1)(n+2)(n+3)/4$
14) $2(e^{-2}+1)$
15) a) $7m^3+3m^2-10m-6$ b) 50
17) $6[1-(\frac{2}{3})^n]-n$
19) $n^2[n^2-2n-2]$
23) a) $\frac{r}{1-r}$ b) $\frac{r}{(1-r)^2}$
c) $\frac{r[n-(n+1)r+r^{n+1}]}{(1-r)^2}$
24) 2, 3, -1; $6e+1$
26) a) $\frac{n}{3}(4n^2-1)$
b) $|a|<1$; $\frac{1}{(1-a^2)(1-a)}$
27) $\frac{r^2-r+2}{2}$, $\frac{r^2+r}{2}$; 22 155
28) b) $\frac{1-x^{n+1}}{1-x}$; $\frac{1-(n+1)x^n+nx^{n+1}}{(1-x)^2}$; $2+(n-1)2^{n+1}$
30) $\frac{4n-1}{4}-\frac{2n+1}{2n(n+1)}$ 31) $2e-2$
32) a) $\frac{n}{6}(n+1)(n+2)$ b) $3e-1$

Exercise 9a – p. 372

1) a, b, c, d (for $x>0$), f, g, h, i (for $|x|\leqslant a$), j (for $x\geqslant 0$), m
2) e) $x=n\pi$; infinite k) $x=n$; finite
l) $x=0$; infinite
3) d) $x\leqslant -1$ e) $x=n\pi$
g) $x=1$ h) $x=0$
i) $|x|>a$ j) $x<0$
k) $x=n$ l) $x=0$
4) a) a, h, i, m
b) b, e, l
5) a) (ii), (iii), (v), (vii)
b) (vi), (vii) c) (iii), (iv)
6) d, e (in two parts)
7) 1 8) 1 9) 4
10) 1, 2 11) 0 12) 0, 1
13) a) $f(x)$ and $f'(x)$ discontinuous when $x=0$
b) $f(x)$ continuous; $f'(x)$ discontinuous when $x=n\pi+\pi/2$
c) $f(x)$ continuous; $f'(x)$ discontinuous when $x=0$

Exercise 9b – p. 378

4) a) 3 b) 3 c) 16
5) a) $f(x)\equiv 2-x$ for $0<x\leqslant 4$
$f(x)\equiv 1$ for $4<x\leqslant 5$
$f(x+5)=f(x)$ for all values of x
b) $f(x)\equiv 2x$ for $0<x\leqslant 1$
$f(x)\equiv 5-2x$ for $1<x\leqslant 2$
$f(x)\equiv -x$ for $2<x\leqslant 3$
$f(x+3)=f(x)$ for all values of x

Exercise 9c – p. 384

5) 0.6931 6) 0.095 31 7) 0.182 32

Exercise 9d – p. 389

1) $1+2x^2+\frac{2}{3}x^4$
2) $\frac{x}{2}+\frac{x^3}{48}+\frac{x^5}{3840}$
3) $1+x+\frac{1}{2}x^2$
4) $1+x^2+\frac{1}{3}x^4$
5) $4x^2+\frac{16}{3}x^4+\frac{128}{45}x^6$
6) $1+\frac{x^2}{2}+\frac{x^4}{24}$
7) $\frac{x^2}{2}-\frac{x^4}{12}+\frac{x^6}{45}$
8) $1+x+\frac{1}{2}x^2$
13) $\sinh(x+y)\equiv\sinh x\cosh y+\cosh x\sinh y$
14) $\cosh(x-y)\equiv\cosh x\cosh y-\sinh x\sinh y$
15) $\sinh 3x\equiv 3\sinh x+4\sinh^3 x$
16) $\cosh x+\cosh y\equiv 2\cosh\frac{x+y}{2}\cosh\frac{x-y}{2}$

17) $\sinh 2x \equiv \dfrac{2\tanh x}{1-\tanh^2 x}$

18) $\cosh 2x \equiv \dfrac{1+\tanh^2 x}{1-\tanh^2 x}$

19) $\frac{4}{5}, \frac{3}{5}, \frac{15}{17}$

20) $\frac{12}{5}, \frac{13}{12}, \frac{65}{72}, \frac{65}{97}$

21) a) $\operatorname{sech} x$, $\sinh^2 x$
   b) $\coth x$, $\operatorname{cosech} x$, $\tanh(-x)$, $\sinh(-x)$
   c) $\coth x$, $\operatorname{sech} x$, $\cosh x + \sinh x$, $\sinh^2 x$, $\tanh(-x)$, $\sinh(-x)$

Exercise 9e – p. 391

1) $0, \ln\frac{5}{3}$ 2) $\ln 4$
3) $\ln 2, -\ln 4$ 4) $\ln 2, -\ln 3$
5) $\operatorname{arcosh}\frac{3}{2}$, $\operatorname{arcosh} 2$
6) $\operatorname{arcosh}\frac{4}{3}$ 7) $0, -\ln 2$
8) 0, $\operatorname{arcosh} 4$ 9) $\ln\frac{2}{5}$
10) $\ln 5, -\ln 4$ 11) $r = 3$, $\tanh y = \frac{4}{5}$
12) 12

Exercise 9f – p. 393

1) $-\operatorname{sech} x \tanh x$ 2) $2\operatorname{sech}^2 2x$
3) $-\operatorname{cosech}^2 x$ 4) $-2\operatorname{sech}^2 x \tanh x$
5) $4\cosh 4x$ 6) $6\cosh^2 2x \sinh 2x$
7) $\sinh x + x\cosh x$
8) $\sinh x\,(1 + \operatorname{sech}^2 x)$
9) $e^x(\cosh x + \sinh x) = e^{2x}$
10) $\dfrac{5\sinh 5x}{2\sqrt{\cosh 5x}}$
11) $2x\tanh 3x(3x\operatorname{sech}^2 3x + \tanh 3x)$
12) $e^x \operatorname{cosech} 2x(1 - 2\coth 2x)$
13) $\sinh x\, e^{\cosh x}$ 14) $\coth x$
15) $\frac{1}{2}\operatorname{sech}^2 x\sqrt{\coth x}$
16) $2\tanh x \operatorname{sech}^2 x\, e^{\tanh^2 x}$
17) $\frac{1}{5}\cosh 5x + K$ 18) $\frac{1}{4}\sinh 4x + K$
19) $\frac{1}{2}\ln(K\cosh 2x)$ 20) $\ln(K\sinh x)$
21) $\frac{1}{4}(\sinh 2x - 2x) + K$
22) $\frac{1}{12}(\sinh 3x + 9\sinh x) + K$
23) $\frac{1}{4}(2x\cosh 2x - \sinh 2x) + K$
24) $K - \operatorname{sech} x$
25) $\frac{1}{4}(e^{2x} + 2x) + K$
26) $\frac{1}{8}(\sinh 4x + 2\sinh 2x) + K$
27) $\frac{1}{16}(\cosh 8x + 4\cosh 2x) + K$
28) $\frac{1}{3}\sinh 12$
29) $\frac{1}{4}(e^2 - 2)(e^2 + 1)$ 30) $\frac{1}{4}\ln(\cosh 4)$
31) $\frac{1}{5}(\tanh 20 - \tanh 5)$
32) $bx\cosh u - ay\sinh u = ab$,
$ax\sinh u + by\cosh u = (a^2 + b^2)\sinh u\cosh u$
33) 24, minimum
34) $y = 2\cosh 2x + \sinh 2x$

Exercise 9g – p. 397

1) $\pm\ln(3 + \sqrt{8})$, $\ln(\sqrt{2} - 1)$, $\ln\sqrt{3}$
2) $x \geqslant \frac{1}{3}$ 3) $-2 \leqslant x \leqslant 2$
4) $y = \frac{1}{2}\ln\dfrac{1+x}{1-x}$ 5) $\pm\ln(2 + \sqrt{3})$
7) a) and b) $\pm\ln\left(\dfrac{1 + \sqrt{1-x^2}}{x}\right)$;
   c) $\ln(x^2 - 1 + \sqrt{x^4 - 2x^2 + 2})$
8) $x = \frac{1}{2}$ 9) $x = \frac{1}{3}$
10) $x = \pm 2$

Exercise 9h – p. 401

2) a) $\operatorname{arsinh}\dfrac{x}{2} + K$ b) $\operatorname{arcosh}\dfrac{x}{3} + K$
   c) $\frac{1}{2}\operatorname{arcosh}\dfrac{x}{2} + K$ d) $\frac{1}{2}\operatorname{arsinh}\dfrac{2x}{3} + K$
3) a) $\pm\dfrac{1}{\sqrt{(x^2 + 2x)}}$ b) $\dfrac{2}{\sqrt{(1 + 4x^2)}}$
   c) $\dfrac{1}{1 - x^2} + K$
   d) $\operatorname{arcosh} x \pm \dfrac{x}{\sqrt{(x^2 - 1)}}$
   e) $\dfrac{-1}{x\sqrt{(x^2 + 1)}}$ f) $\dfrac{\pm 1}{2\sqrt{(x^2 - x)}}$
   g) $\dfrac{\pm 1}{x\sqrt{(1 - x^2)}}$
   h) $\dfrac{1}{(1 - x)\sqrt{(x^2 - 2x + 2)}}$
   i) $\dfrac{\pm 2\cosh 2x}{\sqrt{(\sinh^2 2x - 1)}}$
   j) $\dfrac{\pm 1}{\sqrt{(x^2 - 1)}}\, e^{\operatorname{arcosh} x}$

Add K to each answer in questions 4–15.

4) $\frac{2}{3}\operatorname{arsinh}\dfrac{3x}{2} + \dfrac{x}{2}\sqrt{9x^2 + 4}$
5) $\dfrac{x}{8}(8x^2 - 5)\sqrt{4x^2 - 1} + \frac{3}{16}\operatorname{arcosh} 2x$
6) $\frac{1}{2}\{\operatorname{arsinh}(x + 2) + (x + 2)\sqrt{x^2 + 4x + 5}\}$
7) $\frac{1}{2}\{(x + 2)\sqrt{x^2 + 4x + 3} - \operatorname{arcosh}(x + 2)\}$
8) $\operatorname{arsinh}\left(\dfrac{x\sqrt{2}}{2}\right)$
9) $\sqrt{2}/2\operatorname{arcosh} x\sqrt{2}$
10) $\operatorname{arsinh}(x + 1)$
11) $\operatorname{arcosh}(x + 1)$
12) $\frac{1}{2}(x + 2)\sqrt{x^2 + 4x} - 2\operatorname{arcosh}\left(\dfrac{x + 2}{2}\right)$

13) $\frac{1}{2}$ arsinh $(8x - 1)$
14) $\sqrt{x^2+1}$ + arsinh x
15) $\sqrt{x^2-1}$ − arcosh x
16) $\frac{1}{3}$(arcosh 6 − arcosh 3)
$= \pm\frac{1}{3}\ln\left(\frac{6+\sqrt{35}}{3+\sqrt{8}}\right)$
17) arsinh $\frac{1}{2}$
18) arcosh 2 − arcosh $\frac{3}{2}$
$= \pm\ln\left(\frac{4+2\sqrt{3}}{3+\sqrt{5}}\right)$
19) $\sqrt{5}$ + arcosh $\frac{3}{2}$
20) $2\sqrt{2}$ + arsinh 1
21) $2\sqrt{5} - \frac{8}{3}$ arcosh $\frac{3}{2}$
Add K to each answer in questions 22–33
22) $\frac{1}{4}\arcsin 2x + \frac{x}{2}\sqrt{1-4x^2}$
23) $\arcsin\frac{x}{2}$ 24) $\operatorname{arcosh}\frac{x}{3}$
25) $\frac{1}{6}\ln\frac{3x-1}{3x+1}$ 26) $\frac{1}{3}\arctan 3x$
27) $\frac{1}{3}$ arsinh $3x$
28) $\sqrt{x^2-4} + 2\operatorname{arcosh}\frac{x}{2}$
29) $\ln(x-2)$ 30) $\frac{1}{2}\ln(x^2-4)$
31) $\sqrt{x^2-4}$ 32) $2\sqrt{x-1}$
33) $\sqrt{x^2+1}$ − arcosech x
34) 3 35) 1 36) ∞ 37) 0

Miscellaneous Exercise 9 – p. 404

2) $\frac{5}{6}$ 3) both continuous
6) $n\pi$; (i) $\frac{1}{2}(\pi^2 - 4)$ (ii) $\frac{1}{8}(5\pi^2 - 24)$
11) 4 12) 0, ln 4
13) a) $\frac{1}{3}$arsinh $3x + K$
b) $\frac{1}{12}\sinh 6x - \frac{x}{2} + K$
15) $\pm\frac{1}{2}$ 16) a) $\frac{1}{2}\ln\frac{1}{5}$
17) coth θ 18) $(-\frac{3}{4}, \ln\frac{1}{2})$
19) a) $\frac{e^x - e^{-x}}{e^x + e^{-x}}$;
$-\frac{1}{2}$ (Note. $\tanh y \neq -2$)
b) arcosh $\frac{5}{3}$; $1 - \sqrt{2}$ + arsinh 1
20) $\cosh x \cosh y + \sinh x \sinh y$;
$\sinh x \cosh y + \cosh x \sinh y$;
$2\cosh^2 x - 1$; $4\cosh^3 x - 3\cosh x$
21) $\{\ln(1+\sqrt{2}), -1\}$, minimum
22) $\frac{1}{10}(e^5 - e^{-5}) + \frac{1}{6}(e^3 - e^{-3}) - (e - e^{-1})$
23) $\arctan(\tanh x) + K$
25) $\ln\left(\frac{1+\sqrt{x^2+1}}{x}\right)$; $\sqrt{3}$
27) $x = \ln\frac{1}{2}, y = \ln\frac{3}{2}$

28) a) $\operatorname{arsinh}\left(\frac{x-1}{3}\right) + K$
b) $\arctan\left(\frac{x-1}{3}\right) + K$
29) $\ln(7 + 5\sqrt{2})$
31) $\frac{3}{4}$; $\frac{\pi}{32}(16\ln 2 + 15)$
32) $x = \ln(1+\sqrt{2}), y = \ln(5 + 2\sqrt{6})$;
$x = \ln\left(\frac{1+\sqrt{37}}{6}\right)$,
$y = \ln\left(\frac{25+\sqrt{481}}{12}\right)$

Exercise 10a – p. 418

1) $x = 0$ 2) $y + 3x = 0$
3) $y = x$ 4) $2y = 3x$
5) $x = 0$ 6) $y = 0$
7) $y = \pm x$ 8) $2y = \pm x$
9) (1, 0) 10) none
11) none 12) (0, −1)
13) (0, 0) 14) $(a, 0)$; $(a, -8a)$
15) a) $y^3 = x$ b) $y^3 = x^2$
c) $y^3 = x$ at (0, 0); $x = 0$
16) $y = x^3 - 6x^2 - 3x$
17) $4y = 4x^2 + 5x$ 18) $3y = 8x$

Exercise 10b – p. 432

23) $\begin{cases} x^2 + y^2 \leqslant 4 \\ y < 1 \end{cases}$
24) $\begin{cases} x^2 + y^2 \leqslant 9 \\ 4x^2 + 9y^2 \geqslant 36 \end{cases}$
25) $y \geqslant |x|$
26) $1 \leqslant (x-2)^2 + (y-1)^2 \leqslant 4$
27) $\begin{cases} x^2 + y^2 \geqslant 9 \\ x^2 + 4y^2 \leqslant 16 \end{cases}$

Multiple Choice Exercise 10 – p. 439

1) d 2) e 3) b
4) e 5) b 6) a
7) d 8) b 9) a
10) d 11) e 12) d
13) a 14) a, b, c 15) b, d
16) a 17) a, c 18) b
19) a, c, d 20) b, d 21) b, c
22) a, b 23) a 24) *E*
25) *A* 26) *B* 27) *B*
28) *A* 29) *E* 30) *A*
31) *A* 32) *F* 33) *T*
34) *T* 35) *F* 36) *T*
37) *F* 38) *F* 39) *T*
40) *F*

Miscellaneous Exercise 10 – p. 444

1) $(\pi, \frac{5}{6})$ 2) $y = 4x$
3) $(0, 0), (\sqrt{3}, \frac{1}{4}\sqrt{3}), (-\sqrt{3}, -\frac{1}{4}\sqrt{3})$
4) $(0, 1), (-1, 1); (-\frac{1}{2}, \frac{3}{2})$
5) $(n\pi, n\pi), n = 0, 1, 2, 3, 4$
6) $(0, 1), (-3, 0), (-6, -1)$
7) a) -1 b) $\frac{1}{2}$ c) $-\frac{5}{2}$ or 0
8) $\pm \dfrac{3x - 4}{2\sqrt{x - 2}}$ 9) $\frac{13}{6}a^2$
10) $\frac{5}{6} + 2 \ln 2$ 14) a) 1 b) $-\sqrt{2}$
15) a) $y = \pm x$ b) $y = \pm 1$
16) $a = 1;\ b = 2.7$
17) $A = 5.13 \times 10^7;\ n = -3.89 \times 10^3$
18) 2.3, 3.0
19) $\log y = -0.155x - 0.4$

Exercise 11a – p. 455

1) $-x - 4$ 2) 13
3) $p = 3, q = 5$
4) $a = 9, b = -2, c = -11$
5) $-7x + 10$ 6) $m = \frac{9}{2}, n = -\frac{7}{2}$
7) a) no b) $(x - 3), (x + 3)$
c) $(x - 1)$ d) no
8) $-5, -4$ 9) 3
10) 2 13) 3 14) $p = \frac{3}{2}, q = 2$
16) $(q^2 - p)(p^2 - q) = (pq - 1)^2$
17) -2

Exercise 11b – p. 458

1) 1, homogeneous and cyclic
2) 2, homogeneous
3) 2, homogeneous and cyclic
4) 2, homogeneous
5) 3, homogeneous
6) 3, cyclic
7) 3, homogeneous and cyclic
8) 2, homogeneous
9) 2, homogeneous and cyclic
10) 2, homogeneous and cyclic
11) a) $\alpha(\alpha^2 - \beta^2) + \beta(\beta^2 - \gamma^2) + \gamma(\gamma^2 - \alpha^2)$
b) $xy^2 + yz^2 + zx^2$
c) $\alpha(\beta^2 - \gamma^2) + \beta(\gamma^2 - \alpha^2) + \gamma(\alpha^2 - \beta^2)$
d) $ab^2 + bc^2 + cd^2 + da^2$
12) a) $\sum x^2(y^2 + z^2)$ b) $\sum \alpha^2(\beta + \gamma)$
c) $\sum \alpha$ d) $\sum a^2b$

Exercise 11c – p. 461

1) $2a(a^2 + 3b^2)$ 2) $(x + y + z)^2$
3) $-(x - y)(y - z)(z - x)$
4) $(a - b)(b - c)(c - a)$
5) $3(a - b)(b - c)(c - a)$
6) $-(x - y)(y - z)(z - x) \times (x^2 + y^2 + z^2 + xy + yz + zx)$
7) $(a - b)(a + b)(a^2 + ab + b^2) \times (a^2 - ab + b^2)$
8) $(x - 2)(x + 2)(x^2 + 2x + 4) \times (x^2 - 2x + 4)$
9) $-(p - q)(q - r)(r - p)$
10) $(a + b + c)(a^2 + b^2 + c^2 + 2bc - ab - ac)$
11) $\dfrac{a^n - x^n}{a - x}$;
$(x - 2)(x^4 + 2x^3 + 4x^2 + 8x + 16)$;
$(a - b)(a^4 + a^3b + a^2b^2 + ab^3 + b^4)$
13) $(p - q)^2(p + q)$
14) $(a + 2b + 3c) \times (a^2 + 4b^2 + 9c^2 - 2ab - 6bc - 3ac)$

Exercise 11d – p. 464

1) $-1, -1, \frac{1}{5}, 5$ 2) $-3, -\frac{1}{3}, \frac{1}{2}, 2$
3) $\frac{1}{9}, \frac{1}{7}, 7, 9$ 4) $-\frac{1}{4}, -4, \pm i$
5) 4 6) $6 \pm 2\sqrt{3}$
7) 5 8) $1, 8, -8$
9) $(\sqrt{2}, \sqrt{2}/2), (-\sqrt{2}, -\sqrt{2}/2)$
10) $(2, 3), (-2, -1)$
11) $(-3, 4), (-6, 3)$
12) $4, 1, -2, -2$

Exercise 11e – p. 473

1) $-\frac{3}{4}, -\frac{3}{4}, \frac{2}{5}$ 2) $\frac{2}{3}, \frac{2}{3}, \frac{1}{5}$
3) $\frac{3}{2}, \frac{3}{2}, \frac{2}{5}$ 4) $-\frac{3}{4}, -\frac{3}{4}, \frac{1}{7}$
5) $9, 9, -3 \pm 4i$
6) a) $\frac{1}{4}; \frac{1}{2}; \frac{7}{4}$ b) $0; -3; -1$
c) $0; 0; \frac{1}{8}$ d) $0; -1; 0$
e) $-4; 0; 5$
7) a) $7; 31; 18; -\frac{1}{5}$ b) $\frac{25}{9}; \frac{14}{9}; \frac{13}{9}; 4$
c) $-\frac{3}{2}; \frac{9}{16}; \frac{21}{4}; -\frac{3}{7}$
d) $1; 10; 6; -\frac{2}{3}$ e) $0; 0; 3; 0$
f) $0; 0; 3; 1$
g) $1; 0; 0;$ undefined
8) a) $x^3 + 8x^2 + 15x + 7 = 0$
b) $x^3 - 5x^2 + 2x + 1 = 0$
c) $x^3 + 4x^2 - 20x + 8 = 0$
d) $x^3 - 14x^2 + 21x - 1 = 0$
e) $x^3 + 5x^2 + 2x - 1 = 0$
f) $x + 4x^2 - x - 11 = 0$
g) $x^3 - 12x^2 + 29x - 7 = 0$
9) $\frac{33}{7}; \frac{32}{49}; 7\sum \alpha^5 = 11 \sum \alpha^2 + 4 \sum \alpha^3$
10) a) $b^3d = c^3a$

b) $4abc = b^3 + 8a^2d$
c) $9ad = bc$ and $b^2 = 3ac$
11) $x^3 - 2x^2 - 5x + 6 = 0; -2, 1, 3$
12) $-\frac{4}{3}, -\frac{7}{3}, -\frac{5}{3}, -1;\ \frac{58}{9}, \frac{5}{3};$
$48x^4 + 32x^3 - 28x^2 + 10x = 3$
13) $2, 2, \frac{1}{3}(-2 \pm \sqrt{10})$
15) $X^4 - 5X^3 - 7X^2 + 5X + 6 = 0;$
$x = -3, -3, -1, 4$

Exercise 11f – p. 481

1) 1 2) 2 3) 1 4) 2
5) 1 6) 2 7) 2 8) 3
9) 3 and 4; 1 and 2, 0 and 1; – 4 and – 3, 0 and 1; – 1 and 0, 0 and 1, 2 and 3
11) $4 < \lambda < 5$

Exercise 11g – p. 492

1) 1.6503 2) 0.5879
3) a) 0.4848 b) 0.5543 c) 0.0175
4) 0.20 5) 0.22 6) 0.16
7) 2.07 8) 4.15 9) 5.15

In questions 10–15 the degree of accuracy depends upon the accuracy of the reader's first graphical approximation.

10) 1.1656 (correct to 4 d.p.)
11) 0 (exact), 0.746 85 (correct to 5 d.p.)
12) 0 (exact), 1.2564 (correct to 4 d.p.)
13) 0.510 97 (correct to 5 d.p.)
14) – 2.6691 (correct to 4 d.p.)
15) 0.374 82 (correct to 5 d.p.)
16) 0 (exact), 1.904 17) 1 (exact)
18) 0 19) 0.540, – 1.454
20) 3.104

Multiple Choice Exercise 11 – p. 494

1) b 2) b 3) a
4) b 5) d 6) b
7) b, c 8) b, c 9) a, b, c
10) a 11) b, c 12) *B*
13) *B* 14) *E* 15) *B*
16) *A* 17) *C* 18) *A*
19) *A* 20) *B* 21) *E*
22) c, d 23) *b* 24) *I*
25) *F* 26) *T* 27) *F*
28) *T* 29) *T* 30) *F*
31) *F*

Miscellaneous Exercise 11 – p. 497

1) $a = 1, b = -37$
2) $m = -1, n = -2$
3) $2x^2 - 5x - 3;\ (x-1)^2(2x^2 - 5x - 3)$
4) $[(R_1 - R_2)x + aR_2 - bR_1]/(a - b)$
5) $3, -1;\ 0, -3$
6) $a = \frac{1}{2}f(1), b = -f(2), c = \frac{1}{2}f(3);\ -90$
8) a) 0 b) $a^{n-1}x - a^n$
9) $\sqrt{7}, \frac{1}{2}(-\sqrt{7} \pm \sqrt{5})$
10) $(x-y)(y-z)(z-x)(x+y+z)$
11) $(a+b+c)(a-b+c)(a-b-c) \times (a+b-c)$
13) $(a+b+c)(a-b+c)(a+b-c) \times (b+c-a)$; positive
14) $-(a-b)(b-c)(c-a)$;
$-(a-b)(b-c)(c-a)(a^2+b^2+c^2)$
15) $-\frac{3}{4}, \frac{1}{4}, \frac{1}{2}$ 16) $2x^3 + 3x^2 + 8 = 0$
17) $-\frac{4}{3}, -\frac{1}{2}, \frac{1}{3}$
18) a) $x^3 - 2x^2 - 16x + 40 = 0$
b) $5x^3 - 4x^2 - x + 1 = 0$
c) $x^3 - 2x^2 - 3x - 1 = 0;\ 10$
19) 98 20) – 216
21) $p = 10;\ q = 31$
22) $3, \frac{3}{2}, -\frac{1}{2}, -2$
24) $n = 2, a_2 = 2.20, a_3 = 2.19$
25) $-2pq;\ 0;\ x = -2, 3, 6$
26) 0.7431, 0.7907, 0.8361
27) 2.4, 2.49 28) 0.405
29) 1.79
30) a) – 1; 1 b) $\frac{1}{2}, 2, \frac{5}{6} \pm i\sqrt{11}/6$
31) 1.26 32) 2.908
33) (0, 0) min; $(-\frac{1}{4}, \frac{5}{256})$ max;
(– 2, – 4) min
a) $\frac{1}{2} \pm i\sqrt{3}/2$
b) – 3; 1 and 2 c) 7
34) 0.39
35) $a = 25$, $b = -54$; $1 - 3i$,
$2 \pm \sqrt{3}i$
36) The point where $x = 7.2$ is too near to the tangent at the point where $x = 7$
37) (c) is suitable

Exercise 12a – p. 511

1) a) $\cos 7\theta + i \sin 7\theta$
b) $\cos(-3\theta) + i\sin(-3\theta)$
$\equiv \cos 3\theta - i \sin 3\theta$
c) $\cos(\frac{1}{2}\theta) + i\sin(\frac{1}{2}\theta)$
d) $\cos\pi + i\sin\pi \equiv -1$
e) $\cos\left(-\frac{\pi}{2}\right) + i\sin\left(-\frac{\pi}{2}\right) \equiv -i$
f) $\cos\frac{\pi}{3} + i\sin\frac{\pi}{3} \equiv \frac{1}{2} + i\frac{\sqrt{3}}{2}$
2) a) $(\cos\theta + i\sin\theta)^5$
b) $(\cos\theta + i\sin\theta)^{-2}$
c) $(\cos\theta + i\sin\theta)^{\frac{1}{3}}$

d) $(\cos\theta + i\sin\theta)^{-\frac{1}{2}}$
3) $\cos 7\theta + i\sin 7\theta$
4) $\cos 9\theta + i\sin 9\theta$
5) $\cos 2\theta + i\sin 2\theta$
6) $\cos 3\theta - i\sin 3\theta$
7) $\cos 7\theta + i\sin 7\theta$
8) $\cos 3\theta + i\sin 3\theta$
9) $\cos\left(\frac{10\pi}{3}\right) + i\sin\left(\frac{10\pi}{3}\right) = -\frac{1}{2} - \frac{\sqrt{3}}{2}i$
10) 1   11) $\cos\theta + i\sin\theta$
12) $\frac{\sqrt{3}}{2} + \frac{1}{2}i$
13) $\cos\frac{\pi}{5} + i\sin\frac{\pi}{5}$
14) $\cos\frac{\theta}{n} + i\sin\frac{\theta}{n}$
21) $-1.00, 0.268, 3.73$
22) a) $\frac{1}{4}\sin 4\theta + 2\sin 2\theta + 3\theta + K$
b) $\frac{3}{2}\sin 4\theta + \frac{15}{2}\sin 2\theta + 10\theta + K$
c) $2\sin 2\theta - 3\theta + K$

Exercise 12b – p. 519

1) a) $\pm 2^{\frac{1}{4}}\left\{\cos\frac{\pi}{8} - i\sin\frac{\pi}{8}\right\}$
b) $\pm\sqrt{5}\{\cos(0.46^c) + i\sin(0.46^c)\}$
c) $\pm\sqrt{13}\{\cos(0.98^c) + i\sin(0.98^c)\}$
d) $\pm\sqrt{2}\ \cos\frac{\pi}{12} + i\sin\frac{\pi}{12}$
e) $\pm 8^{\frac{1}{4}}\left\{\cos\left(\frac{5\pi}{8}\right) + i\sin\left(\frac{5\pi}{8}\right)\right\}$
$\equiv \pm 8^{\frac{1}{4}}\left\{\cos\frac{\pi}{8} - i\sin\frac{\pi}{8}\right\}$
f) $\pm\left(\cos\frac{\pi}{4} - i\sin\frac{\pi}{4}\right) = \pm\frac{1}{\sqrt{2}}(1+i)$
g) $\pm\left(\cos\frac{\pi}{2} + i\sin\frac{\pi}{2}\right) = \pm i$

2) a) $2^{\frac{1}{6}}\left\{\cos\left(\frac{2}{3}n\pi - \frac{\pi}{12}\right) + i\sin\left(\frac{2}{3}n\pi - \frac{\pi}{12}\right)\right\}, n = -1, 0, 1$
b) $5^{\frac{1}{3}}\{\cos(\frac{2}{3}n\pi + 0.31^c) + i\sin(\frac{2}{3}n\pi + 0.31^c)\}, n = -1, 0, 1$
c) $(13)^{\frac{1}{3}}\{\cos(\frac{2}{3}n\pi + 0.65^c) + i\sin(\frac{2}{3}n\pi + 0.65^c)\}, n = -1, 0, 1$
d) $2^{\frac{1}{3}}\left\{\cos\left(\frac{2}{3}n\pi + \frac{\pi}{18}\right) + i\sin\left(\frac{2}{3}n\pi + \frac{\pi}{18}\right)\right\}, n = -1, 0, 1$
e) $\sqrt{2}\left\{\cos\left(\frac{2}{3}n\pi - \frac{\pi}{4}\right) + i\sin\left(\frac{2}{3}n\pi - \frac{\pi}{4}\right)\right\}, n = -1, 0, 1$
f) $\cos\left(\frac{2}{3}n\pi + \frac{\pi}{6}\right) + i\sin\left(\frac{2}{3}n\pi + \frac{\pi}{6}\right)$, $n = -1, 0, 1$
g) $\cos\left(\frac{2}{3}n\pi + \frac{\pi}{3}\right) + i\sin\left(\frac{2}{3}n\pi + \frac{\pi}{3}\right)$, $n = -1, 0, 1$

4) $-1, \cos\frac{\pi}{3} + i\sin\frac{\pi}{3}$,
$\cos\left(-\frac{\pi}{3}\right) + i\sin\left(-\frac{\pi}{3}\right)$
5) $\sqrt{2}\left(\cos\frac{\pi}{12} \pm i\sin\frac{\pi}{12}\right)$,
$\sqrt{2}\left(\cos\frac{3\pi}{4} \pm i\sin\frac{3\pi}{4}\right)$,
$\sqrt{2}\left(\cos\frac{-7\pi}{12} \pm i\sin\frac{-7\pi}{12}\right)$
6) $\pm\sqrt{5}(\cos 0.46^c \pm i\sin 0.46^c)$
7) $\pm 1, \pm\left(\cos\frac{\pi}{3} \pm i\sin\frac{\pi}{3}\right)$
8) $3, 2 + \left(\cos\frac{2\pi}{3} \pm i\sin\frac{2\pi}{3}\right)$
10) 2^{13}

Exercise 12c – p. 522

1) a) $\sqrt{2}\,e^{i\pi/4}$ b) $e^{i\pi/2}$
c) $4\,e^{-i\pi/3}$ d) $\sqrt{2}\,e^{i3\pi/4}$
e) $4\,e^{0i}$ f) $5\,e^{0.93i}$
g) $e^{i2\pi/3}$
2) a) $\frac{1}{2} - \frac{\sqrt{3}}{2}i$ b) $-\sqrt{3} + i$
c) -5 d) $-i$ e) -4
3) $e^{i\pi/3}, e^{-i\pi/3}, e^{i\pi}$
4) $2\,e^{-i3\pi/4}, 2\,e^{-i7\pi/20}, 2\,e^{i\pi/20}, 2\,e^{i9\pi/20}, 2\,e^{i17\pi/20}$
5) $e^{0i}, e^{\pm i2\pi/7}, e^{\pm i4\pi/7}, e^{\pm i6\pi/7}$
6) $\sqrt{2}\,e^{-0.18i}, \sqrt{2}\,e^{1.75i}, \sqrt{2}\,e^{3.32i}, \sqrt{2}\,e^{4.89i}$
7) $\frac{1}{2}\{1 - e^{-\pi/2}\}$
8) $\frac{1}{5}\{2\,e^{\pi/4} - 1\}$

Exercise 12d – p. 527

1) a) a circle, centre (4, 0), radius 3
b) the part-line from (4, 0) at $\frac{\pi}{3}$ to Ox
c) a circle, centre 0, radius 5

d) a circle, centre (4, 0), radius 1
e) the part-line from (4, 0) at arctan 2 to Ox
f) a parabola $v^2 = 4(u - 4)$

2) a) $u^2 - v^2 = 4$
b) $\arg(w) = -\frac{\pi}{2}$, i.e. the negative v axis
c) $|w| = 8$ d) $v = 2u - 6$

3) a) $|w| = \frac{1}{6}$ b) $\arg(w) = \frac{\pi}{4}$
c) $u^2 + 4v^2 = 4(u^2 + v^2)^2$
d) $4(u^2 + v^2)^2 + uv = 0$
e) $3u + v + 4(u^2 + v^2) = 0$

4) a) $(u^2 + v^2)^2 = 4uv$
b) the u axis ($v = 0$)
c) the line $u + v = 0$

Miscellaneous Exercise 12 – p. 528

1) -1 2) $9, 6\theta$ 5) -2^{11}

6) $\cos\left\{\frac{p(\theta + 2n\pi)}{9}\right\} + i\sin\left\{\frac{p(\theta + 2n\pi)}{9}\right\}$, $(n = -4, -3, \ldots, 3, 4)$; $\pm i, \frac{1}{2}(\pm\sqrt{3} \pm i)$

10) $\cos\left(\frac{p\theta}{q} + \frac{2pn\pi}{q}\right) + i\sin\left(\frac{p\theta}{q} + \frac{2pn\pi}{q}\right)$;
$\theta = \frac{2}{9}\pi, \frac{4}{9}\pi, \frac{8}{9}\pi, \frac{10}{9}\pi, \frac{14}{9}\pi, \frac{16}{9}\pi$

11) $\cos(\frac{2}{9}n\pi) + i\sin(\frac{2}{9}n\pi)$, $n = 1, 2, 4, 5, 7, 8$;
$\theta = \frac{2n\pi}{9}, n = 1, 2, 4, 5, 7, 8$

12) a) $|u| = 1$ or $\arg u = 0$ or π
b) $(-\frac{1}{2} + i\sqrt{3}/2), (-\frac{1}{2}, -i\sqrt{3}/2)$

13) $2, \pi; 2, \pm\frac{3\pi}{5}; 2, \pm\frac{\pi}{5}; z^2 - 4\cos\frac{3\pi}{5} + 4$,
$z^2 - 4\cos\frac{\pi}{5} + 4$

14) $\pm(1/\sqrt{2} + i/\sqrt{2}); \pm(1 \pm i)$;
$(x^2 + 2x + 2)(x^2 - 2x + 2)$

15) a) 3 or 0

16) a) $\sqrt{2}\left(\cos\frac{\pi}{4} + i\sin\frac{\pi}{4}\right)$; $32(-1 + i)$
b) $\pm 3 \pm 4i$

17) $1, -\frac{1}{2} \pm \frac{\sqrt{3}}{2}i$

18) $-7 - 24i; \frac{1}{25}(3 + 4i); \pm(2 - i)$

19) $32\cos^6\theta - 48\cos^4\theta + 18\cos^2\theta - 1; \frac{1}{60}$

20) $-1, \frac{1}{2} \pm i\sqrt{3}/2$

21) a) (i) 0 (ii) 3 b) $(-\frac{5}{4}, 0), \frac{3}{4}$

22) $\tan\frac{r\pi}{15}, r = 4, 7, 13$

23) $\sqrt{3}; -1$ 24) $2\,e^{-i\pi/6}$

26) $0, 3; 4\omega^2, 4\omega$

27) $2\,e^{i\pi/6}, 2\,e^{i5\pi/6}, e^{-i\pi/2}$

28) $x^2 + x - 1 = 0; 1.62$

29) $8\,e^{i\pi/6}$
b) $2\,e^{i\pi/18}, 2\,e^{i13\pi/18}, 2\,e^{-i11\pi/18}$

30) a) $z = 1, e^{\pm i2\pi/5}, e^{\pm i4\pi/5}$

31) a) $u = \frac{2x}{x^2 + (y + 1)^2}$,
$v = \frac{x^2 + y^2 - 1}{x^2 + (y + 1)^2}$
b) $\frac{1}{3}(2 \pm \sqrt{5}i), \frac{1}{2}(-1 \pm \sqrt{3}i)$

32) $x = u^2 - v^2, y = 2uv$

33) $v^2 = 4(4 - u)$

Exercise 13a – p. 536

1) $nI_n = (n - 1)I_{n-2} - \cos x\sin^{n-1}x$
2) $(n - 1)(I_n + I_{n-2}) = \tan^{n-1}x$
3) $2I_n = (x + 1)^n e^{2x} - nI_{n-1}$
4) $2_nI_n = 2(n - 1)I_{n-2} + \sin 2\theta\cos^{n-1}2\theta$
5) $aI_n = x^n e^{ax} - nI_{n-1}$
6) $2I_n = x^2(\ln x)^n - nI_{n-1}$
7) $nI_n = \sinh x\cosh^{n-1}x + (n - 1)I_{n-2}$
8) $(n - 1)I_n = \tan x\sec^{n-2}x + (n - 2)I_{n-2}$
9) $\frac{1}{6}\sin x\cos^5 x + \frac{5}{24}\sin x\cos^3 x$
$+ \frac{5}{16}\sin x\cos x + \frac{5}{16}x + K$
10) $K - \frac{1}{7}\sin^6 x\cos x - \frac{6}{35}\sin^4 x\cos x$
$- \frac{8}{35}\sin^2 x\cos x - \frac{16}{35}\cos x$
11) $e^{4x}\{\frac{1}{4}(1 - x)^3 + \frac{3}{16}(1 - x)^2$
$+ \frac{3}{32}(1 - x) + \frac{3}{128}\} + K$
12) $\frac{1}{4a}\sin(ax + b)\cos^3(ax + b)$
$+ \frac{3}{8a}\sin(ax + b)\cos(ax + b) + \frac{3x}{8} + K$
13) $\frac{1}{12}\cos\left(\frac{\pi}{4} - 3\theta\right)\sin^3\left(\frac{\pi}{4} - 3\theta\right)$
$+ \frac{1}{8}\cos\left(\frac{\pi}{4} - 3\theta\right)\sin\left(\frac{\pi}{4} - 3\theta\right)$
$+ \frac{3}{8}\theta + K$
14) $\frac{1}{4}\tan^4 x - \frac{1}{2}\tan^2 x + \ln(\sec x) + K$
15) $K - x^3\cos x + 3x^2\sin x$
$+ 6x\cos x - 6\sin x$
16) $\frac{1}{5}\cosh x\sinh^4 x - \frac{4}{15}\cosh x\sinh^2 x$
$+ \frac{8}{15}\cosh x + K$
17) $\frac{1}{3}\tan x\sec^2 x + \frac{2}{3}\tan x + K$
18) $-e^{-x}(x^6 + 6x^5 + 30x^4 + 120x^3$
$+ 360x^2 + 720x + 720)$
19) $x^2\{\frac{1}{2}(\ln x)^3 - \frac{3}{4}(\ln x)^2$
$+ \frac{3}{4}(\ln x) - \frac{3}{8}\} + K$
20) $\frac{1}{12}\sinh 3x\cosh^3 3x + \frac{1}{8}\sinh 3x\cosh 3x$
$+ \frac{3}{8}x + K$

21) $\frac{1}{216}(8x^3-1)(1+x^3)^8+K$
22) $\ln(\cot x-\operatorname{cosec} x)+\cos x$
$+\frac{1}{3}\cos 3x+\frac{1}{5}\cos 5x+K$

Exercise 13b – p. 545

1) $\dfrac{256}{693}$ 2) $\dfrac{63\pi}{512}$ 3) 0
4) $\dfrac{5\pi}{16}$ 5) $\dfrac{105\pi}{256}$ 6) $\dfrac{8}{15}$
7) $\dfrac{3\pi}{4}$ 8) $\dfrac{231\pi}{512}$ 9) $8!$
10) $\dfrac{13}{15}-\dfrac{\pi}{4}$ 11) $\dfrac{7\pi}{512}$ 12) $\dfrac{8}{35}$
13) 0 14) $\dfrac{3\pi}{64}$
15) $I_n=e-nI_{n-1}$; $120-44e$
16) $\dfrac{3^n n!}{(3n+2)(3n-1)\ldots(8)(5)(2)}$
17) a) $37\sinh 1-28\cosh 1$
b) $7\sinh 1-9\cosh 1+6$
18) $(\frac{18}{21})(\frac{16}{19})\ldots(\frac{4}{7})(\frac{2}{5})(\frac{2}{3})$
19) $\left(\frac{\pi}{2}\right)^6-30\left(\frac{\pi}{2}\right)^4+360\left(\frac{\pi}{2}\right)^2-720$
20) $\frac{1}{5}\sinh^5 1-\frac{1}{3}\sinh^3 1+\sinh 1$
$-2\arctan e+\dfrac{\pi}{2}$
21) $\dfrac{5!\,6!\,2^{12}}{12!}$

Exercise 13c – p. 550

1) $xy=e^x+A$
2) $3y\cos x=x^3+A$
3) $2x\ln y=x^2+2x+A$
4) $y=Ax-x\cos x$
5) $y\,e^x=2x+A$
6) $x\,e^y=e^x+A$
7) $4y\ln x=2x^2\ln x-x^2+A$
8) $4y(1+x)=x^4+A$
9) $x\tan y=\ln\sec x+A$
10) $2e^{(x+y)}=e^{2x}+A$
11) $y\,e^{3x}=x+A$
12) $y\sin x=x+A$
13) $xy=x+\ln Ax$
14) $y=(x+1)^3(x^2+2x+A)$
15) $2y\sin x=e^x(\sin x-\cos x)+A$
16) $(2v-1)\,e^{t^2}=A$
17) $x^2y=A-\cos x$
18) $y=x(e^{-x}+A)$
19) $2r\sin\theta=\theta-\sin\theta\cos\theta+A$
20) $2y(x-1)+x\,e^{-x^2}=Ax$
21) $y=x(\sin x-\cos x-1)$
22) $(1-y)\,e^{\frac{1}{2}x^2}=1$

Exercise 13d – p. 555

1) $y=A\,e^x+B\,e^{2x}$
2) $y=A\,e^x+B\,e^{\frac{4x}{3}}$
3) $y=e^x(A+Bx)$
4) $y=A\,e^{-\frac{1}{2}x}\cos\left(\dfrac{\sqrt{3}x}{2}+\epsilon\right)$
5) $y=A\,e^{4x}+B\,e^x$
6) $y=A\,e^{2x}+B\,e^{-2x}$
7) $y=A\cos(2x+\epsilon)$
8) $y=A\,e^{-\frac{x}{4}}\cos\left(\dfrac{\sqrt{15}}{4}x+\epsilon\right)$
9) $y=A+B\,e^{2x}$
10) $y=e^{\frac{1}{3}x}(A+Bx)$

Exercise 13e – p. 561

1) $y=A\,e^x+B\,e^{5x}+\frac{3}{5}$
2) $y=e^x(A+Bx)+e^{2x}$
3) $y=e^x(A+Bx+\frac{1}{2}x^2)$
4) $y=A\,e^{-x}+B\,e^{-2x}$
$+\frac{1}{10}(\sin x-3\cos x)$
5) $y=A^{-\frac{x}{2}}\cos\left(\dfrac{\sqrt{3}x}{2}+\epsilon\right)+x$
6) $y=A\,e^{2x}+(B-\frac{3}{4}x)\,e^{-2x}$
7) $y=A\,e^{\frac{x}{4}}+B\,e^x-\frac{1}{17}(4\cos x+\sin x)$
8) $y=e^{-\frac{x}{3}}(A+Bx)+x^2-10x+45$
9) $y=A\cos(5x+\epsilon)$
$+\frac{1}{629}(50-4i)\,e^{(1+i)x}$
10) $y=A\,e^x+B\,e^{3x}$
$+(8\cos 2x-\sin 2x)$
11) $S=e^{2t}(A+Bt)+\frac{5}{4}$
12) $x=A\,e^{-\frac{1}{2}t}\cos\left(\dfrac{\sqrt{7}t}{2}+\epsilon\right)$
$-\frac{1}{226}(1-15i)\,e^{(2-3i)t}$
13) $y=A\,e^{-x}+B\,e^{-2x}+e^x$
$+\frac{1}{10}(\sin x-3\cos x)$
14) $\theta=A\cos(t+\epsilon)+e^t(\sin t-2\cos t)$
15) $y=2\,e^{5x}-e^{-2x}$
16) $y=\sqrt{37}\,e^{-4x}\cos(3x+\epsilon)+2\cos x$
where $\tan\epsilon=\frac{1}{6}$
17) $f(x)=x\,e^{-x}$; $y=e^{2x}-e^{3x}+x\,e^{-x}$
18) C.F. is $e^{-2x}(A+Bx)$; P.I. is $-\cos 2x$;
$y=2\,e^{-2x}(1+2x)-\cos 2x$

Miscellaneous Exercise 13 – p. 563

1) $\frac{16}{105}$ 2) $\frac{22}{65}e^{\pi}+\frac{6}{65}$ 4) $\frac{1}{4}\ln 3+\frac{1}{3}$
5) $I_m=\dfrac{x^4}{4}(\ln x)^m-\dfrac{m}{4}I_{m-1}$;

$\frac{x^4}{128}\{32(\ln x)^3 - 24(\ln x)^2 + 12\ln x - 3\}$

6) $I_n = \frac{1}{n-1}\tan^{n-1}\theta - I_{n-2}$

7) a) $37\sinh 1 - 28\cosh 1 = \frac{1}{2e}(9e^2 - 65)$

b) $7\sinh 1 - 9\cosh 1 + 6 = 6 - \frac{1}{e}(e^2 + 8)$

8) $(2m-1)J_m = 2^{m-1} + 2(m-1)J_{m-1}$; $\frac{28}{15}$

9) $\frac{1}{2}\pi$

10) $2y = \sin x(\sin 4x + 2\sin 2x + 2x + A)$; $24y\sin x = 3\sin 4x - 2\sin 6x + A$

11) $8y\sin x = 2\cos 2x - \cos 4x + K$

12) $x = Ae^{2t} + Be^{-t} - 3\sin t + \cos t$; $x = 3e^{-t} - 3\sin t + \cos t$

13) a) $y = \tan\{\ln(1+\cos^2 x)\}$
b) $xy + x + 1 = A\,e^x$

14) $x^2 y = 2(x^2-1)^2$

15) a) $\tan y = 2\arcsin\left(\frac{2x}{3} - 1\right) + A$

b) $4y = 2(1+x) + (1-x^2)\ln\frac{A(1-x)}{(1+x)}$

16) $x = \cos t - 3\sin t$ in each case

17) $y = x\cos x - \sin x\cos^2 x + \cos x$

18) a) $\sqrt{(1+y^2)} = \frac{1}{2}\ln\frac{A(1+x^2)}{x^2}$
b) $y = 4e^x - x - 1$

19) a) $y = \sin x(1 + \sin x)$
b) $2y\,e^x = 1 - \cos 2x - \sin 2x$

20) $3y\sin^2 x = \sin^3 x + A$

21) a) $y = 2x^2 - 1$
b) $2y = e^{2x}(1 - \cos 2x - \sin 2x)$

22) a) $2\sqrt{2}\cos x = 1 + e^{-y}$
b) $y\,e^x = A(1+x) - 1$

23) a) $2xy = x^2 - 1$
b) $y = e^x\cos^2 x$

Exercise 14a – p. 574

1) π 2) $c\sinh\frac{x}{c}$

3) $\frac{1}{27}(13\sqrt{13} - 8)$

4) $\frac{1}{2}(\sqrt{2} + \text{arsinh}\,1)$ 5) $\frac{1}{2}\ln 3$

6) $2a\sqrt{2}$

7) a arsinh $t + t\sqrt{(t^2+1)}$

8) $2\pi a$ 9) arcsin (tanh 1)

10) $\frac{k}{2}\text{arsinh}\,\frac{\pi}{2} + \frac{\pi}{4}\sqrt{(\pi^2+4)}$

11) $8a$ 12) $\frac{1}{2}\pi a$ 13) $a\sqrt{2}(e^\pi - 1)$

Exercise 14b – p. 581

1) $2\pi a^2$ 2) $\frac{12}{5}\pi a^2$

3) $4\pi\sqrt{5}$ 4) $4\pi a^2$

5) $4\pi a^2$ 6) $\frac{8\pi}{3}(17^{\frac{3}{2}} - 1)$

7) $\frac{2\pi\sqrt{2}}{5}(2e^\pi + 1)$

8) $\frac{1}{2}\pi a^2\{4\sqrt{7} - 2 + \sqrt{2}(\text{arcosh}\sqrt{2} - \text{arcosh}\,2\sqrt{2})\}$

9) $\frac{1}{2}\pi(2 + \sinh 2)$

10) $\pi\{\text{arsinh}\,e - \text{arsinh}\,1 + e\sqrt{(1+e^2)} - \sqrt{2}\}$

11) $\frac{6\pi}{5}\{247\sqrt{13} + 64\}$

12) a) $2^{10}\pi$ b) $\frac{2^{10}\pi}{5}$

13) $\frac{16\pi}{105}$ 14) $5\pi^2 a^3$

15) $\frac{128\pi\sqrt{2}}{15}$

16) a) $\pi(31 - 32\ln 2)$
b) $\pi(129 - 32\ln 2)$
c) 127π d) 29π

17) $\frac{1}{2}$ 18) $\sqrt{\left(\frac{4}{\pi} - 1\right)}$

19) $\sqrt{\left(\frac{3}{2} + \frac{4}{\pi}\right)}$ 20) $\frac{e\sqrt{2}}{2}\sqrt{(e^2-1)}$

21) $\frac{2\sqrt{6}}{3}$ 22) $\frac{1}{2}\sqrt{(\sinh 4 - \sinh 2 - 2)}$

23) 0.813 24) 0.434 25) $\sqrt{\frac{943}{15}}$

Exercise 14c – p. 588

1) $\sqrt{2}$ + arsinh 1 2) $\frac{1}{12}(17\sqrt{17} - 5\sqrt{5})$

3) $\frac{17}{32}$ arsinh $2 + \frac{7\sqrt{5}}{16}$

4) a) 16 b) $\frac{128}{5}$

5) a) $\frac{1}{4}\pi$ b) $\frac{1}{4}(e^4 - 1)$
c) $\frac{8a^3}{105}$ d) $\frac{1}{4}$

6) a) 2 b) $e^2 - 1$
c) $\frac{3\pi a^2}{32}$ d) $1 - \ln 2$

7) a) $\frac{1}{8}\pi$ b) $\frac{1}{4}(e^2 + 1)$
c) $\frac{256a}{315\pi}$ d) $\frac{1}{4(1 - \ln 2)}$

8) a) $\frac{1}{2}\pi^2$ b) $\frac{1}{2}\pi(e^4 - 1)$
c) $\dfrac{16\pi a^3}{105}$ d) $\dfrac{\pi}{2}$

9) $2; \pi; \dfrac{2}{\pi}; 4\pi$

10) $\frac{8}{3}\pi$; $\left(0, \dfrac{4}{3\pi}\right)$ 11) $(2, \frac{3}{2})$; 15π

Exercise 14d – p. 596

1) a) $\frac{16}{3}ab^3$ b) $\frac{16}{3}a^3b$
2) $\frac{1}{12}l^4$ 3) $\frac{4}{9}$
4) $\dfrac{2^7}{21}$ 5) $\frac{1}{9}(e^3 - 1)$
6) $4 \ln 4 - 3$
7) a) $\frac{864}{35}$ b) $\frac{324}{7}$
8) $\frac{1}{2}\pi a^4 l$ 9) $\frac{8}{15}\pi a^5$
10) $\frac{1}{10}\pi r^4 h$ 11) $\dfrac{128\pi}{315}$
12) a) $\dfrac{\pi}{2}(9e - 24)$
b) $\dfrac{\pi}{24}(-3 + 16e - 36e^2 + 48e^3 - 13e^4)$
13) $\dfrac{3\pi^2}{32}$
14) a) $2\pi r^3 l$ b) $\frac{1}{16}\pi l^4$

Exercise 14e – p. 604

1) $\dfrac{17\sqrt{17}}{4}$ 2) $\dfrac{7\sqrt{7}}{4}$ 3) $2\sqrt{2}$
4) $\dfrac{5\sqrt{5}}{2}$ 5) $4a\sqrt{2}$ 6) $\sqrt{2}$
7) 1 8) $\cosh^2 1$ 9) $4\sqrt{2}$
10) $\dfrac{(a^2 + 3b^2)^{\frac{3}{2}}}{8ab}$ 11) $\dfrac{17c\sqrt{17}}{16}$
12) $\dfrac{3a}{2}$ 13) ∞ 14) a
15) $\dfrac{6\sqrt{13}}{169}$ 16) $\dfrac{\sqrt{5}}{100}$ 17) $\dfrac{2}{a}$
18) $\dfrac{14\sqrt{5}}{25a}$ 19) $\dfrac{3}{4a}$
20) $\dfrac{3\sqrt{(\cos 2\theta)}}{a}$
21) $\arctan(-\frac{1}{2}\cot 2\theta)$
22) $\arctan\left(\dfrac{\sin\theta - 1}{\cos\theta}\right)$ 23) $\dfrac{\pi}{4}$
24) $\arctan(\frac{1}{3}\tan 3\theta)$

Miscellaneous Exercise 14 – p. 604

1) $\dfrac{49\pi}{3}$
3) $\frac{3}{4}$; a) $3 + \ln 2$ b) $\dfrac{\pi}{2}(\ln 2 + \frac{15}{16})$
6) $8a$; $\frac{64}{5}\pi a^2$
7) a) $\dfrac{12\pi}{5}$ b) $2\cosh^2 1$; $4 \sinh 1$
8) $\dfrac{27\pi}{2}$ 9) $\dfrac{6\pi a^2}{5}$ 10) $\dfrac{58a}{35}$
11) $\dfrac{32\sqrt{2}}{7}$ 12) $\dfrac{335a}{27}$
13) $\pi + 2\sqrt{2}$; $\frac{1}{3}\pi$ 14) $\dfrac{\sqrt{2}}{2}$; $\dfrac{2r}{\pi}$
15) $\left(0, \dfrac{4b}{3\pi}\right)$; $\dfrac{2}{3}ab^3$
16) a) 21π b) 84π
17) $(0, 0)$, (π, π), $(2\pi, 2\pi)$, $(3\pi, 3\pi)$, $(4\pi, 4\pi)$; $\frac{1}{6}(32\pi^2 + 15)$

INDEX

Adjoint matrix 267
Adjugate matrix 267
Angle between two lines 160, 203
  between two planes 173, 225
  between line and plane 172, 226
  between tangent and radius vector 600
Arbitrary constant 547
  calculation of 548, 559
Arc, length of 568
  centroid of 584
Area of surface of revolution 575, 585
Argand diagram, transformations in 525
Astroid 411
Asymptotes 123
  of hyperbola 95
Augmented matrix 279
Auxiliary circle 82
Auxiliary quadratic equation 552
Axio 305

Basis vectors 183, 251, 259
Biconditional statement 314

Cartesian base vectors 142, 184
Centroid 583
Chord of contact 130
Cofactor 41
Column vector 1, 19
Common factors of polynomials 454
Complementary function 557
Complex number 504
  roots of 512, 516
  exponential form of 520
Conditional statement 310
Conic sections 62
  as loci 63
Conjugate complex roots of equations 464, 494
Conjugate diameters 76, 84
Consistent equations 277, 283
Continuity 366
Continuous functions 366
Contradiction 326
Contrapositive 314
Convergence 353, 355, 483
Converse 311
Counter example 328
Cube roots of unity 515
Cubic equations 465
Curvature 597
  radius of 598
Cyclic functions 457
  factorisation of 458
Cycloid 410

De Moivre's Theorem 505
  applications of 507
Determinants 36, 40
  expansion of 37, 41
  factorisation of 47
  properties of 44
  simplification of 44
  solution of equations by 40
Diameter 66
  of parabola 66
  of ellipse 75
  conjugate 76, 84
Difference method 344
Differentiable function 369
Differential equations 546
  exact 547
  product type 547
  second order linear 551, 554, 556
Direct proof 318
Direction cosines 145, 184
  ratios 145, 184
  vector 186
Director circle 77
Directrix 64
Discontinuity 367
Distance between parallel planes 172, 223
  between skew lines 216
  of plane from origin 170, 219
  of point from line 215
  of point from plane 171, 227

Eccentric angle 82
  circle 82
Eccentricity 64
Echelon matrix 280
Ellipse 63, 69
  conjugate diameters of 76
  facal distance property of 74
  major axis of 69
  minor axis of 69
Enlargement 8, 250

Equations
  conjugate complex roots of 464, 494
  cubic 465
  hyperbolic 390
  number of real roots of 477
  polynomial 462
  quartic 466
  with repeated roots 469
Even function 364
Exact differential equation 547
Exponential form for complex functions 525
  for circular functions 525

Factorisation of determinants 47
Factor theorem 451
First moment 583
Focal chord 65
  of parabola 65
Focus 64
Free vector 141
Function continuous 366
  cyclic 457
  discontinuous 367
  even 364
  hyperbolic 385
  inverse 402
  inverse hyperbolic 394
  logarithmic 379
  odd 365
  periodic 374

Graph paper log-linear 434, 436
  log-log 434, 437
  Semi-log 434, 436

Homogeneous equation of second order 122
  polynomial 457
Hyperbola 63, 90
  focal distance property of 94
  major axis of 91
  rectangular 96, 111
Hyperbolic functions 385
  calculus of 391
  graphs of 387
  relation to circular functions 523
Hyperbolic cosine 385
  equations 390
  relationships 386
  sine 385

Identity matrix 33
Image 3
Inconsistent equations 277, 238
Indirect proof 325
Induction 318, 350
Inequalities 419
  graphical representation of 422
  simultaneous 426
Inflexion, point of 414

Integrating factor 549
Inverse function 402
Inverse hyperbolic functions 394
  differentiation of 398
  graphs of 395
  logarithmic form of 396
  use in integration 399
Inverse matrix 262, 266, 290
  statements 316
  transformations 262, 266
Iterative methods 481, 485

Latus rectum 65, 84
Length of an arc 568
Line of intersection of two planes 174, 228
Line pair 121
  through the origin 122
  as asymptotes 123
Limit, formal definition of 353
Linear equations 276
  inconsistent 277, 283
  linearly dependent 277, 282
  solution by systematic elimination 279
  unique solution of 276, 282
Linear interpolation 485
Linear transformations 7, 248
Logarithmic function 379
Logarithmic graph paper 434, 436

Map 3
Mathematical proof 317
Matrices 4, 18
  addition of 20
  adjoint 267
  adjugate 267
  augmented 279
  determinant of 37, 40
  echelon 280
  element of 19
  equal 19
  identity 33
  inverse of 262, 266, 290
  multiplication by a scalar 20
  null 33
  products of 23
  reduced 280
  singular 14, 39, 258
  size of 18
  square 19
  transpose 267
  unit 32
Method of differences 344
  of induction 318, 350
Minor 41
Modulus of a vector 141

Negation 307, 312
Newton-Raphson method 488, 489
Normal chord 119

Null matrix 33
Number series 332, 347

Odd function 365
Osborn's Rule 386, 524

Pair of lines 121
through the origin 122
as asymptotes 123
Pappus' Theorems 585, 586
Parabola 63, 64
optical property of 65
Paraplanar 259
Particular integral 556
Period 374
Periodic functions 374
Planes 164, 218
line of intersection of 174, 228
Point of inflexion 414
Polar 130
Pole 130
Polynomials 449
common factors of 454
degree of 449, 456
equations 462
homogeneous 457
of more than one variable 456
of one variable 449
repeated factors of 451
Power series 332
Proof 317
by contradiction 326
by counter example 328
by deduction 318
by induction 318, 350
indirect 325
Proposition 305, 306

Radius of curvature 598
Rectangular hyperbola 96, 111
eccentricity of 96
Reductio ad absurdum 326
Reduction formulae 534, 535
graphical extension to 539
Reduction method of integration 534, 537
Reflection 2, 250
Remainder theorem 449
Repeated factors 451
roots 469
Resultant vector 141
Right-handed set of axes 139
Root mean square value 580
Roots and coefficients, relation between 465, 467
Roots of complex numbers 512, 516
Rotation 7, 11, 250
Row vector 19

Scalar product 200
Second moment of
area 589
a disc 594
a ring 593
a solid of revolution 595
a strip 590
volume 593
Second order linear differential equations 551, 556,
solution of 552, 554, 556
Semi-cubical parabola 411
Sequences 353
Shear 8
Simultaneous equations, nature of solution of 276, 281
Skew lines 158, 197
Statement 306
Straight line in 3-D 153, 191
Surface of revolution, area of 575, 585
Systematic elimination 279

Tangent at the origin 412
Transformations 2, 248
compound 28, 255
in the Argand diagram 525
in 2-D 2
in 3-D 248
inverse 262, 266
linear 7, 248
singular 14, 258
Transpose matrix 267
Trial solution 556, 558
Triple scalar product 214

Unit matrix 32
Unit vector 141, 186
Unity, cub roots of 515
*p*th roots of 516

Vector 1, 141
components of 141
direction 186
free 141
modulus of 141
resultant 141
unit 141, 186
Vector product 200, 207
applications of 212
Volume of revolution 579, 586

$$pa + qb + rc = 1$$

$$p_2 a + q_2 b + r_2 c = 1$$

$$p_3 a + q_3 b + r_3 c = 1$$

$$\frac{q}{q_2} p_2 a + qb + \frac{q r_2}{q_2} c = \frac{q}{q_2}$$

q/q_2 & 2nd

$$\left(p - \frac{q}{q_2} p_2\right) a + \left(r - \frac{q}{q_2} r_2\right) c = D - \frac{q}{q_2}$$

$$\left(p - \frac{q}{q_3} p_3\right) a + \left(r - \frac{q}{q_3} r_3\right) c = D - \frac{q}{q_3} D$$

S is row vector (probs)

C is covariance matrix

P_{ij} = ij both oranges

R_{ij} = Recombination ... ...

PR ...

P_{11} P_{12} P_{13} R_{11} R_{21}

R_{21}

R_{23}

$P_{11}R_{11} + P_{12}R_{12} + P_{13}R_{13}$

P_{11}.